T0399154

Techniques for Corrosion Monitoring

Woodhead Publishing Series in Metals and Surface Engineering

Techniques for Corrosion Monitoring

Second Edition

Edited by

Lietai Yang
Director
Sensors,
Corr Instruments, LLC, Carson City, NV,
United States

WP
WOODHEAD
PUBLISHING
An imprint of Elsevier

Woodhead Publishing is an imprint of Elsevier
The Officers' Mess Business Centre, Royston Road, Duxford, CB22 4QH, United Kingdom
50 Hampshire Street, 5th Floor, Cambridge, MA 02139, United States
The Boulevard, Langford Lane, Kidlington, OX5 1GB, United Kingdom

Library of Congress Cataloging-in-Publication Data
A catalog record for this book is available from the Library of Congress

British Library Cataloguing-in-Publication Data
A catalogue record for this book is available from the British Library

ISBN: 978-0-08-103003-5 (print)

ISBN: 978-0-08-103004-2 (online)

For information on all Woodhead publications
visit our website at https://www.elsevier.com/books-and-journals

Publisher: Matthew Deans
Acquisitions Editor: Christina Gifford
Editorial Project Manager: Rowley Charlotte
Production Project Manager: Anitha Sivaraj
Cover Designer: Victoria Pearson

Typeset by SPi Global, India

Contents

Contributors

C. Sean Brossia Invista Sarl, Houston, TX, United States

Kuangtsan Chiang Southwest Research Institute, San Antonio, TX, United States

Robert A. Cottis University of Manchester, Manchester, United Kingdom

Gustavo A. Cragnolino Southwest Research Institute, San Antonio, TX, United States

Pierangela Cristiani Ricerca sul Sistema Energetico, RSE SPA, Milan, Italy

Christoph Dauberschmidt University for Applied Sciences Munich, Munich, Germany

Frank W.H. Dean Ion Science Ltd., Fowlmere, United Kingdom

Yanxia Du University of Science and Technology Beijing, Beijing, People's Republic of China

Douglas C. Eberle Southwest Research Institute, San Antonio, TX, United States

Christoph Gehlen Technical University of Munich, Munich, Germany

Feng Gui DNV-GL, Dublin, OH, United States

Naeem Khan NK Consulting, LLC, Fort Collins, CO, United States

Ed Kruft American Innovations, Austin, TX, United States

Glenn Light Southwest Research Institute, San Antonio, TX, United States

Leslie Lyon-House American Innovations, Austin, TX, United States

Till Felix Mayer Sensortec GmbH, Munich, Germany

Todd Mintz Southwest Research Institute, San Antonio, TX, United States

Miguel González Núñez (Angel) MISTRAS Group, Inc., Princeton Junction, NJ, United States

Takao Ohtsu Mitsubishi Chemical Corporation, Tokyo, Japan

Sankara Papavinasam CorrMagnet Consulting Inc., Calgary, AB, Canada

Giorgio Perboni CESI SpA, Piacenza, Italy

Hossain Saboonchi MISTRAS Group, Inc., Princeton Junction, NJ, United States

Rich Smalling American Innovations, Austin, TX, United States

Dongmei Sun Polytech, Beijing, People's Republic of China

Huyuan Sun Institute of Oceanology, Chinese Academy of Sciences, Qingdao, China

Xiaodong Sun Corr Instruments, LLC, Carson City, NV, United States

Dale Webb American Innovations, Austin, TX, United States

Kjell Wold Emerson Automation Solutions, Trondheim, Norway

Bo Yang Champion Technologies Inc., Fresno, TX, United States

Lietai Yang Corr Instruments, LLC, Carson City, NV, United States

Introduction

1

Lietai Yang
Corr Instruments, LLC, Carson City, NV, United States

1.1 General

Corrosion is the deterioration process of a material due to reactions with its surroundings. As defined in the *Random House Unabridged Dictionary*: "Corrosion is the act or process of eating or wearing away gradually as if by gnawing, especially by a chemical action." The materials that are subject to corrosion include metals, ceramics, polymers, and even our own teeth. To most corrosion engineers, however, corrosion refers to the oxidation of metals by chemical and/or electrochemical processes. The most common example of metal corrosion relates to its reaction with oxygen or water. Rusting of steel due to exposure to water or humid air is a well-known example of electrochemical corrosion. In this process, the metal reacts with water or oxygen through electrochemical processes and forms iron oxides, eventually causing damage to the steel. The following sections describe the importance of corrosion monitoring and the scope of this book.

1.2 Corrosion cost

Corrosion is a costly worldwide problem. According to a recent study released by NACE International in 2016, the estimated global cost of corrosion was US$ 2.5 trillion in 2013, which is equivalent to 3.4% of the global gross domestic product (GDP) [1]. In the United States, the study commissioned by the Federal Highway Administration (FHWA) indicated that the annual direct cost of corrosion in the United State was 3.1% of the country's GDP in 1998 [2]. In China, a recent study estimated that the annual cost of corrosion in China was 3.4% of the country's GDP in 2015 [3]. Similar studies were also carried out in other countries, such as the United Kingdom, Japan, Australia, and Kuwait. Even though the level of effort varies greatly among these studies, most of them estimated the total annual cost of corrosion as ranging between 1% and 5% of each country's GDP [3].

In addition to the huge cost in economic terms, corrosion is also blamed for many of the disasters that cause loss of life and devastating pollution to the environment. For instance, in April 1992, in Mexico, the Guadalajara Sewer Explosion took the lives of 215 people and caused injury to another 1500 people. The financial loss was estimated at $75 million. The accident was traced to the corrosion of a gas line that caused a leak of the gas into a nearby sewage main [4]. Another example involved the sinking of a tanker, Erika, off the coast of Brittany in France, on 12 December 1999. In this

Techniques for Corrosion Monitoring. https://doi.org/10.1016/B978-0-08-103003-5.00001-1

accident, approximately 19,000 tons of heavy oil was spilled, equal to the total amount of oil spilled worldwide in 1998. Corrosion caused the sinking of the Erika [4].

Because corrosion takes place in many different forms, some of them cannot be eliminated, but others are avoidable by simply applying appropriate corrosion prevention/mitigation technologies. The recent NACE study estimated that by using available corrosion control practices, savings between 15% and 35% of the cost of corrosion which is equivalent to between US$375 and US$875 billion in 2013 on a global basis could be realized [1]. Knowing this, it seems prudent for worldwide industries to use appropriate corrosion prevention and control methods; taking such preemptive measures will not only avert huge economic losses, but also protect the environment and public safety.

1.3 Corrosion monitoring and its importance in corrosion prevention and control

Corrosion monitoring is the practice of acquiring information on the progress of corrosion-induced damage to a material or on the corrosivity of the environment surrounding the material. Corrosion inspection is usually a survey of the material condition at any given time, while corrosion monitoring consists of a series of surveys in a given time period. While test coupons are one of the most widely used and most reliable methods, corrosion monitoring usually relies on the use of electronic corrosion sensors or probes that are exposed to an environment of interest, such as outdoor air or seawater, or inserted into the inner space of a containment system, such as a vessel or a pipe in which a liquid or a gas flows or is contained. On a continuous or semi-continuous basis, the electronic corrosion sensors or probes emit information relating to the corrosion of a metal system.

The study commissioned by the FHWA further pointed out three preventive strategies in technical areas, to lessen or avoid unnecessary corrosion costs and to protect public safety and the environment. They are:

- advance design practices for better corrosion management;
- advance life prediction and performance assessment methods; and
- advance corrosion technology through research, development, and implementation.

These strategies are interrelated. For example, advance design requires better life prediction and performance assessment methods, and better prediction and performance assessment methods require advancement in corrosion technology.

Corrosion monitoring is part of corrosion mitigation technology. In today's electronic age, many of the industrial process parameters—such as temperature, pH, and flow—are controlled by automated feedback controllers. Only after the introduction of these controllers and the associated reliable sensors for these parameters was it possible to precisely manage them and to either improve the product quality or to produce original products. Unfortunately, the control of corrosion in many industries is still in its "stone age" stage. According to the FHWA study, the annual use of corrosion inhibitors in the United States came to over a billion dollars; the annual cost may be over $4 billion worldwide, if the usage of inhibitors is assumed to be proportional to the GDP in each country. It is rather alarming to realize that nearly all of these inhibitors were added into the controlled systems based on parameters that are not the direct measures of corrosion (indirect parameters) or based on historical results acquired from the test coupons that were exposed in the controlled systems several months earlier. Examples of the indirect parameters include the concentration of the inhibitors and the concentration of dissolved oxygen in the controlled systems. Because of the complexity of corrosion, some concentrations of corrosion inhibitor that are shown to be effective under certain conditions may not be effective under others. In addition, the concentration of the inhibitor sampled from the bulk phase may not represent the actual concentration on the metal surface where corrosion takes place.

With the advancement of corrosion monitoring techniques, corrosion sensors may be used in the feedback controllers—as in the control systems for pH, temperature, and other process parameters—to automatically control the addition of corrosion inhibitors. When such a system is implemented, it will not only provide an adequate control of corrosion—which means a better performance and a longer life for the equipment—but will also create tremendous savings in the inhibitor costs by avoiding overdosing. This, in turn, produces a significant risk reduction for our environment, because many corrosion inhibitors are toxic.

Corrosion monitoring also provides performance data and a basis for life prediction; corrosion monitoring is one of the most important components in corrosion prevention and corrosion control.

1.4 Organization of the book

This book is organized into 25 chapters. Chapter 2 presents an overview of corrosion fundamentals and evaluation techniques. It includes detailed discussions on the different forms of corrosion and their electrochemical characteristics.

Chapter 2 is followed by in-depth discussions on the various methods that can be used for corrosion monitoring (i.e., to measure the corrosion damage to metals or the corrosivity of a surrounding environment with an adequate response time so that the measurements can be made at desired time intervals). The response time is relative to the purpose of monitoring. For example, if a corrosion monitor is interfaced with a corrosion inhibitor dosing-controller and the dosing-controller completes the addition of an inhibitor to a system in a few minutes, the response time of the corrosion monitor should be less than 1 minute. However, if the purpose is to monitor the long-term corrosion damage to a metallic structure in the air near a pollution source, 1 month may be considered a sufficient response time.

Chapters 3 through 8 discuss the electrochemical techniques for corrosion monitoring. These techniques include electrochemical polarization techniques (Chapters 3 and 4), electrochemical noise methods and harmonic analyses (Chapter 5), galvanic sensors (Chapter 6), differential flow through cell technique (Chapter 7), and multi-electrode systems (Chapter 8).

Chapters 9 through 14 describe the gravimetric, electrical, and other physical or chemical methods. These methods include gravimetric techniques (Chapter 9), radioactivity tracer methods (Chapter 10), electrical resistance techniques (Chapter 11), nondestructive evaluation techniques based on ultrasonics, eddy current, guided waves, and infrared thermography (Chapter 12), acoustic techniques (Chapter 13), and hydrogen permeation methods (Chapter 14).

The techniques discussed in Chapter 12 were primarily used as corrosion inspection tools; they are now used more and more as monitoring tools with the advancements of the technologies that constantly improve the measurements precision and response time.

Corrosion is an extremely complicated process involving at least two phases, solid and liquid; solid and gas; solid, liquid, and gas; or even solid, first liquid, and second liquid. Corrosion monitoring is a multidisciplinary task. Very often, two or more methods are needed to adequately address the monitoring needs in a given system. In addition, different systems require different methods or a combination of different methods. Chapters 15 through 21 provide detailed discussions regarding the specific monitoring needs in selected environments or under particular conditions. The topics discussed in these chapters include corrosion monitoring in microbial environments (Chapter 15), corrosion monitoring in concrete (Chapter 16), corrosion monitoring in soil (Chapter 17), corrosion monitoring for high-temperature systems in refineries (Chapter 18), corrosion monitoring under coatings (Chapter 19), and corrosion monitoring for structures under cathodic protection or the effect of stray current (Chapter 20).

Data transmission and management is a vital component of corrosion monitoring in the plants or in the fields and an integrated system health management program. Chapter 21 provides detailed discussions on remote monitoring and computer applications.

Finally, Chapters 22 through 25 present applications of corrosion monitoring techniques in specific systems and case studies. These chapters provide readers with quick answers, when they have a need in a particular area or for a specific method. The topics in these chapters include corrosion monitoring in cooling water systems, using a differential flow through cell (Chapter 22), corrosion monitoring in chemical plants using an electrochemical noise method (Chapter 23), corrosion monitoring under cathodic protection conditions using multielectrode sensor techniques (Chapter 24), and corrosion monitoring with the field signature method (Chapter 25).

This suite of in-depth overviews, systematic technique descriptions, and case studies provides a state-of-the-art coverage for corrosion monitoring. Researchers, engineers/operators, and students alike should find this book to be an invaluable resource in meeting their corrosion monitoring needs or advancing research related to corrosion monitoring.

References

[1] NACE International, International Measures of Prevention, Application, and Economics of Corrosion Technologies Study (IMPACT), NACE International, March 2016.

[2] G.H. Koch, M.P.H. Brongers, N.H. Thompson, Y.P. Virmani, J.H. Payer, Corrosion Cost and Preventive Strategies in the United States, FHWA-RD-01-156, Springfield, VA, National Technical Information Service, 2001.
[3] B. Hou, X. Li, X. Ma, C. Du, D. Zhang, et al., The cost of corrosion in China, NPJ Mater. Degrad. 1 (2017) 4, https://doi.org/10.1038/s41529-017-0005-2.
[4] P.R. Roberge, Corrosion Inspection and Monitoring, John Wiley & Sons, Inc., Hoboken, NJ, 2007 (Chapters 1 and 2).

Corrosion fundamentals and characterization techniques

2

Gustavo A. Cragnolino[†]
Southwest Research Institute, San Antonio, TX, United States

2.1 Introduction

Corrosion is defined as the chemical or electrochemical reaction between a material, usually a metal or alloy, and its environment that produces a deterioration of the material and its properties. According to the characteristics of the environment, corrosion processes are classified as chemical or electrochemical.

Chemical corrosion processes are those in which the metal reacts with a nonelectrolyte (e.g., oxidation in high temperature air, dissolution in liquid metals, or dissolution in a carbon tetrachloride solution containing iodine). Electrochemical corrosion processes are those in which the metal dissolves in an electrolyte forming metal cations which implies the transfer of electric charge across the metal/environment interface. Electrochemical corrosion occurs in the large variety of electrolytes found in natural environments and industrial applications ranging from groundwater to molten salts and acids dissolved in organic polar solvents (e.g., hydrochloric acid in methanol).

For practical reasons, corrosions in different natural environments, such as corrosion in soils, atmospheric corrosion, fresh water corrosion, sea water corrosion, etc., are usually considered separately as is also the case for corrosion in different industries (i.e., chemical processing, oil and gas production, power generation, etc.) [1]. However, besides the specificity of the environments and the materials involved, these corrosion processes have fundamental aspects in common related to the thermodynamics and kinetics of the electrochemical reactions at the metal/electrolyte solution interface and the transport of chemical species in a liquid phase to and from the metallic surface [2]. Several recently published books and handbooks are useful sources for general information on corrosion and its electrochemical basis [3–6]. The electrochemical techniques commonly used in corrosion studies are well-presented in Kelly et al. [7] and corrosion tests and standards for materials in a variety of environments and industrial applications are covered in detail in Baboian [8].

According to the morphology of corrosive attack, different modes of corrosion can be identified through visual observation, requiring in many cases sufficiently enlarged magnification by the use of optical or electronic microscopes. The different modes of corrosion will be presented in the following sections and the most common techniques

[†] Deceased.

Techniques for Corrosion Monitoring. https://doi.org/10.1016/B978-0-08-103003-5.00002-3

used to study and characterize corroded metal surfaces and corrosion products will be briefly introduced.

2.2 General corrosion

General corrosion is the most common and benign form of corrosion because it is characterized by a corrosive attack that extends almost uniformly over the whole exposed surface or at least over a large area. Even though the term uniform corrosion is commonly used synonymously with general corrosion, this form of corrosion is seldom completely uniform and the morphology of the corroded surface always exhibits some sort of waviness and roughness. Nevertheless, the average penetration of the attack is practically the same at each point of the corroded surface. Any corrosion process involves at least one anodic (oxidation) reaction and one cathodic (reduction) reaction.

Because charge accumulation cannot occur, the electrons generated by the oxidation reactions must be consumed by the cathodic reactions and, therefore, the total anodic current should be equal to the total cathodic current with the electrical potential at an anodic site equal to that at a cathodic site. This fundamental concept of coupled anodic and cathodic processes was proposed by Wagner and Traud [9]. The hypothesis leads to the definition of the corrosion potential as a mixed potential because the anodic reaction is the dissolution of the metal coupled to the cathodic reaction which could be the reduction of dissolved oxygen molecules, hydrogen ions, or any other reducible species in solution (e.g., Fe^{3+} ions), instead of the reduction of the dissolved metal cations. From an electrochemical point of view, the main characteristic of general corrosion is the fact that metal dissolution takes place without physical separation, even at the microscopic scale, of anodic and cathodic sites.

General corrosion causes by far the largest amount of material losses as a result of corrosion, mostly due to atmospheric corrosion. Leaving aside the costs involved, general corrosion is not of great concern because the corrosion rate and, hence, the expected life of equipment or structures can be accurately estimated by means of relatively simple corrosion tests. The most critical aspect of testing is, however, the correct definition of the environment in which the component will operate and its unexpected variations during operation. Such tests can be immersion tests, in which gravimetric methods are used for measuring the weight loss of a material specimen during a specific testing time to calculate the corrosion rate [10].

Alternatively, electrochemical tests can be used to determine the corrosion rate (CR), expressed as rate of uniform penetration for metals and alloys, from the measured current density by using Faraday laws, according to Eq. (2.1):

$$CR = iE_{w'}/F\rho \quad (2.1)$$

where i is the current density, $E_{w'}$ is the equivalent weight, F is the Faraday constant, equal to 96,485 C/g-equivalent, and ρ is the density of the metal. $E_{w'}$ can be computed using the following expression

$$E_w = 1/\sum (z_j f_j / A_j) \tag{2.1a}$$

where z_j is the oxidation state of the element j in the alloy, f_j is its weight fraction, and A_j is its atomic weight.

In order to calculate the corrosion rate in mm/yr, Eq. (2.1) can be expressed as

$$CR = 3.27 \times 10^{-3} i E_w / \rho \tag{2.2}$$

with i in $\mu A/cm^2$, E_w in g, and ρ in g/cm^3. The corrosion rate is also commonly given in mils per year (mpy) or in units of weight change per unit area per unit time (e.g., mg/cm^2 s) [11]. Values of E_w for many metals and alloys taking into account the various possible oxidation states are also provided in the same standard (see Chapter 3).

The advantage of electrochemical methods over gravimetric methods is that the instantaneous rather than the average corrosion rate can be measured. The linear polarization resistance (LPR) method (see Chapter 3) and electrochemical impedance spectroscopy (EIS) techniques [7, 8] are widely used to obtain real time information on corrosion rates. Corrosion rates at potentials differing from the corrosion potential can be calculated from current density measurements using potentiostatic methods when interferences from cathodic reactions (e.g., cathodic current associated with the oxygen reduction reaction due to the dissolved oxygen in solutions in equilibrium with air) are carefully eliminated.

As noted above, atmospheric corrosion which is the corrosion of materials exposed to air and its pollutants, such as sulfur compounds and NaCl, occurs mostly in the form of general corrosion. The corrosion rate is extremely dependent on relative humidity and the concentration of pollutants in the air, which leads to the distinction of rural, industrial, and sea coastal atmospheres. Prompted by the spectacular development of the electronic industry, the field of atmospheric corrosion has been extended during the last 30 years to include indoor atmospheres to deal with the corrosion problems of electronic devices. Carbon steels, weathering steels, stainless steels (SS), Cu, and certain Cu alloys are typical materials subject to outdoor atmospheric corrosion in the form of general corrosion as a result of their extensive application as structural and architectural materials [1, 12].

The fundamentals of atmospheric corrosion, as well as substantive discussions on outdoor and indoor atmospheric corrosion, are well-covered by Leygraf and Graedel [13]. These authors distinguish three stages of atmospheric corrosion. The initial stage corresponds to the adsorption of molecules of water vapor on the oxide/hydroxide-covered metal surface by surface hydroxylation followed by the formation of several molecular layers. Above 50% relative humidity, the number of water monolayers increases abruptly from two to more than five. The intermediate stages involve the absorption and dissolution of gases (i.e., O_2, CO_2, SO_2, HCl, H_2S) or salts as particulates (i.e., NaCl, $(NH_4)_2SO_4$, $(NH_4)HSO_4$, $(NH_4)Cl$, Na_2SO_4) in the adsorbed water layers followed by chemical reactions and subsequent electrochemical processes of proton and ligand-induced metal dissolution with the nucleation of corrosion products. The final stages are the coalescence, aging, and thickening of corrosion products.

General corrosion is a common occurrence in many industrial applications. In the chemical processing industries in which more corrosion-resistant materials are generally used, the environments include inorganic and organic acids, and also a large number of chemical species, over wide ranges of concentrations and temperatures. A large body of information is usually collected for each chemical (e.g., H_2SO_4, HNO_3, HCl, HF, organic acids, etc.) as isocorrosion maps which are temperature vs concentration plots bound by the respective boiling point curve of the chemical defining regions with different values of the uniform corrosion rate for each alloy of interest [1, 6, 12]. Well-established monitoring techniques are commonly used in these applications [14, 15].

2.3 Passivity and localized corrosion

A metal surface is considered passive when, although it is exposed to an environment under conditions such that dissolution of the metal could be expected from thermodynamic considerations, it remains essentially unchanged with time. The phenomenon of passivity and the breakdown of passivity can be illustrated by considering the schematic anodic polarization curve of metals in an aqueous environment obtained under deaerated conditions, as shown in Fig. 2.1, in which the cathodic reaction is assumed to be the reduction of hydrogen ions.

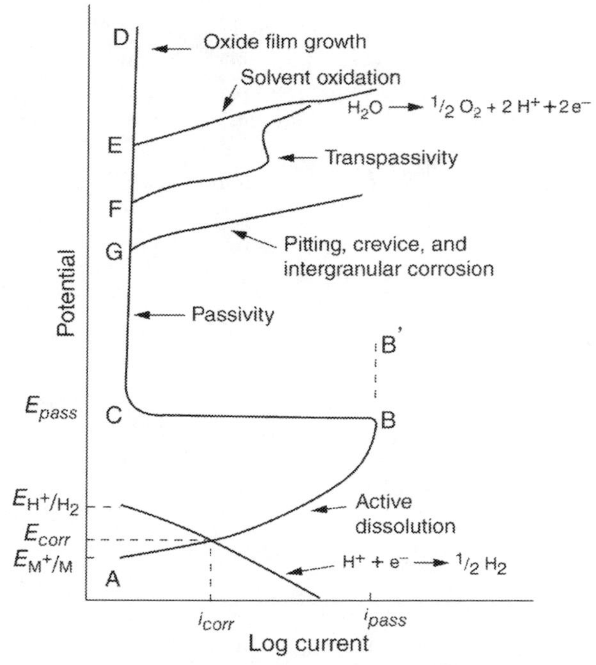

Fig. 2.1 Schematic polarization curve indicating possible variations of the anodic behavior of metals in aqueous solutions.

If the potential is increased in the positive direction from the corrosion potential, E_{corr}, the current will increase initially following a linear relation between the logarithm of the current and the potential (Tafel region) until an approximately constant current is reached due to diffusion limitations or formation of a surface layer of corrosion products. The curve from A to B (only accessible from E_{corr}) is characteristic of the active dissolution of the metal. If the potential is increased further, the current may remain constant and high (B to B′) or decrease significantly by several orders of magnitude (B to C), due to the formation of a passive layer, usually a very thin and protective oxide film. The potential at which this process occurs is called passivation potential, E_{pass}, and the current at B passivation current, i_{pass}. By increasing the potential to even higher values, the thickness of the oxide film will increase, but the current in the passive range, the passive current i_p, will remain essentially constant. If the film is a poor electronic conductor (e.g., Al, Zr, and Ta), very high potentials can be reached, as shown by the curve C to D. If the oxide film is a good electronic conductor (e.g., Fe and Ni), a significant current increase will occur at a certain potential (E) associated with the electrochemical decomposition of the solvent, which in aqueous systems corresponds to the evolution of O_2 as a result of water oxidation. However, if the oxide film is composed by cations which can be oxidized to a higher oxidation state forming soluble products (e.g., Cr, Mo, or alloys such as stainless steels), a current increase may occur at lower anodic potentials (F), accompanied by the dissolution of the metal as a result of transpassivity, in some cases followed by a region of secondary passivity.

Because most of the metals and alloys of industrial importance corrode under passive conditions, the preservation of the passive film in aggressive environments is critical to the life of industrial equipment and components. Under natural corroding conditions, the E_{corr} may vary up to values lower than E as a result of the presence of reducible species in solution, mainly dissolved O_2 in naturally aerated systems. Only in specific, highly oxidizing environments, E_{corr} values as high as F can be attained (e.g., SS in concentrated HNO_3) and transpassive dissolution occurs. In certain applications, the passive condition can be obtained by imposing a potential within the passive range (C to G) through potentiostatic polarization and anodic protection is attained.

For many metals and alloys exhibiting a passive behavior, the passive film becomes locally unstable above a critical potential in solutions containing halide anions, in particular chloride. In the polarization curve shown in Fig. 2.1, a sudden increase in the current is observed above G corresponding to the breakdown of passivity which may lead to various forms of localized corrosion. Depending upon the factors involved in the passivity breakdown and the morphology of the subsequent attack, localized corrosion can be classified as pitting or crevice corrosion if a chemical micro- or macroheterogeneity is developed at the metal/solution interface. If a chemical microheterogeneity exists in the metal, intergranular attack or selective dissolution can occur. Stress corrosion cracking or corrosion fatigue can occur at such potential if dynamic mechanical factors are involved in the passivity breakdown. From an electrochemical point of view, the main characteristic of localized corrosion is the physical separation of anodic and cathodic areas. Under natural corroding conditions, the kinetics of the cathodic reaction on the remaining passive surface, the conductivity of the solution, and the size of the cathodic area are the factors limiting the rate of the localized attack.

Although passivity has been attributed to the formation of an adsorbed layer of dissociated O_2 molecules or to electronic modification of the metal surface, nowadays there is a general agreement that passivity is caused by the formation of a three-dimensional oxide film. Passivity is initiated in certain circumstances by isolated oxide nuclei which grow and spread laterally. Numerous models have been proposed for the growth of passive films. Two of the most well-known models are those developed by Cabrera and Mott [16] and, more recently, the point-defect model [17]. Nevertheless, despite significant experimental progress in recent years, there is still considerable controversy regarding the nature, composition, structure, and long-term stability of passive films in many alloy/environments systems.

2.3.1 Galvanic corrosion

Galvanic corrosion (GC) occurs when two or more metallic materials having different reversible electrode potentials are coupled together and exposed to a corrosive solution or the atmosphere. Due to the metallic contact and the ionic conduction in the electrolyte, a current flows from one metal to the other, resulting in GC of the anodic (negative) member of the couple. Although in many cases GC is morphologically uniform, the most damaging situations occur when it is localized to a limited area. A large cathodic area in contact with a small anodic area is particularly detrimental in promoting this specific form of localized corrosion. In addition to its treatment in corrosion textbooks, important aspects of GC have been reviewed in recent publications by Zhang [18] and Hack [19].

The following factors determine the extent of corrosion experienced by the anodic member of the couple: (a) the E_{corr} of the metals forming the galvanic couple; (b) the nature and kinetics of the cathodic reactions (e.g., O_2 and H^+ reductions) on the more positive metal and the anodic dissolution reactions on the more negative metal; (c) geometrical factors such as the relative areas and spatial positions of the dissimilar metals; (d) the electrolyte solution properties, including concentration of ionic species, pH, temperature, and particularly conductivity.

In many circumstances, the polarity of galvanic couples is different from that expected from the reversible electrode potentials because the corrosion potentials are mixed potentials determined by the kinetics of the anodic and cathodic reactions in the specific environment. In addition, there is more than one anodic oxidation reaction that takes place in the case of alloys. As a result, a specific galvanic series including metals and alloys should exist for each environment instead of the overextended, and sometimes erroneous, use of the galvanic series in sea water for other environments, even though such series can be used with certain confidence in environments similar to sea water.

As illustrated in Fig. 2.2, the following expression provides the relationship between the driving force of the galvanic couple and the galvanic current, I_g

$$\left(E_{corr}^{c} - \eta_c\right) - \left(E_{corr}^{a} + \eta_a\right) = I_g R_s \tag{2.3}$$

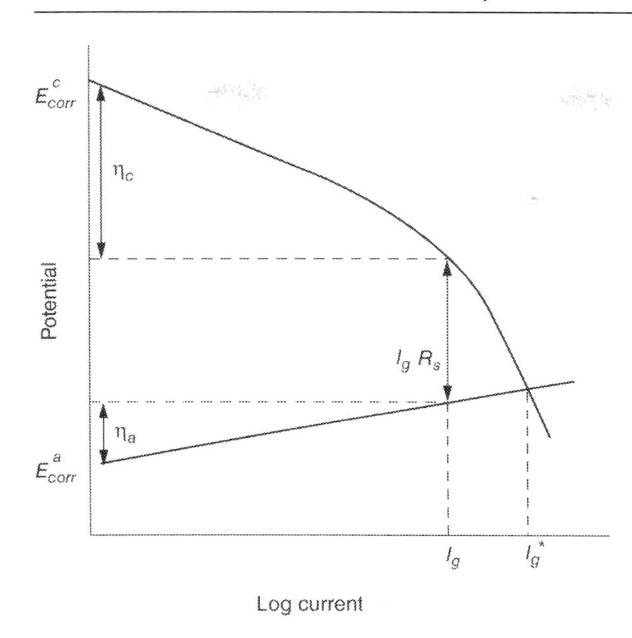

Fig. 2.2 Schematic plot showing the effects of the anodic and cathodic polarization of the members of a galvanic couple and the solution resistance on the galvanic current.

where E^{c}_{corr} and E^{a}_{corr} are the corrosion potentials, and η_c and η_a, the polarization of the cathodic and anodic members of the couple, respectively, and R_s is the resistance of the electrolyte solution in the galvanic circuit. It should be noted that the magnitude of both η_c and η_a depends on the cathodic and anodic polarization curves and on the value of R_s. If R_s decreases, I_g increases until a limiting I_g^* is reached when R_s tends to zero. Obviously, the anodic and polarization curves depend on the kinetics of the electrochemical reactions on each metal in the solution, and hence, are functions of the solution composition, pH and temperature, as well as of the surface conditions of the metals.

When the cathode exhibits strong polarization (large η_c) and the anode does not, the couple is under cathodic control. On the other hand, the couple is under anodic control when the anode displays strong polarization (large η_a). If the resistance controls the current flow, the couple is under resistance control, but most commonly all these three factors affect I_g, and hence, mixed control prevails.

According to Faraday laws, I_g is directly proportional to the corrosion rate of the anode in the galvanic couple. Physical contact between the members of the couple prevents the direct measurement of I_g. However, the I_g can be measured using two isolated specimens of the galvanic couple exposed to the environment of interest and connected through a zero resistance ammeter [20]. Other methods to evaluate GC are discussed by Hack [19].

The galvanic action between two metals is governed by the potential distribution across each electrode surface from which the galvanic current distribution can be calculated. Even though the potential distribution can be obtained by experimental methods (e.g., scanning a reference electrode over the surface of the metallic couple), in most situations this approach is not practical due to the complex configuration of the system. A semiquantitative approach to characterize the current distribution for a given metal/electrolyte combination was proposed by Wagner [21] using a polarization parameter defined as

$$W_p = \sigma[\mathrm{d}\eta_i/\mathrm{d}i_i] \tag{2.4}$$

where σ is the conductivity of the solution, and η_i is the overpotential of the cathode or the anode, and i_i is the current density, and therefore, W_p has dimensions of length. To scale-up a system, either for modeling or experimental simulations, the ratio of W_p to a characteristic length of the system, named the Wagner number, W_n, should be kept constant. Currently, the prevailing approach is based on the development of numerical models for different galvanic couples and geometries, as summarized by Zhang [18]. These models are used to evaluate galvanic corrosion, but also to predict the effect of cathodic protection using sacrificial anodes.

2.3.2 Pitting corrosion

Pitting corrosion (PC) is a form of localized attack morphologically characterized by the development of small cavities usually on openly exposed surfaces of a metal. The surface diameter of these cavities is about the same or less than the depth in most cases, but different pit shapes can be observed depending on the metal or alloy and the environment. Changes in the hydrodynamic conditions or different orientations of the metal surface can alter significantly the pit morphology. The typical appearance of a pit is shown in Fig. 2.3. Under certain conditions, the bottom of the pits may exhibit crystallographic facets, whereas in other cases round electropolished bottoms are observed, generally covered with precipitated corrosion products.

PC occurs by the action of aggressive anions on metals and alloys exhibiting a passive behavior when exposed to environments ranging from slightly acidic to alkaline. Cl^- is the anionic species most commonly associated with PC, in part due to its wide distribution in natural waters. Other anions such as Br^-, I^-, SO_4^{2-}, and ClO_4^- can also promote PC. Equipment failures due to perforation by PC occur without significant weight loss of the entire structure or component because the rate of dissolution in a pit can be 10^4 to 10^6 times faster than that on the rest of the surface [22].

There are numerous papers in the literature and recent conferences in which phenomenological observations as well as mechanistic interpretations of PC have been thoroughly discussed [23–26]. A revised second edition of a book mostly devoted to this subject has been published recently [27].

It has been demonstrated that stable pit growth only occurs above a certain critical potential which depends on the particular metal/anion system. The presence of an aggressive anion in the environment is a necessary, but not a sufficient, condition for the occurrence of PC, as clearly demonstrated by experiments in which the passive film is mechanically disrupted by scratching or straining at potentials below the

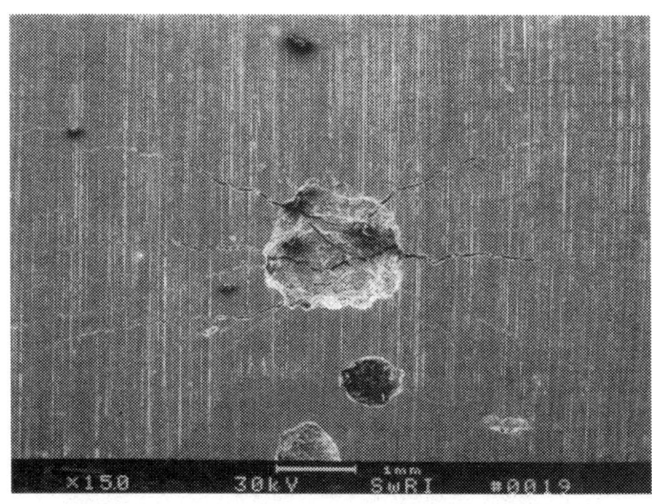

Fig. 2.3 Micrograph showing the appearance of a typical pit of AISI 316 L stainless steel exposed to a hot chloride solution. Cracks are emanating from the pit due to the presence of stress.

critical potential and repassivation occurs. Within such anodic range, metastable pits can be generated, but they do not grow to become stable pits and rapidly die. Only at potentials above the critical potential do metastable pits become stable. It should be noted, however, that pit generation is a stochastic process, and therefore, a distribution of potentials exist above a minimum value that can be considered the "true" pitting potential. This lower potential usually coincides with the protection potential or repassivation potential, which is the potential at which pit growth is arrested.

The pitting initiation potential, E_{pit}, is usually measured potentiodynamically by scanning the potential upward from the corrosion potential or potentiostatically by stepping the potential until a steady state current is attained at each step [28]. E_{pit} corresponds to an abrupt increase in the current density, as illustrated in the schematic anodic polarization curve shown in Fig. 2.4. When the potential is scanned or stepped backward, there is usually an abrupt decrease in the current density corresponding to the arrest of the pit growth at the pitting repassivation potential, E_{rpit}, as also shown in Fig. 2.4.

Under natural corroding conditions, PC occurs when E_{corr} is higher than the "true" E_{pit}. In many metals and alloys (e.g., Al and its alloys), this occurs when the oxidant in the aqueous environment is O_2 in equilibrium with air. In more corrosion-resistant materials, oxidizing cations such as Fe^{3+} are required to attain sufficiently high E_{corr}. Indeed, Fe_3Cl solutions are used in a standard test to evaluate the susceptibility of stainless steel and related alloys to PC [29].

The E_{pit} depends, in addition to the material composition, heat treatment, and surface condition, on the solution composition and temperature. For many metals and alloys, the following relationship exists between the E_{pit} and the concentration

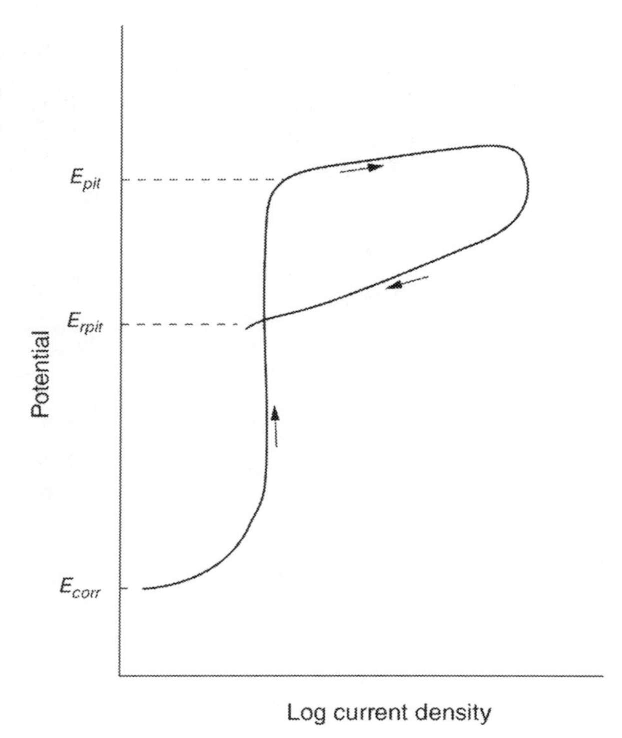

Fig. 2.4 Schematic cyclic potentiodynamic anodic polarization curve showing the pitting initiation potential, E_{pit}, and the pitting repassivation potential, E_{rpit}.

of aggressive anions (e.g., Cl^-, Br^-, I^-) and inhibiting anions (e.g., $CO_3^{2-}, B_4O_7^{2-}, PO_4^{3-}$)

$$E_{pit} = E^0_{pit} - B \log [X^-] + C \log [A^-] \tag{2.5}$$

where E^0_{pit}, B, and C are constants, which depend on the metal/solution system and X^- and A^- are the aggressive and inhibiting anions, respectively. A similar expression exists for E_{rpit}, but with different values for E^0_{rpit}, B, and C.

Some basic aspects of the PC mechanism are well-established. Under natural corroding conditions, rapid anodic dissolution occurs at a pit nucleus, while O_2 reduction takes place on the adjacent passive surface providing the driving force for the anodic reaction. The metal cations react with water forming hydroxo-complexes and H^+ ions, thereby promoting localized acidification inside the pit. The excess of positive charge is counterbalanced by the migration of Cl^- ions creating a very aggressive environment, which stimulates further the local dissolution of the metal in a rather autocatalytic fashion. Other reducible species (i.e., Fe^{3+}, Cu^{2+}, and H_2O_2) can play the same role as O_2.

The lack of dependence of E_{pit} with pH can be explained because the pH inside the pits is strongly acidic, regardless of the bulk pH. In addition, PC inhibitors which are usually the salt of a weak acid or the OH^- anion will increase E_{pit} by consuming H^+ ions and thereby reducing the localized acidification inside the pit, according to the following reaction

$$H^+ + A^- \rightarrow HA \tag{2.6}$$

Galvele [30] developed a mechanistic model based on these concepts and concluded that E_{pit} can be calculated by using the following equation:

$$E_{pit} = E^*_{corr} + \eta + \Phi + E_{inh} \tag{2.7}$$

where E^*_{corr} is the corrosion potential measured in an acidified pit-like solution, η is the overpotential necessary to attain the current density required to maintain the acidic pH inside the pit (which is characteristic of each metal or alloy), Φ is the ohmic drop inside the pit, and E_{inh} the increase in E_{pit} due to the action of the inhibitor (the last term in Eq. 2.5). The current density necessary to maintain the pit actively growing multiplied by the depth of the pit ($i \cdot x$) must be higher than a certain value characteristic of each metal/solution system and the value of η is calculated by assuming Tafel behavior in the pit-like solution. Laycock and Newman [31] developed a rather similar expression (without the last term in Eq. 2.7) for the transition potential between metastable and stable pits by introducing the concept of diffusion-controlled (salt-covered) pit growth. This transition potential is conceived as a "true" E_{pit}.

2.3.3 Crevice corrosion

Crevice corrosion (CC) is a form of localized corrosion that occurs within crevices and other shielded areas where a small volume of a stagnant solution is present. Such crevices can be formed at metal/metal or metal/nonmetal junctions, such as those associated with gaskets, valve seats, rivet and bolt heads and lap joints, as well as under surface deposits (i.e., corrosion products, sand and dirt) or marine biofouling. Crevice corrosion is mainly observed on passive metals and alloys covered with protective oxide films, such as stainless steels, Ti alloys, and Ni-base alloys immersed in aerated aqueous environments containing Cl^- ions (e.g., sea and brackish water).

The morphology of the attack inside the crevice could vary from pitting to a more uniform corrosion. The geometrical dimensions of the crevice are critical, particularly the crevice gap, because the gap must be wide enough to allow entry of the solution, but sufficiently narrow to maintain an occluded environment inside the crevice. In addition, the depth and the exterior/interior area ratio are important factors in determining the severity of the attack. The typical appearance of crevice corrosion is shown in Fig. 2.5.

There are valuable reviews on CC [27, 32] where experimental observations and fundamental concepts are discussed. The development of alloys more resistant to PC led in recent years to place more emphasis on the study of CC. This phenomenon

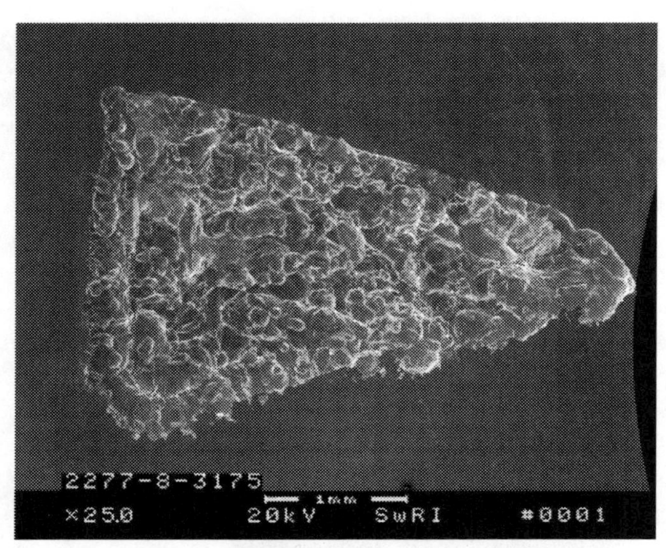

Fig. 2.5 Micrograph showing an example of crevice corrosion. Attack on alloy C-22 under the feet of a crevice formerly exposed to a hot chloride solution.

occurs when the local environment inside the crevice becomes sufficiently aggressive to induce passivity breakdown, either by the generation of pits or a more generalized depassivation. In the occluded environment where mass transport by diffusion and convection is severely limited, hydrolysis of the dissolved metal cations coupled with the buildup of the Cl^- concentration required to maintain electroneutrality leads to the formation of a solution with a critical composition in terms of Cl^- concentration and pH. As soon as the metal in the crevice becomes active, the E_{corr} inside the crevice decreases. This activation is accompanied by a substantial ohmic drop, equal to the product of the current by the solution resistance, due to the difference between E_{corr} inside and outside the crevice. From a mechanistic point of view, there are many similarities between PC and CC and, therefore, the condition for a crevice to be active is that the current density multiplied by the depth of the crevice reaches a certain critical value [33, 34], as originally proposed by Galvele [30].

A thorough review of tests developed to evaluate the CC resistance of alloys was presented by Sridhar et al. [35]. Standard tests in Fe_3Cl solutions in which an oxidizing species (i.e., Fe^{3+} ions) is used in addition to Cl^- to accelerate the initiation of CC are reviewed, as well as electrochemical tests conducted under either open circuit or constant applied potentials. The use of critical potentials and critical temperatures for the initiation of CC as tools for the evaluation and ranking of different alloys is discussed in some detail (Sridhar et al., 2005).

Similarly to PC, a critical potential for the initiation of CC, E_{crev}, exists, but its value exhibits an ill-defined dependence on crevice geometry. As in the case of PC, a distribution of potentials is observed. The most reliable parameter to assess the resistance to CC of passive alloys, particularly SS or Ni-Cr-Mo alloys, in a variety of aggressive environments containing Cl^- anions is the repassivation potential for crevice corrosion, E_{rcrev}, because it is independent of crevice depth [36, 37].

Dunn et al. [38, 39] found that for a given alloy with a specific heat treatment, E_{rcrev} can be expressed as:

$$E_{rcrev} = A_1 + A_2T + (B_1 + B_2T)\log[Cl^-] \tag{2.8}$$

where the coefficients A_1, A_2, B_1, and B_2 are independent of the temperature T. Even though the expression and the coefficients were determined for Alloy 22, similar expressions with the corresponding coefficients can be obtained for other alloys. The expression has been extended to include effect of inhibitors, such as NO_3^-, SO_4^{2-}, and CO_3^{2-}. A semiempirical model has recently been developed by Anderko et al. [40] to account for the effect of Cl^- concentration on E_{rcrev} in the presence of complexing species in solution.

2.3.4 Dealloying

Dealloying is the selective removal by electrochemical dissolution of the most active metal from an alloy under specific environmental conditions. It may occur either as localized corrosion accompanied by perforation (plug type) or as a more uniform attack (layer type). One of the oldest cases reported, the removal of Zn from the brasses (Cu-Zn alloys) is known as dezincification. The phenomenon is also termed selective leaching or parting. Many examples have been reported, such as the removal of Ni from Cu-Ni alloys, Al from Al-bronzes, Sn from Cu-Sn alloys, and Cu from Cu-Au alloys, among others. Dealloying has been extensively studied and there are recent reviews covering the subject such as those by Corcoran [41] and Van Orden [42] in which many alloy/environment systems are discussed.

As noted by Corcoran [41], dealloying can occur in principle in any system where a large difference exists between the reversible electrode potentials of the alloy elements. Fig. 2.6 is a schematic polarization curve for a binary A_pB_{1-p} alloy in which a passivity range is indicated, followed by a substantial increase in the current density at the critical potential for dealloying, E_{crit}. At this potential, selective dissolution of A, the most active element of the alloy, occurs. It should be noted that E_{crit}, even though it is essentially determined by the composition of both the alloy and the electrolyte solution, is affected by the potential scan rate and the prior history of the alloy surface. Above E_{crit}, the morphology of the dealloyed metal surface is characterized by a porous structure (sponge) enriched in the more noble alloying element. The structure consists of metal-void phases with a pore spacing of few nanometers which can be coarsened by annealing at higher temperatures.

Originally, there were two mechanisms proposed to explain dealloying. The first mechanism—the dissolution-redeposition mechanism—suggested that both elements of the alloy dissolve and the more noble elements redeposit on the surface. The second mechanism postulated that only the active element is selectively dissolved from the alloy. A fundamental issue in this case relates to the process by which dealloying

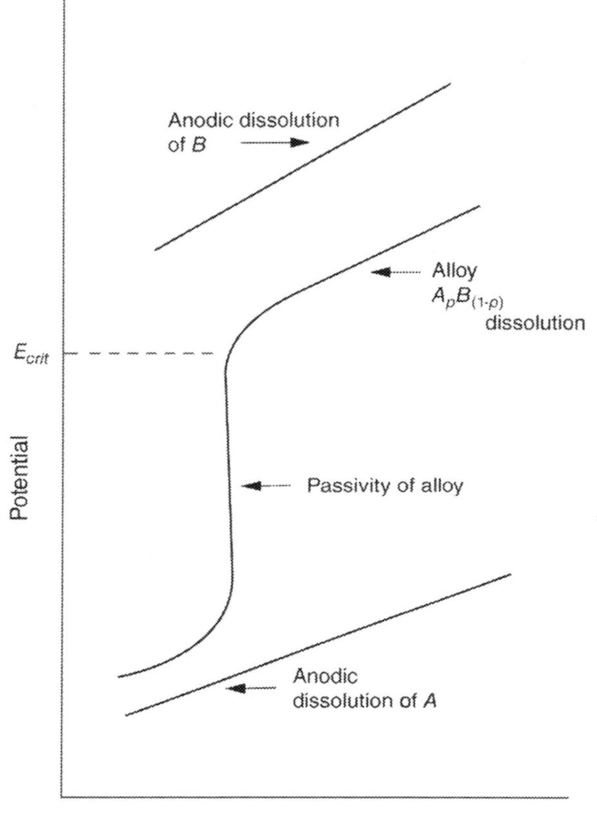

Fig. 2.6 Schematic anodic polarization curve of a binary alloy A_pB_{1-p}, where B is the more noble alloy element, showing the range of passivity and the critical potential for dealloying, E_{crit}.

is sustained over more than a few atomic layers. Some authors suggested that dealloying is supported by solid-state diffusion of both atoms in the alloy to assure the necessary supply of electroactive solute atoms at the reactive layer. Others proposed that enhanced surface diffusion of the noble metal ad-atoms and surface restructuring allow continuous dissolution of the most active metal through a porous layer that progresses in depth. A recent extension of this last mechanism is the percolation model which accounts for preexisting interconnected paths in the binary alloy in order to explain that an alloy compositional threshold exists below which dealloying will not occur for a given alloy system. Van Orden [42] has discussed some techniques used to study and evaluate dealloying, indicating that there are no standards as have been developed for other localized corrosion processes.

2.3.5 Intergranular corrosion

Intergranular corrosion (IGC) is a form of localized corrosion characterized by preferential corrosion at grain boundaries or areas adjacent to them, with little or negligible attack on the grains. Similarly to other forms of localized corrosion, it mainly occurs on passive alloys exposed to specific corrodents. IGC of commercial alloys is generally caused by enrichment or depletion of alloying elements in the area adjacent to the grain boundaries, by intergranular precipitation of second-phase particles or by the presence of alloy impurities segregated at the grain boundaries.

Most alloys, after being submitted to specific heat treatments or as a result of fabrication processes, experience IGC when exposed to an appropriate environment. A large number of cases involve Fe-Ni-Cr alloys, either Fe-based or Ni-based, particularly austenitic stainless steels (SS) due to their widespread use in many industrial applications. Austenitic SSs, such as AISI 304 (UNS S30400), after being slowly cooled through the temperature range of 850 to 550°C, might become susceptible to IGC in relatively benign environments. The phenomenon is called "sensitization" to indicate that the alloy is sensitive to grain boundary attack. Sensitization may occur as a result of various situations: (a) slow cooling from the annealing temperature, which could be the case in heavy section components; (b) stress relieving in the sensitization range, which is possible when ferritic steels that require such treatment are welded to austenitic steel parts that become sensitized; (c) welding operations, which is by far the most common cause of sensitization. The failure of AISI 304 and 316 (UNS S31600) SS components due to IGC in the heat-affected zone (HAZ) of the weld, the so-called weld decay, has been a problem in many industrial applications.

Sensitization of austenitic Fe-Cr-Ni alloys is caused by precipitation of Cr-rich carbides at grain boundaries, accompanied by Cr depletion of the regions adjacent to the carbides to levels below those required for passivation. Within the temperature range of sensitization, C diffuses toward the grain boundaries quite readily, whereas the bulk diffusion of Cr from the austenitic matrix to the depleted region is too slow to allow replenishment. Since at least 12 wt% Cr is necessary to preserve passivity in an acidic medium, depletion of Cr below such level leads to IGC. However, if a sensitized austenitic SS is held long enough at the sensitization temperature, it becomes desensitized because Cr diffusion from the bulk replenishes the Cr-depleted region, even though the carbides are still precipitated. Austenitic SSs become less susceptible to sensitization, and hence intergranular corrosion, by decreasing the C content or by adding alloying elements such as Ti or Nb which are stronger carbide formers than Cr.

Ferritic and duplex SSs are also subject to sensitization, but their susceptibility is quite different. Ferritic SSs are sensitized at temperatures above 925°C by Cr depletion of the matrix in the vicinity of precipitated carbides and nitrides at grain boundaries. Even though the rate of sensitization is faster than that of the austenitic SSs, ferritic SSs are easily desensitized at about 650°C because the diffusion of Cr and C are faster than that in the austenitic phase. Duplex SSs, which contain austenite and ferrite as constituent phases, are far less susceptible to sensitization because typically the C content is lower than 0.03 wt%. These alloys are more prone to exhibit precipitation of intermetallic phases, such as σ and χ, by slow cooling through the

900 to 700°C range due to their relatively high Cr and Mo content. However, the effect of these intermetallic precipitates is more pronounced in terms of impact properties than on corrosion. If the C content is higher, preferential precipitation of carbides at ferrite/austenite boundaries makes the duplex SS less prone to IGC because the depleted zone in the austenite phase can be easily replenished.

In terms of IGC, Ni-based alloys can be divided into two groups: Ni-Cr-Fe alloys such as Alloys 600 (UNS N06600) and 800 (UNS N08800) and Ni-Cr-Mo alloys such as C-276 (UNS N10276) and C-22 (UNS N06022). The first group is also prone to IGC as a result of sensitization. Even though the C content is usually lower (0.02–0.05 wt%) than in the AISI 304 or 316 SS, the solubility of C is also lower in the Ni-base alloys and shorter heat treatments induce sensitization, and therefore, IGC in acidic environments. In the case of Ni-Cr-Mo alloys such as C-22, preferential precipitation of intermetallic phases (μ and P) occurs upon heat treatment at temperatures around 600°C. These intermetallics, rich in Cr and Mo, promote the depletion of both alloying elements in the region around them and facilitate IGC in acidic environments. An example of IGC of Alloy 22 is shown in Fig. 2.7. IGC occurred in this case inside a crevice in which strongly acidic conditions prevailed.

Segregation of impurities such as S or P to grain boundaries is also the cause of IGC. The IGC of austenitic SS in hot concentrated HNO_3 has been attributed to the segregation of P to grain boundaries, whereas the segregation of S to the grain boundaries of Ni is the cause of IGC in H_2SO_4 solutions, in both cases at high oxidizing potentials. Among others, models of sensitization and segregation for Fe-Ni-Cr alloys and steels that account for the most important experimental observations have been developed by Bruemmer [43] and Guttman and McLean [44], respectively.

Aluminum alloys are also susceptible to IGC in Cl^- solutions as a result of certain thermal aging treatments that promote precipitation of intermetallic phases along grain boundaries. The associated depletion of noble alloying elements (e.g., Cu) or the enrichment of the active ones (e.g., Zn) facilitates the preferential dissolution

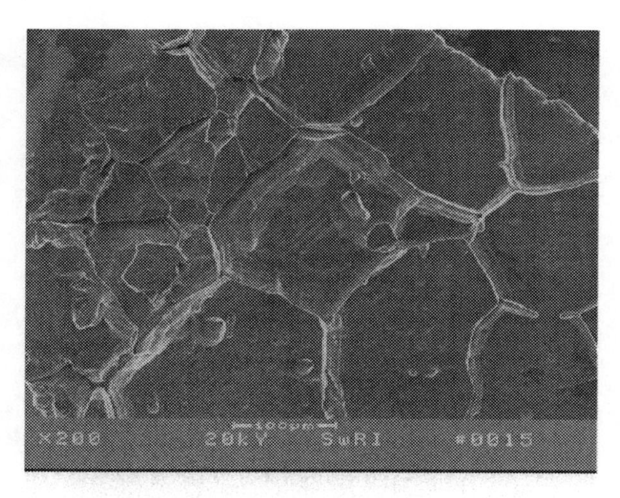

Fig. 2.7 Micrographs showing an example of intergranular corrosion. Attack on alloy C-22 observed inside a crevice where acidified and concentrated chloride solutions exist.

of the depleted area or the intermetallic phases, respectively, at potentials above the E_{pit} of each specific phase.

There are many standard chemical and electrochemical tests for evaluating the susceptibility of alloys to IGC. A good review of the standards and test methods for Fe-Cr-Ni alloys is presented by Streicher [45]. Standards also exist for heat-treatable Al alloys [46, 47].

2.4 Microbially influenced corrosion

Corrosion processes mediated by the action of microbes and their metabolic products are defined as microbially influenced corrosion (MIC). Whereas it is widely recognized that many types of aerobic and anaerobic bacteria are involved in MIC, their action is essentially related to the alteration of the local environmental and surface conditions facilitating the electrochemical reactions responsible for corrosion. Bacteria and other microorganisms, such as algae and fungi, are present in virtually all natural aqueous environments, as well as in industrial fluids and wastewaters.

There is abundant literature on the deterioration of metallic materials since the anaerobic corrosion of cast iron by sulfate reducing bacteria was reported more than 70 years ago [48]. Among recent books, conference proceedings, and handbook chapters, the following ones by Geesey et al. [49], Borenstein [50], Angell et al. [51], Videla [52], Little et al. [53,54], Gu et al. [55], and Dexter [56] deserve to be mentioned. In recent years, it has been recognized that there are types of bacteria that, under certain conditions, can inhibit corrosion [57].

The initial stage of MIC is the formation of a biofilm on the metallic surface exposed to an aqueous environment. Biofilms are complex assemblages of physiologically distinct microbial species that interact to maximize their survival in that environment. A matrix of extracellular polymeric substances (EPS) excreted by the bacteria anchors them to the substratum trapping essential nutrients and buffering any fluctuations in pH, toxic metals, biocides, etc. that may affect the viability and activity of the bacterial colony.

The presence of biofilms and their related heterogeneity creates microenvironments on the exposed surface, thereby affecting the mode and rate of corrosion that in most MIC cases becomes localized. An example of PC promoted by bacteria is shown in Fig. 2.8. According to Dexter [56], bacterial actions promoting MIC comprise: (a) production of organic and inorganic acids as metabolic products; (b) production of sulfur reduced species (e.g., sulfides) under anaerobic conditions; (c) introduction of new redox reactions; and (d) production of oxygen or generation of concentration cells.

It is beyond the scope of this chapter to discuss the specific aspects of all these actions and the type of bacteria involved. The publications listed above provide sufficient information. Nevertheless, it is important to emphasize the role of sulfate reducing bacteria which are able to promote localized corrosion of steels and other alloys under anaerobic conditions, even though the bulk environment could be aerobic because microenvironments depleted of oxygen can exist under biodeposits of

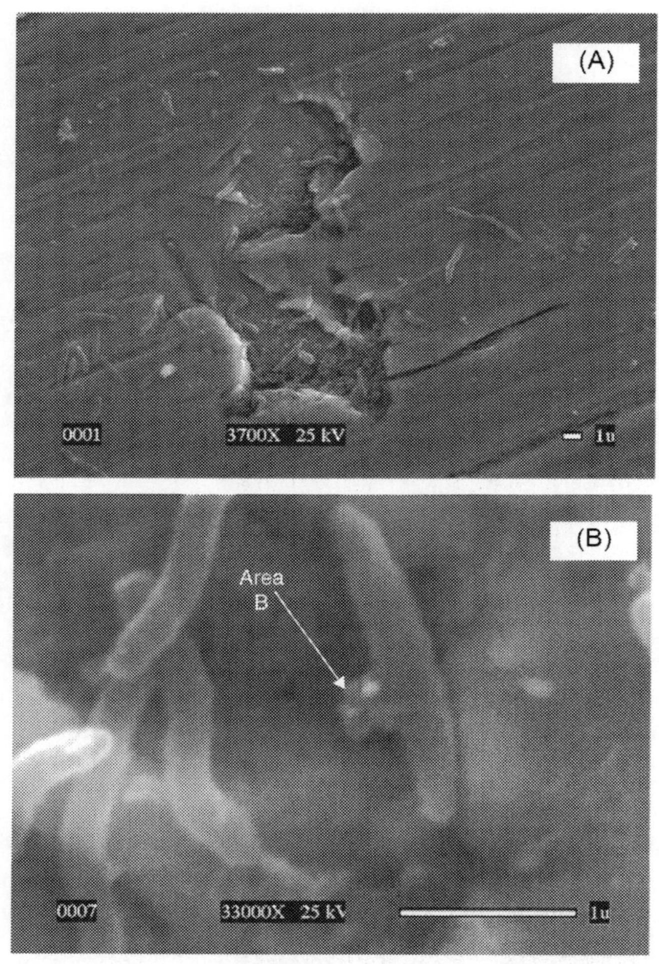

Fig. 2.8 Micrographs showing: (A) Typical pitting corrosion of AISI 304 L stainless steel promoted by sulfate reducing bacteria; and (B) bacteria cells in a close-up view of the lower pit.

anaerobic microorganisms. This example illustrates the complexities of monitoring environmental conditions that may lead to MIC. The use of different techniques for this purpose has been discussed [58]. The application of electrochemical methods and tests to investigate MIC has recently been reviewed by Little and Wagner [59] and by Dexter [57].

2.5 Flow-assisted corrosion and erosion corrosion

These two forms of corrosion are discussed in this section because they are closely related due to the preponderance of hydrodynamic factors on their occurrence. Whereas in some publications [3, 8] no clear distinction is established between these two phenomena and both are considered as erosion corrosion, other authors (e.g., Ref. [60]) emphasize their differences. According to Efird, flow-assisted corrosion (FAC) or flow-influenced corrosion refers to corrosion significantly enhanced by the fast movement of a single-phase fluid. Erosion corrosion (EC) is described as the additional effect exercised by the presence of entrained second-phase particles in a moving fluid, like liquid droplets entrained in a gas or solid particles suspended in a liquid or gas.

In this chapter, both phenomena are considered jointly and the specific differences will be noted. As shown schematically in Fig. 2.9, the rate of corrosion within the FAC regime is accelerated by fluid flow by increasing the rate of mass transport of reactive species to the metal surface or the rate of removal of corrosion products from such surface. At a critical flow velocity, sometimes called the breakaway or erosional velocity, a significant increase in the corrosion rate occurs well beyond the transition from laminar to turbulent flow. The critical flow velocity, V_c, for the occurrence of EC depends on the corrosion behavior of the material in the environment and the hydrodynamic parameters. It should be noted, however, that for many alloy/environment systems the boundary between both types of phenomenon is not as well-defined as suggested by Fig. 2.9.

Below V_c, the corrosion rate increases slowly due to the increase in the rate of transport of electroactive species. The limiting current density, directly related

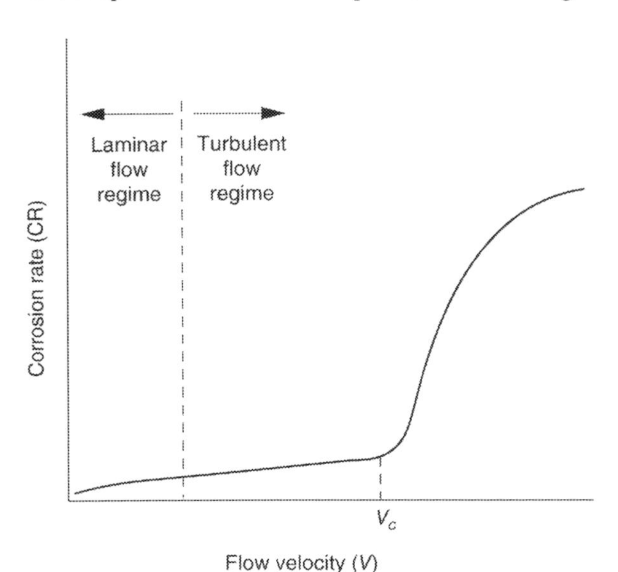

Fig. 2.9 Schematic plot showing the effect of flow velocity on corrosion rate and its significant increase above the critical velocity, V_c, for erosion corrosion.

to the corrosion rate by the Faraday laws, is given by the following general expression

$$i_L = YSc^a \mathrm{Re}^b \tag{2.9}$$

where Y, a, and b are constants, and Sc and Re are the Schmidt and Reynolds number, respectively, as given by

$$Sc = D/v \tag{2.10}$$

$$\mathrm{Re} = dV/v \tag{2.11}$$

where D is the diffusion coefficient, v is the kinematic viscosity, d the characteristic dimension, and V the fluid velocity. Eq. (2.9) is valid for many geometrical configurations (e.g., tubular, annular, rotating disk, rotating cylinder, etc.) under laminar, transitional, and turbulent flow conditions. The values of Y, a, and b for such geometries and conditions are available. The limiting current, and hence, the corrosion rate below V_c usually increases with increasing V according to V^b with $0.33 < b < 0.5$, exhibiting the largest b value in the turbulent regime.

If V is sufficiently high, the shear force at the interface may become sufficiently large to mechanically remove any protective film from the surface, leading to the accelerated corrosion depicted in Fig. 2.9 (i.e., the exponent b becomes higher than 1.0). Although the attack could be localized in regions of high turbulence where reverse flow and eddying occur, corrosion is less localized than in PC and CC which are corrosion processes typical of stagnant fluids.

In addition to the hydrodynamic variables discussed above, the properties of the environment and the composition of the materials have an important influence on FAC and EC. A decrease in pH from neutral values can accelerate significantly the FAC rate, whereas O_2 can have opposite effects. If FAC is under cathodic control and dominated by O_2 reduction, the corrosion rate can be accelerated by increasing the O_2 content. However, an increase in the O_2 content may displace the E_{corr} to a regime in which more protective films can be formed and the corrosion rate will decrease. Very protective passive films, such as those formed on SSs or Ti alloys, are resistant to mechanical disruption even at very high shear forces, and therefore, these materials do not exhibit the abrupt increase in corrosion rate shown in Fig. 2.9, unless the abrasive effects due to the presence of solid particles in the fluid play a role at very high energy transfer rates.

EC should be distinguished from other forms of erosive action such as solid particle and liquid droplet impingement damage and cavitation [61], which can be considered essentially as mechanical forms of metal deterioration because the contribution of corrosion is minimal. Tests to evaluate the effect of these phenomena are discussed by Glaeser [62], whereas Roberge [63] has reviewed test methods for FAC and EC.

2.6 Stress corrosion cracking

Stress corrosion cracking (SCC) is a process by which cracks propagate in a metal or alloy by the concurrent action of a tensile stress (residual and/or applied) and a specific corrosive environment. Depending upon the alloy/environment system, cracks could propagate with velocities ranging from 10^{-12} to 10^{-3} m/s in alloys which do not exhibit significant general corrosion because they are covered with a protective film. For these reasons, SCC is one of the more insidious forms of alloy failure in industrial applications, and therefore, the subject is one of the most researched areas of corrosion science and engineering. Among many books and conferences covering a wide range of alloys, environments, and applications, those by Gangloff and Ives [64], Jones [65], Kane [66], and Shipilov et al. [67, 68] are the most recent ones.

There are two distinctive modes of crack propagation according to the path followed by the cracks. Cracks are intergranular when they propagate along grain boundaries and are transgranular if they run across the grains, in some cases following preferential crystallographic planes (cleavage planes), as illustrated in the fractographs shown in Fig. 2.10. Depending upon the alloy/environment system, either mode of propagation can be observed, but in many cases both modes can occur simultaneously or consecutively. A transition in the mode of propagation could arise from minor changes in the composition of the corrodent (e.g., pH, or concentration of ionic species), the potential, or the composition and microstructure of the alloy.

The three groups of major factors affecting SCC susceptibility are schematically represented in Fig. 2.11. Material variables are uniquely important in determining SCC resistance. Compositional differences even within a class of alloys (e.g., austenitic Ni-Cr-Fe alloys) could have a remarkable effect. Microstructural modifications introduced by heat treatment can alter substantially the SCC resistance (e.g., sensitization of austenitic SS renders them susceptible to intergranular SCC in relatively innocuous environments). Even though variations in surface conditions may not alter substantially the SCC behavior, the initiation time that precedes crack propagation could be affected significantly.

Stress is another major factor required for SCC and various sources of stress are summarized in Fig. 2.11. The time of failure caused by the propagation of cracks under constant load conditions tends to increase with decreasing stress until a threshold stress, σ_{th}, is reached, usually lower than that required for macroscopic yielding as expressed by the yield strength, σ_y, of the material. The effect of applied stress, however, is better represented through the stress intensity factor, K_I, defined by fracture mechanics concepts as

$$K_I = Y\sigma(\pi a)^{1/2} \tag{2.12}$$

where Y is a geometrical factor, σ is the applied stress, and a is the crack length. Crack propagation only occurs above a threshold stress intensity defined as $K_{Iscc.}$ Usually crack growth rate increases rapidly with increasing K_I above K_{Iscc} (Stage 1), until a plateau is reached at which the crack growth rate becomes independent of K_I (Stage 2), as shown schematically in Fig. 2.12.

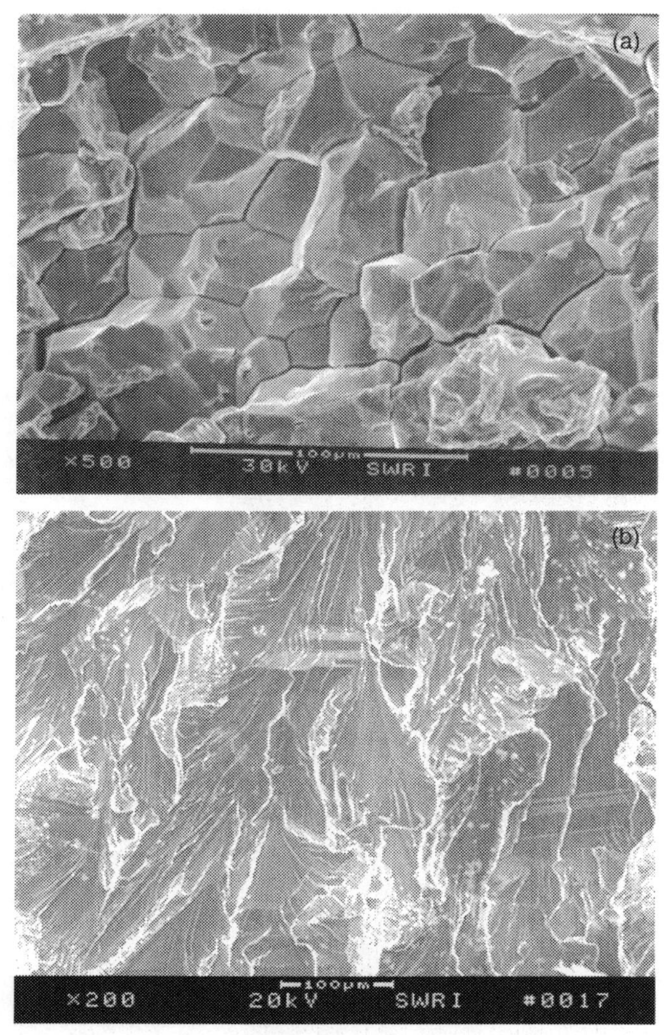

Fig. 2.10 Fractographs showing the typical appearance of: (A) Intergranular stress corrosion cracking; and (B) transgranular stress corrosion cracking. Both crack propagation modes were observed on AISI 316 L strained in a hot $MgCl_2$ solution.

In many cases of service failures, residual stresses caused by manufacturing processes, such as welding, become critical factors, in addition to the effect of stresses due to sustained mechanical loads or dynamic straining during start-up operations (i.e., slow strain rate conditions).

The third major factor influencing SCC is related to the properties of the environment, including variables such as chemical composition, pH, temperature, flow rate,

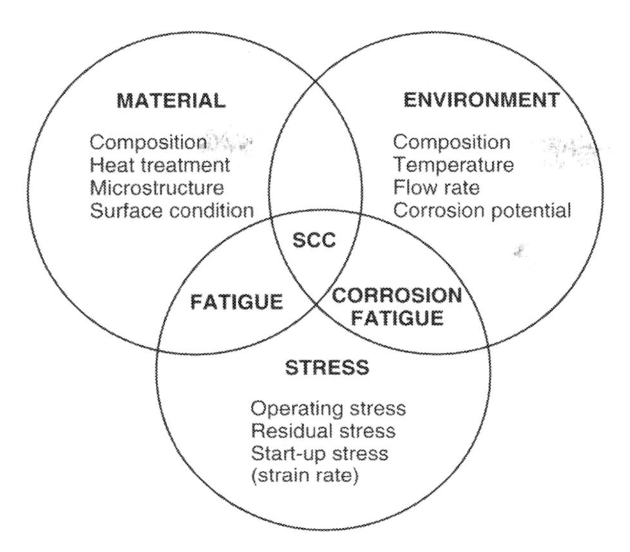

Fig. 2.11 Schematic diagram showing the interplay of the three major influential factors on stress corrosion cracking of metallic materials.

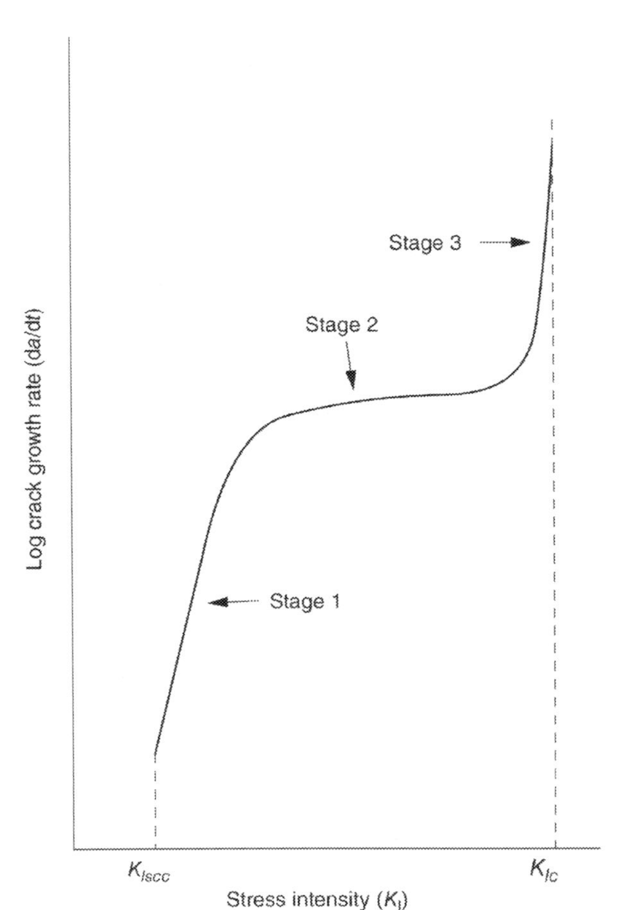

Fig. 2.12 Schematic plot of the logarithm of the crack growth rate, da/dt, as a function of stress intensity, K_I, showing the typical three stages of crack propagation at K_I values above the threshold stress intensity for stress corrosion cracking, K_{Iscc}. The critical stress intensity, K_{Ic}, for fast fracture in air is also indicated.

etc. The concept of a very specific environment required to promote SCC of a given alloy (e.g., NH_3 causing SCC of α-brass) does not hold any more due to the multiplicity of chemical species able to promote SCC under an appropriate set of environmental conditions. In this regard, the potential is recognized as a critical factor because minor modifications of E_{corr} by the presence of reducible species (e.g., O_2, H_2O_2, Fe^{3+}, Cu^{2+}) may promote SCC by displacing the E_{corr} of the metal or alloy to a "window" of susceptibility. For an alloy exposed to the appropriate environment in which anodic dissolution is the predominant crack advance mechanism, the current density at the crack tip should be far faster than the current density at the exposed surface, otherwise general dissolution will occur. Moreover, if the current density at the crack walls is comparable to that at the crack tip, PC will occur instead of SCC. A critical balance must exist between anodic activity at the crack tip and passivity of the overall exposed surface to maintain the high aspect ratio of a crack. A complete passive behavior precludes SCC because any incipient crack nucleus will be soon repassivated.

Parkins [69] proposed that a continuous spectrum of mechanisms can account for the SCC observed in a large variety of alloys and environments and Macdonald and Cragnolino [70] extended such a concept to alloy/environment systems prevailing in the power generating industry. The mechanisms may range from: (a) intergranular SCC (e.g., ferritic steels in caustic solutions); promoted by anodic dissolution along preexisting active paths (i.e., depleted or segregated grain boundaries); (b) transgranular SCC (e.g., austenitic SS in hot chloride solutions), initiated by mechanochemical passivity breakdown in the form of incipient pits just above E_{pit} or E_{rpit} (see Fig. 2.4) and propagated along strain generated paths; and (c) mixed intergranular and transgranular SCC (e.g., high-strength steels in water or steam) induced by absorption of atomic hydrogen and decohesion. Newman [71] proposed that a brittle film on the metal surface, such as a dealloyed layer, can trigger a fast short-life cleavage promoting discontinuous transgranular crack propagation. On the other hand, Galvele [72] after reviewing all the postulated models for SCC has developed a model based on surface mobility. Ionic species in the environments can create a contaminated layer where accelerated self-diffusion of ad-atoms induces the generation of vacancies at the stressed crack tip as the crack advance mechanism. Proposed SCC mechanisms and models have been reviewed and discussed by Jones [73]. The existence of a critical potential for the transgranular SCC of austenitic SSs in hot chloride solutions, E_{SCC}, coincident with E_{rcrev} has been discussed by Cragnolino et al. [74], whereas other authors [75] dispute the existence of such critical potentials on the basis of research conducted in environments such as high temperature water.

The most important parameter for monitoring SCC is the crack growth rate which has been measured in compact tension specimens by using potential drop methods, instrumented loading bolts, and more recently, Micro-Electro-Mechanical Systems (MEMS). Techniques for evaluating SCC are described by Sedriks [76] and Phull [77]. Sensors to measure stress and strain or strain rate are necessary as well as those required for monitoring environmental parameters which are common to other localized corrosion processes, including E_{corr}.

2.7 Corrosion fatigue

Corrosion fatigue (CF) occurs when a metallic component exposed to a corrosive environment is subjected to cyclic stresses. It is distinguished from fatigue (see Fig. 2.11), which takes place in the absence of a corrosive environment (i.e., dry inert gas or vacuum). CF is closely related to SCC and usually jointly considered as environmentally assisted cracking (EAC). However, the effect of the environment is far less specific in CF. As suggested by the lack of superposition with the materials circle in the diagram of Fig. 2.11, almost all metals and alloys, either under active or passive conditions, experience CF in an aqueous environment because their fatigue life is reduced with respect to that in an inert environment. Even humid air is an aggressive environment for many metals and alloys. In alloys exhibiting passive behavior (e.g., ferritic SSs used in turbine blades), transgranular cracks typical of CF—which do not exhibit the branching features characteristic of transgranular SCC—are initiated from pits. Comprehensive reviews of CF can be found in many books, handbook chapters, and conference proceedings. Among those recently published, Magnin [78] and Gangloff [79] provide valuable reviews.

Gangloff [79] describes four successive stages of CF with increasing number of load cycles: (a) cyclic plastic deformation; (b) microcrack initiation; (c) small crack growth to linkup and coalescence; and (d) macrocrack propagation. The initial stages are the subject of high-cycle fatigue (HCF) which is characterized by cyclic loads that lead to small-scale yielding. HCF is studied by using smooth or notched cylindrical specimens and test results are plotted showing the applied cyclic stress amplitude, σ_a, vs the number of cycles to failure, N. The value of σ_a decreases with N until a limiting value named the fatigue strength or endurance limit is reached at $N \approx 5 \times 10^6$ cycles. In an inert environment, this parameter is well-defined and represents the fatigue resistance of the material. In a corrosive environment, however, σ_a tends to decrease continuously with increasing N and the CF strength is usually defined at an arbitrary value of N (e.g., 10^7 cycles). The CF life, as assessed by the σ_a vs N curves, is dominated by microcrack initiation, as evidenced by the fact that notched specimens, where crack initiation is facilitated, fail at substantially lower N than smooth specimens when subjected to the same σ_a values.

Macrocrack propagation is evaluated using fracture mechanics techniques. The CF crack growth rate per cycle, da/dN, is plotted as a function of the stress intensity range, $\Delta K = K_{max} - K_{min}$, as shown in Fig. 2.13A. In vacuum or dry air, a threshold stress intensity range, ΔK_{th}, exists below which da/dN becomes negligible. Above ΔK_{th}, and at intermediate ΔK values, the following Paris relationship is valid

$$\mathrm{d}a/\mathrm{d}N = C(\Delta K)^n \quad (2.13)$$

where C and n are empirical constants. In a corrosive environment, da/dN usually exhibits a more complex dependency with ΔK, $R = K_{min}/K_{max}$ and frequency, f, and the changes in behavior schematically illustrated in Fig. 2.13B–D can be observed. In the first case (Fig. 2.13B), there is a synergism between cyclic loading and the corrosion reactions at the crack tip, leading to what is termed "true corrosion

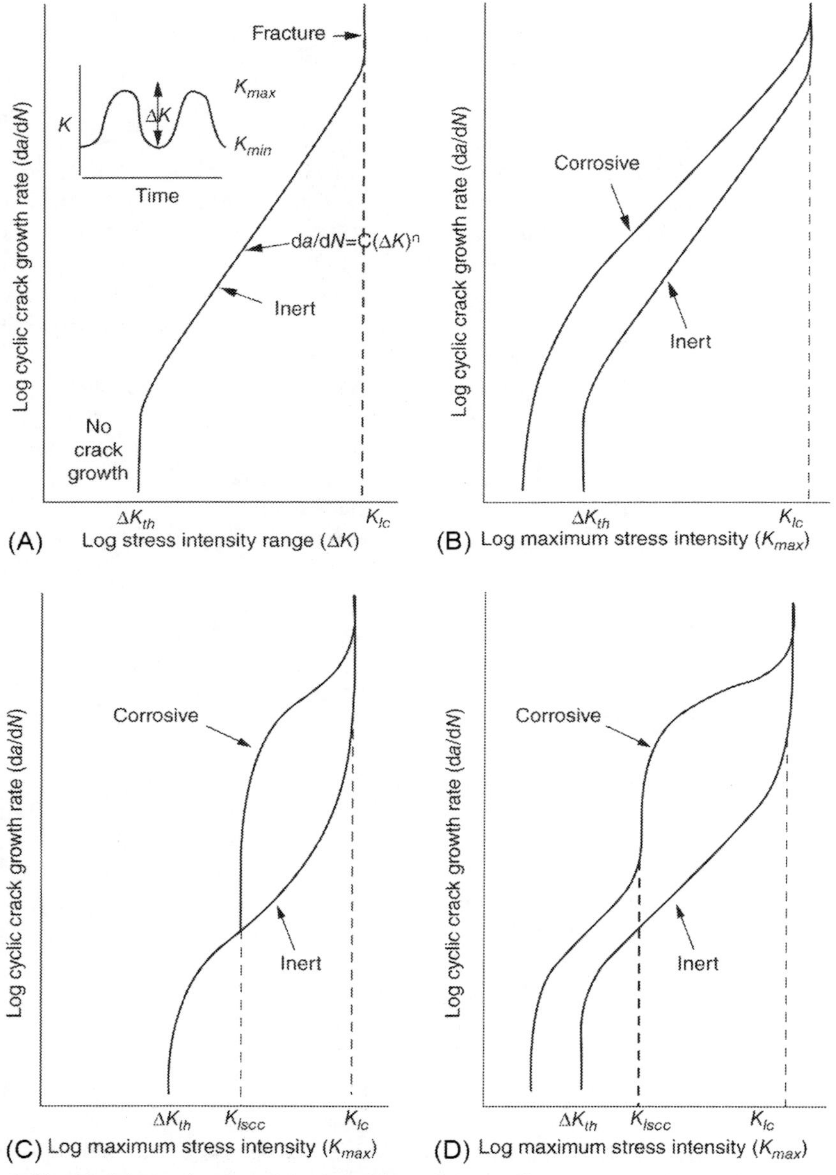

Fig. 2.13 (A) Schematic plot of the logarithm of the cyclic crack growth rate, d*a*/d*N*, as a function of the logarithm of the stress intensity range, ΔK_I, showing the threshold stress intensity range, ΔK_{th}, for fatigue and the critical stress intensity, K_{Ic}, for fast fracture in an inert environment. (B) Schematic plot of the logarithm of the cyclic crack growth rate, d*a*/d*N*, as a function of the logarithm of the maximum stress intensity, K_{max}, showing the effect of a corrosive environment on "true" corrosion fatigue. (C) Schematic plot of the logarithm of the cyclic crack growth rate, d*a*/d*N*, as a function of the logarithm of the maximum stress intensity, K_{max}, showing the effect of a corrosive environment only above the threshold stress intensity for stress corrosion cracking, K_{Iscc}. (D) Schematic plot of the logarithm of the cyclic crack growth rate, d*a*/d*N*, as a function of the logarithm of the maximum stress intensity, K_{max}, showing the effect of a corrosive environment on "true corrosion fatigue" below K_{Iscc} and the superposition of corrosion fatigue and stress corrosion cracking above K_{Iscc}.

fatigue." As a result, da/dN increases with decreasing frequency below 1 Hz because the environment has more time to interact with the crack tip during the rising part of the load cycle. The second case (Fig. 2.13C) is observed in materials/environment systems in which SCC occurs above K_{Iscc} (e.g., high-strength steels in NaCl solution). In this case, the environment has a strong effect on da/dN only within the $K_{Iscc} < K_{max} < K_c$ range. Also, da/dN increases with decreasing frequency and with increasing $K_{mean} = (K_{max} + K_{min})/2$. The third case (Fig. 2.13D) represents a combination of the two previous behaviors (e.g., α/β Ti alloys in NaCl solution), "true corrosion fatigue" below K_{Iscc} and superposition of CF and SCC above K_{Iscc}. This behavior leads to an expression of the crack growth rate for EAC on a time basis given by:

$$(\mathrm{d}a/\mathrm{d}t)_{EAC} = f(\mathrm{d}a/\mathrm{d}N)_{TCF} + (\mathrm{d}a/\mathrm{d}t)_{SCC} \tag{2.14}$$

As indicated in Fig. 2.11, environmental variables play an important role on CF. Although their influence is less specific than in SCC, the localized environment inside the crack which is, among other factors, controlled by the potential determines the rate at which corrosion reactions occur, and hence, affect the CF crack growth rate. Finally, it should be emphasized that the CF behavior is strongly affected by metallurgical variables, including alloy composition and microstructure, and in particular, mechanical properties such as tensile strength and fracture toughness. Comprehensive reviews of CF testing are presented by Phull [80] and Gangloff [79].

2.8 Hydrogen embrittlement

There is a tendency to consider hydrogen embrittlement (HE) as a form of EAC. In this section, however, it is defined separately, taking into account the almost unique physicochemical properties of hydrogen and its specific interaction with metals. Hirth and Johnson [81] in their phenomenological classification of such interaction in nine categories separated three forms of HE from other forms of hydrogen damage. The mode of hydrogen entry in a metal lattice distinguishes hydrogen stress cracking (HSC) from hydrogen environment embrittlement (HEE), both characterized by a loss of tensile stress due to the subcritical growth of cracks at tensile stresses below the yield strength, whereas loss in tensile ductility corresponds to the decrease in elongation or reduction in area in a tensile test. There are many comprehensive books, handbooks, and conference proceedings dealing with HE of different classes of alloys, particularly steels due to their technological significance (e.g., Refs. [82–85]).

The process of greatest interest is HSC because hydrogen entry is caused by a corrosion process in aqueous media, whereas the usual source of hydrogen in HEE is gaseous. The cathodic reduction of H^+ ions by either the chemical desorption mechanism or the electrochemical desorption mechanism leads to the chemical adsorption of atomic hydrogen on the metal surface, according to the following reaction step:

$$\mathrm{M} + \mathrm{H_3O^+} + \mathrm{e^-} \rightarrow \mathrm{MH_{ad}} + \mathrm{H_2O} \tag{2.15}$$

The hydrogen atom can diffuse into the metal lattice, particularly in body-centered cubic (bcc) metals in which the diffusion coefficient is high ($D \approx 10^{-4}$ cm^2/s). The entry of the hydrogen atoms in the lattice is strongly favored by the presence of species in the environment, such as H_2S and other volatile hydrides (e.g., H_2Se, PH_3, AsH_3, and SbH_3), which act as poisons for the recombination of adsorbed hydrogen atoms, limiting the desorption of H_2 molecules, and hence, increasing the available concentration of hydrogen atoms on the metal surface.

There are a large number of atomic, microscopic, and macroscopic defects, where hydrogen atoms may accumulate in a metallic structure. These traps include interstitial and substitutional solute atoms, lattice vacancies, inclusions, voids, and pores. This trapped hydrogen can lead to crack propagation and failure in the presence of a tensile stress of sufficient magnitude. HSC is commonly associated with brittle failure, either in the form of intergranular cracking or transgranular cracking with a quasicleavage appearance, even though in some cases ductile fracture occurs. For steels, the fracture mode depends on the value of K_I. The fracture mode may change from intergranular cracking at low K_I to transgranular tearing with quasicleavage facets at intermediate K_I, and microvoid coalescence, typical of ductile failure at high K_I.

The crack growth rate for a variety of medium- and high-strength steels in environments promoting HE depends on K_I, exhibiting a dependence similar to that shown in Fig. 2.12. above a threshold value defined as K_{IEH}. The crack growth rate in Stage 2 increases with increasing cathodic potentials, suggesting that SCC of high-strength steels in aqueous environments is caused by HE. The crack growth rate also increases in NaCl solutions at anodic potentials above E_{pit} because atomic hydrogen is generated inside the pits as a result of localized acidification and the subsequent reduction of H^+ ions. Test methods for evaluating HE are discussed by Phull [86] and Interrante and Raymond [87].

2.9 Characterization techniques

Techniques used to characterize and analyze corroded and fractured surfaces and features like pits and cracks, as well as corrosion products, are presented and discussed in many papers, books, and other publications. A useful series of articles written by different specialists were published as Surface Analysis Series in many issues of Volumes 41, 42, and 43 of *Corrosion Engineering* (the English translation of Boshoku Gijutsu). Marcus and Mansfeld [88] have recently edited a book in which surface analysis and electrochemical techniques are discussed in some detail. In the following sections, brief descriptions of the different techniques are presented.

2.9.1 Surface characterization

Several techniques are used to examine and characterize topological features of corroded metal and fracture surfaces. They include traditional optical microscopy and electron microscopy. Table 2.1 adopted from Shah [89] summarizes the techniques used to examine the surface topological structure and those to obtain the chemical

Table 2.1 Surface-sensitive analytical techniques, principles, and applications.

Analytical technique	Principle	Target information
Surface morphology		
Optical microscopy	Reflected light is used to generate a magnified image.	Macroscopic surface structure details
Scanning electron microscopy (SEM)	Incident electron beam generates a secondary electron emission that is used to generate the surface image.	Microscopic imaging of the surface structure
Atomic force microscopy (AFM)	Microscopic force sensor (cantilever) is used to sense the force between a sharp tip and the sample surface as the sample is scanned to generate an image.	Imaging of insulated surface structure at atomic resolution
Scanning tunneling microscopy (STM)	Tunneling current is motivated as the probe tip is scanned over a surface of interest in the x-y plane to generate an image.	Imaging of conducting surface structure at atomic resolution
Chemical identification and composition		
Energy dispersive X-ray spectroscopy (EDS)	Incident electron beam generates emission of X-rays characteristic of element present at the surface.	Elemental identification of the surface species
X-ray diffraction (XRD)	Diffraction of the incident X-ray beam from various plans of crystal lattice creates a diffraction pattern characteristic of the sample.	Elemental and phase identity and composition of inorganic corrosion product
Auger electron spectroscopy (AES)	Incident electron beam initiates a multistep process to facilitate the ejection of an outer shell electron. The energy of this ejected electron is characteristic of the surface atoms.	Elemental identity and composition of the surface species and the depth profile
X-ray photoelectron spectroscopy (XPS)	Incident X-rays on the surface eject photoelectrons. The energy of the photoelectrons is characteristic of the surface atoms.	Elemental identity and composition of the surface species, the electronic structure, chemical bonding, and elemental depth profile

Continued

Table 2.1 Continued

Analytical technique	Principle	Target information
Glow discharge optical emission spectrometry (GDS)	Gas discharge with inert gas (Ar) at 1–10 Torr generates plasma and sputtered atoms excited in the plasma will emit light with characteristic wave length.	Elemental identity and composition of surface species and the depth profile
Ion scattering spectroscopy (ISS)	The energy of scattered primary ions from the surface allows identification of surface atoms.	Chemical composition of the surface films at atomic and a molecular level
Secondary ion mass spectrometry (SIMS)	Incident ion beam ejects the surface atoms as ions, and mass of these secondary ions is measured.	Chemical composition of the surface adsorbed species
Extended X-ray absorption fine structure (EXAFS)	The X-ray absorption generates the interference effects between emitted photoelectron waves and backscattering waves characteristic of the local structure.	Composition of adsorbed species, number, and separation distances of surface atoms
Fourier transform infrared absorption spectroscopy (FTIR)	The absorption of the infrared photons results in vibrational excitation that is characteristic of the surface molecules and the environment.	Vibrational structure of molecules and bonding interactions with the surface and the surroundings
Raman and surface-enhanced Raman spectroscopy (Raman, SERS)	Energy shifts of the scattered photons are characteristic of the molecular identity.	Vibrational structure of adsorbed molecules on the surface

Modified from S. Shah, Application of modern analytical instruments in corrosion, in: S.D. Cramer, B.S. Covino Jr. (Eds.), ASM Handbook, vol. 13A, Corrosion: Fundamentals, Testing and Protection, ASM International, Materials Park, OH, 2003, pp. 992–998, with minor modifications.

identity and composition of surface films and layers as well as corrosion products by examining the atomic and molecular structure. More detailed information is available on materials characterization in Volume 10 of the *ASM Handbook*.

Energy dispersive X-ray spectroscopy (EDS) is routinely used in conjunction with scanning electron microscopy (SEM) to examine and analyze corroded and fractured surfaces, whereas it is combined with scanning transmission electron microscopy (STEM) to analyze at higher magnification concentration profiles around precipitates, pits or cracks, intermetallic compositions, etc. X-ray photoelectron spectroscopy

(XPS), Auger electron spectroscopy (AES), and glow discharge optical emission spectrometry (GDS) are commonly used in combination with ion milling or sputtering to study in-depth composition of surface layers or passive films.

Ellipsometry is an optical method used to characterize the optical properties of surface and surface layers using the reflection of polarized light. It has been extensively used to determine the thickness of passive films among other applications.

2.9.2 Corrosion products characterization

Many of the techniques used to characterize and analyze deposited, suspended, or dissolved corrosion products, including ionic species in solution, are listed in Table 2.1. These techniques include, among others, SEM combined with EDS, X-ray diffraction (XRD), Fourier transform infrared absorption spectroscopy (FTIR), laser Raman spectroscopy, and inductively coupled plasma mass spectrometry, as well as traditional methods such as atomic absorption spectroscopy for the analysis of species in solution.

References

[1] S.D. Cramer, B.S. Covino Jr. (Eds.), ASM Handbook, Corrosion: Environments and Industries, vol. 13C, ASM International, Materials Park, OH, 2006.
[2] A.J. Bard, L.R. Faulkner, Electrochemical Methods: Fundamentals and Applications, J. Wiley & Sons, Inc., New York, NY, 1980.
[3] S.D. Cramer, B.S. Covino Jr. (Eds.), ASM Handbook, Corrosion: Fundamentals, Testing and Protection, vol. 13A, ASM International, Materials Park, OH, 2003.
[4] P. Marcus (Ed.), Corrosion Mechanisms in Theory and Practice, second ed., Revised and Expanded, Marcel Dekker, Inc., New York, 2002.
[5] R.W. Revie (Ed.), Uhlig's Corrosion Handbook, second ed., J. Wiley & Sons, Inc., New York, 2000.
[6] L.L. Shreir, R.A. Jarman, G.T. Burstein (Eds.), Corrosion, third ed., Butterworth/Heinemann, London, 1994.
[7] R.G. Kelly, J.R. Scully, D.W. Shoesmith, R.G. Buchheit, Electrochemical Techniques in Corrosion Science and Engineering, Marcel Dekker, Inc., New York, 2003.
[8] R. Baboian (Ed.), Corrosion Tests and Standards: Application and Interpretation, second ed., ASTM International, West Conshohocken, PA, 2005.
[9] C. Wagner, W. Traud, The interpretation of corrosion phenomena by superposition of electrochemical partial reactions and the formation of potentials of mixed electrodes (English translation), Z. Elektrochem. 44 (1938) 391.
[10] ASTM, Standard Practice for Preparing, Cleaning and Evaluating Corrosion Tests Specimens, ASTM International, West Conshohocken, PA, 2017. G1-03(2017).
[11] ASTM, Standard Practice for Calculation of Corrosion Rates and Related Information From Electrochemical Measurements, ASTM International, West Conshohocken, PA, 2015. G102 - 89(2015).
[12] S.D. Cramer, B.S. Covino Jr. (Eds.), ASM Handbook, Corrosion: Materials, vol. 13B, ASM International, Materials Park, OH, 2005.
[13] C. Leygraf, T. Graedel, Atmospheric Corrosion, J. Wiley & Sons, Inc., New York, NY, 2000.

[14] P. Labine, G.C. Moran, Monitoring in Industrial Plants Using Nondestructive Testing and Electrochemical Methods, ASTM STP 908, ASTM International, West Conshohocken, PA, 1986.
[15] P.R. Roberge, R.D. Klassen, Corrosion monitoring techniques, in: S.D. Cramer, B.S. Covino Jr. (Eds.), ASM Handbook, Corrosion: Fundamentals, Testing and Protection, vol. 13A, ASM International, Materials Park, OH, 2003, pp. 514–518.
[16] N. Cabrera, N.F. Mott, Theory of oxidation of metals, Rep. Prog. Phys. 12 (1948–1949) 163–184.
[17] D.D. Macdonald, The point defect model for the passive state, J. Electrochem. Soc. 139 (1992) 3434–3449.
[18] X.G. Zhang, Galvanic corrosion, in: R.W. Revie (Ed.), Uhlig's Corrosion Handbook, second ed., J. Wiley & Sons, Inc., New York, NY, 2000, pp. 137–164.
[19] H.P. Hack, Galvanic, in: R. Baboian (Ed.), Corrosion Tests and Standards: Application and Interpretation, second ed., ASTM International, West Conshohocken, PA, 2005, pp. 233–243.
[20] ASTM, Standard Guide for Conducting and Evaluating Galvanic Corrosion Tests in Electrolytes, ASTM International, West Conshohocken, PA, 2019. G71 - 81(2019).
[21] C. Wagner, Theoretical analysis of the current density distributions in electrolytic cells, J. Electrochem. Soc. 98 (1951) 116–128.
[22] J.R. Galvele, Present state of understanding of the breakdown of passivity and repassivation, in: Passivity of Metals, The Electrochemical Society, Princeton, NJ, 1978, pp. 285–327.
[23] G.S. Frankel, Pitting corrosion of metals. A review of the critical factors, J. Electrochem. Soc. 145 (1998) 2186–2198.
[24] G.S. Frankel, R.C. Newman (Eds.), Critical Factors in Localized Corrosion, The Electrochemical Society, Pennington, NJ, 1992. Proceedings vol. 92-9.
[25] R.G. Kelly, G.S. Frankel, P.M. Natishan, R.C. Newman (Eds.), Critical Factors in Localized Corrosion III, The Electrochemical Society, Pennington, NJ, 1999. Proceedings vol. 98-17.
[26] P.M. Natishan, R.G. Kelly, G.S. Frankel, R.C. Newman (Eds.), Critical Factors in Localized Corrosion II, The Electrochemical Society, Pennington, NJ, 1996. Proceedings vol. 95–15.
[27] Z. Szklarska-Smialowska, Pitting and Crevice Corrosion, NACE International, Houston, TX, 2005.
[28] ASTM, Standard Test Method for Conducting Cyclic Potentiodynamic Polarization Measurements for Localized Corrosion Susceptibility of Iron-, Nickel-, or Cobalt-Based Alloys, ASTM International, West Conshohocken, PA, 2018. ASTM G61 - 86(2018).
[29] ASTM, Standard Test Methods for Pitting and Crevice Corrosion Resistance of Stainless Steels and Related Alloys by Use of Ferric Chloride Solution, ASTM International, West Conshohocken, PA, 2015. G48 - 11(2015).
[30] J.R. Galvele, Transport processes and the mechanism of pitting of metals, J. Electrochem. Soc. 123 (1976) 464–474.
[31] N.J. Laycock, R.C. Newman, Localized dissolution, kinetics, salt films, and pitting potentials, Corros. Sci. 39 (1997) 1771–1790.
[32] P. Combrade, Crevice corrosion of metallic materials, in: P. Marcus (Ed.), Corrosion Mechanisms in Theory and Practice, second ed., Revised and Expanded, Marcel Dekker, Inc., New York, 2002, pp. 349–397.
[33] B.A. Kehler, G.O. Ilevbare, J.R. Scully, Crevice corrosion behavior of Ni-Cr-Mo alloys: comparison of alloys 625 and 22, in: G.S. Frankel, J.R. Scully (Eds.), Proceedings of the

CORROSION 2001 Research Topical Symposium on Localized Corrosion, NACE International, Houston, TX, 2001, pp. 30–64.
[34] N. Sridhar, D.S. Dunn, C.S. Brossia, G.A. Cragnolino, Stabilization and repassivation of localized corrosion, in: G.S. Frankel, J.R. Scully (Eds.), Proceedings of the CORROSION 2001 Research Topical Symposium on Localized Corrosion, NACE International, Houston, TX, 2001, pp. 1–29.
[35] N. Sridhar, D.S. Dunn, G.A. Cragnolino, J.R. Kearns, Crevice corrosion, in: R. Baboian (Ed.), Corrosion Tests and Standards: Application and Interpretation, second ed., ASTM International, West Conshohocken, PA, 2003 (Chapter 19).
[36] D.S. Dunn, G.A. Cragnolino, N. Sridhar, An electrochemical approach to predicting long-term localized corrosion of corrosion-resistant high-level waste container materials, Corrosion 56 (2000) 90–104.
[37] D.S. Dunn, N. Sridhar, G.A. Cragnolino, Long-term prediction of localized corrosion of alloy 825 in high-level nuclear waste repository environments, Corrosion 52 (1996) 115–124.
[38] D.S. Dunn, Y.M. Pan, L. Yang, G.A. Cragnolino, Localized corrosion susceptibility of alloy 22 in chloride solutions. II—Effects of fabrication processes, Corrosion 62 (2006) 3–12.
[39] D.S. Dunn, Y.M. Pan, L. Yang, G.A. Cragnolino, Localized corrosion susceptibility of alloy 22 in chloride solutions. I—Mill-annealed condition, Corrosion 61 (2005) 1078–1085.
[40] A. Anderko, N. Sridhar, D.S. Dunn, A general model for the repassivation potential as a function of multiple aqueous solution species, Corros. Sci. 46 (2004) 1583–1612.
[41] S. Corcoran, Effect of metallurgical variables on dealloying corrosion, in: S.D. Cramer, B. S. Covino Jr. (Eds.), ASM Handbook, Corrosion: Fundamentals, Testing and Protection, vol. 13A, ASM International, Materials Park, OH, 2003, pp. 287–293.
[42] A.C. Van Orden, Dealloying, in: R. Baboian (Ed.), Corrosion Tests and Standards: Application and Interpretation, second ed., ASTM International, West Conshohocken, PA, 2005, pp. 278–288.
[43] S.M. Bruemmer, Quantitative modeling of sensitization development in austenitic stainless steel, Corrosion 46 (1990) 698–709.
[44] M. Guttman, D. McLean, Grain boundary segregation in multicomponents systems, in: W.C. Johnson, J.M. Blakely (Eds.), Interfacial Segregation, ASM International, Materials Park, OH, 1979, pp. 261–350.
[45] M.A. Streicher, Intergranular, in: R. Baboian (Ed.), Corrosion Tests and Standards: Application and Interpretation, second ed., ASTM International, West Conshohocken, PA, 2005, pp. 244–265.
[46] ASTM, Practice for Evaluating Intergranular Corrosion Resistance of Heat Treatable Aluminum Alloys by Immersion in Sodium Chloride + Hydrogen Peroxide Solution, ASTM International, West Conshohocken, PA, 2015. ASTM G110 - 92(2015).
[47] ASTM, Test Method for Determining the Susceptibility to Intergranular Corrosion of 5XXXX Series of Aluminum Alloys by Mass Loss after Exposure to Nitric Acid (NAMLT Test), ASTM International, West Conshohocken, PA, 2018. ASTM G-67(2018).
[48] C.A.H. Von Wolzogen Kuhr, I.S. Van der Vlugt, Graphitization of cast iron as an electrobiochemical process in anaerobic soils (English translation), Water (The Hague) 18 (16) (1934) 147–165.
[49] G.G. Geesey, Z. Lewandowski, H.C. Flemming, Biofouling and Biocorrosion in Industrial Waters, CRC Press, Inc., Boca Raton, FL, 1994.

[50] S.W. Borenstein, Microbiologically Influenced Corrosion Handbook, Industrial Press Inc., New York, 1994.
[51] P. Angell, S.W. Borenstein, R.A. Buchanan, S.C. Dexter, N.J.E. Dowling, B.J. Little, H.G. Ziegenfuss (Eds.), International Conference on Microbially Influenced Corrosion, NACE International, Houston, TX, 1995.
[52] H.A. Videla, Manual of Biocorrosion, CRC Press Inc., Boca Raton, FL, 1996.
[53] B.J. Little, P.A. Wagner, F. Mansfeld, Microbiologically Influenced Corrosion. Corrosion Testing Made Easy, vol. 5, NACE International, Houston, TX, 1997.
[54] Little B.J., Mansfeld F., Arps P.J., Earthman J.C., Microbiologically influenced corrosion testing, in: Cramer S.D., Covino B.S. Jr. (Eds.), ASM Handbook, Corrosion: Fundamentals, Testing and Protection, vol. 13A, ASM International, Materials Park, OH, 2003, pp. 478–486.
[55] J.D. Gu, T.E. Ford, R. Mitchell, Microbiological corrosion of metals, in: R.W. Revie (Ed.), Uhlig's Corrosion Handbook, second ed., J. Wiley & Sons, Inc., New York, NY, 2000, pp. 915–927.
[56] S.C. Dexter, Microbiologically Influenced Corrosion, in: S.D. Cramer, B.S. Covino Jr. (Eds.), ASM Handbook, Corrosion: Fundamentals, Testing and Protection, vol. 13A, ASM International, Materials Park, OH, 2003.
[57] S.C. Dexter, Microbiological effects, in: R. Baboian (Ed.), Corrosion Tests and Standards: Application and Interpretation, second ed., ASTM International, West Conshohocken, PA, 2005, pp. 509–522.
[58] O.H. Tuovinen, G.A. Cragnolino, A review of microbiological and electrochemical techniques in the study of corrosion induced by sulfate-reducing bacteria, in: P. Labine, G.C. Moran (Eds.), Corrosion Monitoring in Industrial Plants Using Nondestructive Testing and Electrochemical Methods, ASTM STP 908, ASTM International, West Conshohocken, PA, 1986, pp. 413–432.
[59] B.J. Little, P.A. Wagner, Application of electrochemical techniques to the study of MIC, in: J.O. Bockris, B.E. Conway, R.E. White (Eds.), Modern Aspects of Electrochemistry, vol. 34, Kluwer Academic/Plenum Publishers, New York, 2001 (Chapter 5).
[60] K.D. Efird, Flow-induced corrosion, in: R.W. Revie (Ed.), Uhlig's Corrosion Handbook, second ed., J. Wiley & Sons, Inc., New York, NY, 2000, pp. 233–248.
[61] J. Postlethwaite, S. Nesic, Erosion-corrosion in single and multiphase flow, in: R.W. Revie (Ed.), Uhlig's Corrosion Handbook, second ed., J. Wiley & Sons, Inc., New York, 2000, pp. 249–272.
[62] W.A. Glaeser, Erosion, cavitation, and fretting, in: R. Baboian (Ed.), Corrosion Tests and Standards: Application and Interpretation, second ed., ASTM International, West Conshohocken, PA, 2005, pp. 273–277.
[63] P. Roberge, Erosion-Corrosion, Corrosion Testing Made Easy, vol. 8, NACE International, Houston, TX, 2004.
[64] R.P. Gangloff, M.B. Ives (Eds.), Environment Induced Cracking of Metals, NACE International, Houston, TX, 1988.
[65] R.H. Jones (Ed.), Stress-Corrosion Cracking. Materials Performance and Evaluation, ASM International, Materials Park, OH, 1992.
[66] R.D. Kane (Ed.), Environmentally Assisted Cracking: Predictive Methods for Risk Assessment and Evaluation of Materials, Equipment, and Structures, ASTM International, West Conshohocken, PA, 2000. ASTM STP 1401.
[67] S.A. Shipilov, R.H. Jones, J.M. Olive, R.B. Rebak (Eds.), Environment-Induced Cracking of Metals: Chemistry, Mechanics and Mechanisms, Elsevier Science, Oxford, 2007.

[68] S.A. Shipilov, R.H. Jones, J.M. Olive, R.B. Rebak (Eds.), Environment-Induced Cracking of Materials: Prediction, Industrial Developments and Evaluation, Elsevier Science, Oxford, 2007.
[69] R.N. Parkins, Stress corrosion spectrum, Br. Corros. J. 7 (1972) 15–28.
[70] D.D. Macdonald, G.A. Cragnolino, Corrosion of steam cycle materials, in: P. Cohen (Ed.), The ASME Handbook on Water Technology for Thermal Power Systems, The American Society of Mechanical Engineers, New York, 1989, pp. 673–1033.
[71] R.C. Newman, Stress-corrosion cracking mechanisms, in: P. Marcus (Ed.), Corrosion Mechanisms in Theory and Practice, second ed., Revised and Expanded, Marcel Dekker, Inc., New York, 2002, pp. 399–450.
[72] J.R. Galvele, Electrochemical aspects of stress corrosion cracking, in: C.G. Vayenas, R.E. White, M.E. Gamboa-Adelco (Eds.), Modern Aspects of Electrochemistry, vol. 27, Plenum Press, New York, NY, 1995, pp. 233–358.
[73] R.H. Jones (Ed.), Chemistry and Electrochemistry of Corrosion and Stress Corrosion Cracking: A Symposium Honoring the Contributions of R. W. Staehle, The Minerals, Metals, and Materials Society, Warrendale, PA, 2001.
[74] G.A. Cragnolino, D.S. Dunn, Y.M. Pan, N. Sridhar, The critical potential for the stress corrosion cracking of Fe-Cr-Ni alloys and its mechanistic implications, in: R.H. Jones (Ed.), Chemistry and Electrochemistry of Corrosion and Stress Corrosion Cracking: A Symposium Honoring the Contributions of R. W. Staehle, The Minerals, Metals, and Materials Society, Warrendale, PA, 2001, pp. 83–104.
[75] P.L. Andresen, T.M. Angeliu, L.M. Young, Immunity, thresholds, and other SCC fiction, in: R.H. Jones (Ed.), Chemistry and Electrochemistry of Corrosion and Stress Corrosion Cracking: A Symposium Honoring the Contributions of R. W. Staehle, The Minerals, Metals, and Materials Society, Warrendale, PA, 2001, pp. 65–82.
[76] A.J. Sedriks, Stress Corrosion Test Methods, Corrosion Testing Made Easy, NACE International, Houston, TX, 1990.
[77] B. Phull, Evaluating stress-corrosion cracking, in: S.D. Cramer, B.S. Covino Jr. (Eds.), ASM Handbook, Corrosion: Fundamentals, Testing and Protection, vol. 13A, ASM International, Materials Park, OH, 2003, pp. 575–616.
[78] T. Magnin, Corrosion fatigue mechanisms in metallic materials, in: P. Marcus (Ed.), Corrosion Mechanisms in Theory and Practice, second ed., Revised and Expanded, Marcel Dekker, Inc., New York, 2002, pp. 451–478.
[79] R.P. Gangloff, Environmental cracking—corrosion fatigue, in: R. Baboian (Ed.), Corrosion Tests and Standards: Application and Interpretation, second ed., ASTM International, West Conshohocken, PA, 2005, pp. 302–321.
[80] B. Phull, Evaluating corrosion fatigue, in: S.D. Cramer, B.S. Covino Jr. (Eds.), ASM Handbook, Corrosion: Fundamentals, Testing and Protection, vol. 13A, ASM International, Materials Park, OH, 2003, pp. 625–638.
[81] J.P. Hirth, H.H. Johnson, Hydrogen problems in energy related technology, Corrosion 32 (1976) 3–16.
[82] R.N. Moody, A.W. Thompson, R.E. Ricker, G.S. Was (Eds.), Hydrogen Effects on Material Behavior and Corrosion Deformation Interactions, The Minerals, Metals, and Materials Society, Warrendale, PA, 2003.
[83] R.A. Oriani, J.P. Hirth, M. Smialowski (Eds.), Hydrogen Degradation of Ferrous Alloys, Noyes Publications, Park Ridge, NJ, 1985.
[84] A.W. Thompson, R.N. Moody (Eds.), Hydrogen Effects on Materials, The Minerals, Metals, and Materials Society, Warrendale, PA, 1996.
[85] A. Turnbull (Ed.), Hydrogen Transport and Cracking of Metals, The Institute of Materials, London, 1995.

[86] B. Phull, Evaluating hydrogen embrittlement, in: S.D. Cramer, B.S. Covino Jr. (Eds.), ASM Handbook, Corrosion: Fundamentals, Testing and Protection, vol. 13A, ASM International, Materials Park, OH, 2003, pp. 617–624.
[87] C.G. Interrante, L. Raymond, Hydrogen damage, in: R. Baboian (Ed.), Corrosion Tests and Standards: Application and Interpretation, second ed., ASTM International, West Conshohocken, PA, 2005, pp. 322–340.
[88] P. Marcus, F. Mansfeld, Analytical Methods in Corrosion Science and Engineering, CRC Press Inc., Boca Raton, FL, 2005.
[89] S. Shah, Application of modern analytical instruments in corrosion, in: S.D. Cramer, B.S. Covino Jr. (Eds.), ASM Handbook, Corrosion: Fundamentals, Testing and Protection, vol. 13A, ASM International, Materials Park, OH, 2003, pp. 992–998.

Part One

Electrochemical techniques for corrosion monitoring

Electrochemical polarization techniques for corrosion monitoring

Sankara Papavinasam
CorrMagnet Consulting Inc., Calgary, AB, Canada

3.1 Introduction

Mechanism of most corrosion taking place in aqueous phase is electrochemical in nature. Therefore, a broad range of electrochemical techniques are used to measure corrosion rates in the laboratory and monitor corrosion in the field. The main advantages of electrochemical techniques are ability to measure low corrosion rates, short measurement duration (corrosion rate can be obtained within as little as 20 s), and well-established theoretical understanding. Electrochemical measurements can be made by applying an external electrochemical signal (potential or current) or without applying any external electrochemical signal. When the external electrochemical signal is applied, the electrode or specimen is polarized.

Commonly used electrochemical polarization techniques include polarization resistance method, Tafel extrapolation method, cyclic potentiodynamic method, potentiostatic method, galvanostatic method, and galvanic current method. The electrochemical polarization methods are used to quantitatively measure and monitor general corrosion. They can also be qualitatively used to monitor localized corrosion (pitting and crevice).

This chapter presents fundamental principles of electrochemical polarization techniques and their advantages and limitations. Several standards have been developed to provide guidelines to measure and monitor corrosion using polarization techniques. These standards are extensively referenced in this chapter. Chapters 6 and 8 discuss electrochemical measurements without polarizing the electrodes.

3.2 Electrochemical nature of corrosion

According to electrochemical mechanism, four elements (ACME) (Fig. 3.1)—anode (A), cathode (C), metallic conductor (M), and electrolytic conductor (E)—are required for corrosion to take place. Three of these elements (anode, cathode, and metallic conductor) are already present within the metals or alloys. The electrolytic conductor is either an ionically conducting liquid (mostly aqueous solution) or ionically conducting solid. Thus, when the metal or alloy is immersed in an ionically conducting solution, corrosion takes place.

Techniques for Corrosion Monitoring. https://doi.org/10.1016/B978-0-08-103003-5.00003-5

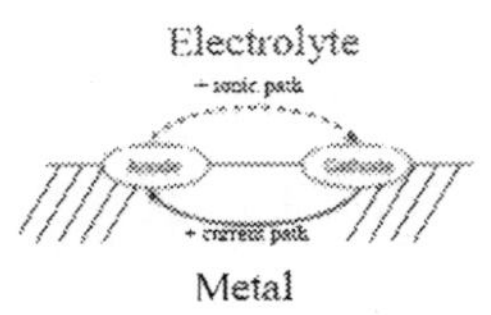

Fig. 3.1 Four elements of a corrosion process.

At the anode (A), metallic ions leave the metal surface (M) and go into the electrolytic conductor (E). In this process, they leave electrons behind on the metal surface. Thus, the metal is oxidized (loss of electrons is oxidation) at the anode (A). This process is corrosion. Anodic (corrosion) process can be written as (Eq. 3.1):

$$M \rightarrow M^{n+} + ne^- \tag{3.1}$$

The metallic ions are carried from the anode (A) to the cathode (C) by the ionically conducting electrolyte (E), whereas the electrons are transferred from anode to cathode by the metallic conductor (M), that is, the metal or alloy itself.

At the cathode (C), metallic ions leave the electrolyte (E) and go back to the metal (M). Thus, the metallic ion is reduced (gain of electrons is reduction) at the cathode (C). This process can be written as (Eq. 3.2):

$$M^{n+} + ne^- \rightarrow M \tag{3.2}$$

It is not necessary for the metallic ion to undergo reduction at the cathode, because, in addition to the metallic ions, the electrolyte may have several other ions that can undergo reduction. Thus, in most cases, corrosion results in loss of metal from the metallic surface, that is, it is deterioration of metals and alloys.

The ions in the electrolyte are of two types: anions and cations. Anions are negatively charged and they migrate towards anode and cations are positively charged and they migrate towards cathode (C). Higher the amounts of ions in an electrolyte, higher its conductivity. Higher the conductivity of the electrolyte, higher the chances for the electrolyte to sustain metallic corrosion.

The most commonly occurring reduction reactions at the cathode are: hydrogen ion reduction (Eq. 3.3) or oxygen reduction (Eq. 3.4):

$$2H^+ + 2e^- \rightarrow H_2 \tag{3.3}$$

$$O_2 + 2H_2O + 4e^- \rightarrow 4OH^- \tag{3.4}$$

In order for electrochemical corrosion to take place, all four elements (ACME) must occur simultaneously. Absence of any one of the four elements prevents corrosion from taking place. In the presence of all four elements, a balance is established, so that the rate of anodic reaction (oxidation) is equal to that of cathodic reaction (reduction).

Electrochemical polarization measurements determine the rate of anodic or cathodic reactions individually or collectively and determine the overall corrosion rate.

3.3 Energy-potential-current relationship

3.3.1 Energy

When you drop a ball (or a bucket of water), it rolls down to a lowest point and ultimately settles down there. It does so because it seeks the state of lowest energy. Under isothermal condition (i.e., at constant temperature), this tendency can be represented by Gibb's free energy change (ΔG). If this energy change, as a result of a process or reaction, is negative, then the result of the process or product of the reaction is at lower energy than the start of the process or starting material. Therefore, the product is more stable and, hence, the process is energetically possible and takes place spontaneously. If the free energy change is positive, then the reaction does not take place spontaneously.

3.3.2 Potential

The free energy change (ΔG) is related to the potential as (Eq. 3.5):

$$\Delta G = +nFE \tag{3.5}$$

where E is the potential, n is the number of electrons transferred, and F is the Faraday constant.

Thus, potential is a measure of reaction (corrosion) tendency. If the potential is negative, the metal would be active, that is, corrosion would take place spontaneously because free energy change (ΔG) is negative.

Potential of a metal in contact with its own ions of concentration equal to unit activity is known as standard potential (E_o). Table 3.1 [1] presents arrangement of selected metals based on their standard potentials. This arrangement of metals is known as Standard Oxidation-Reduction Potential series or redox potential series or Standard Equilibrium Reduction Potential series or Standard Potential or Standard Reversible Potential series or electromotive series or, simply EMF series.

The EMF series can be used as to understand corrosion tendency of metals. For example, when copper and zinc pieces each immersed separately into their own ions of unit activity (Fig. 3.2) and are electrically connected with a high-impedance voltmeter (commonly called open-circuited) [2], a potential of 1.1 V is created. When the two electrodes are short-circuited, the zinc undergoes anodic oxidation (corrosion), whereas copper ion undergoes cathodic reduction at the copper cathode (Note: From Table 3.1, the standard redox potential of copper is +0.337 V and that of zinc is −0.763 V). Thus, EMF series helps to determine corrosion tendency of metals readily and easily.

Table 3.1 Redox potential series (the list is partial—see elsewhere for full list [1]).

Electrode reaction	Standard potential $\phi°$ (in volts) at 25°C
$Au^{3+} + 3e^- = Au$	1.50
$Pt^{2+} + 2e^- = Pt$	ca. 1.2
$Pd^{2+} + 2e^- = Pd$	0.987
$Ag^+ + e^- = Ag$	0.800
$Cu^{2+} + 2e^- = Cu$	0.337
$2H^+ + 2e^- = H_2$	0.000
$Pb^{2+} + 2e^- = Pb$	−0.126
$Mo^{3+} + 3e^- = Mo$	Ca. −0.2
$Ni^{2+} + 2e^- = Ni$	−0.250
$Co^{2+} + 2e^- = Co$	−0.277
$Ti^+ + e^- = Tl$	−0.336
$Cd^{2+} + 2e^- = Cd$	−0.403
$Fe^{2+} + 2e^- = Fe$	−0.440
$Cr^{3+} + 3e^- = Cr$	−0.74
$Cr^{2+} + 2e^- = Cr$	−0.91
$Zn^{2+} + 2e^- = Zn$	−0.763
$Ti^{2+} + 2e^- = Ti$	−1.63
$Al^{3+} + 3e^- = Al$	−1.66
$Mg^{2+} + 2e^- = Mg$	−2.37
$Na^+ + e^- = Na$	−2.71
$Ca^{2+} + 2e^- = Ca$	−2.87
$K^+ + e^- = K$	−2.93
$Li^+ + e^- = Li$	−3.05

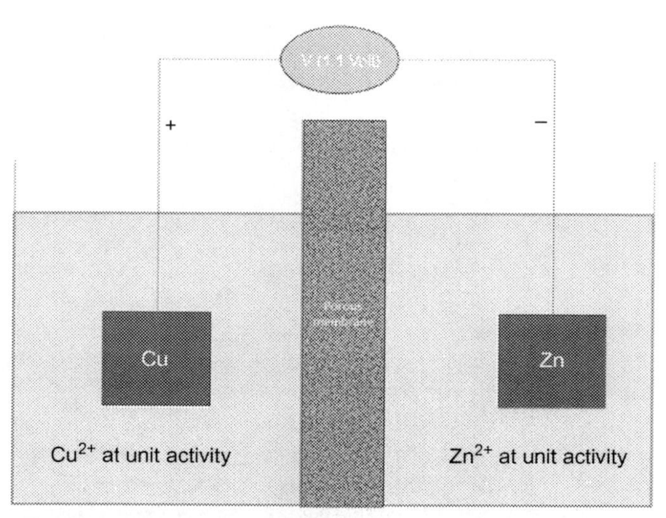

Fig. 3.2 Reversible cell containing copper and zinc electrodes in equilibrium with their ions. *Note:* It is important that the metallic ions must not be allowed to mix by having a porous membrane between them. If they did, copper would deposit on zinc and consequently, the potentials of copper and zinc electrodes would be identical.

In reality, seldom the metal is in contact only with its own ions at unit activity. Therefore, the E_o is only of limited practical application. To determine the potential of a metal in contact with other ions of various concentrations, Nernst derived an equation (Eq. 3.6):

$$E = E_o + 2.303\frac{RT}{nF}\log\frac{a_{oxid}}{a_{red}} \tag{3.6}$$

where E is the potential (corrosion potential or mixed potential), E_o is the standard redox potential, R is the gas constant, T is the absolute temperature, n is the number of electrons transferred, F is the Faraday constant, and a_{ox} and a_{red} are the activities (concentrations) of oxidized and reduced species, respectively. As indicated in Eq. (3.6), potential becomes more positive as the amount of oxidized species increases. For each tenfold increase in the oxidized reactant, at 25°C, the potential increases by 59 mV for a single electron transfer, that is, when $n = 1$.

Measurement of corrosion potential (E) is the primary step in understanding corrosion tendency of metals or alloys in an electrolyte. However, potential of a single electrode can't be directly measured. Only the difference in potential between two electrodes can be measured. For this reason, corrosion potential of an electrode is measured using another electrode called "reference electrode" and the corrosion potential should always be reported with respect to the reference electrode. Some commonly used standard reference electrodes are saturated calomel electrode (SCE), silver/silver chloride (Ag/AgCl) electrode, and copper/copper sulfate (CCS) electrode. The corrosion potential measured against one reference electrode can be converted into against another reference electrode. Table 3.2 presents correction factor to convert the corrosion potential from one reference electrode to another [5, 6].

Metals and alloys can be arranged based on their corrosion potentials in a given environment (electrolyte). Such arrangement of metals and alloys is called "galvanic series" (see Chapter 6 for more information). The tendency of metals or alloys to act as anode or cathode in a given electrolyte can be distinguished from the galvanic series. Table 3.3 presents the difference between galvanic and EMF series.

3.3.3 Current

Corrosion potential indicates the tendency of metals to corrode, but it does not provide the rate of corrosion. Rate of corrosion is proportional to the rate of electron transferred between electrode and electrolyte. Rate of electron transfer is represented as current (I). Amount of current (I) per unit surface area (A) is current density (i).

If a metal (e.g., zinc) is in equilibrium with its ions at unity, it would be at standard redox potential (as given in Table 3.1); the rate of exchange of electrons under this condition is known as exchange current density of zinc electrode ($i_{Zn^{2+}/Zn}$). Similarly, if we consider the hydrogen electrode in equilibrium with H^+ ions at unity, it would be at redox potential ($E_{H^+/H}$); the rate of exchange of electrons under this condition is known as exchange current density of hydrogen electrode ($i_{H^+/H}$). What will happen if we combine these two systems?

Table 3.2 Potentials of standard reference electrodes and conversion factors to convert potentials against one standard reference electrode to another [3, 4].

Electrode	Designation	Potential ^{C} (V)	Thermal temperature coefficient (mV/°C)
(Pt)H_2 (a = 1.0)	SHE	0.000	+0.87
Ag/AgCl/sat. KCl		+0.196	–
Ag/AgCl/1.0 m KCl		+0.235	+0.25
Ag/AgCl/0.1 M KCl		+0.288	+0.22
Ag/AgCl/seawater		+0.25	–
Hg/Hg_2Cl_2/sat. KCl	SCE	+0.241	+0.22
Hg/Hg_2Cl_2/1.0 M KCl		+0.280	+0.59
Hg/Hg_2Cl_2/0.1 M KCl		+0.334	+0.79
Hg/Hg_2SO_4/H_2SO_4		+0.616	–
Cu/sat. $CuSO_4$	CSE	+0.30	+0.90

Table 3.3 Galvanic series vs. EMF series.

Galvanic series	EMF series
List of metals and alloys	List of metals ONLY
Tendency of corrosion of metals and alloys in a given environment (e.g., sea water)	Tendency of corrosion of metals in solution of its own ions at unit activity
Several galvanic series can be developed to represent corrosion tendency of metals and alloys in various environments	Only one series
Associated with "corrosion potential"	Associated with equilibrium "redox potential" or "reversible potential"

Note: There has been some confusion in the corrosion literature on the use of term "open-circuit potential." Some references use the term "open-circuit potential" to represent "corrosion potential" or others to represent "equilibrium potential."

If the zinc is immersed into a solution containing H^+ ions (e.g., hydrochloric acid), the potential of the metal will not be at redox potential of either the zinc or the hydrogen, but the potential will stabilize at a mixed or corrosion potential (E_{corr}). At E_{corr}, the rate of zinc dissolution is equal to the rate of hydrogen reduction to maintain the charge conservation. The current at E_{corr} is known as I_{corr}, which is the rate at which the zinc will corrode when it is immersed in hydrochloric acid. Fig. 3.3 explains how the system moves from redox potentials of zinc and hydrogen towards corrosion potential.

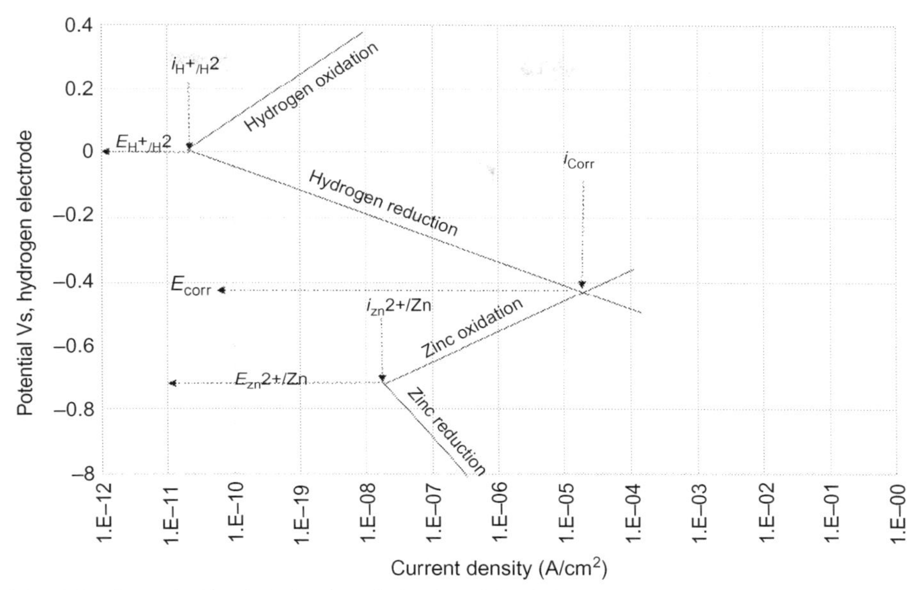

Fig. 3.3 Electrode kinetic behavior of zinc in acid solution—schematic.

3.4 Electrochemical polarization techniques for determining corrosion rates

There are several techniques by which the *i-E* relationship can be measured. Some methods involve application of an electrochemical excitation to an electrode and measure the response of the electrode to that excitation. As a consequence of the excitation, the electrode moves away from the corrosion potential, that is, the electrode is said to be polarized, and hence, these methods are called polarization methods of determining corrosion rates.

If the electrochemical excitation is in the form of controlling potential of an electrode and monitoring the response of current, then the method is called potentiostatic. If the potential of the electrode is varied at a constant rate and the response of the current is continuously monitored, then the method is called potentiodynamic.

On the other hand, if the electrochemical excitation is in the form of controlling the current of an electrode and monitoring the response of potential, then the method is called galvanostatic. If the current of the electrode is varied at a constant rate and the response of the potential is continuously recorded, then the method is called galvanodynamic.

Because of the theoretical relationship between potential and energy, most electrochemical measurements are conducted by controlling potential (potentiostatic or potentiodynamic) rather then by controlling current (galvanostatic or galvanodynamic).

The relationship between applied potential and resulting current is given by (Eq. 3.7):

$$E_{appl.} - E_{corr} = \eta = \pm\beta \log \frac{I}{I_{corr}} \tag{3.7}$$

where E_{corr} is the corrosion potential; $E_{appl.}$ is the applied potential; η is the difference between applied potential and corrosion potential and is commonly known as overpotential; β is a constant, known as Tafel constant; $I_{appl.}$ is the current measured at E_{appl}. and is the measure of the rate of oxidation or reduction; and I_{corr} is the current at corrosion potential. Elucidation of I_{corr}, E_{corr}, and I-E relationship is the underlying principle in measuring corrosion rate by electrochemical polarization techniques. Table 3.4 presents characteristics of different electrochemical polarization techniques.

3.4.1 Polarization resistance method

About 60 years ago, Stern and Geary [3] found that the slope of current-potential plot around corrosion potential is essentially linear; called the slope as polarization resistance (R_p); and defined it mathematically as (Eq. 3.8):

$$R_p = \left(\frac{\Delta V}{\Delta I}\right)_{E_{corr}} \tag{3.8}$$

R_p is related to corrosion current (I_{corr}) as (Eq. 3.9):

$$I_{corr} = \frac{B}{R_p} \tag{3.9}$$

The constant B is defined as (Eq. 3.10):

$$B = \frac{\beta_a \cdot \beta_c}{2.303(\beta_a + \beta_c)} \tag{3.10}$$

where β_a and β_c are anodic and cathodic Tafel constants.

By combining Eqs. (3.8)–(3.10), we get Eq. (3.11):

$$I_{corr} = \frac{1}{R_P}\left(\frac{\beta_a \cdot \beta_c}{2.303(\beta_a + \beta_c)}\right) \tag{3.11}$$

If β_a and β_c values are known, the corrosion rate can be calculated from R_p.

Because only a very small perturbation potential (less than ± 30 mV, typically ± 10 mV) is applied, this technique does not alter corrosion mechanism or change corrosion rate. Fig. 3.4 presents a typical polarization resistance plot [5]. From the slope of the plot, R_p (in ohms/cm^2 if the current density is plotted or in ohms if the current is plotted) is calculated.

Table 3.4 Characteristics of different electrochemical polarization techniques.

Polarization method	Typical measurement	Information obtained	Type of corrosion studied	Relevant standards
Polarization resistance	Application of ±30 mV (typically ±10 mV) around corrosion potential	Corrosion current (I_{corr})	General corrosion	ASTM G3, ASTM G5, ASTM G59, ASTM G102
Tafel extrapolation	Application of an overpotential of +500 mV both in anodic and cathodic directions, from corrosion potential	Corrosion current (I_{corr}), and Tafel slopes (anodic and cathodic)	General corrosion	ASTM G5, ASTM G102
Cyclic potentiodynamic polarization	Application of overpotential from corrosion potential towards noble direction to a potential at which current is 5 mA, where the potential is reversed and scanned until hysteresis loop is completed or until corrosion potential is reached	Critical pitting potential, passive current, transpassive region	Pitting corrosion	ASTM G5, ASTM G61 ASTM G102
Cyclic galvanostatic polarization	Application of current steps (typically in 20 μA/cm^2 increments between 0 and 120 μA) both in anodic and cathodic directions	Protection potential (E_{prot}) and breakpoint potential (E_b)	Pitting corrosion	ASTM G100
Potentiostatic polarization	Application of one potential step (typically to 700 mV vs. SCE)	Change of current with a variable, e.g., temperature (determination of critical pitting temperature)	Pitting corrosion	ASTM G150
	Application of potential step to a more positive potential (above E_b), and stepping it down to a less positive potential (below E_b)	Protection potential and breakpoint potential	Pitting corrosion	ASTM F746
Galvanic corrosion rate	Immersion of two dissimilar metals in an electrolyte and electrically connecting them using zero-resistance ammeter	Galvanic current	Galvanic corrosion	ASTM G71, ASTM G82

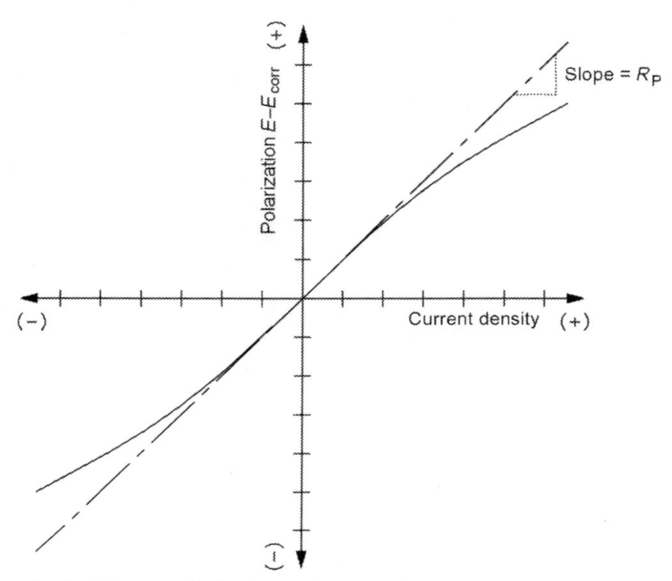

Fig. 3.4 Hypothetical linear polarization resistance plot.
Reprinted from ASTM G3, "Standard Practice for Conventions Applicable to Electrochemical Measurements in Corrosion Testing" Fig. 2, with permission from ASTM International.

It should be noted that the *i-E* curve around corrosion potential may not be linear. Besides, the curve in the anodic and cathodic regions may or may not be symmetrical. Symmetrical *i-E* curve is obtained only when both β_a and β_c are equal.

β_a and β_c values required to calculate corrosion current could either be determined by Tafel extrapolation method (discussed in Section 3.4.2) or could be assumed. Table 3.5 presents typical values of B, β_a, and β_c [5]. For majority of cases, the values of β fall between 60 and 120 mV. Therefore, in many instances, a value of 120 mV is assumed for both β_a and β_c. Consequently, Eq. (3.11) may reduce to (Eq. 3.12):

$$I_{corr} = \frac{26}{R_p} \tag{3.12}$$

Fig. 3.5 presents the error in determining corrosion current using Eq. (3.12) (assuming $B = 26$). From Fig. 3.5, negative error means that actual corrosion rate is lower than the one determined by polarization resistance method and positive error means that actual corrosion rate is larger than the one determined by polarization method. Eq. (3.12) may not be very accurate, but it provides a rapid method of determining corrosion current.

R_p measurements can be obtained by potentiodynamic method or by step-wise potentiostatic polarization method. In both methods, corrosion potential is first measured, typically for one hour (during which time corrosion potentials of most electrodes are stabilized) or until it stabilized. After that, potential step, at increments

Table 3.5 Values of constant "B" for the polarization resistance method [5].[a]

Corroding system	β_a (mV)	β_c (mV)	*B* (mV)
Theoretical (values of *B* calculated from arbitrary β_a and β_c values using formulae given in Eqs. 3.9 and 3.10; β_a and β_c values can be interchanged)	30	30	6.5
	30	60	9
	30	120	10
	30	180	11
	30	∞	26
	60	60	13
	60	90	16
	60	120	17
	60	180	20
	60	∞	26
	90	90	20
	90	120	22
	90	180	26
	90	∞	39
	120	120	26
	90	∞	39
	120	120	26
	120	∞	52
	180	180	39
	180	∞	78

[a] Selected values from Ref. [5].

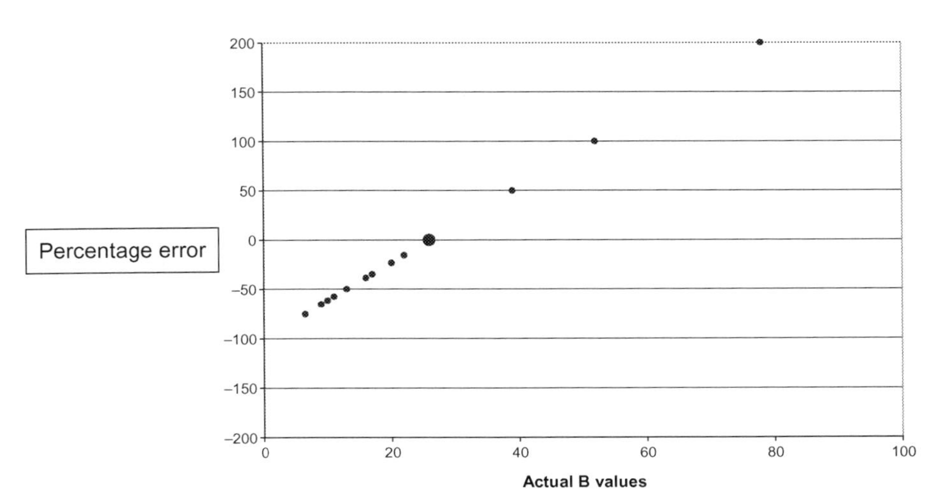

Fig. 3.5 Percentage error in calculating corrosion rate assuming a B value of 26.

of ±5 or ± 10 mV, is applied (potential-step method) or the potential is scanned at a constant rate (typically 0.6 V/h) (potentiodynamic method). In both methods, the experiment is started at negative potential, moving on to positive potential through corrosion potential. From the slope of the plot of potential-current, R_p is determined.

The advantages of polarization resistance method include:

- Corrosion current is measured rapidly, typically within a few minutes, and hence, this technique is widely used as an online monitoring technique in the field.
- Only very small amounts of potential are applied (less than ±30 mV, typically less than ±10 mV); hence, the corrosion rate is not affected due to the measurement.
- This technique can be used to measure low corrosion rates (less than 0.1 mills per year [2.5 μm/yr]).
- Measurements can be made repeatedly.

ASTM G3 [4] provides convention for reporting and displaying electrochemical corrosion data. ASTM G5 [7] provides details of instruments and materials and ASTM G59 [8] describes an experimental procedure required to carry out polarization resistance measurement.

3.4.2 Tafel extrapolation method

About 100 years ago, Julian Tafel found that a linear relationship between E and log I exists if electrode is polarized to sufficiently large potentials, both in anodic and cathodic directions [9]. The regions in which such relationships exist are known as Tafel regions. Mathematically, this relationship is given as (Eq. 3.13):

$$I = I_{corr}\left[\exp\left\{\frac{2.303\,(E - E_{corr})}{\beta_a}\right\}\right] - \left[\exp\left\{-\frac{2.303(E - E_{corr})}{\beta_c}\right\}\right] \quad (3.13)$$

where I is the current, I_{corr} is the current at corrosion potential E_{corr}, E is the applied potential, E_{corr} is the corrosion potential, and β_a and β_c are Tafel constants which are anodic and cathodic slopes of E-log I plots in the Tafel regions, respectively. Corroding metals that show Tafel behavior when polarized will exhibit plot similar to the one shown in Fig. 3.6 [4].

The difference between E-E_{corr} is called overpotential, η. At sufficiently larger values of η (typically between 100 and 500 mV), in the anodic direction, that is, η_a, Eq. (3.13) becomes Eq. (3.14):

$$\eta_a = \beta_a \log \frac{I}{I_{corr}} \quad (3.14)$$

Similarly, at sufficiently large η_c (in the cathodic direction), Eq. (3.13) becomes Eq. (3.15):

$$\eta_c = -\beta_c \log \frac{I}{I_{corr}} \quad (3.15)$$

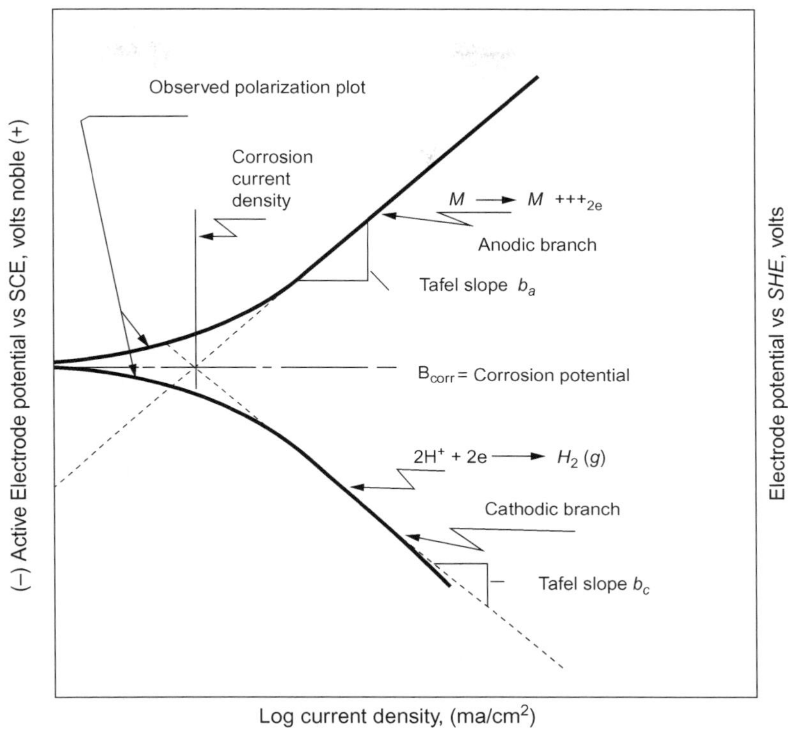

Fig. 3.6 Hypothetical cathodic and anodic Tafel polarization diagram.
Reprinted from ASTM G3, "Standard Practice for Conventions Applicable to Electrochemical Measurements in Corrosion Testing" Fig. 3, with permission from ASTM International.

In cases where Tafel regions are observed, I_{corr} can be determined by the extrapolation of either anodic or cathodic or both of Tafel regions to E_{corr}, as illustrated in Fig. 3.6.

Tafel extrapolation measurements can be performed either by potentiodynamic method or by step-wise potentiostatic polarization method. As in R_p measurement, corrosion potential is first measured, typically for one hour (during which time corrosion potentials of most electrodes are stabilized) or until it stabilized. After that, potential step (typically at increments of ± 25 or ± 50 or ± 100 mV, every 5 min, recording the current at the end of each 5-min period) is applied (potential-step method) or the potential is scanned at a constant rate (typically 0.6 V/h) (potentiodynamic method). In both methods, the experiment is started at corrosion potential and the cathodic polarization is first conducted, by applying an overpotential of approximately 500 mV or until gas (e.g., hydrogen) evaluation occurs at the electrode. After that, corrosion potential is measured again (typically for an hour), and then anodic polarization is conducted by applying an overpotential so that the potential at the end of the anodic polarization is +1.6 V vs. SCE. Tafel plots are generated by plotting both anodic and cathodic data in a semi-log paper as E-log I. From the plot, three values are determined: anodic Tafel slope, cathodic Tafel slope, and I_{corr} (from back-extrapolation of both anodic and cathode curves to E_{corr}).

The main advantage of this method is that it provides a simple, straight-forward method to determine Tafel constants.

This method applies large overpotential to the metal surface, and therefore, is considered as destructive. This is particularly true during anodic polarization during which the metal surface may be permanently changed/damaged. For this reason, it is not used as a monitoring technique in the field.

ASTM G5, however, provides procedure to construct anodic polarization plot. But, ASTM G5 neither provides procedure to construct cathodic polarization plot, nor provides procedure to determine I_{corr} by Tafel extrapolation method [7].

3.4.3 Cyclic potentiodynamic polarization

Cyclic potentiodynamic polarization method is used qualitatively to understand pitting corrosion tendency of metals and alloys.

In this method, the potential is scanned in the noble direction, monitoring the current continuously until the current reaches 5 mA, at which point the scan direction is reversed (i.e., scanned in the active direction) until the hysteresis loop closes or until the corrosion potential is reached. The results are plotted as E-log I, as in the Tafel extrapolation method. Fig. 3.7 presents a representative polarization curve generated by this method.

Initiation of localized corrosion is determined by the potential at which the anodic current increases rapidly. The more noble this potential is, the less susceptible is the alloy to initiation of localized corrosion. Similarly, the more electropositive the potential at which the hysteresis loop is completed, the less likely that the metal or alloy suffers from localized corrosion.

The polarization curve of sample S30400 (Fig. 3.7) indicates that the current starts increasing at less noble potential and does not exhibit hysteresis during reverse sweep; on the other hand, the curve of sample N10276 indicates that the current starts increasing at more noble potential and exhibits hysteresis at more noble potential. These curves indicate that sample N10276 is more resistant to initiation and propagation of localized corrosion than sample S30400.

Some other parameters that can be determined from the potentiodynamic polarization curves (Fig. 3.8) [4] include:

- primary passivation potential (E_{pp}, potential positive to which passive surface layers are formed)
- critical current density (I_{cc}, minimum current required before surface layers are formed)
- breakdown potential (E_b, potential positive to which passive surface layer is destroyed and transpassive region starts)
- protection potential (E_{prot}, potential at which passive layers are stable and protective)
- passive current (I_p, current of the electrode in E_{prot}) and
- area of the hysteresis loop.

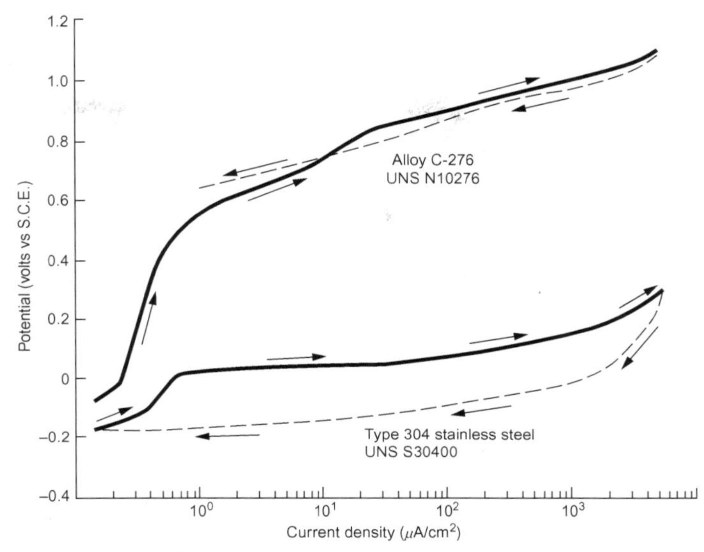

Fig. 3.7 Representative cyclic potentiodynamic polarization curves.
Reprinted from ASTM G61 "Standard Test Method for Conducting Cyclic Potentiodynamic Polarization Measurements of Localized Corrosion Susceptibility of Iron-, Nickel-, or Cobalt-Based Alloys," Fig. 2, with permission from ASTM International.

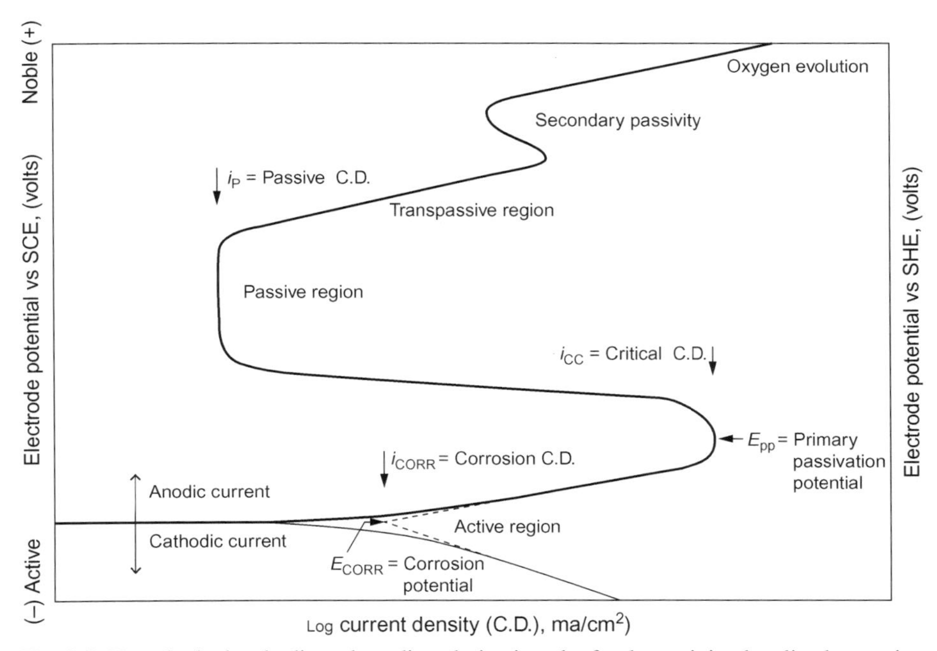

Fig. 3.8 Hypothetical cathodic and anodic polarization plot for determining localized corrosion parameters.
Reprinted from ASTM G3, "Standard Practice for Conventions Applicable to Electrochemical Measurements in Corrosion Testing" Fig. 4, with permission from ASTM International.

Many of these parameters determined are based on empirical observations. The results obtained by cyclic potentiodynamic polarization method depend on scan rate, pit size or depth, polarization curve shape, and specimen geometry. Therefore, the results should be considered as qualitative rather than quantitative.

ASTM G61 provides a procedure for conducting cyclic potentiodynamic polarization measurements [10].

3.4.4 Cyclic galvano-staircase polarization

Susceptibility to localized corrosion of metals is indicated by a protection potential (E_{prot}). More noble the E_{prot} is, the less susceptible is the metal or alloy to initiation of localized corrosion. When the applied potential is more negative than E_{prot}, no pits initiate, and when applied potential is more positive than E_{prot}, pit initiates even when the applied potential is less than breakdown potential (E_b).

Cyclic galvano-staircase polarization (CGSP) method can be used to determine E_{prot} and E_b. In this method, a current density is applied at an increment of 20 μA/cm^2 from 0 to 120 μA/cm^2 and then the current density is decreased at an increment of 20 μA/cm^2 until to the current density reaches 0 μA/cm^2. Two minutes after the application of current density at each step, the potential is measured. The applied current density and resultant potential are plotted (Fig. 3.9) [11]; from the plot, the E_b is obtained by extrapolating of the forward curve to 0 μA/cm^2 and E_{prot} is obtained by extrapolating the reverse curve to 0 μA/cm^2.

The advantage of the method is that it provides a quick method to determine susceptibility to initiation of pit initiation. However, this technique does not provide information on pit propagation and is only qualitative in nature. Because sensitive instruments are required to conduct the experiment, this method is not routinely used.

ASTM G100 [11] provides a procedure for conducting CGSP to determine susceptibility of metals and alloys to localized corrosion. In the round robin experiment conducted (to develop ASTM G100) using aluminium alloy 303-H14 (UNS A93003), it was found that E_b (−636 mV vs. SCE) is more electropositive than E_{prot} (−652 mV vs. SCE).

3.4.5 Potentiostatic polarization

There are several ways to conduct potentiostatic polarization measurements. In one method, only one potential step is applied and the variation of current is monitored. For example, to measure the resistance of stainless steel and related alloys to pitting corrosion, the potential is stepped and held constant at 700 mV (vs. SCE) and the variation of current is monitored as a function of other changes (e.g., temperature variation). ASTM G150 provides the procedure to determine critical pitting temperature (temperature in which these alloys become susceptible to pitting corrosion) of stainless steel and related alloys [12].

In another procedure, the potential is first stepped to a higher value and the current is monitored (this potential is set high enough value to induce and grow stable pits). The potential is then stepped down and the current is continued to be monitored. In

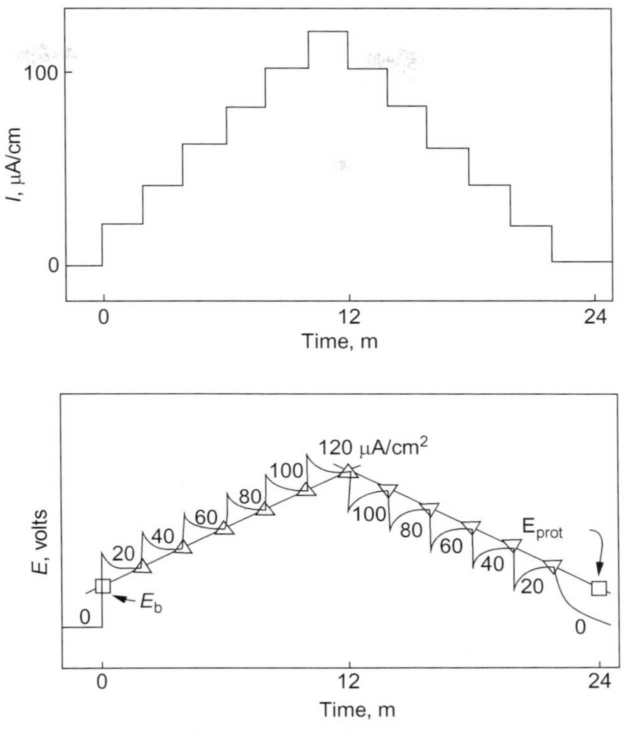

Fig. 3.9 Relationship of a schematic GSCP curve (lower) to the current staircase signal (upper). Reprinted from ASTM G100, "Standard Test Method for Conducting Cyclic Galvano-staircase Polarization" Fig. 3, with permission from ASTM International.

order to induce pits, the initial potential should be stepped more electropositively than E_b and the second potential applied is above E_{prot}, so that pit propagation is sustained. For this reason, this technique cannot be used independently without results from other techniques and prior knowledge of the alloy and metal behavior. ASTM F746 provides a test procedure for this type of measurements [13].

3.4.6 Galvanic corrosion rate

When two dissimilar metals in electrical contact are exposed to a common electrolyte, one of the metals (in the active side of the EMF or galvanic series) will undergo increased corrosion while the other (in the noble side of the EMF or galvanic series) will undergo decreased corrosion. This type of corrosion is known as galvanic corrosion.

A simple method of determining galvanic corrosion rate involves immersing two dissimilar metals in an electrolyte and electrically connecting the metals together using a zero-resistance ammeter and measuring the resulting current as a function of time.

ASTM G71 provides a guide for conducting galvanic corrosion tests both in the laboratory and in the field [14]. ASTM G82 describes procedures to develop galvanic series and predict the effect of one metal on the corrosion behavior of another metal when they are electrically connected while immersed in an electrolyte [15].

3.5 Conversion of I_{corr} into corrosion rate

Corrosion current values obtained from polarization resistance and Tafel extrapolation methods should be converted into corrosion rate. For this reason, it should be assumed that the current distribution is uniform across the area used in the calculation. With this assumption, the current value is divided by the surface area as (Eq. 3.16):

$$i_{corr} = \frac{I_{corr}}{A} \tag{3.16}$$

where i_{corr} is the current density (μA/cm^2), I_{corr} is the current (μA), and A is exposed specimen area, cm^2. Other commonly used units for current are “mA” and “A”.

Based on Faraday’s law, then corrosion rate (CR) (Eq. 3.17) or mass loss rate (MR) (Eq. 3.18) can be calculated as:

$$CR = K_1\left(\frac{i_{corr}}{d}\right)EW \tag{3.17}$$

$$MR = K_2 \cdot i_{corr} \cdot EW \tag{3.18}$$

where CR is given in mm/yr, i_{corr} in μA/cm^2, K_1 is 3.27×10^{-3} in mm g/μA cm y. MR is g/m^2, K_2 is 8.954×10^{-3} in g cm^2/μA m^2, d is the density in g/cm^3, and EW is the equivalent weight. Table 3.6 [16] presents other values for K_1 in other units.

Equivalent weight, EW, is the mass in grams that will be oxidized by the passage of one Faraday (96,489C (amp-sec)) of electric charge and EW of pure elements is calculated as (Eq. 3.19):

$$EW = \frac{W}{n} \tag{3.19}$$

where W is the atomic weight of the element and n is the number of electrons required to oxidize an atom of the element in the corrosion process, that is, the valence of the element.

The EW of alloy can be calculated as (Eq. 3.20):

$$EW = \frac{1}{\sum \frac{n_i f_i}{w_i}} \tag{3.20}$$

where n_i is the valence of the ith element of the alloy; w_i is the atomic weight of the ith element of the alloy; and f_i is the mass fraction of the ith element of the alloy.

Table 3.6 Constants to convert corrosion current into penetration rate and mass loss rate.

A				
Penetration rate unit (CR)	I_{cor} **unit**	ρ **unit**	K_1	**Units of** K_1[a]
mpy	μA/cm^2	g/cm^3	0.1288	mpy g/μA cm
mm/yr[b]	A/m^2[b]	kg/m^3[b]	327.2	mm kg/A m y
mm/yr[b]	μA/cm^2	g/cm^3	3.27×10^{-3}	mm g/μA cm y

B			
Mass loss rate unit	I_{cor} **unit**	K_2	**Units of** K_2[a]
g/m^2 d[b]	A/m^2[b]	0.8953	g/A d
mg/dm^2 d (mdd)	μA/cm^2	0.0895	mg cm^2/μA dm^2 d
mg/dm^2 d (mdd)	A/m^2[b]	8.953×10^{-3}	mg m^2/A dm^2 d

[a] EW is assumed to be dimensionless.
[b] SI unit.
Reprinted from ASTM G102, "Standard Practice for Calculation of Corrosion Rates and Related Information from Electrochemical Measurements" Table 2, with permission from ASTM International.

Calculation of EW of alloys may create two problems:

- It is assumed that the process of oxidation (corrosion) is uniform and does not occur selectively to any component of the alloy. If this not true, then the calculation should be adjusted.
- Valance assignments for elements that exhibit multiple valences can create uncertainty. It is best if an independent technique is used to establish the proper valence for each alloying element.

ASTM G102 provides details of calculating corrosion rate from corrosion current determined by polarization techniques [16]. This standard also provides the *EW* values for selected metals.

3.6 Measurement of corrosion rate by polarization methods in the laboratory

One of the greatest advantages of using electrochemical polarization experiments is that they provide a quick, easy, and relatively straight-forward results in short period of time [17–31]. Before using them, their types should be understood. Electrochemical polarization methods may be broadly divided into three types:

- Electrochemical measurements
- Electrochemical tests
- Electrochemical monitoring

Electrochemical measurements: Measuring corrosion rate in the laboratory requires two basic components: A laboratory methodology, to simulate corrosive conditions and an electrochemical technique to repeatedly measure the corrosion rate simulated by the laboratory methodology. The monitoring techniques by themselves do not (more importantly should not) alter the corrosion rate. Example for this type is polarization measurement (see Section 3.4.1).

Electrochemical tests: These tests use a laboratory methodology to carry out the tests and an electrochemical technique to stimulate corrosion conditions. In these tests, the monitoring techniques indeed alter the corrosion rate by polarization. Example for this type is Tafel extrapolation method (see Section 3.4.2).

Electrochemical Monitoring: Monitoring corrosion rate in the field requires two basic components: A plant (e.g., pipeline, infrastructure, building) containing corrosive conditions and an electrochemical technique to repeatedly monitor the corrosion rate/condition of the plant. The monitoring techniques by themselves do not (more importantly should not) alter the corrosion rate. Example for this type is polarization measurement in the field (see Section 3.6).

Table 3.7 [17, 18] lists various laboratory methodologies in which electrochemical polarization techniques are used. The list in Table 3.7 is not extensive, but is provided only as an illustration. Fig. 3.10 [4] presents a basic laboratory methodology in which electrochemical polarization technique is used as measurement technique.

Irrespective of the type, certain accessories are required to carry out electrochemical polarization techniques. They include working electrode (metal whose corrosion rate to be determined), counter electrode, reference electrode, conducting electrolyte (i.e., environment), potentiostat, potential-measuring instrument, and current-measuring instrument, Luggin capillary with salt-bridge connection to reference electrode.

Before proceeding to understand these components, we should understand why we need three different electrodes to conduct electrochemical polarization measurement. Recollect that potential of a single electrode cannot be measured; only the difference in potentials between two electrodes can be measured. Therefore, we need at least two electrodes to conduct electrochemical polarization measurement. During the polarization measurement, a potential or current is impressed on these electrodes, and as a consequence, the potentials of both the electrodes move from their respective corrosion potentials. The potential difference measured between these two electrodes then includes the values of two overpotentials. But we are interested in the polarization of only one electrode (working electrode). To resolve this issue, another electrode, auxiliary or counter electrode, is used. During electrochemical measurements, this counter electrode gets polarized in the opposite direction of the working electrode, that is, if the working electrode is polarized in the noble direction, the counter electrode is polarized in the active direction. A standard reference electrode (Table 3.1) is connected to the working electrode through a high-impedance voltmeter. This arrangement prevents any current passing through the reference electrode. Therefore, the reference electrode is not polarized.

Table 3.7 Laboratory methodologies in which electrochemical polarization method is used as the monitoring technique.

Laboratory methodology	Variables simulated in the laboratory methodologies	Relevant standards
Electrochemical polarization cell	Composition (of material and environment) and temperature	ASTM G5
Kettle or bubble test	Composition (of material and environment) and temperature	NACE 1D196 [17]
Rotating disk electrode (RDE)	Composition (of material and environment), temperature, and velocity	
Rotating cylinder electrode (RCE)	Composition (of material and environment), temperature, and velocity	ASTM G170 ASTM G185
Jet impingement (JI)	Composition (of material and environment), temperature, and velocity	ASTM G170 ASTM G208
High-pressure versions of RCE and JI	Composition (of material and environment), temperature, pressure, and velocity	ASTM G170 ASTM G184 ASTM G185 ASTM G208

3.6.1 Working electrode

This is the primary electrode whose corrosion rate is being measured. Care should be taken that this electrode is properly prepared and mounted (without any crevice). Fig. 3.11 presents a simple method in which the working electrode is mounted. A leak-proof assembly is obtained by the proper compression fit between the electrode and an insulator gasket.

3.6.2 Counter electrode

Auxiliary or counter electrode is usually made up of inert materials, for example, platinum or graphite. Generally, two auxiliary electrodes or one large sheet of one auxiliary electrode is used. The auxiliary electrode is mounted in the same way as that of working electrode.

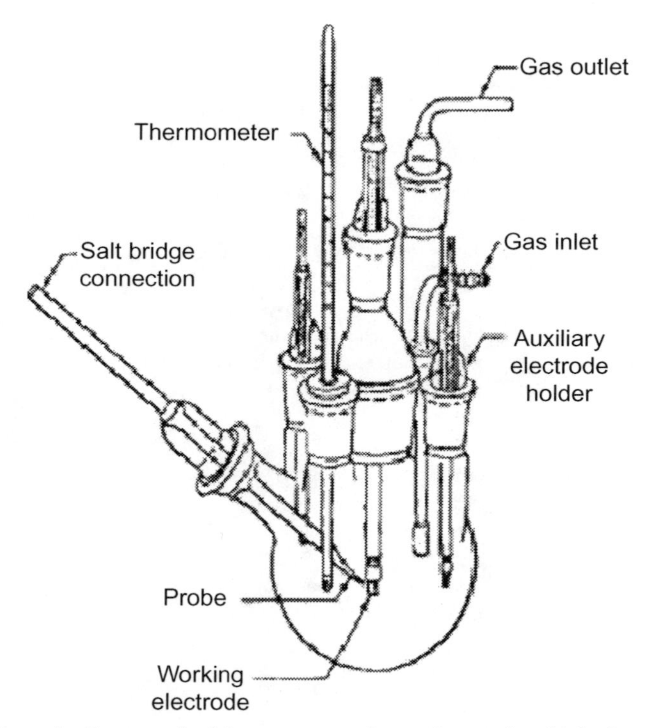

Fig. 3.10 Schematic diagram of a laboratory experimental setup in which electrochemical polarization method is used as the monitoring technique.
Reprinted from ASTM G5, "Standard Reference Test Method for Making Potentiostatic and Potentiodynamic Anodic Polarization Measurements" Fig. 3, with permission from ASTM International.

3.6.3 Reference electrode

Reference electrodes should have stable and reproducible potential. Reference electrodes used are preferably reversible type electrodes. In a reversible electrode, a small cathodic current produces the reduction reaction, while a small anodic current produces the oxidation reaction. Table 3.2 presents several commonly used reference electrodes and their reversible potential relationships. A SCE reference electrode with a controlled rate of leakage (about 3 μL/h) has been used for a long time. The SCE electrode is durable, reliable, and commercially available. However, use of mercury is banned by some governmental agencies. Therefore, SCE may not be available in locations where mercury is banned. In these locations, an alternate reference electrode should be used.

Irrespective of the type of standard reference electrode, the potential of reference electrode should be checked at periodic intervals to ensure the accuracy of the reference electrode.

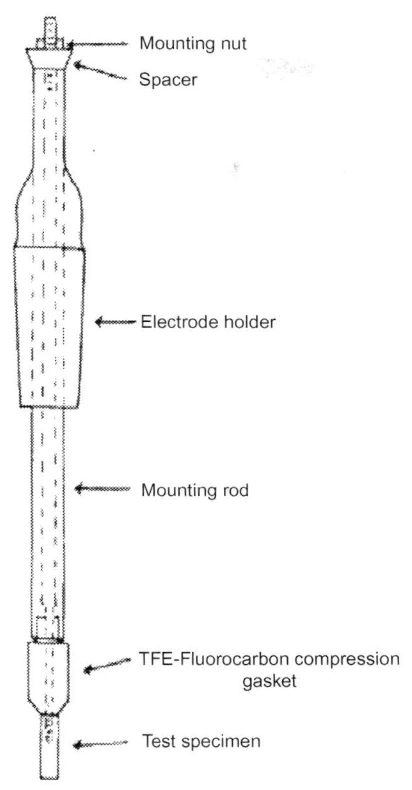

Fig. 3.11 Schematic diagram of an electrochemical specimen mounted on holder. Reprinted from ASTM G5, "Standard Reference Test Method for Making Potentiostatic and Potentiodynamic Anodic Polarization Measurements" Fig. 5, with permission from ASTM International.

Sometimes the reference electrode is placed in a separate vessel and is connected to the working electrode through Luggin capillary with salt-bridge. This arrangement minimizes contamination of the reference electrode from the products of corrosion reactions, or the contamination of the test solution by the solution in the reference electrode (Fig. 3.12). This arrangement requires special care and precaution.

3.6.4 Electrolyte

One basic requirement of any electrochemical measurement is a reasonably conducting (low-resistivity) electrolyte. Electrolyte with low-conductivity produces erroneously low corrosion rate, because their resistance (solution resistance, R_s) is added to the polarization resistance (R_p) during measurement. If in doubt, the resistivity or conductivity of the electrolyte should be determined. ASTM D1125 describes test methods for electrical conductivity and resistivity of water [21]. Fig. 3.13 provides

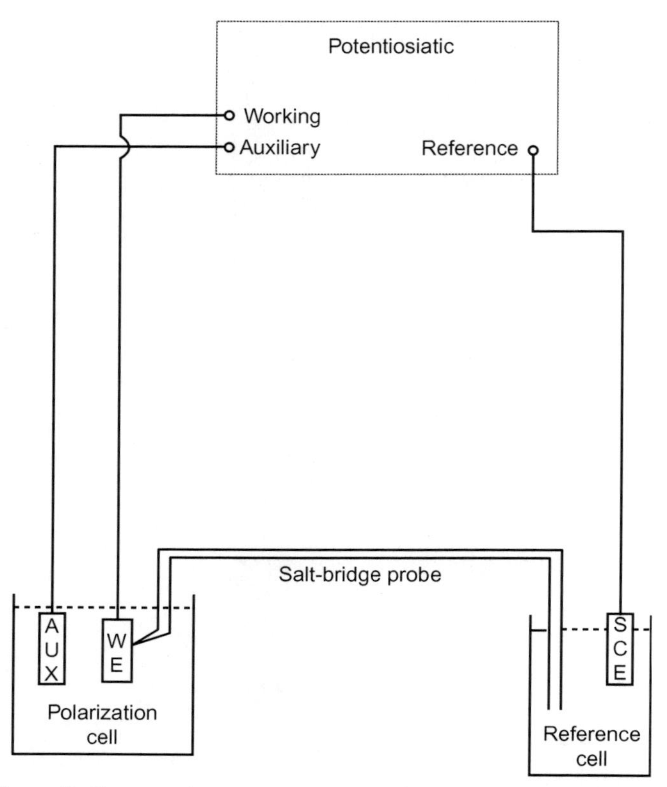

Fig. 3.12 Schematic diagram of a setup to connect reference electrode to an electrochemical cell through salt-bridge.
Reprinted from ASTM G5, "Standard Reference Test Method for Making Potentiostatic and Potentiodynamic Anodic Polarization Measurements" Fig. 4, with permission from ASTM International.

guidelines on operating range of solution conductivity in which electrochemical polarization (R_p) method can be conducted.

3.6.5 Potentiostat

Potentiostat is the instrument that maintains the potential of the working electrode. Most of the corrosion reactions occur between -2 to $+2$ V potential range and between 1 and 10^6 μA current range. Therefore, the instrument should have sufficient range of potential and current to supply them on a standard electrode type and shape. Modern computer-controlled, software-enabled, potentiostat can perform multiple functions including applying potential, monitoring current, storing data, analyzing data, and producing trend.

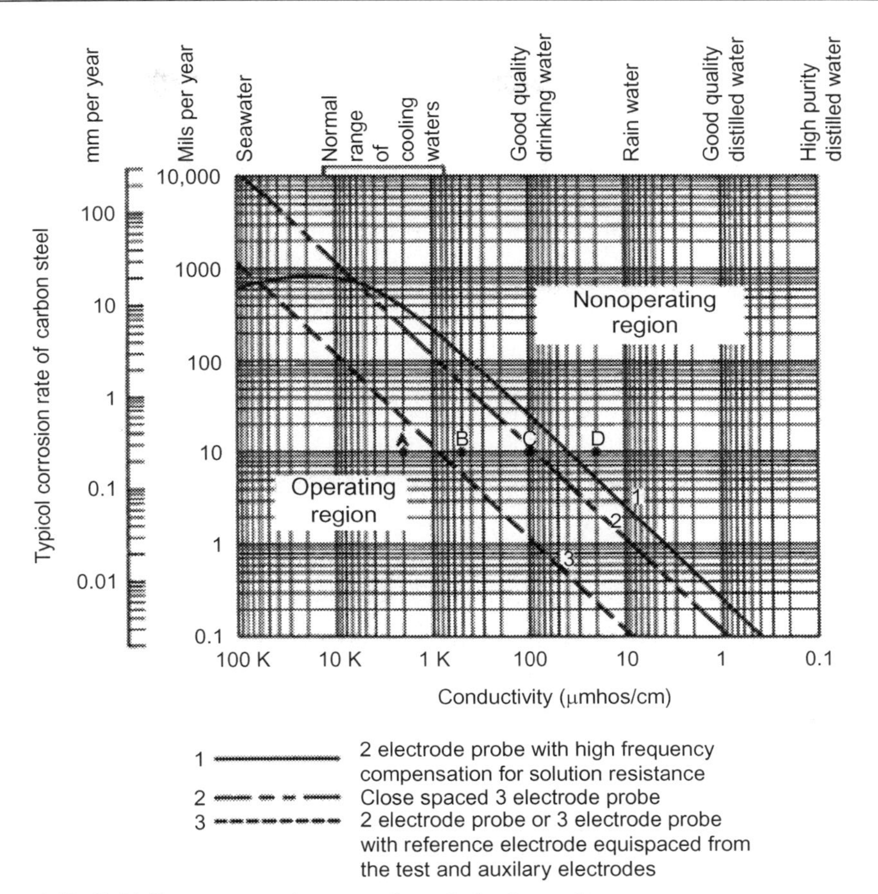

Fig. 3.13 Guidelines on operating range for polarization resistance.
Reprinted from "Standard Guide for On-line Monitoring of Corrosion in Plant Equipment (Electrical and Electrochemical Methods)" Fig. 2, with permission from ASTM International.

3.7 Monitoring of corrosion rate by polarization methods in the field

Because polarization resistance technique (see Section 3.4.1) produces results within minutes and does not affect the corrosion rates, it is the most widely used in the field. All other polarization techniques are invariably used only in the laboratory.

When compared to measuring corrosion rate in the laboratory, monitoring corrosion in the field using polarization resistance technique requires several adjustments. Table 3.8 compares the main differences in measuring corrosion rate in the laboratory and in monitoring corrosion rate in the field.

Several configurations of probes (electrodes for field usage are commonly known as probes) or coupons are used. Tools to introduce and retrieve probes under pressure

Table 3.8 Difference in using polarization techniques in the laboratory corrosion measurement and field corrosion monitoring.

Electrochemical parameters	Laboratory measurement	Field monitoring
Counter electrode	Platinum or graphite—two electrodes	Normally made of the same material as that of working electrode
Reference electrode	Standard reference electrode (reversible)	Normally made of the same material as that of working electrode
Tafel constants	Determined by Tafel extrapolation method	Normally assumed
Luggin capillary	Used	Not used
Corrosion potential	Measured and reported against standard reference electrode	Only stable values are noted

and temperature are available. Figs. 3.14 and 3.15 [19, 20] present some commonly used coupon/electrode designs. It is necessary to place the electrodes/coupons in the most corrosive spot in the system. For instance, referring Fig. 3.15, measurements using probes/coupons 5 and 6 are not relevant, as they may be in oil or gas phase. Placing the electrodes probes in those locations is relatively easy from the operational standpoint, but the results obtained from those probes are irrelevant or erroneous. Recording R_p measurements using electrodes coupons 1 and 2 are most relevant because the electrodes probes are in the brine solution (high-conductivity). When electrodes coupons 3 and 4 are used, one should make sure that they are indeed in the conducting brine phase during the entire time when the R_p measurements are obtained.

Another precaution that should be considered in using the electrochemical polarization probes in the field is the ability to make direct electrical connection between the probe and the potentiostat (instrument). These connections must be insulated; sometimes installation of such a system may be costly.

3.8 General limitations of polarization methods of determining corrosion rate

As with any other techniques, electrochemical polarization techniques also have advantages and disadvantages. In this section, the shortcomings of electrochemical techniques, methods to overcome the shortcomings, and caution exercised when the shortcomings cannot be addressed are discussed [23–26].

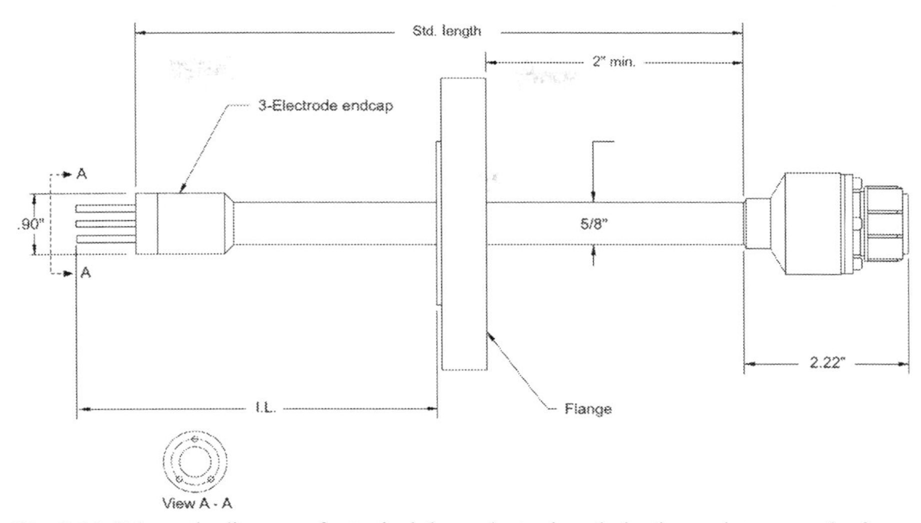

Fig. 3.14 Schematic diagram of a typical three-electrode polarization resistance probe for determining corrosion rate in the field.
Courtesy Metal Samples, A Division of Alabama Speciality Products, http://www.alspi.com/lp6000.htm.

3.8.1 Solution resistance

The data obtained during electrochemical measurements should be directly relevant to corrosion. In all electrochemical polarization methods, either current is applied and potential monitored or potential applied and current monitored. From the measurements, resistance of the electrode is determined from Ohms law, that is, potential, $V =$ current, and I times resistance, R. The measured R is made up of two components: R_s and R_p. It is assumed that the R_s, that is, solution resistance, is low (highly conducting solution), so that the measured $R \sim R_p$. If R_s value is appreciated, the measured corrosion rate will be underestimated (because R_p is inversely proportional to corrosion rate, higher values of R_s exhibit higher value of R). The error due to R_s in the R_p measurement is normally significant in systems with high corrosion rates and low conductivity.

3.8.2 Scan rate

The electrode (i.e., corroding element) can be considered as a capacitor. A capacitor is an electrical device that can store energy in the electric field. When voltage or potential is applied to the capacitor, it builds up electric charge. The amount of charge built on a capacitor is measured in terms of capacitance. When a potential is applied at a higher rate, the electrode's charge build-up is higher. Thus, the current measured will have higher component of capacitance current rather than current resulting from

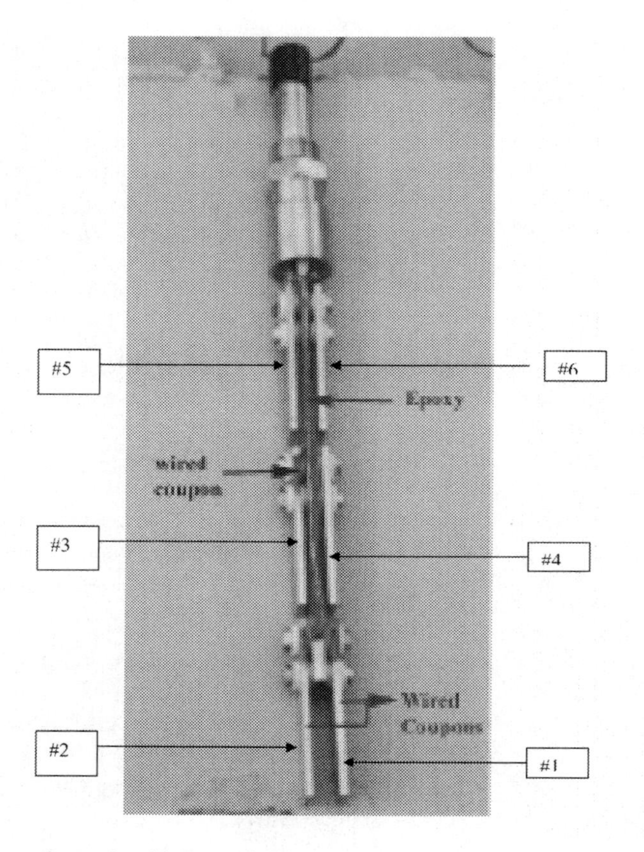

Fig. 3.15 Photo of a retrievable 3-tree holder to mount a 6-coupon assembly in an operating field.

corrosion process. The current associated with corrosion process is known as Faraday current. The relationship between measured current, capacitance current, and Faraday current (i.e., corrosion current) is given as (Eq. 3.21):

$$I_{total} = I_F + C\frac{dV}{dt} \tag{3.21}$$

where I_{total} is the total current measured, I_F is the Faradaic current associated with corrosion rate, C is the electrode capacitance, dV/dt is the scan rate (i.e., rate of change of potential as a function of time).

From Eq. (3.20), the error due to capacitance increases as the scan rate increases. As a result of this error, the corrosion rate is overestimated. To minimize the error due to capacitance current, polarization experiments are conducted at lower scan rate (typically 0.6 V/h).

3.8.3 Electrode-bridging

This error occurs mostly in the field-monitoring probes when the products of corrosion reactions electrically connect the electrodes. This usually happens when the electronic conductivity of the corrosion products is higher (e.g., FeS). As a result, component "*E*" of the ACME is lost. This system behaves as an electronic circuit rather than as an electrochemical corrosion system. When this problem is suspected, the distances between the electrodes should be increased (see Fig. 3.15).

3.8.4 Presence of oxidation-reduction species

If electrochemically active species are present in the electrolyte, they undergo reaction. Under this condition, the current (or potential) change measured in the polarization method is not related to corrosion. Under this situation, measurement of corrosion rates from electrochemical techniques has no meaning.

3.8.5 Variation of corrosion potential

If the corrosion potential varies during electrochemical measurements, the results obtained are not meaningful. Therefore, before conducting any electrochemical measurement, the corrosion potential of the system should be monitored for sufficiently larger duration to establish that it is stable. If corrosion potential does not stabilize, then electrochemical techniques may not be used.

3.8.6 Diffusion-controlled condition

In order to use electrochemical techniques for measuring corrosion, the reaction should be under charge-controlled, that is, the rate of the reaction is proportional to the rate at which electron transfer occurs at the electrode-electrolyte interface. If the charge-transfer rate is high, a situation is developed—usually near the cathode—when the reduction species is fast depleted. At this stage, the reaction rate is controlled by the rate of diffusion of species onto the cathode, that is, diffusion-controlled corrosion, under which conditions, the results from polarization methods are not accurate.

3.8.7 General corrosion only

Most electrochemical polarization parameters measured are only related to general corrosion rate. Initiation and propagation of localized corrosion complicate the response of polarization measurements. Localized corrosion alters chemistry of solution, electrode geometry, and surface area; all of which make quantification of polarization response difficult. These effects may be quantified to some extent by cyclic potentiodynamic polarization (Section 3.4.3) or by CGSP (Section 3.4.4) methods.

3.9 Applications of polarization methods in the field

Various fields in which electrochemical polarization techniques are used include oil and gas field (for determining corrosion inhibitor), chemical processing plants (for monitoring process changes), airplanes (for monitoring crevice corrosion), pulp and paper (for monitoring liquor composition effect on corrosion), and water handling systems (for identifying corrosion upsets) [18, 32].

Electrochemical polarization technique is also used in many standard tests. For example, in cathodic disbondment tests to understand the compatibility between protective polymeric coatings and cathodic protection, the electrode sample is polarized, typically to −1.5 V vs. SCE, and the resultant current is monitored [33].

here.

3.10 Future trends

Computers and computer-based information systems have revolutionized many fields.

The growth of computer software and hardware technologies, communication technologies, and the internet has been a phenomenon in the past decade.

Electrochemical techniques have also benefited from the application of computer-based systems and tools promoting automatic data collection, analysis, and problem solving. Currently, almost all electrochemical monitoring instruments are computer-controlled. Graphical user-friendly interfaces (GUI) and software packages enable electrochemical measurements easy and fast as well as analysis of results instantaneous. Many constants, for example, Tafel slope, equivalent weight of metals, surface area, and conversions of corrosion current density to corrosion rate, can be now selected by the click of a button from the preloaded database. These advances currently make possible to conduct electrochemical polarization methods repeatedly and continuously. As a result, there are now systems available to conduct online, real-time measurements using pocket-size computer.

In future, the electrochemical polarization probes will increasingly be used as online sensors in the field, in addition to their traditional rule as laboratory monitoring tools. The advancements in information technology and web-based knowledge sharing will make the amount of data collected using one sensor phenomena. The data collected will be integrated in the decision-making process online. It is anticipated that the decision-making centers will in future be used to perform four functions:

1. Interpret the data based on the signal provided from electrochemical polarization sensors,
2. Predict future behavior of the system in question,
3. Trigger an action (the action triggered may include taking corrosion control measurement, e.g., addition of corrosion inhibitor if the corrosion rate is high, or adjusting operating parameter, e.g., vary temperature, pressure, or shutting-down the system is necessary),
4. Store the data in a suitable form and place for future reference.

Thus, electrochemical will function as front-end of an intelligent system to model, mitigate, monitor, maintain, and manage corrosion online and in real time. In future, the sensors will thus play a key rule in the Supervisory Control And Data Acquisition (SCADA) systems.

All automotive and fast systems have shortcomings. Caution should be exercised in using electrochemical polarization techniques; their advantageous and limitations should be understood before they are used as sensors. It will serve best for the corrosion professionals to use at least two or more independent measurement techniques, preferably one electrochemical and another nonelectrochemical so that the limitations of one technique is compensated by the other.

3.11 Further information

The purpose of this chapter is to provide an overview of the types of electrochemical polarization techniques that can be used to understand corrosion. The discussions provided are not inclusive. They are meant to provide a flavor of types of techniques available, how to use them, and what information can be obtained.

Several advancements are made in the field of monitoring and new standards are being developed continuously. Most of the advancements are published in corrosion journals (including *Corrosion, Material Performance*, *The Journal of Electrochemical Society*, *British Corrosion Journal*, and *Corrosion Science*), in symposia proceedings (including those organized by ASTM International, NACE International, European Federation of Corrosion (EFC)), in books (including those published by ASTM, ASM, NACE, and EFC), and in standards (including those developed by ASTM G01 committee and NACE STG 62 committees).

The readers are strongly encouraged to refer to the references provided at the end of this chapter and the sources provided in the preceding paragraph.

References

[1] R.W. Revie, H.H. Uhlig, Corrosion and corrosion control, in: An Introduction to Corrosion Science and Engineering, fourth ed., John Wiley and Sons, 2008 978-0-471-73279-2.

[2] M.G. Fontana, Corrosion engineering, in: McGraw-Hill Series in Materials Science and Engineering, third ed., McGraw-Hill Book Company, 2017 978-0070607446.

[3] M. Stern, A.L. Geary, Electrochemical polarization. I. A theoretical analysis of the shape of the polarization curves, J. Electrochem. Soc. 104 (1957) 56.

[4] ASTM G3, "Standard practice for conventions applicable to electrochemical measurements in corrosion testing", Annual Book of ASTM Standards, vol. 03.02, ASTM West Conshohocken, PA.

[5] R. Baboian, C.G. Munger, R.S. Treseder (Eds.), NACE Corrosion Engineer's Reference Book, NACE International, Houston, TX, 2016 9781523106578.

[6] ASTM G215, "Standard guide for potential measurement", Annual Book of ASTM Standards, vol. 03.02, ASTM West Conshohocken, PA.

[7] ASTM G5, "Standard reference test method for making potentiostatic and potentiodynamic anodic polarization measurements", Annual Book of ASTM Standards, vol. 03.02, ASTM West Conshohocken, PA.

[8] ASTM G59, "Standard test method for conducting potentiodynamic polarization resistance measurements", Annual Book of ASTM Standards, vol. 03.02, ASTM West Conshohocken, PA.
[9] Z. Tafel, Über die Polarisation bei kathodischer Wasserstoffentwicklung (Discovery of Tafel Equation), Phys. Chem. 50 (1904) 641.
[10] ASTM G61, "Standard test method for conducting cyclic potentiodynamic polarization measurements for localized corrosion susceptibility of iron-, nickel-, or cobalt-based alloys", Annual Book of ASTM Standards, vol. 03.02, ASTM West Conshohocken, PA.
[11] ASTM G100, "Standard test method for conducting cyclic galvanostaircase polarization", Annual Book of ASTM Standards, vol. 03.02, ASTM West Conshohocken, PA.
[12] ASTM G150, "Standard test method for electrochemical critical pitting temperature testing of stainless steel", Annual Book of ASTM Standards, vol. 03.02, ASTM West Conshohocken, PA.
[13] ASTM F746, "Standard test method for pitting or crevice corrosion of metallic surgical implant materials", Annual Book of ASTM Standards, vol. 13.01, ASTM West Conshohocken, PA.
[14] ASTM G71, "Standard guide for conducting and evaluating galvanic corrosion tests in electrolytes", Annual Book of ASTM Standards, vol. 03.02, ASTM West Conshohocken, PA.
[15] ASTM G82, "Standard guide for development and use of a galvanic series for predicting galvanic corrosion performance", Annual Book of ASTM Standards, vol. 03.02, ASTM West Conshohocken, PA.
[16] ASTM G102, "Standard practice for calculation of corrosion rates and related information from electrochemical measurements", Annual Book of ASTM Standards, vol. 03.02, ASTM West Conshohocken, PA.
[17] NACE Task Group T-1D-34 Technical Committee Report, Laboratory Test Methods for Evaluating Oilfield Corrosion Inhibitors, NACE Publication 1D196NACE International, Houston, TX, 1996, December.
[18] S. Papavinasam, Corrosion Control in the Oil and Gas Industry, Gulf Professional Publication (Imprint of Elsevier), 2013, October. ISBN 978-0-1239-7022-0.
[19] S.W. Dean, Laboratory Corrosion Testing of Metals and Alloys, Materials Technology Institute, 2014 ISBN: 978-1-57698-262-0.
[20] S. Papavinasam, R.W. Revie, M. Attard, A. Demoz, K. Michaelian, Comparison of laboratory methodologies to evaluate corrosion inhibitors for oil and gas pipelines, Corrosion 59 (10) (2003) 897.
[21] ASTM D1125, "Standard test methods for electrical conductivity and resistivity of water" Annual Book of ASTM Standards, vol. 11.01, ASTM West Conshohocken, PA.
[22] S.W. Dean, Corrosion monitoring for industrial processes, in: S.D. Cramer, B.S. Covino (Eds.), ASM Handbook, vol. 13A: Corrosion: Fundamentals, Testing, and Protection, ASM International, 2003, . ISBN 0-87170-705-5. p. 537.
[23] S. Papavinasam, R.W. Revie, M. Attard, A. Demoz, K. Michaelian, Comparison of techniques for monitoring corrosion inhibitors in oil and gas pipelines, Corrosion 59 (12) (2003) 1096.
[24] D.C. Silverman, Practical corrosion prediction using electrochemical techniques (Chapter 68), in: R.W. Revie (Ed.), Uhlig's Corrosion Handbook, third ed., John Wiley and Sons, 2011, ISBN 978-0-470-87285-7. p. 1129.

[25] S. Papavinasam, R.B. Rebak, L. Yang, and N.S. Berke, "Advances in electrochemical techniques for corrosion monitoring and laboratory corrosion measurements", STP 1609, ASTM International, 2019, ISBN 978-8031-76638.
[26] F. Mansfeld, Electrochemical methods of corrosion testing, in: ASM Handbook, vol. 13A, Corrosion: Fundamentals, Testing, and Protection, ASM International, 2003, ISBN 0-87170-705-5. p. 446.
[27] J.R. Scully, Electrochemical tests, in: R. Baboin (Ed.), ASTM Corrosion Tests and Standards: Application and Interpretation, second ed., ASTM International, 2004. ISBN 0803120982.
[28] S. Papavinasam, Advancements in modelling and prediction – pitting corrosion (Chapter 28), in: A.M. El-Sherik (Ed.), Trends in Oil and Gas Corrosion Research and Technologies – Production and Transmission, Woodhead Publishing Limited (Imprint of Elsevier), 2017, ISBN 978-0-08-101105-8, pp. 663–688.
[29] D.A. Jones, Principles and prevention of corrosion, second ed., Prentice-Hall, Upper Saddle River, NJ, 1996 ISBN 0-13-359993-0.
[30] Corrosion Testing and Evaluation, R. Baboian, S.W. Dean (Eds.), ASTM STP 1000, 1989 ISBN 0-8031-1406-0.
[31] J.R. Scully, Polarization resistance methods for determination of instantaneous corrosion rates, Corrosion 56 (2000) 218.
[32] S. Papavinasam, Evaluation and selection of corrosion inhibitors (Chapter 67), in: R. W. Revie (Ed.), Uhlig's Corrosion Handbook, John Wiley and Sons, Inc., 2011, . ISBN 978-0-470-87286-4p. 1121.
[33] S. Papavinasam, M. Attard, R.W. Revie, Modified Cathodic Disbondment Testing of External Polymeric Pipeline Coatings, Paper #7021, NACE Corrosion 2007, Houston, TX(2007).

Electrochemical polarization technique based on the nonlinear region weak polarization curve fitting analysis

4

Huyuan Sun
Institute of Oceanology, Chinese Academy of Sciences, Qingdao, China

4.1 Introduction

As discussed in Chapter 3, electrochemical polarization techniques are widely used for the measurements of corrosion rates that are controlled by the charge transfer processes (activation-controlled processes). The polarization techniques are based on the Butler-Volmer Equation:

$$I = I_{\text{corr}} \cdot \left\{ \exp\left(\frac{2.303\Delta E}{\beta_{\text{a}}}\right) - \exp\left(\frac{-2.303\Delta E}{\beta_{\text{c}}}\right) \right\} \tag{4.1}$$

where I is the current, I_{corr} is the current at corrosion potential E_{corr}, ΔE is the difference between E and E_{corr}, β_{a} and β_{c} are the Tafel constant in the anodic and cathodic directions, respectively.

4.1.1 Measurement in the linear polarization region near the corrosion potential

The derivative of Eq. (4.1) is:

$$\frac{dI}{d\Delta E} = I_{\text{corr}} \cdot \left\{ \frac{2.303}{\beta_{\text{a}}} \cdot \exp\left(\frac{2.303\Delta E}{\beta_{\text{a}}}\right) - \left(\frac{-2.303}{\beta_{\text{a}}}\right) \cdot \exp\left(\frac{-2.303\Delta E}{\beta_{\text{c}}}\right) \right\} \tag{4.2}$$

At the point of $E = E_{\text{corr}}$, the above expression is simplified to:

$$\frac{dI}{d\Delta E}\Big|_{\Delta E=0} = I_{\text{corr}} \cdot \left\{ \frac{2.303}{\beta_{\text{a}}} + \frac{2.303}{\beta_{\text{c}}} \right\} \tag{4.3}$$

That is to say, the derivative of the linear segment of the polarization curve is proportional to I_{corr} and a constant, and this derivative is the reciprocal of the polarization

Techniques for Corrosion Monitoring. https://doi.org/10.1016/B978-0-08-103003-5.00004-7

resistance. Eq. (4.3) is applicable when the ΔE is small (typically in the 0 to ± 10 mV range) and is the fundamental basis for the widely used linear polarization resistance method. Because of the linear relationship between the corrosion current and the reciprocal of the polarization resistance, measurements of the relationship between the current and the overpotential near the corrosion potential would allow the derivation corrosion current. So, this method is widely used for the measurement of corrosion rate in both laboratories and in the field. Readers are encouraged to read Section 4.1 of Chapter 3 for more information about the linear polarization method.

4.1.2 Measurement in the Tafel region

When the ΔE is large (typically between 100 and 500 mV), the Butler-Volmer can be reduced to the following simple expressions:

$$i_{\text{anode}} = i_{\text{corr}} \cdot \exp\left(\frac{2.303\Delta E}{\beta_{\text{a}}}\right) \tag{4.4}$$

for polarizations in the anodic direction, and:

$$i_{\text{cathode}} = -i_{\text{corr}} \cdot \exp\left(\frac{-2.303\Delta E}{\beta_{\text{c}}}\right) \tag{4.5}$$

for polarizations in the cathodic direction.

These relationships were first established by Tafel 100 years ago and they are called Tafel equations. The Tafel Equations mean that the logarithm of *the* I-*to*-i_{corr} ratio is linearly related to the overpotential (or potential) when the polarization curve is measured in a region significantly away from the corrosion potential. Therefore, the corrosion current can be derived by simply extrapolating the linear E-to-log(I) polarization curve. This method is also called Tafel extrapolation method and is widely used in the laboratory. As the large scale of polarization (up to 500 mV from the corrosion potential) may significantly alter the metal surface property, this method has not been widely used in the field. Readers are encouraged to read Section 4.2 of Chapter 3 for more information on Tafel extrapolation.

4.1.3 Measurement in the nonlinear medium polarization region

Fig. 4.1 shows the polarization curve (the thick solid curve). According to the Butler-Volmer equation (Eq. 4.1), this polarization current is equal to the sum of both branch currents (the dotted lines) at the same polarized value ΔE. This thick solid curve is reduced to the linear relationship (Eq. 4.3) near zero overpotential (the dotted line) and to the Tafel relationships (Eqs. 4.4, 4.5) when the polarized values are large (>100 mV) in both the anodic and the cathodic directions (the dashed curves). When the polarized value is positive, the absolute value of branch current of anodic reaction is larger than that of cathodic reaction, so the total polarization current is positive;

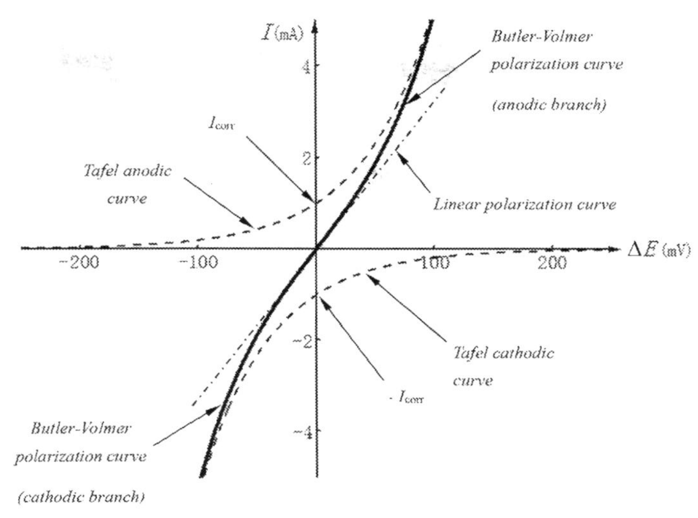

Fig. 4.1 Typical electrochemical polarization curve of uniform corrosion of metal electrode.

on the contrary, the absolute value of branch current of cathodic reaction is larger than that of anodic reaction; the total polarization current is negative.

When the polarized value ΔE is relatively small, the polarization curve is an approximate straight line passing through the origin, and its slope is the reciprocal of the polarization resistance. So, Eq. (4.3) can be used to measure the corrosion current. However, parameters β_a and β_c are required for the calculation of corrosion current. But the parameters β_a and β_c vary from one system to another and they also vary in the same system during the corrosion process because corrosion changes the surface properties and the surrounding electrolyte properties. The determination of these two parameters has been a challenge, especially for field application, and they were often assumed as constants (for example, $\beta_a = \beta_c \equiv 120$ mV), which would inevitably produce errors when used for the calculation of corrosion rate.

When the polarized value ΔE is high, the linearly related Tafel equation can be used to extrapolate the corrosion current. However, the large overpotential during the polarization may significantly alter the metal electrode property and corrosion current measured with the altered metal electrode may not represent the true corrosion current.

A compromised method is to obtain the corrosion current from the polarization curve in the nonlinear medium range. In general, the polarized value ΔE for this range is from -70 to 70 mV, so this is called the weak polarization region. Unlike the linear polarization region and the Tafel region, the polarization curve in the weak polarization region is nonlinear and the corrosion current must be derived by mathematical simulation or curve fitting in this region. This chapter presents the nonlinear curve fitting method for corrosion monitoring.

4.1.4 Characteristics of the curve fitting method

In fact, the curve fitting method is not only applicable to the nonlinear weak polarization region, it also applies to the other regions, including the linear polarization region and the Tafel regions, as long as the corrosion is controlled by the charge transfer processes. With the aid of computer or microprocessor, this calculation method is simple, reasonable, and can be easily adopted for corrosion monitoring in the field. The curve fitting method does not involve the assumption for linear relationship as in the case of linear polarization resistance method and the Tafel extrapolation methods. So, the curve fitting method is more accurate than the linear polarization and the Tafel extrapolation methods.

4.2 Numerical simulation of the polarization curves in the nonlinear region—Weak polarization analysis

4.2.1 Computing software platform

The method of simulation analysis began when computers, especially microcomputers, were available about 50 years ago. Then many researchers had developed some relevant calculation programs. The early programs used to analyze electrochemical parameters were written in FORTRAN, BASIC, and other languages [1, 2].

However, these computer programs were mainly used by the researchers who wrote these programs; very few of them were used by other people in the corrosion community. Particularly, these programs were difficult to use; they were not visual, and there was no visual graphical interface in the calculation process. With the introduction of visual programming languages, user-friendly simulation programs became available. An example is the Fitting 1.0 software written with the Visual Basic software development kit [3]. Fitting 1.0 runs in IBM PC 386 hardware and Windows 3.1 or later platform. The latest version of the Fitting software is 3.0, which can be run on computers with WinXP, Win7, or later versions of Windows.

4.2.2 Mathematical expression of calculation principle of numerical simulation method based on polarization measurement data

In fact, the Butler-Volmer equation (Eq. 4.1) can be used to calculate the corrosion current for all activation-controlled corrosion processes in all regions, including the Tafel region, the linear region, and the nonlinear weak polarization region. The numerical simulation method does not take any approximate calculation in the mathematical model, and it can obtain all of the undetermined parameters in the equation from the measured data in each experiment. The simulation process consists of the following steps:

(1) Impose a voltage pulse E to the metal.
(2) Measure the response current density i.
(3) Repeat the steps 1 and 2 automatically to get arrays of E and i.

(4) Set the initial electrochemical parameters in the Butler-Volmer equation, such as current density i_{corr}, cathodic and anodic Tafel slopes, etc.
(5) Calculate the i value as an objective function for every E value, according to Butler-Volmer equation—this is the key step, which generates a new array i_{calc} based on the array E and electrochemical parameters.
(6) Calculate the total deviation Err by array i_{calc} minus array i, that is, calculate the differences between the i values calculated in step 5 and the measured i values in step 3.
(7) Adjust the initial values and repeat steps 5 and 6 until the electrochemical parameters such as corrosion current density i_{corr} and Tafel slopes and so on are such that the total deviation Err is acceptably small.

In step 1, the stepwise potentiostatic polarization method is used to measure the stable value of the measured polarization current density under different polarization potential values. First of all, an imposed electric pulse voltage is applied to the working electrode (such as a metal) and counter electrode, and then the electric voltage E between the working electrode and the reference electrode and the response electric current density i between the working electrode and counter electrode are measured and recorded. When determining the stable value of the response electric current density i in step 2 under different polarization E values, the polarized value ΔE is taken as the independent variable, and the polarization current density measurement value i is taken as the corresponding variable. Here, ΔE equals to E minus E_{corr}.

Each experiment will obtain a series of measurement data, and all these measurement data are integrated into the array of E and i, as described in step 3.

Next, in step 4, before starting the numerical simulation, i.e., fitting iteration calculation, it is necessary to set initial values for all uncertain electrochemical parameters, which are generally based on experience.

Accordingly, the polarized value ΔE is the independent variable, and the corresponding polarization current density measurement value i is the dependent variable which can also be regarded as the objective function of ΔE based on the Butler-Volmer polarization equation. In this way, the initial values are substituted into the Butler-Volmer Eq. (4.1), and the calculated i value of polarization current density is obtained corresponding to each polarized value ΔE, as described in step 5.

There are differences between the calculated i value and the measured i value, so the total deviation Err between the objective function i value and the measured i value (step 6) needs to be calculated. Ideally, the difference should be zero for a full, 100% fit. In engineering mathematics, it is usually necessary to set a threshold that is acceptably small. The total deviation Err above is calculated by changing the values of electrochemical parameters (step 7), until the total deviation Err between the objective function i value and the measured i value is less than this threshold, then the fitting accuracy is considered to be reached and the calculation will be finished.

4.2.3 Curve fitting software that simultaneously solves for the corrosion current and Tafel slopes

Fig. 4.2 shows the user interface of a typical curve fitting software, Fitting 3.0 (trade name by Huyuan Sun). As advanced options, the software opens the setting of the range of weak polarization region and maximum of fitting cycles (36,000 as default).

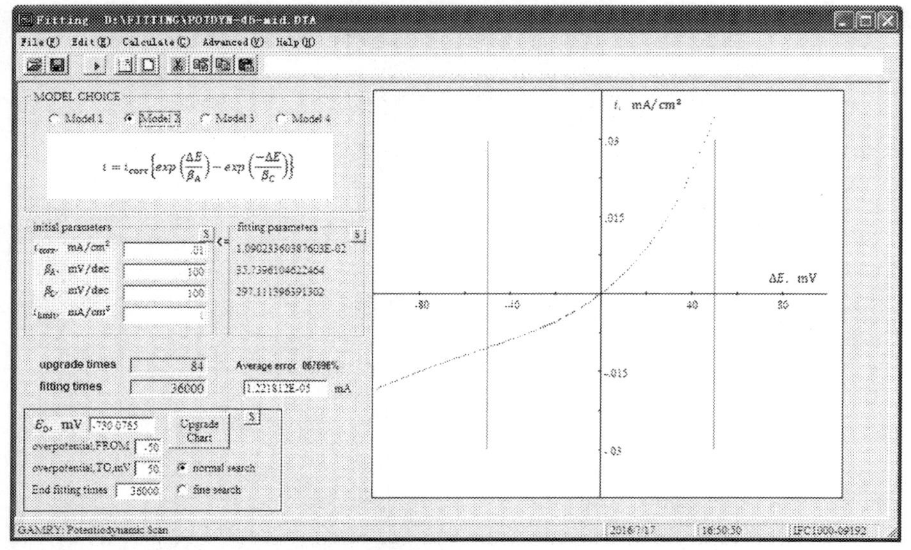

Fig. 4.2 Fitting 3.0 software user interface.

To obtain the better precision, we can set the range of weak polarization region to reasonable zone and set more larger fitting cycles to end the fitting program.

The basic operation flowchart of Fitting 3.0 software is as follows:

- Open the user data file;
- select the fitting calculation model;
- set the initial fitting parameters and fitting calculation parameters (optional, by default if no choice);
- save the fitting results based on each calculation model.

The fitting software can be intervened manually during the process of data processing, including updating the initial fitting parameters, adjusting the search range or end conditions of iterative calculation, and so on. After processing the data, the software will recommend the most suitable fitting model according to the minimum value of the total deviation Err, give the best fitting calculation electrochemical parameters, including the corrosion current density i_{corr}, Tafel cathodic and anodic slopes, and save all the intermediate calculation results in the text file for the user. The fitting software is a perpetual freeware and shareware and can be obtained from the author by email. Specifically, the software is divided into the following functional areas, including:

(A) Model Selection Area for selection of the models. In the example model 2, the Butler-Volmer equation was selected. Model 3 and model 4 are simplified models when cathodic or anode reaction can be ignored, respectively. If concentration polarization of the cathode reaction in corrosion process cannot be ignored, model 1 should be used to obtain reasonable results which will be discussed in next session (see Section 4.2.4 for further information).

(B) Parameters Setting Area for setting initial electrochemical parameters, including corrosion current density, Tafel cathodic slope Tafel anodic slope, and limitation of diffusion current density (model 1 only, see Section 4.2.4). Moreover, the calculation results of electrochemical parameters will be displayed here in real time. During the calculating process, initial electrochemical parameters can be replaced by current electrochemical parameters, according to the user's operation.

(C) Fitting Setting Area for setting the corrosion potential E_{corr}, the starting polarized value and ending polarized value for fitting calculation, and the number of cycles of ending the fitting calculation or the calculation accuracy achieved.

(D) Fitting Display Area for displaying the number of iteration updates of the current calculation, the number of repetitions of the current fitting calculation, and the current calculation error.

(E) Graphic Display Area for displaying the curve image of calculated i value (in red dotted line; gray in print version) and measured i value (in blue dotted line; light gray in print version) relative to the polarized value ΔE, respectively. This function is very important; especially when you choose a wrong model, it gives you a chance to correct the mistakes.

(F) Menu and Button Function Area for data file operation and saving and processing of various calculation results, and even the batch file automation operation.

4.2.4 Advanced model that accounts for diffusion control

From the previous section, we can see that the choice of mathematical model for numerical simulation is very important. If the choice is not suitable, the implied electrochemical meaning will be lost. The Butler-Volmer equation is the most commonly used equation. For the electrochemical process of most metals, the fitting of this equation can generally obtain satisfactory results. However, the Butler-Volmer equation is only applicable for the charge transfer-controlled processes, but some corrosion processes are controlled by both the charge transfer process and the mass transfer process. In this case, if the Butler-Volmer equation is directly used for the fitting calculation, there still will be a relatively large error. To solve this problem, Cao [1] modified the Butler-Volmer equation and incorporated the concept of limiting current density into the calculation. The limiting current density is the maximum diffusion current density that can be achieved when the electrode is sufficiently electrochemically polarized. The modified Butler-Volmer equation is:

$$i = i_{corr} \cdot \left\{ \exp\left(\frac{2.303\Delta E}{\beta_a}\right) - \frac{\exp\left(\frac{-2.303\Delta E}{\beta_c}\right)}{1 - \frac{i_{corr}}{i_L}\left[1 - \exp\left(\frac{-2.303\Delta E}{\beta_c}\right)\right]} \right\} \tag{4.6}$$

where i_L stands for limiting current density. This equation is also known as Cao's equation and it is the Model 1 in the Fitting software (Fig. 4.3). Eq. (4.6) applies to the electrochemical processes that are controlled by both the charger and diffusion processes. When $i_{corr} \ll i_L$, i.e., the concentration polarization of the cathodic reaction in corrosion process can be ignored, then i_{corr}/i_L approaches zero, and Eq. (4.6) is reduced to the Butler-Volmer equation (Eq. 4.1). Eq. (4.1) is the Model 2 in the Fitting software (see Fig. 4.3).

When dealing with the experimental results of polarization curve, we usually need to choose an appropriate fitting mathematical model. The following are the principles and suggestions about the selection of the fitting mathematical model:

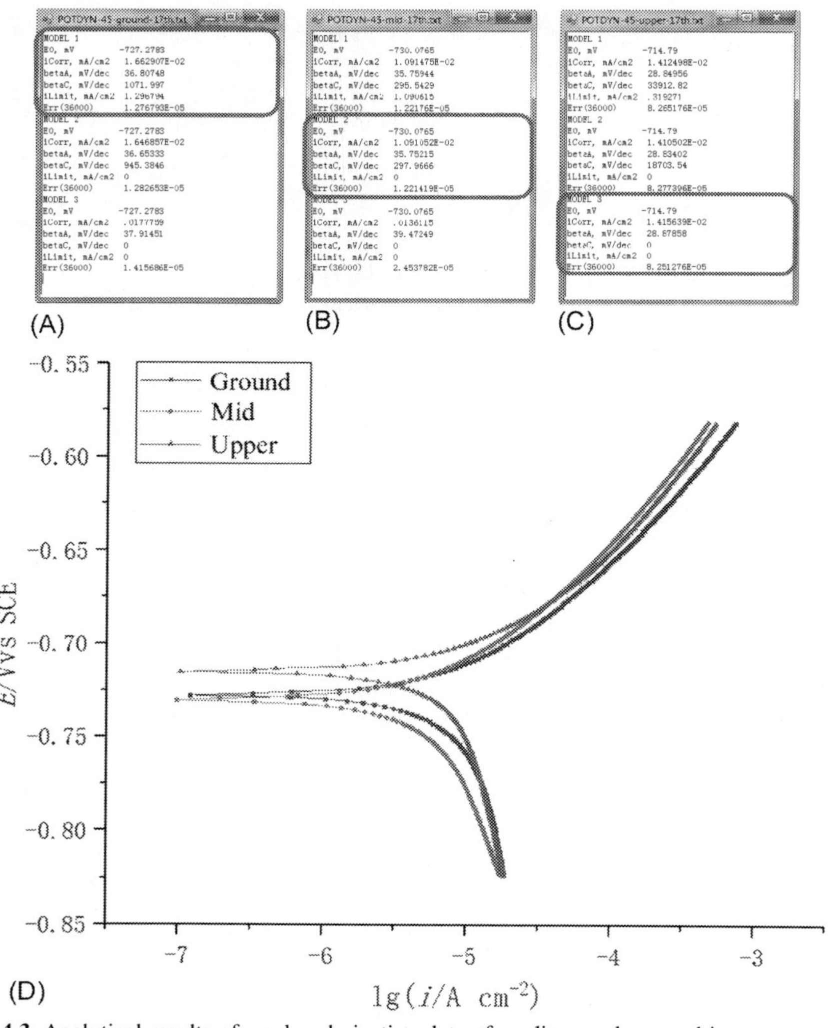

Fig. 4.3 Analytical results of weak polarization data of medium-carbon steel in seawater of Bohai Bay, China, at different depth: (A) Ground layer seawater, −28 m; (B) middle layer seawater, −15 m; and (C) upper layer seawater, −3 m, (D) the three corresponding polarization curves.

Firstly, the electrochemical environment of the corrosion system should be considered. For the general open system, the Butler-Volmer equation (model 2) is good enough. But for some closed systems, or in the case of membrane sealing, or when it cannot be determined, we can use the most complex four parameter Cao's equation to carry out the numerical fitting calculation.

By fitting the calculation results, we get the limiting current density and corrosion current density of the corrosion system, and then compare these two parameters. If the corrosion current density is close to the limiting current density, the system is under diffusion control and the mass transfer process governs the electrochemical reaction in the corrosion environment. If the difference between the limiting diffusing current density and the corrosion current density is more than one or two orders of magnitude, then we recommend using the Butler-Volmer equation to get the more reliable calculation results.

As description above, when $i_{\text{corr}} \ll i_{\text{L}}$, Eq. (4.6) (Model 1) is reduced to the Butler-Volmer equation (Model 2). When the corrosion process is completely controlled by the diffusion step, i.e., $i_{\text{corr}} = i_{\text{L}}$, we have

$$\text{Model 3:} \quad i = i_{\text{corr}} \cdot \left\{ \exp\left(\frac{2.303\Delta E}{\beta_{\text{a}}}\right) - 1 \right\} \tag{4.7}$$

For a passive system, the anodic polarization current stays low [2], i.e., β_{a} approaches infinity and $1/\beta_{\text{a}}$ approaches zero. According to the Butler-Volmer equation, we have

$$\text{Model 4:} \quad i = i_{\text{corr}} \cdot \left\{ 1 - \exp\left(\frac{-2.303\Delta E}{\beta_{\text{c}}}\right) \right\} \tag{4.8}$$

For model 3 and model 4, we should first observe the shape of the polarization curve. From Fig. 4.1, we have seen a typical polarization curve and the shapes of its anodic curve and cathodic curve. In other words, we can see clearly that these two curves based on Eqs. (4.4), (4.5) have different shapes. Obviously, the curve shape of Eq. (4.4) is similar to that of Eq. (4.7), and the curve shape of Eq. (4.5) is similar to that of Eq. (4.8). If the transverse axe of the ΔE is moved to make the anode curve or cathode curve cross the origin respectively, it is the curve shape of model 3 and model 4.

That is to say, the polarization curve fully controlled by anodic reaction is characterized by that with the increase of ΔE—the increase of polarization current is faster and faster and there is no maximum value for the current; while the polarization curve completely controlled by cathodic reaction is characterized by that with the increase of ΔE—the increase of polarization current is slower and slower and there is a maximum value for the current. This is the most significant difference between model 3 and model 4 in curve shape. Similarly, model 3 has a minimum when the ΔE is low enough, while model 4 has no minimum when the overpotential is low enough.

When doing numerical fitting calculation again, we need to carefully observe the shape of the polarization curve. If the curve drawn according to the calculated i value does not have the same characteristics as the polarization curve drawn according to the measured i value in shape, it means that we have chosen the wrong model. Such a calculation will be meaningless and must be stopped immediately to avoid the wrong results and incorrect conclusion.

4.2.5 Electrochemical polarization method for general corrosion monitoring

The corrosion monitoring and detection technology applied in the field is of great value for real-time understanding of the corrosion status, evaluating the effectiveness and safety and preventing the occurrence of corrosion damages. The data from the continuous monitoring of general corrosion of materials in the field are of great guiding value for the evaluation and analysis of the maintenance, repair, and protection of equipment or devices under long-term service conditions. Table 4.1 compares the

Table 4.1 Characteristics of electrochemical polarization techniques for general corrosion measurement.

Polarization method	Measurement polarized value	Information obtained	Influence on corrosion system
Polarization resistance (linear polarization)	±30 mV typical ±10 mV	Corrosion current density (i_{corr})	Tiny
Tafel extrapolation (Tafel polarization)	±500 mV typical >±250 mV	Corrosion current density (i_{corr}) cathodic Tafel slope anodic Tafel slope	Strong
Nonlinear region polarization curves (weak polarization analysis)	±70 mV typical ±50 mV	Corrosion current density (i_{corr}) cathodic Tafel slope anodic Tafel slope limiting current density	Moderate or Weak

main characteristics of polarization resistance method, Tafel extrapolation method, and weak polarization analysis method for general corrosion rate measurements.

In terms of polarization amplitude, the polarization resistance method has the least influence on the corrosion system. Generally, its polarized value ΔE is plus or minus 30 mV, and the typical value is only plus or minus 10 mV. In contrast, the polarization amplitude of Tafel method is plus or minus 500 mV. Therefore, Tafel extrapolation method is not suitable for use in the field as a corrosion monitoring technique, because it has strong interference and influence on corrosion system. With weak polarization analysis, the polarization amplitude is plus or minus 70 mV, and the typical value is plus or minus 50 mV; therefore, the effect on corrosion is moderate or weak. Only electrochemical method which has weak influence on corrosion system can be used for long-term continuous monitoring and detection. Therefore, this method is suitable for long-term corrosion monitoring.

In terms of the accuracy of calculation, the nonlinear curve fitting method (weak polarization analysis) is more accurate because it does not assume a linear relationship to approximate the curves. In addition, it is applicable to all the segments of the whole polarization curve segment.

Usually, there are a series of electrode reactions that may occur on the metal electrode in a field system at different potentials. Because some of these reactions are not related to corrosion of the metal, too large polarization amplitude is not suggested in field monitoring.

4.2.6 Calculation examples

Fig. 4.3 shows a fitting calculation result of the corrosion current density of the medium-carbon steel obtained in the seawater of Bohai Bay, China, at different depth in 2016. Here, all models were tried; however, only models 1 to 3 gave the valid fitting results. During the curve fitting process, models 4 was automatically discarded due to the nonconformity of the curve shape. For the fitting parameters from model 1 to model 3, their total deviation Err is different. According to the principle of minimum value, the calculation results of the optimal model are marked.

4.3 Design of low-power consumption real-time sensor systems for general corrosion monitoring

This section briefly introduces the design of a general purpose low-power real-time corrosion monitoring and acquisition system, which was used for on-site monitoring the corrosion status of metal materials in the environments, which provides fundamental data for the service condition and residual life evaluation of metal structures. The application includes the corrosion monitoring for offshore and onshore oil and gas platforms, drilling platforms, cross-sea highway bridges and railway bridges, and marine science research.

The system has 16 digits A/D conversion data acquisition, built-in temperature unit, real-time clock, 32K bytes ferroelectric storage space, 60K bytes erasable flash storage space, and product identification number which cannot be changed after writing. The system integrates the module that accesses the flash disk through the serial port, which can easily exchange instructions and data with the upper computer. The system is designed with low-power consumption and has been successfully applied in the intelligent corrosion monitoring equipment.

In general, the user can set the sampling interval in advance. The typical requirement is to collect a group of data every minute to several hours. The sampling resolution is required to be 16 bits or higher.

Based on the firmware design written with assembly language, the sensor device will work automatically according to user's default setting and generate binary result file that can be decoded to comma separated text format files (.csv) by special software after they were transferred from the device to user's removable mass storage device, such as a U disk.

The decoded text files are input data files of weak polarization analysis software, which can calculate the simulated fitting results according to four model equations and auto judge the best model with minimum fitting error. Based on these, we can obtain the corrosion current density and Tafel slopes and even limiting diffusion current density.

4.4 Application of corrosion sensors based on weak polarization analysis method

4.4.1 Smart Marine Corrosion Sensor

Marine surveys are important for improving the ability of predicting and alerting marine disasters, improving the ability of monitoring marine pollution and ecological environment, improving the support ability of marine resource development, and improving the acquirement ability of the marine environment-integrated monitoring data so as to benefit the development of marine sciences. In the marine service vessels, there were Conductivity Temperature Depth Profilers (CTD) to measure the real-time temperature, salinity, and depth, and Acoustic Doppler Current Profilers (ADCP) to measure the velocity of seawater. The equipment to measure the corrosivity of seawater was not available until the Smart Marine Corrosion Sensor was developed recently under the financial aid of Chinese 863 research project and NSFC project.

The Smart Marine Corrosion Sensor (SMCS) is based on electrochemical principles. It was designed to measure the corrosion potential and corrosion current in different sea zones, season, day-night, and depth.

4.4.1.1 Measuring the corrosion potential

The corrosion potential of a metal, often termed E_{corr}, is a most useful variable measured in corrosion studies as well as during the corrosion monitoring of complex field situations. Corrosion potential is readily measured by determining the potential difference between a metal immersed in a certain environment and an appropriate reference electrode. Measurement is accomplished by using a high impedance voltmeter which is capable of measuring small voltages accurately without drawing any appreciable current. For measuring metal materials with organic coating, the voltmeter should have an impedance of more than 100M ohms.

4.4.1.2 Measuring the corrosion current

Many electrochemical technologies, such as polarization resistance, weak polarization, electrochemical impedance spectroscopy, Tafel extrapolation, and galvanic cell testing, can be used to estimate the corrosion current density, which is respondent to the corrosion rate. In this study, we chose the popular weak polarization method to calculate corrosion current.

4.4.1.3 Inner design of SMCS

The Model FS-1 SMCS was designed in a pressure tolerance sealed titanium shield. The SMCS is a lithium cell-powered instrument that can real-time measure the corrosion potential and corrosion current of materials in seawater. The lithium cell group was rated to 7.2 V 13.2 Ah. Waterproof cables were used to connect the measured materials and the printed circuit board inside the corrosion sensor. The photograph and the inner block diagram are shown in Fig. 4.4.

The monitored materials such as metals or their alloy or metal with coatings were connected to a copper wire and sealed and then were connected to the corrosion sensor electronics.

4.4.1.4 Characteristics of SMCS

The Model FS-1 SMCS has the following characteristics:

- Real-time measurement of the corrosion potential and corrosion current of up to five kinds of metal materials in seawater or similar aqueous medium.
- Precision: corrosion potential 0.1 mV, corrosion current 0.1 μA
- Range: corrosion potential from −2 to 2 V, corrosion current from 1 μA to 20 mA
- Data collection interval: from 1 to 720 min (user setting)
- Maximum data storage: 2G bytes
- Valid supply voltage: from 5.5 to 11.5 V.
- Power supply: 7.2 V13.2 Ah lithium cell group included.
- Maximum depth: 500 m.
- Shield material: titanium alloy.
- Automatic start while immersed more than 0.4 m depth.

The sensor has five channels to measure the corrosion potentials. A standard cell (model BC-9, Grade 0.005) was used as a voltage signal provider. The voltage of standard cell is almost stable and only change within a narrow gap, from 1018.55 to 1018.68 mV typically. Moreover, its actual voltage was 1018.63 mV under the authorized measurement verification. Based on this, it proves the accuracy of potential results of ground vs positive pole of the standard cell measured by FS-1 SMCS. All of the errors are less than 0.1 mV.

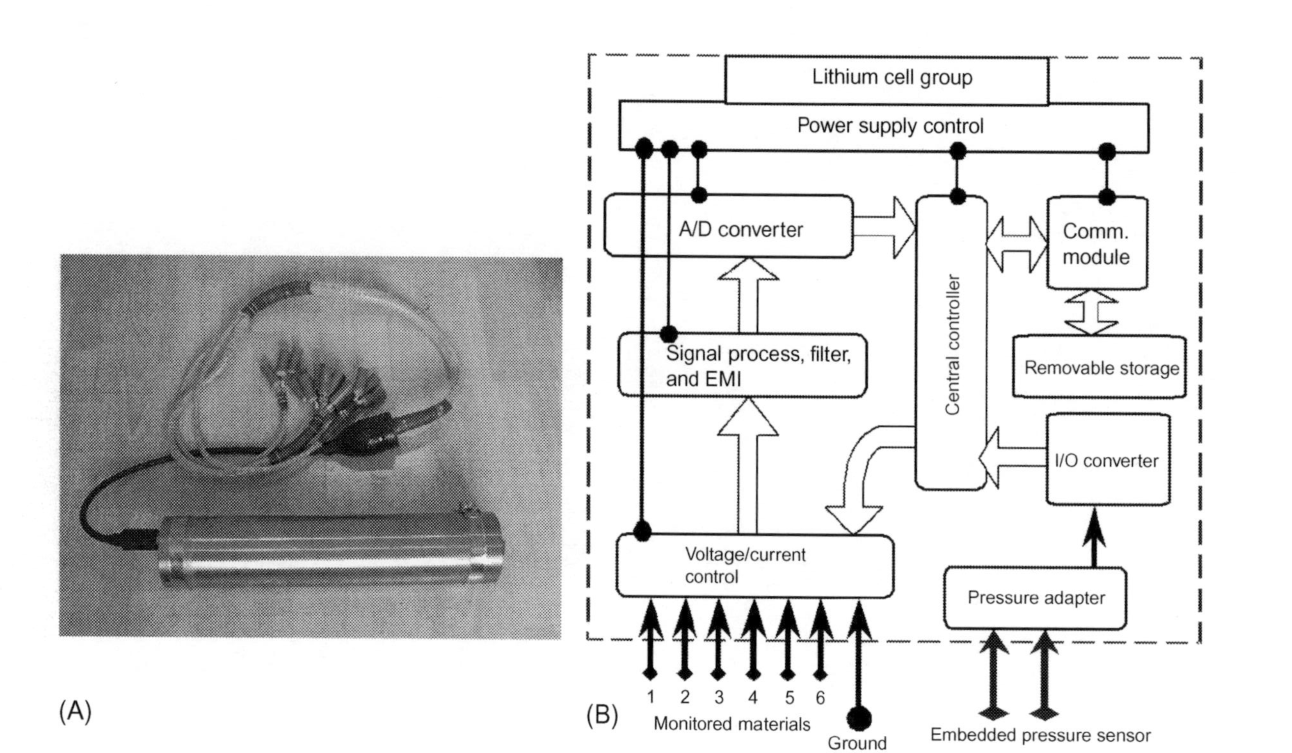

Fig. 4.4 Photograph (A) and inner block diagram (B) of Model FS-1 SMCS.

4.4.1.5 Application of SMCS

The Model FS-1 SMCS is suitable for application in aqueous media, especially in seawater. SMCS does not work if it is not immersed into water or the water depth is less than 0.4 m, because there is an embedded pressure sensor to control the power switch. SMCS mainly includes a complex cover with cables and probes and a titanium shield (Fig. 4.4A). Prior to commission, the lithium cell group can be charged and taken out from the complex cover. The removable storage apparatus such as flash disk is mainly storage media. Control parameters such as record data interval, waiting time, and remark information can be set into a control file, and the measuring results will also be recorded into the flash disk in a binary file. The measured result data file can be opened by the text software to reveal corrosion potential and corrosion current.

For typical application of record interval at 10 min, the SMCS system can work properly for 3 months if it was full filled. And if only corrosion potential need to be recorded, the SMCS can work properly no less than 1 year. Therefore, SMCS is not only suitable for ship carried, but also suitable for fixed application in marine survey.

This corrosion sensor had been successfully used to investigate the seawater's corrosivity to several typical metals in an autumn open cruise as shown in Fig. 4.5, which was first performed on a *Science I* Research Vessel in November 2009.

4.4.2 Deep-sea corrosion rate sensor

4.4.2.1 Technical principle

The deep-sea corrosion sensor adopts the nonlinear range polarization curves (weak polarization analysis) measurement principle. The corrosion current is measured in real time through the three-electrode system, and then the corrosion rate is obtained through Faraday transformation. The three electrodes refer to the working electrode

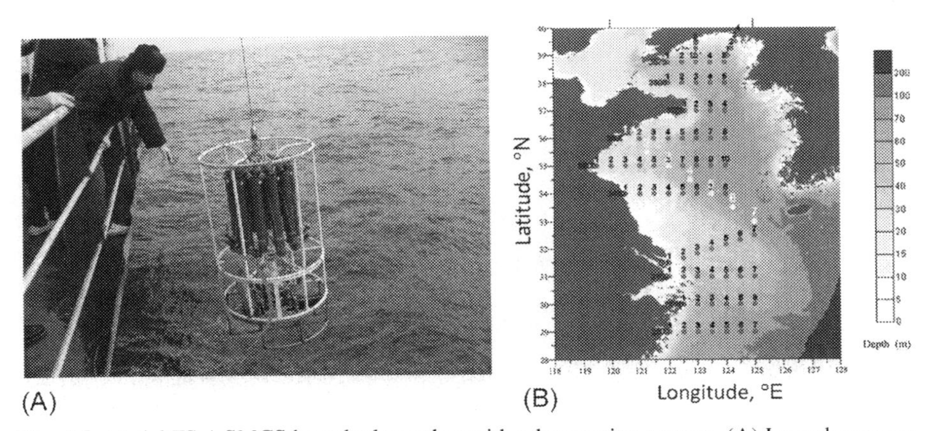

Fig. 4.5 Model FS-1 SMCS launched together with other marine sensors. (A) Launch process, SMCS is in the ellipse ring (B) investigation stations.

(the measured equipment material or component), reference electrode, and auxiliary electrode. The weak polarization analysis technology is a good method which is suitable for the field rapid measurement and accurate solution of corrosion rate.

The deep-sea environment is an area that human beings do not know very well—the deep-sea operating environment is not simply the high-pressure environment that people understand, but also a variety of factors such as submarine hydrothermal, current, and so on; deep-sea equipment may suffer from high-pressure, high temperature, and various chemical substances corrosion. As the corrosion process is a gradual process with increasing risks and accumulating, the damage of corrosion was often underestimated, which often caused huge losses. Therefore, a real-time in situ deep-sea corrosion monitoring technology is urgently needed to predict and study the possible corrosion risks.

4.4.2.2 Research status, latest progress, and development prospect

The Institute of Oceanography of Chinese Academy of Sciences has developed the first remotely operated vehicle (ROV) for underwater corrosion detection in China [4, 5]; under the support of the National 863 project, the FS-1 type marine corrosion sensor has been developed, as described in the previous session, with a design water depth of 500 m. In November 2009, the ROV carried out the first sea test at a depth of 83 m in the Yellow Sea (Fig. 4.5), and then it has carried out the project of the National Natural Science Foundation of China and the marine science of Qingdao. It has been used in the East China Sea and the Yellow Sea for many times with the support of the National Laboratory. The latest development is the production of a 3000 m deep-sea corrosion sensor (Fig. 4.6), which has passed the pressure test.

The National Institute of Marine Technology of India conducted corrosion tests in the Indian Ocean, with the mooring system deployed by submerged buoy [6]. The corrosion rates of test coupons were calculated using weight loss method, rather than electrochemical method and the maximum test depth was 5100 m. Similar technologies are based on methods such as on-site hanging. By installing the sample, only the final corrosion state data can be obtained. The deep-sea corrosion sensor of our production can automatically measure and continuously record the corrosion current data in situ in real time to obtain the change of corrosion process of deep-sea materials, which is an important technical measure to obtain the fundamental data of deep-sea key materials in situ corrosion.

4.4.2.3 Performance of the present deep-sea corrosion rate sensor

Model JF-1 deep-sea corrosion rate sensor is shown in Fig. 4.6; the bottom of the Fig. 4.6A is the physical photograph and the middle of Fig. 4.6A is its accessory. Fig. 4.6B is its breakdown view, which is in data retrieving state, including a flash memory reader, a sensor body with corrosion probe inside, as shown in Fig. 4.6C, and an internal battery. Through this sensor, the corrosion data of X80 steel at 3000 m depth in South China Sea are obtained for the first time, which is shown in

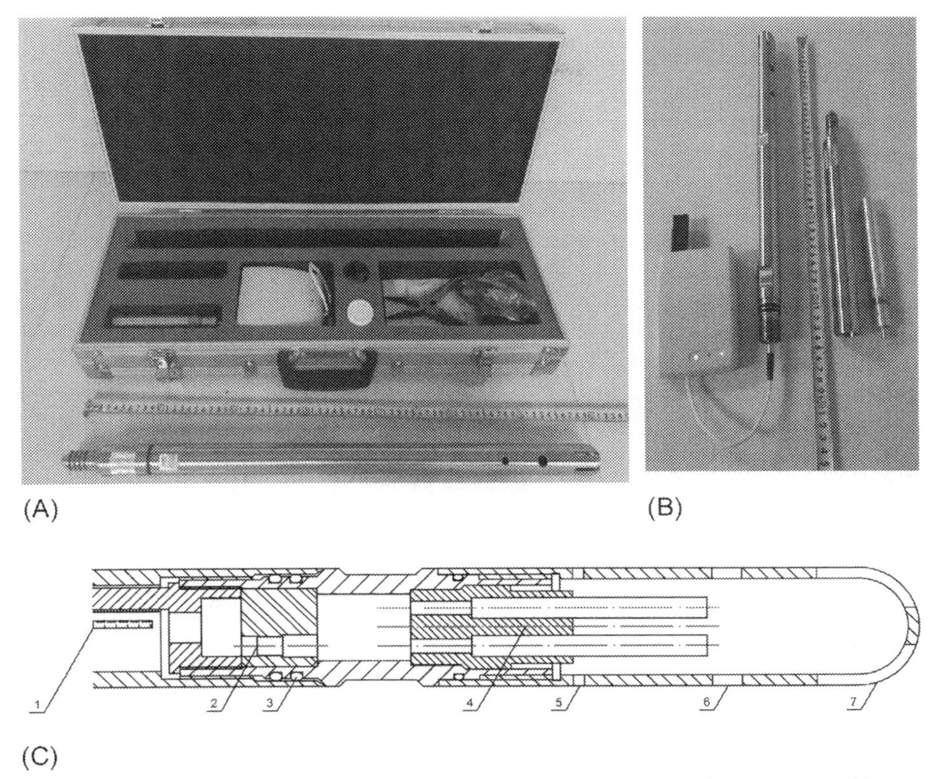

Fig. 4.6 Model JF-1 deep-sea corrosion rate sensor. (A) Photograph of the sensor and its accessories; (B) Photograph for the breakdown view of the sensor; (C) Partial details of the sensor include (1) print circuit board, (2) seal combination, (3) O-type ring, (4) 3-electrode combination (1 behind), (5) overflow hole, (6) mounting hole, (7) elongated guide hole.

Fig. 4.7. This is a continuous observation curve of corrosion rate, which can clearly find the trend of corrosion rate of X80 steel under different water depths (Fig. 4.7B).

The 3000 m marine test in South China Sea is shown in Fig. 4.7B, which is performed by R/V Xiangyanghong 18 supported by the 2018 autumn sea trial voyage of China National Laboratory for Marine Science and Technology.

4.4.3 Challenges of deep-sea corrosion sensors

4.4.3.1 Communication

The current deep-sea corrosion sensor is a self-contained instrument, without real-time communication interface. To meet the real-time access requirements of the platform equipment, bottom access of the equipment is urgently needed to be solved, especially RS-232 and RS-485 communication ports need to be added and the communication protocol should be modbus compatible.

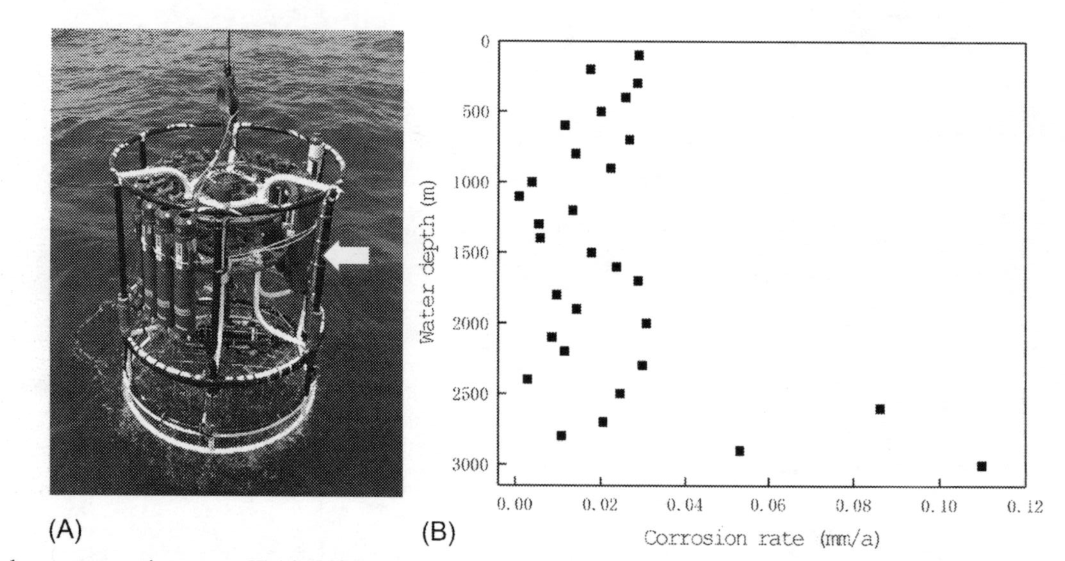

Fig. 4.7 Smart deep-sea corrosion sensor JF-1 in 3000 m marine test. (A) JF-1 sensor to be launched as arrow figured. (B) The measured corrosion rate in different depth.

4.4.3.2 Deeper depth

The design of the deep-sea corrosion sensor at the depth of 6000 m for short-term use and long-term use is of great value in marine exploration, especially the three-electrode system used for corrosion current measurement need to be improved. The three electrodes include the working electrode made of the measured equipment material or component, which must be exposed to the seawater at one end and connected to the core print circuit board at the other end through a robust high-pressure seal into the sensor cabin.

4.4.3.3 Power supply

The power supply of deep-sea corrosion sensor is another major issue, particularly under long-term service. Seawater battery should be considered for the power source, although it has different requirements for sealing. The seawater battery power supply can achieve a greater capacity and a higher water-pressure resistance relatively due to less consideration of sealing design.

4.4.3.4 Underlying software

User-level corrosion risk classification is based on the existing accumulated data analysis of corrosion big data and using the mathematical model to classify and draw the evaluation conclusion, which is pushed to the user in the way of underlying software or bottom layer middleware algorithm.

Acknowledgments

The work has been supported by National Natural Science Foundation of China (Nos. 41476067, 41076046, 40776044), China National 863 Project (2006AA09Z181), and high-end user project of R/V Kexue of Chinese Academy of Sciences (KEXUE2018G13). The acquisition of relevant scientific data is strongly supported by the R/V Kexue, R/V Science I, R/V Dongfanghong 2, R/V Xiangyanghong 18, under the Shared Voyage Plans of National Natural Science Foundation of China, Qingdao Pilot National Laboratory for Marine Science and Technology, and Center for Ocean Mega Science, Chinese Academy of Sciences.

References

[1] C. Cao, Estimation of electrochemical kinetic parameters of corrosion processes by weak polarization curve fitting, J. Chin. Soc. Corr. Prot. 5 (3) (1985) 155–165.

[2] J. Zhang, Y. Niu, W. Wu, Electrochemical evaluation of corrosion resistance of metals and alloys in molten salts—polarization curve fitting, J. Chin. Soc. Corr. Prot. 10 (3) (1990) 207–215.

[3] H. Sun, X. Li, X. Sun, The application of visualization technology in fitting of weak polarization curve and the electrochemical mechanism analysis of hot dip coating in seawater, J. Basic Sci. Eng. 5 (3) (1997) 268–275.

[4] Huyuan Sun, A submarine underwater detector. Chinese patent, 200420030218.3, 2005.8.3, 2004.
[5] H. Sun, L. Sun, Z. Wang, B. Hou, RM-1 ROV for marine corrosion detection, in: 16th International Corrosion Congress. Beijing, China, Sep. 2005, p. 690.
[6] R. Venkatesan, M.A. Venkatasamy, T.A. Bhaskaran, et al., Corrosion of ferrous alloys in deep sea environments, Br. Corros. J. 37 (4) (2002) 257–266.

Electrochemical noise for corrosion monitoring

Robert A. Cottis
University of Manchester, Manchester, United Kingdom

5.1 Introduction to electrochemical noise

5.1.1 What is electrochemical noise?

Electrochemical noise (EN) is a generic term used to describe the fluctuations in potential and current that occur on a corroding electrode. EN is produced by the processes causing the corrosion (or other electrochemical reactions), and it has been a hope of corrosion researchers that its interpretation would provide an understanding of the corrosion process that cannot be obtained by other means. So far, this hope has not been completely realized, but some progress has been made, and the method has been used for corrosion monitoring. This chapter will review the development of our current understanding of EN and the methods that have been used in corrosion monitoring.

5.1.2 History of EN measurement

Any measurement of the potential of a corroding electrode or of the current to an electrode at a controlled potential will implicitly measure EN, in addition to whatever is being measured deliberately. However, it was only in 1968 that noise was seen as a possible source of information, rather than an "error" in the measurement. The first deliberate measurement of EN for a corrosion system was made by Iverson [1], who recorded the electrochemical potential noise (EPN) and concluded "Investigations of these voltage fluctuations appear to offer much promise for the detection and study of the corrosion process and for the study of corrosion inhibitors." At about the same time, Tyagai [2] examined EN from an electrochemical engineering perspective and presented a relatively advanced interpretation of the expected characteristics of the noise. While it was not generally described as EN at the time, several workers, including Stewart and Williams [3] recorded the fluctuations in current associated with the phenomenon of metastable pitting. This was thus probably the first measurement of electrochemical current noise (ECN). The next major advance came when Eden et al. [4] realized that three electrodes could be used to make a simultaneous measurement of potential and current noise. By dividing the potential noise by the current noise (as the standard deviations), a parameter with units of resistance, and consequently termed the electrochemical noise resistance and commonly referred to as R_n, was obtained. The details of the derivation of this parameter and its

Techniques for Corrosion Monitoring. https://doi.org/10.1016/B978-0-08-103003-5.00005-9

significance are discussed further below. The three-electrode measurement configuration has become the standard for EN measurements, although a number of attempts have been made to try to make an EN measurement on a single electrode, thereby avoiding some of the compromises that are inherent in the analysis of the conventional measurement. These are discussed further below.

5.2 Measurement of EN

At first sight, the measurement of EN is simple, and this has been promoted as an advantage of the method. However, considerable care is needed to obtain reliable measurements; early results should be viewed with some suspicion, as they were often seriously contaminated with noise from other than electrochemical sources, which, even at the time of writing poor quality data, occasionally finds its way into the peer-reviewed literature.

5.2.1 Electrochemical potential noise

The measurement of EPN can be made either by recording the potential difference between a corroding electrode and a low-noise reference electrode or the potential difference between two corroding electrodes. The latter technique has advantages for practical corrosion monitoring, although the results may be slightly more difficult to interpret as it is not possible to determine unequivocally which of the two electrodes is the source of the noise.

5.2.2 Electrochemical current noise

The measurement of ECN is normally made by measuring the current between two nominally identical electrodes. Alternatively, it can be made by measuring the current drawn by a single electrode held at a fixed potential. The first method is simpler, as it avoids the requirement for a low-noise reference electrode and potentiostat and also avoids questions about the effect of holding the working electrode at a fixed potential, rather than allowing it to vary naturally.

5.2.3 Simultaneous measurement of potential and current noise

If the ECN is measured as the current between two nominally identical working electrodes, the potential of this coupled working electrode pair can be measured with respect to a reference electrode or a third working electrode. This permits the measurement of R_n as the standard deviation of potential divided by the standard deviation of current and has become the conventional method of measurement of EN (see Fig. 5.1). There are differences between the method using a reference electrode or a third working electrode; the former method is scientifically somewhat better, since it avoids complications associated with uncertainty as to whether the noise emanates from the working electrode used as a pseudo-reference electrode, but it requires a reliable

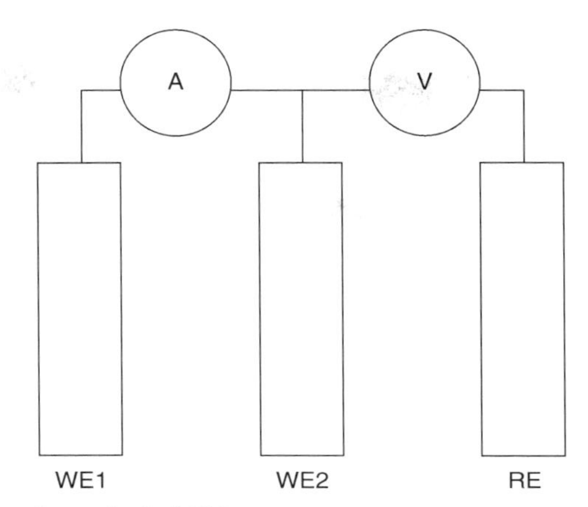

Fig. 5.1 Three-electrode method of EN measurement.

reference electrode, which may be problematic in practical monitoring applications. Thus, two working electrodes and a reference electrode are normally used in the laboratory, whereas the three similar electrode methods are more common in corrosion monitoring.

5.2.4 Instrumental requirements

It is beyond the scope of this chapter to present a detailed description of the instrumentation aspects of the measurement of EN (see Ref. [5] for further information), but it is important that users are familiar with the requirements of reliable measurements, as there are many artifacts that can lead to erroneous results.

5.2.4.1 Potential measurement

Most electrical measuring devices are designed to measure voltage, and at first sight, the measurement of EPN is simple. However, the amplitude of EPN is typically relatively low (<1 mV), and if a true reference electrode is used, the average potential may be several hundred mV. Thus, the measuring system should have a high sensitivity, and it may be beneficial to offset the average dc level, either by subtracting a predefined value or by using a very low frequency high-pass filter to remove the dc. Any real voltage amplifier (whether the input of the voltage measuring system or the input of a signal-conditioning circuit used before the voltage measurement) will have a number of error and noise sources. The influence of these on the measurement will depend on the source impedance of the system being measured, and it is recommended that the performance of the measuring system should be checked using a dummy cell with properties similar to those expected for the real measurement (see Fig. 5.2).

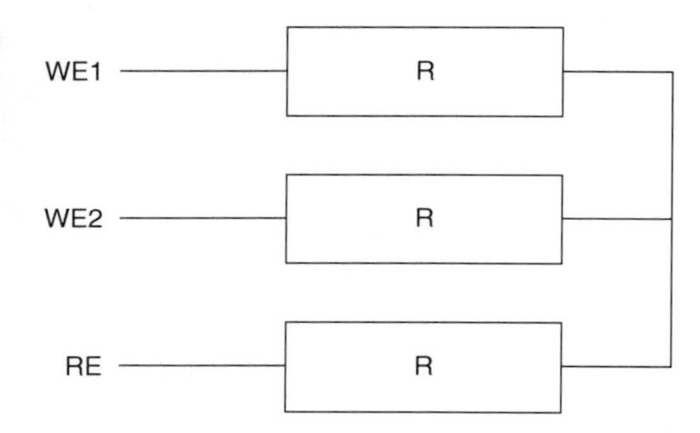

Fig. 5.2 Dummy cell for measurement of instrument noise levels—all resistor values can be the same and should be comparable to the impedance of the source being measured.

5.2.4.2 Current measurement

In general, current is measured as the voltage across a resistor. This resistor can be placed directly in the current path, but for the conventional ECN measurement this will inevitably lead to a voltage difference between the two working electrodes. While the resistor, and hence the voltage drop, can be kept small, this in turn leads to difficulties in the measurement of the voltage. Therefore, it is normal to use a current amplifier that produces a voltage output that is proportional to a current input, with (nominally) no voltage drop between the input terminals. Because of the latter property, the device is normally called a *zero resistance amplifier* (ZRA) in the corrosion community. As with the voltage amplifier, a real ZRA has a number of error sources, and the performance of the system should be checked using a dummy cell.

5.2.4.3 Filtering

EN is almost invariably analyzed using computing techniques (though for monitoring application, the computer may be embedded in the measuring instrument). Consequently, it is necessary to sample the continuous, analogue signal before digitizing it to produce a sequence of numbers that can be handled by the computer. It is beneficial to remove some signal frequencies before the analogue signal is sampled. A low-pass filter (one that allows frequencies below a specified value, known as the cut-off frequency, to pass, while removing higher frequencies) should normally be used to remove frequencies above the Nyquist frequency (half the sampling frequency, the maximum frequency that can be represented by the digitized output) to avoid the problem of aliasing. A high-pass filter (one that allows frequencies above a specified value to pass, while removing lower frequencies) can be used to remove the dc component of a signal and low frequency drift, although it is important to be aware that a high-pass filter with a low cut-off frequency can take a very long time to respond to step changes in the input, so a long settling time is needed after the filter is first connected to a signal source.

5.2.4.4 *Error sources*

Besides the noise and other error sources associated with the potential and current amplifiers, additional errors may arise as a result of artifacts in the signal-conditioning or digitization process.

Aliasing

Aliasing occurs when the input signal contains frequencies above the Nyquist frequency for the analogue-to-digital converter. Once the signal has been sampled, it is impossible to distinguish components produced by aliasing of higher frequencies from real lower frequency information, so it is important to prevent aliasing from occurring by appropriate filtering or by the use of a digitization method that automatically removes higher frequencies. The most common problem with aliasing that can be easily detected is due to power-line interference, which will manifest itself as a peak in the power spectrum. However, the more insidious result of aliasing is erroneous, but apparently "normal" data.

Quantization

Quantization is the inevitable result of representing a value as a number with a fixed number of decimal places. This leads to a step between successive possible values, producing a form of noise. Quantization noise is apparent in a plot of the time record as a set of clear discrete steps in the data (see Fig. 5.3). Quantization noise is not too serious if the steps are small compared with the overall signal, but if the digital signal consists of only a few steps, then the resolution of the data acquisition system needs to be improved.

Interference

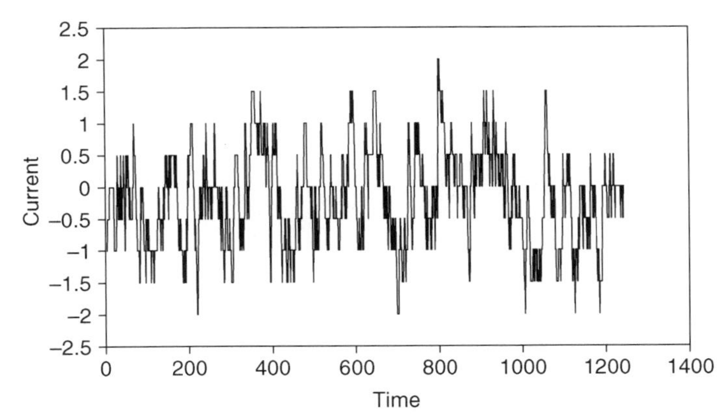

Fig. 5.3 Simulated time record showing quantization noise with a quantization step size of 0.5.

Interference is a result of voltages or currents produced by the coupling of electromagnetic radiation with the measuring circuit. Much interference will be induced by power cables and will, therefore, consist of a sine wave (often quite distorted and hence containing higher harmonics) at the power-line frequency (normally 50 or 60 Hz). Other sources of interference include inductive spikes due to switching of large currents (refrigerator on-off cycling is a common source in laboratories) and high frequency ac interference due to radio frequency induction heaters and the like. Interference typically has a higher frequency than the regions of interest for EN measurements, and it can therefore often be removed by filtering. However, it is best minimized by shielding the measurement system (e.g., by using a Faraday cage), by paying careful attention to the layout of signal cables, and by avoiding ground loops.

Validation

Owing to the scope for measurement errors and artifacts, EN measurement systems should be validated when starting work on a new system, A process for achieving this is described in Ref. [6].

5.3 Alternative EN measurement methods

While the conventional three-electrode method is widely used, a number of modifications of this method have been proposed. All too often these methods have been developed on an ad-hoc basis, with no analysis of the underlying theory.

5.3.1 Methods using asymmetric electrodes

One class of method uses asymmetric electrodes, either through necessity (e.g., when making measurements on a straining electrode during stress corrosion cracking studies), accident (as when studying painted samples, where it is almost inevitable that the samples will differ somewhat) or design.

The theoretical analysis of EN generation with asymmetric electrodes had been reported by Bautista et al. [7]. A key conclusion of this analysis is that it is not possible to determine all of the properties of the working electrodes (the impedance and current noise of each of the electrodes), as there are four unknowns and only two measured parameters. The solution to this problem is to arrange for the properties of one electrode to be known, so that the properties of the other electrode can be determined. However, this still does not provide a complete solution, as it tends to be difficult to measure both properties of the "unknown" electrode. Thus, one variant of the method uses a working electrode coupled to a low-noise electrode that is nominally used to sense the current noise. In practice, the sensing electrode is polarized by the potential noise of the working electrode and the resultant current provides a measure of the impedance of the sensing electrode, and the impedance of the working electrode cannot be determined.

Benish et al. [8] used working electrodes that were nominally identical, but they were maintained with a potential difference between them using a modified ZRA. This was intended as a method of ensuring that all anodic transients emanated from one electrode in order to simplify the analysis. The method may achieve this, but a more important potential advantage may be the tendency to encourage pitting corrosion on the more positive electrode, thereby providing an early warning of possible problems.

Another method using deliberately asymmetric electrodes is that of Chen and Bogaerts [9]. This uses a single working electrode, coupled to a platinum "micro-cathode," with the potential of the coupled electrode pair being measured against a conventional reference electrode. An analysis of the system appears to show that this configuration can measure both the impedance and current noise of the working electrode, but it has been shown [10] that the analysis is flawed and the configuration is not recommended.

A similar method has been used by Klassen and Roberge [11], who used a graphite electrode coupled to nickel aluminium bronze. This work uses a heuristic analysis to estimate the polarization resistance; based on the work of Bautista et al. [7], it is expected that the majority of the current noise will be produced on the metal electrode, and consequently the potential noise will be similar to that for the uncoupled metal electrode, but modified by the galvanic connection to graphite, while the current noise will be the result of the potential noise acting on the impedance of the graphite electrode. Thus, the measured polarization resistance should be that of the graphite, and the net result is that the coupling to graphite serves only to degrade the potential measurement, without providing information of any relevance to the corrosion of the metal.

5.3.2 Switching methods

A problem with the conventional three-electrode method is the need to assume that the two working electrodes are similar. A number of workers have attempted to make near-simultaneous measurements of current and potential noise by switching between a current and a potential measuring configuration. These attempts have often been flawed by the treatment of the corroding metal electrode as a pure resistor, without taking capacitive and diffusional effects into account. Thus, initial attempts switched rapidly between current and potential control. In this configuration, the current-measuring part of the cycle, which necessarily imposes a controlled potential on the working electrode, effectively swamps any potential fluctuations. Subsequent attempts have recognized this problem and switch more slowly (of the order of minutes per cycle or longer). In the latter case, the measurements are valid (providing a sufficient settling time is allowed following the switch in the measurement configuration), but there is a question about the stationarity of the system, especially in the presence of significant changes in the connection of the system.

Two implementations of switching systems have been used commercially:

- The CorrElNoise method [12] was first reported in about 1996. It uses rapid switching (the actual switching frequency is unclear from the manufacturer's website, but appears to be of the order of 10 Hz) between current and potential measurement between two nominally identical electrodes. The analysis presented is based on the treatment of the corroding system as a voltage noise source in series with a purely resistive impedance, with no account taken of the capacitive character of the electrodes. Based on the analysis above, it seems questionable whether the measured potential noise is of any value.

- In 2005, a method of switching between current noise and potential noise measurement was patented by Gill et al. [13]. This uses a relatively long switching cycle (of the order of 30 min), with an EPN time record for a single electrode first being monitored in open-circuit conditions, and then switched to potentiostatic control at the free corrosion potential (at the time of switching—this will include a contribution of the potential noise) to measure an ECN time record. The method appears sound from a theoretical viewpoint, although the current and potential noise will not be correlated in time, so it is difficult to see that it offers a major advantage over the conventional method.

5.3.3 Combined noise and impedance measurement

An alternative approach to the problem of making noise measurements on a single electrode is to recognize that the only reason for measuring both current and potential noise is to permit the estimation of the impedance of the electrode. Consequently, an alternative approach is to measure just the potential noise (which will provide an indication of the nature of the corrosion, or the current noise can be inferred from the potential noise and the impedance), but at the same time to measure the impedance of the electrode at a single frequency by applying a low frequency current signal. This method was patented by Cottis [14], but has yet to be used in practical corrosion monitoring.

5.3.4 Testing EN instrumentation

In most cases, EN instrumentation will be purchased from a commercial supplier and should be of a high quality. However, there is a relatively complex relationship between the source of the noise and the errors induced by the measuring instrument, and this cannot easily be deduced from instrument specifications (and it is rare for the specification to give sufficient information). Thus, it is recommended that EN-measuring instrumentation should be tested to determine the level of instrument noise [6]. This should be done using a dummy cell with impedance comparable to that expected in the real measurement. Even resistors produce a certain amount of thermal or Johnson noise—the dummy cell will produce a potential noise power spectral density (see below) of 4 kTR V^2/Hz (where k is Boltzmann's constant, T the temperature in K, and R the resistance) and a thermal current noise power spectral density of 4 kT/R A^2/Hz. These are the lowest levels of noise that can be detected—good instruments should be able to achieve these noise levels for high source impedances, but will probably not be able to do it for the low source impedances that will characterize corroding electrodes. Most electronic devices also exhibit $1/f$ noise at low frequencies (typically below 1 Hz), so the instrument noise usually also becomes more of a problem at the low frequencies that are often of interest for the interpretation of EN.

5.4 Interpretation of EN

5.4.1 Introduction

The fluctuations that are measured as EN have been produced by the corrosion process without any external influence, and it might therefore be supposed that it would be possible to obtain a lot of information about the corrosion process from the EN data. There have been many attempts to extract useful information from EN data. These have been successful in part, and it is usually possible to get a reasonably good indication of the rate of corrosion. A more challenging objective is to determine the type of corrosion occurring; some progress has been made in this, with parameters that give a reasonable indication of the tendency to localize corrosion having been developed, but only very limited success has been obtained in discriminating between different types of localized corrosion—see Ref. [15] for further information.

This section will present the various methods that have been used, with an emphasis on the techniques that are sufficiently mature, and sufficiently simple to interpret, for use in practical corrosion monitoring.

5.4.2 Direct examination of time records

The simplest method of interpretation of EN data (at least in terms of computational complexity) is direct examination of the potential and current time record. A number of properties can be derived from this:

- Metastable pitting will show as transients in potential and current. If the measurement uses a true reference electrode, all of the transients will be produced on one or other of the working electrodes, so the transients in current may go in either direction (depending on which working electrode the pit occurs), but the potential transient will always go in the negative direction, since the anodic transient from the pit will polarize the cathodic reaction of the two working electrodes. Every current transient should be matched by a potential transient (if not, then the transient may be due to interference rather than being true EN). If the three working electrode method is used, then the behavior is a little more complex. Pits on one or other of the two working electrodes should give a corresponding negative-going potential transient with a corresponding current transient. Pits on the electrode used as a reference electrode will give a positive-going potential transient, but no current transient.
- The initiation period of crevice corrosion will give similar behavior to metastable pitting (indeed, it can be argued that it is metastable pitting). Stable crevice corrosion will typically lead to a significant drop in the potential of the electrode with the crevice, and consequently, a relatively large coupling current if the electrode is one of the current-measuring electrodes. Little work has been done on stable pitting corrosion, but this might reasonably be expected to exhibit similar behavior.
- The initiation of stress corrosion cracks also gives transient events in current and potential, but these tend to diminish in amplitude as the cracks become longer and hence more effective at shielding the currents and potential fluctuations at the crack tip.

While direct examination of time records provides a relatively simple interpretation of the behavior for systems showing clear independent transients, it becomes much less effective as the frequency of the corrosion events increases and transient events start to overlap. It is also dependent on interpretation by a skilled human, which makes it impractical for routine use in corrosion monitoring. However, it is commonly the primary method used to validate computational methods of interpretation. This may be considered as a weakness of most work on EN, in that it is quite rare for EN interpretation methods to be compared with anything other than other EN interpretation methods, notably direct examination of the time record. While unfortunate, this weakness is understandable, in that there are currently few alternative methods that provide the necessary sensitivity for the instantaneous detection of localized corrosion.

5.4.3 Statistical methods

A large class of methods (and almost all that are currently used in practical corrosion monitoring) are based on statistical analysis of the EN data. These methods treat the potential and current noise data as an unordered sample of values from a population. Note the word unordered, which means that the position of a particular value in a time record is not given any significance, nor is the correlation between the potential and current measured at a given time. It might be expected that the loss of the information present in the ordering of the data would limit the capability of these methods, but this is compensated for by the very effective data reduction (usually to only a few values), which makes subsequent interpretation relatively easy.

5.4.3.1 Mean current and potential

It is a moot point whether or not the mean potential or current can truly be regarded as EN, and some measurement methods explicitly remove this by the use of a high-pass filter. However, the mean potential and current are usually measured, and they may contain useful information. Thus, the mean potential of steel in concrete can provide an indication of whether or not corrosion is likely, and a mean current with a large absolute amplitude can be used as an indication that one electrode of a coupled pair is suffering from stable crevice or pitting corrosion.

5.4.3.2 Standard deviation of current and potential

The standard deviation is a direct indication of the amplitude of the fluctuations associated with the noise, and it is by far the most common parameter used in EN applications for corrosion monitoring, at least as the first step in the analysis process.

A number of features of the standard deviation need to be appreciated:

- The magnitude of the standard deviation is dependent on the range of frequencies included in the measurement (usually described in terms of the bandwidth), with a larger bandwidth leading to a larger standard deviation. The exact dependence will be influenced by the shape of the power spectrum.

- Drift in the time record will also contribute to the standard deviation, and it may be appropriate to remove this, either by high-pass filtering before sampling, or by digital filtering or linear trend removal applied to the digital time record.
- The standard deviation of current is normally expected to be proportional to the square root of the specimen area, while the standard deviation of potential is expected to be inversely proportional to the square root of the specimen area. These relationships are, however, not guaranteed to be correct and it is recommended that the specimen area should be quoted, but no normalization for the area should be performed. This is not usually a significant issue for corrosion monitoring, where the probe area is generally fixed.

5.4.3.3 *Noise resistance*

Eden et al. [4] proposed the division of the standard deviation of potential times the specimen area by the standard deviation of current to obtain a parameter with units of resistance times area, known as the electrochemical noise resistance, R_n:

$$R_n = \frac{\sigma_E A}{\sigma_I}$$

where σ_E and σ_I are the standard deviations of potential and current, respectively, and A the area of the sample. Several series of experiments have shown that R_n is comparable with linear polarization resistance, R_p (see also Section 3.1 of Chapter 3). This can also be demonstrated theoretically, providing it is assumed that the response of the metal-solution interface to the noise current can be described by R_p.

Accepting the equivalence of R_p and R_n, it is possible to determine the corrosion current density, i_{corr}, from the Stern-Geary equation:

$$i_{corr} = \frac{B}{R_n}$$

where B is the Stern-Geary coefficient. In the first edition of this work, it was stated that "*this method is arguably the only established EN method, and it has been widely used in corrosion monitoring*." Much work has been published since that was written, but it is probably still valid.

5.4.3.4 *Skewness of current and potential*

The skew or skewness is an indicator of the extent to which the distribution of values is skewed in one direction or another. It is normalized with respect to the standard deviation, and so is dimensionless. An unskewed distribution has a skew of 0. A time record exhibiting a moderate number of transients in one direction can be expected to have a nonzero skew, and the skew has been used as an indicator of metastable pitting events. However, bidirectional transients will tend to give a skew of zero, so this parameter can be expected to be less useful for corrosion monitoring using three similar electrodes.

5.4.3.5 *Kurtosis of current and potential*

The kurtosis is an indicator of the "peakedness" of the distribution of values. As conventionally defined, the kurtosis of a normal distribution is 3, and it is always positive. It is common to subtract 3 from the measured value, in which case it is recommended by this author to use the term "normalized kurtosis" to avoid confusion, although this is not always done. A distribution showing a moderate number of bidirectional transients can be expected to have a kurtosis that is greater than 3, and this has been used as an indicator of localized corrosion [16].

5.4.3.6 *Coefficient of variation*

The coefficient of variation is the standard deviation divided by the mean. It clearly only makes sense for the current, as the mean potential will vary according to the reference electrode used. The coefficient of variation is an indicator of the relative scatter of the values, and it was initially suggested that a large value would indicate localized corrosion. However, it is now recognized that it has a major problem; since the mean current can be zero (indeed, the expected value is zero), the value of the coefficient of variation may go to infinity (in principle, it can also go to minus infinity, as the mean may be positive or negative, but the absolute value is normally taken). This problem is a manifestation of a more fundamental difficulty of the coefficient of variation, as it is only meaningful for a one-sided distribution (one where all values are either all positive or all negative). For this reason, the coefficient of variation of current is not now in common use.

5.4.3.7 *Localization index*

The localization index was developed as a replacement for the coefficient of variation in order to avoid the problem of the possibility of very large values. It is defined as the standard deviation of current divided by the root mean square (rms) current. Since the rms is necessarily greater than the mean, the value can never exceed one. Unfortunately, it can readily be shown that the localization index is a simple transformation of the coefficient of variation, and it suffers exactly the same fundamental limitation.

5.4.3.8 *Pitting factor*

The fundamental problem with the coefficient of variation and the localization index is that the measured mean current is actually the difference between the currents from the two working electrodes, whereas the standard deviation of current will be derived from the sum of the individual standard deviations (strictly the variances add). An indication of the sum of the currents is often available as the estimated corrosion current, and a better parameter can, therefore, be obtained by dividing the standard deviation of current by the corrosion current. This method has been developed by Kane et al. [17] and is used in commercial monitoring systems. In this implementation, the corrosion current is obtained from an independent measurement using harmonic analysis. There is a dimensional problem with the parameter, in that the standard deviation of current (which is expected to be proportional to the square root of area) is divided by the corrosion current density (which will be independent of area) times the area, so the pitting factor is inversely proportional to the square root of the area. However, this is not significant in corrosion monitoring, where the probe size will remain constant. A similar parameter, termed the true coefficient of variation, may also be computed using the corrosion current, I_{corr}, derived from R_n, in which case it can be shown [18] that the true coefficient of variation is the standard deviation of potential divided by the Stern-Geary coefficient:

$$\text{True coefficient of variation} = \frac{\sigma_I}{I_{corr}} = \frac{R_n \sigma_I}{B} = \frac{\sigma_E}{B}$$

Thus, the standard deviation of potential provides a simple indicator of the tendency to localize corrosion, but note that it is also a function of specimen area.

5.4.3.9 Shot-noise parameters

Many of the parameters used in the interpretation of EN data are heuristic, which essentially means that they seem to work, but don't have a theoretical basis. This is a rather disappointing aspect of the use of EN in corrosion studies, as there are relatively simple theoretical analyses available for noise processes. One of the simplest approaches is based on the theory developed originally by Schottky [19] for noise in vacuum electronic devices and is known as shot noise.

The basic assumptions of the shot-noise theory are:

- The current is comprised exclusively of packets of charge of a fixed size. In the case of electronic noise, the packet of charge is the electron.
- The passage of individual packets of charge is independent of other packets; that is to say that the probability of a packet passing in a particular time interval is not influenced by when a packet last passed.
- The packets of charge pass instantaneously (this implies an infinite current, but passing for an infinitesimal time). This is known as a Dirac delta function, a transient of zero width, but finite area.

With these assumptions, it can be shown that the standard deviation of the current will be given by

$$\sigma_I = \sqrt{2qIb}$$

where q is the charge in each packet, I is the average current, and b is the bandwidth of the measurement.

This theory can be applied to corrosion processes if it is assumed that the corrosion is produced by a series of "events" of short duration and constant charge. The basic theory assumes a Dirac delta function for the packets of charge, but if the measurement bandwidth is restricted to low enough frequencies that the events are short compared with the period of the highest frequencies considered, this requirement is effectively met. If it is also assumed that only one of the anodic or cathodic processes is producing noise, and all of that current is produced as packets of charge, then the current I will be the corrosion current, while the charge q will be the charge produced by each event. Hence, the standard deviation of current can be calculated.

If it is further assumed that the corrosion current can be derived from R_n, then it is possible to estimate both I_{corr} and q:

$$I_{corr} = \frac{B}{R_n} = \frac{B\sigma_I}{\sigma_E}$$

$$q = \frac{\sigma_I \sigma_E}{Bb}$$

where B is the Stern-Geary coefficient.

Note that q and I_{corr} are, respectively, the charge and the current resulting from the passage of packets of charge q. Thus, it is also possible to calculate the average frequency of the corrosion events, f_n (termed the characteristic frequency):

$$f_n = \frac{I_{corr}}{q} = \frac{B^2 b}{\sigma_E^2}$$

As it is reasonable to assume that the corrosion current and the frequency of events are proportional to the sample area, it is reasonable to normalize these by dividing by area.

Note that f_n is similar in character to the pitting factor, although it provides a better normalization with respect to area, while q should be independent of area. Also, note that the shot-noise theory depends on the frequencies included in the measurement of standard deviation being low enough to include many complete transient events. A better way of making this measurement is to use the power spectral density (PSD) at a given, low frequency, but the standard deviation forms are given here as these are probably simpler for corrosion monitoring applications (the PSD form can be obtained by using the relationship PSD $\equiv \sigma^2/b$).

5.4.3.10 Coulomb counting

The "coulomb counting" or CoulCount method was developed by Schmitt et al. [20] as a heuristic method, although the underlying theory is reasonably accessible [21]. It depends on the recording of the current noise only. The signal is filtered with a high-pass filter with a cut-off frequency of 0.01 Hz. The filter may be either analogue (although the implementation of a good-quality 0.01 Hz analogue filter is difficult) or digital. The absolute value of the measured current samples is then summed over time. A steep slope is taken as an indication of rapid corrosion, although it is not possible to calibrate this in terms of an estimated corrosion rate.

It can be shown that this method is similar to integrating the standard deviation of current over time [20]. The use of the integrating plot provides a form of low-pass filtering that may make it easier to see trends, but this can also be accomplished by digital filtering. The use of only the current noise restricts the value of the technique, as it does not allow for the estimation of the noise resistance (and hence the corrosion rate), and it is difficult to see what real advantages the method has over the conventional three-electrode method, although the use of integrating plots may be of value.

5.4.4 Spectral methods

In spectral methods, the noise data are transformed from the time domain (i.e., potential or current versus time) into the frequency domain, in which the power present at different frequencies is plotted as a function of frequency. The power is normally plotted as V^2/Hz or A^2/Hz, termed the power spectral density (PSD), and the plot of PSD against frequency is known as a power spectrum (usually plotted on log-log axes). The process of transforming from the time to the frequency domain is known as spectral estimation. Two methods are commonly used, the Fast Fourier Transform (FFT) and the Maximum Entropy Method (MEM) [22]. It should be appreciated that there is a close relationship between the variance (the square of the standard deviation) and the power spectral density. In effect, the PSD can be thought of as the variance measured over a narrow frequency range and normalized to a bandwidth of 1 Hz. The integral of the PSD over the full frequency range included in the measurement is equal to the variance.

Spectral methods generally produce an output that is too complex for use in monitoring, but they may be useful in research. They also provide improved parameters compared with statistical measures for the calculation of noise resistance, pitting factor, and f_n, since the most appropriate frequency range can be chosen, rather than using the arbitrary collection of frequencies present in the standard deviation. One potentially useful extension of spectral methods is the computation of the electrochemical noise impedance by dividing the PSD of potential by the PSD of current and then taking the square root (the calculation being performed at each frequency). The result is also known as the spectral noise resistance, on the basis that only the modulus of the impedance can be obtained, with no phase information.

5.4.5 Wavelet methods

Wavelet methods may be regarded as a form of spectral method, in that wavelets of finite duration are fitted to the time record, rather than a series of continuous sine waves. A nominal advantage of the use of wavelet methods is the avoidance of assumptions about stationarity of the system. It can be shown that wavelet methods have a close relationship to conventional spectral estimation [16]. It is commonly stated that certain wavelet "crystals" (essentially the energy present in a specific frequency range) correspond to specific processes such as pitting corrosion. However, this is something of an oversimplification, since all real transient events will produce power over a wide range of frequencies. Thus, wavelet methods are complex to interpret, and while they are of theoretical interest, and the subject of ongoing study, they are, as yet, inappropriate for corrosion monitoring.

5.4.6 Time-frequency methods

Since the publication of the first edition, a considerable body of work has been concerned with the analysis of the variation of the frequency content of the EN time records as a function of time. The first work in this area was a by-product of wavelet analysis methods, but, more recently, a wide range of methods have been used, including both the discrete and continuous wavelet transform (DWT [23] and CWT [24]), the short-time Fourier transform (STFT) [25], and the Huang-Hilbert transform (HHT) [26]. All such methods produce a time-frequency spectrogram (though other names may be used), a three-dimensional plot of power as a function of time and frequency (Fig. 5.4). While these methods may be of interest in the laboratory, they all suffer from the fundamental problem for corrosion monitoring that they do not reduce the quantity of data (indeed some increase it) and so are not suitable for corrosion monitoring, which needs a small number of simple parameters if it is to be useful in a real-world situation.

5.4.7 Chaos methods

The methods of analysis of chaotic systems are concerned with the detection of deterministic behavior in apparently random signals. There is some evidence that localized corrosion processes have a chaotic character, and parameters derived from chaos analysis have been used as inputs to classifier methods.

5.4.8 Classifier and neural network methods

A general class of methods for the interpretation of measured data can be described as classifiers, in which a range of measured parameters are used to predict outcomes based on prior experience. Essentially, there are two approaches to such analyses:

(1) Classifiers, in which mathematical functions are fitted to the measured data, a simple example being the fitting of polynomial functions to data. Such methods are deterministic, in that the same function will always be obtained with a given classifier when starting from the same initial data.
(2) Heuristic methods, such as neural networks, where mathematical functions are fitted by "training" a function starting from randomly selected control variables. These methods are nondeterministic, in that different functions will be obtained, depending on the randomly selected starting condition.

In essence, both classifiers and neural networks provide a mechanism for the fitting of complex functions to measured data, without the need for detailed knowledge of the functional relationship between the input and output variables [see [27] for a review of the use of neural network methods in corrosion]. In the case of EN, neural network methods have been used to construct a model of the relationship between the statistical properties of potential and current noise and the type of corrosion [28]. As the type of corrosion was assessed by the examination of the EN data by a human expert, the neural network was effectively being used to emulate a human. This is potentially a viable approach to the production of simple outputs that are suitable for corrosion

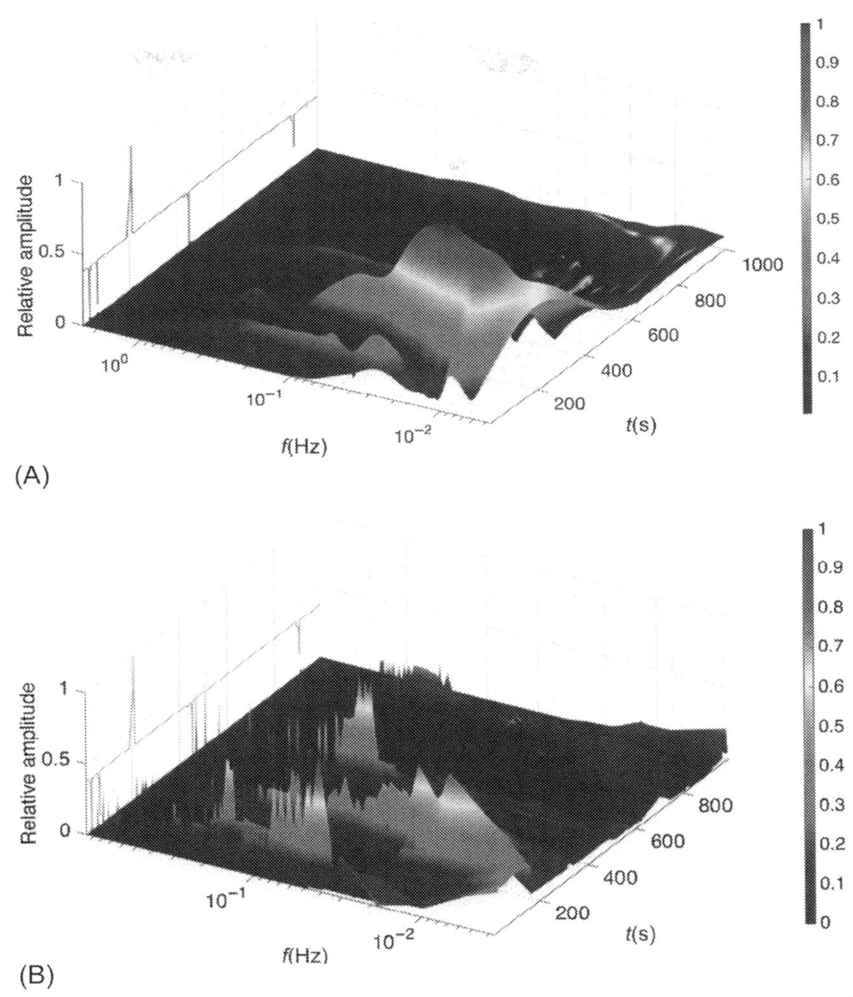

Fig. 5.4 CWT (A) and Hilbert (B) spectrum of the ECN signal of AISI304 exposed in a 10^{-4} M HCl solution for a duration of 1000 s.
From A.M. Homborg, R.A. Cottis, J.M.C. Mol, An integrated approach in the time, frequency and time-frequency domain for the identification of corrosion using electrochemical noise, Electrochim. Acta 222 (2016) 627–640.

monitoring, and it merits further investigation, especially if combined with some of the more effective conventional analysis methods, such as R_n and pitting factor or f_n.

Recently, Alves et al. [29] have used a variety of machine learning methods to analyze data on the corrosion of AISI 1040 carbon steel in a variety of solutions. They found that the best discrimination was achieved with SVM (Support Vector Machines). It is expected that further work on this and similar approach will be forthcoming in the near future, and this seems a promising area for monitoring applications.

However, it seems probable that effective models will be system-dependent, so considerable development will be necessary to use these methods in practice.

5.5 Comparison of EN and polarization resistance for the estimation of corrosion rate

Both R_n and R_p can be used to estimate the corrosion rate using the Stern-Geary equation, and both can be applied reasonably easily in a corrosion monitoring situation. Relatively few detailed comparisons of the merits of the two methods have been reported, although several workers have demonstrated a reasonable correlation between them. Even fewer have tested the "correct" corrosion rate by gravimetric methods, so it is difficult to say which of the two methods is better in respect of the accuracy of measurement. However, we can compare a number of claimed advantages of EN.

5.5.1 Claimed advantages of noise resistance

- It is often suggested that R_n is simpler to measure than R_p. This is rather questionable, especially when considering dedicated instruments for corrosion monitoring; the electronic requirements of the two methods are very similar, and the major difference between the two measurements is liable to be the program in the control microprocessor, rather than the electronic hardware. Furthermore, R_n typically requires a more sensitive measurement and is more easily contaminated by interference.
- The measurement of R_n is claimed not to perturb the system being measured. This can be demonstrated by the thought experiment of considering the two working electrodes in the conventional measurement as being two halves of the same piece of metal. Some workers have been concerned about the "alien" influence of the ZRA used to measure the current, but provided this is well-designed, it is difficult to see how it can influence the behavior of the two electrodes. A more difficult question is whether the measurement of R_p affects the behavior of the electrodes. Another thought experiment is to record the potential noise of a corroding electrode, and then use this as a control signal to measure the properties of a new working electrode—this would effectively measure the impedance of the working electrode, and hence R_p; is this then any different from an R_n measurement? Thus, it is arguable that measurements of R_p can be made in such a way that they do not influence the behavior of the electrodes any more than an R_n measurement (although this is not the case for conventional LPR or impedance measurements).
- In some early work, it was suggested that R_n was not affected by the solution resistance. This is now known to be incorrect, and R_n has just the same dependence on solution resistance as R_p. Similarly, R_n and R_p both measure the properties of the most rapid electrochemical reactions, which may not be the corrosion reactions; they both assume that the reactions are far from equilibrium, and they both depend on at least one of the reactions being activation-controlled.

At present, it is reasonable to say that the jury is still out on the relative merits of the two methods, but it seems that R_p is rather less noisy and possibly a little more reliable.

5.5.2 Use of EN for the identification of the type of corrosion

Where EN really has a "unique selling point" is in its potential ability to give an indication of the type of corrosion. For corrosion monitoring, the method used for identification must be relatively simple, ideally just providing a uniform/localized green-amber-red indication. However, the best method of obtaining this information remains to be determined; some of the methods that have been proposed are:

- The coefficient of variation of current was proposed in early work. It does tend to have a larger value when localized corrosion is occurring, but it is also very sensitive to the value of the mean current, and it is therefore little used now. The Localization Index is similar.
- The Pitting Factor is used in commercial systems. It has some theoretical justification, other than its rather questionable dimensionality, and the performance in service seems to be relatively reliable.
- The characteristic frequency, f_n, derived from a shot-noise analysis has also been found to correlate reasonably well with the occurrence of localized corrosion. It is similar to the pitting factor, but with the minor advantage of a well-defined area dependence. Note that f_n is a function of the standard deviation of potential, and the latter could also be used directly (and very simply) as an indicator. A method of presentation that may be useful is to present the measurements on a map of R_n against f_n [30], which should map different types of corrosion to different parts of the map. This could be considered as a simple classifier, and classifier methods seem promising for this application.
- The slope of the frequency-dependent part of the power spectrum has been suggested as being indicative of the type of corrosion. However, while this may be valid for a small set of experiments, comparison of a wider range gives conflicting results [21], and the method is probably not appropriate for corrosion monitoring.

While the measurement of R_n is relatively insensitive to other sources of noise, such as flow fluctuation, it should be appreciated that the parameters used to identify the type of corrosion are generally very sensitive to such interference. Thus, in work that used a peristaltic pump to change the corrosive solution, it was found that the measured characteristic frequency dropped by several orders of magnitude to approximately the pump pulsation frequency when the pump was switched on [31]. In contrast, the measurement of R_n was hardly affected.

5.6 Practical applications

A number of practical applications of EN to corrosion monitoring have been reported, many at Symposia on Electrochemical Noise that have been held at the Annual NACE Corrosion Conference, or more recently, in related sessions (see www.nace.org for past conference papers).

One application that has been thoroughly reported is the use of EN in the monitoring of the nuclear waste storage tanks at the Hanford site. This is a difficult monitoring problem. As a result of the required safety assessments and decontamination procedures, the installation of probes is extremely expensive, the composition of the solutions stored in the tanks is not known exactly (and they tend to be inhomogeneous due to precipitation of salts), and there is no way of obtaining an independent check on the validity of the measurements. The solutions stored in the tanks are designed to passivate the tanks, so the rate of general corrosion is low and the main concern is the possibility of localized corrosion. For this reason, the use of EN provides the only realistic option for corrosion monitoring, despite uncertainties over the optimum interpretation of the data. As far as it is possible to judge, the programme has been successful [32].

5.7 Harmonic distortion analysis

Any methods that use the Stern-Geary equation to estimate corrosion rate, including the use of electrochemical noise resistance, depend on the value of the Stern-Geary coefficient. This is not a constant and presents an element of uncertainty in the estimation of corrosion rate. There are a number of solutions available for this problem, including the use of an arbitrary value (usually of the order of 25 mV), calibration against corrosion rate measured by another method for the system in question, or measurement of Tafel slopes in a separate polarization experiment. None of these solutions are guaranteed to provide a correct value, since the behavior of the real system is obviously not known exactly (otherwise there would be no need to monitor it), and the method of harmonic distortion analysis (often termed just harmonic analysis) provides the only method of directly measuring the Stern-Geary coefficient on the actual system being monitored. The method relies on the determination of the Tafel coefficients by analysis of the distortion of a sine wave applied to a corroding probe (see Ref. [33] for more details of the method). The distortion leads to the production of harmonics of the original sine wave, and the amplitudes of the harmonics can be used to estimate the value of the Stern-Geary coefficient. This is illustrated in Fig. 5.5, which shows the response of a simulated metal-solution interface to an applied potential sine wave. For this simulation, the anodic Tafel slope was set at 60 mV/decade, while the cathodic Tafel slope was set at 1 V/decade (simulating a mass-transport limited reaction). The peak to peak amplitude of the potential sine wave was the 60 mV, and this gives a severely distorted signal. Fig. 5.6 shows the power spectrum calculated for this signal. At the same time, the amplitude of the response at the fundamental frequency provides a measure of the linear polarization resistance (although it is implicitly measured in somewhat nonlinear conditions). Thus, harmonic distortion analysis should provide a more reliable estimate of the corrosion rate than other methods that have to use a less direct measure of the Stern-Geary coefficient. Note, however, that the analysis makes similar assumptions to those used in the derivation of the Stern-Geary equation, and it is not guaranteed that these assumptions will be valid in all cases:

(1) It is assumed that both the anodic and the cathodic reactions follow Tafel's Law. This is only true if the reactions are activation-controlled; while this is often the case for rapid corrosion processes, such as active metals corroding in acid, it is by no means certain that it is valid for real corroding systems, where factors such as mass-transport limitation and solution/film resistance effects may be important.

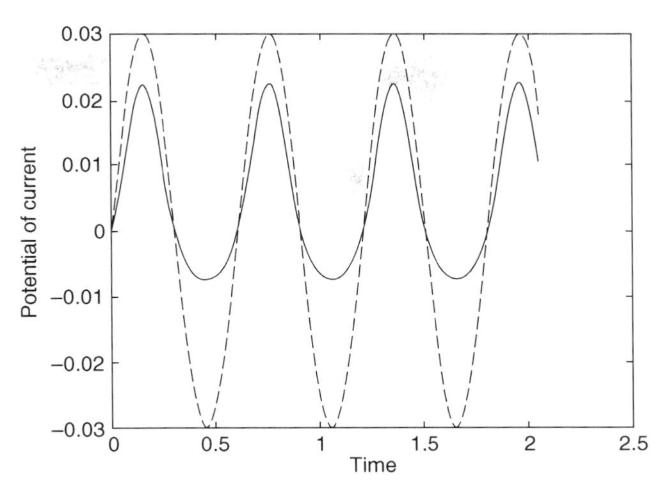

Fig. 5.5 Simulated potential (—) and current (–) signals for a nonlinear metal-solution interface (arbitrary units).

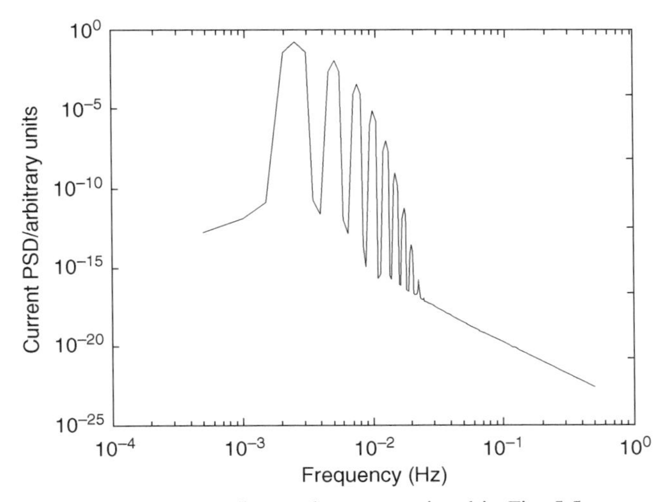

Fig. 5.6 Power spectrum corresponding to the current signal in Fig. 5.5.

(2) It is assumed that both anodic and cathodic reactions are far from equilibrium, so that the rates of the reverse reactions can be ignored. While this is often true, there are situations where it is not. Thus, in hydrogenated water systems, such as are used in the primary circuit of pressurized water reactors, the hydrogen-water reaction is essentially in equilibrium, and electrochemical measurements essentially just measure the exchange current density of that reaction. Similarly, for copper corrosion processes, it is likely that copper will be close to equilibrium with copper ions in the solution.

(3) It is assumed that the frequency of measurement is low enough for capacitive currents to be ignored. This implies a very low frequency of measurement, and in some systems, it is questionable whether a valid measurement can be achieved. Thus, iron sulfide deposits formed in sour (H_2S-containing) environments can be electronically conductive and have a very large surface area by virtue of their porosity, such that the apparent capacitance is many mF/cm^2, leading to great difficulty in making valid measurements.

Despite these potential problems, harmonic distortion analysis has proved commercially successful. It has been used in conjunction with electrochemical noise measurement to provide an indicator of localized corrosion in the form of the pitting factor (see above).

5.8 Electrochemical frequency modulation

A further extension of the harmonic distortion method uses two simultaneous sine waves. When applied to a nonlinear corroding interface, these interact to produce sum and difference frequencies. The method has been termed electrochemical frequency modulation by the developers [34] (the process is known as "intermodulation" in electrical engineering). Analysis of these components also allows determination of the Stern-Geary coefficient, and additionally, allows the production of two "quality indicators" which can be used to check that the measurement is valid. As far as the author is aware, the method has not been used in commercial corrosion monitoring, although it may have promise for the future.

References

[1] W.P. Iverson, Transient voltage changes produced in corroding metals, J. Electrochem. Soc. 115 (1968) 617–618.
[2] V.A. Tyagai, Faradaic noise of complex electrochemical reactions, Electrochim. Acta 16 (1971) 1647–1654.
[3] J. Stewart, D.E. Williams, The initiation of pitting corrosion on austenitic stainless steel: on the role and importance of sulphide inclusions, Corros. Sci. 33 (1992) 457–474.
[4] D.A. Eden, D.G. John, J.L. Dawson, Corrosion monitoring, UK Patent 8611518 (1986), US Patent 5139627 (filed 1987, granted 1992).
[5] S. Turgoose, R.A. Cottis, Corrosion Testing Made Easy: Electrochemical Impedance and Noise, NACE, Houston, TX, 2000.
[6] S. Ritter, F. Huet, R.A. Cottis, Guideline for an assessment of electrochemical noise measurement devices, Mater. Corros. 63 (4) (2012) 297–302.
[7] A. Bautista, U. Bertocci, F. Huet, Noise resistance applied to corrosion measurements: V. Influence of electrode asymmetry, J. Electrochem. Soc. 148 (10) (2001) B412–B418.
[8] M.L. Benish, J. Sikora, B. Shaw, S. Sikora, M. Yaffe, A. Krebs, G. Martinchek, A New Electrochemical Noise Technique for Monitoring the Localized Corrosion of 304 Stainless Steel in Chloride-Containing Solutions, Corrosion 98, Paper 370, 1998.
[9] J.F. Chen, W.F. Bogaerts, Electrochemical emission spectroscopy for monitoring uniform and localized corrosion, Corrosion 52 (10) (1996) 753–759.
[10] R.A. Cottis, The significance of electrochemical noise measurements on asymmetric electrodes, Electrochim. Acta 52 (27) (2007) 7585–7589 (available online from Jan 2007).

[11] R.D. Klassen, P.R. Roberge, Self Linear Polarization Resistance, Corrosion/2002; NACE, Paper 02, 2002.
[12] B. Röseler, C.A. Schiller, Strom-Potential-korrelierte Rauschmessung (CorrElNoise)—Ein neues Verfahren zur elektrochemischen Rauschanalyse, Mater. Corros. 52 (6) (2001) 413–417.
[13] R.P. Gill, V. Jovancicevic, W.Y. Mok, P. Hammonds, Quantitative, real time measurements of localized corrosion events, US Patent publication number 20060144719 (filed 31 October 2005, published 6 July 2006).
[14] R.A. Cottis, Method and apparatus for monitoring corrosion, GB2407169 (filed 6 June 2003, granted 16 November 2005).
[15] R.A. Cottis, Sources of electrochemical noise in corroding systems, Russ. J. Electrochem. 42 (5) (2006) 497–505.
[16] R.A. Cottis, A.M. Homborg, J.M.C. Mol, The relationship between spectral and wavelet techniques for noise analysis, Electrochim. Acta 202 (2016) 277–287.
[17] R.D. Kane, D.C. Eden, D.A. Eden, Evaluation of Potable Water Corrosivity using Real Time Monitoring Methods, Corrosion/2003, Paper 03271, 2003.
[18] R.A. Cottis, Parameters for the identification of localized corrosion: theoretical analysis, in: Electrochemical Society Proceedings PV 2001-22, 2001, pp. 254–263.
[19] W. Schottky, Über spontane Stromschwankungen in verschiedenen Elektrizitätsleitern, Ann. Phys. 362 (23) (1918) 541–567.
[20] G. Schmitt, K. Moeller, P. Plagemann, A New Service Oriented Method for Evaluation of Electrochemical Noise Data for Online Monitoring of Crevice Corrosion, Corrosion/2004, Paper 04454, 2004.
[21] R.A. Cotta, R.A. Cottis, Methods for the Visualisation of Electrochemical Noise Data, Corrosion/2007, Paper 07363, 2007.
[22] R.A. Cottis, The interpretation of electrochemical noise data, Corrosion 27 (3) (2001) 265–285.
[23] A. Aballe, M. Bethencourt, F.J. Botana, M. Marcos, Using wavelets transform in the analysis of electrochemical noise data, Electrochim. Acta 44 (26) (1999) 4805–4816.
[24] A.M. Homborg, R.A. Cottis, J.M.C. Mol, An integrated approach in the time, frequency and time-frequency domain for the identification of corrosion using electrochemical noise, Electrochim. Acta 222 (2016) 627–640.
[25] J. Smulko, K. Darowicki, A. Zielinski, Detection of random transients caused by pitting corrosion, Electrochim. Acta 47 (8) (2002) 1297–1303.
[26] A.M. Homborg, E.P.M. van Westing, T. Tinga, X. Zhang, P.J. Oonincx, G.M. Ferrari, J.H. W. de Wit, J.M.C. Mol, Novel time–frequency characterization of electrochemical noise data in corrosion studies using Hilbert spectra, Corros. Sci. 66 (2013) 97–110.
[27] R.A. Cottis, L. Qing, G. Owen, S.J. Gartland, I.A. Helliwell, M. Turega, Neural networks for corrosion data reduction, Mater. Des. 20 (4) (1999) 169–178.
[28] S. Reid, G.E.D. Bell, G.L. Edgemon, The use of Skewness, Kurtosis and Neural Networks for Determining Corrosion Mechanism from Electrochemical Noise Data, Corrosion 98, Paper 176, NACE, 1998.
[29] L. Alves, R. Cotta, P. Ciarelli, Identification of types of corrosion through electrochemical noise using machine learning techniques, in: Proceedings of the 6th International Conference on Pattern Recognition Applications and Methods (ICPRAM 2017), 2017, pp. 332–340, ISBN: 978-989-758-222-6.
[30] H. Al-Mazeedi, R.A. Cottis, Parameter Maps for the Assessment of Corrosion Type from Electrochemical Noise Data, NACE Corrosion/2004, Paper 04460, 2004.

[31] R.A. Cottis, H.A. Al-Mazeedi, S. Turgoose, Measures for the Identification of Localized Corrosion from Electrochemical Noise Measurements, NACE Corrosion/2002, Paper 02329, 2002.
[32] G.L. Edgemon, Design and Performance of Electrochemical Noise Corrosion Monitoring Systems at the Hanford Site, NACE Corrosion 2004, Paper 04448, 2004.
[33] C. Gabrielli, M. Keddam, H. Takenouti, An assessment of large amplitude harmonic analysis in corrosion studies, Mater. Sci. Forum 8 (1986) 417–427.
[34] R.W. Bosch, J. Hubrecht, W.F. Bogaerts, B.C. Syrett, Electrochemical frequency modulation: a new electrochemical technique for online corrosion monitoring, Corrosion (USA) 57 (1) (2001) 60–70.

Galvanic sensors and zero-voltage ammeter

Lietai Yang
Corr Instruments, LLC, Carson City, NV, United States

6.1 Introduction

When two dissimilar metals are in contact with a conductive solution, a potential difference develops between them. This difference produces a current when the two metals are connected by an electronic conductor or a metal wire and this current is called galvanic current. Of the two metals, the more corrosion-prone one undergoes anodic reaction and releases electrons. The electrons flow through the metal wire to the less corrosion-prone metal and are consumed by cathodic reactions. The cathodic reactions are usually oxygen reduction or hydrogen evolution, neither of which consumes metal. The metal that undergoes anodic reaction is termed the anode; the metal that supports the cathodic reaction is termed cathode.

The term, galvanic, is named after Luigi Galvani. In the mid-1780s, Galvani was studying the effects of lightning with a dissected frog. He accidently observed that the frog's legs twisted wildly when he fastened the frog between a brass hook and an iron railing. This phenomenon occurred even when there was no lightning and the sky was calm. His finding ignited research among scientists of Europe and led to the discovery of the Galvanism concept, which refers to the production of electrical current from the contact of two metals in a moist environment. In Galvani's experiment, the frog's body was the electrolyte to permit the galvanic current to pass between the brass and iron [1].

6.2 Galvanic current and corrosion current

As discussed above, galvanic current is the current that flows between two coupled dissimilar metals in contact with a conductive solution or an electrolyte. Corrosion current is the transfer of all the charge released on the surface of one metal in a solution. The two terms are different and they are discussed below.

6.2.1 Galvanic current

Fig. 6.1 shows the galvanic current flow produced by two dissimilar metals in contact with an electrolyte and coupled by a metal wire. Metal #1 is the more corrosion-prone metal (more active metal) and acts as the anode; Metal #2 is the less corrosion-prone metal (less active metal) and acts as the cathode. When the two metals are immersed in

Techniques for Corrosion Monitoring. https://doi.org/10.1016/B978-0-08-103003-5.00006-0

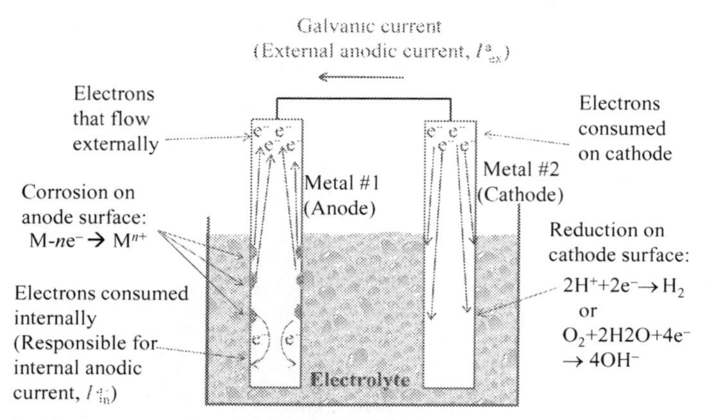

Fig. 6.1 Galvanic current generated by two dissimilar metals in contact with an electrolyte and coupled by a metal wire. It is the current that flows through the external wire due to the corrosion of the anode and it is represented as I^a_{ex}.

an electrolyte and coupled by a metal wire, the anode undergoes corrosion reaction and some of the more active metal, M, on the surface becomes the corrosion products, M^{n+}:

$$M - ne \rightarrow M^{n+} \tag{6.1}$$

where n is the valance of the metal in the corrosion product.

Under the galvanic corrosion condition, the majority of the electrons released at the anode due to the corrosion reaction flow through the external wire from the anode to the cathode and are consumed by the cathodic reaction on the cathode. In some cases, however, small amount of the electrons produced by the corrosion reaction may also flow to the nearby cathodic sites on the same anode and consumed by the cathodic reaction internally. If the electrolyte is acidic, the cathodic reaction is usually the evolution of hydrogen:

$$2H^+ + 2e^- \rightarrow H_2 \tag{6.2}$$

If the oxygen is present, however, cathodic reactions may also include the reduction of oxygen:

$$O_2 + 2H_2O + 4e^- \rightarrow 4OH^- \tag{6.3}$$

All metals produce an electrode potential when it is immersed in an electrolyte. The electrode potential reaches the equilibrium electrode potential when the metal is not connected to any other metals in the solution (under open-circuit condition) after a certain period of time. The equilibrium electrode potential is also called corrosion potential of the metal if the potential involves the corrosion of the metal in the solution; it is also called the open-circuit potential because the potential is measured under open-circuit conditions. The driving force for the galvanic current is the difference between the open-circuit potentials of the two metals in the electrolyte. The galvanic series is a relative ranking of the equilibrium potentials of selected different metals and alloys in seawater electrolyte [2].

As shown in Fig. 6.1, galvanic current is the anodic current produced by the electrons that flow externally through the external wire and internal anodic current is the current produced by the electrons that are consumed internally. The galvanic current is represented as I^a_{ex} and the internal anodic current is represented as I^a_{in} in this chapter to differentiate them from the corrosion current (see below).

6.2.2 Corrosion current

In addition to the open-circuit potential differences of the two metals, the galvanic current also depends on the dynamic kinetic behaviors of the anodic reaction on the anode (more active metal) and the cathodic reaction on the cathode (less active metal). Fig. 6.2 shows how the galvanic current (I^a_{ex}) is related to the differences of the open-circuit potential (or corrosion potentials) of the two electrodes and the polarization curves (kinetic behaviors) of the two electrodes for a general case where the corrosion rate is controlled by both charge-transfer and mass-transfer processes [3, 4]. The thick red (dark gray in print version) solid curves represent the dissolution and reduction polarization behaviors of the anode, respectively. The blue (light gray in

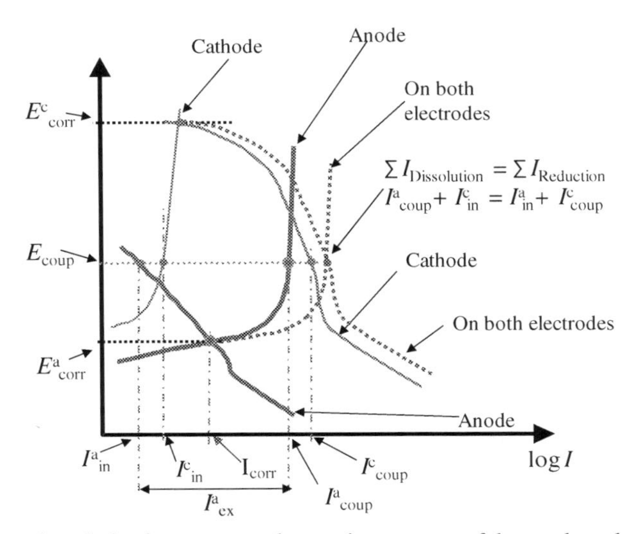

Fig. 6.2 Schematic polarization curves and corrosion currents of the anode and cathode and the galvanic current, corrosion current, and internal anodic current of the anode after it is coupled to the cathode.
Modified from L. Yang, Multielectrode systems, in: L. Yang (Ed.), Techniques for Corrosion Monitoring, Woodhead Publishing, Sussex, UK, 2008, Figure 8.12.

print version) solid curves represent the dissolution and reduction polarization behaviors of the cathode, respectively. The thick blue (light gray in print version) dashed curves represent the total reduction current and the total dissolution current on the combined electrodes (both anode and cathode), respectively. The less active metal (cathode) has a higher corrosion potential, E^{c}_{corr}, and the more active metal (anode) has a lower corrosion potential, E^{a}_{corr}. The current at the corrosion potential of the anode (E^{a}_{corr}) is the corrosion current of this more active metal, I^{a}_{corr}. Similarly, the current at the corrosion potential of the cathode (E^{a}_{corr}) is the corrosion current of the less active metal, but this current is not important as far as galvanic corrosion is concerned.

When the two electrodes are coupled by an external wire, the more active metal acts as the anode and the less active metal acts as the cathode and the potentials of the two electrodes are forced to the same value, E_{coup}, which is called the coupling potential (or mixed potential). At the coupling potential, the corrosion current of the more active metal increased from I^{a}_{corr} to I^{a}_{coup} and the total anodic dissolution current equals the total cathodic current (see the thick dashed curves in Fig. 6.2):

$$I^{a}_{coup} + I^{c}_{in} = I^{a}_{in} + I^{c}_{coup} \quad (6.4)$$

where

I^{a}_{coup} = Corrosion current (dissolution current) on the anode at the coupling potential
I^{c}_{in} = Dissolution (anodic) current on the cathode at the coupling potential
I^{a}_{in} = Reduction current on the anode which is also the internal anodic current on the anode (see Fig. 6.1) at the coupling potential because the same amount of current from the cathodic sites flows to the anodic sites within the same anode. When the anode is not coupled to the cathode and its potential is at E^{a}_{corr} (under open-circuit condition), the I^{a}_{in} is the natural corrosion current on the anode, I^{a}_{corr}
I^{c}_{coup} = Reduction current on the cathode at the coupling potential.

As indicated in Fig. 6.1, the galvanic current (I^{a}_{ex}) is the current that flows through the external wire from the cathode to the anode and the internal anodic current (I^{a}_{in}) is produced by the electrons that are consumed internally. Because the galvanic current flows through the external wire, it is also called the external anodic current on the anode. As shown in Fig. 6.2, at the coupling potential, the corrosion current (dissolution current) on the anode, I^{a}_{coup}, is equal to the sum of the external anodic current, I^{a}_{ex} and the internal anodic currents, I^{a}_{in}:

$$I^{a}_{coup} = I^{a}_{ex} + I^{a}_{in} \quad (6.5)$$

For metal pairs that have sufficient large difference between their corrosion potentials, I^{a}_{in} on the anodic electrode is often much smaller than its I^{a}_{ex} at the coupling potential (note that the axis in Fig. 6.2 is on log scale); the external anodic current on such anode is often used to estimate the corrosion current at the coupling potential:

$$I^{a}_{coup} \approx I^{a}_{ex} \quad (6.6)$$

Eq. (6.6) provides the theoretical basis for why the galvanic current, I^a_{ex}, can be used to assess the corrosion damage to the more active metal (I^a_{coup}) by the coupling of the metal to a less active metal. An example of such cases is mild steel structures that have stainless steel bolts exposed to atmosphere in the presence of salt deposits. Here, the stainless steel is the less active metal and serves as the cathode and the salt deposits serve as the electrolyte when there is moisture in the air and the stainless steel causes the surrounding area of the mild steel to corrode at an accelerated corrosion rate.

ASTM standard G71 is a widely used international standard that provides the detailed procedures for the evaluation of galvanic corrosion currents of two dissimilar metals in contact with an electrolyte both under laboratories conditions and in the fields [5]. ISO standard **7441:2015** is another international standard specifically for the evaluation of the corrosion of two dissimilar metals exposed to atmospheric environments [6].

6.2.3 Galvanic current from two pieces of same metals

The above described Galvanism concept can also be extended to two interconnected pieces of a same metal that have different microstructures in contact with an electrolyte. The piece that exhibits more anodic properties can act as the anode and the piece that exhibits more cathodic properties can act as the cathode. The variations in microstructures on a metal are often the causes for localized corrosion of the metals because the more anodic areas and the more cathodic areas are interconnected by the metal itself. For example, localized corrosion often initiates at the site where there is sulfide microstructures on the surface of a stainless steel. The concept of using a coupled multielectrode array sensor (CMAS) to measure localized corrosion is based on this principle [7, 8] (see Chapter 8 for more information).

Galvanic phenomena also occur between two pieces of the same metal and have the same microstructure if the two metals are surrounded by different local solutions. For example, if one metal has been corroded and covered by a corrosion product, the solution underneath the corrosion product will be different (usually more aggressive) from the bulk solution and there will be a galvanic current if the two metals are connected by a metal conductor. The flow-through sensor is based on this principle [9] (see Chapter 7 for more information).

6.3 Measurement of galvanic current and zero-voltage ammeter

As shown in Fig. 6.1, the galvanic current is the current that flows from Metal #2 to Metal #1 through the external metal wire with both metals at the same electrode potential. To measure the galvanic current, an ammeter must be inserted between Metal #1 and Metal #2. This ammeter should not impose any significant voltage because such voltage would cause a potential difference between the two metals and change the coupling condition for the two metals. According to ASTM 217 [10], a zero-voltage ammeter (ZVA) is a device that imposes a negligibly low voltage drop when inserted

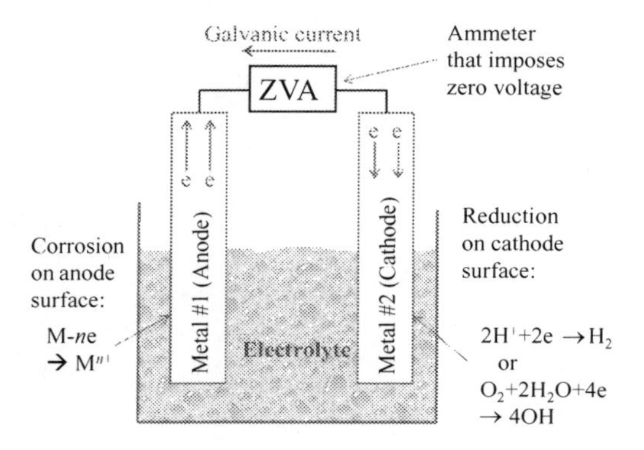

Fig. 6.3 A zero-voltage ammeter (ZVA) is used to measure the galvanic current.

into a circuit for measurement of current. Therefore, ZVA can be used to measure the galvanic current and ensure that the potentials of the two metals are essentially the same during the measurement (Fig. 6.3).

6.3.1 Zero-voltage ammeters formed with operational amplifiers

The circuit of the ZVAs that are built with an operational amplifier (Op-Amp) is shown in Fig. 6.4 [11]. This type of ZVA is often called zero-resistance ammeter (ZRA) and it is often perceived to have near zero input resistance as the term implies. As a matter of fact, the ZRA term is only applicable if the Op-Amp is ideal in which case the potential at the inverting terminal (V_n) is equal to the potential at the non-inverting terminal (V_p). Hence, the input resistance which is the ratio of the voltage drop ($V_p - V_n$) to the measured current (I) is also zero.

In reality, however, Op-Amps are not ideal and do have input voltage. For example, every Op-Amp has an offset voltage (V_{OS}) which is defined as the small differential voltage that must be applied to the input of an Op-Amp ($V_p - V_n$) to produce a zero output (V_{out}). As a matter of fact, the offset voltage is one of the most important parameters that are reported in manufactures' product sheets.

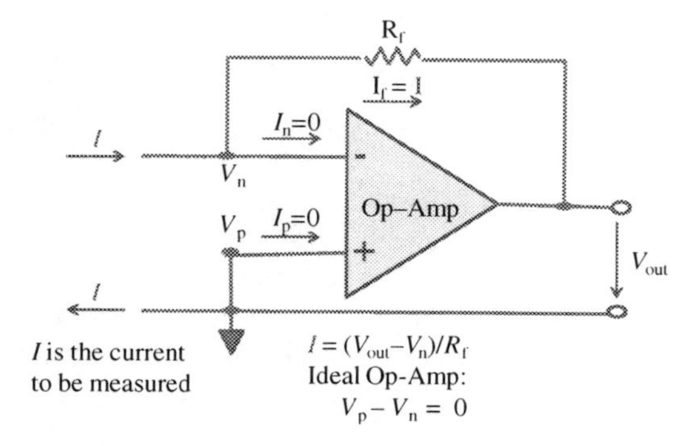

Fig. 6.4 Basic circuit of a zero-voltage ammeter (ZVA) built with an ideal Op-Amp.
Reproduced with permission from L. Yang, A.A. Yang, On zero-resistance ammeter and zero-voltage ammeter, J. Electrochem. Soc. 164 (2017) C819–C821.

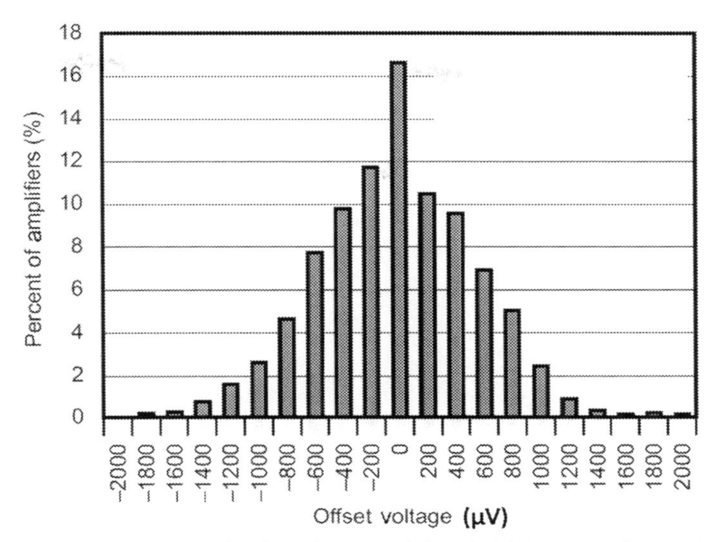

Fig. 6.5 Typical production distribution of packaged OPA2134 Op-Amp of Texas Instruments. Reproduced with permission from L. Yang, A.A. Yang, On zero-resistance ammeter and zero-voltage ammeter, J. Electrochem. Soc. 164 (2017) C819–C821. Copyright 2017, The Electrochemical Society.

Fig. 6.5 shows the production distribution of offset voltage of a typical high-performance Op-Amps by Texas Instruments [11]. The offset voltage of this type of Op-Amps varies between −2 and +2 mV. In addition, the offset voltage may drift with time and temperature. When they are used to build the ZVA, the input voltage may be as high as 2 mV when the measured current is zero. Special Op-Amps with extremely low offset voltage are available, but the lowest offset voltage is still 5 μV. Because there are other integrated components in this type of assembled ZVA and the offset voltage also varies with time and temperature, it is difficult to achieve 5 μV for this type of ZVA. A survey of manufacturers conducted in 2016 indicated that the lowest specified input voltage of this type of ZVA is 50 μV and they are called ZRA by the manufacturers [11]. When they are used to measure the current at 1 nA, the input resistance is still 50 kohm. They are excellent zero-voltage ammeters because 50 μV (or even 1 mV) can be considered as a negligibly low voltage when used for galvanic current measurements (see Section 6.3.4), but it is misleading to call them zero-resistance ammeter because 50 kohm is by far not close to zero resistance.

6.3.2 Zero-voltage ammeters formed with a potentiostat

Potentiostats were often used as ZVAs for galvanic corrosion measurements and for other corrosion studies such as electrochemical current noise measurements (see Chapter 5 for more information). In these measurements, the potentiostat's counter

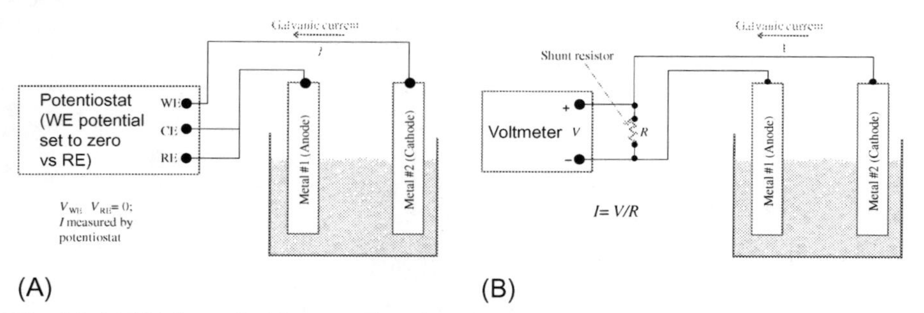

Fig. 6.6 A ZVA formed with a potentiostat (A) and a ZVA formed with a simple voltmeter and a shunt resistor (B).

electrode terminal (CE) is joined to the reference electrode terminal (RE) and both are connected to one of the electrodes in the galvanic couple, and the working electrode terminal (WE) is connected to the other electrode (see Fig. 6.6A). Theoretically, by setting the working electrode potential (relative to the potential of the reference electrode) to zero, the potentiostat measures the galvanic current without imposing a voltage between the two metals.

A survey of the manufacturers of potentiostats conducted in 2016 indicated that the specified accuracy of the potentiostat's working electrode potentials were between ± 1 and ± 2 mV, but no manufacturers claimed for better than ± 1 mV accuracy. Therefore, potentiostats can also be used as the ZVAs for the measurements of the galvanic current for most galvanic sensors because the difference between the two metals' open-circuit potentials is often quite large compared to the 2 mV that may be imposed by the potentiostat (see Section 6.3.4).

Like the ZVAs formed by operation amplifiers, the ZVAs formed with potentiostats are also often called ZRAs in the corrosion community. However, this term is also misleading because the commercial potentiostats with the best ability to control potential may still impose a voltage between -1 and $+1$ mV and the corresponding resistance may be as high as 1 Mohm if the measured current is 1 nA which is often the case for galvanic sensors.

6.3.3 Zero-voltage ammeters formed with a low-cost voltmeter and shunt resistor

The device built with a simple shunt resistor and a low-cost voltmeter has also been used by many researchers in their work with CMAS probes [7] and this device may also be used to measure the galvanic current as shown in Fig. 6.6B. With the shunt-resistor approach, the voltmeter is used to measure the small voltage drop across the shunt resistor to derive the current according to Ohm's Law. As long as the voltage drop across the shunt resistor is negligibly low (<0.5 mV, for example), the device built with a shunt resistor can also be satisfactorily used as the ZVA for galvanic sensors. Compared with the commercial ZVAs formed with operational amplifiers and the potentiostats, a ZVA built with a shunt resistor and a thermocouple-grade voltmeter may be 10 times cheaper. Yet, the low-cost ZVA composed of a shunt resistor and a voltmeter is as effective as the high-cost ZVAs formed with operational amplifiers and potentiostat.

As mentioned in Sections 6.3.1 and 6.3.2, the ZVAs formed with operational amplifiers or a potentiostat are often called ZRA, but their input resistances are often extremely large (10 kohm to 10 Mohm) when they are used to measure the low level currents (0.1 μA to 0.1 nA) as often found in galvanic sensors.[a] So, this term (ZRA) is misleading because most users often consider that they do not have any resistance as the term implies. In the ASTM standard for electrochemical current noise (ECN) measurements (G199) [12], a ZRA is defined as an electronic device used to measure current without imposing a significant IR drop by maintaining close to $0-V$ potential difference between the inputs. This definition is essentially the same as the definition of the ZVA and it does not mention anything about resistance. Therefore, the resistance of these ZRAs is irrelevant and the only criterion that matters is how low the imposed voltage is. In addition, the ZRA term is limited to the ammeters that are formed with the operational amplifiers or a potentiostat and the low-cost ZVAs formed with shunt resistors cannot be called a ZRA. As the current-measuring devices for galvanic measurements described in some international standards [5, 12] and research papers are called ZRA, the low-cost ZVAs formed with shunt resistors are often excluded for consideration, even though they are as effective. It is the author's opinion that the ZRA term in these standards should be changed to ZVA because such a change would avoid discouraging users from using the low-cost ammeters formed with shunt resistors.

6.3.4 *Effect of the voltage imposed by ZVA on galvanic current measurements*

As discussed above, a true zero-voltage ammeter is not practically obtainable. A ZVA formed with shunt-resistor imposes a small voltage drop in the measuring circuit governed by Ohm's Law. A ZVA formed by operational amplifiers or a potentiostat also imposes a small voltage in the measuring circuit because the commercially available units usually have an offset voltage of 0.05–2 mV. The following sections discuss what voltage imposed by the ZVA is acceptable. For simplicity, the discussions omit the effect of the voltage drop caused by the resistance of the solution between the electrodes.

Fig. 6.7 shows a simplified version of Fig. 6.2 when the corrosion process is controlled by only the charge-transfer process [13]. Similar to Fig. 6.2, the thick red (dark gray in print version) solid lines represent the dissolution and reduction polarization behaviors on the anode, respectively. The thin blue (light gray in print version) solid

[a] Strictly speaking, the ZRA type of ZVAs often has near zero dynamic resistance, but extremely large static resistance when they are used to measure low-level of currents. Because static resistance also causes the voltage drop which is the main concern for galvanic measurements, this chapter does not differentiate the two types of resistances.

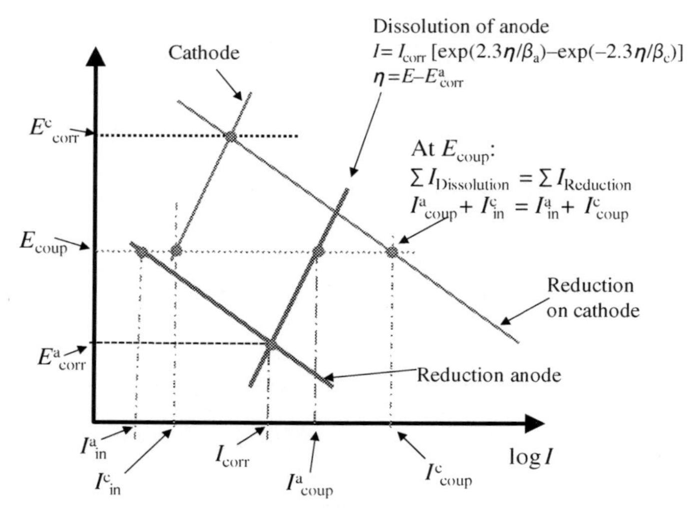

Fig. 6.7 A variation of Fig. 6.2 when the corrosion rate is controlled by only the charge-transfer process.

lines represent the dissolution and reduction polarization behaviors on the cathode, respectively. The E_{coup} is the mixed potential when the two electrodes are coupled and E^a_{corr} is the corrosion potential of the anode before the anode is connected to the cathode. Because the corrosion is controlled only by the charge-transfer processes, the polarization behavior can be described by the Butler-Volmer equation. For dissolution of the anode, the Butler-Volmer equation is as follows:

$$I = I_{corr}\left[\exp(2.3\eta/\beta_a) - \exp(-2.3\eta/\beta_c)\right] \tag{6.7}$$

where β_a and β_c are Tafel slopes in the anodic and cathodic directions, respectively. η is the overpotential of the anode:

$$\eta = E - E^a_{corr}$$

The corrosion current at E_{coup} may be represented by:

$$\begin{gathered} I^a_{coup} = I_{corr}\left[\exp\left(2.3\eta_{coup}/\beta_a\right) - \exp\left(-2.3\eta_{coup}/\beta_c\right)\right] \\ \eta_{coup} = E_{coup} - E^a_{corr} \end{gathered} \tag{6.8}$$

where η_{coup} is the overpotential of the anode after it is coupled to the cathode.

Fig. 6.8 shows that when the voltage of the ZVA is not zero [14], the actual coupling potential of the anode E^a_{coup} will be slightly lower than E_{coup} and the actual coupling potential of the cathode E^c_{coup} will be slightly higher than E_{coup}. The difference between the two actual coupling potentials is the voltage of the ZVA, ΔE:

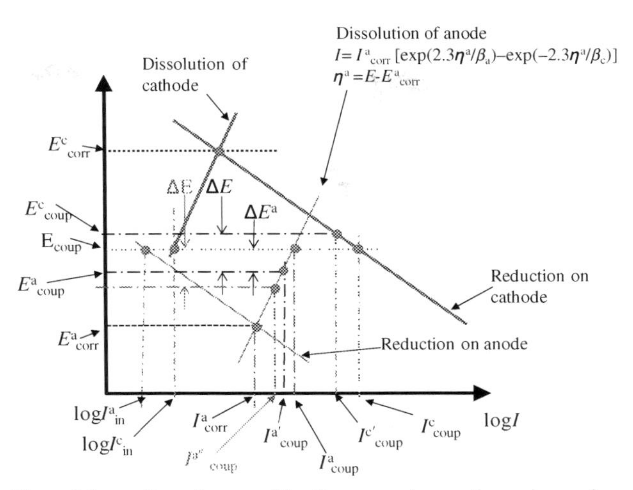

Fig. 6.8 The effect of the voltage imposed by the ammeter on the measured corrosion current. The voltage of the ammeter lowered the coupling potential from E_{coup} to $E^{\text{a}}_{\text{coup}}$ and corrosion current from $I^{\text{a}}_{\text{coup}}$ to $I^{\text{a}'}_{\text{coup}}$.

$$\Delta E = E^{\text{c}}_{\text{coup}} - E^{\text{a}}_{\text{coup}} \tag{6.9}$$

Under this condition, the anode will be corroding at $E^{\text{a}}_{\text{coup}}$, instead of at E_{coup} and the corrosion current at $E^{\text{a}}_{\text{coup}}$ may be represented by:

$$\begin{gathered} I^{\text{a}'}_{\text{coup}} = I_{\text{corr}}\left[\exp\left(2.3\eta^{\text{a}}_{\text{coup}}/\beta_{\text{a}}\right) - \exp\left(-2.3\eta^{\text{a}}_{\text{coup}}/\beta_{\text{c}}\right)\right] \\ \eta^{\text{a}}_{\text{coup}} = E^{\text{a}}_{\text{coup}} - E^{\text{a}}_{\text{corr}} \end{gathered} \tag{6.10}$$

The relative error caused by the shift of the electrode potential from E_{coup} to $E^{\text{a}}_{\text{coup}}$ because of the nonzero voltage of the ZVA is:

$$\begin{aligned} I_{\text{error}} &= \left| \left(I^{\text{a}'}_{\text{coup}} - I^{\text{a}}_{\text{coup}}\right)/I^{\text{a}}_{\text{coup}} \right| \\ &= \left| \left\{\left[\exp\left(2.3\eta^{\text{a}}_{\text{coup}}/\beta_{\text{a}}\right) - \exp\left(-2.3\eta^{\text{a}}_{\text{coup}}/\beta_{\text{c}}\right)\right] - \left[\exp\left(2.3\eta_{\text{coup}}/\beta_{\text{a}}\right)\right.\right.\right. \\ &\quad \left.\left.\left. - \exp\left(-2.3\eta_{\text{coup}}/\beta_{\text{c}}\right)\right]\right\}/\left[\left[\exp\left(2.3\eta_{\text{coup}}/\beta_{\text{a}}\right) - \exp\left(-2.3\eta_{\text{coup}}/\beta_{\text{c}}\right)\right]\right| \end{aligned} \tag{6.11}$$

As shown in Fig. 6.9, $\Delta E = (E^{c}_{coup} - E^{a}_{coup})$ which is larger than $\Delta E^{a} = (E_{coup} - E^{a}_{coup}) = \eta_{coup} - \eta^{a}_{coup}$. Hence,

$$\Delta E = \left(E^{c}_{coup} - E^{a}_{coup}\right) > \Delta E^{a} = \left(E_{coup} - E^{a}_{coup}\right) = \eta_{coup} - \eta^{a}_{coup}$$
$$\Delta E > \eta_{coup} - \eta^{a}_{coup}$$
$$\eta^{a}_{coup} > \eta_{coup} - \Delta E$$

When the η^{a}_{coup} in Eq. (6.10) is substituted with $(\eta_{coup} - \Delta E)$ which is lower than η^{a}_{coup}, it would produce a current $(I^{a''}_{coup})$ that is even more away from the I^{a}_{coup} than the $I^{a'}_{corr}$ is from the I^{a}_{coup}. The value of $I^{a''}_{coup}$ is also shown in Fig. 6.8. Therefore, the following term is the upper bound for I_{error} after substituting η^{a}_{coup} in Eq. (6.11) with $(\eta_{coup} - \Delta E)$:

$$\left| \left[\exp\left(2.3\left(\eta_{coup} - \Delta E\right)/\beta_a\right) - \exp\left(-2.3\left(\eta_{coup} - \Delta E\right)/\beta_c\right)\right] - \left[\exp\left(2.3\eta_{coup}/\beta_a\right) - \exp\left(-2.3\eta_{coup}/\beta_c\right)\right] / \left[\exp\left(2.3\eta_{coup}/\beta_a\right) - \exp\left(-2.3\eta_{coup}/\beta_c\right)\right] \right| \quad (6.12)$$

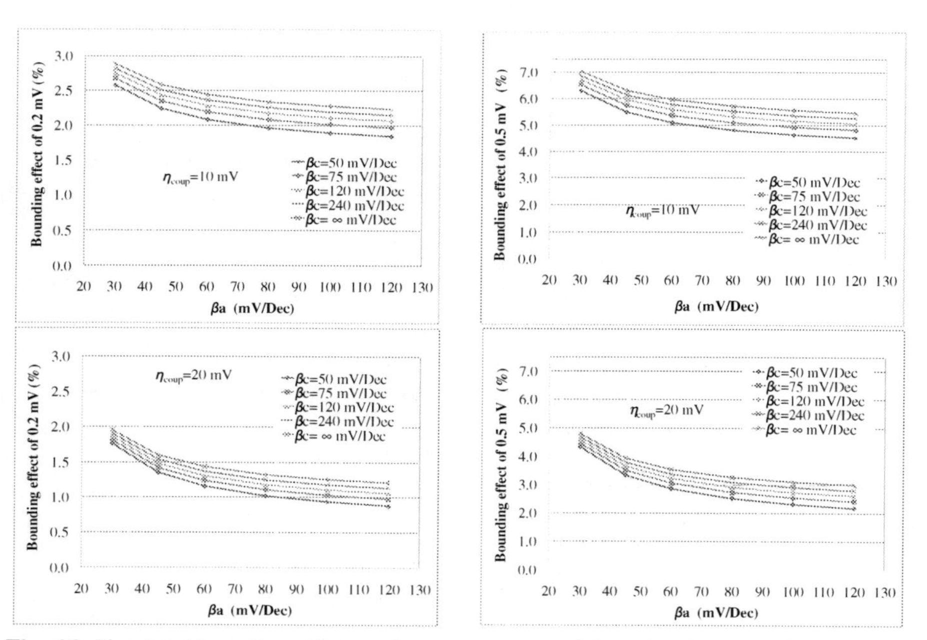

Fig. 6.9 Simulated bounding effect on the measurement of the galvanic current by ZVA voltages of 0.2 and 0.5 mV when the overpotential after coupling is 10 and 20 mV. Reproduced with permission from NACE International, Houston, TX. All rights reserved. L. Yang, Effect of voltage between electrodes of a coupled multielectrode array sensor on corrosion rate measurement, CORROSION/2016 Conference Proceedings, Vancouver, BC, Canada. Paper No. 7911, Houston, TX, NACE, 2016. © NACE International 2016.

According to the review by Mansfeld, [15] the data measured at or near room temperature were all from 30 to 120 mV/Dec for β_a and mostly 50 mV/Dec to infinity (diffusion-controlled process) for β_c. The percentage bounding values for the effect of voltage of ZVA were calculated using Eq. (6.12) for ΔE of 0.2 and 0.5 mV, β_a from 30 to 120 mV/Dec, and β_c from 50 mV/Dec to infinity. Fig. 6.9 shows the calculated results for $\eta_{coup} = 10$ and 20 mV, respectively. When $\eta_{coup} = 20$ mV, the effects by a 0.5-mV ZVA voltage are less than 5% for all the considered cases.

Similarly, the Ohmic voltage drop caused by solution resistance also affects the measurements of the galvanic current and such effect may be much more significant than that by the ZVAs when the conductivity of the solution is extremely low or if the ionic flow path is long or extremely narrow (such as the atmospheric condensate film formed on metal surface). The effect of the ohmic voltage caused by solution resistance is similar to the effect of the voltage caused by the ZVA. The analysis of this effect, however, is beyond this chapter and the above analyses do not include the effect of ohmic voltage drop caused by solution resistance.

6.4 Galvanic sensors

A galvanic sensor is a simple corrosion monitoring device consisting of a pair of dissimilar metals and works on the principle of galvanic corrosion. For corrosion monitoring, the real goal is to measure the natural corrosion rate or natural corrosion current, I_{corr}, for a metal equipment or metal structure in an environment of interests at its natural corrosion potential. As indicated by the thick red (dark gray in print version) curve in Fig. 6.2, however, at the natural corrosion potential (E^a_{corr}) of the metal, I_{corr} equals to I^a_{in} which is the internal anodic current and I^a_{ex} equals to zero. So, 100% of I_{corr} flows to the cathodic sites on the same anode internally at the corrosion potential (E^a_{corr}); none of the current flows externally. Hence, the I_{corr} is not measurable.

In a typical galvanic corrosion sensor, a less active metal must be connected to the active metal of interests to raise the potential of the active metal to a higher potential (E_{coup} in Fig. 6.2) so that there is a nonzero current that is measurable in the external circuit. As described above, this current is the galvanic current, I^a_{ex}. Clearly, I^a_{ex} is not the natural corrosion current of the active metal (I_{corr}) that we want to measure, but it is the only current that we can measure from a galvanic sensor. If the cathode is properly selected and it can raise the coupling potential (E_{coup}) to a sufficiently high value, I^a_{ex} can be used to approximate I^a_{coup}, (Eq. 6.6). Obviously, I^a_{coup} is an artificially created current, and its value is higher (often much higher) than the true natural corrosion current (I_{corr}). Because the I_{corr} is always lower than the I^a_{coup}, I^a_{coup} can be used to estimate the bounding corrosion rate for I_{corr}. Hence, the corrosion rate measured from a galvanic sensor is often the accelerated corrosion rate and it bounds the true corrosion rate. For this reason, galvanic sensors are often used as qualitative sensors for detecting corrosion, rather than a quantitative sensor for measuring the corrosion rate.

Depending on the applications, the anode of a galvanic sensor is not always made of the same metal whose corrosion rate is of interest. For example, cadmium and silver have been used as the anode of some galvanic sensors for monitoring the atmospheric corrosivity [16]. In these sensors, the galvanic currents are due to the corrosion of cadmium or silver, but not necessarily the corrosion of the metal of interests. The galvanic current measured from the cadmium or silver is used to characterize the corrosivity of the environments to metals of general interests. The measurements reflect the level of humidity, contents of aerosols, and pollutants in the air.

6.5 Applications of galvanic sensors

6.5.1 Galvanic sensors for monitoring corrosion in industrial processes

Fig. 6.10 shows a typical galvanic sensor for application in high-pressure processes. The anode is made of steel and the cathode is made of copper. When installed in the process stream, the galvanic current measured from the steel is used to detect the corrosion of the steel pipe wall in contact with the process fluid.

Fig. 6.11 shows typical responses of a galvanic sensor made of copper cathode and steel anode and installed at the 6-o'clock position of a high-pressure crude oil pipeline with an outside diameter of 36 in. (914 mm) [17]. The pipeline was used to transport the crude oil from ships on a nearby dock. Corrosion inhibitors were injected into the pipeline from time to time to control the corrosion. In Fig. 6.11, the two sharp increases (one from 0.5 to 14 μA and the other from 3 to 30 μA) in the galvanic current coincide with the arriving of the crude oil from different ships, indicating that the crude oil from some ships were more corrosive than the others. The galvanic probe in the pipeline was very useful in seeing the changes in the system. The surface area of the anode was not specified because the purpose of the monitoring was to detect the trend of corrosion, rather than to measure the corrosion rate. As discussed in Section 6.4, galvanic sensors are not meant for quantitative measurement of corrosion rate because the anode operates at artificially raised potentials.

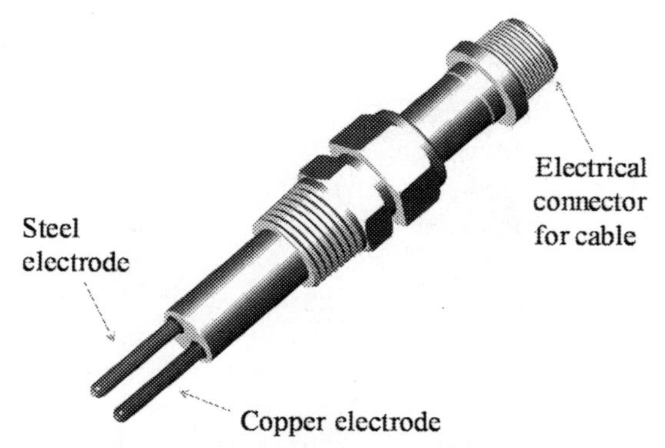

Fig. 6.10 A typical galvanic probe for applications in pressurized process streams. Courtesy of Metal Samples Co, Alabama, United States.

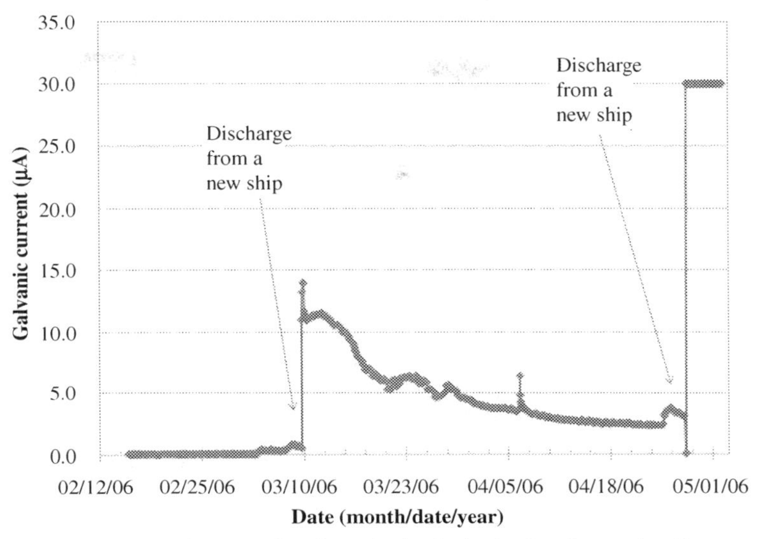

Fig. 6.11 Typical signals from a galvanic probe in dock pipeline for crude oil.
Data from T. Pickthall, V. Morris, H. Gonzalez, Corrosion monitoring of a crude oil pipeline a comparison of multiple methods, CORROSION/2007, paper no. 07340, (Houston, TX: NACE), 2007.

6.5.2 Galvanic sensors for monitoring atmospheric corrosion

Galvanic sensors have been used to measure the atmospheric corrosivity since the late 1950s [18]. Unlike the corrosion in a liquid solution, the electrolyte that causes atmospheric corrosion is often the thin water film with dissolved pollutants or aerosol formed on the metal surface. The anode and cathode in the galvanic sensors for atmospheric corrosion monitoring should be as close as possible to ensure the two electrodes are both in contacts with the thin film electrolyte. Interdigitized finger designs for the anode and cathode are often used in these sensors. Fig. 6.12 shows a typical such sensor [1] where the finger anode was cadmium film and the finger cathode was gold film, both plated onto a Kapton insulator. The width of the finger films was as small as 150 μm and the gap between the 2 electrodes was as thin as 120 μm. The sensor surface, including the insulator gap, was sensitized for greater adsorption of water moisture from the air [19].

6.5.3 Galvanic sensors for corrosion monitoring in other systems

Galvanic sensors are also widely used for monitoring corrosion in other systems. These applications include the rebar in concrete, pipelines and tanks buried in soil, and metal substrate under coatings. Readers are encouraged to read Chapter 16 for monitoring corrosion in concrete, Chapter 17 for monitoring corrosion of pipeline in soil, and Chapter 19 for monitoring corrosion under coatings and insulations.

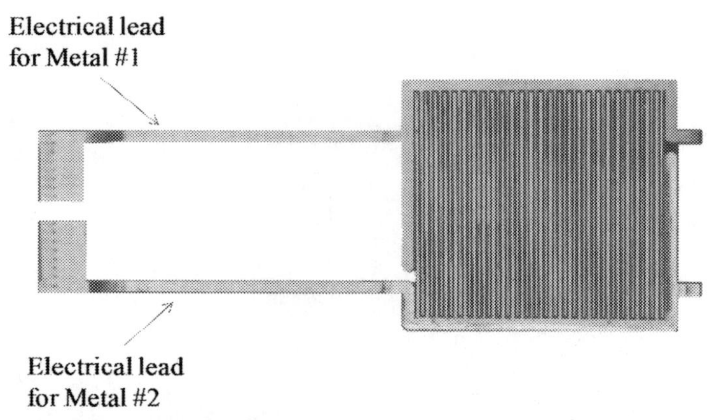

Fig. 6.12 Photo of the interdigitized galvanic element of a galvanic probe used for monitoring atmospheric corrosivity. Cadmium film was used as the anode and gold film was used as the cathode. Both films were plated on a Kapton substrate.
From R.D. Klassen, P.R. Roberge, Zero resistance ammetry and galvanic sensors, in: L. Yang, (Ed.), Techniques for Corrosion Monitoring, Woodhead, Sussex, UK, 2008, Figure 5.3.

6.6 Advantages and limitations of galvanic sensors

A galvanic sensor is easy to build with only two metal electrodes and only requires the measurement of DC current with a ZVA. The ZVA built with voltmeters and shunt resistors is one of the most economic corrosion monitoring instruments. Hence, galvanic sensors can be deployed economically in large numbers to form a network of monitoring system. This is important because the corrosion in many systems (such as in concrete and soil) varies from one location to another and a large number of sensors are required to capture the worst case corrosion. Galvanic sensors are highly effective for qualitatively detecting both general corrosion and the localized corrosion.

The main limitation of the galvanic sensors is that there is not a simple way to derive the quantitative rate of corrosion from the signal of a galvanic sensor. So, it is difficult to use the data from galvanic sensors to predict the service life of metal equipment or metal structures in a given environment. In systems where the solution resistance is not significant, the corrosion rate from the galvanic sensors usually overestimates the true corrosion rate because the anode of the galvanic sensors operates at artificially raised electrode potential and corrodes at an accelerated rate.

6.7 Summary

When dissimilar metals are in contact with a conductive solution, a potential difference develops between them. When the two metals are coupled by an electronic conductor (i.e., metal wire), a current is produced and flows through the conductor. This current is called galvanic current. Of the two metals, one is more active and the other one is less active; the more active metal undergoes corrosion. The corrosion current under the coupling condition is usually much higher than its natural corrosion current which occurs when it is not coupled to any other metals. The galvanic current is often a good approximation of the corrosion current for the more active metal under the coupling conditions if the less active metal has a sufficient power to raise the potential of the more active metal. Because the natural corrosion current is difficult to measure and the galvanic current is easy to measure, galvanic current which is close to the corrosion current under coupling conditions is often used to estimate the bounding corrosion rate of the more active metal in a solution.

A simple device that is consisted of two dissimilar metals and an ammeter that measures the galvanic current to evaluate the corrosion of the active metal or the corrosivity of the environment surrounding the two dissimilar metals is called a galvanic sensor. If the active metal is made of the metal whose corrosion rate is to be evaluated, the corrosion rate from the galvanic sensor is the accelerated corrosion rate and it bounds the natural corrosion rate of the metal. If the more active metal is made from a relatively noble metal such as cadmium or silver, the galvanic sensor can be used for evaluating the corrosivity of the environment to metals of general interests. An example is the monitoring of atmospheric corrosion.

Because galvanic sensors are easy to build and only require the measurement of DC current, they are one of the most economic sensors for corrosion monitoring and suitable for deployment in large numbers to form a network of monitoring system. Galvanic sensors are highly effective for qualitatively detecting both general corrosion and the localized corrosion. The main limitation of the galvanic sensors is that it cannot reliably measure the quantitative rate of corrosion. So, the data from galvanic sensors cannot be used to predict the service life of metal equipment or metal structures.

To minimize the voltage caused by the current-measuring device between the pair of dissimilar metals, the ammeters that impose near zero voltage (zero-voltage ammeters or ZVA) should be used to measure the galvanic current. Both the ammeters that are formed with operational amplifiers and commonly referred to as zero-resistance ammeters (ZRA) and many of the low-cost ammeters formed with a voltmeter and a shunt resistor are effective ZVAs for galvanic sensors. Contrary to the common perception, many of the ZRAs do exhibit a significant resistance (10–1000 kohm) when they are used in galvanic sensors, but the high resistance usually does not affect their ability to be used as effective ZVAs. The only criterion to determine whether an ammeter can be used to measure the galvanic current is how the imposed voltage is close to zero.

References

[1] R.D. Klassen, P.R. Roberge, Zero resistance ammetry and galvanic sensors, in: L. Yang (Ed.), Techniques for Corrosion Monitoring, Woodhead, Sussex, UK, 2008, pp. 111–126.

[2] ASTM G82, Standard Guide for Development and Use of a Galvanic Series for Predicting Galvanic Corrosion Performance, ASTM, West Conshohocken, PA, 2014. Reapproved.

[3] L. Yang, Multielectrode systems, in: L. Yang (Ed.), Techniques for Corrosion Monitoring, Woodhead Publishing, Sussex, UK, 2008, pp. 199–203.

[4] L. Yang, K.T. Chiang, P.K. Shukla, N. Shiratori, Internal current effects on localized corrosion rate measurements using coupled multielectrode array sensors, Corrosion 66 (2010) 115005–115017.

[5] ASTM G71, Standard Guide for Conducting and Evaluating Galvanic Corrosion Tests in Electrolytes, ASTM, West Conshohocken, PA, 2014. Reapproved.

[6] ISO 7441:2015, Corrosion of Metals and Alloys—Determination of Bimetallic Corrosion in Atmospheric Exposure Corrosion Tests, The International Organization for Standardization, Geneva, Switzerland, 2015.

[7] L. Yang, N. Sridhar, O. Pensado, D. Dunn, An in-situ galvanically coupled multielectrode array sensor for localized corrosion, Corrosion 58 (2002) 1004–1014.

[8] L. Yang, N. Sridhar, Coupled multielectrode online corrosion sensor, Mater. Perform. 42 (9) (2003) 48–52.

[9] B. Yang, Differential flow through cell technique, in: L. Yang (Ed.), Techniques for Corrosion Monitoring, Woodhead, Sussex, UK, 2008 (Chapter 6).

[10] ASTM G217, Standard Guide for Corrosion Monitoring in Laboratories and Plants with Coupled Multielectrode Array Sensor Method, ASTM, West Conshohocken, PA, 2017. Reapproved.

[11] L. Yang, A.A. Yang, On zero-resistance ammeter and zero-voltage ammeter, J. Electrochem. Soc. 164 (2017) C819–C821.

[12] ASTM G199, Standard Guide for Electrochemical Noise Measurement, ASTM, West Conshohocken, PA, 2014. Reapproved.

[13] H. Cong, F. Bocher, N.D. Budiansky, M.F. Hurley, J.R. Scully, Use of coupled multielectrode arrays to advance the understanding of selected corrosion phenomena, J. ASTM Int. 4 (10) (2007). Paper ID JAI101248.

[14] L. Yang, Effect of voltage between electrodes of a coupled multielectrode array sensor on corrosion rate measurement, in: CORROSION/2016 Conference Proceedings, Paper No. 7911, Houston, TX, NACE, 2016.

[15] F. Mansfeld, The polarization resistance technique for measuring corrosion currents, in: M.G. Fontana, R.W. Staehle (Eds.), Advances in Corrosion Science and Technology, vol. 6, Plenum Publishing, New York, 1976, pp. 185–187.

[16] V.S. Agarwala, S. Ahmad, Corrosion detection and monitoring—a review, in: CORROSION/2000, paper no. 271 (Houston, TX: NACE International, 2000.

[17] T. Pickthall, V. Morris, H. Gonzalez, Corrosion monitoring of a crude oil pipeline a comparison of multiple methods, in: CORROSION/2007, paper no. 07340, (Houston, TX: NACE), 2007.

[18] P.J. Sereda, American Society for Testing and Materials, Bulletin No. 228 (February), (1958), pp. 53–55.

[19] V.S. Agarwala, In-situ corrosivity monitoring of military hardware environments, in: CORROSION/1996, Paper No. 632, (Houston, TX: NACE International), 1996.

Differential flow cell technique

7

*Bo Yang**
Champion Technologies Inc., Fresno, TX, United States

7.1 Introduction

In this chapter, a corrosion measurement technique, i.e., the differential flow cell method, designed for online and real-time determination of the rates of localized corrosion (such as pitting, crevice/under-deposit corrosion, microbiologically influence corrosion or MIC, and galvanic corrosion) of metals used in water containing systems (such as cooling water systems, engine cooling systems, oil field production systems, etc.) is described. The chapter begins with a detailed discussion of the operating principles of the method. The localized corrosion rate calculation methods, typical flow cell and monitoring instrument design, verification of the results by other known reliable measurement methods, and guidance on data interpretation are provided. The usefulness of the methods is also discussed. The chapter ends by pointing future developmental needs of the method to broaden its application potential. Examples of laboratory and field applications of the methods are described in Chapter 22.

7.2 Principles of the differential flow cell (DFC) method

7.2.1 *The problem the method designed to solve*

Corrosion can be classified into two categories: general corrosion and localized corrosion. Even though the term general corrosion is often used synonymously with uniform corrosion, this form of corrosion is seldom completely uniform (see Chapter 2 for more information). Whether a corrosion attack is uniform is determined by the degree of inhomogeneities on the substrate surface (e.g., surface active sites such as the presence of defects or impurities, or steps and kinks, or mechanical stresses), and its immediate environment (e.g., concentration or temperature gradients), or in other words, by the degree of inhomogeneities of the solid-electrolyte interfacial region. The more inhomogeneous electrochemically the solid-electrolyte interfacial region is, the more severe the localized attack. One of the most notable examples of uniform corrosion is the corrosion of Zinc (Zn) amalgam in acid solutions. In this case, a homogeneous surface can be realized down to the atomic level because of the high mobility of the liquid metal surface. In addition, there is no significant inhomogeneity in the solution side (e.g., oxygen concentration cells are absent). The well-known mixed potential theory for metal corrosion proposed by Wagner and Traud in 1938 was based on the results obtained

* Current affiliation: Prestone Products Corporation, Danbury, CT, United States.

Techniques for Corrosion Monitoring. https://doi.org/10.1016/B978-0-08-103003-5.00007-2

from such as a system [1]. According to the mixed potential theory, the assumption of the physically identifiable local anodes and cathodes (i.e., the local cell theory) is not necessary to explain corrosion phenomena because the cathodic and anodic partial reactions can occur at the metal-electrolyte interface in constant change, with random distribution of location and time of the individual reaction.

In practice, corroding metal surfaces of an engineering component is seldom homogeneous. It is well-known that majority of corrosion phenomena encountered in engineering applications are nonuniform or localized. In cooling water systems, available industry survey data suggest that the most commonly observed forms of localized corrosion for components made of carbon steel, cast iron, galvanized steel, and aluminum are pitting, crevice/under-deposit corrosion, and microbiologically influenced corrosion [2–5]. For stainless steel components, stress corrosion cracking is the predominant cause of corrosion failures [4–8]. Pitting and under-deposit corrosion, often associated with microbial activity, are also frequently observed. Furthermore, pitting and under-deposit corrosion on stainless steel components (e.g., heat exchangers) often become the initiation sites leading to the subsequent stress corrosion cracking failures. For components made of copper and copper alloys, corrosion failures are often attributed to under-deposit corrosion, microbiologically influenced corrosion, selective dissolution (e.g., dezincification), ammonia corrosion, and erosion corrosion [5, 9]. Similarly, localized corrosion such as pitting, crevice/under-deposit corrosion, galvanic corrosion, and cavitation-erosion corrosion are known to be the major types of corrosion commonly observed in vehicle engine cooling systems [10]. Under-deposit/crevice corrosion, pitting, galvanic corrosion, microbiologically influenced corrosion, stress corrosion cracking, and erosion corrosion are also known to be the major types of corrosion commonly observed in petroleum production and refining processes, as well as other petrochemical operations [11–13].

Ferrous metals such as carbon steel, cast iron, and galvanized steel are among the most widely used structure materials in industrial cooling water systems. Transfer lines or pipes, heat exchanger shells, tubes, baffles, and water boxes, pump components, valves, screens, plumbing fixtures, and cooling tower basins, and other tower components are often made of these ferrous metals. Localized corrosions, such as pitting, crevice/under-deposit corrosion, and microbiological influenced corrosion, are usually the main limiting factors in determining their service life in cooling water systems [2–4]. The complex and dynamic nature of industrial scale operations, combined with the lack of effective monitoring tools, mean little is known about the impact of variations in process conditions and water chemistry on localized corrosion. As a result, premature equipment failures occur, in spite of the use of relatively large design safety factors. A recent survey of the Japanese petrochemical industry shows that more than 70% of cooling water failures of carbon steel heat exchangers are due to localized corrosion [2, 3]. In addition, 59% of industrial heat exchangers have lifetimes of less than 10 years. Obviously, reduction of localized corrosion will lead to the extension of equipment service life. Hence, the ability to reliably monitor and control localized corrosion is a major industrial need.

Many direct or indirect measurement techniques are available for corrosion assessment [14, 15]. Indirect measurement techniques, such as pH, conductivity, corrosive ion or corrosion product ion concentrations (e.g., chloride or ferrous ions), and temperature and flow rate measurements, measure parameters that may greatly influence, or are influenced by, the extent of corrosion. While these indirect measurement methods are generally useful for predicting or assessing the potential risk of corrosion and may also be useful for process control, they do not yield quantitative corrosion rate results. To obtain quantitative corrosion information, direct corrosion measurement techniques are used.

The commonly used methods that measure the extent of corrosion directly in industrial water systems include weight (mass) loss methods such as coupons (see Chapter 9), polarization resistance-based methods (see Chapter 3), electrical resistance-based methods (see Chapter 11), and plant equipment inspection or failure analysis [2, 14, 16–18]. Although both coupons and plant equipment inspection or failure analysis can provide useful information about corrosion, obtaining quantitative corrosion rate data (especially localized corrosion data, such as maximum pit depth) from these methods is often time-consuming and costly. They usually require equipment shutdown or sample extraction. Furthermore, these two methods usually yield the average corrosion rate over the period of exposure, typically ranging from several months to several years. Since many important factors influencing corrosion (e.g., temperature, flow rate, solution composition, inhibitor and/or chemical treatment dosages, pH, etc.) in an industrial or commercial system are time-dependent variables, it is usually very difficult (if not nearly impossible) to identify clearly the root causes for the observed high corrosion rate based on equipment inspection and/or coupon analysis results. In addition, equipment inspection methods do not permit remedial measures to be taken before significant damage to the system components has occurred. On the other hand, polarization resistance and electrical resistance-based methods cannot be used to determine localized corrosion rate because the corrosion rate obtained is averaged over the whole surface area of the test probe that is generally not corroded uniformly.

To obtain more timely information about localized corrosion, several techniques, including the pitting index (or imbalance technique, similar to electrochemical noise) provided by some polarization resistance-based general corrosion monitors, electrochemical noise techniques (see Chapter 5), scanning vibrating electrode techniques, and zero-resistance ammeter-based occluded cell techniques, have been used to indicate or study localized corrosion [14–16]. However, none of these techniques can be used to obtain accurate quantitative localized corrosion rate under normal corrosion conditions because the anodic corrosion and cathodic corrosion reactions do not occur at well-separated sites and time (see also Section 7.4). They remain largely qualitative in measuring localized corrosion.

To overcome these limitations, the differential flow cell (DFC) method was developed recently for the real-time, online, and reliable measurement of localized corrosion rates (e.g., pitting, crevice/under-deposit corrosion, microbiological influenced corrosion, and galvanic corrosion) of metals in industrial water systems [2, 16–20]. In the rest of the chapter, the term localized corrosion is narrowly defined to include only pitting, crevice/under-deposit corrosion, microbiologically influenced corrosion, and galvanic corrosion.

7.2.2 The physical model

The DFC method was initially developed for measuring pitting and crevice or under-deposit corrosion of carbon steel in cooling water systems [18]. In cooling water systems, propagating pitting and crevice or under-deposit corrosion on carbon steel has similar characteristics. They can all be considered as a form of "concentration cell" corrosion. Concentration cell corrosion occurs when metal surfaces encounter very different physical and chemical conditions within the same water system. These differences predispose specific surface sites to localized corrosion. At these sites, an increase in the corrosive nature of the process conditions or water chemistry can lead to the formation of a localized corrosion cell (a pit). Examples of metal surfaces that are predisposed to pitting corrosion are:

- a surface beneath a deposited particle (mineral scale, corrosion product, silt)
- a surface beneath a biofilm deposit
- a surface between rifled tube grooves
- a surface immediately after a weld or flange
- a surface behind a baffle, within a "dead-leg," or other low-flow areas
- a surface that is touching the surface of a different material
- a surface with microstructure defects or impurities (e.g., sulfide inclusion)

These metal surface differences allow for the creation of anodes and cathodes (regions differing in electrochemical potential) that are localized within a micro-environment. Differences in dissolved ion (or other chemical) concentration, such as hydroxyl ion, oxygen, chloride, and sulfate, eventually develop between predisposed and nearby regions. In this way, localized corrosion cells are established. The predisposed sites become anodic, resulting in loss of metal to the solution (e.g., $Fe \rightarrow Fe^{2+} + 2e^-$). The nearby "open" sites become cathodic, accelerating reduction reaction (e.g., $H_2O + 1/2O_2 + 2e^- \rightarrow 2\ OH^-$).[a] The generation of hydroxide results in highly alkaline localized surface conditions, which can result in the formation of a mixed deposit, containing corrosion products and mineral scales [e.g., $Fe(OH)_2$, $Fe(OH)_3$, $FePO_4$, $Mg(OH)_2$, $Ca_3(PO_4)_2$, $Zn(OH)_2$, etc.]. The intense localized attack at the site leads to the formation of a deposit layer that is porous and not as protective as the inhibited deposit layer formed at other part of the surface. The presence of relatively high concentrations of corrosion product metallic ions (such as Fe^{2+}, Fe^{3+}, Cu^{2+}, and Al^{3+}) near the corrosion sites is generally known to have a significant negative effect on the performance of scale inhibition and particle dispersing ability of dispersant polymers typically included in cooling water treatments and would be one of the contributing

[a] The predominant cathodic reaction in the corrosion of carbon steel or copper alloys is normally oxygen reduction in cooling water systems with a pH value typically between 6.5 and 9.5. Hydrogen evolution occurring in pits or under-deposit corrosion sites generally contributes only a small part to the overall cathodic corrosion current [2]. It should be noted that when low pH upset operating conditions (e.g., acid overfeed, acid cleaning or excessive growth of acid producing bacteria)are present in the system, hydrogen evolution reaction may become the major contributor to the corrosion processes. Similarly, halogen reduction reaction may become the major contributor to the corrosion processes when overfeed of an oxidizing biocide occurs.

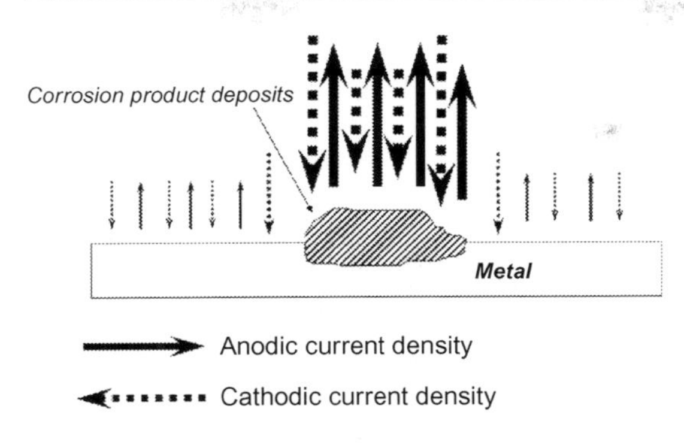

Fig. 7.1 Physical model for the DFC method—distribution of anodic and cathodic currents around a localized corrosion site.

factors in forming the porous, loosely attached, and less protective deposit layers on the corrosion sites.

As localized corrosion proceeds, a crust (called a "tubercle") can form over the top of the localized corrosion site, creating an under-deposit localized corrosion environment. Biofilm or biofouling may also start to accumulate around the deposits covering the localized corrosion site (the pit) under suitable conditions.

Systematic investigation carried out by Yang and coworkers [2, 16–20] shows that the anodic and cathodic current distribution around a localized corrosion (e.g., pitting, crevice or under-deposit corrosion) site can be represented by a model shown in Fig. 7.1. In other words, the mixed potential corrosion theory is found to be applicable to the typical localized corrosion conditions. According to this model, the anodic reaction and cathodic reaction of the corrosion process can occur at the same surface location at the same time around a localized corrosion site.

The anodic current density is proportional to the rate of metal dissolution partial reaction in the corrosion process. The cathodic current density is proportional to the cathodic partial reactions in the corrosion process. In cooling water systems, the cathodic partial reactions are commonly comprised of oxygen reduction, halogen (commonly used as oxidizing biocide) reduction, and hydrogen evolution (e.g., inside the localized corrosion sites for carbon steel, other ferrous metals, aluminum alloys, etc.). Experimental studies demonstrate that the model depicted in Fig. 7.1 can be used to describe the localized corrosion processes of carbon steel, admiralty brass, aluminum alloys, and magnesium alloys in aqueous fluids [2, 13, 20–22], such as cooling water, process water in paper mill and food processing plant, boiler steam condensate, vehicle engine coolant [20], and oil field production stream [13, 22].

7.2.3 The DFC method to obtain localized corrosion rate

The DFC method uses an electrolytic cell assembly to simulate the localized corrosion conditions of interest and an electrical instrument/data acquisition system to measure the localized corrosion rate in real-time from the simulation cell [16, 18, 23]. A unique combination of linear polarization resistance (LPR) and zero-resistance measurements is usually used to obtain the rate of localized corrosion (such as pitting, crevice/under-deposit, microbiologically influenced corrosion, and galvanic corrosion) for metals in aqueous solutions from the simulation electrolytic cell.

7.2.3.1 Typical electrolytic cell assembly

Several methods have been used to configure the electrolytic cell assembly to simulate the localized corrosion conditions [2, 18, 20, 24]. In a typical application in cooling water systems, the electrolytic cell assembly has a differential flow cell configured to simulate the essential features of the propagating localized corrosion process. The flow cell generally contains two small anodes commonly placed in a slow flow condition and one large cathode placed in a faster flow condition. The anodes are used to simulate the preferential or localized attack areas of the metal in a system. The large cathode simulates the nonpreferential attack area of the metal. To simulate faithfully the localized corrosion conditions, the anodes and the cathode are normally connected electrically together via a zero-resistance ammeter (ZRA) in the electrical instrument assembly to allow the occurrence of galvanic interaction among them. Fig. 7.2 shows a flow cell design for use in field applications in commercial or industrial cooling water systems.

The flow cell showed in Fig. 7.2 can be used to simulate the localized corrosion conditions found in tube-side heat exchanger and pipeline, or shell-side heat exchanger flow conditions. When the flow cell is used to simulate localized corrosion in a shell-side heat exchanger, the cathode tube A placed in the upper glass tube is used as the cathode and the flow rate in cathode flow channel is set at a higher flow rate than the one at the lower flow channel where the anodes are located. When the flow cell is used to simulate localized corrosion condition in a tube-side heat exchanger or in a pipeline, the cathode tube B placed in the lower glass tube is used. In this case, the anodes are usually recessed into the ports to create slower flow conditions than the one experienced at the cathode tube surface. The flow rates in the two flow channels and the distance between anodes and cathode tube B may be adjusted to better simulate the localized corrosion conditions in the cooling water system of interest. It should be noted that in the DFC design shown in Fig. 7.2, the separation distance between the anodes and cathode is usually larger in the case used to simulate shell-side heat exchanger corrosion than the one used to simulate tube-side heat exchanger corrosion. The flow rates used in the type of setup shown in Fig. 7.2 were typically between 45 and 1130 L/h or about 0.04–2.3 m/s. More information about using Tube A or Tube B will be described in the Chapter 22 where the application examples of the method are provided.

A number of other different designs of the electrolytic cell were also used, including the use of various occluded cell (e.g., lead-in-pencil type setup commonly used in many pitting and crevice corrosion studies) designs, anode or cathode or both under different heat-rejection conditions, and various anode/cathode configuration designs for studying galvanic corrosion. Wide ranges of exposed anode and cathode surface areas can be selected to simulate the application conditions. Exposed anode surface areas used in the DFC method were often between 0.07 and 2 cm^2. The cathode/anode

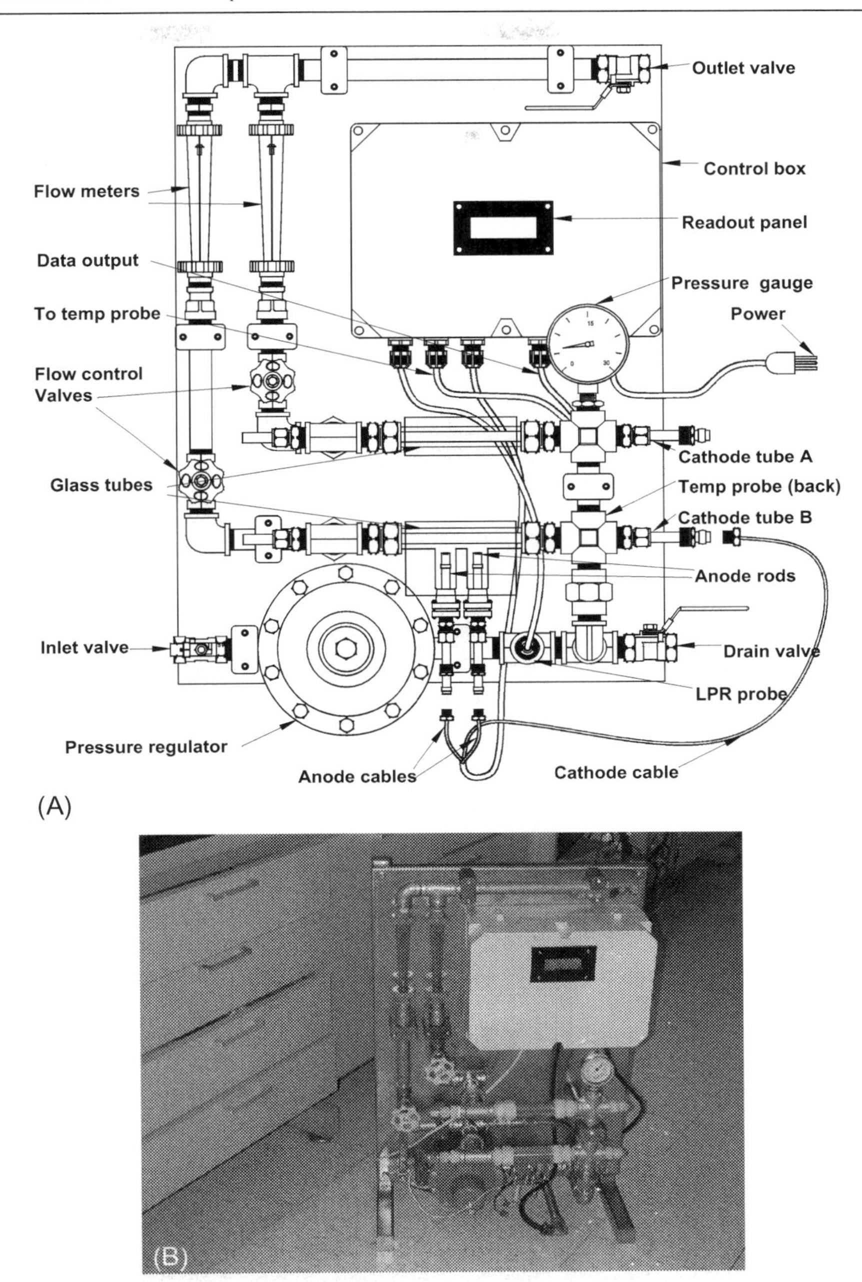

Fig. 7.2 (A) Schematics of a localized corrosion monitor (LCM) based on the DFC method for use in cooling water systems. (B) A photo of the DFC-based localized corrosion monitor (LCM) with optional stand.

surface area ratio was typically between 1 and 1000. In practice, one should choose the most appropriate design to simulate realistically the localized corrosion process under study.

7.2.3.2 Electrical instrument assembly

The electrical instrument assembly typically consists of a ZRA, a LPR-based general corrosion monitor with IR-drop compensation capability, instrument control, and data acquisition system. In addition, relay switches, display or read-out panel, data output (e.g., as 4–20 mA signals), power source conditional circuit, instrument enclosure, etc. may also be included in the instrument assembly. In the design shown in Fig. 7.2, the electrical instrument assembly is housed in the control box. The ZRA, LPR, relay switches, data acquisition and control, and data analog outputs are all integrated in a circuit board placed inside the control box. The control box also has an enclosure [rated NEMA (National Electrical Manufactures Association) 12 ×] to protect against circulating dust, falling dirt, and dripping noncorrosive liquids. In addition to the LPR probe for measuring general corrosion, the monitor design shown in Fig. 7.2 also has a sensor to measure the temperature of the test fluid. Sensors to detect other parameters (e.g., pH, flow rate, conductivity, corrosion and scale inhibitor concentrations, biocide concentration, and corrosive ion or corrosion product concentrations, such as Cl^- and Fe^{3+}, etc.) important to the corrosion process may also be included in the assembly.

7.2.3.3 Methods to obtain localized corrosion rates

Three methods can be used to determine the localized corrosion rates from a DFC cell [16, 18, 24]. One method is based on the Tafel extrapolation of the anodic polarization curve of the anode (or the slow flow electrode) to the corrosion potential of the DFC cell (i.e., the corrosion potential recorded when the anodes and cathode are connected together). The second method is based on measuring the polarization resistance of the anodes and cathode connecting together, and then measuring the polarization resistance of the cathode when it is temporarily disconnected from the anode. A detailed description of these two methods is given in previous publications [18, 24]. In theory, the Tafel extrapolation should have yielded the most accurate localized corrosion rate because it gives the corrosion rate of the anode at the mixed potential raised by the cathode. The second method (polarization resistances only) would have underestimated the localized corrosion rate because the corrosion potential of the cathode alone was more positive than the corrosion potential of anodes and cathode connected together. It should be noted that the Tafel extrapolation method is generally not suitable for use to monitor localized corrosion rate as a function of time, because it is considered as a destructive method and the surface conditions of the anode generally change irreversibly after the measurement.

The third method of determining localized corrosion rates is commonly used to determine localized corrosion rates from a DFC-based localized corrosion monitor (e.g., a typical design is shown in Fig. 7.2). The localized corrosion current density.

($I_{localized\ corrosion}$) of the anode is calculated according to the following equation: [16, 18, 24].

$$I_{localized\ corrosion} = I_{ZRA} + I_{LPR} \tag{6.1}$$

where I_{ZRA} represents the galvanic coupling current density measured between the anode and the cathode from the differential flow cell; I_{LPR} represents the LPR (or linear polarization resistance) corrosion current density measured from the anode alone when the anode is temporarily disconnected from the cathode (Note: To obtain a more accurate result, a solution ohmic drop compensated instrument should be used to measure the polarization resistance). The Stern-Geary constant for the metal used for the anode is calculated from either the Tafel slopes measured experimentally or the Tafel slopes recommended in literatures or obtained in previous studies under similar corrosion conditions. The localized corrosion rate of the anode is calculated from the localized corrosion current density, Faraday constant, equivalent weight, and density of the anode. I_{LPR} reflects the contribution of cathodic reactions (e.g., oxygen reduction, hydrogen evolution and halogen reduction, etc.) occurring on the anodes toward their own corrosion. I_{ZRA} reflects the contribution of cathodic reactions occurring on the cathode toward the corrosion of anodes or the contribution of the galvanic interaction among the anodes and cathode toward corrosion of the anodes.

In a DFC cell design shown in Fig. 7.2, two localized corrosion rates may be obtained as a function of time from the cell according to the Eq. (6.1) listed above. In some earlier designs of DFC method-based localized corrosion monitor, both localized corrosion rates from the two anodes were reported. In a newer design, only the maximum value of the two readings from the two anodes at a given time was reported. In addition, the time average of the localized corrosion rate and the general corrosion rate [the rate averaged over the surface area of the electrodes of the linear polarization resistance (or LPR) probe] is also saved and displayed in the DFC design shown in Fig. 7.2.

In theory, Eq. (6.1) or the third method yields an approximate localized corrosion rate. It was deduced under two assumptions: the corrosion attack at the anodes was uniform and the corrosion potential (E_{corr}) of the anodes alone was not much different from the E_{corr} of anodes and cathode connected together. For electrodes made of carbon steel, the first assumption can be met in most cases. It has been observed in >20 field applications of the DFC-based localized corrosion monitor that the corrosion attack on the anodes was quite uniform, especially after exposure longer than several days since the anodes usually were covered completely by corrosion products and other deposits. It should also be noted that in several laboratory tests using aluminum alloys as the anode in engine coolants, the corrosion attack on the anode was also uniform based on posttest visual examination. In bench-top tests using two rotators to create the differential flow conditions, E_{corr} of the anode alone was generally more negative than the E_{corr} of the anode and cathode connected together. Depending on the difference in the extent of corrosion on the carbon steel anode and the carbon steel cathode, the difference in E_{corr} values was as much as 50–60 mV several minutes after disconnecting the anode from the cathode. In theory, this means that Eq. (6.1) would have tended to overestimate the localized corrosion rates. However, as shown in previous studies, the overestimation was not significant. The differences between localized corrosion rates determined using Eq. (6.1) and the rate determined from Tafel extrapolation of the anode-only anodic polarization curve to the E_{corr} of the anode and cathode connected together were <20%. The high cathodic Tafel slope values typically observed in cooling water system applications and the fact that the main cathodic reaction (i.e., oxygen reduction) in the anode corrosion process is often close to diffusion control in the observed E_{corr} region may be the likely reasons for the relatively small errors observed. The potential errors of using Eq. (6.1) in DFC-based localized corrosion monitors (e.g., Fig. 7.2) due to the differences in E_{corr} appeared to be fewer than the bench-top differential flow cell setup using two rotators. One reason is that the cathode/anode surface area ratio in the DFC-based field localized corrosion monitors (i.e., >250) shown in Fig. 7.2 was much greater than the one in the bench-top setup (i.e., 20). Another reason is that the low-flow entrance and exit regions in the field DFC localized corrosion monitor cathode also will corrode preferentially. These factors ensure that there would be many more corrosion sites on the field localized corrosion monitor than on the bench-top setup. Thus, the relatively high corrosion rates of the anodes tended to have less of an effect on the E_{corr} in the DFC-based field localized corrosion monitor setup than in the bench-top setup.

In cases where the corrosion attack on the anode is not uniform, a surface area correction factor should be used in Eq. (6.1) to account for the percentage of actual surface suffering corrosion attack. One alternative solution is to change the design of the electrolytic cell used to simulate the localized corrosion. Experiments showed that the use of occluded cell type of designs (in this case, the anodes are located in an occluded site similar to the ones found in a crevice. The flow velocity in the anode, or slow flow electrode, can be considered to be negligibly small or close to zero) would be helpful to ensure that the corrosion attack on the anode remains largely uniform.

Another factor to consider is the separation distance between anodes and cathode in an electrolytic cell shown in Fig. 7.2. When the DFC setup shown in Fig. 7.2 is used to simulate shell-side heat exchanger localized corrosion, the cathode is located in the top flow channel. In this case, the anodes and cathode could be separated by a relatively large distance (e.g., ~10 cm). In cooling water with relatively high conductivity, the additional solution ohmic drop introduced by the separation is not expected to affect the measured corrosion rate substantially after the corrosion sites in the electrodes are covered with a layer of corrosion products. In bench-top experiments using two rotators to create the differential flow conditions, the solution ohmic drop in the anode corrosion products was 10 times higher than the ones in the solution, even with a solution conductivity as low as 80 μS/cm. Nevertheless, in the absence of corrosion product deposits (e.g., in the initial stage of immersion of a sand-paper polished anode), the localized corrosion rate obtained from the lab setup was reduced by the separation of anode and cathode, especially in cases where the solution conductivity was low (e.g., <200 μS/cm). No systematic investigation has been conducted to study the effect.

7.2.3.4 Validation of the technique

Similar to other commonly used electrochemical techniques for measuring corrosion, the DFC technique is a simulation measurement method. The advantage is that the DFC-based localized corrosion monitor (LCM) is an electrochemical measurement device providing a time-averaged localized corrosion rate that is quantitative to that measured independently on the same anodes used in the instrument. In Fig. 7.3, the time-averaged and steady state localized corrosion rates determined by the DFC-based LCM are compared to the results of penetration depth measurements on the DFC anode surfaces [18]. In principle, only the time-averaged corrosion rates should be directly comparable to the corrosion rates determined from penetration depth measurements. However, under well-controlled steady state conditions, the instantaneous corrosion rates closely approximate the time-averaged rates.

In addition, the localized corrosion rates of carbon steel, admiralty brass, and aluminum alloys determined from Eq. (6.1) were also shown to be in agreement with the ones determined by the Tafel extrapolation of the anodic polarization curves of the anodes (Using the fist method as discussed in Section 7.2.3), providing further confirmation of the validity of the method [18, 20]. Under properly chosen simulation conditions, the corrosion rate results (both localized and general rates) obtained from the DFC-based localized corrosion monitor in the field applications were also in agreement with independent physical measurement (i.e., coupons) and heat exchanger inspection (e.g., ultrasonic inspection) or operation results (e.g., service life or leakage frequency) obtained under comparable conditions [2, 16, 21, 23].

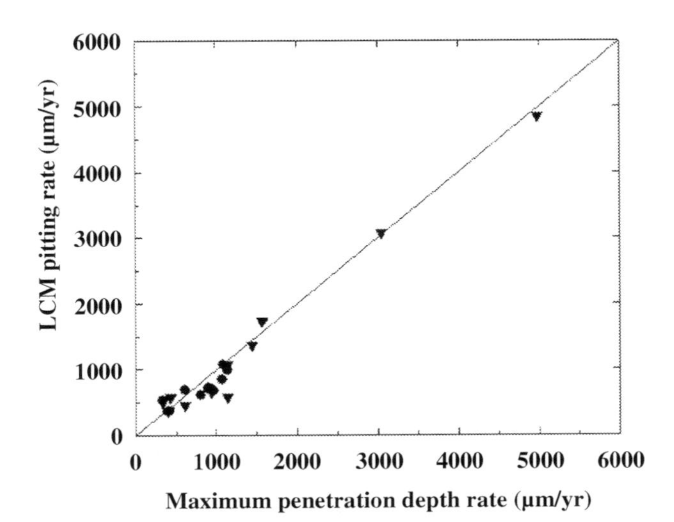

Fig. 7.3 Comparison of localized corrosion rates obtained from penetration depth measurements and DFC-based LCM readings.

7.2.3.5 How to use a DFC-based localized corrosion monitor (LCM) for field applications

Because of its ability to obtain quantitative, reliable, and real-time localized corrosion rate, the DFC method offers great potential for use in localized corrosion research and localized corrosion monitoring and control in industrial systems. The extensive use of this method in the laboratory and in field applications has been very fruitful for developing new insight into the localized corrosion process, for aiding troubleshooting and failure analysis of cooling water system corrosion-related problems, for developing novel, highly effective, environmentally friendly cooling water corrosion inhibitors, and for performance-based optimization and control of chemical treatments (i.e., corrosion inhibitors, scale inhibitors, and biocides) in industrial and commercial cooling water systems [2, 16–23, 25–35]. In the past, the effective use of the method for performance-based optimization and control of chemical treatments in field applications has been the most challenging, because it not only requires substantial corrosion training and knowledge, but also in-depth understanding of modern cooling water chemical treatment technology, as well as detailed knowledge about the operation of the cooling water systems. Thus, guidance is provided below to help a potential user to obtain greater benefits of applying the technique.

The DFC-based LCM (shown in Fig. 7.2) measures the localized corrosion rate, general corrosion rate, and water temperature in real-time. The readings are updated every 10–15 min. The lower the localized corrosion rate and general corrosion rate, the better corrosion protection the treatment is providing to the system.

Most importantly, one should be less concerned about the localized corrosion rate reading if the general corrosion rate is unacceptably high. In this case, the large amount of corrosion product generated by the high general corrosion rate is likely to result in widespread under-deposit corrosion and scaling problems in other parts of the system. Thus, one should first perform program optimization practices to reduce the general corrosion rate to an acceptable level. Then, further program optimization can be performed to address any localized corrosion issues.

Corrosion in an industrial system is often nonuniform and is a function of time and location. Thus, one usually observes that corrosion failure in a plant occurs individually at a specific time. For this reason, one should never expect to be able to equate one single corrosion rate reading obtained at a specific location to all the observable corrosion rates in a plant or even within a single heat exchanger. This is true for direct corrosion measurement methods and for side-stream simulation devices.

Therefore, even though the LCM has many distinguished capabilities in measuring localized corrosion, the readings obtained from it do not yield all the corrosion rates occurring throughout an industrial water system. There is only one way to obtain all the actual corrosion rates of a system—to measure them directly from every location in the system. Obviously, this is not an economically sensible approach. The next preferred solution is to install as many LCMs as needed, with each unit setup to obtain localized and general corrosion rate information from well-simulated conditions at each specific location. It is likely that one may need tens or hundreds of LCMs for simulating the different operating environments in a chemical processing plant, paper mill, utility, office building, or refinery. Thus, this approach may also not be economically feasible.

Often only one or a few LCMs can be made available for a given plant, leaving two primary options:

1. Simulate a specific plant condition at a particular location of concern. Then, monitor and optimize the treatment for that specific location and condition (for instance, a critical heat exchanger previously known to have corrosion problems).
2. Choose a typical system condition. Then, monitor and optimize the treatment for that specific condition (for instance, water from the cooling water return line).

Based on well-established engineering principles, one can assume that when corrosion under the chosen condition is under good control, other parts of the system are also most likely be under good control, unless the conditions differ drastically. The benefit of option 1 is that by selecting a critical, highly stressed system condition and by minimizing localized corrosion at this location, it is very likely that localized corrosion at all other locations will be similarly improved. The challenge of option 1 is that it may be logistically very difficult to plumb-in the LCM sufficiently close to the actual plant equipment of interest (for instance, a critical heat exchanger located 30 m above the ground in the middle of a refinery).

Checking corrosion control at various parts of the system from time to time and making required changes would ensure the success of these approaches. In this way, the LCM can be used to monitor changes in the corrosivity of the water or the effectiveness of the treatment program to control the propagation of pitting and general corrosion. The data obtained will be most effective in program selection, optimization, control, and diagnosis of failure mechanisms.

One should always remember that LCM is a simulation device. It is not a substitution for plant equipment inspection. It does provide new capability in the battle against corrosion. The more techniques one can afford to use to monitor or measure the extent of corrosion, the greater the confidence one will have on the results obtained, and the greater the chance one will be able to keep corrosion under control in an industrial system.

7.3 Data interpretation and use

7.3.1 General considerations for effective carbon steel corrosion control

In order to interpret the LCM localized corrosion data correctly, one must first understand that many factors that influence corrosion are time-dependent variables.

Some of the factors known to have a major effect on corrosion are listed below:

Material factors

(a) Composition of the alloy
(b) Structure of the alloy (single phase, multiple phases, and presence of precipitates)
(c) Mechanical state of the alloy (degree of cold work, presence of defects)
(d) Equipment design features (stagnant regions, galvanic contact, etc.)

Operational factors

(a) Upset conditions (e.g., process leaks, extended shutdown, etc.)
(b) Chemical treatments (type and dosages)
(c) Improper treatment program selection or application
(d) Poor control of water chemistry and operating parameters
(e) Economic or environmental restrictions on chemical use or proper dosages
(f) Presence of mechanical stresses
(g) Debris blockage

Chemical/environmental factors

(a) Water chemistry (O_2, pH, Cl^-, SO_4^{2-}, HCO_3^-, Ca^{2+}, $Fe^{2+/3+}$, PO_4^{3-}, SiO_2, etc.)
(b) Temperature
(c) Flow velocity
(d) Bacteria
(e) Coating

Owners of industrial system generally expect a chemical treatment vendor to provide the most cost-effective and reliable solutions to their cooling water system operation problems. One of the problems the chemical treatment needs to provide solutions for is carbon steel corrosion control. Many methods are available for controlling corrosion:

- switch to a more corrosion-resistant material
- apply an inert barrier or coating
- use cathodic protection (e.g., sacrificial anodes, such as Zn, Mg alloy, or Al)
- use anodic protection (e.g., H_2SO_4 production plant)
- adjust water chemistry
- use corrosion inhibitors

Use of corrosion inhibitors in the form of a cooling water treatment program is often the most cost-effective solution.

7.3.2 How corrosion inhibitors work

Based on inhibition mechanisms, corrosion inhibitors may be classified into three types: oxidation, deposition, and adsorption inhibitors. Chromate (CrO_4^{2-}) and nitrite (NO_2^-) are oxidation inhibitors, which inhibit corrosion by oxidizing the metal surface, promoting the formation of a protective passive film on the metal surface. Tolyltriazole, benzotriazole, and many organic corrosion inhibitors used in iron-oxide cleaning treatments, such as thiourea and acetylenic alcohols, may be considered adsorption inhibitors. Adsorption inhibitors reduce corrosion by forming a thin adsorbed layer on the metal surface. Zn^{2+}, phosphates, pyrophosphate, and phosphonates [e.g., AMP (aminotrimethylenephosphonic acid), HEDP (1-hydroxy ethylidene-1,1-diphosphonic acid), etc.] may be considered as deposition inhibitors. Deposition inhibitors function by forming a compact deposit layer on the metal surface with Ca^{2+} in the solution, Fe^{2+}/Fe^{3+} generated from the corrosion (metal dissolution) process (for phosphates or phosphonates), or OH^- (for Zn^{2+}) generated from the corrosion process (oxygen reduction). Usually, the protective layers formed by deposition inhibitors are thicker than the ones formed by either oxidation or adsorption inhibitors.

Deposition corrosion inhibitors are most commonly used for mild steel corrosion protection in open-recirculating industrial cooling systems today. Major types and the typical components of the most widely used modern cooling water treatments for corrosion and scale control are shown in Table 7.1 [21, 24, 30, 35–40].

Water-soluble inorganic phosphates such as orthophosphates and pyrophosphates are commonly used as the phosphate components in the treatments. Phosphonates typically used in the treatments include PBTC (2-butane phosphono-1,2,4-tricarboxylic acid), HEDP, AMP, HPA (2-hydroxyphosphono acetic acid, or Belcor 575), and PCAM (phosphono carboxylate acid mixture, or Bricorr 288, a mixture of sodium salts of organophosphonic acid, $H\text{-}[CH(COONa)CH(COONa)]_n\text{-}PO_3Na_2$, where $n < 5$ and $n_{mean} = 1.4$). The phosphinate used in the chemical treatments is PSO (phosphinic acid oligomers) which is a mixture of mono, bis, and oligomeric phosphinosuccinic acid adduct described in a recent US patent [36]. Acrylate-based polymers frequently used include Good-Rite K797 and K798 (both are acrylate/sodium styrene sulfonate/2-acrylamido-2-methylpropane sulfonic acid terpolymers) from Noveon (Ohio, USA); Aquatreat AR-540 (an acrylic acid/2-propenoic acid, 2-methyl, methyl ester/benzenesulfonic acid, 4-[(2-mthyl-2-propenyl)oxy]-, sodium salt/2-propene-1-sulfonic acid, 2-methyl-, sodium salt quad-polymer) and AR-545 (an acrylic

Table 7.1 Components in modern cooling water chemical treatments.

Classification	Typical active components	Use
All Organic	Phosphonates, Acrylate-based copolymer or terpolymer, Azole	Open-recirculating systems
Low Zn	Zn^{2+}, Phosphate, Phosphonate, Acrylate-based copolymer or terpolymer, Azole	Open-recirculating systems
Stabilized Phosphate	Phosphates, Phosphonate or phosphinate, Acrylate-based copolymer or terpolymer, Azole	Open-recirculating systems
Alkaline-Stabilized Phosphate	Phosphate, Phosphonate or phosphinate, Acrylate-based copolymer or terpolymer, Azole	Open-recirculating systems
Zn-Phosphonate	Zn^{2+}, Phosphonate, Acrylate-based copolymer or terpolymer, Azole	Open-recirculating systems
Molybdate	Molybdate, Phosphonate, Acrylate-based copolymer or terpolymer, Azole	Open-recirculating systems
Silicate-Molybdate	Silicate, Molybdate, Acrylate-based polymer, Azole	Open-recirculating or closed systems
Nitrite-Molybdate	Nitrite, Molybdate, Azole, Polymer dispersant	Closed systems
Silicate-Nitrite	Silicate, Nitrite, Borate, Azole, Polymer dispersant	Closed systems

acid/2-acrylamido-2-methylpropane sulfonic acid copolymer) from Alco Chemical (Tennessee, USA); Acumer 2000 (an acrylic acid/2-acrylamido-2-methylpropane sulfonic acid copolymer) and Acumer 3100 (an acrylic acid/t-butyl acrylamide/2-acrylamido-2-methylpropane sulfonic acid terpolymer) from Rohm and Haas (Pennsylvania, USA); acrylic acid/acrylamide/acrylamidomethanesufonic acid terpolymers, acrylate/hydroxypropyl acrylate copolymers, acrylic acid/allyloxy-2-hydroxypropylsufonic acid copolymers, and acrylic acid/allyloxy-2-hydroxypropylsufonic acid/polyethyleneglycol allyl ether terpolymers. Azole compounds suitable for use include benzotriazole and tolytriazole. Polymer dispersants suitable for use are generally polyacrylate or acrylate-based polymers. In some cases, PESA, a polyepoxysuccinic acid acting mainly as a calcium carbonate scale inhibitor, is used to substitute the phosphonate or phosphinate used in the phosphate or zinc containing chemical treatments listed in Table 7.1. The benefits of using PESA instead of some commonly used phosphonates such as HEDP, PBTC, or AMP may include the reduction of polymer demand, and in some cases, an enhancement of halogen stability of the treatments if the phosphonate being replaced is not halogen-stable. In addition to the corrosion and scale control treatments, a biocide treatment (using either an oxidizing biocide or a nonoxidizing biocide or both) is also generally required for microbial growth control in a cooling water system [38, 40].

The formation of a compact and protective layer on metal surface using a deposition type corrosion inhibition chemical treatment depends on a number of factors. The important factors include Ca^{2+}, phosphate, Zn^{2+}, calcium carbonate scale inhibitors (e.g., phosphonates, PSO or PESA), dispersant polymer concentrations, pH and/or alkalinity concentration, water temperature, corrosive ion (e.g., Fe^{3+}, Cu^{2+}, Cl^{-}, and SO_4^{2-}) concentrations, flow rate, and microbial activity level in the system. Generally, the chemical treatment dosages have to be selected based on the water chemistry and system operating conditions. In order to make correct treatment program optimization decisions, one needs to measure the water chemistry, product feed rates, and residual active concentrations and relates to the system operational and corrosion performance information.

Cooling water treatments are typically developed by conducting bench-top screening tests of the candidate components, followed by pilot cooling system tests of candidate formulations, and finally, field testing to confirm the treatment performance.

Tables 7.2 and 7.3 show the examples of screening test results on commonly used corrosion inhibitors under typical cooling water operating conditions.

7.3.3 Performance issues in cooling water treatments

Corrosion control, scale control, and microbial control (or biofouling control) are the main performance parameters used to judge whether the application of a cooling water program is meeting the system needs [2]. In addition, compliance of applicable regulations and discharge limits are also the required performance parameters. As in all engineering applications, a safety margin needs to be included in the selection and application of cooling water treatment chemicals and their use dosages. In this regard,

Table 7.2 Screening test results on the effect of corrosion inhibitors on maximum pitting penetration rate of a 25 μ Fe foil in a synthetic plant water containing 1.2 mM CaCl2, 0.68 mM MgSO4, 3 mM NaHCO3, 2.1 mM NaCl, pH = 8.4 at room temperature. Solution aerated and under magnetic stirring with a Teflon-coated bar.

Inhibitor treatment	Maximum pitting penetration rate on a 25 μm Fe (99.5% pure) foil at room temperature, μm/y	Mean maximum pitting rate on a 25 μm Fe foil, μm/y
Blank		1841.1
Run 1	4142.7	
Run 2	1775.5	
Run 3	1414.8	
Run 4	1282.7	
Run 5	1348.7	
Run 6	1082.0	
1 mg/L AA/AM copolymer		2788.0
Run 1	2519.7	
Run 2	3056.4	
5 mg/L AA/AM copolymer		1832.6
Run 1	2415.5	
Run 2	1249.7	
5 mg/L Pyrophosphate as PO_4		2165.4
Run 1	2933.7	
Run 2	1397.0	
2 mg/L Pyrophosphate as PO_4 + 1 mg/L AA/AM copolymer		807.7
Run 1	1082.0	
Run 2	533.4	
5 mg/L Orthophosphate as PO_4 + 1 mg/L AA/AM copolymer		1225.6
Run 1	2115.8	
Run 2	335.3	
5 mg/L HEDP +1 mg/L AA/AM copolymer		2642.9
Run 1	3208.0	
Run 2	2077.7	
5 mg/L HIB		1607.2
Run 1	2677.2	
Run 2	1280.2	
Run 3	1191.3	
Run 4	1280.2	
5 mg/L MoO4		5936.0
Run 1	4183.4	

Continued

Table 7.2 Continued

Inhibitor treatment	Maximum pitting penetration rate on a 25 μm Fe (99.5% pure) foil at room temperature, μm/y	Mean maximum pitting rate on a 25 μm Fe foil, μm/y
Run 2	7688.6	
5 mg/L CrO4		628.7
Run 1	800.1	
Run 2	457.2	
5 mg/L Belcor 575 (HPA) as active		932.8
Run 1	1275.1	
Run 2	807.7	
Run 3	995.7	
Run 4	652.8	
5 mg/L Bricorr 288 as Solid		660.4
Run 1	701.0	
Run 2	619.8	
5 mg/L PBTC as active	4582.2	4582.2
2 mg/L Zn^{2+}	1003.3	1003.3

Table 7.3 An example of corrosion inhibitor screening test results for general corrosion protection of carbon steel (The corrosion inhibitor labeled as Example 1 in the table is PSO (phosphinic acid oligomers) [36].)

Corrosion inhibitor screening test results: Hard water 360 ppm $CaCl_2$, 200 ppm MgS04, 100 ppm $NaHC0_3$, pH = 8.4, 120°F., 160 rpm; 16 h immersion			
Compound	**Dosage as active in acid form or stated**	**MS general corrosion rate (mpy)**	
Blank	None	43.50	
Orthophosphate	15 ppm as PO_4	17.35	
Pyrophosphate	30 ppm as PO_4	11.99	
HPA	15 ppm	2.13	
	30 ppm	2.47	Av. of 3 tests
PCAM	15 ppm	15.43	
PCAM	20 ppm	6.26	
PCAM	30 ppm	23.68	
PCAM	40 ppm	15.28	Av. of 6 tests
Example 1	15 ppm	2.78	
Example 1	20 ppm	2.16	

Table 7.3 Continued

Corrosion inhibitor screening test results: Hard water 360 ppm $CaCl_2$, 200 ppm MgS04, 100 ppm $NaHC0_3$, pH = 8.4, 120°F., 160 rpm; 16 h immersion			
Compound	**Dosage as active in acid form or stated**	**MS general corrosion rate (mpy)**	
Example 1	30 ppm	0.97	
Example 1	40 ppm	2.17	Av. of 2 tests
HEDP	15 ppm	13.05	
HEDP	30 ppm	8.09	
AMP	15 ppm	9.25	
AMP	30 ppm	16.11	
PBTC	15 ppm	6.17	
PBTC	30 ppm	10.49	
Polyacrylate (MW = 2000)	15 ppm	11.84	
Polyacrylate (MW = 2000)	30 ppm	34.49	
Molybdate	15 ppm as MoO_4	19.89	
Vanadate	15 ppm as VO_3	17.43	
Nitrite	30 ppm as NO2	9.38	
Zn^{2+}	5 ppm as Zn	37.24	
Gluconate	30 ppm	5.69	
Phosphonosuccinic acid	15 ppm	6.80	
Blow results are obtained at 100°F			
Blank	None	21.5	Av. of 2 tests
Example 1	40 ppm	0.98	Av. of 2 tests
Example 1 +	40 ppm Example 1 +	0.92	Av. of 3 tests
Polymer 1	10 ppm Polymer 1		
PCAM +	40 ppm PCAM +	19.50	Av. of 3 tests
Polymer 1	10 ppm Polymer 1		

Note: HPA = 2-hydroxy-phosphonacetic acid.
PCAM = phosphonocarboxylic acid mixture, H—[CH(COOH)CH(COOH)]$_n$—PO_3H_2, where $n < 5$ and $n_{mean} = 1.4$.
HEDP = l-hydroxyethylidene-1,1-diphosprionic acid.
AMP—Amino tri(methylene phosphonic acid).
PBTC = 2-phosphono-butane-1,2,4-tricarboxylic acid.
Polymer 1 = Acrylic acid (50–60 mol%)/acrylamide (20–36 mol%)amino methane sulfonate (14%–20%) terpolymer.

past application experience and treatment active monitoring are essential in ensuring the operation of the system in the safe region without excessive chemical feeds. These factors and the cost of the treatment program are then used to establish the cost vs. performance relationships. The cost vs. performance relationships can be used as the objective criteria for choosing a particular treatment for a given system.

Generally, determining whether a chosen treatment program will meet the requirements of applicable regulations and discharge limits is relatively easy to accomplish. It typically requires one to choose the appropriate treatments and/or their dosages to avoid or limit the use of some specified restricted chemicals. In some cases, the effluent can be treated to remove the harmful substances before discharge to the environment. For example, in utilities industry, Na_2SO_3 may be used to react with the residual NaOCl or NaOBr in the blowdown water before discharging it to the river. In this case, only periodical monitoring of NaOCl or NaOBr in the blowdown water to control the Na_2SO_3 feed may be needed.

Controlling scale formation can be easily accomplished via pH control, cycle control, effective use of scale inhibitors, and a combination of these methods. Monitoring of scale formation on equipment such as exchangers can also be accomplished with well-established temperature, flow, and/or pressure drop measurements. In fact, many chemical processing plants, refineries, and power plants are already conducting real-time total (including the effects of both water side and process side) heat transfer efficiency measurements on their critical exchangers and/or reactors. Prediction of the likelihood of scale formation under specified conditions based on water analysis results, temperature, and flow rate data is also well-established, e.g., Langelier Saturation Index [38] and French Creek WaterCycle[b] software. Additions of biocides, biodispersants, and side-stream filtration to remove nutrients are the proven methods to control excessive microbiological growth and biofouling. In addition, the damaging effects of scaling and biofouling can often be reversed via cleaning. In contrast, the only options of reversing a corrosion failure, e.g., perforation of an exchanger tube, is to replace it with new tube or cease usage via plugging.

Generally, cooling chemical treatments and their application control parameters are developed based on the general corrosion control performance obtained from laboratory tests and a small number of controlled field tests. In a typical application of the chemical treatments in a cooling water system, general corrosion control performance is also often monitored by periodic testing of corrosion product concentrations, such as Fe^{3+} and Cu^{2+}, in the water. In addition, corrosion coupons are also often used in many systems. Thus, chemical treatment technologies and their application methods to control general (average) corrosion rate to an acceptable level, e.g., <3 mpy (75 μm/y), are generally well-developed and widely practiced in many cooling water systems. However, localized corrosion remains unpredictable. Often, the extent of localized corrosion under a given application condition has been the key remaining unknown in the cost vs. performance relationships [2, 21].

[b] Trade name of French Creek Software, PA, United States.

7.3.4 Integrated solutions needed to improve cooling water treatment performance

Controlling corrosion, scale formation, and biofouling (or microbiological growth) is required for the effective operation of cooling water systems. The problems caused by corrosion, scale formation, and biofouling are interrelated (see Fig. 7.4) and can present serious challenges for the normal operation of cooling water systems.

The primary function of a cooling water system is to satisfy the heat transfer needs of the plant. Thus, preventing and/or controlling scale and biofouling to ensure adequate heat transfer efficiency is one of the primary requirements a chemical treatment program must fulfill. However, although scaling and biofouling may be controlled to provide effective process cooling needs, they may still pose a serious threat to localized corrosion control. For instance, the tendency of stress corrosion cracking of a stainless steel heat exchanger in a CPI (or chemical processing industry) plant usually increases as scaling and fouling increase. One reason for the need of controlling general corrosion to an acceptable level is to prevent the deposition of corrosion products downstream on other equipment, which may initiate new concerns about reduction of heat transfer efficiency and promotion of further corrosion damage. Under cooling water conditions, a high carbon steel localized corrosion rate is usually observed when the metal is partially protected and/or there is high microbial activity.

Following the operating guidelines provided by the chemical treatment supplier, the carbon steel localized corrosion rate (or pitting rate) in a synthetic plant water containing 450 ppm $CaCl_2$ as $CaCO_3$ was found to vary according to the following chemical program sequence in bench-top tests [18]:

$$\text{Stabilized Phosphate} = \text{Low Zinc} < \text{Alkaline} - \text{Stabilized Phosphate} < \text{All Organic}$$

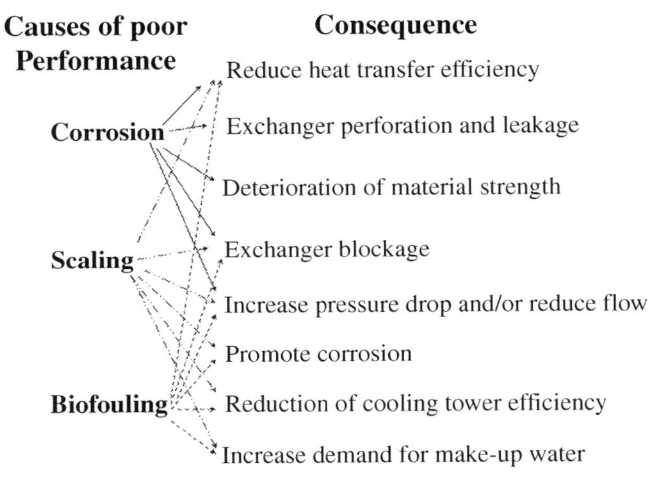

Fig. 7.4 Problems caused by corrosion, scaling, and biofouling in cooling water systems.

While all of these programs can be optimized to give low localized corrosion rates, the treatments that operate at the higher pHs require more attention due to the greater likelihood of under-deposit corrosion by deposition of calcium carbonate and calcium phosphate in isolated regions of heat exchanger surfaces.

Depending on the operating conditions, the following have been shown to be effective in reducing the carbon steel localized corrosion rate [2, 21]:

- adjusting the corrosion inhibitor feed
- adjusting dispersant polymer feed
- adjusting the pH
- feeding a more effective biocide or a more proper biocide dosage
- increase the blowdown rate

Examples of using these methods to reduce pitting and general rates are discussed in Chapter 22.

7.3.5 Factors to consider in interpreting LCM readings

7.3.5.1 Time-dependence of corrosion rate measurements

A carbon steel corrosion rate is time-dependent even under constant conditions (e.g., water flow, water temperature, chemical treatment type and dosages, cycle of concentration, absence of process leaks, absence of biofouling, etc.). This can be mostly attributed to the effect of corrosion product and inhibitor film build-up on the metal surface, creating a barrier to restrict mass transfer and increase solution resistance. For this reason, a minimum exposure time is recommended in order to report steady state corrosion rate values. The following equation is recommended by ASTM for use to obtain consistent general corrosion rate information using coupons [41].

$$\text{Minimum coupon exposure time (hour)} = 2000/\text{expected general corrosion rate in mpy } (25.4\,\mu m/\text{yr})$$

For the LPR general corrosion rate reading obtained from the LCM, the steady state value under constant operating conditions should be reached in at most ~1/10 of the time needed for the coupon general rate measurements [2, 21, 23, 25, 29, 30, 32, 33]:

$$\text{Minimum LCM (LPR) exposure time (hour)} = 200/\text{expected general corrosion rate in mpy } (25.4\,\mu m/\text{yr}).$$

In field applications, the time needed to reach the steady state general corrosion rate value is often shorter than the time given by the above equation, assuming a ~15% error is allowed.

Like the general corrosion rate, the localized corrosion rate requires time to reach a steady state condition. Meanwhile, changes in corrosivity may be reliably monitored even before the steady state condition is reached. Because the LCM is sensitive to even slight changes in system corrosivity, such as the swings in the daily water temperature from night-time to day-time, reliable detection of variations in localized corrosion is possible prior to reaching the steady state localized corrosion value. In this way, the LCM is quite different from a device that only monitors the general corrosion rate.

One critical phenomenon needs to occur in order for the LCM to become fully reliable. This involves the formation of a tubercle over each anode surface. In some cases, during the first week of operation, considerable fluctuations in the localized corrosion rate may be observed which are specifically due to partial formation of a tubercle, followed by partial sloughing off the tubercle. In some cases, little or no signal fluctuations are observed during the first week. Once a stable tubercle covering has developed over each anode, any further fluctuations in the localized corrosion rate can be attributed to actual corrosivity events in the cooling water system.

Because of the time required for tubercle growth to develop and the presence of complete corrosion product deposits on the localized corrosion sites, the time-dependence of the localized rate is different from that of the general rate. Depending on the corrosivity of the water, it normally takes about 2–3 weeks for the localized corrosion rate to reach a quasi-steady state value. It should be noted that even after the initial 2–3 week period, the localized corrosion rate would continue to decrease if the system conditions remain largely the same.

For example, in Great Lakes of United States drinking water, the maximum carbon steel pitting rate was found to be given by the following Eq. [38].

$$\text{Max .pitting rate (mpy)} = \text{constant}{*}t^{-2/3}.$$

t is time in year, or

Max. Pitting Depth = constant * $t^{1/3}$.

The constant is a function of water temperature and Ryznar Index [38]. It should be pointed out that these equations are obtained in the absence of chemical treatments except bleach. It should be used as a rough indicator for trends only.

Based on these expected time-dependence of localized and general corrosion rates, the time-averaged localized and general corrosion rates are expected to decrease continuously with time, assuming other conditions, e.g., temperature, flow, chemical treatments, biocide treatments, and water chemistry, remain largely the same.

If there is an interruption in this general trend, then it indicates that a major corrosivity change in the system has occurred.

7.3.5.2 Effect of water temperature

The rate of corrosion is dependent on temperature, as are the rates of most other chemical reactions. In the absence of an insulating, nonporous layer of scale on the metal surface, the corrosion rate will increase with increasing temperature. Since the LCM measures the water temperature as well as the corrosion rates, the effect of temperature on general and localized corrosion can be easily observed and evaluated. Fig. 7.5 clearly shows the sensitivity of the localized corrosion rate compared to the general

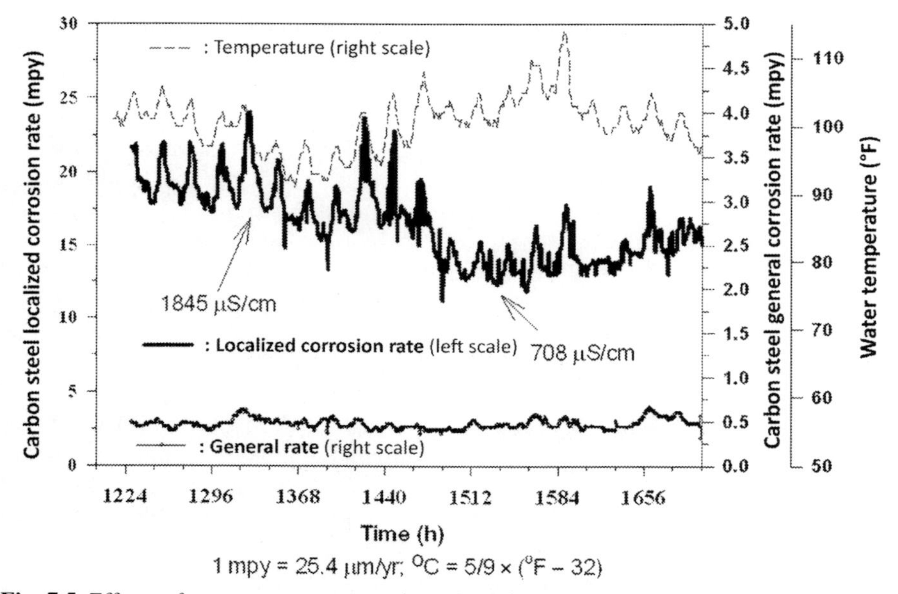

Fig. 7.5 Effects of water temperature and conductivity on carbon steel corrosion rates.

corrosion rate, with respect to temperature fluctuations. In this figure, each temperature increase, followed by decrease, corresponds to the daily temperature fluctuation in the cooling system. As can be seen, in this system, a 5°F (2.8°C) change causes a 5 mpy (127 μm/yr) change in the localized corrosion rate. Little or very slight correlation can be seen in the general corrosion rate readings. In addition to temperature, other factors (e.g., conductivity, as shown in Fig. 7.5) can also influence corrosion rates.

7.3.5.3 *Attributing causes of corrosion rate variations and treatment optimization strategy*

After the initial tubercle growth and presteady state corrosion period has passed (2–3 weeks), if the LCM localized corrosion rates continue to decrease, it indicates either:

- Water temperature decrease
- Blowdown rate increase
- Better corrosion control due to the following
  - Adjustment to a more appropriate pH set point
  - Use of a more effective corrosion inhibitor (or more appropriate inhibitor dosage)
  - Use of a more effective polymer (or more appropriate polymer dosage)
  - Use of a more appropriate biocide or increase of biocide feed if there is an existing MIC problem
- Formation of a large amount of nonconducting scales (e.g., $Ca_3(PO_4)_2$, $CaCO_3$)

If the LCM localized corrosion rate is increasing, it indicates that one or more of the following events may be happening:

- Water temperature increase
- Increase in cycles of concentration (or increase of total dissolved solid concentration)
- Poor pH control (e.g., acid overfeed, or incorrect pH set point)
- Insufficient corrosion inhibitor feed
- Overfeed corrosion inhibitor (ortho-PO_4, pyro-PO_4, phosphonate, and/or Zn)
- Insufficient polymer feed
- Oxidizing biocide overfeed
- Biocide underfeed, causing MIC problems
- Process leaks
- Instrument malfunctions: Need to check flow and/or electronics.

Since a typical LCM measures water temperature at the same time as the corrosion rates, the effects of temperature changes should be the easiest to check first before considering other factors as the possible cause of corrosion rates changes. The localized corrosion rate is usually more sensitive to all corrosivity changes than the general rate. Therefore, the localized corrosion rate may show significant increasing or decreasing trends, while the general corrosion rate may remain unchanged.

An example is shown in Fig. 7.5. The effect of cycle change can be clearly distinguished from the daily water temperature variation (due to daily change of sunlight and/or atmospheric weather conditions) [16]. In this cooling system, a decrease in conductivity from 1845 to 708 μS/cm resulted in a decrease in the localized corrosion rate from 15 to 20 to 10–15 mpy (381–508 to 254–381 μm/y). Meanwhile, the general corrosion rate fluctuated between 0.4 and 0.6 mpy (10 and 15 μm/yr) during this study and did not show a correlation with the conductivity change.

In general, all of the major changes in LCM readings can be correlated to one or a combination of the events listed above. However, developing reliable correlation may not be an effort-free task. Reliable correlation can generally be developed by comparing the corrosion rate trends to other readily measured parameters, like water temperature, conductivity, pH, halogen feed rate or ORP (oxidation-reduction potential) readings, changes in biocide feed strategies, process operating conditions, changes in program feed rates, changes in the chemical program, changes in the residual program actives levels, microbial activity changes, and the changing quality of the make-up water. Additionally, one needs to check that the flow rates for the water passing through the LCM and other side-stream monitoring devices are maintained and the equipment is properly calibrated. Although many changes can occur in actual cooling water system, rarely do all or several of them occur simultaneously at the same time or within a relatively short time period, e.g., 1–2 h. Because of the capability of the LCM to provide reliable, sensitive, real-time, and continuous measurement of corrosion rates and water temperature, using it to establish correlation between localized corrosion changes and process variations has become possible. It should be stressed that the more one knows about the system operating conditions, water chemistry parameters, and chemical treatment residuals, the more capable one will be able to establish meaningful correlation of localized corrosion rate changes with the events occurring in the system. One should also be aware that since many of the system changes listed above can be controlled and easily repeated at will, a correlation that may not be established with high confidence at its first occurrence can be easily repeated and confirmed.

7.3.5.4 Maximum localized corrosion rate vs. length of localized corrosion events

To provide true cost-effective solutions to industrial cooling water treatment problems, one must have a sense of proportion. For instance, the actual corrosion damage done to a cooling system depends not only on how high the instantaneous corrosion rate is, but more importantly on how long a high corrosion rate event lasts. For example, a localized corrosion rate of 200 mpy (5.08 mm/yr) lasting one day will only produce a penetration depth of 0.54 mils (13.7 μm). This amount of attack by itself is not significant in reducing the service life of a carbon steel exchanger with a thickness of 1/8 in. (3.18 mm), assuming that the localized corrosion rate is reduced to an acceptable level soon afterwards. On the other hand, one should be more concerned if the localized corrosion rate increases from 10 to 50 mpy (from 254 to 1270 μm/yr) for one year since 40% of the tube thickness could be penetrated. Thus, in general, one needs to worry more about sustained and prolonged corrosivity changes than short-life, transient spikes of corrosion rates.

Some process variations in cooling water systems are unavoidable, such as cycle changes due to fluctuating make-up water quality, daily or seasonal temperature fluctuations, pH variations due to poor acid feed control, occasional process leaks, etc. One objective of using the LCM is to optimize the chemical treatment (both in terms of dosage and application methodology) so that the damaging effects of these process variations can be minimized.

7.4 Comparison with ZRA-based occluded cell measurement results

Occluded cell is a type of artificial pit simulation test method that has been widely used to study pitting, crevice/under-deposit corrosion in a variety of corrosion systems [13, 42]. Since an occluded cell can also be considered as a differential flow cell with the anode having a flow rate approaching zero, based on the results described in previous studies using the DFC-based method, one could expect that a ZRA-based occluded cell technique can only measure a small portion of the current related to localized corrosion rate. In order to further examine the capability of the technique, an occluded cell test device similar to the ones described earlier [42] was built. The occluded cell consists of a section of carbon steel (C1010, length = 16.5 cm, diameter = 1.27 cm, area = 65.8 cm^2) tubing as the cathode and four carbon steel disks of the same material (diameter = 0.9 cm, exposed area = 0.2 cm^2) as the anodes. These anodes were placed at the bottom of the crevices of various depths (i.e., 5 mm, 15 mm, 25 mm, and 40 mm) in a Plexiglas block. The diameter of the crevices was 1 mm. The cathode/anode area ratio was 82. The occluded cell was connected to a

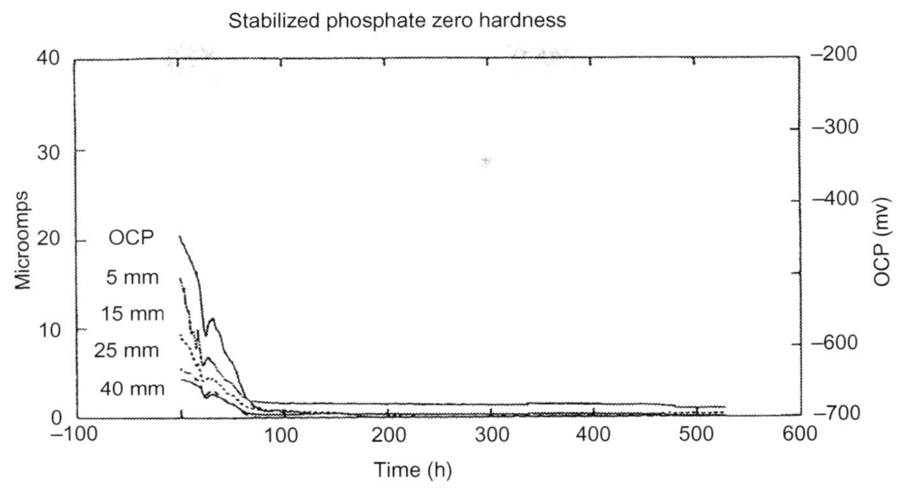

Fig. 7.6 Short circuit currents from various occluded anodes obtained from a ZRA and cathode potential (open-circuit potential [OCP] as a function of time for the stabilized phosphate program operated at zero-hardness conditions: pH = 7.2, 18 mg/L inorganic PO_4).

laboratory circulator. The setup provided provisions for automatic pH and temperature control (100°F at the basin), holding time simulation, chemical treatment addition, and heat transfer simulation, similar to a pilot cooling water system. The cathode and the anodes were connected electrically through a high sensitivity zero-resistance ammeter. The current between the anodes and cathode was monitored and recorded by a computerized data acquisition system. The corrosion potential of the cathode vs. a saturated calomel reference electrode was also monitored and recorded. A more detailed description of the setup and experimental procedures was described in a previous publication [16, 42].

The occluded cell device was used to test the performance of the stabilized phosphate program used in the refinery against mild steel localized corrosion. Fig. 7.6 shows the current flowing from various occluded anodes to the cathode as a function of time for a test carried out in a zero-hardness water containing 360 ppm $Na2SO_4$, 200 ppm $MgSO_4$, 400 ppm $NaHCO_3$ (initial concentrations, all as $CaCO_3$), and the recommended dosages of the stabilized phosphate program (e.g., 18 ppm Inorganic PO_4). The solution pH was controlled at 7.2 by feeding 1% H_2SO_4. The current from all occluded anodes dropped below 1 μA (or less than 2.3 mpy) after ~75 h immersion. At the same time, the corrosion potential of the system also dropped from −430 mV/SCE to −690 mV/SCE. Visual inspection of the cathode tube showed the decrease in current coincided with the formation of tuberculation-type corrosion products on the mild steel tube surface. Thus, the current decreases did not indicate the deactivation of the occluded anodes. Rather, it indicated that the conditions at the tube surface (the cathode) were changing toward those found in the occluded anode. This leads to the occluded discs and the tube corroding at roughly equally high rate (localized rate or pitting rate greater than 50–100 mpy were observed under similar conditions). Under these conditions, the cathodic corrosion reactions occurring on the anodes themselves were supplying close to 100% of current needed to balance the Fe dissolution reaction on the anodes, in agreement with the results obtained from the LCM at the refinery [16]. Thus, it is clear that a ZRA-based occluded cell measurement alone cannot be used to measure accurately the localized corrosion rate or pitting rate.

7.5 Applications

The DFC-based LCM technology has been used in many cooling water systems (the author had been involved in about 30 field applications), including once-through, open-recirculating, and closed-loop cooling water systems in petroleum refineries, nuclear and fossil fuel power plants, chemical processing industry plants, pulp and paper mills, and food processing plants. Several examples of the applications of this technique in cooling water systems are described in Chapter 22. The DFC-based LCM was demonstrated to be very useful for localized corrosion monitoring, for failure cause analysis, and for the performance-based optimization and control of chemical treatments (i.e., corrosion inhibitors, scale inhibitors, polymer dispersants, and biocides) and their application methodologies in these systems [2, 16–19, 21, 23, 25–30, 32–35]. The LCM technology has also been adapted for use to study localized corrosion in oil field systems [22], iron-oxide cleaning process [31], boiler condensate corrosion, and aluminum corrosion in engine coolants [20]. The interest readers may consult the references provided in the end of this chapter.

7.6 Future trends and additional information

To the author's knowledge, the DFC-based LCM technology is the first real-time localized corrosion monitoring method based on the mixed potential theory of corrosion initially developed by Wagner and Traud [1]. Although the mixed potential theory has been widely accepted for use to interpret general corrosion, previously the more phenomenological local cell theory has been commonly assumed to be valid for localized corrosion. Before the development of DFC-based LCM technology, the local cell theory has been universally used as the theoretical basis for various real-time localized corrosion monitoring technologies, including the more widely used methods based on electrochemical noise measurements and various ZRA-based occluded cell methods. The success of the DFC-based LCM technology in cooling water applications suggests that the mixed potential theory may also be generally applicable to localized corrosion processes in other environments. New real-time methods based on the mixed potential theory for localized corrosion are expected to be developed in the future to allow a more accurate and reliable measurement of localized corrosion in these environments.

References

[1] C. Wagner, W. Traud, On the interpretation of corrosion processes through the superposition of electrochemical partial processes and on the potential of mixed electrodes, Z. Elektrochem. 44 (1938) 391. English Translation by M.O. Friedman, Corrosion, Vol. 62, 844 (2006).
[2] B. Yang, Minimizing localized corrosion via new chemical treatments and performance-based treatment optimization and control, in: NACE Corrosion/99, Paper no. 307, 1999.
[3] Field usage data on soft-steel heat exchangers in cooling water environments, edited by Corrosion Subcommittee of Chemical Equipment Materials Committee, Japanese Society of Chemical Engineers, October 1990.
[4] Heat Exchanger Management and Residual Life Prediction, Edited by the Task Group for Heat Exchanger Residual Life Prediction, Corrosion and Anticorrosion Subcommittee, Japanese Material Society, September 1996.
[5] Handbook of Corrosion and Corrosion Protection – Corrosion and Protection of Production Equipment of Chemical Industry, Edited by Research Institute of Chemical Engineering Machinery, P.R. China, Chemical Industry Publishing House, Beijing, 1991.
[6] D.R. McIntyre, MTI Publication No. 27, Experience Survey Stress Corrosion Cracking of Austenitic Stainless Steels in Water, in: Materials Technology Institute of the Chemical Process Industries, St. Louis, MO, 1987.
[7] S. Hayruyama, "Stress Corrosion Cracking by Cooling Water of Stainless Steel Shell and Tube Heat Exchangers", Materials Performance, March, 1982, pp 14–19.
[8] Results of Second Field Survey on Stress Corrosion Cracking in Multi-tube Stainless Steel Heat Exchangers and Life Analysis, Japanese Society of Chemical Engineers, 1984.
[9] M.O. Speidel, A. Atrens (Eds.), Corrosion in Power Generating Equipment, Plenum Press, New York, 1984.
[10] B. Yang, P. Woyciesjes, and A. Gershun, "Comparison of Extended Life Coolant Corrosion Protection Performance," SAE Technical Paper 2017-01-0627, 2017, https://doi.org/10.4271/2017-01-0627.
[11] R.D. Kane, Corrosion in Petroleum Production Operations. in: S.D. Cramer, B. S. Covino Jr. (Eds.), ASM Handbook, Vol. 13C, Corrosion: Environments and Industries, 2006, pp. 922–966, https://doi.org/10.31399/asm.hb.v13c.a0004210.
[12] R.D. Kane, Corrosion in Petroleum Refining and Petrochemical Operations. in: S. D. Cramer, B.S. Covino Jr. (Eds.), ASM Handbook, Vol. 13C, Corrosion: Environments and Industries, 2006, pp. 967–1014, https://doi.org/10.31399/asm.hb.v13c.a0004211.
[13] J. Han, B.N. Brown, S. Nesic, Investigation of the galvanic mechanism for localized carbon dioxide corrosion propagation using the artificial pit technique. Corrosion 66 (9) (2010). https://doi.org/10.5006/1.3490308. 095003.
[14] NACE Publication 3T199-2012-SG, Techniques for Monitoring Corrosion and Related Parameters in Field Applications, NACE, Houston, TX, 2012.
[15] R. Baboian (Ed.), Corrosion Tests and Standards: Application and Interpretation, ASTM International, West Conshohocken, PA, 2005.
[16] B. Yang, Real-time localized corrosion monitoring in industrial cooling water systems, Corrosion 56 (2000) 743.
[17] B. Yang, "Localized Corrosion Monitoring in Cooling Water Systems", Corrosion/95, Paper no. 541, NACE, Houston, TX (1995).
[18] B. Yang, Method for on-line determination of Underdeposit corrosion rates in cooling water systems, Corrosion 51 (1995) 153.

[19] M. Enzien, B. Yang, Effective use of monitoring techniques for use in detecting and controlling MIC in cooling water systems, Biofouling 17 (1) (2001) 47.
[20] B. Yang, F. J. Marinho, and A.V. Gershun, "New electrochemical methods for the evaluation of localized corrosion in engine coolants," J. ASTM Int., Vol. 4, No. 1, (2007). Available online at *www.astm.org*.
[21] B. Yang, Corrosion control in industrial water systems, in: Presented at NACE Central Area Conference, Corpus Christi, TX, October 2001.
[22] S.L. Fu, A.M. Griffin, J.G. Garcia and B. Yang, "A New Localized Corrosion Monitoring Technique for the Evaluation of Oilfield Inhibitors", Corrosion/96, Paper No. 346, NACE, Houston, TX (1996).
[23] B. Yang, Advances in localized corrosion control in cooling water systems, Power Plant Chem. GmbH 2 (6) (2000) 321.
[24] B. Yang, "Method and Apparatus for Measuring Underdeposit Localized Corrosion Rate or Metal Corrosion Rate under Tubercles in Cooling Water Systems", US5,275,704.
[25] B. Yang, Real time localized corrosion monitoring in industrial cooling water systems, Corrosion Rev. 19 (3–4) (2001) 315.
[26] M. Enzien and B. Yang, "On-line Performance Monitoring of Treatment Programs for MIC Control", Corrosion/2001, paper no. 279, NACE, Houston, TX (2001).
[27] B. Yang, Advances in localized corrosion control in cooling water systems, in: Proceedings of 9th European Symposium on Corrosion Inhibitors, vol. 2, Univ. of Ferrara, Ferrara, Italy, 2000, p. 821.
[28] M. Enzien and B. Yang, "Effective Use of Biocide for MIC Control in Cooling Water Systems", Corrosion/2000, paper no. 384, NACE, Houston, TX (2000).
[29] E.R. Hale, B. Yang, M.V. Enzien, Corrosion control in cooling water systems: recent experience using a new corrosion monitor, in: Proceedings of AIM 2000, Venice, Italy, 2000.
[30] D.A. Meier, E.B. Smyk, and B. Yang, "Advances in Zinc Free Alkaline Cooling Water Treatment", Corrosion/99, paper no. 304, NACE, Houston, TX (1999).
[31] K. Sotoudeh and B. Yang, "On-line Cleaning of Year-Round Building HAVC Loops", Corrosion/99, paper no. 102, NACE, Houston, TX (1999).
[32] E. Hale, B. Yang, Real-time, on-line localized corrosion rate measurements for monitoring and control of cooling water corrosion, in: EPRI Corrosion and Degradation Conference, June 12–13, St. Pete Beach, FL, 1999.
[33] E. Hale, B. Yang, Evaluation of various application regimes to control mild steel corrosion in a nuclear power service water system using a localized corrosion monitor, in: EPRI Service Water System Reliability Improvement Seminar, July 13, Biloxi, MS, 1999.
[34] B. Yang, "Real Time Localized Corrosion Monitoring in Refinery Cooling Water Systems", Corrosion/98, paper no. 595, NACE, Houston, TX (1998).
[35] B. Yang, "A novel Method for On-line Determination of Underdeposit Corrosion Rates in Cooling Water Systems", Corrosion/94, Paper No. 335, NACE, Houston, TX (1994).
[36] B. Yang, P.E. Reed, and J.D. Morris, "Corrosion inhibitors for aqueous systems", US6,572,789B1.
[37] B. Yang and J. Tang, "Hydroxyimino alkylene phosphonic acids for preventing corrosion and scale in aqueous systems", US5,788,857A1.
[38] R.W. Lane, Control of Scale and Corrosion in Building Water Systems, McGraw-Hill, New York, 1993.
[39] B. Yang and W. Chen, "Water treatment chemicals and addition method thereof", CN1,483,684.

[40] K.D. Demadis, B. Yang, P.R. Young, D.L. Kouznetsov, D.G. Kelley, Rational development of new cooling water chemical treatment programs for scale and microbial control, in: Z. Amjad (Ed.), Advances in Crystal Growth Inhibition Technologies, Kluwer Academic/Plenum Publishers, New York, 2000, p. 215.
[41] ASTM Standard G4-01, Standard guide for conducting corrosion tests in field applications, in: ASTM Standards on Disc, vol. 03.02, ASTM International, West Conshohochen, PA, 2003.
[42] S. H. Shim, and D. A. Johnson, "Studies of Localized Corrosion Using a Occluded Cell Device," Corrosion/91, paper no. 86 (Houston, TX: NACE, 1991).

Multielectrode systems

8

Lietai Yang
Corr Instruments, LLC, Carson City, NV, United States

8.1 Introduction

Multielectrode systems have been used for electrochemical and corrosion studies for at least three decades. Coupled multielectrode systems, in which the multiple electrodes are connected together through an external circuit and all electrodes are essentially at the same potential, were first used as high-throughput devices for the measurements of the stochastic behavior of pitting corrosion in 1977. Similar concepts were reported for corrosion detection in concrete in 1991 and for crevice corrosion measurements in 1993.

Multielectrode arrays, in which the multiple electrodes are arranged in a given pattern and each electrode is addressable, have been used since 1984. The early work on the multielectrode arrays appears to have been for the development of electronic devices. The concept of multielectrode arrays for corrosion studies, in the uncoupled form, was first reported in 1991. The first coupled multielectrode array that was used to simulate a one-piece metal for studying the spatiotemporal electrochemical behaviors and corrosion processes of iron in sulfuric acid solutions was published in 1996. Because the electrodes in a multielectrode array can be arranged in any given pattern and each of the electrodes is addressable, the coupled multielectrode arrays have been widely used by researchers to study the corrosion processes, especially the localized corrosion processes, of metals.

More recently, coupled multielectrode arrays have been used as sensors [called coupled multielectrode array sensors (CMASs)] for online and real-time monitoring of corrosion in laboratories and industrial fields. Because a CMAS does not require the presence of bulk electrolytes, CMAS probes have been used to quantitatively measure the rate of nonuniform corrosion of metals not only in aqueous solutions, but also in wet gases, oil/water mixtures, salt deposits, biodeposits, soil, concrete, and undercoatings. CMAS probes are also used for real-time monitoring of the performance of cathodic protection systems. Nonuniform corrosion includes localized corrosion, such as pitting corrosion and crevice corrosion, and most types of general corrosion experienced by most of the industrial metal equipment and field metal structures. This chapter presents a review for the development and applications of the multielectrode systems, especially the coupled multielectrode array sensors, for corrosion monitoring and corrosion studies.

Techniques for Corrosion Monitoring. https://doi.org/10.1016/B978-0-08-103003-5.00008-4

8.2 Earlier multielectrode systems for high-throughput corrosion studies

The first published use of a multielectrode system appears to have been by Shibata and Takeyama [1] concerning the evaluation of the stochastic behavior of pitting corrosion. Fig. 8.1 shows the schematic diagram of the multielectrode system. Twelve stainless steel specimens were assembled onto a specimen holder and each of them was individually connected, through independent ammeters and relays, to a common joint which was in turn connected to a potentiostat. The potential of the specimens was increased at a given constant rate (mV/s) in the anodic direction by the potentiostat. A timer was used for each specimen to trigger the opening of the relay if the current from that specimen exceeded a given value, which indicated the initiation of pitting corrosion for that specimen. Because the potential was increased linearly from a known value, the potential at which each relay was triggered was known from the timer. In this way, the authors were able to obtain 12 pit initiation potentials in one experiment, which greatly increased the efficiency of the pitting potential measurements and enabled them to study the stochastic behavior of the pitting corrosion of stainless steel.

Because the multiple electrodes were connected to one common joint and all the electrodes were at the same electrode potential during the test (before the trigger was initiated by the timer), this system can also be categorized as a coupled multielectrode system.

Similar high-throughput measurements were also reported by Tan and Xu in 1987 in the evaluation of coating performance on steel [2]. Multiple specimens coated with different corrosion-prevention oils were assembled in an electrochemical cell and connected to a common joint through individual manual switches (Fig. 8.2). The common

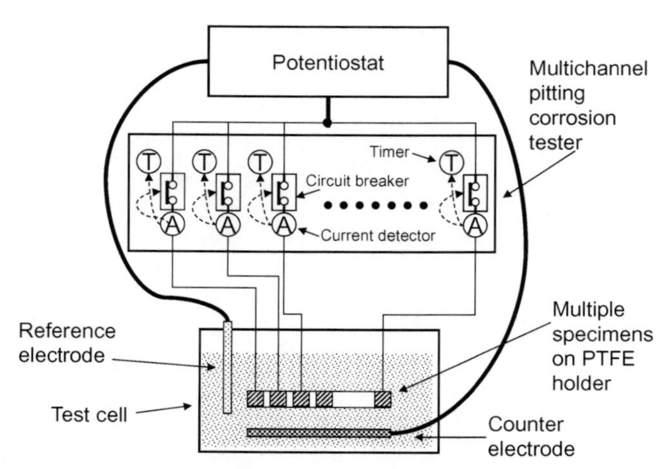

Fig. 8.1 Schematic diagram of a multielectrode system for high-throughput pitting potential measurements [1].

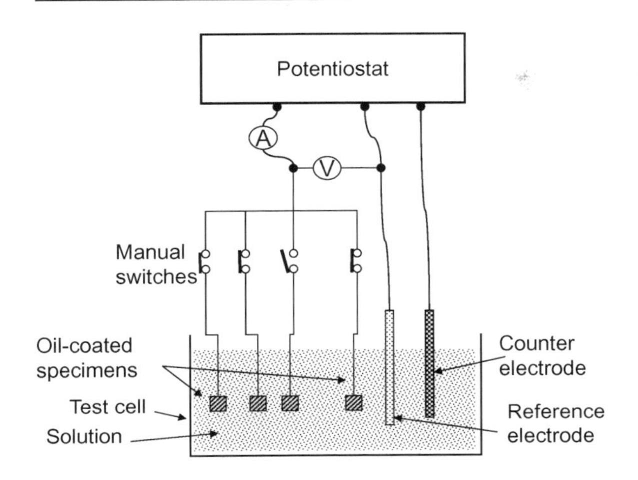

Fig. 8.2 Schematic diagram of a multielectrode system for high-throughput evaluation of coatings [2].

joint was connected to a potentiostat; an ammeter and a voltmeter were used to measure the total current flowing through all the specimens (electrodes) and the potential of the electrodes (at the common joint), respectively. During the measurements, a constant anodic current was applied to the multiple specimens (measured as the total current as compared with the individual current in the study by Shibata and Takeyama). Because the specimens with poor coatings allowed the anodic current to pass easily and the specimens with high-quality coatings did not allow the current to flow, the authors were able to identify the quality of the different coatings by interrogating the switches. For example, if the potential changed significantly only after opening the switch connected to the first specimen, the coating quality on the first specimen was poor. On the other hand, if the potential did not change before and after opening the first switch, the quality of the coating on the first specimen was at least not worse than one of the others that were connected to the common joint.

8.3 Uncoupled multielectrode arrays

The first published multielectrode array appears to be the multiple gold microelectrode system on a single-crystal silica substrate developed by the group at Massachusetts Institute of Technology (USA) [3–5] using microfabrication techniques. In one example [4], the array was comprised of eight individually addressable gold microelectrodes, each being about 0.12 μm thick, 3 μm wide, and 140 μm long, and separated from each other by a distance of 1.4 μm (Fig. 8.3). The authors coated electroactive polymers on the microelectrode arrays and electrochemically characterized the semiconducting properties of the arrays as molecule-based electronic devices such as diodes. A similar microelectrode array was also used by Wang to study the electrochemical behavior of the high-resistance lubricant by coating the closely packed electrodes with ionically conducting polyethylene oxide film [6].

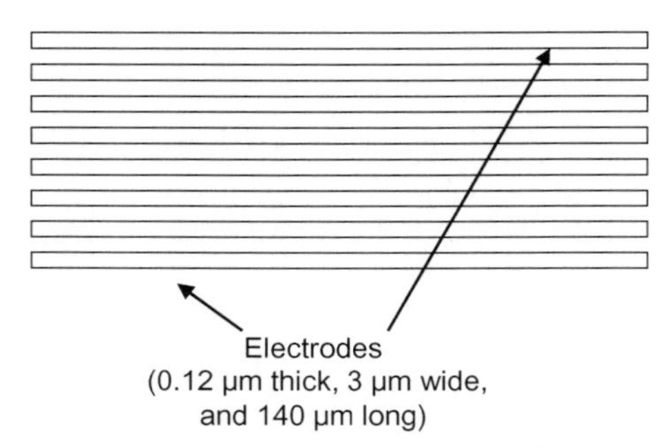

Fig. 8.3 Schematic diagram of a micro multielectrode array device [4].

The first published uncoupled multielectrode array for corrosion studies (also called wire beam electrode, WBE) appears to be the one described by Tan [7] and Tan and Yu [8] in 1991 for measuring the corrosion behavior of carbon steel under oil-based coatings. Later, similar multielectrode arrays were used to study nonuniform corrosion under protective coatings and salt deposits [9–16]. In a typical uncoupled multielectrode array, a large number of wires (0.5 to 2 mm in diameter) are flush-mounted in epoxy and arranged in a square configuration with the cross-section exposed to an electrolyte. The spatial behavior of corrosion or the electrochemical heterogeneity is characterized by the measurement of the open-circuit potential map from each electrode and by the measurement of the current map or electrical resistance map between pairs of selected wires.

8.4 Coupled multielectrode systems for corrosion detection

One of the devices described by Schiessl in a US patent [17] issued in 1991, and in a patent application [18] initially filed in Germany in 1988, appeared to be the first coupled multielectrode system for corrosion monitoring (Fig. 8.4). This coupled multielectrode system consisted of multiple steel anodes that were composed of materials similar to the reinforcement material and a corrosion-resistance cathode. Both the anodes and the cathode were embedded in a concrete structure. The steel anodes were separated from each other at different distances from the external surface of the concrete structure, and each electrode was independently connected (coupled) to the cathode through a resistor. The coupling current from each anode to the cathode as a result of corrosion was measured by a voltage-measuring system that was connected to both ends of the resistor to establish the temporal course of the penetration of substances (e.g., chloride) that were capable of damaging the reinforcement. Similar applications in concrete were reported later in other publications by Schiessl and coworkers and by other investigators [19–23].

In 1993, Steinsmo and coworkers reported on similar galvanically coupled multielectrode systems used to study the crevice corrosion of stainless steel materials in seawater [24–27]. Fig. 8.5 shows the galvanically coupled multielectrode system used

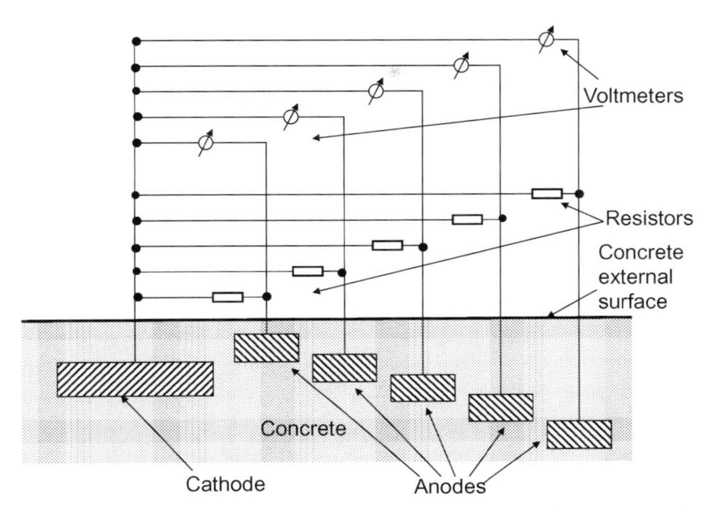

Fig. 8.4 Schematic diagram of a galvanically coupled multielectrode system used to detect the onset of corrosion of carbon steel reinforcing materials at different locations. Note: The anodes were made of carbon steels similar to the reinforcing material; the cathode was made of special steels or alloys [17].

by Steinsmo and coworkers. In this system, several creviced stainless steel specimens were independently coupled to a large noncreviced specimen (similar to the creviced specimens in composition) through resistors. The corrosion process taking place in the crevices made the electrode potential of the creviced specimen more negative (or anodic) than the large noncreviced specimen (cathode) and produced a coupling current from the large noncreviced specimen to the creviced specimen. The coupling current was measured across its associated resistor and used to indicate the degree of crevice corrosion. The reference electrode in Fig. 8.5 was used to measure the potential of the coupled multielectrode system.

Fig. 8.6 shows typical results obtained with the coupled multielectrode system by Steinsmo et al. [26] The measurement was conducted in a seawater test loop at 15°C, a condition under which the specimen is normally not subject to crevice corrosion. The purpose of this measurement was to study the effect of temporary upset conditions on crevice corrosion of a welded UNS S31254 material in seawater. Prior to the start of the measurement in Fig. 8.6, the crevice specimens were temporarily aged at high temperatures and high electrochemical potentials to initiate crevice corrosion. Fig. 8.6 shows that the corrosion attack continued to propagate at 15°C for 10 to 20 days before repassivation took place.

The use of resistors between the coupling joint and each electrode in Figs. 8.4 and 8.5 might cause potential variations among the different electrodes. As long as the currents flowing through the individual electrodes were low, or the resistors were small, such variations would not be significant compared with the potential changes required to cause significant current changes on a typical polarization curve (see Chapter 6 for additional information). Therefore, Figs. 8.4 and 8.5 may be considered

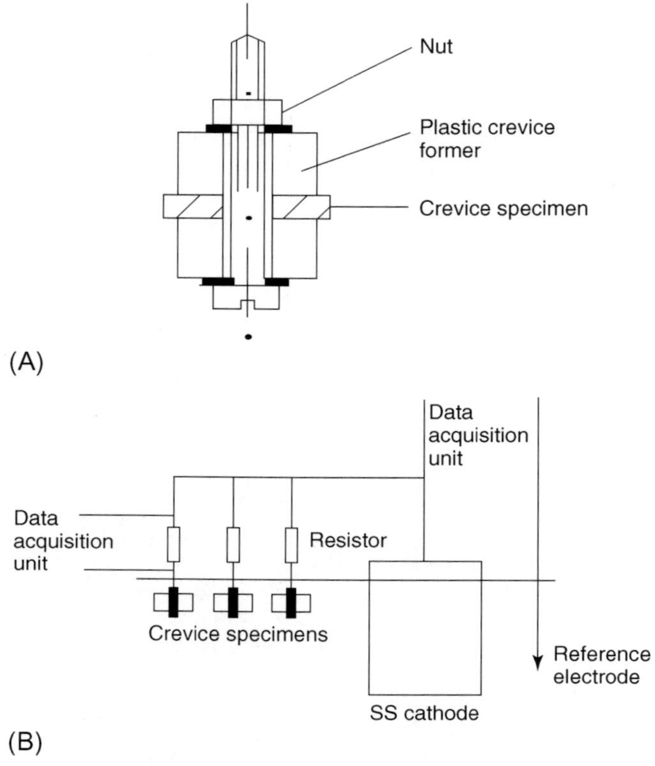

Fig. 8.5 Schematic diagram of a galvanically coupled crevice cell system. (A) Crevice assembly and (B) galvanic coupling of crevice assemblies and stainless steel cathode [26]. © NACE International 1997.

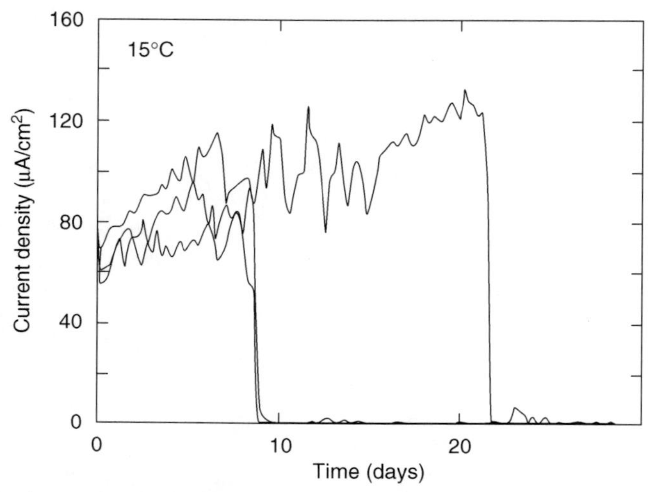

Fig. 8.6 Galvanic current densities between three creviced specimens of welded UNS S31254 and a cathode as a function of time in seawater at 15°C [26]. © NACE International 1997.

as the coupled multielectrode systems according to the definition at the beginning of this chapter (all electrodes are essentially at the same potential).

The multielectrode technique shown in Figs. 8.4 and 8.5 requires only straightforward voltage measurements and yet is capable of detecting the onset of general and localized corrosion under service conditions. Depending on its accuracy and resolution, the voltage-monitoring instrument may be able to detect corrosion in the early stages of degradation. The disadvantage of this technique is that it cannot be used to determine the true rate of general or localized corrosion because of the large surface area of the creviced specimen (5.5 cm^2 in Fig. 8.5).

8.5 Coupled multielectrode arrays for spatiotemporal corrosion and electrochemical studies

The first published coupled multielectrode array for corrosion and electrochemical studies appears to be the one described by Fei et al. in 1996 [28]. The coupled multielectrode array was used to simulate a one-piece metal electrode for the study of the spatiotemporal electrochemical behavior of iron metal in a sulfuric acid solution. The electrode arrays consisted of 16 (a 2×8 rectangle or a 4×4 square) or 61 (a hexagonal-shaped bundle) electrodes (cross-section of 0.5 mm diameter wires) embedded in epoxy (Fig. 8.7) [28, 29]. The individual wires were coupled to a common joint using a multichannel zero-resistance ammeter (ZRA) box and the current flowing

through each individual electrode was independently measured. The common coupling joint was connected to the working electrode jack of a potentiostat so that the electrode array could be polarized to study the electrochemical spatiotemporal pattern of the electrode array at different potentials. With the 61-electrode array, it was shown (Fig. 8.8) that: (a) the activation (corrosion) of metal started at the center electrodes and propagated from the center to the edge; and (b) the passivation started from the edge electrodes and propagated in the opposite direction (from the edge to the center). By comparing the behavior of the total current from the electrode array with the behavior of the current from a one-piece electrode, the authors concluded that the array of electrodes behaved similarly to a one-piece electrode of the same shape and total area. Thus, the electrochemical spatial patterns observed with the electrode array are representative of patterns on a one-piece metal electrode. By using the coupled multielectrode array, the authors were able directly to determine the spatial pattern of the active-passive electrochemical oscillations and how the oscillation wave front travels on a large iron electrode.

Similar coupled multielectrode arrays were also used extensively by others to study the corrosion mechanism and spatial interactions among the localized corrosion sites on different metals [30–42], and the localized corrosion behavior of aluminum alloys [43–49], copper alloy [50], nickel [51], the characteristics of chromate coatings [52] and atmospheric corrosion [50, 53].

A typical use of the coupled multielectrode arrays for studying the propagation of localized corrosion was demonstrated by work conducted at the University of

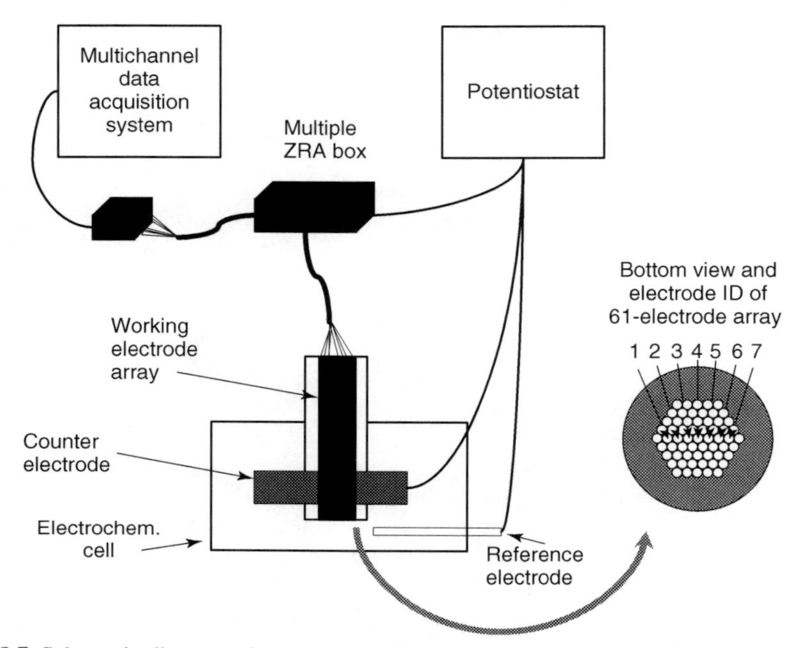

Fig. 8.7 Schematic diagram of a coupled multielectrode array system used to study the spatiotemporal pattern of the nonuniform corrosion of iron in sulfuric acid solution. Modified according to Z. Fei, R. G. Kelly, J. L. Hudson, Spatiotemporal patterns on electrode arrays. J. Phys. Chem. 100 (1996) 18986–18991, with permission from Z. Fei and J.L. Hudson.

Virginia. For example, Fig. 8.9 shows the spreading of the depassivated surface when only a group of the coupled electrodes were held at an elevated potential [35]. The array used to obtain these results was a 5×20 array. The potential of each electrode is presented as a 3D cylinder chart. The currents measured on the array are presented as a grid with each square representing a single electrode. The current value is monochromatically color-coded in shades of grey (from 0 to 2 mA/cm^2). The sign in each square indicates if the current is anodic or cathodic. The time scale is in seconds. The results show that corrosion spreads rapidly across the surface from active sites on carbon steel electrodes held at $0.2V_{SCE}$, while stainless steel demonstrated complete resistance to spreading from preferentially initiated sites held at $1V_{SCE}$ onto surrounding electrodes. Moreover, the preferentially active sites resisted initiation and did not fully activate until approximately 11,000 s. It can be seen from these results that the spreading of corrosion along the electrode surface is greatly influenced by the material composition.

Coupled multielectrode arrays are also used for corrosion studies of galvanic coupling between dissimilar metals [54–56] or the galvanic corrosion between weld zones

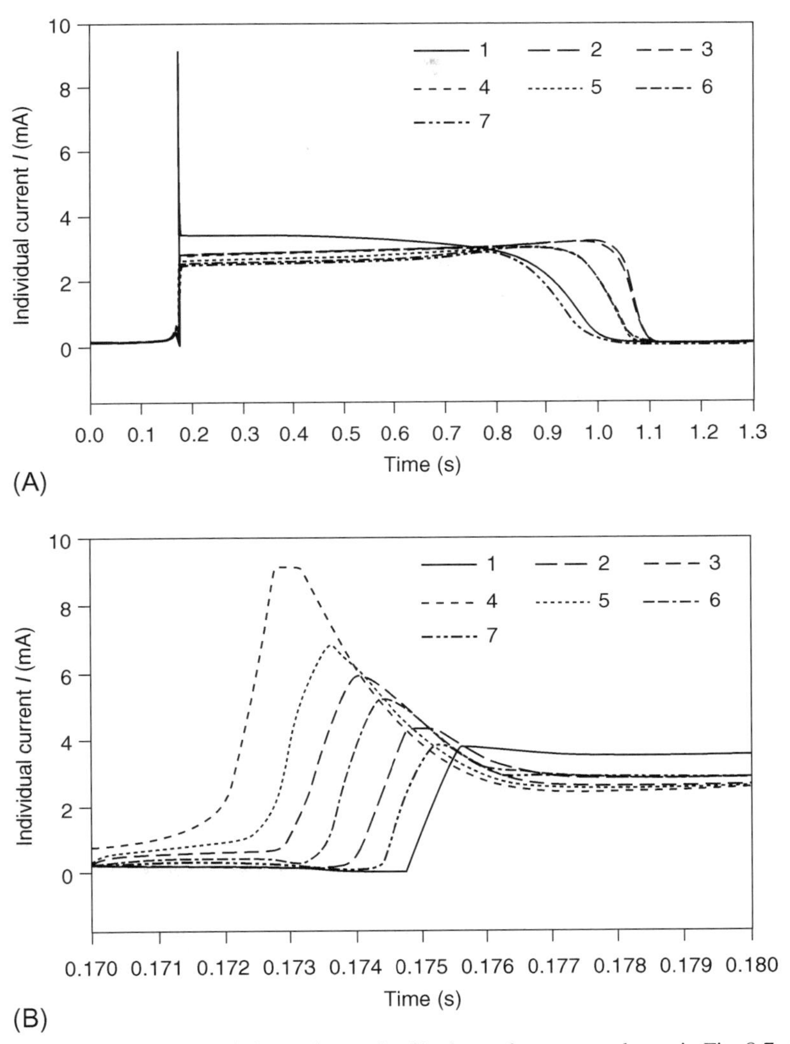

Fig. 8.8 Currents of individual electrodes on the 61-electrode array, as shown in Fig. 8.7, during (A) activation and (B) passivation. Note: The electrode numbers are shown in Fig. 8.7. *Source*: Z. Fei, Spatiotemporal Behavior of Iron and Sulfuric Acid Electrochemical Reaction System', PhD Thesis, University of Virginia, Charlottesville, Virginia, Diss. Abstr. Int., B 1997, 58, 3, 1402, May, 1997, with permission from the American Chemical Society.

[57]. An example is the magnesium alloy, AZ31B-H24, joined by tungsten inert gas welding that exhibits characteristic of anodic zones and cathodic activation in Mg alloys. In addition, coupled multielectrode systems were also used to study the deposition of metals [58, 59].

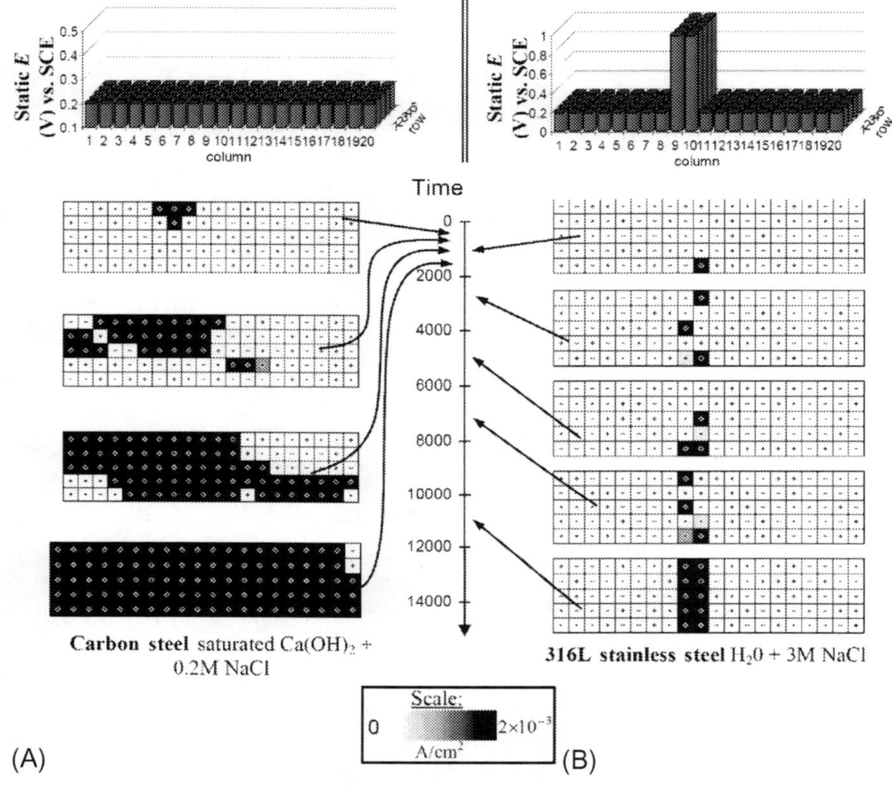

Fig. 8.9 Schematic diagrams of a CMA with closely-packed electrodes for studying the spreading of depassivated surface when all electrodes were held at an elevated potential (A) and only the middle columns were held at an elevated potential (B) [38].
Reproduced with permission from ASTM International, Cong, H., Bocher, F., Budiansky, N. D., Hurley, M. F., Scully, J. R., Use of coupled multi-electrode arrays to advance the understanding of selected corrosion phenomena. J. ASTM Int. 4(10) (2007), Fig 12.

8.6 Coupled multielectrode arrays for spatiotemporal corrosion measurements

Coupled multielectrode arrays have been extensively used to measure the spatiotemporal patterns of the nonuniform corrosion [60–76]. The multielectrode array is also called wire beam electrodes (WBE). Fig. 8.10 shows a typical multielectrode array (or WBE) system used by Tan et al. in their studies [63]. In Fig. 8.10, multiple electrodes were directly connected to a common joint. The current flowing through each electrode was measured by momentarily decoupling the electrode from the common joint and inserting a ZRA between the electrode and the common joint. The potential of the individual electrodes and the coupling joint was measured with a voltmeter (V) and a

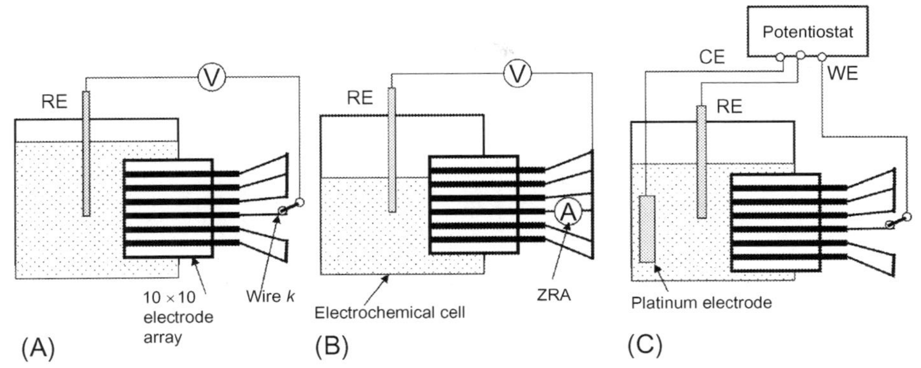

Fig. 8.10 Schematic diagram of the coupled multielectrode array system for measurement of: (A) open-circuit potentials of individual wires; (B) galvanic currents between individual wires and the system; and (C) Tafel slopes of individual wires [67]. © NACE International 1998.

reference electrode (RE). Because the potential of each electrode was measured under open-circuit conditions, this system was not operating under truly coupled conditions. The decoupling of electrodes for the measurement of the open-circuit potential may affect the ability of this system to simulate the behavior of a one-piece metal.

In a nonuniform corrosion environment, each mini-electrode was considered to corrode uniformly and behave as an ideal electrochemical system because of the small size (three to four orders of magnitude smaller than the electrode in a typical conventional corrosion probe such as a linear polarization resistance probe). Therefore, the following formulae were derived on the basis of the Butler-Volmer equation to calculate the corrosion current on each electrode [63]:

$$I_{ka} = I_{kcouple} / \left\{ 1 - \exp\left[-(2.3/b_{ka} + 2.3/b_{kc}) \left(E_{coup} - E_{kopen} \right) \right] \right\} \tag{8.1}$$

where I_{ka} is the anodic current (or corrosion current) from electrode k; $I_{kcouple}$ is the coupling current from electrode k measured with the ZRA; E_{kopen} is the open-circuit potential of electrode k; E_{coup} is the coupling potential of all the electrodes; and b_{ka} and b_{kc} are the anodic and the cathodic Tafel slopes, respectively. It was proposed to use a separate *RE* to measure the values for E_{kopen}, when electrode k is disconnected from the coupling joint, and E_{coup}. Linear polarization measurements were conducted to obtain the Tafel slopes.

It was hypothesized that if the value of $E_{coup} - E_{kopen}$ for an electrode is greater than 100 mV, the exponential term in Eq. (8.1) will vanish, and the corrosion current from electrode k could be estimated by the coupling current from the electrode:

$$I_{ka} = I_{kcouple} \tag{8.2}$$

The system used to experimentally measure the coupling currents, coupling potential, open-circuit potential, and Tafel slopes is shown in Fig. 8.10A–C.

This method has been applied to mapping the localized corrosion behavior (coupled currents) of carbon steel materials in aqueous solutions [62], crevice [69], water/gas interface [70], and soil [72]. Tan and coworkers also used the multielectrode arrays in conjunction with the electrochemical noise technique to measure the localized corrosion rate distributions [66, 73]. With this approach, the electrochemical noise resistances between paired electrodes in the multielectrode array were measured, and the localized corrosion rate for each electrode was calculated based on the electrochemical noise theory. Because the noise resistance was measured in pairs, the multielectrode array that operated under this mode was not a coupled multielectrode array.

8.7 Ammeters used for the measurements of coupling currents

As discussed in Chapter 6, the coupling currents can either be measured with a voltmeter and a shunt-resistor as shown in Fig. 8.5 [26] or a zero-resistance ammeter (ZRA) as shown in Fig. 8.7 [28] and Fig. 8.10 [67]. The coupling current can also be measured with a potentiostat by connecting the potentiostat's counter electrode and reference electrode terminals to one of the electrodes and the working electrode terminal to the coupling joint. An ideal current-measuring device for this purpose should not impose any voltage between the electrodes, so that all the electrodes are at the same electrode potential during the measurement. Such a device is called zero-voltage ammeters (ZVA) [77–79] (see Chapter 6 and Section 8.8 of this Chapter for information) which includes ZRAs, potentiostats, as well as the shunt-resistor with a voltmeter. Ideal ZVAs may not be in existence because all ZVAs, including the ZRA, may impose a small voltage in the measuring-circuit. As long as the current-measuring device imposes a near-zero voltage (e.g., 0.5 mV [77, 79]), it can be called a ZVA and can be used for the measurement of the current in a coupled multielectrode systems.

8.8 Coupled multielectrode array sensors with simple output parameters for corrosion monitoring

The coupled multielectrode array systems have been used as sensors (called coupled multielectrode array sensors or CMAS) for real-time monitoring the rate of non-uniform corrosion, particularly localized corrosion [80–142]. As described in Sections 8.4 through 8.6, coupled multielectrode systems can provide not only temporal, but also spatial information on corrosion, especially localized corrosion in two dimensions. It is an excellent tool for the studies of corrosion phenomena in laboratories. However, the data from a multielectrode system are often huge and incomprehensible by ordinary plant or facility operators. The data must be greatly reduced to one or to a few simple parameters so that the multielectrode systems can be used as a sensor to provide real-time data for the operation of a plant or facility in the field. This section describes the working principle of the coupled multielectrode array sensors that are suitable for corrosion monitoring and the methods used to derive the simple parameters.

8.8.1 Principle of coupled multielectrode array sensors for corrosion monitoring

Fig. 8.11 shows the schematic diagram of a typical CMAS [87, 116, 121]. When a metal is exposed to a corrosive electrolyte, electrons are produced by corrosion reaction at the anodic sites and flow to the cathodic sites (see the upper section). If the metal is divided into small pieces, some of them have more of the anodic sites and some of them have more of the cathodic sites (see below). When all the small pieces are coupled to a common joint through a multichannel zero-voltage ammeter (ZVA) that imposes near-zero voltage in the circuit (see Chapter 6 for more information about ZVA), electrons produced on the more anodic pieces (also called anodic electrodes) flow through the ZVA as represented by the anodic external current, I^{a}_{ex}, to the more cathodic pieces (also called more cathodic electrodes) as represented by the cathodic external current, I^{c}_{ex}. On each electrode, some of the electrons may also flow to the local cathodic sites on the same electrode as represented by the internal currents (I^{a}_{in} on the net anode or I^{c}_{in} on the net cathode in Fig. 8.11), which are usually small if the electrode is significantly more anodic or cathodic than the others (see Section 8.9 for more details). Because the I^{a}_{ex} is driven by the difference between the open-circuit potential of the more anodic electrode and the open-circuit potential of the remaining electrodes connected to the coupling joint and such difference is the characteristic of nonuniform corrosion, especially localized corrosion, I^{a}_{ex} represents the degree of nonuniform corrosion or nonuniform corrosion current.

This concept of dividing a corrodible metal into multiple small pieces and coupling them together through an external circuit so that the nonuniform corrosion current on the most anodic electrode (the electrode that gives the highest anodic current) which simulates the most anodic area on the corrodible metal can be measured was independently proposed by the author and coworkers in 2000. This concept is the base for the CMAS that measures localized corrosion and does not require a reference electrode which may have a limited service life and does not need the externally applied polarization which usually affects the measurements, so that the CMAS is suited for monitoring the corrosion of the nuclear waste containers to be disposed off in nonaccessible and hot drifts during the initial 100 years of emplacement.

As discussed in Chapter 6, the ZVA can be a zero-resistance ammeter or a simple voltmeter with a shunt-resistor as long as it does not impose a noticeable voltage drop so that all the multiple electrodes in the CMAS probe are at the same electrode potential and simulate the behavior of a one-piece corrodible metal. When the corrosion currents vary in a large range, variable shunt resistors whose value can be automatically changed by a controller during the measurements may be needed to minimize the potential differences across the shunt resistors [126].

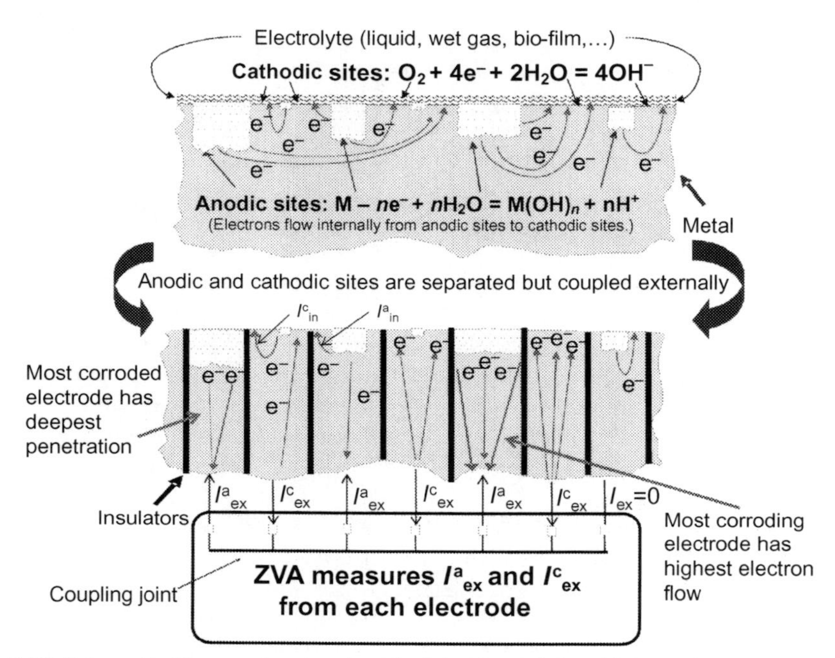

Fig. 8.11 Schematic diagram showing the principle of coupled multielectrode array sensors for localized corrosion monitoring [116]. Note: I^a_{ex} and I^a_{in} represent the currents on the net anodic electrodes; I^c_{ex} and I^c_{in} represent the currents on the net cathodic electrodes; the most corroded electrode always has the highest cumulative electron flow, but may not be the most corroding electrode at a given time.

The CMAS probes can be made in many configurations and sizes, depending on the applications. They may be used for high-temperature and high-pressure applications if the electrodes are properly sealed and insulated with high temperature insulators. Fig. 8.12 shows a typical commercial CMAS monitor (nanoCorr®) and some typical commercial probes for real-time corrosion monitoring in plants or in fields. (nanoCorr is the trade name of Corr Instruments, LLC, Carson City, Nevada, USA.).

Fig. 8.13 shows the theoretical basis for the CMAS probe, assuming that one electrode on the probe is anodic and all the other electrodes are cathodic. Because localized corrosion often involves small areas of corroded anodic sites accompanied by large areas of cathodic sites, such assumption is often reasonable under many environments. The thin solid curves represent the dissolution and reduction polarization behaviors on the anodic electrode. The thick solid curves represent the combined dissolution and reduction polarization behaviors on the rest of the electrodes (the cathodic electrodes) if these cathodic electrodes are coupled as a single electrode. The thick dashed lines represent the reduction curve for all electrodes, or the dissolution behavior for all electrodes on the CMAS probe. For a passive metal, in the cathodic area (or the cathodic electrodes in a CMAS probe) where no localized corrosion has been initiated, the anodic current is usually extremely low due to the

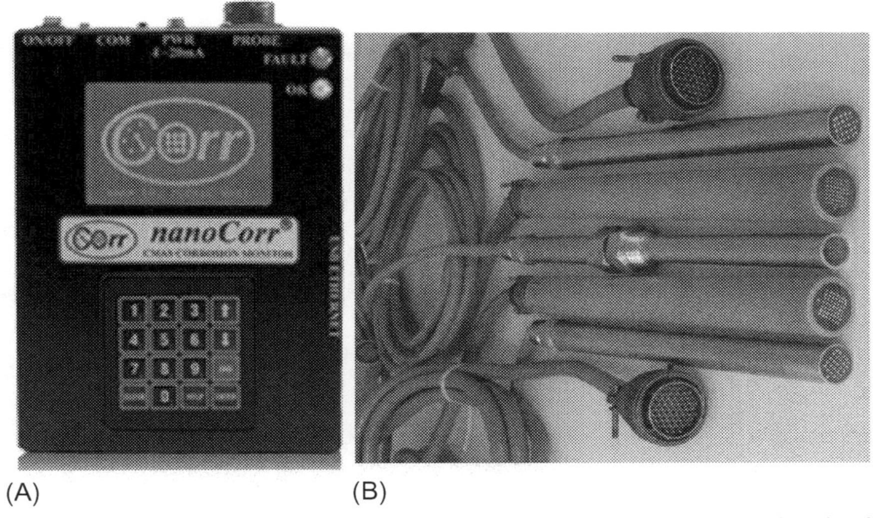

Fig. 8.12 A typical commercial CMAS monitor (A) and typical commercial CMAS probes for real-time corrosion monitoring in plants or fields (B).
Courtesy of Corr Instruments, LLC.

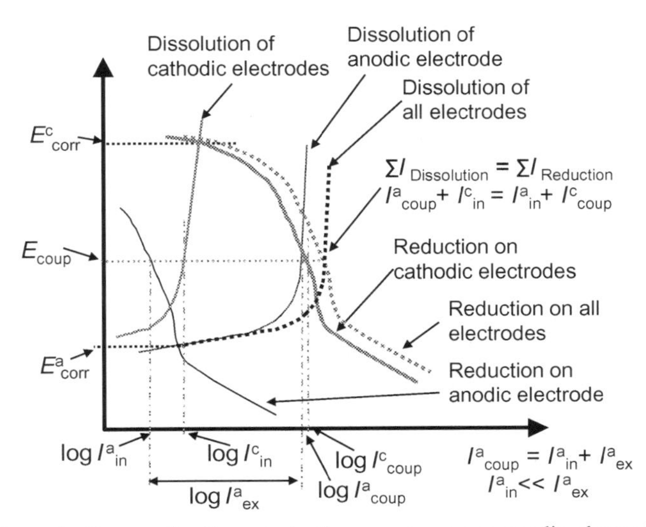

Fig. 8.13 Schematic diagram for the polarization curves on one anodic electrode and several cathodic electrodes that are connected together on a coupled multielectrode sensor. Note: because the internally flowing anodic current is solely supported by the internally flowing cathodic current on any given electrode, the internal reduction current on the anodic electrode, I^{a}_{in}, is also the internally flowing anodic current on the anodic electrode.

protective layer of the oxide formed on the metal (see Chapter 2) and the corrosion potential for the cathodic electrodes, E^{c}_{corr}, is high (or noble). For the anodic electrode where localized corrosion has been initiated and the protective layer has been compromised, however, the anodic current is usually high and the corrosion potential for the anodic electrode, E^{a}_{corr}, is low (or active). Note in Fig. 8.13, the cathodic current on the combined cathodic electrodes is significantly higher than that on the anodic electrode. This is because we have assumed that the surface area on the anodic electrode is significantly smaller than that of the cathodic electrodes (one anodic electrode versus many cathodic electrodes). In addition, the cathodic reactions deep in an anodic pit on the anodic electrode require more effort for the reactants (O_2 or H^+) to overcome the mass transfer barrier.

When the anodic electrode and the combined cathodic electrodes are coupled, the corrosion potential changes to a new value, E_{coup} (or E_{corr} for all coupled electrodes), and the total anodic dissolution currents equal the total cathodic reduction currents (see the thick dashed lines in Fig. 8.13):

$$I^{a}_{coup} + I^{c}_{in} = I^{a}_{in} + I^{c}_{coup} \tag{8.3}$$

Where $I^{a}_{coup}$is the corrosion current or dissolution current on the anodic electrode, I^{c}_{in} is the internal dissolution (anodic) current on all the cathodic electrodes (anodic current that flows within all the cathodes), I^{a}_{in} is the internal reduction current on the anode (the cathodic current that flows within the anode), and I^{c}_{coup} the cathodic current on the combined cathodes. Because the internally flowing anodic current is solely supported by the internally flowing cathodic current on any given electrode, the internal reduction current on the anodic electrode, I^{a}_{in}, is also the internal anodic current on the anodic electrode. Visa versa, the internal dissolution current on the combined cathodic electrodes, I^{c}_{in}, is also the internal cathodic current on combined cathodic electrodes.

On the anodic electrode, the corrosion current (total dissolution current), I^{a}_{coup}, is equal to the sum of the externally flowing anodic currents, I^{a}_{ex}, and the internally flowing anodic currents which is equal to I^{a}_{in}. Therefore,

$$I^{a}_{coup} = I^{a}_{ex} + I^{a}_{in} \tag{8.4}$$

Because the I^{a}_{in} on the anodic electrode, especially when the anodic electrode is the most anodic electrode of the CMAS probe (see Section 8.9), is often much smaller than its I^{a}_{ex} at the coupling potential in a nonuniform corrosion environment, the externally flowing current from such anodic electrode of the probe can often be directly used to estimate the localized corrosion current:

$$I^{a}_{coup} \approx I^{a}_{ex} \tag{8.5}$$

In a less corrosive environment, however, there would be less separation between the anodic electrodes and the cathodic electrodes. The behavior of even the most anodic electrode may be similar to the other electrodes in the CMAS probe. In this case, the I^{a}_{in}

for the anodic electrode would be close to I^{a}_{coup}, and I^{c}_{in} would be close to the I^{c}_{coup}. The most anodic electrode may still have significant cathodic sites available, and the electrons from the anodic sites would flow internally to the cathodic sites within the same electrode. Therefore, I^{a}_{in} in Eq. (8.4) cannot be ignored in the calculation of the corrosion current.

Eq. (8.4) may also be represented as [82, 87, 105]:

$$I^{a}_{ex} = \varepsilon I^{a}_{coup} \tag{8.6}$$

where ε is a current distribution factor that represents the fraction of electrons resulting from corrosion that flows through the external circuit. The value of ε may vary between 0 and 1, depending on parameters such as surface heterogeneities on the metal, the environment, the electrode size, and the number of sensing electrodes. If an electrode is severely corroded and significantly more anodic than the other electrodes in the probe, the ε value for this corroding electrode would be close to 1 ($I^{a}_{in} = 0$), and the measured external current would be equal to the localized corrosion current [82, 87]. More discussion of ε is given in Section 8.9.

8.8.2 Maximum localized corrosion rate

In a corrosion management program for engineering structures, field facilities, or plant equipment, the most important parameter is the remaining life (often the remaining wall thickness) of the systems. If localized corrosion is a concern, the remaining wall thickness in the most corroded area or site is often used to evaluate the remaining life. Therefore, the corrosion depth (the corrosion-induced wall thinning) or the maximum localized corrosion depth (the corrosion-induced wall thinning at the most corroded area) for localized corrosion is often the most important parameter in an operator's mind. Because the corrosion depth is a parameter that takes a long time (often many years) to accumulate, the corresponding parameter that is important to the day-to-day operation would be the corrosion rate or the maximum localized corrosion rate if localized corrosion is a concern. Because the anodic electrodes in a CMAS probe simulate the anodic sites on a metal surface, the maximum anodic current (the current from the most anodic electrode) may be considered as the corrosion current from the most corroding site on the metal. Therefore, the maximum anodic current should be used as one of the most import parameters for the CMAS probes [82, 87].

The value, based on three times the standard deviation of currents from a CMAS probe, was sometimes used to represent the maximum anodic current [82, 87]. Because the number of electrodes in a CMAS probe is always limited and usually far fewer than the number of corroding sites on the surface of a metal coupon, the use of the value based on the statistical parameter, such as three times the standard deviation of current, to represent the maximum anodic current may sometimes be a better choice than the true maximum anodic current [82, 87]. The standard deviation value may be derived from the anodic currents or from both the anodic and the cathodic currents [82, 107].

Accordingly, the maximum localized corrosion rate (or maximum localized corrosion penetration rate) may be derived from the maximum anodic current [82, 87]. The following equation has been used to calculate the maximum localized corrosion rate:

$$CR_{max} = (1/\varepsilon)I_{max}W_e/(F\rho A) \tag{8.7}$$

where CR_{max} is the calculated maximum penetration rate (cm/s), I_{max} is the maximum anodic current, or the most anodic current, F is the Faraday constant (96 485 C/mol), A is the surface area of the electrode (cm^2), ρ is the density of the alloy or electrode (g/cm^3), W_e is the equivalent weight (g/mol) (see Chapter 3 for details). Eq. (8.7) assumes that corrosion on the most corroded electrode is uniform over the entire surface. Because the electrode surface area is usually between 0.01 and 0.0003 cm^2, which is approximately two to four orders of magnitude less than that of a typical linear polarization resistance (LPR) probe or a typical electrochemical noise (EN) probe, the prediction of penetration rate or localized corrosion rate by assuming uniform corrosion on the small electrode is realistic in most applications [82, 87].

The corrosion depth or penetration is related to the total damage accumulated in a given time period. The corrosion depth of the ith electrode may be derived from the cumulative charge that can be obtained by integrating the corrosion current through the electrode from time zero to time t:

$$Q_i = \int I_i(t)\mathrm{d}t \tag{8.8}$$

where Q_i is the cumulative charge of the ith electrode. Similar to the maximum localized corrosion rate, the following equation has been used to calculate the maximum cumulative localized corrosion depth or penetration (cm):

$$CD_{max} = (l/\varepsilon)Q_{max}W_e/(F\rho A) \tag{8.9}$$

where Q_{max} is the maximum of the cumulative charges (coulomb) from all the electrodes. The cumulative charge of each electrode is calculated individually using Eq. (8.8).

Fig. 8.14A shows typical responses of the standard deviation of the currents measured from a 25-electrode probe made of Type 304 stainless steel (UNS S30400) [82, 87]. The following order of increasing corrosiveness was observed: deionized water $<$ saturated KCl $<$ 0.0025 M $FeCl_3$ $<$ 0.25 M $FeCl_3$. Fig. 8.14A also shows that $NaNO_3$ is an effective corrosion inhibitor for stainless steel in $FeCl_3$ solution.

Fig. 8.14B shows typical responses of the maximum localized corrosion rate (derived from the most anodic current) of a low carbon steel CMAS probe measured by a commercial instrument [120]. The maximum localized corrosion rate in air was close to the instrument theoretical detection limit (10 nm/year). The initial maximum localized corrosion rate was 10 μm/year in distilled water. The maximum localized corrosion rate in simulated seawater was approximately 1 mm/year. When 10 mM H_2O_2 was added to the simulated seawater, the corrosion rate was 10 mm/year. It

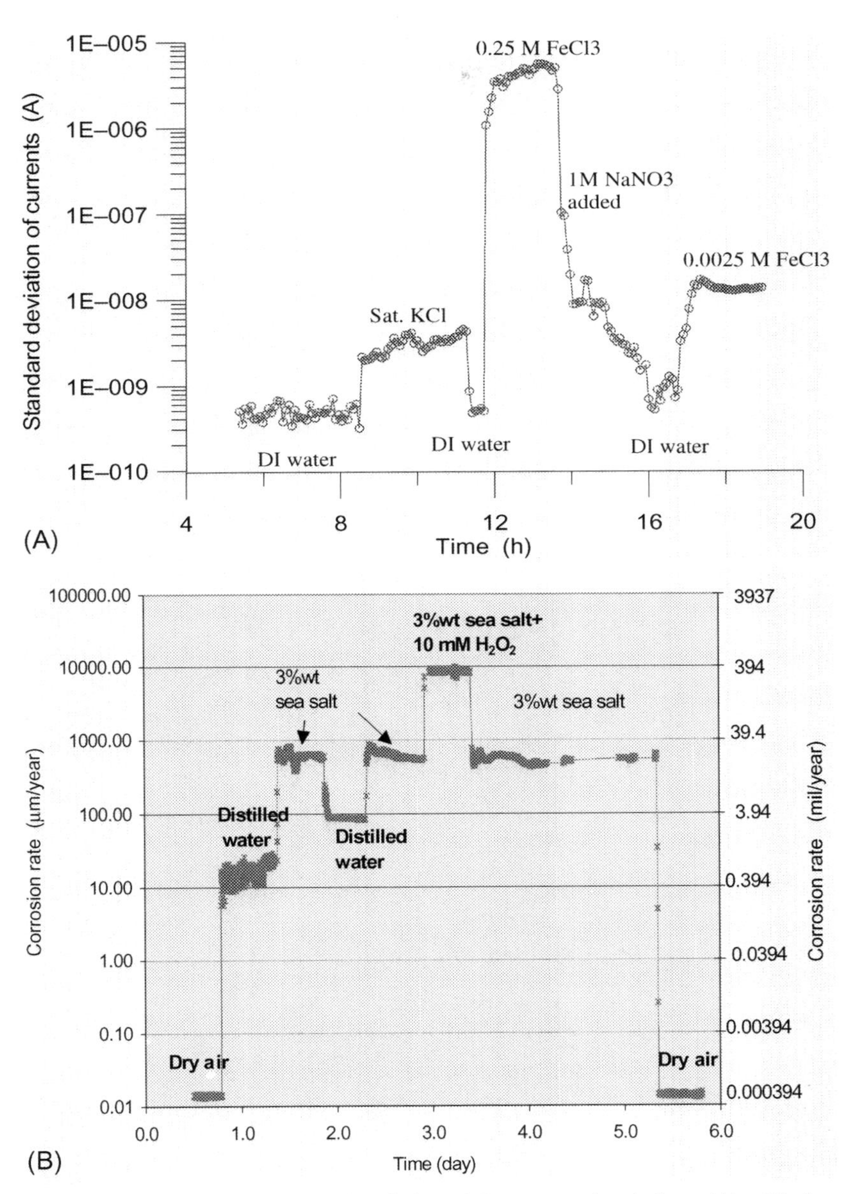

Fig. 8.14 Typical response of standard deviation of the current signals from (A) a 25-electrode type 304 stainless steel probe and (B) a 16-electrode carbon steel probe to the changes in solution chemistry [87, 120]. © NACE International 2003/2004.

was later verified that the corrosion of the electrode in the distilled water and in the simulated seawater was mainly in the form of pitting corrosion [116]. Therefore, the maximum corrosion rate in Fig. 8.14B represents the maximum pitting rate.

8.8.3 Estimation of general corrosion rate using coupled multielectrode array sensors and localized corrosion rate factor

The maximum localized penetration rate and cumulative maximum localized penetration depth are, no doubt, the two most important parameters for the assessment of localized corrosion. However, the maximum penetration rate data for alloys are rarely available, and the measured localized corrosion rates by the CMAS probes cannot be easily compared with the general corrosion rates commonly reported in literature. In addition, maximum penetration rate is difficult to measure, whereas general corrosion rate can be measured by many methods, such as the LPR probes (see Chapter 3) and ER probes (see Chapter 11), or weight loss methods.

In most cases, localized corrosion is associated with some degree of general corrosion. When a metal is undergoing corrosion, the corroding metal is usually at an electrochemical potential (corrosion potential) that is significantly higher than the metal's deposition potential. Thus, the cathodic currents do not directly contribute to the metal loss or the metal gain (electroplating), and therefore, can be ignored in the corrosion rate calculation. For this reason, the average corrosion penetration rate may be calculated using the average value of the anodic currents from the CMAS probe [109].

$$I_{avg} = \left(\sum I_i^a\right)/n, i \text{ from } 1 \text{ to } n \tag{8.10}$$

where I^a_i is the anodic current from the ith electrode, n is the number of electrodes in the probe, and I_{avg} is the average of the anodic currents. If I^a_i is cathodic, it is set to zero in the summation. Fig. 8.15 shows the current from each electrode, the maximum anodic current, and the average of the anodic currents from a typical 16-electrode stainless steel probe in simulated seawater [131].

Thus, the average corrosion rate, CR_{avg}, may be calculated by [109]

$$CR_{avg} = (1/\varepsilon) I_{avg} W_e/(F\rho A) \tag{8.11}$$

The general corrosion penetration rate obtained using the weight loss method or by an electrochemical method using relatively large electrodes is essentially the corrosion rate averaged over the sample surface area. Therefore, Eq. (8.11) may be used to estimate the general corrosion rate. It should be noted that the general corrosion rate estimated with Eq. (8.11) should be used with caution because some of the corrosion currents on the corroding electrodes flow internally (see Sections 8.9), especially on those that are not corroding more significantly than the others (in contrast to the

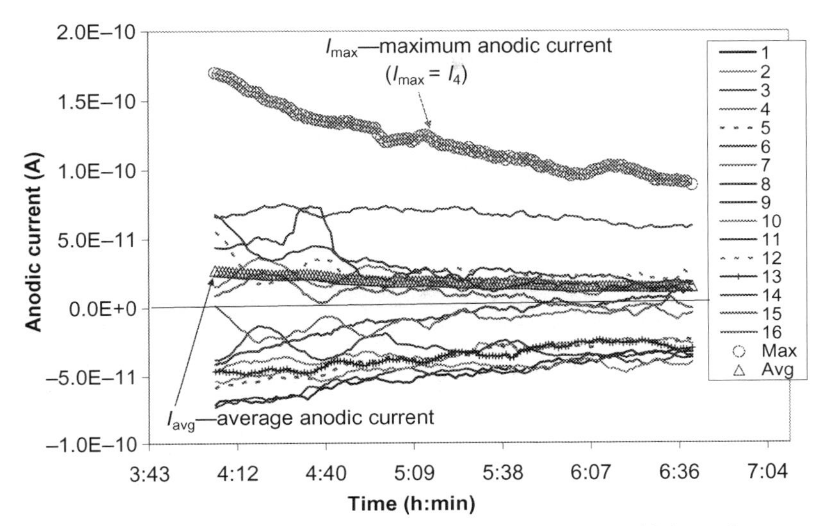

Fig. 8.15 Typical currents measured from a stainless steel coupled multielectrode array sensor probe in simulated seawater. Maximum anodic current was due to one electrode (Electrode #4) throughout the measurement period. Note: the numbers in the key indicate the ID of the sensor electrodes; a negative value of the anodic current means that the current is actually cathodic [131]. © NACE International 2007.

most corroding electrode). The general corrosion rate estimated using Eq. (8.11) may be lower than the true general corrosion rate.

The localized corrosion rate factor, f_{rate}, was defined as the ratio of the maximum localized corrosion rate to the average corrosion penetration rate. It can be expressed by [109]

$$f_{rate} = CR_{max} / CR_{avg} \tag{8.12}$$

The localized corrosion rate factor indicates how much higher the localized corrosion rate (e.g., the penetration rate of the fastest growing pit on the surface of a coupon in the case of pitting corrosion) is than the average corrosion rate (e.g., the average penetration rate on the surface of a coupon). It should be noted that the maximum localized corrosion rate may not always be found on one electrode. In a CMAS probe, one electrode may have the highest corrosion rate at one time, but another electrode may corrode at the highest rate at another time (see Section 8.8.5). Fig. 8.16 shows the maximum localized corrosion rate, the general corrosion rate calculated using Eq. (8.11), and localized corrosion rate factor for a typical freshly polished Type 1008 carbon steel (UNS G10080) probe in simulated seawater at room temperature. The data were obtained by the author using a probe with 16 electrodes (1 mm in diameter) and the equipment described by Sun and Yang [109].

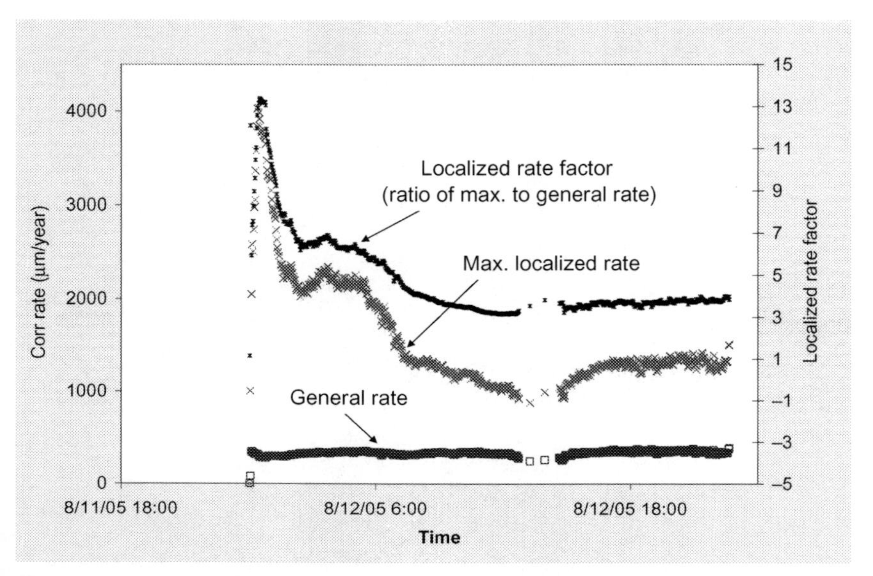

Fig. 8.16 Maximum localized corrosion rate, average corrosion rate, and localized corrosion rate factor for a typical freshly polished Type 1008 carbon steel probe in simulated seawater at room temperature. Note: Number of electrode: 16; electrode diameter: 1 mm.

8.8.4 Estimation of general corrosion depth using coupled multielectrode array sensors and localized corrosion depth factor

Similar to the average corrosion rate, the average anodic charge, Q^{a}_{avg}, may be calculated by [109]

$$Q^{a}_{avg} = \left(\sum Q^{a}_{i}\right)/n, i \text{ from } 1 \text{ to } n \tag{8.13}$$

where Q^{a}_{i} is the anodic charge from the *i*th electrode, and *n* is the number of electrodes in the coupled multielectrode probe. If the value of Q^{a}_{i} is cathodic, it is set to zero in the summation. Fig. 8.17 shows the charge from each electrode, the maximum anodic charge, and the average of the anodic charge for the data shown in Fig. 8.15 [131].

Thus, the average corrosion penetration depth, CD_{avg}, may be calculated by [109]

$$CD_{avg} = (1/\varepsilon)\left(Q^{a}_{avg}\right)W_{e}/(F\rho A) \tag{8.14}$$

Similar to the average corrosion rate, Eq. (8.14) may be used to estimate the general corrosion penetration depth. It should be noted, however, that the general corrosion depth estimated with Eq. (8.14) may be lower than the true general corrosion depth because some of the corrosion currents on the corroding electrodes flow internally

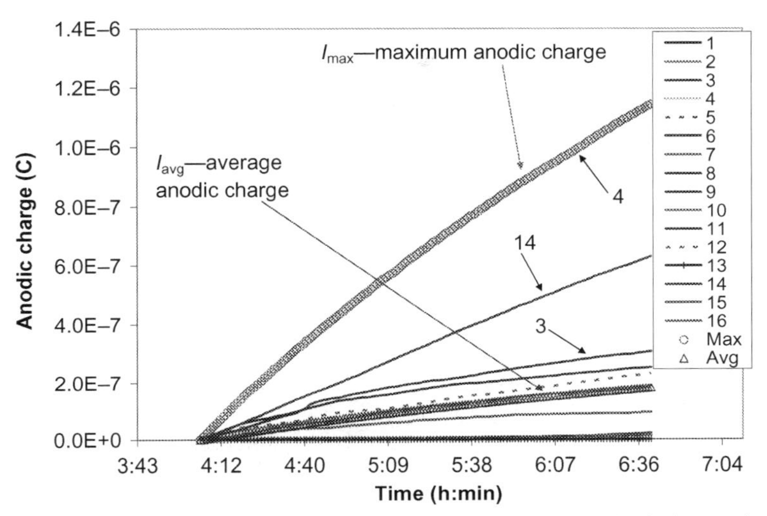

Fig. 8.17 Anodic charges corresponding to Fig. 8.15. Maximum charge is the integration of the current from one electrode (Electrode #4) [131]. © NACE International 2007. Note: the numbers in the legend indicate the ID of the sensor electrodes.

(see Sections 8.9), especially those electrodes that are not corroding more significantly than the others (in contrast to the most corroding electrodes).

The localized penetration depth factor, f_{depth}, may be defined as the ratio of the maximum localized penetration depth to the average corrosion penetration depth [109]

$$f_{depth} = CD_{max}/CD_{avg} \tag{8.15}$$

The localized corrosion penetration depth factor indicates the severity of localized corrosion depth (e.g., the deepest penetration of the most corroded pit on the surface of a coupon in the case of pitting corrosion) compared to the general corrosion penetration or the average corrosion penetration (e.g., the average loss in thickness on the surface of a coupon). Similar to the localized corrosion rate, the maximum penetration may not always be found on the same electrode. It also might not be found on the electrode that has the highest penetration rate at a given time. Instead, the electrode that has the highest cumulative penetration has the maximum penetration depth (see Section 8.8.5 for details). Fig. 8.18 shows the maximum localized corrosion depth, average corrosion depth, and localized corrosion depth factor for the data shown in Fig. 8.17. Even though localized corrosion depth factor was high initially (close to 11), it decreased quickly to 6 in about 3 h and reached 4 in about another 7 h. If the trend continued, the ultimate results would be general corrosion. This result is consistent with the general observations for carbon steel in seawater—general corrosion is the main mode of attack in most cases.

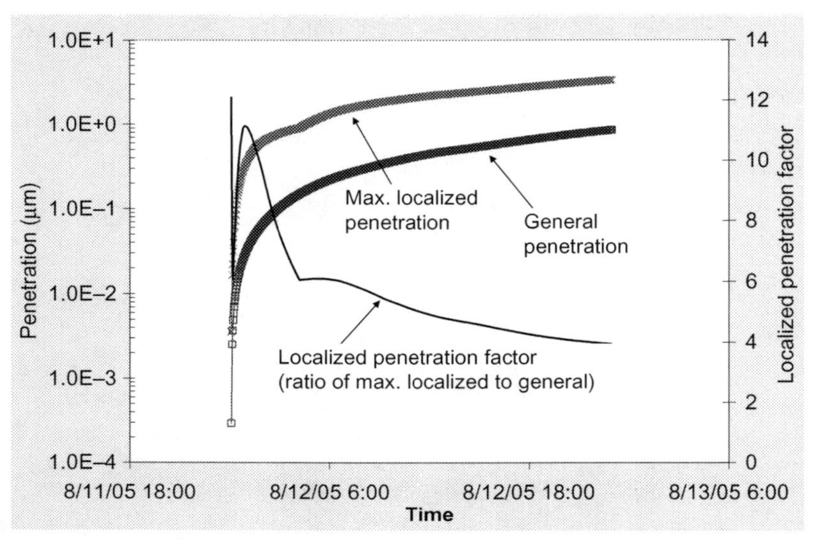

Fig. 8.18 Maximum localized corrosion depth, average corrosion, and localized corrosion depth factor for the data derived from the same experiment for Fig. 8.17.

8.8.5 Cumulative maximum localized corrosion rate

As shown in Fig. 8.15, the maximum anodic current was due to the current from a single electrode (Electrode #4). By applying Eqs. (8.7) through (8.9) to the data shown in Figs. 8.15 and 8.17, and assuming $\varepsilon = 1$, the maximum localized corrosion rate and maximum corrosion depth were obtained and shown in Fig. 8.19. In Fig. 8.19, the maximum localized corrosion depth curve is the integration of the maximum localized corrosion rate curve. Because the maximum localized corrosion rate curve in Fig. 8.19 was from one electrode, the maximum localized corrosion rate curve and the maximum localized corrosion depth curve are directly related to each other.

Fig. 8.20 shows typical currents measured from an aluminum coupled multi-electrode array sensor probe in simulated seawater [131]. The aluminum probe had 16 electrodes made of annealed Type 3003 (UNS A93003) aluminum wire. In Fig. 8.20, the maximum anodic current was from different electrodes (Electrodes #5, #13, #12 and #16) at different times. The anodic charges corresponding to Fig. 8.20 are shown in Fig. 8.21. By applying Eqs. (8.7) through (8.9) to the data shown in Figs. 8.20 and 8.21, the maximum localized corrosion rate and the maximum localized corrosion depth were obtained and are shown in Fig. 8.22. Unlike Fig. 8.19, the maximum localized corrosion depth curve is not the direct integration of the maximum localized corrosion rate curve in Fig. 8.22. The integration of the maximum localized corrosion curves would produce a much higher depth than the true maximum localized corrosion penetration depth as shown in Fig. 8.22 (dashed line). Therefore, there is no direct relationship between the maximum localized corrosion rate and the maximum localized corrosion penetration depth.

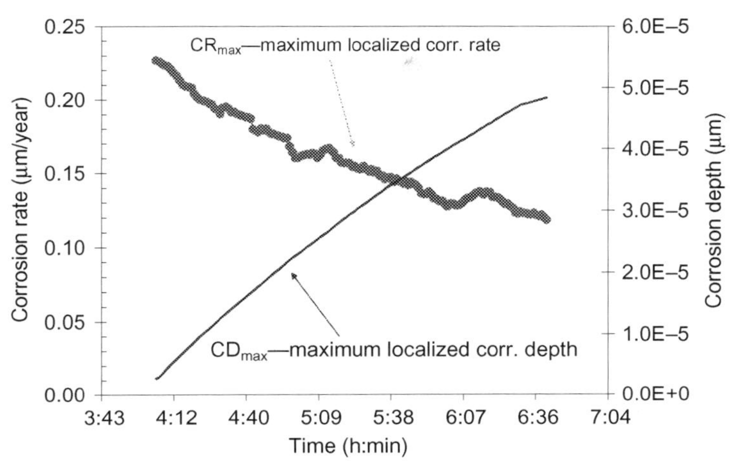

Fig. 8.19 Maximum localized corrosion rate and maximum localized corrosion depth (penetration) for the data shown in Figs. 8.15 and 8.17. The maximum localized corrosion depth is a direct integration of the maximum localized corrosion rate [131]. © NACE International 2007.

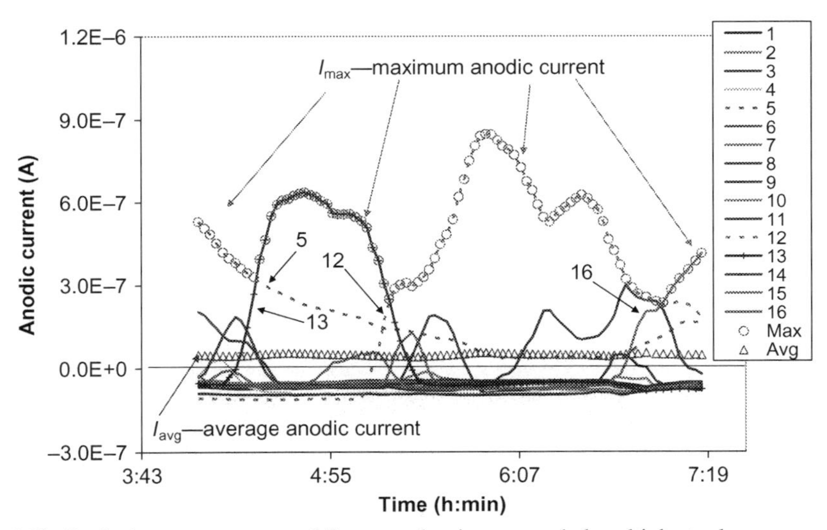

Fig. 8.20 Typical currents measured from an aluminum coupled multielectrode array sensor probe in simulated seawater. Maximum anodic current was due to different electrodes (Electrodes #5, #13, #12, and #16) at different times [131]. Note: the numbers in the legend and figure indicate the ID of the sensor electrodes; a negative value of the anodic current means that the current is actually cathodic. © NACE International 2007.

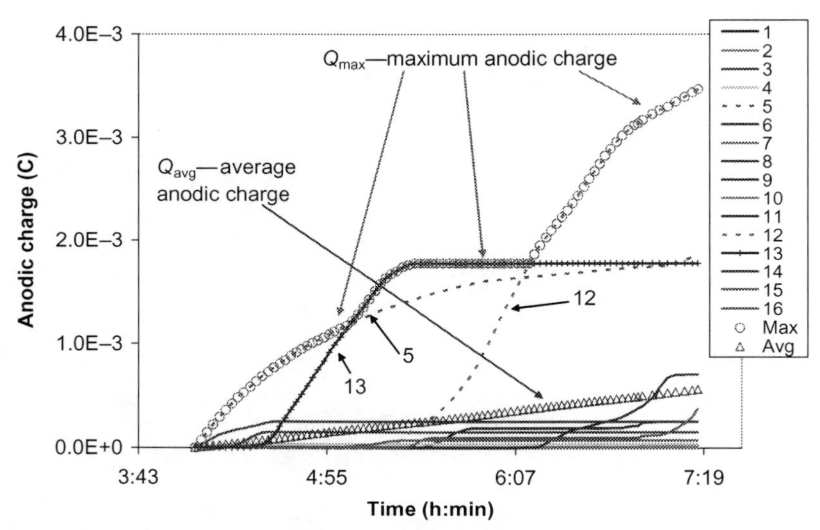

Fig. 8.21 Anodic charges corresponding to Fig. 8.20. Maximum charge is the integration of the currents from different electrodes (Electrode #5, #13 and #12) [131]. © NACE International 2007.

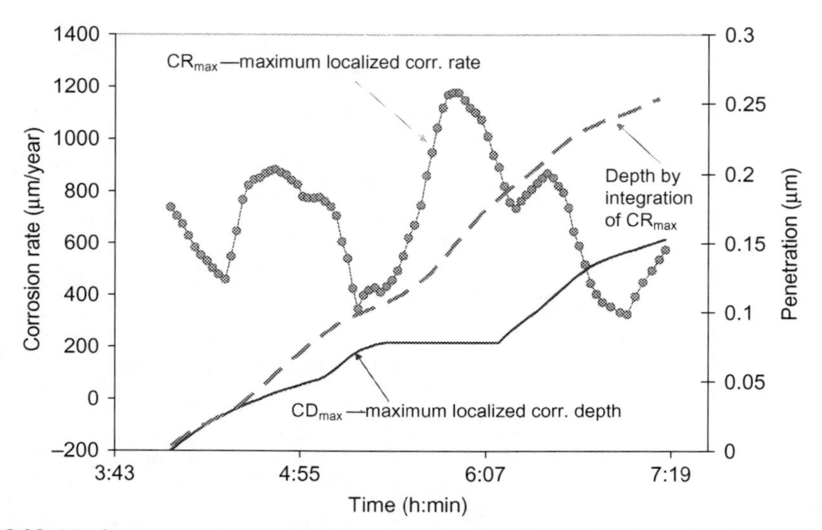

Fig. 8.22 Maximum corrosion rate and maximum localized corrosion depth for the data shown in Figs. 8.20 and 8.21. The dashed line was obtained by the integration of the maximum localized corrosion rate, CR_{max}, and is much higher than the actual maximum localized corrosion depth, CD_{max} [131]. © NACE International 2007.

To solve the problem of discrepancy between the maximum localized corrosion rate and maximum localized corrosion depth, as shown in Fig. 8.22, a new parameter called cumulative maximum localized corrosion rate, CR_{cmax}, was introduced [131]. It was defined as the derivative of the maximum localized corrosion depth curve:

$$CR_{cmax}(t) = d[CD_{max}(t)]/dt \tag{8.16}$$

The maximum localized corrosion depth, $CD_{max}(t)$, is proportional to the maximum anodic charge, Q_{max}, as shown in Fig. 8.21 (see Eq. 8.9). The maximum anodic charge in Fig. 8.21 was obtained by integrating the currents from the electrodes that had the maximum anodic charge since time zero (or had been corroded the most at time t). Therefore, the derivative of $CD_{max}(t)$ is simply a function of the current from the electrode that has the highest anodic charge at time t, I_{max_charge}:

$$CR_{max}(t) = (1/\varepsilon)I_{max\_charge}W_e/(F\rho A) \tag{8.17}$$

The CR_{cmax} values for the data shown in Figs. 8.20 and 8.21 are calculated and shown in Fig. 8.23. The maximum localized corrosion depth, CD_{max}, and the calculated corrosion rates for the electrodes that exhibited the highest corrosion currents at certain times during the test (see Fig. 8.20) are also plotted in Fig. 8.23. In the specified three time periods, the maximum localized corrosion depth was the value measured from Electrodes #5, #13, and #12, respectively (see Fig. 8.21). Therefore, the cumulative maximum localized corrosion rate, CR_{cmax}, was equal to the corrosion rates of these

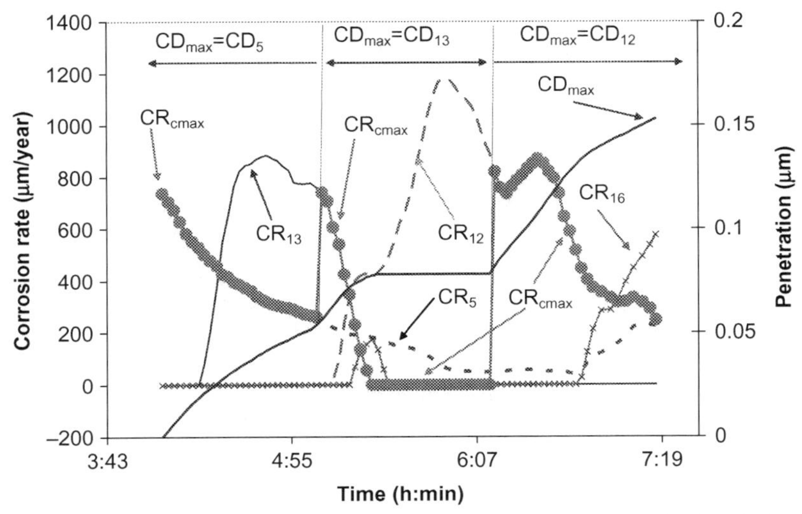

Fig. 8.23 Cumulative maximum localized corrosion rate, corrosion rates of selected electrodes, and maximum localized corrosion depth for the data shown in Figs. 8.21 and 8.22. Note, in the specified three time periods, the maximum localized corrosion depth was due to the corrosion of Electrodes #5, #13, and #12, respectively (see Fig. 8.21). © NACE International 2007.

three electrodes during the different corresponding time intervals. Another way to look at the cumulative maximum localized corrosion rate is that it is the corrosion rate of the cumulatively most corroded electrode at any given time (or the electrode with the deepest pit, if the mode of localized corrosion is pitting corrosion). Because of the nature of localized corrosion, the deepest pit may be repassivated under certain conditions and the corrosion rate in this pit may drop to zero. The zero value of CR_{cmax} in the second time interval indicates that the most corroded electrode (Electrode #13) was repassivated, and the corrosion on it was stopped from approximately 5:20 to 6:20 a.m. (Fig. 8.23).

Because the cumulative maximum localized corrosion rate is simply the corrosion rate of the electrode that passed the maximum amount of charge, a sorting algorithm may be built into a real-time corrosion monitoring software to track the most corroded electrode and obtain the cumulative maximum localized corrosion rate. The software may give the maximum localized corrosion rate, the cumulative maximum localized corrosion rate, and the average corrosion rate, CR_{avg}, which is calculated from the average of the anodic currents. The average corrosion rate may be used to represent the general corrosion rate [109]. Fig. 8.24 shows the maximum localized corrosion rate, cumulative localized corrosion rate, and average corrosion rate from a commercial software, CorrVisual (CorrVisual is a trade name of Corr Instruments, LLC, Carson City, NV, USA.)

Because the cumulative maximum localized corrosion rate is defined as the derivative of the maximum localized corrosion depth, the CR_{cmax} and CD_{max} values are directly related to each other. One can solve for CR_{cmax} from CD_{max} by differentiation or solve for CD_{max} from CR_{cmax} by integration. Unlike the maximum localized corrosion rate, CR_{max}, whose integration would produce an imaginary number that may

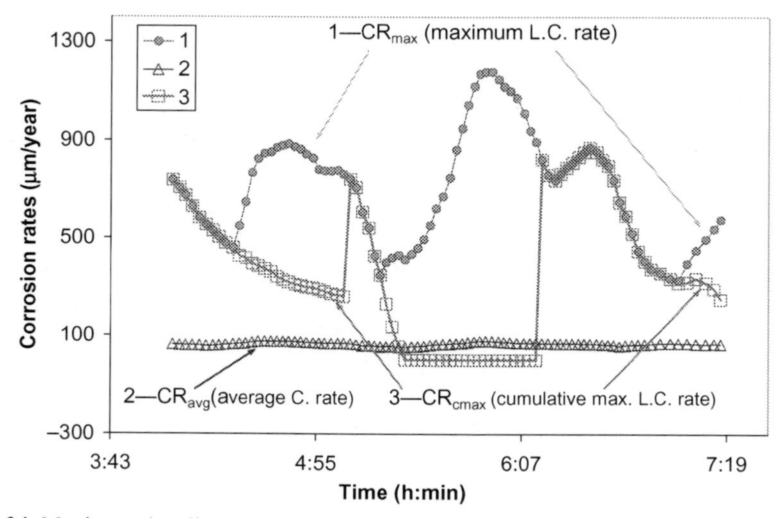

Fig. 8.24 Maximum localized corrosion rate, cumulative maximum localized corrosion rate, and average corrosion rate for the data shown in Fig. 8.20 [131]. © NACE International 2007.

be higher than the actual maximum localized corrosion depth, the integration of the cumulative maximum localized corrosion rate would produce exactly the maximum localized corrosion depth.

Maximum localized corrosion rate is a measure of localized corrosion rate at a given moment of time. Cumulative localized corrosion rate is a measure that relates to the cumulative damage of localized corrosion to a metal. If the maximum localized corrosion rate is high, but the cumulative maximum localized corrosion rate is low, the corrosion rate at one electrode is high at one time, but remains low at other times and there is always one electrode that has a high corrosion rate. The cumulative effect of this kind of high maximum localized corrosion rate in a given time period is a high general corrosion rate (i.e., every electrode is significantly corroded after a given period of time). When the maximum localized corrosion rate is significantly different from the cumulative maximum localized corrosion rate, the cumulative maximum localized corrosion rate should be used to evaluate the persistence of localized corrosion.

Because the cumulative maximum localized corrosion rate is the corrosion rate on the most corroded electrode and the corrosion rate on the most corroded electrode may be lower than the corrosion rate taking place on the most corroding electrode (the maximum localized corrosion rate) at certain times, the cumulative maximum localized corrosion rate is always lower than or equal to the maximum localized corrosion rate.

Therefore, for process control applications (i.e., inhibitor dose control), the maximum localized corrosion rate, CR_{max}, should be used. However, for the evaluation of the effect of localized corrosion on the cumulative damage to a metal component, the cumulative maximum localized corrosion rate should be used. It should be noted that at a given time in a certain environment, the cumulative maximum localized corrosion rate may be very low, even zero, but the maximum localized corrosion rate may still be very high. This means that the localized corrosion on the most corroded electrode has stopped, but the localized corrosion on the other electrodes is still high.

8.9 Effects of internal currents on CMAS and minimization of the internal effect

8.9.1 *Internal current effects on nonuniform corrosion rate measurement using coupled multielectrode array sensors*

As discussed in Section 8.8.1, in a uniform corrosion case, there would be a lower degree of deviation between the anodic electrodes and the cathodic electrodes. Fig. 8.25 shows the electron flow paths on the different kinds of electrodes in a CMAS probe [105]. If the most anodic electrode on a probe in a given environment is a partially corroded electrode (Fig. 8.25B) rather than a totally corroded active electrode (Fig. 8.25A), it is more likely that the electrode still has the cathodic sites to accept electrons from the neighboring anodic sites on the same anode [82, 87], and the I^{a}_{in} in Eq. (8.4) or the ε in Eq. (8.6) cannot be ignored when calculating the corrosion current, I^{a}_{coup}.

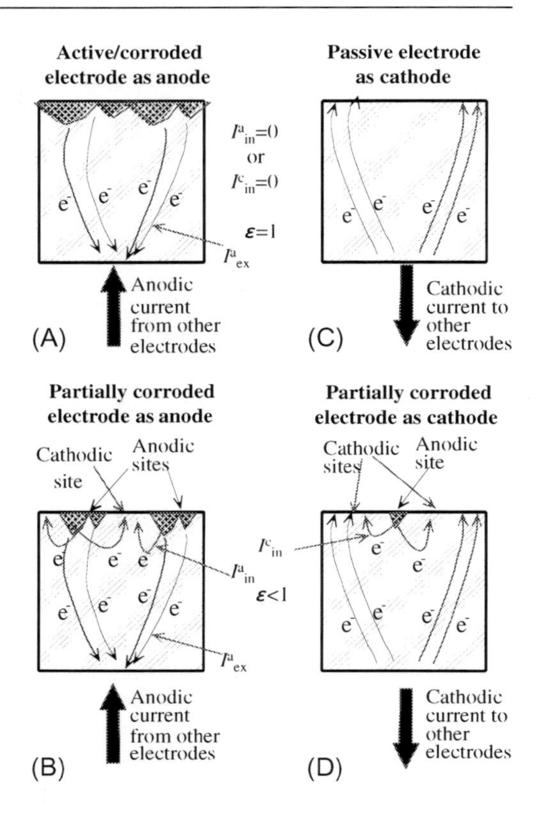

Fig. 8.25 Flow of electrons on (A) a totally corroded active electrode as anode, (B) a partially corroded electrode as anode, (C) a passive electrode as cathode, and (D) a partially corroded electrode as cathode in a CMAS probe [105]. © NACE International 2005.

In Fig. 8.25A, the electrode is totally active and fully covered by the corrosion products. Because the corrosion products may act as a diffusion layer for O_2 or H^+ to reach the metal surface, the I^a_{in} (also see Fig. 8.13) on this electrode is expected to be small. If the other electrodes on the CMAS probe are not corroded, all or most of the corrosion electrons from this totally corroded electrode would flow to the other electrodes through the external circuit and the corresponding ε in Eq. (8.6) would be equal or close to unity. On the other hand, in Fig. 8.25B, the electrode is only partially corroded; it is likely that a portion of its corrosion electrons would flow to the local cathodic sites, and the corresponding ε is less than 1. Therefore, in the case of Fig. 8.25B, there may be a large uncertainty in localized corrosion rate measurements using the CMAS probe if ε is assumed to be unity (Eq. 8.7).

8.9.2 Evaluation of Internal current effects on nonuniform corrosion rate measurement

The author and coworkers conducted experiments to evaluate the effect of the internal current on the measurement of nonuniform corrosion rate using the CMAS probe [110].

As discussed in Chapter 3, Tafel extrapolation method can be used to measure the corrosion current of a metal if it is under activation control. The Tafel extrapolation line is based on the polarization curve obtained when the overpotential is between 100 mV and 500 mV. As mentioned in Section 8.8, the internal anodic current equals to the internal cathodic current on the electrodes at the coupling potential; Tafel extrapolation based on the cathodic curve from the most anodic electrode can be used to measure the internal anodic current on the most anodic electrode ($I^{a'}_{in}$), as shown in Fig. 8.26A, if the corrosion is under activation control. If the corrosion is controlled by diffusion or by both diffusion and activation processes, the internal corrosion current should be lower than that under purely activation control. Therefore, the internal anodic current derived from the Tafel method ($I^{a'}_{in}$ in Fig. 8.26A) is the bounding value of the actual internal anodic current (I^{a}_{in} in Fig. 8.26A). It is a conservative approach by using the value of $I^{a'}_{in,}$ as the value of $I^{a}_{in,}$ in the evaluation of the internal current effect.

Fig. 8.26B shows typical cathodic polarization curves for 16 electrodes of a carbon steel CMAS probe in simulated seawater (3 wt% NaCl) and the Tafel extrapolation current from the most anodic electrode ($I^{a'}_{in,\ MAE}$). It also shows the externally measured coupling current from the most anodic electrode before the polarization ($I^{a}_{ex,\ MAE}$), which is the same as the maximum anodic current or the most anodic current (I_{max}) used for deriving the maximum nonuniform corrosion rate in Section 8.8.2.

Table 8.1 shows the externally measured coupling current ($I^{a}_{ex,\ MAE}$) and the estimated internal current ($I^{a}_{in,\ MAE}$) from the most anodic electrode for an aluminum CMAS probe and a carbon steel CMAS probe exposed to simulated seawater (3% NaCl) and hydrochloric acid solutions. The corrosion currents on the most anodic electrode ($I^{a}_{coup,\ MAE}$) are calculated according to Eq. (8.4) ($I^{a}_{coup,MAE} = I^{a}_{ex,MAE} + I^{a}_{in,MAE}$). The ratio of $I^{a}_{ex,MAE}/I^{a}_{coup,MAE}$ represents the percentage of the corrosion current

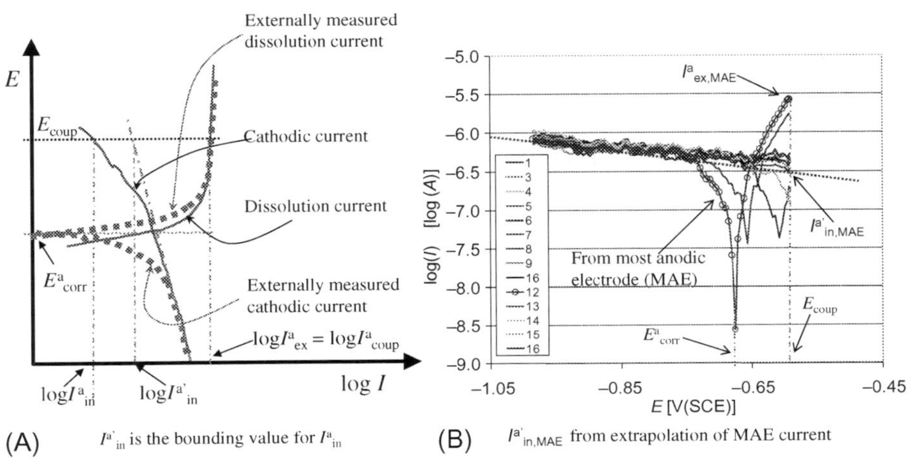

Fig. 8.26 Determination of the bounding value of the internal corrosion current on an anodic electrode (A) and typical bounding internal anodic current on the MAE of a carbon steel CMAS probe in simulated seawater (B). Note: The numbers shown in (B) are the identification numbers of the electrodes of the probe.

Table 8.1 Parameters for evaluation of internal current effects for aluminum (Al) and carbon steel (CS) CMAS Probes in seawater and hydrochloric acid solutions.

	Al in 3% NaCl	CS in 3% NaCl	Al in 0.2% HCl	CS in 2% HCl	CS in 0.2 wt% HCl
$I^a_{ex,MAE}$[a] (A)	4.27E−07	2.60E−06	4.36E−07	1.10E−06	7.94E−07
$I^a_{in,MAE}$	<2.50E−08	<2.10E−07	<1.90E−06	<3.90E−06	<6.30E−07
$I^a_{coup,MAE}$[b] (A)	4.52E−07	2.81E−06	2.34E−06	5.00E−06	1.42E−06
$I^a_{ex,MAE}/I^a_{coup,MAE}$	0.94	0.93	0.19	0.22	0.56
Effect ($I^a_{in,MAE}/I^a_{coup,MAE}$)	6%, Insignificant	7%, Insignificant	81%, Severe	78%, Severe	44%, Significant
Corrosion mode	Localized	Nonuniform	Uniform	Uniform	Uniform

[a] $I^a_{ex,MAE}$—Externally measured anodic current from most anodic electrode (MAE).
[b] $I^a_{coup,MAE}$—Corrosion current on MAE from CMAS ($I^a_{coup,MAE} = I^a_{ex,MAE} + I^a_{in,MAE}$).

measured by the CMAS probe in the presence of the internal current. The ratio of $I^a_{in,MAE}/I^a_{coup,MAE}$ represents the percentage of the effect of the internal current on the measurements of corrosion rate with the CMAS probe. All of these values are also included in Table 8.1

In all cases in which the metals exhibited typical localized corrosion or nonuniform general corrosion, the $I^a_{in,\ MAE}$ values are much smaller than the $I^a_{ex,\ MAE}$ values and the effects of $I^a_{in,\ MAE}$ on the measure of $I^a_{coup,MAE}$ or maximum nonuniform corrosion rate are less than 10%. Table 8.1 also shows that the effect is only significant (44% or higher) when the corrosion is uniform when the two metals were exposed to hydrochloric acid. Even for carbon steel in 0.2% wt hydrochloric acid, the CMAS probe still measures 56% of the corrosion on the most anodic electrode. It is noted that the error of 44% is exceptionally large for the measurement of most types of industrial process parameters such as temperature and pressure, but it is not a large number when corrosion rate is concerned. The large variation is true, especially when localized corrosion is concerned. Evaluation of pitted metal specimens after exposure in a corrosive industrial environment revealed that the maximum pit depth often varies from one coupon to another by multiple factors [98].

uniform corrosion such as those on the metals exposed to diluted acids is rare for industrial metal equipment and field metal structure; the vast majority of the corrosion found in the industry are nonuniform general corrosion and localized corrosion. So, the CMAS method is an effective tool for monitoring corrosion in the industry because the effect of the internal current is negligible for localized corrosion and nonuniform general corrosion.

8.9.3 CMAS that measures the internal current effect during the measurement of corrosion rate

As mentioned above, the internal effect is not significant if the mode of corrosion is localized or nonuniform corrosion, but the effect cannot be ignored if the mode of corrosion is uniform or dominated by uniform corrosion. A simple and convenient way to determine the effect during the measurement of the corrosion rate was introduced by Yang and Sun [111].

8.9.3.1 Estimation of the internal current based on overpotential measurements

Because active and uniform type of corrosion is usually under activation control, Volmer-Butler equation can be used to describe the polarization behaviors. Fig. 8.27 is similar to Fig. 8.13, but it is only applicable if the corrosion is activation-controlled and it is suitable for the present analysis. The thick solid lines represent the dissolution and reduction polarization behaviors on the MAE if the potential is significantly away from the respective reversible reaction potentials (reversible potentials for oxygen reduction reaction and for metal dissolution reaction, for example) when the Tafel relationship applies. The thin solid curves represent the dissolution and reduction polarization behaviors on the remaining coupled electrodes without the MAE when the potential is such that the Tafel relationship applies. The dashed thin lines represent the combined dissolution and reduction polarization behaviors on the MAE and on the remaining coupled electrodes.

Because the MAE is more active than the rest of the electrodes, its corrosion potential (E^{a}_{corr}) is lower than the corrosion potential of the rest of the electrodes when they are coupled as one electrode (E'_{coup}). When the MAE is connected to the remaining electrodes, the potential changes to a new value, E_{coup} (or E_{corr} for all coupled electrodes), such that the total anodic dissolution currents equal the total cathodic reduction currents (as shown by the thin dashed lines in Fig. 8.27). The corrosion current on the MAE changes from $I^{a}_{corr,MAE}$ to $I^{a}_{coup,MAE}$ and the internal anodic current on the

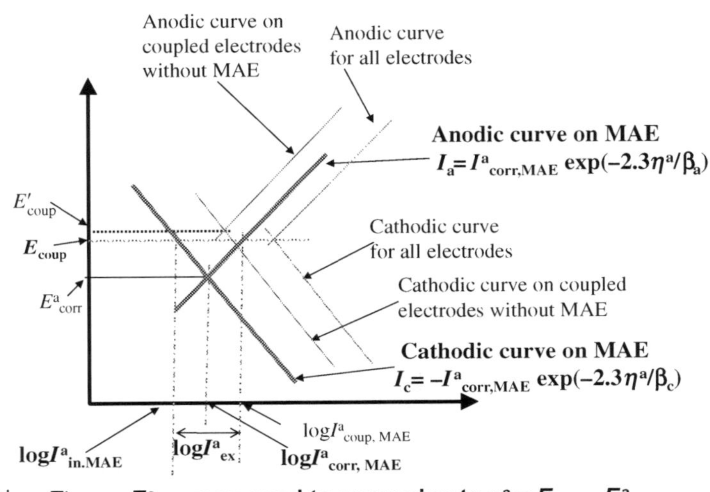

Fig. 8.27 Schematic diagram for the polarization curves on the MAE and on the remaining electrodes that are connected together on a CMAS, assuming activation control.

MAE changes from $I^a{}_{corr,MAE}$ to $I^a{}_{in,MAE}$. The difference between $I^a{}_{coup,MAE}$ and $I^a{}_{in,MAE}$ is the external current that flows from the coupling joint (rest of the electrodes) to the MAE ($I^a{}_{ex,MAE}$) and is measured by the CMAS instrument. The external current to the MAE can be described by the following equation:

$$I^a{}_{ex,MAE} = I^a{}_{corr,MAE}\left[\exp\left(-2.3\eta^a/\beta_a\right) - \exp\left(-2.3\eta^a/\beta_c\right)\right] \tag{8.18}$$

where η^a is the overpotential for the MAE ($\eta^a = E_{coup} - E^a{}_{corr}$), β_a and β_c are the Tafel slopes in the anodic direction and the cathodic direction, respectively. Eq. (8.18) is also called Tafel equation (See Chapter 3). The internal anodic current on the MAE is equal to the absolute value of the second term of the right-hand side of Eq. (8.18):

$$I^a{}_{in,MAE} = I_c = I^a{}_{corr,MAE}\left[\exp\left(-2.3\eta^a/\beta_c\right)\right] \tag{8.19}$$

The first term of the right-hand side of Eq. (8.18) is the total corrosion current ($I^a{}_{coup,MAE}$) on the MAE at η^a:

$$I^a{}_{coup,MAE} = I^a{}_{corr,MAE}\left[\exp\left(-2.3\eta^a/\beta_a\right)\right] \tag{8.20}$$

As mentioned in Section 8.9.2, the ratio of $I^a{}_{in,MAE}/I^a{}_{coup,MAE}$ represents the percentage of the effect of the internal current on the measurements of the nonuniform corrosion rate with the CMAS probe if $I^a_{ex,MAE}$ is used to calculate the maximum nonuniform corrosion rate.

The E'_{coup} has the same meaning as the E^c_{corr} in Fig. 8.13. Here, the emphasize is that it is the coupling potential of all the rest electrodes without the MAE. In fact, it is also the corrosion potential of the rest electrodes connected together without the MAE.

In the review by Mensfeld [112], the data measured at or near room temperatures were all from 30 to 120 mV/dec for β_a and mostly from 50 mV/dec to infinity (diffusion-controlled process) for β_c. For a given set of β_a and β_c, there is a minimum overpotential, η^a, for the externally measured current, $I^a_{ex,MAE}$ to be equal or more than the corrosion current, $I^a_{corr,MAE}$, and this overpotential can be calculated according to Eq. (8.18). Although $I^a_{corr,MAE}$ is not the $I^a_{coup,MAE}$, it is the corrosion current if the linear polarization (LPR) method (see Chapter 3) is used to determine the corrosion current on the MAE and it is not too far from the $I^a_{coup,MAE}$ when the corrosion is dominated by activation control. Fig. 8.28A shows the minimum overpotential required for the externally measured current, $I_{ex,MAE}$, to be equal to or greater than the corrosion current, $I^a_{corr,MAE}$, of a CMAS probe for wide ranges of β_a and β_c. Fig. 8.28B shows the corresponding maximum internal current effect. For the vast majority values of β_a and β_c, the minimum overpotential is between 7 and 29 mV and the corresponding effect ($I^a{}_{in,MAE}/I^a{}_{coup,MAE}$) is less than 50% which is the extreme case of β_c being infinity. Therefore, by measuring the overpotential, one can estimate the effect of the internal current on the measurement of the maximum nonuniform corrosion rate with the CMAS probe.

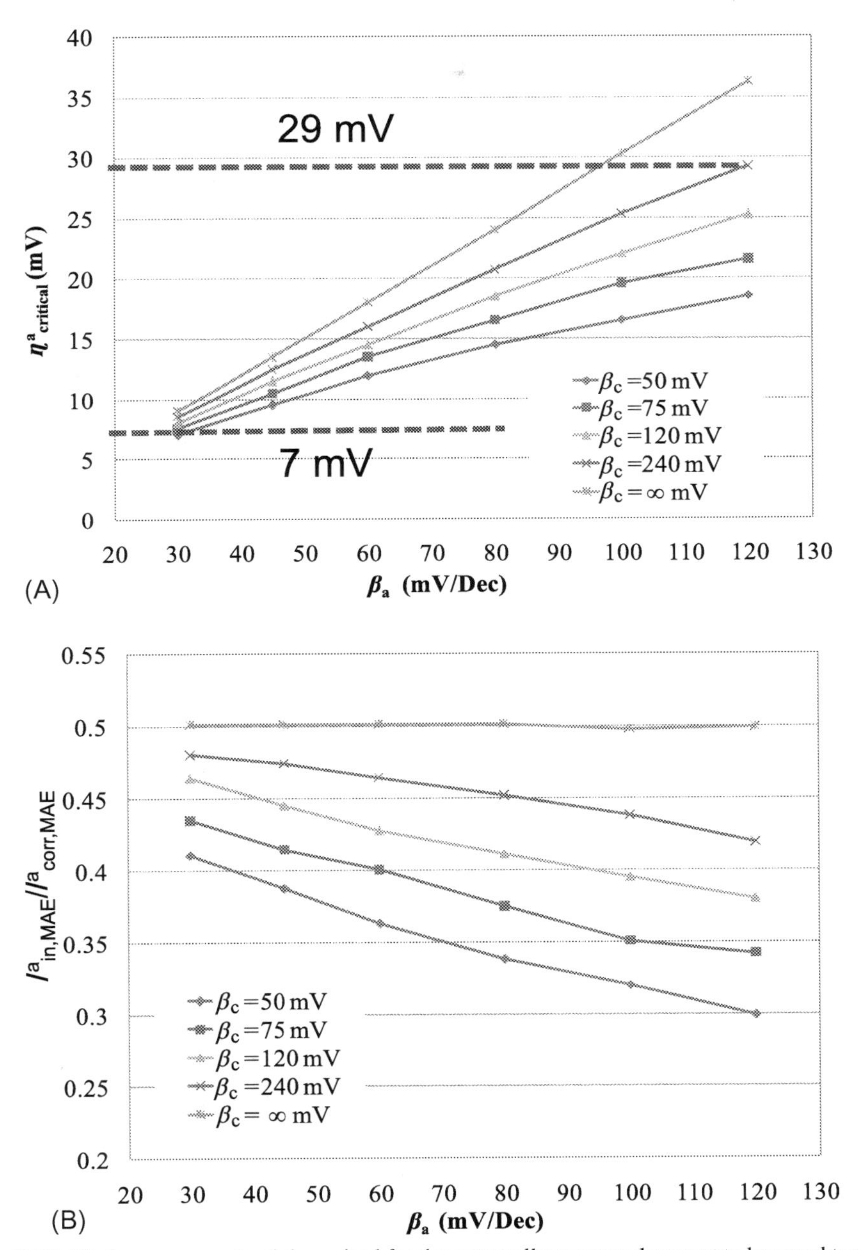

Fig. 8.28 Minimum overpotential required for the externally measured current to be equal to or greater than the corrosion current on the MAE of a CMAS probe for wide ranges of β_a and β_c (A) and the corresponding maximum internal current effect (B).

The above analysis assumes that the corrosion is activation-controlled. If the corrosion is controlled by both activation and diffusion as is the case for most nonuniform, especially localized, corrosion, the above estimation is conservative (see discussion for Fig. 8.26A). The effects in these cases should be smaller than the that of activation-controlled corrosion processes.

8.9.3.2 Method to determine the overpotential without using a reference electrode

The method of determining the effect of the internal current requires the measurement of the overpotential and this measurement normally requires the use of a reference electrode. Because the CMAS probe is intended for use in harsh environments and for long-term measurements, the requirement for a reference electrode which often requires frequent maintenance and has limited service life presents a challenge for field applications. Yang and Sun proposed to use the differences between the corrosion potential of the most anodic electrode ($E^{a}_{corr.MAE}$) and the coupling potential of the CMAS probes after the most anodic electrode was decoupled (E'_{coup}) to approximate the overpotential [111]. The measurement of this difference ($\eta^{a'} = E'_{coup} - E^{a}_{corr.MAE}$) does not need the reference electrode. Because the normal coupling potential (E_{coup}) is the potential when the most anodic electrode is connected to all other electrodes, the coupling potential after the most anodic electrode is decoupled (E'_{coup}) may be slightly higher than the normal coupling potential (E_{coup}), but the difference should be small if the corrosion is dominated by activation control because the CMAS probes usually have nine or more electrodes and the removal of only one electrode may not significantly change the coupling potential. If the corrosion is controlled by diffusion or both diffusion and activation processes, E'_{coup} may be significantly higher than the E_{coup}, but the higher E'_{coup} makes the difference ($\eta^{a'} = E'_{coup} - E^{a}_{corr.MAE}$) even larger. The larger $\eta^{a'}$ means that the most anodic electrode is significantly more anodic than the rest of the electrodes, which is a typical sign of severe nonuniform or localized corrosion for which case the effect of internal current is negligible (see Section 8.9.1). Therefore, in all cases, the measurement of $\eta^{a'}$, without the need for a reference electrode, is an effective way to determine the internal current. If it is larger than the minimum value required for the externally measured current to equal the corrosion current, the internal current effect is less than 50% in all cases.

Fig. 8.29 shows real-time maximum nonuniform corrosion rates (CR_{max}) and overpotentials ($\eta^{a'} = E^{a}_{corr.MAE} - E'_{coup}$) from a CMAS instrument that measures two probes simultaneously. The absolute values of $\eta^{a'}$ for carbon steel and aluminum in NaCl solution are all in the range from 30 to 60 mV which is large compared to the abovementioned minimum overpotential of 7 to 29 mV (more precisely, 16 mV for carbon steel in seawater with $\beta_a = 57$ and $\beta_c = \infty$ [112]), indicating that there were no significant internal current effect. On the other hand, the absolute values of $\eta^{a'}$ for carbon steel and aluminum in hydrochloric acid solution are only 10 mV, indicating that there was significant internal current effect. These results are consistent with the results presented in Section 8.9.2.

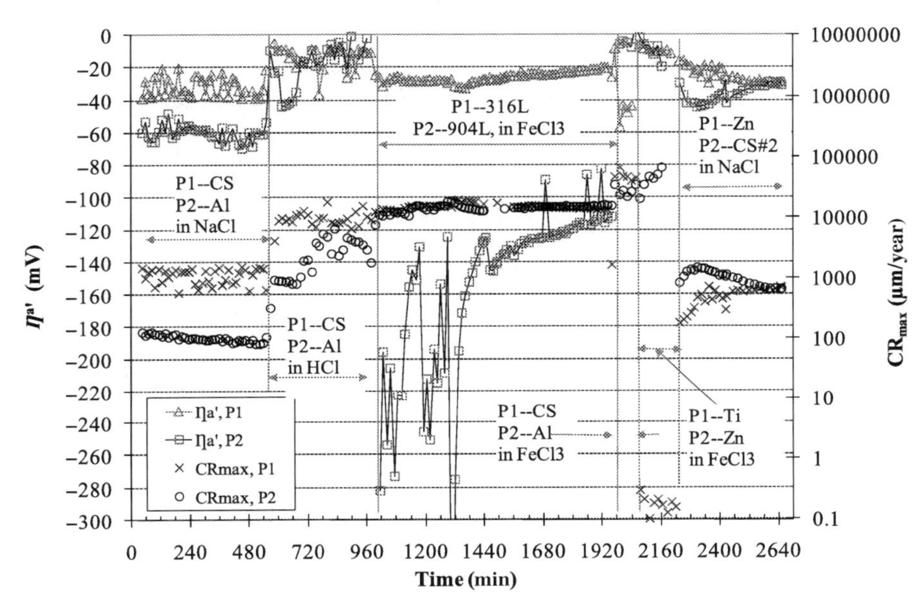

Fig. 8.29 Real-time maximum nonuniform corrosion rates and overpotentials from a CMAS instrument that measures two probes (P1 and P2) simultaneously. Note: CS, Al, 316L, 904L, Zn, and Ti represent probes for carbon steel, aluminum, 316L stainless steel, 904L stainless steel, zinc, and titanium, respectively.

In addition, the values of $\eta^{a'}$ are large ($\geq$20 mV) for Zn in simulated seawater, 316 SS, 904L SS, Zn, and CS in FCl_3 solutions, which is an indication of nonuniform corrosion. The value of $\eta^{a'}$ is small (<10 mV) for aluminum in $FeCl_3$, indicating that the corrosion is dominated by uniform corrosion. Even though there were significant effect of internal current for aluminum and carbon steel in 2% HCl, and aluminum in $FeCl_3$, CMAS probes still measured substantially high corrosion rates (5000, 7500, and 16,000 μm/year, respectively).

8.9.4 Minimization of internal current effects on localized corrosion rate measurement using coupled multielectrode array sensors

To represent the case as depicted in Fig. 8.25B (coexistence of cathode on the most anodic electrode), Fig. 8.13 is modified into Fig. 8.30. In Fig. 8.30, the anodic and cathodic behaviors are only shown for two electrodes: one is the most anodic electrode, the other is the most cathodic electrode (rather than for many cathodic electrodes as shown in Fig. 8.13 because there may not be many cathodic electrodes for one anodic electrode in the case as shown in Fig. 8.25B). In addition, the cathodic curve on the most anodic electrode is deliberately shifted to the right due to the

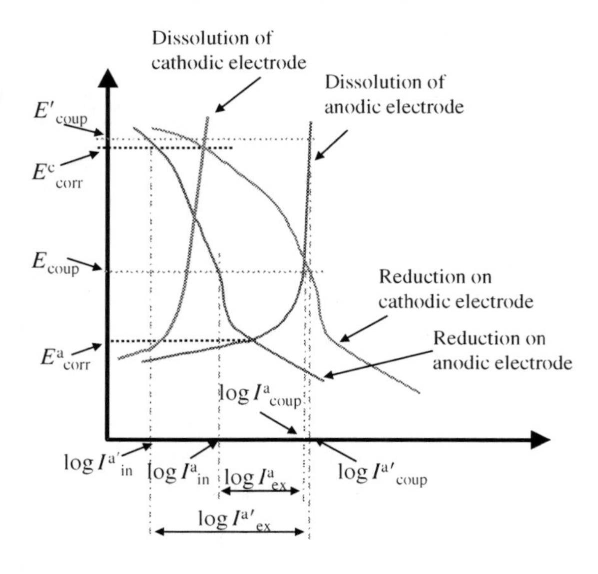

Fig. 8.30 Schematic diagram for the polarization curves on one anodic electrode and one cathodic electrode of a coupled multielectrode sensor.

existence of the uncorroded sites on the most anodic electrode that may support a higher cathodic current. At the coupling potential, E_{coup} (assuming only these two electrodes are coupled), the internal current, which is equal to $I^a{}_{in}$, is large and cannot be ignored. However, if the coupling potential can be altered to E'_{coup}, the corresponding $I^{a'}{}_{in}$ would be much smaller than the externally measured anodic current, $I^{a'}{}_{ex}$. In this case, Eq. (8.5) can be used to calculate the corrosion current. In the work by Yang et al. [105], E'_{coup} was defined as a potential at which the current from the most cathodic electrode is about to become zero. In other words, E'_{coup} is the highest open-circuit potential measured on all the electrodes of a CMAS probe (after the electrodes are decoupled). Because E'_{coup} is the open-circuit potential of one of the electrodes which are made of identical metals, it is not an unreasonable potential for the metal in the solution of interest.

From a statistical point of view, the multiple electrodes in a CMAS probe simulate the different cathodic and anodic sites of a metal coupon; the highest potential (usually the potential measured from the most cathodic electrode) can be considered to statistically represent the potential of the most cathodic site on the metal (after it is electrically isolated from the other sections of the metal). The potential measured from the most cathodic electrode may thus be considered as the highest bounding potential for all the cathodic sites on the most anodic electrode. Therefore, if the coupling potential of a CMAS probe is raised to a value that is slightly higher than the potential measured from the most cathodic electrode, there should be no cathodic reaction on the most anodic electrodes, even though some areas are still uncorroded. Under these conditions, $\varepsilon \rightarrow 1$ and $I^a{}_{in} \rightarrow 0$, the uncertainty in the measured corrosion rate from the CMAS probe due to $I^a{}_{in}$ can be ignored or eliminated.

A polarization unit, as shown in Fig. 8.31, was used for a CMAS instrument to dynamically adjust the potential of the coupling joint of the probe such that the current from the most cathodic electrode of the CMAS probe is zero or slightly anodic (lower than zero if an anodic current is recorded as a negative current by the instrument or higher than zero if a cathodic current is recorded as a negative current by the instrument). Under this condition, the coupling potential is higher than the open-circuit potential of the most cathodic electrode.

Because raising the coupling potential also increases the current from the active sites, the corrosion rate measured at the raised coupling potential may be higher than the actual corrosion rate. Thus, the corrosion rate measured at the raised coupling potential may be considered the upper bound of the corrosion rate. The corrosion rate measured at the natural coupling potential, E_{coup}, may be considered the lower bound of corrosion rate because of the possible nonzero internal flow of electrons on the most corroding electrode at E_{coup}.

Preliminary measurements were conducted in a 0.5-M NaCl solution using the improved CMAS method as shown in Fig. 8.31 [105]. The upper bounds of the maximum localized corrosion rates were found to be 2.2 to 2.7 times higher than the lower bounds for the probes made of Alloy 276 (UNS N10276), Type 316 stainless steel (UNS S31600), and Type 1008 carbon steel (UNS G10080), respectively. Therefore, the maximum localized corrosion rate values measured at the natural coupling potential were close to the actual maximum localized corrosion rates for these three metals in 0.5 M NaCl solution.

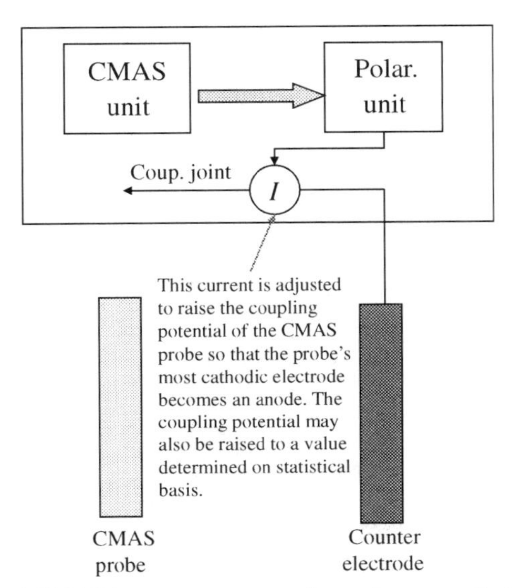

Fig. 8.31 An improved CMAS instrument system that reduces or eliminates the effect of internal electrons on the most anodic electrode of a CMAS probe [105]. © NACE International 2005.

8.10 Electrode spacing effect on corrosion rate measurement with CMAS

As discussed by Cong et al., the ability for a coupled multielectrode array to effectively simulate the electrochemical behavior of a one-piece metal for studying the special behavior depends on two types of couplings [38]. One is coupling through chemical interaction and the other is coupling through electrical potential field. The effective coupling through potential fields requires that both the voltage drop caused by the solution resistance and the voltage imposed by the ZVA must be negligibly low. The effective coupling through the chemical interaction requires that the spacing between the neighboring electrodes must be as small as reasonably achievable. As the solution resistance also depends on the spacing of the electrodes, the effective coupling of electrical potential fields also requires that the electrode spacing be reasonably small.

When the coupled multielectrode array is used as a sensor, namely the CMAS, the spacing must not be too small in some applications. An example is the special CMAS probes designed for the systems containing H_2S that must have a large spacing to avoid the bridging effect (see below) that may be caused by the electron-conducting sulfide corrosion products deposited between the electrodes [113, 114].

Yang and Sun conducted experiments on the effect of electrode spacing of the CMAS probes on corrosion rate measurements for carbon steel in simulated seawater [115]. Three CMAS probes with 16 functional electrodes made of Type 1018 carbon steel (UNS G10180) were tested in simulated seawater (0.5 M sodium chloride [NaCl] solution saturated with air). The electrodes were made of 1.0-mm-diameter wires. The electrodes of the first CMAS probe (Probe A) were closely packed together in a random pattern as shown in Fig. 8.32A, with the spacing between the adjacent electrodes being 0.05 to 0.25 mm. The electrodes of the second CMAS probe (Probe B) were packed in a regular four-by-four pattern as shown in Fig. 8.32B, with the spacing between the adjacent electrodes being 1.0 mm. The electrodes of both Probe A and

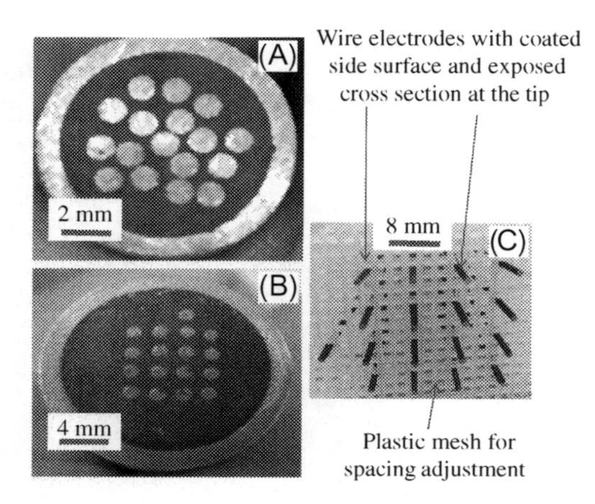

Fig. 8.32 Configuration of the electrodes on the CMAS probes for studying spacing effects [115]. Note: Spacing was 0.05 to 0.25 mm for Probe A (A), 1.0 mm for Probe B (B), and 8.0 mm for Probe C (C).

Probe B were flush-mounted in an epoxy. The electrodes of the third CMAS probe (Probe C) were spaced out in a regular four-by-four pattern as shown in Fig. 8.32C, with the spacing between the adjacent electrodes being 8 mm. The side surfaces of all of the electrodes of Probe C were coated with an epoxy, and the wire cross-sectional areas at the tips were exposed to act as the active surfaces of the electrodes. The electrode arrangement of Probe C simulates the CMAS probes with finger-like electrodes that are for application in systems containing a hydrogen sulfide (H_2S) environment. The finger-like electrodes effectively prevent the formation of contiguous electron-conducting corrosion products such as iron sulfide (FeS) in the presence of H2S between adjacent electrodes [114]. The electron-conducting corrosion products between adjacent electrodes had been blamed for causing short-circuiting of the electrodes for electrochemical sensors (see Chapter 3).The cross-sections of all the electrodes in the aforementioned probes were polished with 240-grit silicon carbide paper before the experiment, and the exposed surface area of each electrode was 0.785 mm^2.

Fig. 8.33 shows typical maximum nonuniform corrosion rates from the probes with different spacing (0.05–8 mm). The rates from the three probes varied between 0.5 mm/year and 2 mm/year, which is not significant considering the stochastic nature of nonuniform corrosion, especially localized corrosion which may vary by a factor of five or more for the same metal immersed in similar solutions (see Section 8.13.). Probe C had the largest spacing (8 mm) and it should measure a lower corrosion rate if there is the distance effect because it has the highest solution resistance between the electrodes. On the contrary, Fig. 8.33 shows that nonuniform corrosion rate from Probe C was mostly higher than those from the other two probes. Similarly, Probe

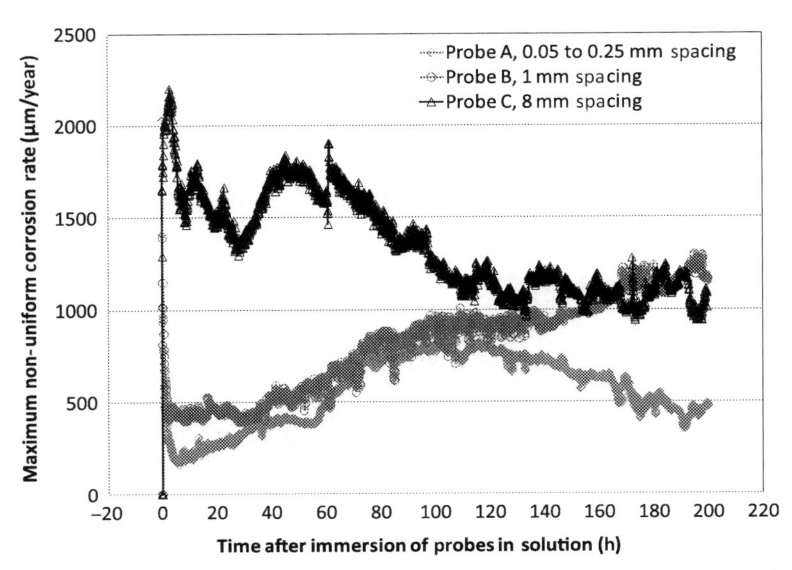

Fig. 8.33 Maximum nonuniform corrosion rate of carbon steel in simulated seawater from the CMAS probes with different spacing [115].

A had the smallest spacing (0.05–0.25 mm) and the nonuniform corrosion rate from this probe should be higher than those from the other probes if there is any spacing effect because it has the lowered solution resistance. In fact, the data from Fig. 8.33 show that the corrosion rate from Probe A was for the most part slightly lower than the corrosion rates from Probe B and Probe C. Therefore, the data in Fig. 8.33 does not indicate any effect of the spacing (from 0.05 to 8 mm) on the measurement of nonuniform corrosion rate for carbon steel in simulated seawater.

8.11 Minimization of the effects by corrosion products formed in H_2S-containing environment on localized corrosion rate measurement using coupled multielectrode array sensors

As discussed in Chapters 3 and 11, both linear polarization resistance probes (LPR) and the electrical resistance (ER) probes are subject to the effect of electron-bridging caused by the corrosion products formed in H_2S-containing environments. The sulfide corrosion products, such as FeS, are semiconductors and have similar behaviors as metal conductors for conducting electrons. In an ER probe, the corrosion measurement is based on the increase in the resistance of the sensing element that is caused by corrosion. If the electron-conducting corrosion products are deposited onto the sensing element and cause the changes in the measured resistance, it is not difficult to understand that such changes would give false readings for the ER probe.

Fig. 8.34A shows the effect of the electron-conducting deposits on the measurement of corrosion using an LPR probe. An external potential or current perturbation source (I source in Fig. 8.34A) is required to cause a slight polarization to the sensing electrode, and the change in current or voltage (measured by V in Fig. 8.34A) that is caused by the polarization is used to derive the corrosion rate (see Chapter 3). Normally, the conductivity between the two electrodes in the solution is provided purely by the ionic species in the solution, in which case, the number of electrons flowing through the solution, m', is zero, and the externally measured or supplied number of electrons, n, equals to the number of electrons that are as a result of the dissolution or corrosion reaction at the metal-solution interface, m. However, if an electron-conducting deposit is formed between the two electrodes in the solution, a portion of the electrons supplied or measured in the external circuit, m', would go through the electron-conducting deposits. Because this portion of electrons does not take part in the corrosion reaction at the metal-solution interface, the number of electrons that take part in the corrosion reaction, m, does not equal the externally supplied or measured number of electrons, n. Because there is no way of knowing the value of m', sulfide corrosion products severely affect the performance of an LPR probe. It also affects any other electrochemical methods that require an external perturbation source.

On the other hand, as shown in Fig. 8.34B, the corrosion current from a coupled multielectrode array sensor probe is obtained by measuring the currents generated

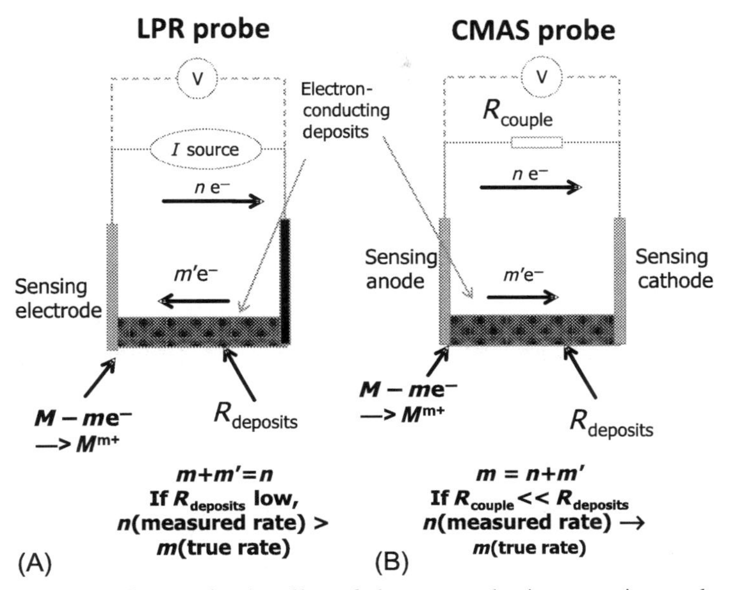

Fig. 8.34 Schematic diagram for the effect of electron-conducting corrosion products on (A) linear polarization resistance (LPR) probe and (B) coupled multielectrode array sensor (CMAS) probe.

by the coupling of an anode in the probe with the cathodes on the same probe. There is no external perturbation needed during the measurements. The coupling current is measured by a coupling resistor (or a zero-resistance ammeter which also has a defined internal resistance, especially when the current is low) that is in parallel with the resistance formed by the corrosion products. As long as the value of the coupling resistor (R_{couple}) is much smaller than the values of the electronic-conducting deposits ($R_{deposits}$), the number of electrons going through the electron-conducting deposit would be much smaller than that going through the coupling resistor ($m' \ll n$), and the current measured in the external circuit would be close to the current involved in the corrosion reaction ($n = m$).

Fig. 8.35 shows the results obtained with a carbon steel CMAS probe in a solution saturated with 100% H_2S at ambient temperature and pressure using two CMAS analyzers [118]. The first analyzer used larger resistors, and the second analyzer used smaller resistors as the shunt-resistors (also called coupling resistors) in the ZVA (see Fig. 8.11). The measured maximum localized corrosion rate was high initially, but decreased gradually with time when the probe was connected to the instrument unit that had large resistors. Because the corrosivity of the solution should not decrease after introducing the H_2S, the decrease in the measured corrosion rate was apparently caused by the bridging effect. At the end of the experiment, the DC resistance measurement between each electrode and the coupling joint showed a significant decrease (from several megaohms to less than 100 ohms) which indicates the bridging effect. However, when the maximum localized corrosion rate for the same

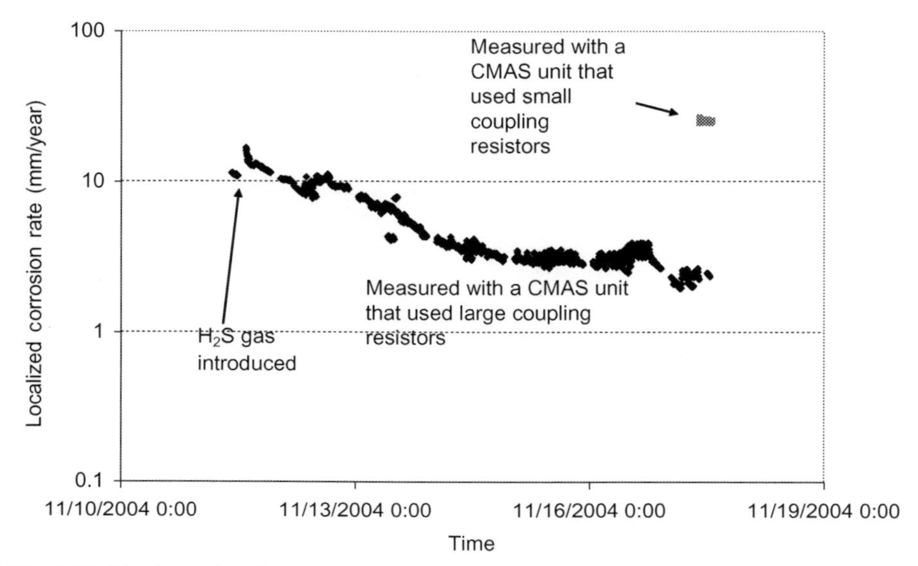

Fig. 8.35 Maximum localized corrosion rate measured from a carbon steel CMAS probe in a solution saturated with pure H_2S at ambient temperature and pressure using two CMAS instrument units [118]. © NACE International 2006.

probe was measured with the second analyzer that had lower values of coupling resistors, the localized corrosion rate was close to the value measured initially when there were no corrosion products. Therefore, the CMAS probe works in H_2S environments as long as the coupling resistors are low.

An important criterion for the coupled multielectrode array sensor probes to perform properly in systems that contain H_2S is that the value of the coupling resistor must be substantially smaller than that of the corrosion products that bridge the anodes and the cathodes. A mechanism built in a coupled multielectrode array sensor instrument that measures the bridging resistance on a continuous basis would be ideal for this application.

8.12 Minimization of the effect by crevice on corrosion rate measurement using coupled multielectrode array sensors

Similar to all electrochemical methods in corrosion rate measurements, the electrode surface area must be well-known to apply Faraday's law to calculate the corrosion rate. For highly corrosion-resistant alloys or for corrosion in less corrosive environments, CMAS probes are sensitive to the presence of crevices that may be formed between each sensing electrode and the surrounding insulator. This is because the

CAMS probes have a much smaller designated sensing surface area than the probes based on other electrochemical techniques (see Section 8.8). In addition, the crevice between the sensing electrode and the surrounding insulator, if formed, may also promote localized corrosion. If the objective is to measure localized corrosion, including crevice corrosion, such as in the case of steam generator tube corrosion under the tube sheet, measurement with a probe that has a crevice would not be a concern. But for cases where no crevice exists, the measurement with a probe that has a crevice would give false results. It should be mentioned that the concern only exists for highly corrosion-resistant alloys or for corrosion in less corrosive environments. If the solution is highly corrosive or the metal is highly active, such as in the case of carbon steel in seawater [108], most of the corrosion reactions would take place on the boldly exposed surfaces rather than deep in the crevice because of the limitation by mass transfer through the relatively thin crevice. If the solution is not corrosive or the metal is highly corrosion-resistant, however, mass transfer would not be a limiting factor and corrosion reactions may take place anywhere the metal is in direct contact with the solution, even if the location is deep inside the crevice. In this case, the effective corrosion area is the total area of the metal in contact with the solution.

At low temperatures (<80°C), this requirement can be satisfied by using a proper sealing material such as an epoxy as the insulator or applying a proper coating around the electrodes to achieve good bonding between the electrode and the insulating material or the coating. At elevated temperatures (e.g., 150°C), however, it is difficult to find a proper coating or sealing material that would resist a harsh chemical environment. In the measurements of localized corrosion rate of Alloy 22 (UNS N06022) material in a NaCl-$NaNO_3$-KNO_3 brine at 150°C, Chiang and Yang [132] successfully coated the sensing electrode with a diamond-like carbon coating using a chemical vapor deposition process. It was demonstrated that the electrical-insulating diamond-like carbon coating formed an excellent bonding with the Alloy 22 substrate. Fig. 8.36 shows a comparison for the sensing electrode surface between a probe that was uncoated and a probe that was coated with the diamond-like carbon (both were supported by an outer layer of epoxy) after the probes were exposed to a low pH NaCl-$NaNO_3$-KNO_3 brine at 150°C for 2 weeks. Clearly, a crevice was formed between the outer layer epoxy and the Alloy 22 electrode, but no degradation was observed at the interface between the diamond-like carbon coating and the Alloy 22 electrode. Measurements with electrochemical impedance spectroscopy (EIS) showed that the diamond-like carbon coating was highly protective before and after the exposure to the aggressive NaCl-$NaNO_3$-KNO_3 brine at 150°C [132].

8.13 Stochastic nature of localized corrosion and variability of localized corrosion rates of metals

Corrosion is a complex process and it is affected by the intrinsic properties of the metal such as composition and microstructure, the environments such as the concentrations of aggressive or inhibiting species, and the conditions experienced by the metal such as mechanical loading. In addition, localized corrosion processes are often governed

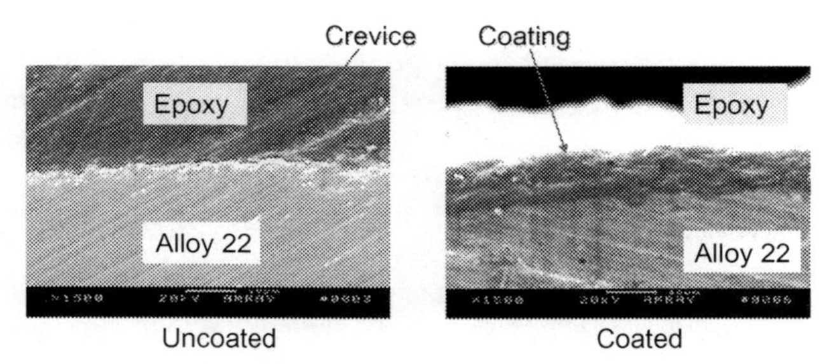

Fig. 8.36 Posttest appearance of the electrode with and without the diamond-like carbon coating [132]. © NACE International 2007.

by the above-mentioned properties or conditions on the local scale, even at the lattice level, which are impossible to control or to predict. Because of these unknown or uncontrollable effects, corrosion, especially localized, is stochastic in nature and difficult to reproduce. Unlike the measurements of process parameters such as temperature and pressure for which one can specify an accuracy of 5% or even 1%, the reproducibility of localized corrosion rate cannot be expected to be within 5% or even 50%. This is not because of the accuracy of the monitoring instruments; it is rather because of the variability of the corrosion of metals.

For these reasons, reliable localized corrosion rate should only be described by a statistical probability distribution function based on multiple measurements. The localized corrosion rate from a single sensor installed at a particular location can trend the process conditions near that location. However, the evaluation of the performance of a metal in a particular system or the evaluation of a corrosion mitigation strategy should be based on a statistical analysis of the data from multiple sensors and/or from multiple measurements.

Southwell and coworkers conducted series of long-term experiments to evaluate the corrosion of a range of metals exposed to fresh water as well as marine water [99, 100]. In their experiments, large metal plates (9 by 9 in. or 229 by 229 mm) were exposed to the water environments. At the end of year 1, year 2, year 4, year 8, and year 16, two of the plates were removed for measurement of general corrosion rate by weight loss method and the measurement of the pitting corrosion penetration. Fig. 8.37 shows the depth of the deepest pit found on the two sides of each of the two plates at different time intervals for carbon steel and copper exposed to seawater. In Fig. 8.37A, the maximum depth on the two carbon steel plates removed after 4 years was less than the maximum depth on the two carbon steel plates removed after 2 years. This is clearly an indication of the variability of the pitting corrosion for carbon steel on the identical plates, but exposed to the same environment. The variations may be shown by the upper and lower bounding curves. At year 2, the upper curve is 2.6 times higher than the lower bounding curve, which corresponds to a variation of 260% in localized corrosion rate. Similarly, the data for copper (Fig. 8.37B) indicated a variation of 620%.

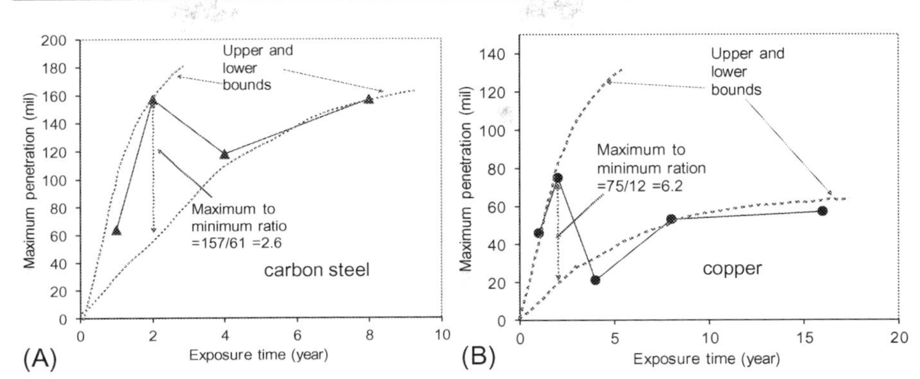

Fig. 8.37 Variations of maximum penetration depth of pitting corrosion obtained from large carbon steel coupons (A) [99] and copper coupons (B) [100] exposed to seawater. Note: 1 mil = 25.4 μm; each data point was measured from the deepest pit found on the 2 sides of two 229 × 229 mm plates.

8.14 Validation of corrosion rate measurement using coupled multielectrode array sensors

An important assumption in the derivation of the maximum localized corrosion rate using a CMAS probe is that if the number of electrodes is large enough, there is at least one electrode (the most corroding electrode) that would not have significant internal current flow (see Section 8.9). In addition, the insulator that separates the electrodes from electrically contacting each other and the spacing among the electrodes in typical CMAS probes would not have a significant effect on the measurements of the CMAS probe. Therefore, validations are required to compare the measured corrosion rates against those measured with coupons or other types of probes that measure the wall loss as a direct signal.

8.14.1 Comparison with literature coupon data

Sun and Yang measured the corrosion rates of the following alloys in simulated seawater using coupled multielectrode array sensors [109]:

Type 1008 carbon steel (UNS G10080)
Type 304L stainless steel (UNS S30403)
Type 316L stainless steel (UNS S31603)
Type 904L stainless steel (UNS N08904)
Type 1100 aluminum (UNS A91100)
Type 3003 aluminum (UNS A93003)

Fig. 8.38 shows the comparison between the maximum localized corrosion rates averaged over the testing periods for the different types of alloys and the average corrosion rates averaged over the testing periods from the CMAS probes, and the pitting and general corrosion rates of relevant alloys from the literature.

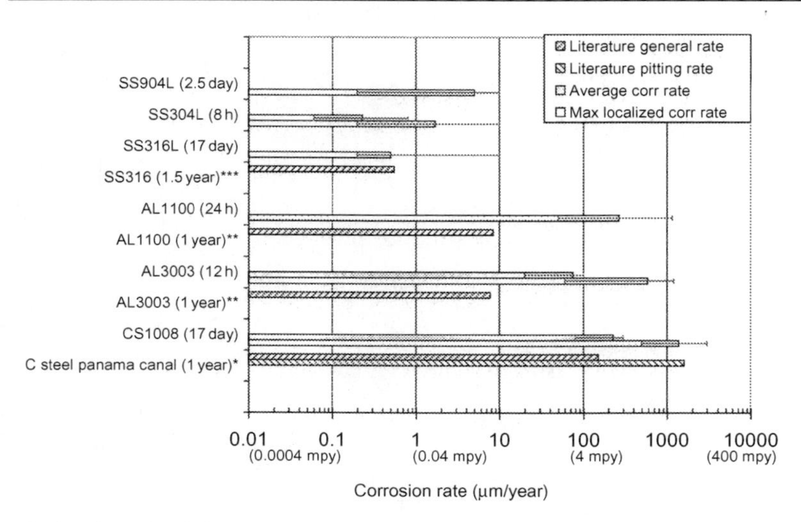

Fig. 8.38 Comparison between the corrosion rates measured with coupled multielectrode array sensors and literature data [109]. © NACE International 2006.

The measured maximum localized corrosion rate (1380 μm/year) and average corrosion rate (228 μm/year) for carbon steel 1008 are in good agreement with the reported maximum pitting rate (1600 μm/year) and general corrosion rate (228 μm/year), respectively [109]. These reported corrosion rates for carbon steel were obtained from one-year immersion test in Panama Canal seawater. The average corrosion rates for the stainless steels (~0.25 μm/year) are close to the reported general corrosion rates for stainless steel 316 (0.55 μm/year). The reported stainless steel general corrosion rate was obtained in an 18-month test in seawater. The average corrosion rate for Al3003 (20–100 μm/year) is slightly higher than the reported value (7.8 μm/year). The reported general corrosion rate for Al3003 was obtained in a one-year immersion test in seawater.

8.14.2 Comparison with penetration probes with multilayer, multithickness elements or multiwires

Penetration probes were also used to validate the corrosion rates measured with CMAS probes [129]. Fig. 8.39A shows a schematic diagram of a penetration probe with multilayer foils for measuring the penetration rate for a metal in a corrosive environment. Multiple electrodes were embedded inside one or more layers of metal foils in tube shape which were formed by spirally winding a piece of large foil onto a cylindrical bar. Two layers of acid-free paper were placed between two layers of the foils and each of the electrodes was placed between the two layers of paper inside a given layer of the foil so that no electrode was electrically contacting the metal foil. The bottom end of the foil tubes was sealed with epoxy. One side of the foil tube was cut and filled with epoxy so that the space within each layer of the foil was sealed

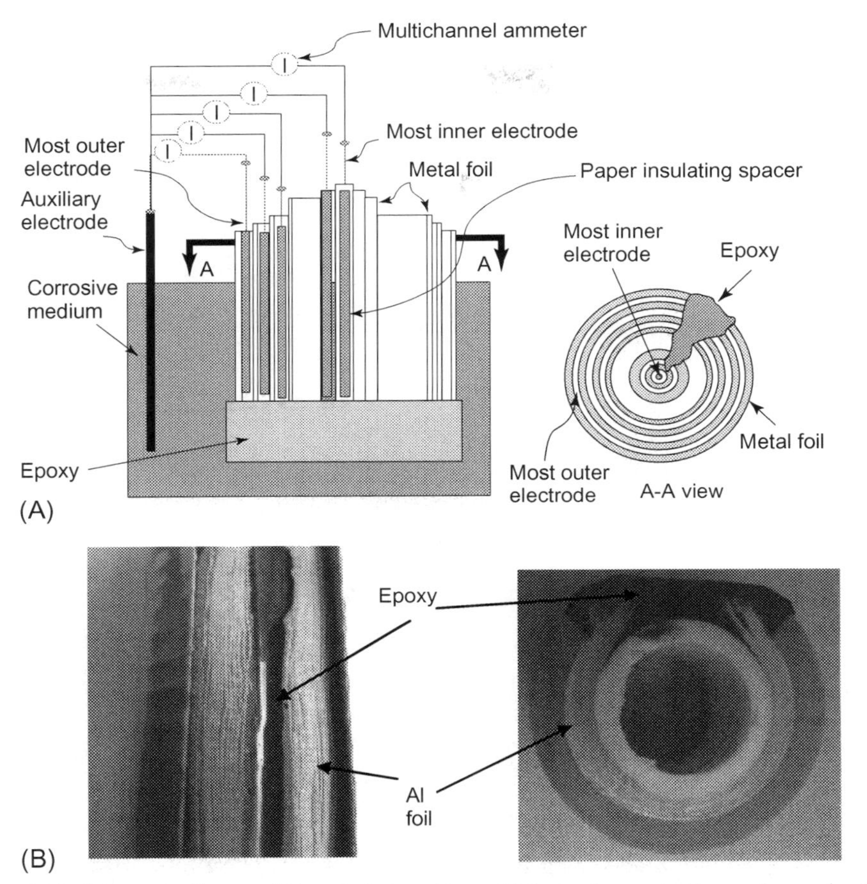

Fig. 8.39 Schematic diagram (A) and photos of a multielectrode penetration probe made of aluminum foil (B) [129]. © NACE International 2007.

(separated) from the neighboring layers, to avoid having the electrolyte migrate from the space inside one layer of the foil tube to the space inside another layer of the foil tube. The assembled system containing multiple electrodes and multiple foil tubes was then placed in a polyvinylchloride (PVC) protection tube to form an integrated multielectrode penetration probe. The protection tube had an opening (window) for the exposure of the foil to a corrosion environment. Fig. 8.39B shows the horizontally cut and vertically cut sectional views of the multielectrode penetration probe made of an aluminum foil. During the measurements, the probe was exposed to a corrosive liquid and each electrode was connected to an auxiliary electrode placed in the same electrolyte through an ammeter. When a layer of foil and its outside layers of foil were penetrated by corrosion (both general corrosion and localized corrosion), the corrosive medium would wet the electrode inside that layer of foil and form an electrical path between the electrode and the auxiliary electrode and a galvanic current would flow through the electrode.

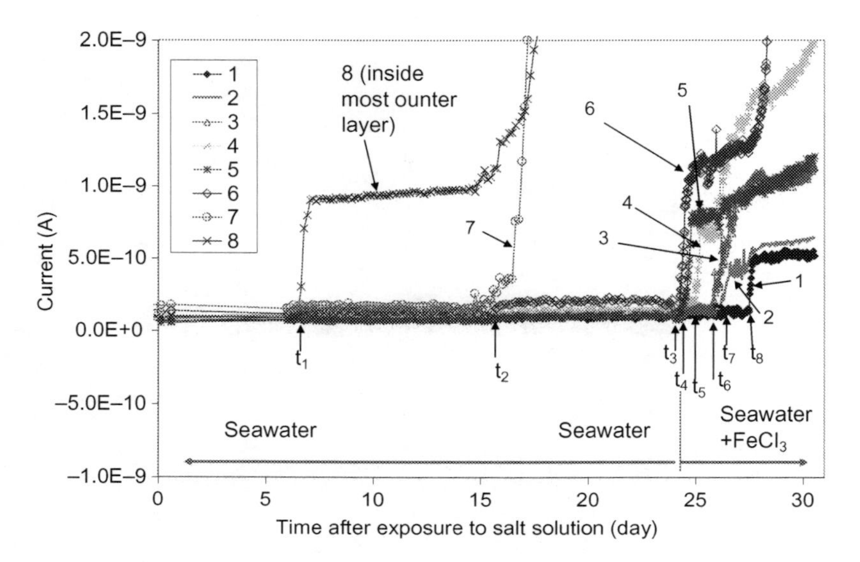

Fig. 8.40 Typical responses of currents and breakthrough times measured from a multielectrode penetration probe [129]. © NACE International 2007.

Fig. 8.40 shows the anodic currents from the different electrodes in an eight-electrode multilayer penetration probe made of aluminum foil to a Type 316L stainless steel auxiliary electrode in simulated seawater and a simulated seawater plus 10 mM ferric chloride solution. The breakthrough times (t_1, t_2, ….) are the times the current from the corresponding electrodes started to increase due to the perforation of the foils. Fig. 8.41 shows the penetration rates calculated using the breakthrough times shown in Fig. 8.40 and the thickness of the foil (16 μm in thickness). The average penetration rates for the aluminum foil were approximately 0.73 mm/year (29 mil/year) in simulated seawater and 12.5 mm/year (490 mil/year) in the simulated seawater plus ferric chloride solution, respectively. The aluminum foil was Type 8111 aluminum with a chemical composition of 98.5 wt% aluminum and balance of iron and silicon. The maximum localized corrosion rates measured with the CMAS probes were 0.2 to 1.1 mm/year for Types 3003 aluminum (98.6% Al, 0.12% Cu, and 1.2%Mn) and 0.2 to 0.5 mm/year for Type 1100 aluminum (99% minimum Al and 0.12% Cu) in simulated seawater, and 7 to 40 mm/year for Type 1100 aluminum in simulated seawater plus 10 mM $FeCl_3$ solution. These results from the CMAS probes are in good agreement with the results from the penetration probes.

In summary, the use of the penetration probe with multilayer, multithickness elements or multiwires [130] is an effective way to verify the performance of the CMAS probe. If these probes are installed in parallel with the CMAS probes, the verification is obtained at certain time intervals during the service without the need to remove the probes for examination. This is especially important for system under high pressures or with hazardous materials such as H_2S near the oil and gas wells where the installation and removal of the probes require special procedures.

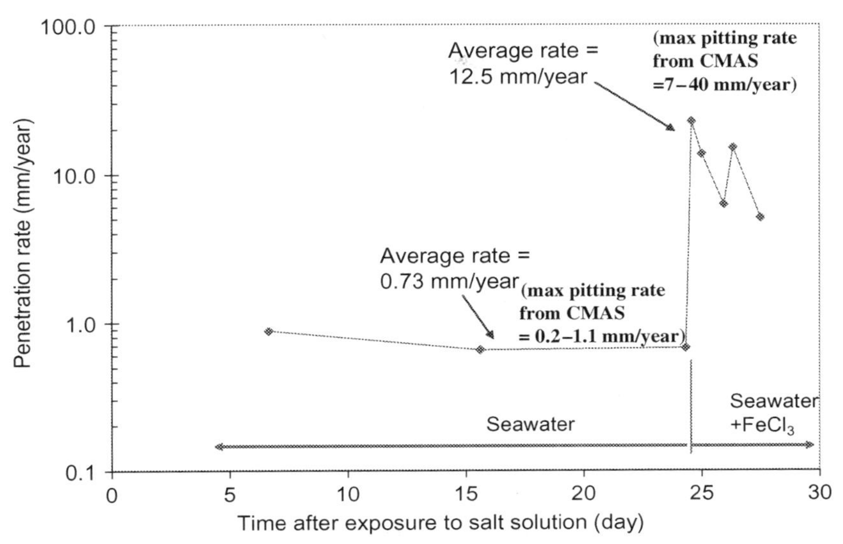

Fig. 8.41 Comparison between the penetration rates calculated using the breakthrough times shown in Fig. 8.40 and penetration rates measured with coupled multielectrode array sensors [129]. © NACE International 2007.

8.14.3 Comparison with parallel coupons and with the corroded depth of the electrodes on the CMAS probes

8.14.3.1 Corroded depth of the electrodes on the CMAS probes

Dorsey et al. used the coupled multielectrode array sensors to measure the localized corrosion rates of low carbon steel material in a chemical plant [98]. Fig. 8.42 shows the real-time maximum localized corrosion rates from two carbon steel CMAS probes during a 5-month test. The 5-month average of the maximum localized corrosion rate was 145 mil/year (3.68 mm/year). Fig. 8.43A shows the appearance of the sensing electrodes of a CMAS probe after the test. The sensing electrodes of the probes were covered by a thick layer of deposits when they were removed from the monitoring station. Bacteria were found in the water samples and in the corrosion products, and microbially influenced corrosion was attributed to the high corrosion rate of carbon steel in the system. As indicated in Fig. 8.43A, the penetration rate calculated from the corroded depth on the most corroded electrode [~149 mil/year (3.73 mm/year)] represents the cumulative maximum corrosion rate (see section 8.8.5) and it is close to the average maximum localized corrosion rate from the probe [145 mil/year (3.7 mm/year), see Fig. 8.42A)].

Similar comparisons were conducted by Ohtu and Miyazawa [101] and Nikam et al. [102].

The authors of both teams measured the corroded depth of the electrodes on the CMAS probes after services using laser profilometry. In the work by Ohtu and Miyazawa, the CMAS probe with carbon steel electrodes showed the maximum

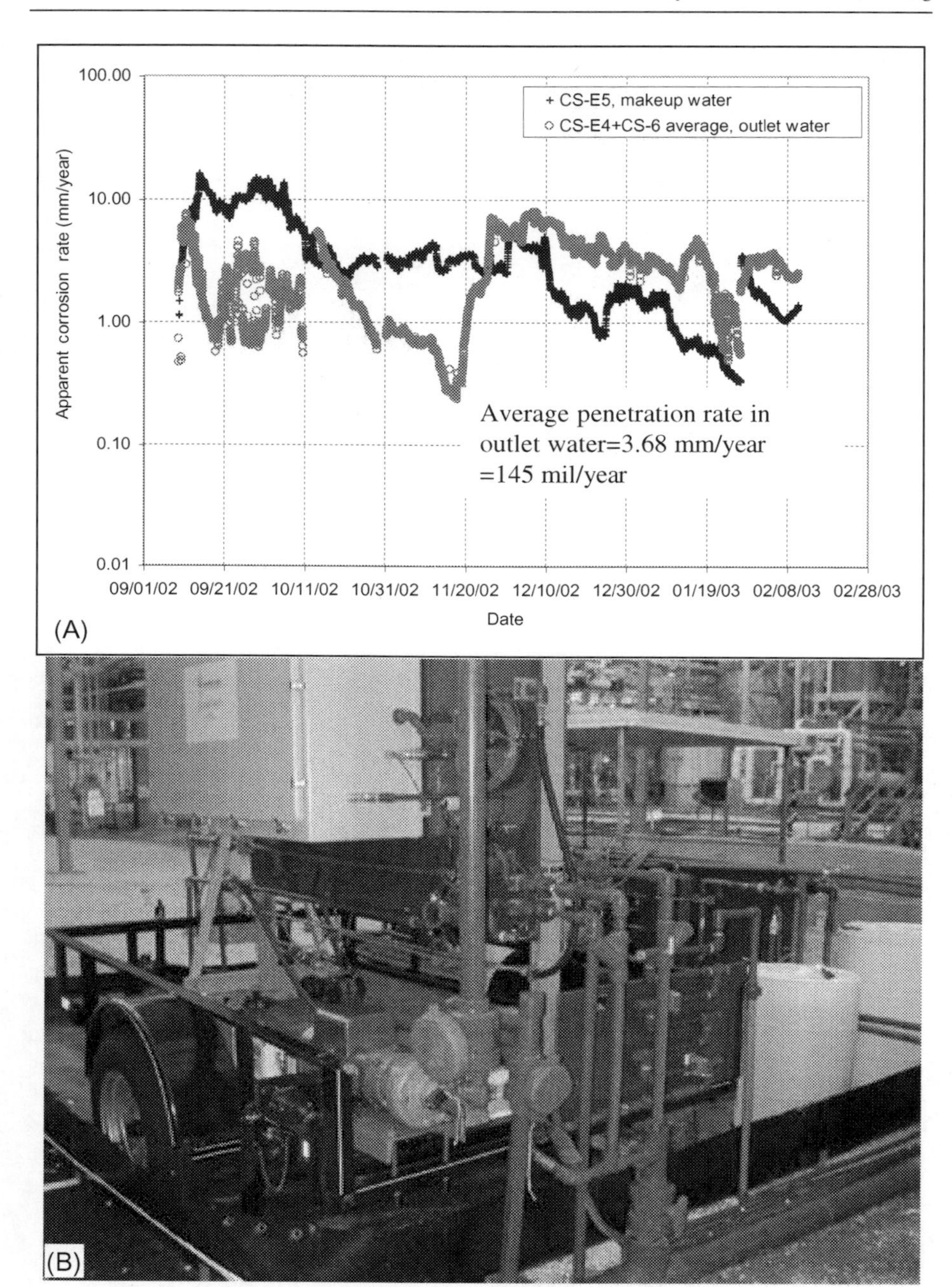

Fig. 8.42 Typical maximum localized corrosion rates of CMAS carbon steel probes (A) measured in a cooling water monitoring station of a chemical plant (B) [98]. © NACE International 2004.

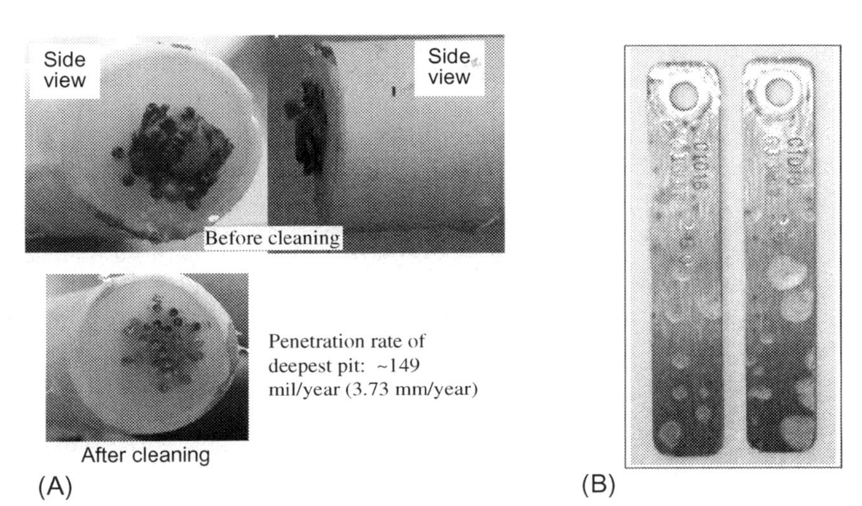

Fig. 8.43 Comparison of the posttest appearances of the CMAS carbon steel probes (A) and the coupon test specimens (B) in the cooling water system of a chemical plant [98]. Note: The pitting rates calculated from the deepest pit depth on different coupons varied between 100 to 200 mil/year (2.5 to 5 mm/year).

localized corrosion rate of 4 mm/year and the laser measurement showed a maximum pitting rate of 3 mm/year (difference of 33%) after the exposure to a contaminated river water. In the work by Nikam for the pitting corrosion of 316 SS stainless steel in 6% ferric chloride solutions at 40°C, the difference between the cumulative depth from the CMAS instrument and the posttest laser measurements varied from 2% to 20% for two different experiments. Because of the variability nature of localized corrosion as discussed in Section 8.13, the difference of 20% to 33% should be considered excellent agreement.

8.14.3.2 Comparison with parallel coupons

ASTM standard recommended that multiple coupons be installed in parallel with the CMAS probes for validation of the probe signals whenever possible [77]. In the work conducted by Dorsey et al. as discussed above, multiple coupons were installed with the CMAS probes for comparison [98]. Fig. 8.43B shows the pitting corrosion on typical coupons. The penetration rates calculated from the depth of the deepest pit on each of the coupons varied from 100 to 200 mil/year (2.5–5 mm/year). Compared with the real-time average maximum localized corrosion rate from the probe [145 mil/year (3.68 mm/year), see Fig. 8.42A], the difference is within 50% and this range of difference should be considered good agreement.

8.15 Applications of coupled multielectrode array sensor for real-time corrosion monitoring

CMAS probes have been extensively used for monitoring localized corrosion of a variety of metals in the following environments:

- Cooling water [83, 87, 98, 107, 134]
- Simulated seawater [109, 134]
- Salt-saturated aqueous solutions [88]
- Concentrated chloride solutions [84]
- Concrete [121, 135, 136]
- Soil [122]
- Low-conductivity drinking water [116]
- Process streams of chemical plants at elevated temperatures [91, 95, 104]
- Coatings [119, 123]
- Deposits of sulfate-reducing bacteria [90, 97]
- Deposits of salt in air [89, 93]
- High pressure simulated natural gas systems [118]
- H_2S systems [118, 138]
- Oil/water mixtures [106, 128, 137, 138]
- High temperature solutions [132, 133]
- Engine coolant [127] and engine exhaust [139]
- Formation of oil field [140]
- Gas turbine compressor components [141]

The CMAS probes were also used in the evaluation of corrosion inhibitors [83, 102, 142] and monitoring the effectiveness of cathodic protections [120]. In addition, CMAS probes have been used to measure the propagation rate of metals in crevices [108, 137].

Chapter 24 describes in details the application of the CMAS probes in real-time monitoring of cathodic protection in simulated seawater, drinking water, soil, and concrete.

8.16 Limitations of multielectrode systems

All electrochemical techniques have limitations. Coupled electrode systems, depending on their operating principles, have their own limitations. Corrosion engineers need to understand these limitations in order to use them effectively for research or monitoring. As previously discussed, an obvious limitation of the CMAS probe is a possible underestimation of actual dissolution rates because of concurrent cathodic reactions occurring on the most anodic electrode of the array. The corrosion rate from an unpolarized CMAS probe may be lower than the actual corrosion rate and is the lower bound of the corrosion rate. The proposed method to overcome this limitation is to slightly polarize the probe to or near the potential of the most cathodic electrode. This polarization method gives the upper bound of the corrosion rate. Although the tests with selected systems showed that the lower bound values were close to the upper bound values (factors of 2.2 to 2.7), the lower and upper bound values may be significantly different for other systems. An alternative approach is to use small electrodes, because the smaller the electrode, the greater the probability for anodic reactions to completely cover the most anodic electrode surface. This, however, may not be practical for all situations. In addition, if the electrode is too small, it may no longer represent the corrosion behavior of a large metal. The electrode size effect should be determined before any attempt is made to use extremely small electrodes.

A second limitation, especially for purposes of research in localized corrosion phenomena, is the crevice between each electrode and the surrounding insulator. Although different methods and different types of epoxy or coatings have been used to reduce or eliminate the formation of crevices, the fabrication of a crevice-free multielectrode probe for certain environments, especially under elevated temperature conditions, has been a challenge. For a less corrosion-resistant metal in a corrosive environment, such as carbon steel in seawater, this may not be an issue because corrosion takes place mostly on the boldly exposed metal surfaces. If a crevice forms, the area of corrosion is less defined and the calculation of corrosion rates may be difficult.

Finally, because the electrode areas are small and low levels of current are measured, precautions must be taken to minimize noise from thermal junctions, dissimilar electrode contacts, and other noise sources. Care must also be taken in the electrode preparation to ensure that the metallurgical state of the actual material of interest is adequately represented by the electrodes. Often, wires of an alloy do not have the same microstructure as other wrought forms. Special machining techniques have been used to fabricate the electrode from exactly the same kind of wrought forms for the probes, but the cost for such fabrication is high.

8.17 Summary

Coupled multielectrode systems have been used for qualitative corrosion detection in concrete for nearly two decades and for crevice corrosion detection for more than 25 years. Coupled multielectrode arrays with spatially patterned electrodes have been used for electrochemical studies for at least 22 years. The coupled multielectrode arrays that can be arranged in any given patterns have been used by many researchers for studying the spatial patterns and the electrochemical behaviors of the corrosion processes, especially the localized corrosion processes of metals.

With the advancement of the coupled multielectrode arrays and multichannel instrumentation, real-time CMAS probes have been developed. These probes give simple parameters, such as maximum localized corrosion rate, maximum localized corrosion penetration depth, and estimation for average corrosion rate. The CMAS probes have been extensively used for online and real-time corrosion monitoring in laboratories and industrial fields. Because CMAS probes do not require the presence of bulk electrolytes, they have been used not only in aqueous solutions, but also in wet gases, oil/water mixtures, salt deposits, biodeposits, soil, concrete, and undercoatings. CMAS probes also have been used for real-time monitoring of the effectiveness of cathodic protection in cathodically protected systems. In addition to the real-time measurement of the quantitative rate of localized corrosion, such as pitting and crevice corrosion, CMAS probes have also been used to estimate the general corrosion rate based on the average corrosion rate. With proper design in seals and access fittings, CMAS probes also have been used for corrosion monitoring at high-pressure and elevated temperatures. Another important feature of CMAS probes is that the life of these probes is extremely long because there is no limit on the length of the sensing electrodes embedded in the insulators. The probes can be reused after the sensing surface is repolished or after the severely corroded sensing tip is cut off.

At the present time, the costs of the multichannel high-resolution multielectrode instruments are rather high. As the demand for the multielectrode systems increases and the instruments as well as the probes can be mass produced, the costs of the multielectrode systems will decrease and the multielectrode systems will play an important role in corrosion monitoring.

References

[1] T. Shibata, T. Takeyama, Stochastic theory of pitting corrosion, Corrosion 33 (12) (1977) 243–251.

[2] Y.-J. Tan, X. Xu, Evaluation of corrosion protective oil using poly working electrode system, Mater. Protect. (Chin.) 20 (2) (1987) 38–40.

[3] H.S. White, G.P. Kittlesen, M.S. Wrighton, Chemical derivatization of an array of three gold microelectrodes with polyrrote: fabrication of a molecule-based transistor, J. Am. Chem. Soc. 106 (1984) 5375–5377.

[4] G.P. Kittlesen, H.S. White, M.S. Wrighton, Chemical derivatization of a microelectrode arrays by oxidation of polyrrote and N-methylpyrrole: fabrication of molecule-based electronic device, J. Am. Chem. Soc. 106 (1984) 7389–7396.

[5] E.W. Paul, A.J. Rico, M.S. Wrighton, Resistance of polyaniline film as a function of electrochemical potential and the fabrication of polyaniline-based microelectronic devices, J. Phys. Chem. 89 (1985) 1441–1447.

[6] S.S. Wang, Microelectrode arrays for lubrication studies, J. Electrochem. Soc. 136 (3) (1989) 713–715.

[7] Y.J. Tan, The effect of inhomogeneity in organic coatings on electrochemical measurements using a wire beam electrode, part 1, Prog. Org. Coat. 19 (1991) 89–94.

[8] Y.J. Tan, S.T. Yu, The effect of inhomogeneity in organic coatings on electrochemical measurements using a wire beam electrode, part 2, Prog. Org. Coat. 19 (1991) 257–263.

[9] Y.J. Tan, A new crevice corrosion testing method and its use in the investigation of oil stain, Corrosion 50 (4) (1994) 266–269.

[10] Q. Zhong, Electrochemical technique for investigating temporarily protective oil coatings, Prog. Org. Coat. 30 (4) (1997) 213–218.

[11] Q. Zhong, Wire beam electrode: a new tool for investigating electrochemical inheterogeneity of oil coating, Prog. Org. Coat. 30 (4) (1997) 279–285.

[12] Z. Qingdong, A novel electrochemical testing method and its use in the investigation of underfilm corrosion of temporary protective oil coating, Corrosion 56 (7) (2000) 722–726.

[13] Q. Zhong, Potential variation of a temporarily protective oil coating before its degradation, Corros. Sci. 43 (2) (2001) 317–324.

[14] Q. Zhong, Study of corrosion behaviour of mild steel and copper in thin film salt solution using the wire beam electrode, Corros. Sci. 44 (5) (2002) 909–916.
[15] Q. Zhong, A novel electrochemical testing method and its use in the investigation of the self-repairing ability of temporarily protective oil coating, Corros. Sci. 44 (6) (2002) 1247–1256.
[16] Q. Zhong, Z. Zhao, Study of anti-contamination performance of temporarily protective oil coatings using wire beam electrode, Corros. Sci. 44 (12) (2002) 2777–2787.
[17] P. Schiessl, Corrosion Measuring Cell, United States Patent 5,015,355, 1991.
[18] P. Schiessl, Corrosion Measuring Cell, Federal Republic of Germany Patent Application 3,834,628, 1988.
[19] P. Schiessl, M. Raupach, Monitoring system for the corrosion risk for steel in concrete, Concr. Int. 7 (1992) 52–55.
[20] P. Schiessl, M. Raupach, New approach for monitoring of the concrete risk for the reinforcement-installation of sensors, in: Proceedings, International Concrete: Across Borders, Odense, Denmark, 1994, pp. 65–78.
[21] R. Bassler, J. Mietz, M. Raupach, O. Klinghoffer, Corrosion risk and humidity sensors for durability assessment of reinforced concrete structures, in: Proc. EUROCORR 2000, London, 2000. article 100805.
[22] R. Bassler, J. Mietz, M. Raupach, O. Klinghoffer, Corrosion monitoring sensors for durability assessment of concrete structures, in: Proc. SPIEs 7th International Symposium on Smart Structures and Materials, Newport Beach, 2000. article 398806.
[23] O. Klinghoffer, P. Goltermann, R. Bassler, Embeddable sensors for use in the integrated monitoring systems of concrete structures, in: First International Conference on Bridge Maintenance, Safety and Management IABMAS 2002, Barcelona, 14–17 July, 2002.
[24] U. Steinsmo, The effect of temperature on propagation of crevice corrosion of high alloyed stainless steel in natural seawater, in: J.M. Costa, A.D. Mercer (Eds.), Progress in the Understanding and Prevention of Corrosion, 10th Eur. Corros. Congr. Paper, Vol. 2, Institute of Materials, London, UK, 1993, pp. 974–982.
[25] U. Steinsmo, T. Rone, J.M. Drugli, Aspects of Testing and Selecting Stainless Steels for Sea Water Applications, CORROSION/94, Paper No. 492, NACE International, Houston, TX, 1994.
[26] U. Steinsmo, T. Rogne, J. Drugli, Aspect of testing and selecting stainless steels for seawater applications, Corrosion 53 (12) (1997) 955–964.
[27] U. Steinsmo, Test method for localized corrosion of stainless steel in sea water, Pitture Vernici Eur. 74 (14) (1998) 15–18.
[28] Z. Fei, R.G. Kelly, J.L. Hudson, Spatiotemporal patterns on electrode arrays, J. Phys. Chem. 100 (1996) 18986–18991.
[29] Z. Fei, Spatiotemporal Behavior of Iron and Sulfuric Acid Electrochemical Reaction System, PhD Thesis, University of Virginia, Charlottesville, Virginia, Diss. Abstr. Int., B 1997, 58, 3, 1402, 1997.
[30] T.T. Lunt, V. Brusamarello, J.R. Scully, J.L. Hudson, Interactions among localized corrosion sites investigated with electrode arrays, Proceedings—Electrochemical Society, 99–27, in: Passivity and Localized Corrosion, 1999, pp. 414–424.
[31] T.T. Lunt, V. Brusamarello, J.R. Scully, J.L. Hudson, Interactions among localized corrosion sites investigated with electrode arrays, Electrochem. Solid-State Lett. 3 (6) (2000) 271–274.
[32] T.T. Lunt, J.R. Scully, V. Brusamarello, A.S. Mikhailov, J.L. Hudson, Spatial interactions among localized corrosion sites: experiments and modeling, Proceedings—

Electrochemical Society, 2000–25, in: Pits and Pores II: Formation, Properties, and Significance for Advanced Materials, 2001, pp. 115–125.

[33] T.T. Lunt, J.R. Scully, V. Brusamarello, A.S. Mikhailov, J.L. Hudson, Spatial interactions among localized corrosion sites: experiments and modeling, J. Electrochem. Soc. 149 (5) (2002) B163–B173.

[34] N.D. Budiansky, J.L. Hudson, J.R. Scully, Origins of persistent interaction among localized corrosion sites on stainless steel, J. Electrochem. Soc. 151 (4) (2004) B233–B243.

[35] N.D. Budiansky, F. Bocher, H. Cong, M.F. Hurley, J.R. Scully, Use of coupled multi-electrode arrays to advance the understanding of selected corrosion phenomena, Corrosion J. 63 (6) (2007) 537–554.

[36] M.F. Hurley, J.R. Scully, Corrosion propagation behavior of new metallic rebar materials in simulated concrete environments and engineering implications, ECS Trans. 3 (13) (2007) 53–70.

[37] H. Cong, N.D. Budiansky, H. Michels, J. Scully, Use of coupled electrode arrays to elucidate copper pitting as a function of potable water chemistry, ECS Trans. 3 (31) (2007) 531–544.

[38] H. Cong, F. Bocher, N.D. Budiansky, M.F. Hurley, J.R. Scully, Use of coupled multi-electrode arrays to advance the understanding of selected corrosion phenomena, J. ASTM Int. 4 (10) (2007).

[39] F. Bocher, F. Presuel-Moreno, N.D. Budiansky, J.R. Scully, Coupled multi-electrode investigation of crevice corrosion of 316 stainless steel, Electrochem. Solid-State Lett. 10 (3) (2007) C16–C20.

[40] F. Bocher, F. Presuel-Moreno, N.D. Budiansky, J.R. Scully, Investigation of crevice corrosion of AISI 316 stainless steel compared to Ni–Cr–Mo alloys using coupled multi-electrode arrays, J. Electrochem. Soc. 155 (5) (2008) C256–C268.

[41] H. Cong, J.R. Scully, The use of coupled multi-electrode arrays to elucidate the pH dependence of copper corrosion as a function of water chemistry, J. Electrochem. Soc. 157 (1) (2010).

[42] F. Bocher, J.R. Scully, Stifling of crevice corrosion and repassivation: cathode area versus controlled potential decreases assessed with a coupled multi-electrode array, Corrosion 71 (9) (2015) 1049–1063.

[43] D. Battocchi, J. He, G.P. Bierwagen, D.E. Tallman, Emulation and study of the corrosion behavior of Al Alloy 2024-T3 using a wire beam electrode (WBE) in conjunction with scanning vibrating electrode technique (SVET), Corros. Sci. 47 (2005) 1165–1176.

[44] N. Missert, R.G. Copeland, F.D. Wall, C.M. Johnson, J.C. Barbour, P. Kotula, Aluminum corrosion at systems of engineered copper particles, in: Corrosion Science, vol. 2002–13, Proceedings–Electrochemical Society, 2002, pp. 307–313.

[45] F.D. Wall, M.A. Martinez, C.M. Johnson, J.C. Barbour, N. Missert, R.G. Copeland, Does anything pin the pitting behavior of aluminum? in: Proceedings–Electrochemical Society, vol. 2003–23, 2004, pp. 1–11.

[46] F.D. Wall, M.A. Martinez, Using microelectrodes to determine the availability and behavior of pit initiation sites in aluminum, in: Corrosion and Corrosion Prevention of Low Density Metals and Alloys, vol. 2000–23, Proceedings–Electrochemical Society, 2001, pp. 229–238.

[47] F.D. Wall, M.A. Martinez, A statistics-based approach to studying aluminum pit initiation—intrinsic and defect-driven pit initiation phenomena, J. Electrochem. Soc. 150 (4) (2003) B146–B157.

[48] F.D. Wall, C.M. Johnson, J.C. Barbour, M.A. Martinez, The effects of chloride implantation on pit initiation in aluminum, J. Electrochem. Soc. 151 (2) (2004) B77–B81.

[49] F.D. Wall, M.A. Martinez, J.J. Van den Avyle, Relationship between induction time for pitting and pitting potential for high-purity aluminum, J. Electrochem. Soc. 151 (6) (2004) B354–B358.
[50] F.D. Wall, Applications of Multi-electrode Techniques to Aqueous and Atmospheric Corrosion Testing, CORROSION/06, paper no. 06672, NACE, Houston, TX, 2006.
[51] K.R. Cooper, M. Smith, J.R. Scully, N.D. Budiansky, Development of a Multielectrode Array Impedance Analyzer for Corrosion Science and Sensors, CORROSION/06, paper no. 06674, NACE, Houston, TX, 2006.
[52] W. Zhang, B. Hurley, R.G. Buchheit, Characterization of chromate conversion coating formation and breakdown using electrode arrays, J. Electrochem. Soc. 149 (8) (2002) B357–B365.
[53] F.D. Wall, M.A. Martinez, The effect of electrode thickness and inter-electrode spacing on electrochemical signals at low humidity, in: Abs. 480, 204th Meeting, The Electrochemical Society, Inc, 2004.
[54] A.D. King, J.S. Lee, J.R. Scully, Galvanic couple current and potential distribution between a Mg electrode and 2024-T351 under droplets analyzed by microelectrode arrays, J. Electrochem. Soc. 162 (1) (2015) C12–C23.
[55] A.D. King, J.S. Lee, J.R. Scully, Finite element analysis of the galvanic couple current and potential distribution between Mg and 2024-T351 in a Mg rich primer configuration, J. Electrochem. Soc. 163 (7) (2016) C342–C356.
[56] E. Schindelholz, L.-k. Tsui, R.G. Kelly, Hygroscopic particle behavior studied by interdigitated array microelectrode impedance sensors, J. Phys. Chem. A 118 (1) (2013) 167–177.
[57] L.G. Bland, B.C. Rincon Troconis, R.J. Santucci Jr., J.M. Fitz-Gerald, J.R. Scully, Metallurgical and electrochemical characterization of the corrosion of a Mg-Al-Zn Alloy AZ31B-H24 tungsten inert gas weld: galvanic corrosion between weld zones, Corrosion 72 (10) (2016) 1226–1242.
[58] Y.J. Tan, Studying non-uniform electrodeposition using the wire beam electrode method, Int. J. Modern Phys. B (World Sci.) 16 (2002) 144–150. No. 1&2—special issue for proceedings of International Conference on Materials for Advanced Technologies, Singapore (1–6 July 2001).
[59] Y.J. Tan, K.Y. Lim, Characterising nonuniform electrodeposition and electrodissolution using the novel wire beam electrode method, J. Appl. Electrochem. 34 (2004) 1093–1101.
[60] G. Hinds, A. Turnbull, Novel multi-electrode test method for evaluating inhibition of underdeposit corrosion. Part 1: sweet conditions, Corrosion 66 (4) (2010), 046001.
[61] G. Hinds, A. Turnbull, Novel multi-electrode test method for evaluating inhibition of underdeposit corrosion—part 2: sour conditions, Corrosion 66 (5) (2010) 056002.
[62] Y.J. Tan, Wire beam electrode: a new tool for localized corrosion studies, in: Proceedings of Australasian Corrosion Association, Corrosion & Prevention 97, Australasian Corrosion Association, Australia, 1997. Paper no. 52.
[63] Y.J. Tan, Monitoring localized corrosion processes and estimating localized corrosion rates using a wire-beam electrode, Corrosion 54 (5) (1998) 403–413.
[64] H. Eren, A. Lowe, Y.J. Tan, S. Bailey, B. Kinsella, An auto-switch for multi-sampling of a wire beam electrode corrosion monitoring system, IEEE Trans. Instrum. Meas. 47 (1998) 1096–2001.
[65] Y.J. Tan, Wire beam electrode: a new tool for studying localized corrosion and other inheterogeneity electrochemical processes, Corros. Sci. 41 (2) (1999) 229–247.

[66] Y.J. Tan, S. Bailey, B. Kinsella, A. Lowe, Mapping corrosion kinetics using the wire beam electrode in conjunction with electrochemical noise resistance measurements, J. Electrochem. Soc. 147 (2) (2000) 530–540.
[67] Y.J. Tan, Method and Apparatus for Measuring Localized Corrosion and Other Heterogeneous Electrochemical Processes, United States of America Patent No. 6132593, 2000, p. 24.
[68] Y.J. Tan, S. Bailey, B. Kinsella, Mapping non-uniform corrosion in practical corrosive environments using the wire beam electrode method (I)—multi-phase corrosion, Corros. Sci. 43 (2001) 1905–1918.
[69] Y.J. Tan, S. Bailey, B. Kinsella, Mapping non-uniform corrosion in practical corrosive environments using the wire beam electrode method (II)—crevice corrosion, Corros. Sci. 43 (2001) 1919–1929.
[70] Y.J. Tan, S. Bailey, B. Kinsella, Mapping non-uniform corrosion in practical corrosive environments using the wire beam electrode method (III)—water-line corrosion, Corros. Sci. 43 (2001) 1930–1937.
[71] Y.J. Tan, Measuring localized corrosion using the wire beam electrode method, in: Corrosion Science: A Retrospective and Current Status, The Electrochemical Society 201st Meeting, Philadelphia, USA, Special Electrochemical Society publication, 2002, pp. 377–384.
[72] N.N. Aung, Y.J. Tan, A new method of studying buried steel corrosion and its inhibition using the wire beam electrode, Corros. Sci. 46 (2004) 3057–3067.
[73] Y.J. Tan, An experimental comparison of three wire beam electrode based methods for determining corrosion rates and patterns, Corros. Sci. 47 (2005) 1653–1665.
[74] Y. Tan, Sensing electrode inhomogeneity and electrochemical heterogeneity using an electrochemically integrated multielectrode array, J. Electrochem. Soc. 156 (6) (2009) C195–C208.
[75] Y. Tan, Sensing electrode inhomogeneity and electrochemical heterogeneity using an electrochemically integrated multi-electrode array, J. Electrochem. Soc. 156 (2009) C195–C208.
[76] F. Mahdavi, M. Forsyth, M.Y.J. Tan, Understanding the effects of applied cathodic protection potential and environmental conditions on the rate of cathodic disbondment of coatings by means of local electrochemical measurements on a multi-electrode array, Prog. Org. Coat. 103 (2017) 83–92.
[77] ASTM G217, Standard Guide for Corrosion Monitoring in Laboratories and Plants With Coupled Multielectrode Array Sensor Method, ASTM, West Conshohocken, PA, 2017.
[78] L. Yang, A.A. Yang, On zero-resistance ammeter and zero-voltage ammeter, J. Electrochem. Soc. 164 (2017) C819–C821.
[79] L. Yang, Effect of voltage between electrodes of a coupled multielectrode array sensor on corrosion rate measurement, in: CORROSION/2016 Conference Proceedings, Paper No. 7911, NACE, Houston, TX, 2016.
[80] K.T. Chiang, L. Yang, A review of recent patents on coupled multielectrode array sensors for localized corrosion monitoring, Rec. Pat. Corrosion Sci. 1 (2011) 108–117.
[81] L. Yang, N. Sridhar, O. Pensado, Development of a multielectrode array sensor for monitoring localized corrosion, in: Presented at the 199th Meeting of the Electrochemical Society, Abstract #182, Extended Abstract Volume I, 2001.
[82] L. Yang, N. Sridhar, O. Pensado, D. Dunn, An in-situ galvanically coupled multielectrode array sensor for localized corrosion, Corrosion 58 (2002) 1004.
[83] L. Yang, D. Dunn, Evaluation of Corrosion Inhibitors in Cooling Water Systems Using a Coupled Multielectrode Array Sensor, Corrosion/2002, paper no. 004, NACE International, Houston, TX, 2002.

[84] L. Yang, N. Sridhar, G. Cragnolino, Comparison of Localized Corrosion of Fe-Ni-Cr-Mo Alloys in Concentrated Brine Solutions Using a Coupled Multielectrode Array Sensor, NACE CORROSION/2002, paper no. 545, NACE International, Houston, TX, 2002.
[85] C.S. Brossia, L. Yang, D.S. Dunn, N. Sridhar, Corrosion sensing and monitoring, in: Proceedings of the Tri-Service Corrosion Conference, Jan. 14–18, San Antonio, TX, USA, 2002.
[86] L. Yang, N. Sridhar, Monitoring of localized corrosion, in: S.D. Crammer, B.S. Covino Jr. (Eds.), ASM Handbook, Volume 13A—Corrosion: Fundamentals, Testing, and Protection, ASM International, Materials Park, OH, 2003, pp. 519–524.
[87] L. Yang, N. Sridhar, Coupled multielectrode online corrosion sensor, Mater. Perform. 42 (9) (2003) 48–52.
[88] L. Yang, R.T. Pabalan, L. Browning, G.A. Cragnolino, Measurement of Corrosion in Saturated Solutions Under Salt Deposits Using Coupled Multielectrode Array Sensors, CORROSION/2003, paper no. 03426, NACE International, Houston, TX, 2003.
[89] L. Yang, R.T. Pabalan, L. Browning, D.S. Dunn, Corrosion behavior of carbon steel and stainless steel materials under salt deposits in simulated dry repository environments, in: R.J. Finch, D.B. Bullen (Eds.), Scientific Basis for Nuclear Waste Management XXVI, Vol. 757, Materials Research Society, Warrendale, PA, 2003, pp. 791–797. M. R. S. Symposium Proceedings.
[90] C.S. Brossia, L. Yang, Studies of Microbiologically Influenced Corrosion Using a Coupled Multielectrode Array Sensor, CORROSION/2003, paper no. 03575, NACE International, Houston, TX, 2003.
[91] A. Anderko, N. Sridhar, C.S. Brossia, D.S. Dunn, L. Yang, B.J. Saldanha, S.L. Grise, M. H. Dorsey, An Electrochemical Approach to Predicting and Monitoring Localized Corrosion in Chemical Process Streams, CORROSION/2003, paper no. 03375, NACE International, Houston TX, 2003.
[92] V. Jain, S. Brossia, D. Dunn, L. Yang, Development of sensors for waste package testing and monitoring in the long term repository environments, Ceram. Trans. 143 (2003) 283–290.
[93] L. Yang, R.T. Pabalan, D.S. Dunn, The study of atmospheric corrosion of carbon steel and aluminum under salt deposit using coupled multielectrode array sensors, in: Presented at the 204th Meeting of the Electrochemical Society, Abstract #465, Extended Abstract Volume 2003-II, 2003.
[94] L. Yang, N. Sridhar, Sensor Array for Electrochemical Corrosion Monitoring, US patent 6,683,463, 2004.
[95] L. Yang, N. Sridhar, S.L. Grise, B.J. Saldanha, M.H. Dorsey, H.J. Shore, A. Smith, Real-Time Corrosion Monitoring in a Process Stream of a Chemical Plant Using Coupled Multielectrode Array Sensors, CORROSION/2004, paper no. 04440, NACE International, Houston, TX, 2004.
[96] L. Yang, N. Sridhar, D.S. Dunn, C.S. Brossia, Laboratory Comparison of Coupled Multielectrode Array Sensors With Electrochemical Noise Sensors for Real-Time Corrosion Monitoring, CORROSION/2004, paper no. 04429, NACE International, Houston, TX, 2004.
[97] L. Yang, G.A. Cragnolino, Studies on The Corrosion Behavior of Stainless Steels in Chloride Solutions in the Presence of Sulfate Reducing Bacteria, CORROSION/2004, paper no. 04598, NACE International, Houston, TX, 2004.
[98] M.H. Dorsey, L. Yang, N. Sridhar, Cooling Water Monitoring Using Coupled Multielectrode Array Sensors and Other On-Line Tools, CORROSION/2004, paper no. 04077, NACE International, Houston, TX, 2004.

[99] C.R. Southwell, B.W. Forgeson, A.L. Alexander, Corrosion of metals in tropical environments, Corrosion 16 (10) (1960) 512t–518t.
[100] C.R. Southwell, C.W. Hummer Jr., A.L. Alexander, Corrosion of Metals in Tropical Environments, Part 7—Copper and Copper Alloys—Sixteen Years' Exposure, NRL Report 6452, Naval Research Laboratory, Washington, DC, 1966, p. 4.
[101] T. Ohtsu, M. MiyazawaM, Application of Analysis and Control for Corrosion Damage in Cooling Water Systems Using Corrosion Monitoring, CORROSION/2009, paper no. 09439, NACE International, Houston, TX, 2009.
[102] V.V. Nikam, et al., Evaluation of CMAS Probe in Field Simulated MIC Conditions, CORROSION/2014, paper no. 4083, NACE International, Houston, TX, 2014.
[103] L. Yang, N. Sridhar, C.S. Brossia, D.S. Dunn, Evaluation of the coupled multielectrode array sensor as a real time corrosion monitor, Corros. Sci. 47 (2005) 1794–1809.
[104] A. Anderko, N. Sridhar, L. Yang, S.L. Grise, B.J. Saldanha, M.H. Dorsey, Validation of a Localized Corrosion Model Using Real-Time Corrosion Monitoring in a Chemical Plant, Corros. Eng. Sci. Technol. 40 (2005) 33–42 (formerly British Corrosion J.).
[105] L. Yang, D.S. Dunn, G. Cragnolino, An Improved Method for Real-Time and Online Corrosion Monitoring Using Coupled Multielectrode Array Sensors, CORROSION/2005, paper no. 05379, NACE International, Houston, TX, 2005.
[106] L. Yang, D.S. Dunn, Y.-M. Pan, N. Sridhar, Real-time Monitoring of Carbon Steel Corrosion in Crude Oil and Brine Mixtures Using Coupled Multielectrode Sensors, CORROSION/2005, paper no. 05293, NACE International, Houston, TX, 2005.
[107] M.H. Dorsey, D.R. Demarco, B.J. Saldanha, G.A. Fisher, L. Yang, N. Sridhar, Laboratory Evaluation of a Multi-Array Sensor for Detection of Underdeposit Corrosion and/or Microbially Influenced Corrosion, CORROSION/2005, paper no. 05371, NACE International, Houston, TX, 2005.
[108] X. Sun, L. Yang, Real-Time Measurement of Crevice Corrosion With Coupled Multielectrode Array Sensors, CORROSION/2006, paper no. 06679, NACE International, Houston, TX, 2006.
[109] X. Sun, L. Yang, Real-Time Monitoring of Localized and General Corrosion Rates in Simulated Marine Environments Using Coupled Multielectrode Array Sensors, CORROSION/2006, paper no. 06284, NACE International, Houston, TX, 2006.
[110] L. Yang, K.T. Chiang, P.K. Shukla, N. Shiratori, Internal current effects on localized corrosion rate measurements using coupled multielectrode array sensors, Corrosion 66 (2010) 115005.
[111] L. Yang, X. Sun, A Method to Determine the Internal Current Effect on Corrosion Rate Measurements with Coupled Multielectrode Array Sensors, CORROSION/2012, paper no. C2012-0001707, NACE International, Houston, TX, 2012.
[112] F. Mansfeld, The polarization resistance technique for measuring corrosion currents, in: M.G. Fontana, R.W. Staehle (Eds.), Advances in Corrosion Science and Technology, vol. 6, Plenum Publishing, New York, 1976, pp. 185–187.
[113] X. Sun Yang, L. Yang, Electrochemical Probes for Corrosion Monitoring in Hydrogen Sulfide Systems and Methods of Avoiding the Effect of Electron-Conducting Deposits, US Patent, #8,298,390, United States Patent and Trademark Office, 2012.
[114] L. Yang, X. Sun, R. Barnes, Coupled Multielectrode Array Sensors with Solid Electrolyte-Coated Finger-Like Electrodes for Applications in Oil-Water Mixtures and Natural Gas Systems Containing Hydrogen Sulfide, CORROSION/2014, paper no. 4406, NACE International, Houston, TX, 2014.
[115] L. Yang, X. Sun, S. Papavinasam, R.B. Rebak, L. Yang, N.S. Berke (Eds.), Experimental studies on the effect of electrode spacing in coupled multielectrode array sensors on

corrosion rate measurements, in: Advances in Electrochemical Techniques for Corrosion Monitoring and Laboratory Corrosion Measurements, ASTMSTP1609, ASTM International, West Conshohocken, PA, 2019, pp. 180–195, https://doi.org/10.1520/STP1609/20170233.
[116] X. Sun, L. Yang, Real-Time Monitoring of Localized and General Corrosion Rates in Drinking Water Utilizing Coupled Multielectrode Array Sensors, CORROSION/2006, paper no. 06094, NACE International, Houston, TX, 2006.
[117] K.T. Chiang, L. Yang, Monitoring Corrosion Behavior of a Cu-Cr-Nb Alloy by Multielectrode Sensors, CORROSION/2006, paper no. 06676, NACE International, Houston, TX, 2006.
[118] N. Sridhar, L. Yang, F. Song, Application of Multielectrode Array to Study Dewpoint Corrosion in High Pressure Natural Gas Pipeline Environments, CORROSION/2006, paper no. 06673, NACE International, Houston, TX, 2006.
[119] X. Sun, Online Monitoring of Undercoating Corrosions Utilizing Coupled Multielectrode Sensors, CORROSION/2004, paper no. 04033, NACE International, Houston, TX, 2004.
[120] X. Sun, Online Monitoring of Corrosion Under Cathodic Protection Conditions Utilizing Coupled Multielectrode Sensors, CORROSION/2004, paper no. 04094, NACE International, Houston, TX, 2004.
[121] X. Sun, Online and Real-Time Monitoring of Carbon Steel Corrosion in Concrete, Using Coupled Multielectrode Sensors, CORROSION/2005, paper no. 05267, NACE International, Houston, TX, 2005.
[122] X. Sun, Real-Time Corrosion Monitoring in Soil with Coupled Multielectrode Sensors, CORROSION/2005, paper no. 05381, NACE International, Houston, TX, 2005.
[123] X. Sun, Online monitoring of undercoating corrosion using coupled multielectrode sensors, Mater. Perform. 44 (2) (2005) 28–32.
[124] P. Angell, Use of the Multiple-Array-Sensor to Determine the Effect of Environmental Parameters on Microbial Activity and Corrosion Rates, CORROSION/2006, paper no. 06671, NACE International, Houston, TX, 2006.
[125] L. Yang, N. Sridhar, Sensor Array for Electrochemical Corrosion Monitoring, US Patent, #6,987,396, United States Patent and Trademark Office, 2006.
[126] X. Sun Yang, Electronic System for Multielectrode Sensors and Electrochemical Devices, US Patent, #7,180,309, United States Patent and Trademark Office, 2007.
[127] B. Yang, F.J. Marinho, A.V. Gershun, New electrochemical methods for the evaluation of localized corrosion in engine coolants, J. ASTM Int. 4 (1) (2007).
[128] T. Pickthall, V. Morris, H. Gonzalez, Corrosion Monitoring of a Crude Oil Pipeline A Comparison of Multiple Methods, CORROSION/07, paper no. 07340, NACE International, Houston, TX, 2007.
[129] X. Sun, L. Yang, Multielectrode Penetration Sensor for Monitoring Localized and General Corrosion, CORROSION/2007, paper no. 07391, NACE International, Houston, TX, 2006.
[130] X. Yang, L. Yang, Multihole and Multiwire Sensors for Localized and General Corrosion Monitoring, US Patent, #7,508,223, United States Patent and Trademark Office, 2009.
[131] L. Yang, X. Sun, Measurement of Cumulative Localized Corrosion Rate Using Coupled Multielectrode Array Sensors, CORROSION/2007, paper no. 07378, NACE International, Houston, TX, 2007.
[132] K.T. Chiang, L. Yang, Development of Crevice-Free Multielectrode Sensors for Elevated Temperature Applications, CORROSION/2007, paper no. 07376, NACE International, Houston, TX, 2007.

[133] K.T. Chiang, L. Yang, High-temperature electrochemical sensor for online corrosion monitoring, Corrosion 66 (2010) 095002.
[134] D. Duke, L. Yang, Laboratory and Field Studies of Localized and General Corrosion Inhibiting Behaviors of Silica in Zero Liquid Discharge (High TDS Cooling Water) Using Real Time Corrosion Monitoring Techniques, CORROSION/2007, paper no. 07626, NACE International, Houston, TX, 2007.
[135] L. Yang, et al., Threshold chloride levels for localized carbon steel corrosion in simulated concrete pore solutions using coupled multielectrode array sensors, Corrosion 70 (2014). pp. 850-315.
[136] R.T. Pabalan, L. Yang, K.-T. Chiang, Boric acid corrosion of concrete rebar, in: EPJ Web of Conferences, Vol. 56, EDP Sciences, 2013, p. 06005.
[137] R. Colbert, R. Reich, Corrosion Monitoring of a Water Based Rolling Facility With Coupled Multielectrode Array Sensors and the Correlations With Other Process Variables: Conductivity, pH, Temperature, Dissolved Oxygen and Corrosion Potential, CORROSION/2008, paper no. 08295, NACE International, Houston, TX, 2008.
[138] N.M. Alanazi, et al., Corrosion of pipeline steel X-60 under field-collected sludge deposit in a simulated sour environment, Corrosion 71 (2015) 305–315.
[139] A.A. Yang, X. Sun, Vehicle emission testing with corrosion sensors, Mater. Perform. 53 (3) (2014) 2–7.
[140] Z. Guan, M. Du, Z. Cui, Study pitting corrosion of P110 steel by electrochemical frequency modulation technique and coupled multielectrode array sensor, Corrosion 3 (2017) 998–1006.
[141] R. Hefner, J.P. Czapiewski, System and Method for Online Monitoring of Corrosion of Gas Turbine Components, US Patent 8,475,110, 2013.
[142] G. Tormoen, J. Dante, N. Sridhar, Correlation of In-Situ VCI Adsorption Monitoring with Real-Time Corrosion Rate Measurements, CORROSION/2007, paper no. 07356, NACE International, Houston, TX, 2007.

Part Two

Other physical or chemical methods for corrosion monitoring

Gravimetric techniques

9

Kuangtsan Chiang[†] *and Todd Mintz*
Southwest Research Institute, San Antonio, TX, United States

9.1 Introduction

Measuring weight is one of the fundamental techniques of corrosion sciences. Measurements of mass changes (weight loss or weight gain) of test specimens using analytical balances with various capacities and sensitivities have been widely used to evaluate materials degradation as a function of exposure times or temperatures in corrosive environments. In this chapter, two gravimetric techniques for corrosion monitoring will be described: thermogravimetric analysis (TGA) and quartz crystal microbalance (QCM). This chapter will examine the operating principles of TGA and QCM and will review types of corrosion that can be evaluated with the two techniques. Additionally, the various types of TGA and QCM measuring and monitoring systems that are currently available on the market and future trends for applying these techniques will be discussed throughout the text.

9.2 Thermogravimetric analysis technique

The thermogravimetric analysis (TGA) technique requires continuous weighing of a sample in a specific environment, either as a function of temperature or at a constant temperature. This technique is widely used to study oxidation and high-temperature corrosion. Gulbransen and Andrew [1] developed vacuum microbalance to study and analyze the reaction rates of metals and alloys in controlled atmospheres of O_2, N_2, H_2, and H_2O. Such analysis relies on a high degree of precision in four measurements: specimen weight, temperature, rate of temperature change, and exposure time. An automatic recording microbalance [2] with precision of 0.1 micrograms ($\mu g = 10^{-6}$ g) permits long duration runs at high temperatures and has been commercially available since the 1960s. A schematic diagram of an automatic recording microbalance is shown in Fig. 9.1. The specimen is freely suspended from a precision microbalance located outside the furnace chamber. The microbalance mechanism records weight change due to chemical reactions in a controlled environment, along with specimen temperature and elapsed time. The technique has been used to study the kinetics and mechanisms of engineering alloys in high-temperature oxidation [3, 4]. The technique was also used for laboratory study of high-temperature corrosion kinetics of nickel- and cobalt-based turbine alloys [5–10]. The effects

[†] Deceased.

Techniques for Corrosion Monitoring. https://doi.org/10.1016/B978-0-08-103003-5.00009-6

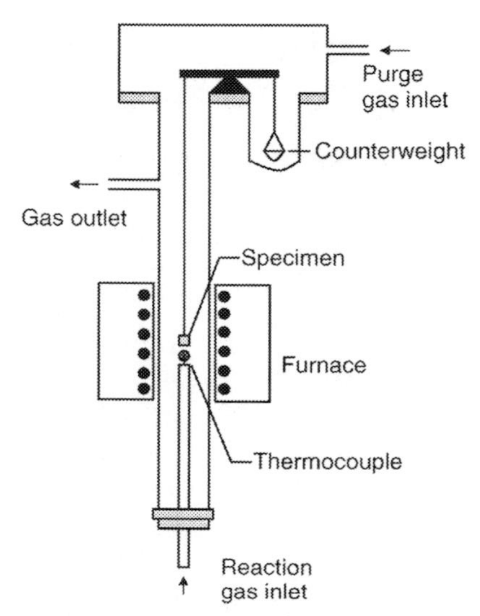

Fig. 9.1 Schematic diagram of a thermogravimetric apparatus with automatic recording microbalance.

of alloy composition, salt composition, gas composition, and temperature on the corrosion rates and mechanisms were reported.

9.2.1 Examples of microbalance applications

Multiple processes can lead to mass changes that would correspond to chemical reactions as a function of temperature. The types of processes that can be studied using microbalance systems include but are not limited to:

- Thermal stability and decomposition
- Surface adsorption/desorption
- Evaporation
- Oxidation rates of metals, alloys, polymers, ceramics, composites, and coatings
- Reaction rates in CO, CO_2, H_2O, N_2, H_2, and mixed gases
- Halogen corrosion
- Salt deposit corrosion.

Examples of microbalance applications are given next.

9.2.1.1 Thermal decomposition of calcium oxalate hydrate

Fig. 9.2 shows the TGA weight change curve for calcium oxalate hydrate ($CaC_2O_4{\cdot}H_2O$) compound when the sample is heated from 20°C to 900°C in a furnace with a heating rate of 10°C min^{-1}. Each downward step in the weight change curve corresponds to a decomposition process for calcium oxalate hydrate compound. The three steps correspond to losing water (H_2O), carbon monoxide (CO), and carbon dioxide

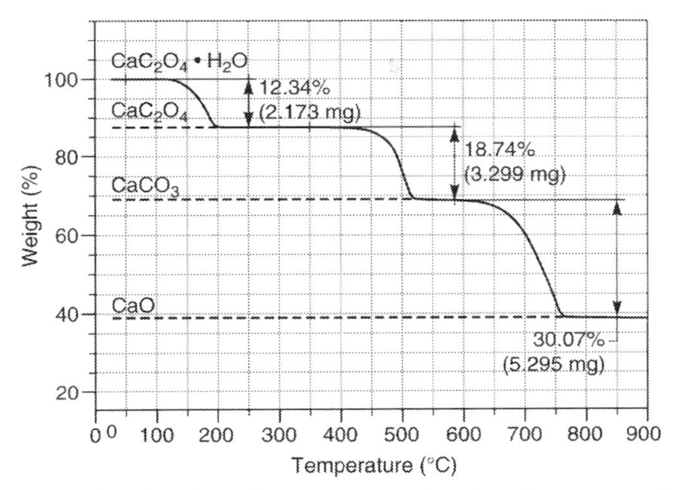

Fig. 9.2 Thermogravimetric weight change curve as a function of temperature for calcium oxalate hydrate.

(CO_2). The specimen weight data were converted to percentage weight with respect to the original weight of calcium oxalate hydrate [11].

9.2.1.2 Oxidation kinetics of nanostructured coatings

TGA technique was used to study the oxidation kinetics of a nanostructured Cu-Cr coating deposited on copper alloy substrate [12]. The nanostructured coating was deposited using an ion-beam-assisted deposition (IBAD) method [13]. The oxidation rate of uncoated and IBAD single-layer Cu-Cr-coated NARloy-Z (Cu-3 wt%Ag–0.5 wt% Zr) at 650°C is shown in Fig. 9.3A as mass gain as a function of exposure time. Both specimens exhibited high oxidation rates during an initial transient period of approximately 6 min, which was reduced to lower rates at longer times. Oxidation rate decreases markedly with the coating. The data are also plotted in Fig. 9.3B as mass gain versus square root of time. The data fall along a straight line, indicating that the oxidation kinetics after the initial transient period obey parabolic rate law:

$$\Delta m/A = \left(K_p t\right)^{1/2} + C \tag{9.1}$$

where

$\Delta m/A$ = mass change per unit area
K_p = parabolic rate constant
t = exposure time
C = a constant

The parabolic rate constant (square of the slope in Fig. 9.3B) for the uncoated NARloy-Z is $4.5 \times 10^3\ \mu g^2\ cm^{-4}\ min^{-1}$. In comparison, the IBAD Cu-Cr-coated

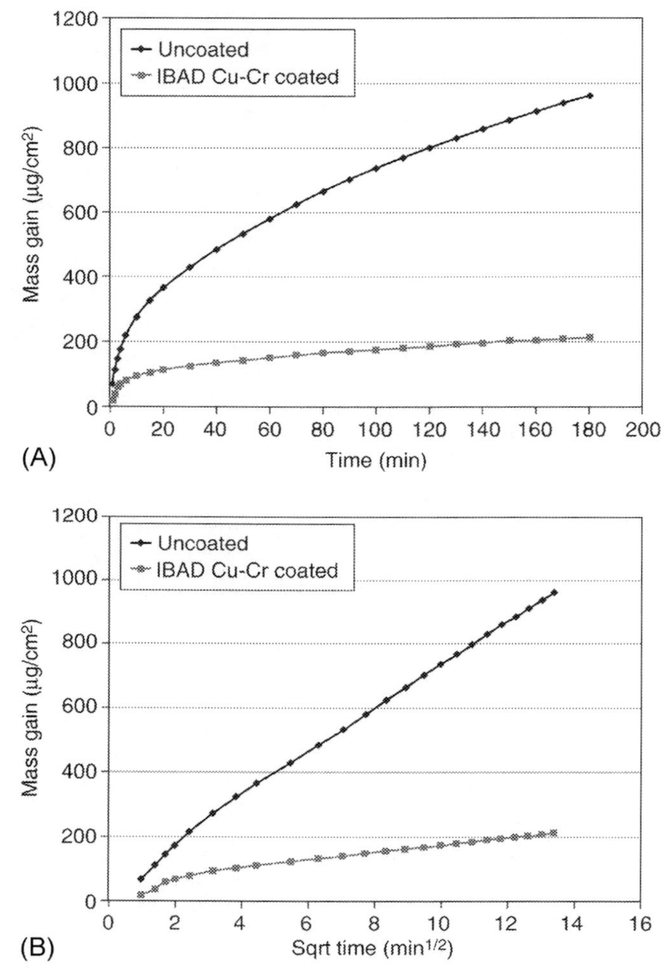

Fig. 9.3 Oxidation kinetics of uncoated and IBAD Cu-Cr-coated NARloy-Z (Cu-3 wt%Ag–0.5 wt%Zr) in air at 650°C. (A) Mass gain versus time plot and (B) mass gain versus square root of time plot [12]. (IBAD: ion-beam assisted deposition).

specimen has a parabolic rate constant of 1.3×10^2 μg^2 cm^{-4} min^{-1}, which is about 35 times smaller than the parabolic rate constant obtained for the uncoated specimen. The oxidation rate of NARloy-Z is controlled by outward diffusion of Cu ions to form external Cu oxide. In contrast, oxidation of the Cu-Cr coating surface is controlled by outward diffusion of chromium ions to form protective Cr_2O_3 scales.

Most commercial TGA systems are supplied with data acquisition and analysis software to display the test in progress on a monitor, store the data, and perform analyses of the data.

9.3 QCM technique

The QCM was initially used to measure mass changes during thin film deposition in a vacuum system and to study adsorption processes [14–16]. This technique can measure mass changes on the order of nanograms (10^{-9} g) [17], and with some new equipment, may be able to measure picogram (10^{-12} g) changes. Within the last 45 years, the QCM has become a highly useful instrument when in situ detection and monitoring of phenomena occurring on the surface of materials at the microscopic level are required [18].

9.3.1 Principle of QCM

The fundamental properties of quartz have been studied for over a century. QCM is based on the inverse piezoelectric effect, in which a shift in the resonance frequency of the quartz crystal is used for measuring shifts in mass of the crystal and any surface film. QCM consists of a thin plate of single-crystal quartz with electrodes affixed to each side of the plate. Synthetic quartz is composed of silicon and oxygen in the form of silicon dioxide (SiO_2) and is normally made in autoclaves under high temperature and pressure. These quartz crystals exhibit piezoelectric properties, which means that they can generate an electric potential when pressure is applied to the crystal surfaces. If a current-measuring circuit is attached to the quartz crystal and compressive stresses are applied, current will begin to flow through the circuit in one direction. If tensile stresses are applied to the same quartz crystal, the current will flow in the reverse direction. Conversely, when an electric potential is applied to the surface of the crystal, mechanical deformation or vibrations are generated [19].

The main technology behind QCM is the piezoelectric properties of the quartz crystal sandwiched between two electrodes. In 1880, the Curie brothers discovered the piezoelectric effect on a quartz crystal [19]. Their first test included placing a weight on the surface of a quartz crystal while measuring the electrical charge that appeared on the surface. It was discovered later that if an alternating current (AC) is applied to the electrodes, the quartz crystal will start to oscillate at its resonance frequency due to the piezoelectric properties. The resonance frequency of the crystal will be a function of its physical properties and environment. The theoretical resonance frequency for a bar can be determined by Eq. (9.2) [19].

$$f_R = \frac{1}{2l}\sqrt{\frac{Y_o}{\rho}} \tag{9.2}$$

where

f_R = resonance frequency
l = length of the bar
Y_o = Young's modulus along the bar
ρ = density

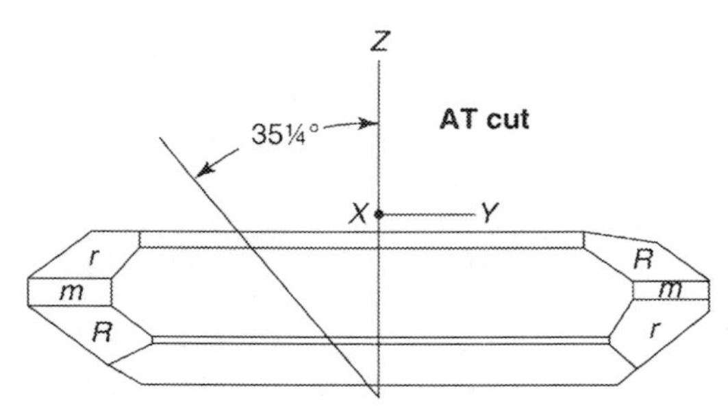

Fig. 9.4 Quartz crystal AT cut.
Modified from R.A. Heising, Quartz Crystals for Electrical Circuits, D. Van Nostrand Company, New York, 1946.

Some factors affecting the frequency can be controlled, such as the density and viscosity of the phases adjacent to either side of the quartz crystal, the pressure differences across the crystal, and the temperature. Other factors that affect the frequency that are not necessarily controllable are the mass of the attached electrode, an absorbed film or a thin deposited film [20].

The preparation of the quartz crystal will affect the crystal device's properties. The crystal's oscillation direction will depend upon the orientation of the crystal lattice. Quartz is typically a six-sided prism that ends in a six-sided pyramid, as seen in Fig. 9.4. Because of the quartz crystal structure, various types of crystal cuts can be made that will affect the piezoelectric properties of the crystal. One of the more important properties is the frequency stability: the amount that the frequency deviates from the ambient temperature frequency over the operating temperature range. As shown in Fig. 9.4, a quartz crystal that is cut so that the plate contains the X-axis and makes an angle of about 35 degrees with the optic or Z-axis is called an AT cut [21]. Because of the frequency stability, this is one of the most common quartz crystal cuts, with fundamental resonance frequencies predominately in the range of 5–30 MHz [22]. An AT cut will lead to mechanical shear oscillations and displacements that are parallel to the wafer surface. Other types of cuts are available, but the AT cut is most often used for QCM. The frequency stability and the operating temperatures required by the type of testing will determine the necessary angle of the cut. Generally, the oscillations of quartz crystals are very stable due to the high quality of the oscillation (Q-factor). The ability to measure the resonant frequency for particular quartz crystals can be so accurate that it is possible to measure the adsorption of a monolayer of hydrogen in a vacuum system.

The QCM system utilizes the inverse piezoelectric effect to measure the changes in mass with time. The quartz crystal is coated by the material of interest, which is considered the working electrode. The material used for the working electrode will depend upon the type of test being performed. Any change in the mass that is rigidly attached to the working electrode will result in a change of the quartz crystal's resonance frequency. The mass change can be measured by comparing the frequency of the quartz/working electrode to the frequency of a standard reference quartz crystal and computing a differential frequency [23]. The mass change is related to the differential frequency by the Sauerbrey equation, shown in Eq. (9.3).

$$\Delta f = \frac{-2\Delta m f_o^2}{A\rho_q v_q} \tag{9.3}$$

where

Δf = change in oscillation frequency of a piezoelectric quartz crystal
Δm = change in mass of quartz crystal/working electrode system
f_o = resonant frequency of the crystal
A = active area of the crystal
ρ_q = density of quartz
v_q = shear wave velocity in the quartz

Sometimes the Sauerbrey equation is written as Eq. (9.4).

$$\Delta f = -C_f \cdot \Delta m \tag{9.4}$$

where

$$C_f = \frac{2f_o^2}{A\rho_q v_q} = \text{Sensitivity factor for the quartz crystal}$$

In this equation, the constant C_f (sensitivity factor) depends only upon the thickness of the quartz slab and on the intrinsic properties of the quartz. C_f will increase proportionally with the overtone number and depends upon the square of the fundamental frequency. For an AT-cut 10 MHz quartz crystal, C_f equals 0.444 cm^2 Hz ng^{-1} if changes of mass on each face of the quartz crystal are equal [24].

Limitations to the Sauerbrey equation include high dissipation (i.e., damping) that affects the accuracy of the calculation when the mass that is added to the electrode surface (a) is not rigidly deposited, (b) slips on the surface, or (c) is not deposited evenly on the surface in the QCM system being evaluated. The damping effect can be measured and accounted for with more elaborate QCM testing equipment, which will be described in Section 9.3.3.

The Sauerbrey equation is also only useful for thin solid films, where the frequency shift is less than 1% of the initial quartz resonator frequency [25]. To deal with this limitation, Miller and Bolef [26] developed the composite acoustic resonator model to account for thicker films. Miller and Bolef developed an equation that maintains Sauerbrey-like results while treating the quartz resonator and deposited film as a composite resonator. Lu and Lewis [27] further developed this technique for a more accurate equation.

An additional limitation to the Sauerbrey equation is that it assumes that a rigid mass is attached to the crystal in air and not in a liquid environment. When using the QCM system in a liquid, the viscosity of the liquid will decrease the amplitude of the oscillation and affect the resonance frequency [28]. In the 1980s, both Kanazawa and Gordon [29] and Nomura and Okuhara [30] studied the use of QCM in a liquid environment to evaluate the effect of liquids on the QCM. Kanazawa and

Gordon showed that an exponentially damped shear wave develops at the surface of a QCM when it is submerged in a solution. The change in resonant frequency as a function of the liquid environment's viscosity can be calculated by Eq. (9.5).

$$\Delta f = -f_o^{3/2}\left(\frac{\eta_l \rho_l}{\pi \rho_q \mu_q}\right)^{1/2} \tag{9.5}$$

where

η_l = viscosity of the liquid
ρ_l = density of the liquid
ρ_q = density of quartz
μ_q = shear modulus of quartz

Once it was found that the excess viscosity loading of a liquid did not affect the sensitivity of QCM to measurements in mass changes, QCM was utilized in direct contact with liquids and/or viscoelastic films. With this development, QCM could be used as a tool to measure corrosion processes in a range of aqueous environments.

9.3.2 QCM experiments and equipment

QCM was first used in 1959, when Sauerbrey published a paper showing that the change in frequency of a quartz crystal can be directly proportional to the change in mass. Since then, many different uses for this technology have been developed, some of which are described next.

9.3.2.1 QCM equipment

Many QCM systems are currently available on the market, with various features. A typical QCM system consists of a few components, as shown in Fig. 9.5. These include a controller, oscillator electronics, frequency counter, high-precision digital voltmeter, crystal holder, and quartz crystals with electrodes attached to each side. The controller is attached to the oscillator and the frequency counter. The controller sets the inputs for the experiment and outputs the measured frequency to the frequency counter. The oscillator is the electronic circuit that specifies the output signal, and as such, controls the resonance frequency of the crystal. It receives input from the controller and then transmits the electric signal to the quartz crystal holder. The frequency counter measures the frequency from the output of the controller. The selection criteria for a frequency counter will include resolution speed, time-base stability, computer interfaces, and software drivers. Careful selection is required; otherwise, the quality of the mass measurement may be degraded.

Depending upon the experiments planned, a variety of quartz crystal coatings can be used to cover a wide range of physical and chemical properties, bonding properties, and various interaction mechanisms. The most commonly used electrode for QCM is pure gold, which is chemically inert and can be chemically modified to allow users to optimize the test setup. For example, gold can be chemically modified to be

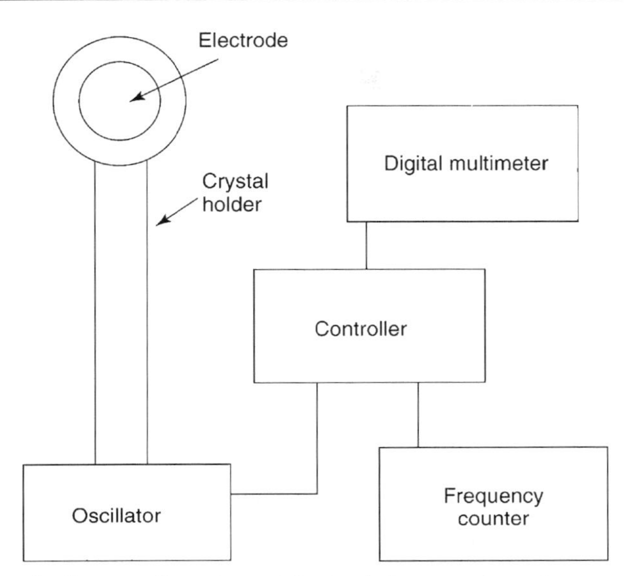

Fig. 9.5 Schematic diagram of quartz crystal microbalance equipment.

hydrophobic rather than the customary hydrophilic surface. In addition to the gold sensor, there are silica-, stainless steel-, polystyrene- and titanium-coated surfaces that can be easily purchased. Silica surfaces are very hydrophilic and are mainly used to form lipid bilayers. Stainless steel surfaces have been used to examine biofilm formation. Polystyrene can be used to evaluate the interaction between polymers and various different biomolecules. Titanium-coated samples have been used for implant research. While these types of sensors can be purchased off-the-shelf, surface deposition techniques can also be used to develop specific sensors if there are none in stock that fit specific requirements.

In today's computer-based world, the newer systems have integrated some of the components shown in Fig. 9.5 into a peripheral component interconnect (PCI) board/computer. The system is interfaced to a personal computer (PC) by means of a digital counter PCI board, which replaces the controller and frequency counter in the conventional setup. This type of system allows for simultaneous control of multiple quartz crystals. In addition, very low noise electrogravimetric QCM systems have been developed that can measure picogram level changes in mass.

9.3.2.2 QCM applications

QCM experiments can be used to measure various types of phenomena. The working electrode is typically gold-deposited onto the quartz crystal, unless specified as something other than gold. The working electrode is a thin film that is placed onto one side of the quartz crystal. The quartz crystal oscillates nominally in the shear mode due to the AT-cut crystal. Any change in the mass rigidly attached to the working electrode results in a change in the quartz crystal oscillation frequency, given by the Sauerbrey equation. Differential frequency measurements are recorded, which means the frequency of a reference quartz crystal is subtracted from the frequency of the working crystal.

There are multiple processes that can lead to mass changes which would correspond to an effective frequency change. The types of process that can be studied using QCM systems include, but are not limited to,

- Corrosion and corrosion protection
- Surface oxidation
- Surfactant research
- Adsorption/desorption
- Metal/alloy plating
- Oxidation
- Etching
- Moisture accumulation.

9.3.3 Dissipation technique

As mentioned previously, the Sauerbrey equation is based upon having a rigid film form on the surface of the working electrode. However, not all processes such as adsorption or corrosion may lead to a rigid film being attached to the electrode surface. When certain layers on the surface of the working electrode have some degree of structural flexibility or viscoelasticity, another technique will be needed. An alternative method that has been developed for studying these types of surface properties using QCM is called the dissipation technique.

The viscoelasticity of a surface film can be measured by analyzing the energy loss, or dissipation, of the shear movement of the quartz crystal. One means to measure the dissipation is to drive the quartz crystal at its resonant frequency by an oscillator that is intermittently disconnected, causing the crystal oscillation amplitude to decay exponentially. The decay of the frequency is recorded which provides a method for measuring not only the frequency, but also the dissipative properties of the film. This data provides a wealth of knowledge on the film's viscoelastic properties. The dissipative properties can be measured by the dissipation factor [24, 31], which can be calculated using Eq. (9.6).

$$D = \frac{E_{\text{dissipated}}}{2\pi E_{\text{stored}}} \tag{9.6}$$

where

D = dissipation factor
$E_{\text{dissipated}}$ = energy dissipated during one period of oscillation
E_{stored} = energy stored in the oscillating system

The total dissipation factor is the sum of all the losses due to damping in the system. Measurement of the dissipation factor can provide some quantitative analysis of the viscous nature of the film. For example, if an absorbed film is somewhat viscous, it will add to the total dissipation factor.

9.3.3.1 Dissipation monitoring equipment

There are no significant differences between the traditional QCM and equipment that allows for a dissipation study. The main difference is the addition of a relay which can pulse the quartz crystal A/C driver on and off.

9.3.4 Electrochemical quartz crystal microbalance

Because QCM was shown to be reliably operated in liquid environments, it was possible to develop a new technique called electrochemical quartz crystal microbalance (EQCM) [20]. EQCM has been used for electrochemical studies including potentiostatic- and galvanostatic-controlled experiments. The typical configuration for these electrochemical studies is to place the quartz crystal between two O-rings. One side of the quartz crystal, including the deposited electrode, contacts the solution in the electrochemical cell. The electrode contacting the solution serves as both part of the QCM oscillator circuit and as the electrochemical working electrode. By combining the concepts from Sauerbrey's work with the frequency change due to a solution damping, the EQCM system can measure mass changes in solutions.

9.3.4.1 EQCM equipment

The EQCM is utilized to study electrochemical processes. The typical equipment is not much different from the standard QCM equipment. The main difference is that an electrochemical cell and some accessories are needed. A QCM crystal is mounted into a holder, which is connected to the chemical cell so that only one of the electrodes is exposed to the conductive solution. The holder is connected to a crystal oscillator, and the QCM crystal electrode exposed to the conductive solution is connected to the working electrode lead of a potentiostat. A standard reference electrode can be used in conjunction with a counter electrode (e.g., platinum mesh). A QCM controller provides independent frequency and conductance outputs. This data is collected and displayed by a frequency counter and digital multimeter, respectively. An example of the EQCM setup is shown in Fig. 9.6.

9.3.4.2 Applications

There are many potential applications for EQCM [20]. One is to measure the mass changes in thin films on the electrode. Because of this measurement capability, it may be possible to use EQCM as a monitor in plating or stripping baths. In addition to its use as a monitor, EQCM may be used to measure mass changes associated with the reduction or oxidation reactions. The EQCM technique can also be used with dissipative techniques to quantitatively treat the influence of viscoelasticity. EQCM has also been used to measure the formation of a monolayer on the electrode surface. One study by Quan et al. [32] utilized QCM to examine the self-assembled monolayer

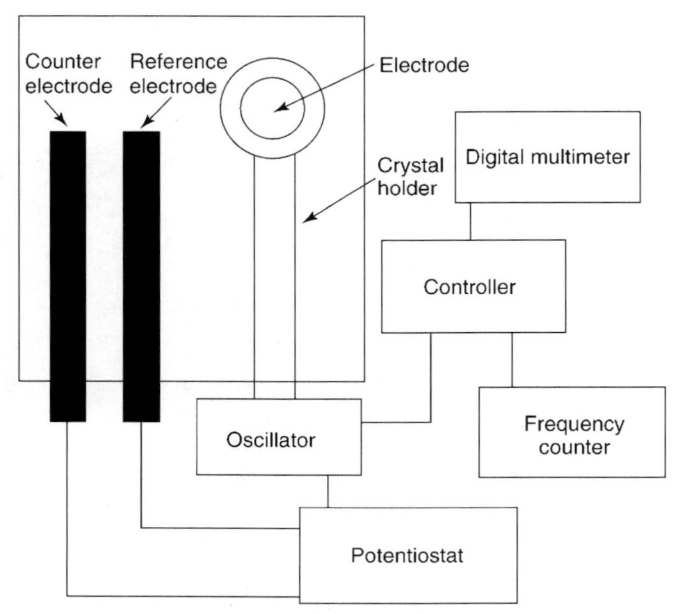

Fig. 9.6 Schematic diagram of electrochemical quartz crystal microbalance equipment.

of Schiff bases on copper surfaces in ethanol. AT-cut quartz crystal disks were precoated with a silver underlayer. Copper was then electroplated onto one side of the silver underlayer, and this side was used as the working electrode in the experiments. Once the copper was electroplated, it was initially rinsed with double distilled water, quickly rinsed with absolute ethanol, and immediately transferred to the QCM cell containing absolute ethanol. The adsorbate solutions were injected into the ethanol to start the process of forming the self-assembled monolayer. Two Schiff bases were used in this QCM study: *N,N′-o*-phenylen-bis-(3-methoxy-salicylindenimine) (V-*o*-Ph-V) and *N*-2-hydroxyphenyl-(3-mthoxy-salicyli-denimine) (V-bso). Fig. 9.7 shows the QCM results for the different self-assembled monolayers. As the Figure shows, after injection of either base, the frequency decreased very rapidly and reached a constant value within 15 and 30 min for V-*o*-Ph-V and V-bso, respectively. It was assumed that a monolayer of material had formed onto the surface once the frequency reached a steady state. A mass associated with the change in frequency can be calculated from the frequency results. QCM tests determined that the V-*o*-Ph-V formed a more densely packed film on the surface of copper than did the V-bso. This information along with additional corrosion data can be used to make conjectures about the inhibiting effect of the individual monolayer. This and other types of experiments are possible with the use of an EQCM system.

In addition, the EQCM system has been combined with other surface analysis techniques, such as X-ray photoelectron spectroscopy, infrared spectroscopy, and electrochemical impedance spectroscopy to study the mechanisms of various corrosion reactions, including growth and dissolution of passive oxide films of alloys in corrosion environments [33].

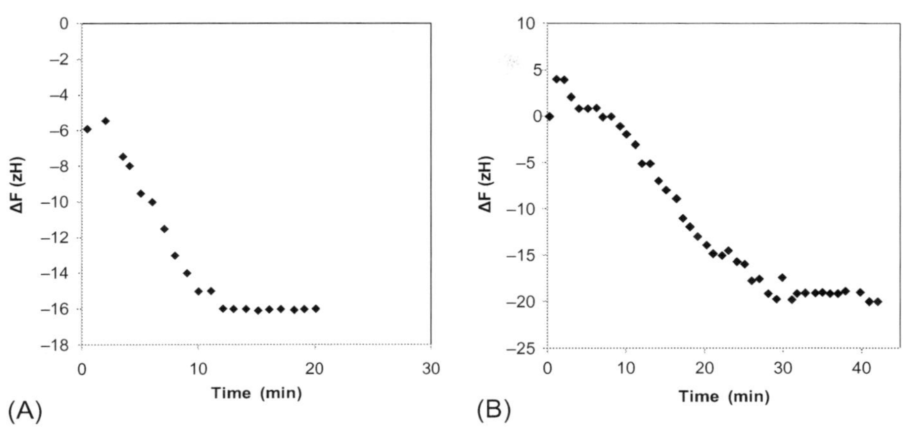

Fig. 9.7 A plot of frequency change (ΔF) against time during the self-assembly of (A) V-o-Ph-V and (B) V-bso. The final concentration of the solution in both tests was 10^{-3} M with a total volume of 20 mL. Frequency change of 1 Hz corresponds to mass change of 18 ng cm^{-2}. Data from Z. Quan, X. Wu, S. Chen, S. Zhao, H. Ma, Self-assembled monolayers of Schiff bases on copper surfaces, Corrosion 57 (2001) 195–201.

9.4 Gravimetric techniques summary

Two gravimetric techniques are examined in this chapter: thermogravimetric analysis (TGA) and quartz crystal microbalance (QCM). The fundamental theories behind the techniques for corrosion monitoring are covered next, including some of the typical experiments that would be conducted and the equipment that would be used. Gulbransen and Andrew [1] developed vacuum microbalance to study gas-metal reactions in a controlled environment at high temperatures. Automatic recording microbalance [2] with precision of 0.1 μg permits long duration runs at high temperatures and has been commercially available since the 1960s. The gravimetric method is widely used to study the kinetics of oxidation and other forms of high-temperature corrosion [9], surface reaction, and protective coatings.

QCM is based on the inverse piezoelectric effect, where a shift in resonance frequency can be used to measure changes in mass. QCMs consist of a thin plate of single-crystal quartz with electrodes affixed to each side of the plate. The crystal can be made to oscillate at its resonant frequency when an external potential is applied. The frequency of the quartz crystal will depend upon several factors, including physical properties of the quartz, properties of the environment, and changes in the mass attached to the electrode. The quartz crystal can be prepared in different ways, but the most common type of quartz used for QCM is an AT-cut quartz, which will produce a mechanical shear oscillation with displacements parallel to the wafer surface.

In 1959, Sauerbrey derived an equation (Eq. 9.3) that relates the change in resonance frequency to a change in mass. This equation was used to conduct early QCM experiments in vacuum and air. However, this equation assumed that the films on the QCM electrodes were thin, were rigidly deposited, did not slip on the surface, and were deposited evenly. The Sauerbrey equation does not account for any of these circumstances. Therefore, QCM was not initially used in a liquid environment due to the viscoelastic nature of the solution.

Kanazawa and Gordon [29] evaluated the use of QCM in liquids, and in the mid-1980s, an equation was developed (Eq. 9.5) to account for the frequency change due to contact with a solution. These authors have conducted experiments that compared the variation in the QCM frequency to changes in solution density or viscoelasticity. Further experiments have been expanded to look at the electrochemical properties for different systems.

QCM experiments can examine many different types of phenomena. These include, but are not limited to, adsorption/desorption, oxidation, metal/alloy plating, corrosion, etching, corrosion protection, surfactant research, and moisture accumulation. EQCM can be used to study the electrochemical properties of a system. One of the electrodes contacts the solution and is attached to a potentiostat as the electrochemical working electrode. EQCM can measure the formation of monolayers or mass changes in thin films. It can also measure mass changes in plating or stripping baths, or mass transport associated with redox reactions.

Finally, there is a dissipation technique that can be used to measure films that are not fully rigid and bonded to the electrode surface. A dissipative factor can be measured by intermittently disconnecting the driving potential to the QCM crystal holder. The crystal oscillation amplitude decay curve provides information about the dissipative properties of the film on the surface of the electrodes. QCM techniques offer a wide range of potential applications for conducting in situ experiments in various environments. Depending upon the properties that are being evaluated, different QCM techniques that have been developed over the last half-century may be applicable.

References

[1] E.A. Gulbransen, K.F. Andrew, An enclosed physical chemistry laboratory—the vacuum microbalance, in: M.J. Katz (Ed.), Vacuum Microbalance Techniques, vol. 1, Plenum Press, New York, 1961, pp. 1–21.

[2] L. Cahn, H.R. Schultz, The Cahn gram electrobalance, in: R.F. Walker (Ed.), Vacuum Microbalance Techniques, vol. 2, Plenum Press, New York, 1962, pp. 7–18.

[3] N. Birks, G.H. Meier, F.S. Pettit, Introduction to the High-Temperature Oxidation of Metals, second ed., Cambridge University Press, New York, 2006.

[4] G.Y. Lai, High Temperature Corrosion of Engineering Alloys, ASM International, Materials Park, OH, 1990.

[5] N.S. Cheruvu, K.T. Chiang, Isothermal and cyclic oxidation behavior of turbine blade alloys, in: Proceedings of Turbo Expo 2006: Power for Land, Sea and Air. Paper No. GT2006-90756. ASME. New York, 2006.

[6] K.T. Chiang, F.S. Pettit, G.H. Meier, Low temperature hot corrosion, in: R.A. Rapp (Ed.), High Temperature Corrosion: NACE-6, National Association of Corrosion Engineers, Houston, TX, 1983, pp. 519–530.

[7] K.T. Chiang, G.H. Meier, R.A. Perkins, The effect of deposits of CaO, $CaSO_4$, and MgO on the oxidation of several Cr_2O_3-forming and Al_2O_3-forming alloys, J. Mater. Energy Syst. ASM 6 (2) (1984) 70–86.

[8] J.A. Goebel, F.S. Pettit, Na_2SO_4-induced accelerated oxidation (hot corrosion) of nickel, Metall. Trans. 1 (1970) 1943–1954.
[9] E.N. Fedorova, M. Braccini, V. Parry, C. Pascal, M. Mantel, F. Roussel Dherbey, D. Oquab, Y. Wouters, D. Monceau, Comparison of damaging behavior of oxide scales grown on austenitic stainless steels using tensile test and cyclic thermogravimetry, Corros. Sci. 103 (2016) 145–156.
[10] T. Huang, E.A. Gulbransen, G.H. Meier, Hot corrosion of Ni-base turbine alloys in atmospheres in coal-conversion systems, J. Metals 31 (3) (1979) 28–35.
[11] W.W. Wendlandt, Thermal Methods of Analysis, John Wiley & Sons, Inc, New York, 1964.
[12] K.T. Chiang, R. Wei, J.H. Arps, Development of nanostructured Cu-Cr coatings for liquid rocket engine applications, in: Proceedings of the 43rd AIAA/ASME/SAE/ASEE Joint Propulsion Conference, Paper No. 2007-5583. AIAA. Reston, Virginia, 2007.
[13] K.T. Chiang, J.H. Arps, R. Wei, Oxidation resistance of multilayered Cu-Cr coatings produced by ion-beam assisted deposition, in: N. Padture, L. Francis, J. Jampikian, N. Dahotre (Eds.), Proceedings of Coatings 2005 Symposium. Materials Science & Technology Conference. Pittsburgh, Pennsylvania, 2005, pp. 7–10.
[14] H.L. Eschbach, E.W. Kruidhof, A direct calibration method for a crystal oscillator film-thickness monitor, in: K.H. Behrndt (Ed.), Vacuum Microbalance Techniques, vol. 5, Plenum Press, New York, 1966, pp. 207–216.
[15] C.D. Stockbridge, Effect of gas pressure on quartz-crystal microbalances, in: K.H. Behrndt (Ed.), Vacuum Microbalance Techniques, vol. 5, Plenum Press, New York, 1966, pp. 147–178.
[16] A.W. Warner, C.D. Stockbridge, Mass measurement with resonating crystalline quartz, in: K.H. Behrndt (Ed.), Vacuum Microbalance Techniques, vol. 3, Plenum Press, New York, 1963, pp. 55–73.
[17] D.A. Buttry, M.D. Ward, Measurement of interfacial processes at electrode surfaces with the electrochemical quartz crystal microbalance, Chem. Rev. 92 (1992) 1355–1379.
[18] C. Lu, A.W. Czanderna, Applications of Piezoelectric Quartz Crystal Microbalances, Elsevier, Amsterdam, 1984.
[19] R.A. Heising, Quartz Crystals for Electrical Circuits, D. Van Nostrand Company, New York, 1946.
[20] M.R. Deakin, D.A. Buttry, Electrochemical applications of the quartz crystal microbalance, Anal. Chem. 61 (1989) 1147A–1154A.
[21] W.G. Cady, Piezoelectricity, Dover Publications, New York, 1964.
[22] Z. Lin, C.M. Yip, I.S. Joseph, M.D. Ward, Operation of an ultrasensitive 30-MHz quartz crystal microbalance in liquids, Anal. Chem. 65 (1993) 1546–1551.
[23] G. Sauerbrey, Verwendung von Schwingquarzen zur Wägung dünner Schichten und zur Mikrowägung, Zeitschrift für Physik 155 (1959) 206–222.
[24] M. Rodahl, F. Höök, A. Krozer, P. Brzezinski, B. Kasemo, Quartz crystal microbalance setup for frequency and Q-factor measurements in gaseous and liquid environments, Rev. Sci. Instrum. 66 (1995) 3924–3930.
[25] V.M. Mecea, J.O. Carlsson, R.V. Bucur, Extensions of the quartz-crystal-microbalance technique, Sens. Actuators A 53 (1996) 371–378.
[26] J.G. Miller, D.I. Bolef, Acoustic wave analysis of the operation of quartz-crystal film-thickness monitors, J. Appl. Phys. 39 (1968) 5815–5816.
[27] C.S. Lu, O. Lewis, Investigation of film-thickness determination by oscillating quartz resonators with large mass load, J. Appl. Phys. 43 (1972) 4385–4390.

[28] B.A. Martin, H.E. Hager, Velocity profile on quartz crystals oscillating in liquids, J. Appl. Phys. 65 (1989) 2630–2635.
[29] K.K. Kanazawa, J.G. Gordon, The oscillation frequency of a quartz resonator in contact with liquid, Anal. Chim. Acta 175 (1985) 99–105.
[30] T. Nomura, M. Okuhara, Frequency shifts of piezoelectric quartz crystals immersed in organic liquids, Anal. Chim. Acta 142 (1982) 281–284.
[31] M. Rodahl, F. Höök, C. Fredriksson, C.A. Keller, A. Krozer, P. Brzezinski, M. Voinova, B. Kasemo, Simultaneous frequency and dissipation factor QCM measurements of biomolecular adsorption and cell adhesion, Faraday Discuss. 107 (1997) 229–246.
[32] Z. Quan, X. Wu, S. Chen, S. Zhao, H. Ma, Self-assembled monolayers of Schiff bases on copper surfaces, Corrosion 57 (2001) 195–201.
[33] C.-O.A. Olsson, D. Landolt, Electrochemical quartz crystal microbalance, in: P. Marcus, F. Mansfeld (Eds.), Analytical Methods in Corrosion Science and Engineering, CRC Press, Taylor and Francis Group, Boca Raton, FL, 2006, pp. 733–751.

Radioactive tracer methods

10

Douglas C. Eberle
Southwest Research Institute, San Antonio, TX, United States

10.1 Principle and history

The radioactive tracer method was first conceived by George de Hevesy in the early 1900s. According to Van Houten [1], de Hevesy was employed at Ernest Rutherford's lab in Manchester, England, between 1910 and 1913. Rutherford was interested in studying the properties of a substance named radium-D. The problem was that the radium-D he had obtained was mixed with lead. Rutherford had been unable to separate it, so he challenged de Hevesy to the task. De Hevesy attempted the separation unsuccessfully for more than a year before he had the idea that if this substance was inseparable from lead, it might be used as a tracer to follow lead through chemical reactions by making radioactive measurements. This was the birth of the radioactive tracer method and it eventually led to de Hevesy winning the Nobel Prize for his work on the use of isotopes as tracers in the study of chemical processes in 1943. As a side note, it was eventually discovered that radium-D was actually an isotope of lead, namely ^{210}Pb.

Since de Hevesy's initial conception, radioactive tracer methods (sometimes known as radiotracers) have been used by industry to measure a wide variety of corrosion, erosion, and wear mechanisms. The technique offers several advantages over conventional techniques. First of all, it is highly sensitive. Radioactive material can often be detected at quantities much lower than that can be detected through other means such as chemical analysis or gravimetric or dimensional means. Material loss rates in the order of micrograms per minute (μg/min) or nanometers per minute (nm/min) can be measured with time resolutions in the order of several minutes. Under certain circumstances, the technique can make measurements in near real-time without having to periodically take apart an assembly. This is a significant advantage when the component is difficult to get to and disassembly means having process downtime. It also means that measurements can be made under changing conditions. This allows for cause and effect analysis, such as the effect of temperature on corrosion rate. Lastly, the technique can quantify material losses from discrete areas or surfaces of interest. This allows for tracing of corrosion, erosion, or wear rates from a discrete component within an assembly, such as wear from piston rings in an engine, without the confounding effect of other sources of wear material from the engine.

The basic principle of operation is that a surface or volume of material to be tested is first labeled with a radioactive isotope or isotopes. Ehmann and Vance [2, p. 8] defines isotopes as "forms of an element with the same number of protons, but different masses due to different numbers of neutrons in their nuclei." Isotopes are often an

Techniques for Corrosion Monitoring. https://doi.org/10.1016/B978-0-08-103003-5.00010-2

unstable form of the element and these unstable forms decay by various means at a known rate. As isotopes decay, they often emit emanations such as gamma rays, alpha particles, or beta particles. These emanations are known as radioactivity and can be measured with different types of detectors.

Once a material has been radio-labeled, it is exposed to an environment (either a wearing or corroding condition, usually in a fluid) for a period of time. As the part wears or corrodes, labeled radioactive material is removed at the same rate as the unactivated parent material. Quantification can be made of this surface loss through either measuring the increase in radioactivity of the fluid or by measuring the decrease in radioactivity from the activated surface, after accounting for natural radioactive decay.

To give an illustration of this technique, Treuhaft and Eberle [3] describe a method often employed to measure real-time wear from components within internal combustion engines. In this case, the components of interest (piston rings, rod bearings, etc.) are made radioactive by exposing the parts to a thermal neutron flux (bulk or neutron activation) near the core of a nuclear reactor. The high energy neutrons penetrate the nucleus of a small number of atoms within the test component. For each interaction, there exists a probability that the free neutron will be captured. This results in an increase in the mass of the nucleus, transmuting the atom into a heavier isotope, which is often radioactive. The now radioactive component of interest is then installed in the engine. The engine is configured with a special sump oil loop for interrogation with a gamma spectroscopy system, see Fig. 10.1. As the engine is operated and the

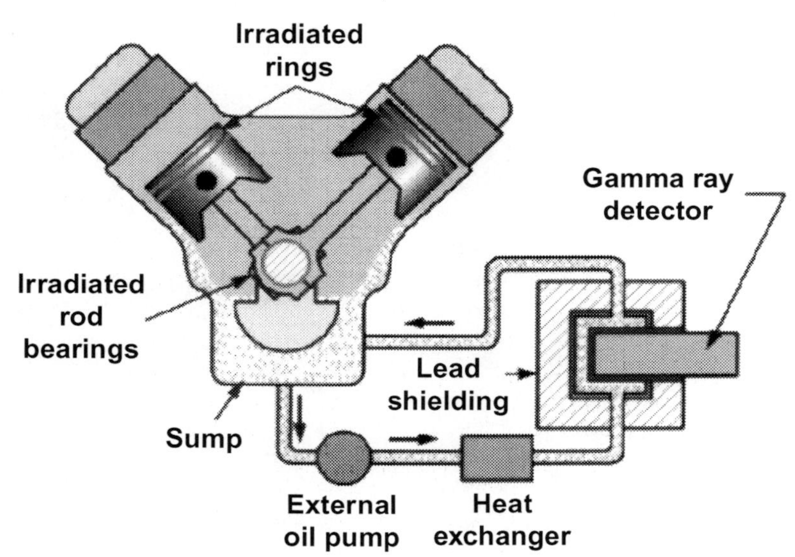

Fig. 10.1 Wear measurement flow loop schematic.
Courtesy Southwest Research Institute.

component wears, wear particles are abraded from the parent part and are transported into the lubricating oil. Since the wear debris consists of very small particles, the debris is readily suspended and distributed throughout the oil system. Oil is drawn from the sump and circulated past a gamma ray detector, which measures the buildup of radiation from the wear particles. The activity of the wear particles is mathematically proportional to the concentration of radioactive wear particles present, after making corrections for natural radioactive decay.

10.2 Assumptions

The most important assumption that must be made is that the process of labeling a material with a radiotracer does not alter its mechanical or chemical properties. This is almost always valid for radiotracer studies. The amount of radioactively transmuted material is typically in the order of 1 ppm to 1 ppb, and thus, the gross material properties are not affected. The amount of radiation produced is also typically at a low enough level that the radioactivity itself does not significantly affect the system. In most cases, the process of activation produces isotopes that are chemically the same as the parent isotopes. These isotopes contain the same number of protons as the parent, with only an additional neutron. The electron structure is not affected. There is a slight difference in the mass of the nucleus caused by the additional neutron, which may have a minor effect on diffusion type studies of lighter elements. According to Ehmann and Vance [2, p. 314], for heavy isotopes (such as ^{51}Cr or ^{59}Fe), the mass differences are minor and the isotopes behave so similarly in chemical and physical processes that they are very difficult to separate by chemical means. Some types of activations produce isotopes that contain an additional proton. While these isotopes are chemically different from the parent, they reside within the same grain structure as the parent isotope in the metallurgy. In wear and corrosion studies, the assumption is made that, as the grain corrodes, the labeled atom is transported with neighboring atoms.

Although not explicitly necessary for qualitative evaluations, quantitative evaluations require a few additional assumptions. When monitoring the buildup of radioactivity in a fluid, it must be assumed that the entrained radioactive material is evenly distributed and is not removed from the fluid by means of other processes. In many cases, this is a good assumption; however, this must be considered as a source of potential error when designing an experiment, as it can have a significant effect on measurements. Examples of processes which can remove radioactive material from fluids include filtration, particle settling, and plating of the radioactive materials onto other contacting surfaces.

When monitoring an activated surface for loss of activity through direct measurement, it must be assumed that all of the material that has been removed from the surface has been carried away from the immediate area such that it does not significantly affect the measurement. If material has replated on the surface, or has settled through another means near the detector, it can significantly skew the results.

10.3 Labeling methods

10.3.1 Bulk or thermal neutron activation

Thermal neutron activation is sometimes referred to as bulk activation. This technique (as well as thin layer activation, described below) creates radioactive isotopes from naturally abundant isotopes within the parent material. The component to be activated is placed in a neutron flux near the core of a nuclear reactor, where it is bombarded with thermal neutrons. There is a small probability (referred to as a cross-section) that a free neutron will directly interact with and be absorbed by a nucleus within the component. This causes a nuclear reaction to take place, sometimes resulting in the formation of a radioisotope. Since neutrons are uncharged particles, they can penetrate relatively large distances through matter without interaction. Thus, this method allows the creation of radioisotopes that are homogeneously distributed throughout the part.

There are several limitations to this technique. The physical size of components is limited by the space available near the core of the nuclear reactor. Typically, research reactors are used that are available at certain governmental and university campuses around the world. Depending upon the design of the vessels used for activation, the size of components that can be activated is often limited to 15–20 cm in diameter. Some reactor setups have much smaller limitations.

Since the entire mass of the component is activated, the total induced radioactivity in the part is higher than in other methods. This higher activity often requires special licensing for increased radiation levels and additional safety precautions and facilities. In addition, to be able to obtain high sensitivity, measurement of corrosion or wear rates is accomplished by monitoring the buildup of radioactive material in the transport fluid. Monitoring for the loss of activity from the surface directly is typically not possible as the change in activity is generally too small to detect. In most cases, this requires a recirculating flow system in order to allow the cumulative radioactive corrosion product to build up to detectable levels. Certain processes do not lend themselves to recirculating systems. Sensitivity is also dependent upon the total volume of the recirculating fluid. If the system volume is too large, the radioactive material will be too dilute to detect.

Despite these limitations, there are a lot of good reasons to use bulk activation. For small components, the cost of irradiation is often much lower than other methods for the same specific activity. Multiple components can be activated at the same time for a lower cost and activating to a higher specific activity level increases the sensitivity of the experiment. Bulk activation can produce different isotopes from parent materials than other methods. This increases the choices of available radiotracers. In some cases, bulk activation is the only way to produce useful isotopes from certain metallurgies. Since the available choice of isotopes is increased, bulk activation is sometimes combined with other techniques to produce multiple isotopes within an assembly from the same parent metallurgy. This is useful to be able to trace processes from multiple discrete components simultaneously. In the case of internal combustion engines, it has been used to measure piston ring wear and wear from localized areas of the engine bores in the same measurement.

10.3.2 *Thin or surface layer activation*

Thin layer activation (TLA) is sometimes referred to as surface layer activation (SLA) or as charged particle activation. Laguzzi et al. [4] give a good description of this technique. To summarize, in TLA, a thin layer of material at the surface of a component is activated using a beam of charged particles (ions) from either a linear accelerator or a cyclotron. Typical ions used for activation include protons and deuterons, although sometimes heavier ions such as ^{3}He are used. As in neutron activation, there is a small probability (cross-section) that a charged particle will directly interact with a parent nucleus, causing a nuclear reaction that can result in a radioactive isotope. Charged particles do not penetrate through matter without interaction as easily as neutrons. This characteristic limits the depth of activation to a very shallow layer on the surface of the component. Typically, activation depths range from 50 to 150 μ and can be tailored by adjusting the energy of the particle beam or the angle of incidence.

One of the drawbacks of the technique is that activation is much more complicated. The charged particle beam is typically narrower than the area to be activated. This requires that either the beam be rastered or the part itself be accurately moved through either mechanical or electromechanical means. In order to readily quantify the loss of material, the material near the surface must be homogeneously and uniformly activated, without a surface dead-band. If a surface dead-band exists, the part will lose material initially without showing any loss of activity. This characteristic can be taken advantage of in certain applications as an indication, for instance, a surface coating has been compromised. If the surface is not homogeneously or uniformly activated, the specific activity of the material changes with depth or position. Quantifying the loss of material requires that the specific activity be constant or significant errors can result.

In an activity to depth profile, a linear curve represents a homogeneous activation as shown in Fig. 10.2. While no profile is perfectly linear, an experienced activation facility should know how to generate a nearly linear profile with errors of less than a few percent over most of the useful depth. This curve can be generated by either empirical modeling or physical measurement as described in Section 10.5.

10.4 Potential isotopes

Table 10.1 contains a sampling of available isotopes, related production methods, half-lives, and gamma ray energies. For SLA production methods, the beam type is listed as p for proton or d for deuteron.

10.5 Calibration and conversion to corrosion units

In order to quantify radiotracer measurements into physical units of either mass or thickness, a calibration must be accomplished. The method used for determining the calibration factor is dependent upon the method used to activate the sample. For bulk-activated materials, since the activity is homogeneously distributed through

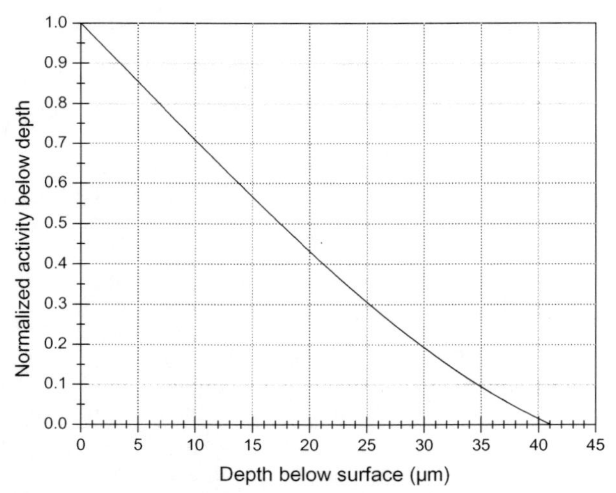

Fig. 10.2 TLA activity to depth profile.
Courtesy Southwest Research Institute.

Table 10.1 A sampling of available isotopes, related production methods, half-lives, and gamma ray energies.

Parent element	Nuclide made	Production method	$t_{1/2}$ (days)	Gammas (keV)
Fe	^{59}Fe	Bulk	44.5	1099, 1291
Fe	^{56}Co	SLA, p	77.2	846
Fe	^{57}Co	SLA, d	272	122
Cr	^{51}Cr	Bulk	27.7	320
Cr	^{52}Mn	SLA, p	5.6	744, 935, 1434
Cr	^{54}Mn	SLA, p	312	835
Sn	^{113}Sn	Bulk	115	255
Sn	^{124}Sb	SLA, p	60.2	603, 1691
Ni	^{58}Co	Bulk	70.8	811
Zn	^{65}Zn	Bulk	244	1115
In	^{114m}In	Bulk	49.5	190
Cu	^{65}Zn	SLA, p	244	1115
Co	^{60}Co	Bulk	1925	1173, 1332
Sb	^{124}Sb	Bulk	60.2	602.7, 1691

For SLA production methods, the beam type is listed as p for proton or d for deuteron.

the material, calibration to correlate measured activity to mass is relatively straightforward. Samples of the test material are activated simultaneously with the test article to ensure that the calibration samples have the same specific activity as the labeled part of interest. These calibration samples are accurately weighed, then dissolved into an acidic solution to form a base calibration solution. This base solution is then diluted through several orders of magnitude into a series of known concentrations. The detector is then sequentially exposed to this series of calibration solutions in the same or identical vessel as to be used during the experiment. Counts are accumulated for each sample and corrected for time decay to a fixed but arbitrary reference time, usually chosen as the time of activation. This series of counts is plotted as a function of concentration. If performed properly, a linear curve is developed relating measured counts for each isotope to solution concentration. A nonlinear curve is an indication that the radioactive isotope has partially come out of solution as either a precipitate or as plating on the solution container walls. A sample calibration curve is shown in Fig. 10.3.

Surface layer activation calibrations require a less direct approach, since the surface layer cannot easily be separated from the unactivated material below and the homogeneity of the activation must be checked. Several options are possible. If thin foils of known thickness made from the same or very similar parent material are available, the foils can be stacked and activated with the same accelerator setup as used for the primary activation. Care must be used to ensure that the accelerator parameters are held constant to produce the same activation profile in the foil stack as in the primary target. Following activation, the foil stack is exposed to the detector. Successive measurements are made in which one layer of foil at a time is stripped away. By performing time decay correction and plotting this series of measurements, a curve representing activity to depth is developed.

If adequate foils are not available, an ablation technique can be used. In the ablation technique, a calibration coupon made from a similar alloy material is activated using the same parameters as during the primary component activation. A calibration profile is developed by measuring the activity of the coupon between sequential abrasive

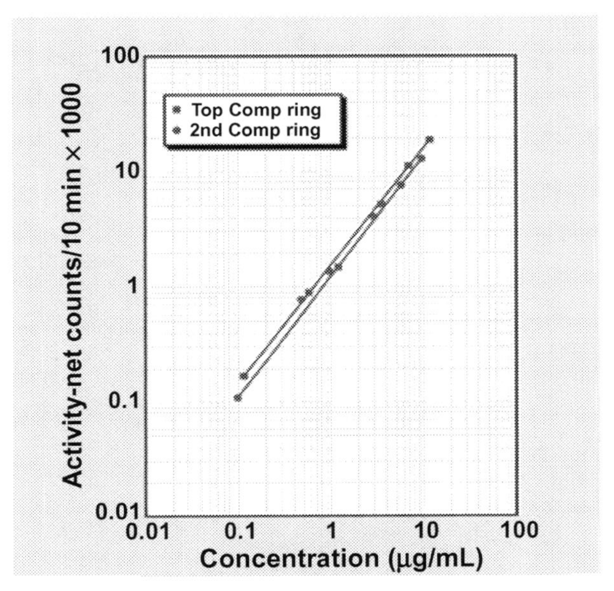

Fig. 10.3 Bulk activation calibration curve.
Courtesy Southwest Research Institute.

ablation steps which remove a measured amount of material. The method is more prone to error than the foil technique as it is difficult to evenly remove material by abrasion and to accurately quantitate the material removed.

Finally, if similar material activations have been previously accomplished and are well-known, an empirically calculated activity profile can be used. While most commonly employed, this method is the least preferred as there is no confirmation of the actual depth profile.

10.6 Applications and limitations

10.6.1 Example applications

10.6.1.1 Real-time crude oil corrosivity measurement

The corrosivity of crude oils has been a long-term area of interest for refineries and crude oil buyers and sellers. Crude oils with higher total acid numbers (TAN) are becoming more prevalent on the market. Due to the perceived higher corrosivity of these high TAN crude oils, the crude oils are often sold at a discount. Though TAN has historically been used as an indicator of higher corrosivity, the correlation between TAN and corrosivity is poor. Crude oil corrosivity is a difficult parameter to measure under high temperature conditions, since the corrosive agents (typically a combination of naphthenic acids and sulfur compounds) can undergo thermal degradation. Traditionally, crude oil corrosion is measured by coupon weight loss techniques that range from 48 to 96 h, or longer, exposure under various conditions. Weight loss has several confounding issues in that the oil properties can change in this time period and passivating or other types of films can build up on the surface of the coupon, potentially leading to coupon weight gain if not properly removed.

In order to obtain better sensitivity than loss-in-weight techniques, and to make the measurement prior to thermal degradation and without confounding film buildup, two techniques have been developed by Eberle et al. [5] using radiotracers to measure crude oil corrosivity at elevated temperatures. One technique uses a heated, stirred autoclave, while the second uses a high shear stress flow loop arrangement. The flow loop technique allows for real-time measurement of corrosion with 10-min resolution. In both techniques, 1018 carbon steel coupons were radio-labeled for demonstration studies with ^{59}Fe by bulk activation.

The first technique involves exposure of the activated coupon to heated oil in a stirred autoclave. A radio-labeled disk-shaped coupon is mounted between low conductivity ceramic spacers onto the autoclave's stirring rod to prevent galvanic corrosion. Test oil is added and the autoclave is sealed and pressurized with nitrogen gas to minimize oil phase change. The coupon is rotated, while the oil is heated to the test temperature. The autoclave is held at this temperature for a period of 30 min and then is allowed to cool. Following disassembly, the coupon is removed from the test oil. The test oil is measured with a gamma spectrometer for the accumulation of radioactive soluble corrosion products. The test coupon is next tumbled in abrasive media to remove any adhered products on the coupon surface. The abrasive media is measured with a gamma spectrometer to measure the products which were removed during the tumbling process. In both cases, measurements are first converted to mass units by applying appropriate time decay and calibration factors. Mass units can be converted to average thickness loss over the exposed coupon surface area. A chart showing soluble, adhered, and total corrosion product loss resulting from testing is shown in Fig. 10.4.

In the flow loop technique, a rod-shaped coupon is installed in a special test fixture designed to create a constant shear stress along the coupon surface when oil flows through. As in the autoclave technique, the coupon is mounted between low conductivity ceramic isolators to prevent galvanic corrosion. The fixture is installed in a heated recirculating flow loop containing the test oil. An image of the flow loop is shown in Fig. 10.5. The flow loop is pressurized with nitrogen gas to minimize oil phase change. Shear stress is measured as pressure drop over the length of the coupon. Typically, oil flow rate is adjusted during testing to maintain a constant shear stress as the oil is heated. As the coupon corrodes, corrosion products are entrained in the oil stream and carried past a calibrated gamma spectrometer. The spectrometer accumulates detected counts over a fixed time period. At the end of each period, the accumulated counts are saved, the spectrometer is zeroed, and the counts restarted. The counts for each period are time-decay-corrected and converted using calibration factors to accumulated corrosion product in the oil. Fig. 10.6 shows a sample corrosion curve for a mineral oil spiked to a TAN of 3.0 mg KOH. In this curve, corrosion rate increases as temperature increases and stabilizes once the system reaches a steady state temperature.

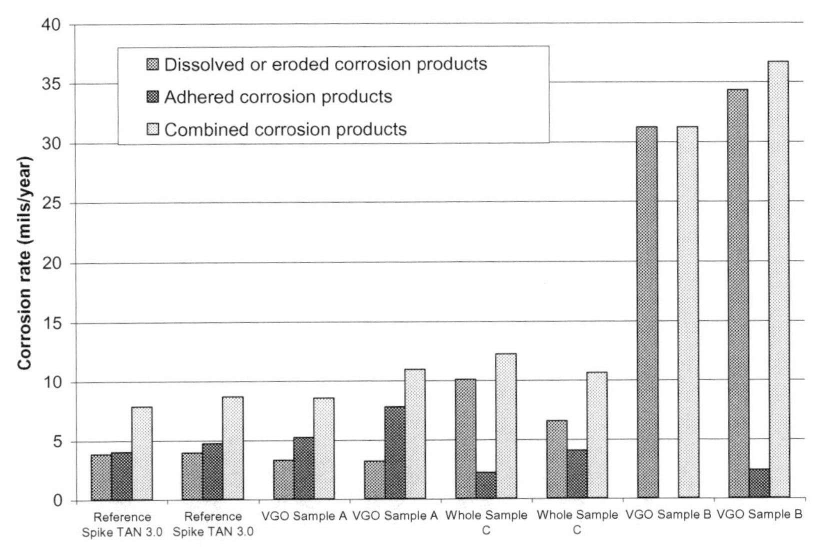

Fig. 10.4 Grouped repeats of autoclave corrosivity tests.
Courtesy Southwest Research Institute.

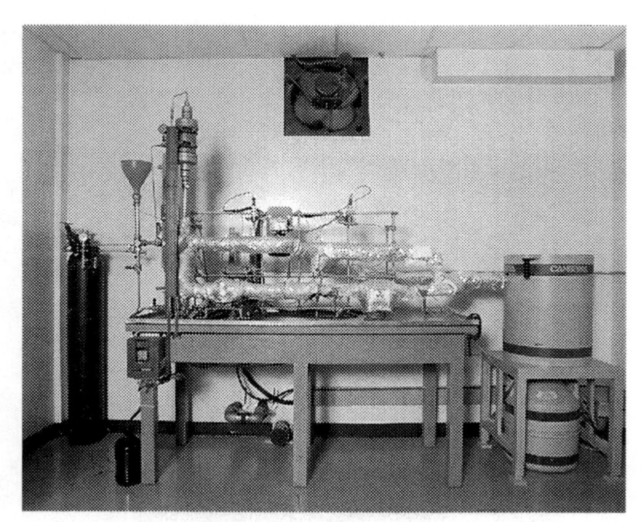

Fig. 10.5 Real-time corrosivity measurement flow loop.
Courtesy Southwest Research Institute.

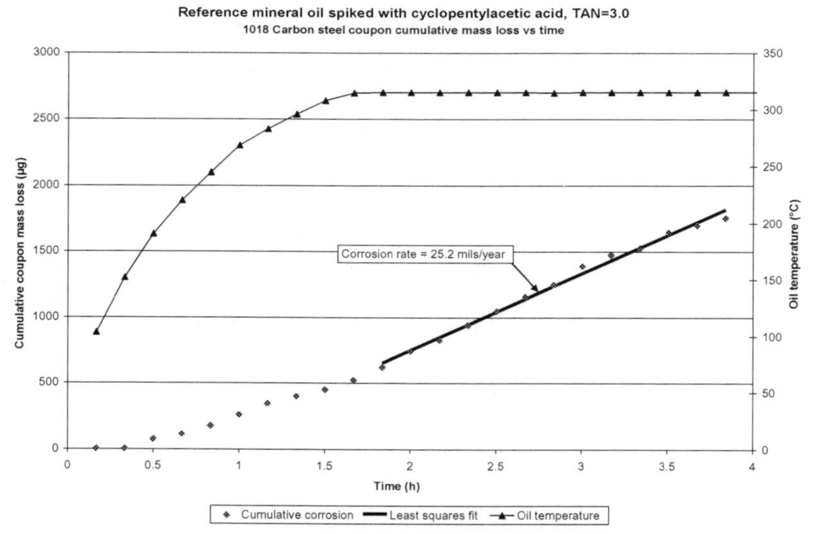

Fig. 10.6 Sample flow loop test cumulative corrosion as a function of time and temperature. Courtesy Southwest Research Institute.

10.6.1.2 Real-time measurement of in-cylinder corrosion in operating internal combustion engine

Corrosion of components in internal combustion engines under exposure to certain fuels, use of Exhaust Gas Recirculation (EGR), or under certain operating conditions is of concern to automotive manufacturers for durability and warranty considerations. In some areas of the world, fuels with high sulfur content are still in use, although regulations restricting use of sulfur are becoming more prevalent. Radioactive tracers have been used in order to quantify corrosion and wear effects in engines for many years.

Fukawa et al. [6], followed up by Ohmiya et al. [7], report using radioactive tracers to measure the effects of ethanol and sulfur concentrations in fuel, surface temperature, and engine oil base number on corrosive wear. In these studies, a radioisotope was used to label a specific area of the cylinder wall via SLA to depths of 100 μm (Fukawa) or 50 μm (Ohmiya). In both studies, the engines were configured with gamma spectroscopy systems to be able to measure real-time buildup of radioisotopes in the engine oil sumps. Additionally, each engine exhaust system was fitted with a particle filter trap to capture radioactive particles which pass through the exhaust valves and enter the exhaust stream. The exhaust filter traps were measured for buildup of radioactivity in real-time.

As each engine was operated across a test matrix of conditions, corrosive wear rates were measured over a range varying from tens of μg/h up to thousands of μg/h. Increased sulfur level in fuel showed higher levels of corrosive wear that peaked at a specific engine coolant temperature and then decreased at higher temperatures. Correlation was also shown between corrosive wear and oil base number.

10.6.2 Discussion of limitations

Although radiotracer corrosion measurement techniques have many significant advantages over other techniques, there are limitations as well. First of all, the technique involves the use of radioactive materials. Although the relative activities are low and can be handled safely with appropriate precautions, radioactive materials are regulated materials. Laws vary in almost every country. In most cases, specific licenses to handle radioactive materials are required. Licenses can be costly both to obtain and to maintain.

Another significant limitation to the technique is that in order to make the measurement, the corrosion products must be physically separated from the originating surface. In many cases, this is accomplished by transporting the corrosion products from the surface in the corroding fluid. In other cases, the surface can be cleaned either mechanically or chemically to remove the products.

Measurement equipment costs are relatively high for this technique. Depending upon the instrumentation used, detector costs can range from several thousand to tens of thousands of US dollars or more.

10.7 Sources of further information

For information on radiation measurement equipment, refer to Knoll, G 2000, *Radiation Detection and Measurement*, *Third Edition*, Wiley, New York.

References

[1] J. Van Houten, A century of chemical dynamics traced through the Nobel Prizes, 1943: George de Hevesy, J. Chem. Educ. 79 (3) (2002) 301–303.
[2] W.D. Ehmann, D.E. Vance, Radiochemistry and Nuclear Methods of Analysis, Wiley, New York, 1991.
[3] M.B. Treuhaft, D.C. Eberle, The Use of Radioactive Tracer Technology to Measure Real-Time Wear in Engines and Other Mechanical Systems, (2007). SAE paper no. 2007-01-1437.
[4] G. Laguzzi, L. Luvidi, N. De Cristofaro, M.F. Stroosnijder, Corrosion monitoring of different steels by thin layer activation, J. Radioanal. Nucl. Chem. 262 (2) (2004) 325–330.
[5] D.C. Eberle, C.M. Wall, M.B. Treuhaft, Applications of radioactive tracer technology in the real-time measurement of wear and corrosion, Wear 259 (2005) 1462–1471.
[6] K. Fukawa, K. Mase, Y. Ohmiya, H. Moritani, Y. Kamiyama, Experimental Analysis of Cylinder-Bore Corrosive Wear by Utilizing a Radioisotope Tracer (First Report), (2018). JSAE paper no. 20185416.
[7] Y. Ohmiya, K. Fukawa, K. Mase, H. Moritani, M. Tohyama, T. Kittaka, K. Obara, J. Kobayashi, T. Goto, Experimental Analysis of Cylinder-Bore Corrosive Wear by Utilizing a Radioisotope Tracer (Second Report), (2018). JSAE paper no. 20185417.

Electrical resistance techniques 11

C. Sean Brossia
Invista Sarl, Houston, TX, United States

11.1 Introduction and background

Corrosion monitoring probes based on detecting and measuring changes in electrical resistance have been in use since the 1950s [1, 2]. These electrical resistance (ER) probes, sometimes referred to as electronic coupons [3], monitor the electrical resistance of a sensing element that is exposed to the environment of interest. As the sensing element experiences corrosion, its cross-sectional area will decrease and the measured resistance will increase. ER probes are utilized in a wide range of application and industries including oil and gas production, pipeline, chemical process, pulp and paper, power generation, and archeological preservation, to name a few.

ER probes are often selected because of their relatively simple operations, relatively low maintenance, ease of data interpretation, real-time data collection, and reliability. Because the probes essentially are exposed to the environment, they are relatively passive in nature and are no more difficult to install or utilize than traditional weight loss coupons. Thus, the main issues in utilizing ER probes include ensuring that the probe element is exposed to the environmental conditions of interest (e.g., the same chemistry, temperature, flow regime, etc. as the structure being monitored), safety considerations associated with ensuring that penetrations into a high pressure system are done correctly, and that provisions are made for eventual retrieval for maintenance and replacement. Thus, ER probes are commercially available for service operations up to and exceeding pressures of 100 bar and temperatures of 500°C. Higher pressure and temperature probes are often also available for specialized cases and can be custom-produced without significant difficulties.

Maintenance of the probes usually arises due to deposit build-up and precipitation. In cases where scaling formation is of interest, the probe can still be used with caution since some scale deposits can be electronic-conducting (note, ionic-conducting generally does not affect the measurements) and others are not. Thus, with conducting scales, the overall resistance will tend to decrease with the time of exposure. Should corrosion occur simultaneously with scale formation, interpretation of the results may be difficult. Data interpretation and decision-making are generally fairly easy. As the resistance increases, the loss of metal can be directly determined which will be discussed below. Because the effect of actual metal loss is measured, this can then be translated to a corrosion rate and provides a reliable indication of possible metal loss experienced by system components. It has been reported that the nominal minimum thickness loss that can be detected by ER probes is on the order of 1 μm [4]; however, the level of sensitivity is strongly dependent on the probe design. Thin film ER probes

Techniques for Corrosion Monitoring. https://doi.org/10.1016/B978-0-08-103003-5.00011-4

have been shown to be sensitive to metal losses as small as 5–10 nm when the starting ER sensing element is on the order of 25 μm [5]. Others have used sensing elements as thin as a few μm and were able to resolve corrosion rates as low as 1.3 μm/year [6]. Though an indication of metal loss is obtained from ER probes, they do not measure the direct metal loss experienced by system components.

Unlike corrosion coupons, ER probes can be used to provide real-time corrosion rate information (see Fig. 11.1). For corrosion coupons, the corrosion rate is generally determined after a prescribed exposure period has been reached and the weight change of the specimens is then determined. The assumption in using these coupons is that the corrosion rate throughout the exposure period is relatively uniform. As a result, any process system upsets that dramatically increase the corrosion rate will be difficult to detect and tend to be minimized using coupons. In the case of ER probes, the corrosion rate can be determined either through periodic or continuous resistance change measurements. In the case of periodic measurements, the periodicity can be much more frequent than that employed with corrosion coupons. For example, typical coupon exposures might call for weight loss measurements after a period of several weeks or months, whereas the ER probe can be interrogated as often as desired. Thus, the corrosion rate can be determined daily or more frequently. When monitored continuously, both the instantaneous effects of corrosion and the integrated time effects of corrosion can be determined. For example, a linear polarization resistance probe (see Chapter 3 for more information) can provide a direct measure of the instantaneous corrosion rate, but the cumulative effects of corrosion can only be determined by the integration of a sequence of continuous measurements with time. In the case of the ER probe, the cumulative effect of corrosion is always known and the instantaneous effects of corrosion, like the effects of a process upset, can be determined by comparing subsequent or a sequence of measurements over a short period of time. That is, if the resistance is noted to increase dramatically over a short period of time and then continues to increase at a slower rate consistent with the preevent rate, then an upset or some other event can be concluded to have occurred. This approach was successfully utilized by Abdulhadi et al. [7] to verify that sufficient corrosion inhibitor injection was reducing the corrosion rate in a gas gathering system. The authors also

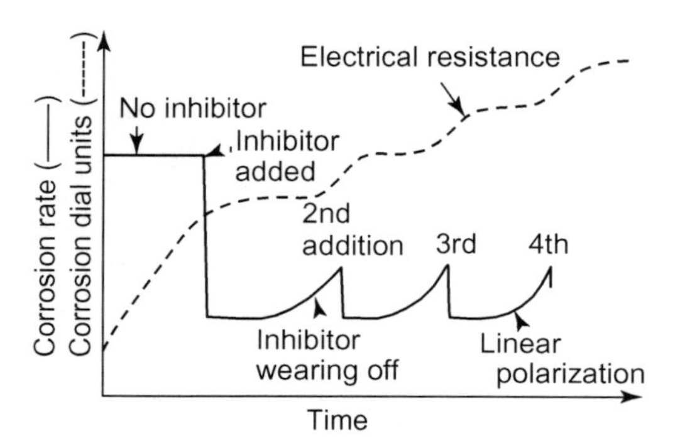

Fig. 11.1 Illustration and comparison of ER probe data and linear polarization data as a function of time.

installed a suite of 29 ER probes to monitor various locations in a sour water system as a part of a continuous monitoring and alarming system.

Because of their relative simplicity, ER probes are generally quite reliable and cost effective. Issues and concerns that need to be considered include probe replacement due to corrosion and the resultant effect on useful life, the effects of nonuniform corrosion on rate measurements, and installation and manufacturing defects. Because the probe element corrodes away during exposure, eventually the probe will corrode to the point that unreliable measurements are obtained. These errors typically arise due to nonuniformity in corrosion of the sensing element after approximately half of the original thickness (one-quarter of the diameter for wire elements) has been consumed [8]. Nonuniform corrosion, such as pitting, crevice corrosion, and stress corrosion cracking, can lead to errors in measurement. These errors can be both in the form of underestimation and overestimation of the true corrosion rate, thus visual inspection of the probe element on replacement is required. Installation and manufacturing defects can also be a source of errors, and thus, spurious readings should not always be associated with process system changes.

Another complication with using ER probes is the possibility that conductive deposits and films form on the sensing element. In such cases, deposition can lead to a metal loss reading that is smaller than actual, and thus, underreporting the effective corrosion rate. In some cases, the rate of deposition can be high enough that negative corrosion rates are observed from the probe. One of the most commonly encountered situations where this occurs is where carbon steel probes are installed in systems where H2S and other sulfur species may be present. Jarragh et al. [9] observed highly variable corrosion rates using ER probes in high pressure crude oil lines which was attributed to the presence of H_2S. The ER probes also often underreported the corrosion rates observed using coupons that were also installed for comparison purposes. Pitting was also noted on both the ER probe sensing elements and coupons. This was not apparent from the ER probe data, highlighting the previously mentioned challenges of using ER probes where localized pitting corrosion may take place.

11.2 Sensing probe designs

In its basic form, an electrical resistance (ER) probe consists of a metal sensing element whose resistance is measured while the element is exposed to the environment of interest. As corrosion takes place and metal loss occurs, the cross-sectional area of the sensing element, shown generically in Fig. 11.2 will decrease and the electrical resistance will increase via Eq. (11.1):

$$R = \frac{\rho l}{A} \tag{11.1}$$

where ρ is the resistivity of the metal (element), l is the length of the element, and A is the cross-sectional area. Utilizing Ohm's law, the resistance can thus be measured by

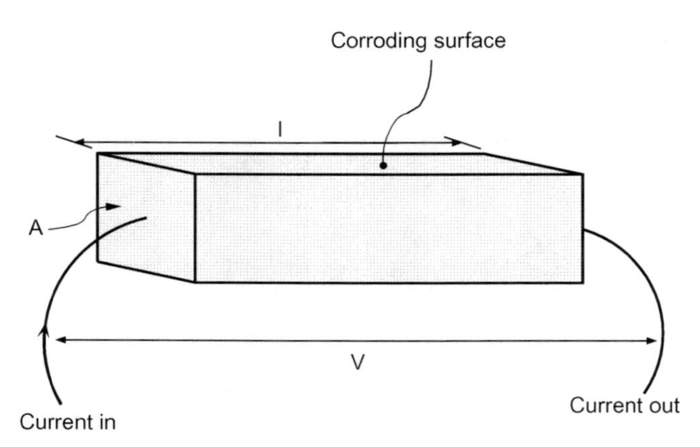

Fig. 11.2 Schematic diagram of ER probe sensing element.

determining the potential (voltage) drop across the sensing element when a small current is applied. The resistance of the sensing element is then compared to the resistance of a separate reference element that is not exposed to the environment. Changes in the ratio of the two resistances as a function of time (the slope of the resistance ratios with time) then yield the corrosion rate. Because resistivity is also a function of temperature, compensation for temperature effects separate from the effect of temperature on corrosion rate needs to be considered. To accomplish this, the reference element is allowed to reach the same temperature as the sensing element, thus any changes in the inherent resistivity due to temperature will be subtracted out.

Sensing probes come in a wide range of forms including wires and loops and tubular and thin film/sheets and strips. Examples of some of these probe designs are shown in Fig. 11.3. The choice of probe element design is often driven by the application and conditions being monitored. For example, thinner elements such as thin films/sheets and small diameter wires can provide very high sensitivity and improved response times for detecting corrosion. However, these elements also will be perforated and will experience significant metal loss as a fraction of the total original much sooner than other forms. Thus, their life expectancy in service will be shorter than thicker forms. The relationship between ER probe response time, probe life, and the expected corrosion rate is illustrated in Fig. 11.4. Because the temperature of the environment will change the resistance of the element and thus introduce an error in the corrosion rate measurement, a thicker probe, especially the tubular form where the reference element can be inside the tube and shows the same temperature as the sensing element, tends to counteract this and improves temperature compensation. By improving temperature compensation, the response time to corrosion rate changes also improves.

It is important also to note that probes can be designed to be inserted into the process stream as well as to be flush-mounted. The use of flush-mounted probes is important when the fluid dynamic properties play a significant role in performance. This could arise in multiphase systems such as mixed product pipelines where liquid petroleum, liquid water, and various gases are present. Under stratified flow conditions, the

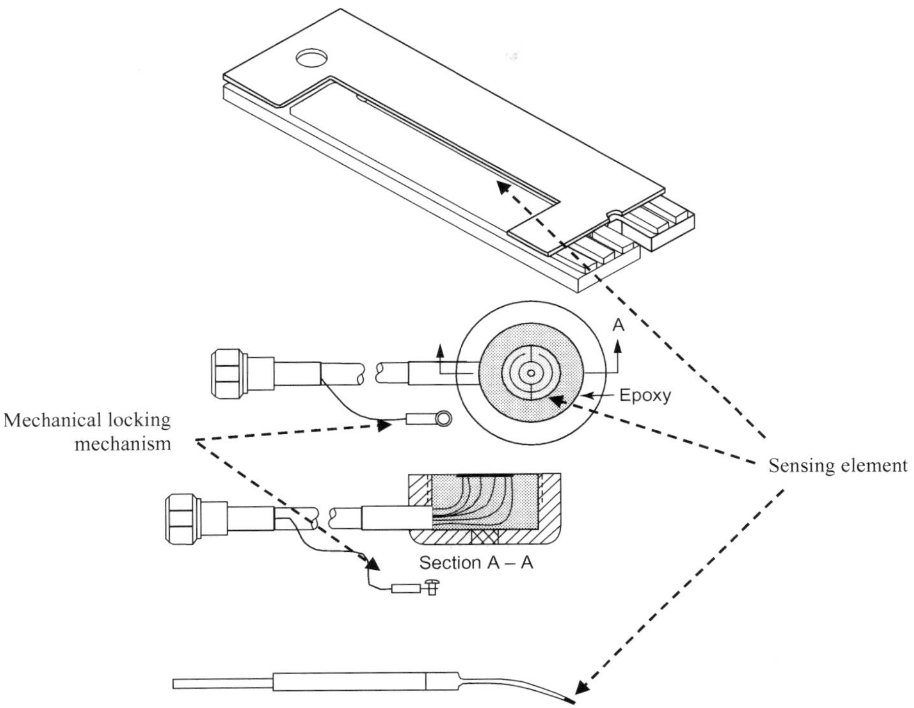

Fig. 11.3 Examples of different commercially available ER probes including illustrations of mechanical locking mechanisms to prevent ejection of the probe when installed in high pressure systems.

water phase will settle to the bottom of the pipeline and thus corrosion monitoring for inhibitor effectiveness at these locations is critically important. If the ER probe is located in the middle of the stream, it will predominantly see the oil phase which may be of lesser interest. Similarly, if entrained solids lead to erosion-corrosion, placement of the ER probe at locations within the system where erosion is most likely (e.g., bends, elbows) would be in order. Thus, not only the design of the probe element itself but how it is inserted and located within the system is important.

11.3 Examples of application and use

For the most part, all of these probe designs discussed previously are suitable for immersion service in any form of electrolyte including aqueous, nonaqueous (e.g., organic solvents, alcohols), concrete, and soils. For some cases, such as atmospheric corrosion, the use of the thin film probe is preferred due to increased sensitivity. Examples of the use and application of ER probes for corrosion monitoring are described below.

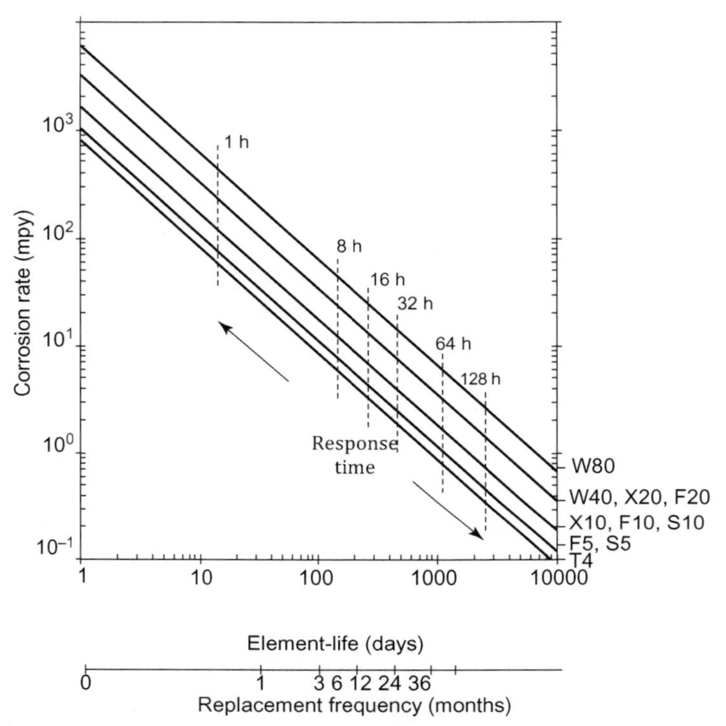

Fig. 11.4 Expected life and response times for ER probes as a function of corrosion rate. Note: 1 mpy = 25.4 μm/year. Different codes indicators (e.g., W80) designate the response of different ER probe sensing element shape and size.

11.3.1 Chemical process and oil and gas industries

Bergstrom [10] described results using an ER probe to monitor the corrosion rate in an acidified organic solvent process stream where attempts were made to neutralize the acid by caustic soda additions. The system had been experiencing a series of unexplained leaks that were diagnosed as caused by a faulty pH probe that was part of the caustic addition control system such that the pH remained very low and the corrosion rate high. In a separate example, Bergstrom [10] showed where an ER probe was able to detect the failure of fluid velocity controlling baffles in a concentrated brine solution (Fig. 11.5). The resultant increase in corrosion rate experienced by the ER probe element was a direct result of the baffles failure leading to increased flow rates.

Torres et al. [11] utilized ER probes to aid in understanding biodiesel plant piping and equipment failures. As a part of their investigation, these authors installed a set of ER probes with 316 stainless steel sensing elements at different locations near an HCl injection point. The HCl was used as a neutralizing agent for the biodiesel-glycerin process where many of the observed failures had occurred as well as in other locations of the process system. Based upon the results obtained using the installed ER probes,

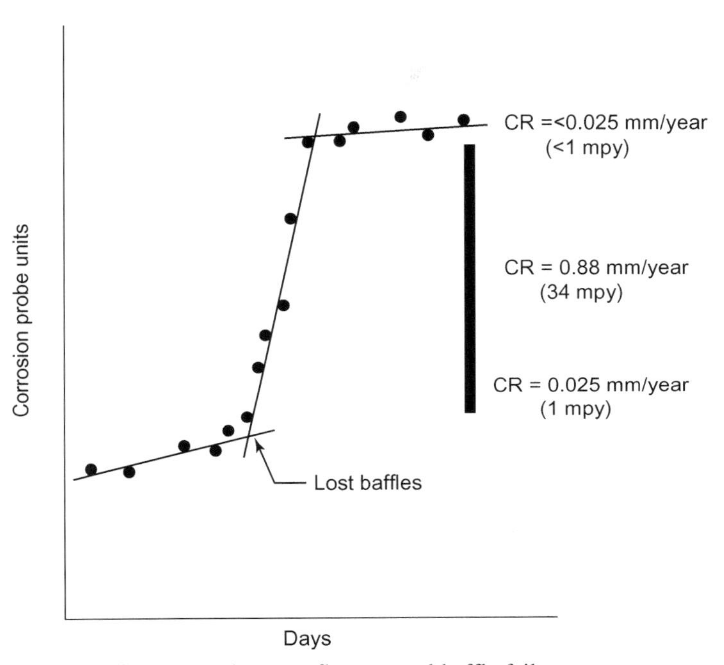

Fig. 11.5 Response of an ER probe upon flow control baffle failure.

the authors were able to determine that one of the primary drivers for failure was a lack of homogenization immediately downstream of the HCl injection points and that modification to promote better mixing might be helpful.

Jarragh et al. [9], as discussed previously, installed ER probes in several locations in three process streams consisting of crude oil, effluent water, and brackish water. In their study, the authors also installed coupons and linear polarization resistance (LPR) probes in an attempt to compare the results of these three methods of corrosion monitoring. Though the installed coupons showed pitting which could be characterized postexposure, the ER probes results did not indicate any pitting even when pits were observed on the sensing elements. The LPR probes were not found to be viable in the crude oil systems because of the highly resistive nature of crude oil. The authors did point out that short-term corrosion rate determination using their ER probe systems was more problematic and less accurate than the long-term corrosion rate determination. Though not fully discussed by the authors, this observation could be a result of the probe configuration selected which can have a significant effect on sensitivity.

11.3.2 Concrete structures

Legat et al. [12] demonstrated that ER probes could detect the addition of corrosion inhibitors to concrete and measure the cumulative effects of corrosion damage. In their study, they also examined the effects of probe depth and the presence of salt on the

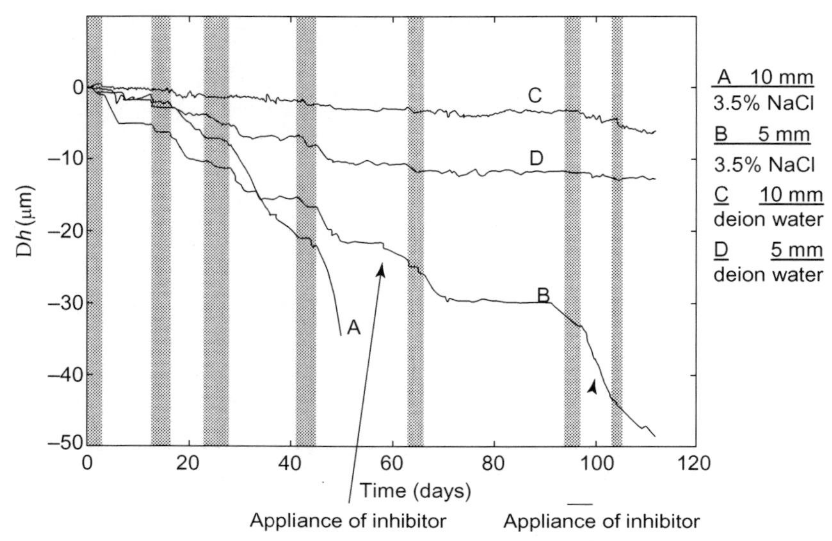

Fig. 11.6 ER probe results embedded in concrete at different depths and exposed to different wetting environments.

corrosion rate. As shown in Fig. 11.6, the probes where deionized water (DI) water was used to wet the concrete showed consistently lower corrosion rates than the concrete wetted with a 3.5% NaCl solution (as would be expected). The thickness loss at locations closer to the surface was also noted to be more dramatic than deeper into the concrete. On the addition of inhibitor to the chloride solution wetted concrete, the thickness loss of the ER probe was noted to stop until the inhibitor was consumed.

11.3.3 Atmospheric

The thin film/strip probe can also be used for atmospheric corrosion monitoring. For example, McKenzie and Vassie [13] used a thin film ER probe and compared the resultant corrosion to that measured using weight loss coupons over a period of 12 months. They demonstrated very good correlation (Fig. 11.7) in total metal loss experienced by the weight loss coupons and the corrosion loss measured using the ER probe at intervals of 3, 6, 9, and 12 months. They concluded that the ER probe results were comparable to the traditionally used standard weight loss coupon and thus could be used to provide better short-term atmospheric corrosion rate estimates, and importantly, can provide an early indication of the onset of corrosion due to higher sensitivity. In addition, they found that the ER probe was able to more effectively determine and quantify the effects from directional wind/ rain exposure than weight loss coupons.

Jung et al. [14] constructed a thin film ER probe to monitor atmospheric corrosion and compared its performance to a traditional time-of-wetness sensor. They were able to demonstrate that the ER probe was much more sensitive to alternating wet/dry

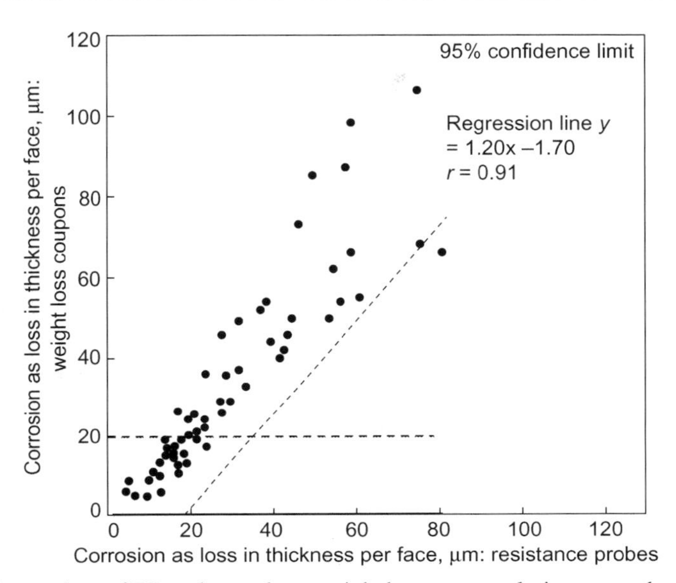

Fig. 11.7 Comparison of ER probe results to weight loss coupons during atmospheric exposure.

conditions and could clearly show the elevated corrosion rates that result from salt concentration effects and increased oxygen transport during the dry out periods. In contrast, the time-of-wetness sensors were unable to clearly distinguish these episodic events.

Though not strictly corrosion in the atmosphere, Li et al. [6] used ER probes to monitor corrosion in the dead space in steel bridge girders. As a part of the corrosion management program for this particular bridge, a dehumidification system was installed to minimize the risk of corrosion inside the box girders. Thin film ER probes to monitor corrosion were deployed as a methodology to provide in situ corrosion rate measurements in relatively real-time. The probes used were custom-fabricated as thin films of steel approximately 1 μm thick. As a result, very high sensitivity was achieved which in the case where corrosion rates are expected to be low, a reasonable probe life can then be expected. Their results indicated that the corrosion rate inside the girder was less than 4 μm/year, while outside the girder the rate was approximately 68 μm/year. Based on their observations, the authors concluded that the dehumidification system was performing as intended and that minimal corrosion was likely to occur inside the girder.

Madden and Baker [15] used ER probes to monitor corrosion under tank bottoms. The primary focus of this study was to monitor the efficacy of vapor phase corrosion inhibitors that were injected into the void spaces to protect the portions of the tank bottoms not in contact with the soil and thus not subject to protection by the cathodic protection system. They were able to show over a period of several years that the corrosion rate with active vapor phase corrosion inhibitor injection averaged just under 0.01 mm/year (0.36 mpy) across 30 tanks. After a period of several years, the vapor phase corrosion inhibitors were consumed and additional injections were not performed and the corrosion rates measured using the ER probes increased to 0.15 mm/year (5.7 mpy). Using ER probes in this way enabled the authors to determine when the inhibitors had been consumed and were no longer providing corrosion protection. The ER probes also provided some quantitative assessment of likely corrosion rates on the tank bottoms that otherwise would only be measurable using internal remaining metal thickness measurements using UT measurements requiring the tank to be taken out of service.

11.3.4 Corrosion in soils

Thin film ER probes have also been used to monitor the onset of microbially influenced corrosion in soils [16]. Fig. 11.8 shows the response of three individual probes that were exposed to soils containing sulfate-reducing bacteria. Some differences in performance were noted and, based on postexposure examination, showed that all the probe elements had formed FeS films, but had experienced ruptures on two of the probes.

Thus, where the ruptures were present, the corrosion rate was substantially higher and the remaining thickness considerably decreased.

Horton et al. [17] described a program to monitor possible corrosion of ductile iron pipe in the Florida Everglades, United States. The Florida Everglades environment is highly corrosive because of its high concentration of decaying organic matter, brackish water chemistry, and high microbial concentrations. Because these pipes traditionally were encased in polyethylene, a determination of the effectiveness of this encasement for this environment was needed. To aid in that determination, both bare and polyethylene-encased ER probes were installed in several locations. The authors were able to distinguish corrosion rate differences between the encased and nonencased ER probes, but also where the probes were located with respect to the pipe. For example, they noted that the corrosion rate measured on top of the pipe (12 o'clock

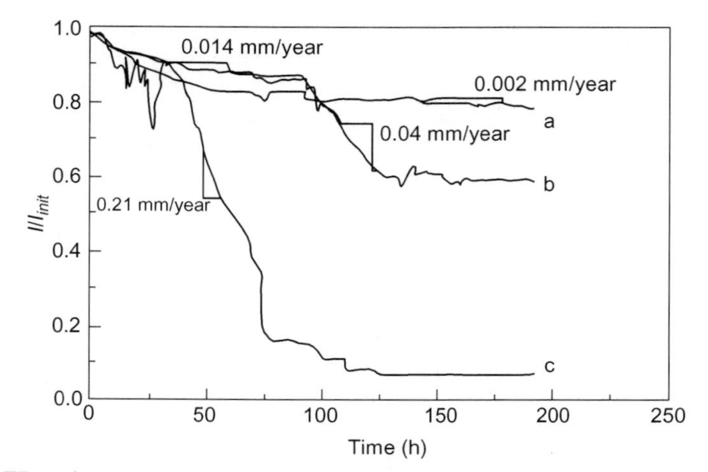

Fig. 11.8 ER probe response in soils containing sulfate-reducing bacteria.

position) showed virtually no corrosion, whereas underneath the pipe (6 o'clock position) corrosion rates on the order of a few mils per year (mpy) (or 0.025–0.1 mm/year) were observed. This work helps to demonstrate not only that corrosion and where it occurs on a structure is highly spatially dependent, but also the utility in using ER probes to help make these determinations.

Martin [18] described efforts to better understand and mitigate potential telluric current (current induced in buried metallic structures due to the earth's magnetic field) corrosion effects on buried pipelines in Australia. To accomplish this, he attached a series of ER probes to the cathodic protection (CP) system of the pipeline and then compared the pipe potential and CP system current draw with the ER probe data. He was able to show that in some cases telluric currents did in fact result in appreciable changes in the corrosion rate. This work illustrates an important advantage that ER probes have over many other monitoring methods. If a system is protected using cathodic protection, the ER probe can likewise be polarized and the resultant corrosion rate while under CP can be determined. In contrast, other methods such as linear polarization resistance (see Chapter 3) cannot easily be modified to enable this capability.

Bell et al. [19] reported on the development and use of ER probes to monitor the corrosion of ductile iron wastewater lines. In previous attempts to monitor corrosion using ER probes, carbon steel sensing elements were utilized and higher corrosion rates were measured compared to the actual corrosion rates experienced by the ductile iron. The authors attributed this difference in corrosion behavior to the presence of a more protective oxide on the ductile iron surface. As a result, the authors developed a set of ER probes that were constructed by machining out sections of a ductile iron pipe that had been thermally aged to create this protective oxide. In their work, Bell et al. [19] clearly showed that matching the sensor element to the chemistry, metallurgical condition, and other factors to the actual structure of interest was a key component to obtain corrosion rate measurements that could be reliably acted upon as a part of an overall corrosion management program.

ER probes have also been used in an attempt to better understand long-term corrosion of copper for use as a containment material for nuclear waste disposal. Though used as part of a laboratory study and not field-monitoring, Marja-aho et al. [20] were able to demonstrate that thin film ER probes provided a more sensitive measurement of the corrosion rate of copper in simulated groundwater environments in the presence and absence of sulfate-reducing bacteria. Their results led them to conclude that the planned use of copper in this application was suitable, but also that the thin film ER probes would serve as an effective tool to monitor the integrity of the waste storage packages once in service.

11.4 Sensing probe electronics and instrumentation

The electronics used for ER probe measurements are shown schematically in Fig. 11.9 and essentially consist of a Wheatstone bridge configuration. A Wheatstone bridge is particularly well-suited to determine the value of an unknown resistance (in this case, the resistance of the sensing element, R_X). Because the values for the other resistances are known (R_1 and R_2 are in the meter and R_3 is the unexposed reference probe that has

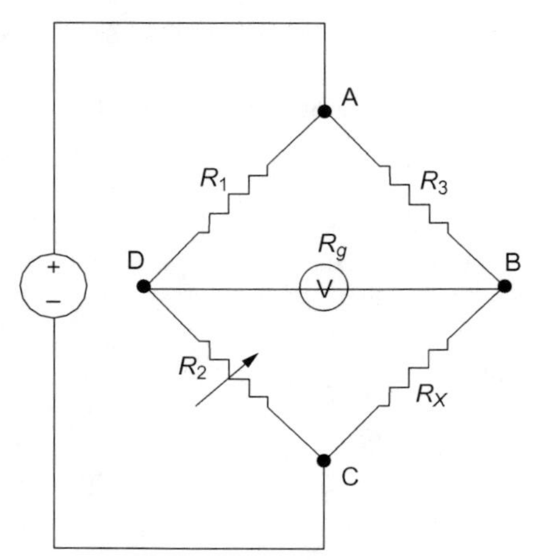

Fig. 11.9 Wheatstone bridge.

the initial resistance of the sensing element without experiencing any corrosion), if the ratio of R_X/R_3 equates R_2/R_1, then the current flow and voltage between points B and D will be zero. Using a sensitive galvanometer, the current between these two midpoints can be measured very accurately and can be nulled out by adjusting the resistance of R_2. With no current flow between the midpoints, the total resistance of the entire circuit is given by Eq. (11.2):

$$R_T = \frac{(R_1 + R_2) \cdot (R_3 + R_X)}{R_1 + R_2 + R_3 + R_X} \tag{11.2}$$

Since R_T can be measured and R_1, R_2, and R_3 are known, R_X can be solved. Essentially all commercial instruments automatically determine the sensing element resistance and either provide that resistance value on the meter or take the additional steps to compare previous readings to determine a corrosion rate.

Several different instrumentation designs are available. Some of the simpler systems are battery-powered and consist of a dial to adjust (null) out the sensing probe resistance, thus providing its value. Other systems are more complex, but still in the form of battery-powered portable units, that automatically convert the probe measurement response into a corrosion rate by comparing previous measurements and are displayed as a digital readout. Other systems are available that are intrinsically safe for applications in fire-hazardous locations and are permanently mounted in place. These permanently mounted systems are then read using a portable reader or automatically by a data logger.

11.5 Variations on the ER theme

Modifications and variations on the ER theme have also been developed that encompass changes to how the measurements are made to increase sensitivity and stability and to allow the application of ER measurement principles to the actual structure.

Each of these advancements provides an improvement upon the overall ER probe concept, but the underlying principles and applications still generally apply.

11.5.1 Inductance method

The use of inductive resistance to detect metal loss in the sensing element instead of directly measuring the resistance is an example of one of these changes. Patented under the trade name Microcor, changes in the inductive resistance of a coil in proximity to the sensing element are monitored [21]. Because changes in the magnetic permeability of the sensing element will be more pronounced than changes in its resistance, these magnetic changes result in potentially significant changes in the inductive resistance of the coil. As a result, it has been reported that this configuration results in decreases in response times by factors of between 100 and 2500. In addition to improvements in response times, improved temperature sensitivity (less sensitive to temperature changes) was also reported. (Microcor was originally developed and patented by Cortest Instrument Systems (Ohio, United States). Microcor is now available from Rohrback Cosasco Systems (California, United States).)

When compared to linear polarization resistance method (see Chapter 3) and traditional ER methods, the inductive resistance method was shown to provide comparable results (Figs. 11.10 and 11.11). In Figs. 11.10 and 11.11, the linear polarization resistance probe measurements are the same and the difference in response between the inductive resistance probe and the traditional ER probe can be seen. When the environment corrosivity changed from less than 5 mpy (0.125 mm/year) for steel to variable corrosion rates between 15 and 30 mpy (0.375–0.75 mm/year), the inductive resistance probe was able to detect the change in the corrosivity as evidenced by the change in slope. In contrast, the changes in corrosivity were not detected by the standard ER probe over the same time period. In addition, careful examination of the

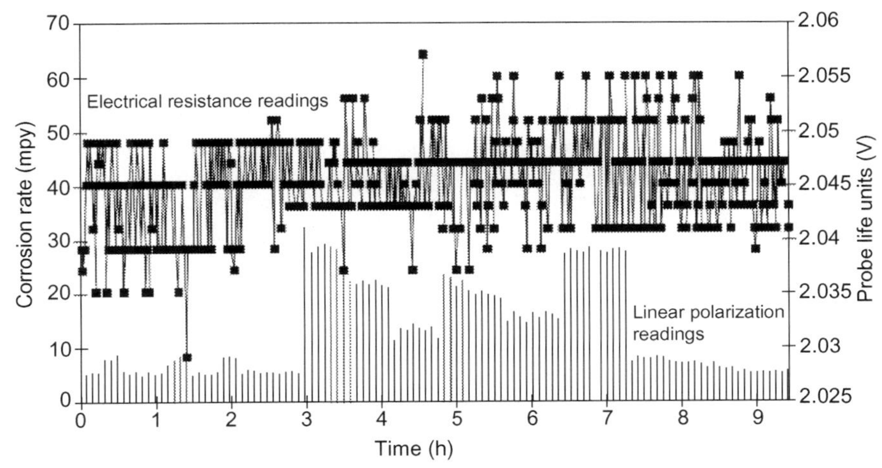

Fig. 11.10 Comparison of standard ER probe to linear polarization resistance under the same conditions as shown in Fig. 11.9. Note: 1 mpy = 25.4 μm/year.

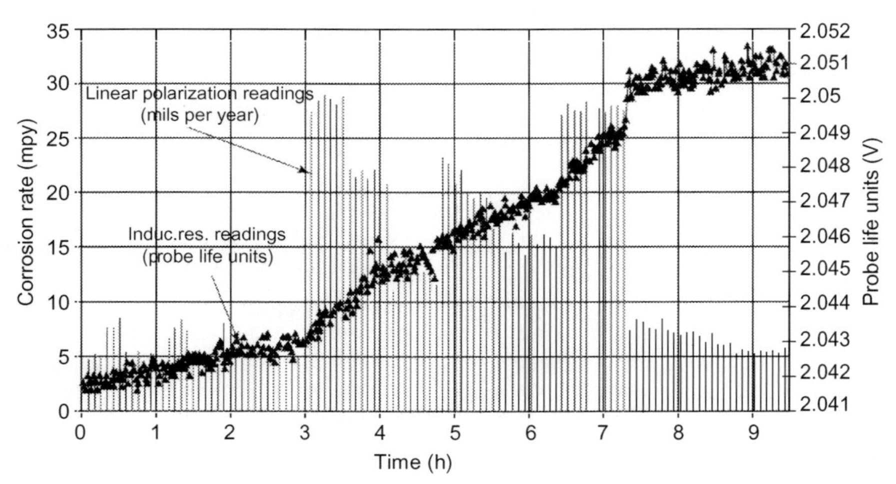

Fig. 11.11 Comparison of linear polarization resistance to inductive resistance ER probe. Note: 1 mpy = 25.4 μm/year.

inductive resistance probe results also shows that some of the corrosion rate changes measured by linear polarization resistance method could also be detected as small changes and deviations in the slope of the inductive resistance probe data.

The inductive resistance probe, however, has some limitations. For example, because it relies on changes in the magnetic permeability of the sensing element as corrosion occurs, only highly magnetic materials can be used. Thus, carbon steel (and other ferritic steels) can be used; however, nonmagnetic materials cannot and weakly magnetic materials may not work as well. As a result, monitoring systems constructed from aluminum, copper, and nickel alloys as well as austenitic stainless steels will not be possible with this probe system. For comparison purposes, the magnetic permeabilities of aluminum and copper are relatively similar and are only 0.1% of the value of steel. Other reported limitations with the inductive resistance probe include a short probe life and the influence of mechanical stresses on magnetic properties resulting in erroneous readings.

CEION, produced and marketed by Cormon Ltd. (United Kingdom), is also supposed to be based on an ER probe type concept. The probe, which appears to be in the form of a spiral ring embedded in epoxy, has been used in a wide range of oil field applications and has been shown to provide reasonable agreement with corrosion rate measurements found using other methods (Fig. 11.12). Though this technology appears promising and seems to be based in some fashion on ER principles, the precise functionality and background of the method are unclear. Based on what is available in company-provided information, the resistance of the element is apparently determined using alternating current methods rather than direct current. It has also been claimed that the resolution of the probe and its associated electronics is on the order of 1 nm [22, 23]. Others who report using ER probes that appear to be similar report resolution that are sub-μm [5]. Because of the limited amount of independent

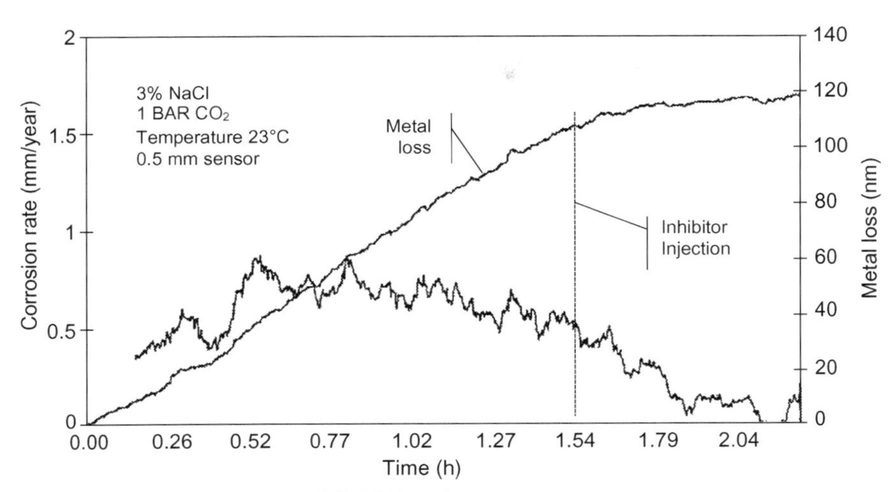

Fig. 11.12 Corrosion response of CEION probe.

information available and the lack of details concerning the operating principles, it is difficult to determine the accuracy of these claims.

11.5.2 Field signature method

A slightly different variation on the ER probe concept is the use and application of the same operating principles, but applied to the entire structure of interest. The Field Signature Method (FSM, see Chapter 25 for additional information), developed by the University of Oslo in Norway in the mid-1980s and presently marketed by CorrOcean, is capable of detecting metal loss and other defects by observing the changes induced in the way that current flows through the structure. To accomplish this, a series of pins are attached to the structure in an array spaced at approximately two to three times the wall thickness. When a current is applied to the structure, the current flow and resultant potential drop due to the metal's resistance are measured. When corrosion occurs, this resistance between a given set of sensing pins will increase (similar to an ER probe) and the potential drop across that region will also increase. When the original potential distribution "signature" is compared to subsequent readings, the effects of corrosion can be determined. This method has a nominal sensitivity reported to be in the order of 0.1% of the wall thickness and has the significant advantage of monitoring the actual component. That is, instead of monitoring a sensing element, the structure in question is itself the sensing element. The other advantage of FSM is that it is nonintrusive, which is an important aspect for application in internal corrosion monitoring, especially if the system is pressurized.

FSM has been widely used for monitoring corrosion in subsea pipelines, onshore pipelines, in gas plants, and in the refinery industry. Strommen [24] described the results of a 7-year internal corrosion monitoring program using FSM on subsea pipelines. A comparison of the FSM response as compared to independent corrosion rate measurements in the well fluids showed excellent agreement (0.11 mm/year by FSM compared to 0.12 mm/year). Scanlan et al. [25] used the FSM technique to monitor possible corrosion at a series of locations within a refinery and noted that the method could provide a three-dimensional image of metal loss at bends and elbows and could distinguish between pitting and general corrosion.

Kawakam et al. [26] successfully used FSM to monitor fatigue cracks on steel-constructed bridges. Multiple steel decks were noted to have experienced fatigue cracks and FSM was deployed to monitor crack growth. Because FSM has been shown to be able to detect cracking, its success in this application was not surprising. What was unique in this study was that the authors examined the effects of progressively increasing the distance between adjacent sensing pins. If close sensing pin spacing was required to effectively monitor crack growth, the installation of such a system to monitor an entire bridge would be impractical. The authors were able to successfully place pins on 10–12 m spacing and still be able to detect fatigue crack growth.

Identification and detection of pits can be challenging using FSM. Gan et al. [27], however, conducted research that indicates that detection of small pitting defects may be more readily possible using FSM depending on how the data was analyzed. By conducting a systematic study of pits of different diameters and depths, the authors were able to demonstrate that the traditional approach used gave erroneous measurements, particularly for small diameter pits. By analyzing the data differently, it was shown that better resolution and sizing of small pits was possible. The authors did, however, state that additional work was necessary to make the approach more robust and to account for a broader set of conditions and pit densities.

11.6 Advantages and limitations

ER probes are often chosen due to their relatively simple operation, low cost, and ease of operation and interpretation. Because the ER probe method does not rely on electrochemistry, it can be used in nearly any condition including atmospheric and high resistivity environments where electrochemistry-based monitoring tools often experience difficulties or cannot be used. In addition, ER probes can be connected to a cathodic protection system which then enables the monitoring of structures that utilize CP for corrosion mitigation.

ER probes, however, do have some limitations. First, the sensitivity of the ER technique is generally too low to show rapid changes in corrosion conditions. As shown in Fig. 11.4, if a corrosion rate on the order of 1 mpy (25 μm/year) is expected, the response time for detecting that rate is between 1 and 10 days. On the other hand, if the expected corrosion rate is 100 mpy (2.5 mm/year), the response time is significantly diminished to a matter of less than a few hours. Increased sensitivity can be achieved using thinner/ smaller elements, but this comes at the sacrifice of probe life.

Limitations also arise when corrosion mechanisms other than uniform, general corrosion are expected or encountered. The reduction in the cross-sectional area that results in an increase in probe resistance is assumed to be uniform. Depending on the probe size and geometry, the presence of a pit or stress corrosion cracking

(SCC) may or may not be detected since the reduction in the cross-sectional area associated with a pit or a crack can be quite small.

Additional limitations arise when the possibilities of conductive corrosion products or scales are precipitated on the probe. For example, sulfide environments are particularly difficult to interpret using ER probes because the iron sulfide corrosion product that forms is conductive which will tend to underestimate the amount of metal loss experienced. Iron sulfide precipitation has been heavy enough in some cases to indicate that metal gain had taken place. This type of response, then, somewhat limits the applicability of ER probes in sour oil and gas systems.

11.7 Summary and conclusions

ER probes have been in use since the mid-1950s and provide a valuable tool for the corrosion engineer. They are easy to use, interpret, and can provide highly valuable information regarding the propensity for a system to experience corrosion. Improvements and modifications have also been developed that help enhance the sensitivity and responsiveness of the overall ER probe concept. As research continues, it is expected that the sensitivity to corrosion, the response time to corrosion events (e.g., process upsets and transient corrosion events), and the ability to detect localized corrosion events will improve.

References

[1] A. Dravnieks, H.A. Cataldi, Corrosion 10 (1954) 224.
[2] A.J. Freedman, E.S. Troscinski, A. Dravnieks, Corrosion 14 (1958) 175.
[3] G.K. Brown, Pipeline and Gas Industry, (1996) 53April.
[4] M.A. Clarke, Ind. Corrosion 9 (1991) 11.
[5] B. Ridd, T.J. Blakset, D. Queen, Corrosion98, paper no. 78(1998).
[6] S.Y. Li, S. Jung, K.-W. Park, Corrosion2008, paper no. 08297(2008).
[7] A. Abdulhadi, M.F. Al-Subaie, A.M. Al-Zahrani, M.M. Al-Qarni, Corrosion2007, paper no. 07265(2007).
[8] R.D. Davies, Appita, vol. 39, (1986) 472.
[9] A. Jarragh, S. Al-Sulaiman, A.R. Al-Shamari, B. Lenka, M. Islam, S. Prakash, Corrosion2014, paper no. 3934(2014).
[10] D.R. Bergstrom, Materials Performance, (1981) 17September.
[11] C.E. de Almeida Souza Torres, C.G.F. Costa, A.P. Pereira, M. das Mercês Reis de Castro, V. de Freitas Cunha Lins, Eng. Fail. Anal. 66 (2016) 365.
[12] A. Legat, V. Kuhar, M. Leban, A. Vernekar, Corrosion2003, paper no. 03390(2003).
[13] M. McKenzie, P.R. Vassie, Br. Corros. J. 20 (1985) 117.
[14] S. Jung, Y.-G. Kim, H.-S. Song, S.Y. Li, S.-M. Le, Y.-T. Kao, Corrosion2003, paper no. 03392(2003).
[15] C. Madden, K. Baker, Corrosion2018, paper no. 11540(2018).
[16] S.Y. Li, Y.G. Kim, K.S. Jeon, Y.T. Kho, T. Kang, Corrosion 57 (2001) 815.
[17] A.M. Horton, D. Lindemuth, G. Ash, Materials Performance, (2006) 50May.
[18] B.A. Martin, Corrosion 49 (1993) 343.
[19] G.E.C. Bell, C.G. Moore, S. Williams, Corrosion2007, paper no. 07335(2007).

[20] M. Marja-aho, P. Rajala, E. Huttunen-Saarivirta, A. Legat, A. Kranjc, T. Kosec, L. Carpén, Sens. Actuators A Phys. A274 (2018) 252.
[21] A.F. Denzine, M.S. Reading, Corrosion97, paper no. 287(1997).
[22] B.J. Hemblade, J.R. Davies, J. Sutton, Corrosion99, paper no. 225(1999).
[23] G.K. Brown, J.R. Davies, B.J. Hemblade, Corrosion2000, paper no. 00278(2000).
[24] R.D. Strommen, Corrosion2002, paper no. 02251(2002).
[25] R.J. Scanlan, R.M. Boothman, D.R. Clarida, Corrosion2003, paper 03655(2003).
[26] Y. Kawakam, H. Kanaji, K. Oku, Proc. Eng. 14 (2011) 1059.
[27] F. Gan, G. Tian, Z. Wan, J. Liao, W. Li, Measurement 82 (2016) 46.

Nondestructive evaluation technologies for monitoring corrosion

12

Glenn Light
Southwest Research Institute, San Antonio, TX, United States

12.1 Introduction

Corrosion begins and increases over time, and in most cases, is hidden from view either inside a component such as a pipe or under a surface covering such as thermal insulation or concrete. Unfortunately, corrosion does not grow predictably. The present practice is to estimate a period of time over which corrosion will not cause a catastrophic failure and conduct an inspection to measure remaining wall thickness at specific intervals. This is usually done by marking certain areas on the component and attempting to collect data at the same location each time so that any changes in the component wall can be detected. This approach sounds like a logical one, but many times data are collected at specific spatial locations—for example, on 50-mm centers. There is no guarantee that the corrosion effect is occurring at one of the sites where data are collected. In fact, there can be a situation where a throughwall corrosion pit can grow next to a region that is being periodically inspected and the defect never detected. So, this type of process has uncertainties and can lead to unexpected failure.

In addition, corrosion can occur on the inside or outside. The product flow or product process may lead to corrosion on the inside of the pipe or component. Other times, the corrosion is caused by the environment. For example, corrosion on the outside surface of a pipe under the insulation is caused by the moisture condensed between the outside surface of the pipe and the insulation.

A better way to determine the amount and severity of damage being caused by corrosion is to use sensor systems that can remain in place to monitor a component while the process that causes corrosion is ongoing. This process is called "condition monitoring." This monitoring can be either active or passive.

12.2 NDE methods for corrosion monitoring

There are a number of nondestructive evaluation (NDE) methods that can be used to monitor the condition of a component for defects such as corrosion. These include ultrasonics, eddy current, acoustic emission, guided waves, and infrared thermography.

Techniques for Corrosion Monitoring. https://doi.org/10.1016/B978-0-08-103003-5.00012-6

In the following sections, typical applications of these techniques and limitations are discussed, including probes, electronics, and software.

For condition monitoring programs, it is necessary to establish the baseline condition of any component or installation. This may be done at any time, but should preferably be implemented directly after construction. In an early stage, one should consider using a type of inspection that may be applied in an equal fashion in a later stage during service. Periodic condition assessment allows trending and prediction of the remaining mean-time-to-failure (MTTF). Operational lifetime extension may be agreed upon with authorities and increased plant availability is the result [1]. Lifetime calculation models may predict the remaining service time of any asset. These calculations require input data provided by the asset's history and available nondestructive testing (NDT) data. The outcome can be only as reliable as the input. High probability of detection (POD) is required to ensure reliable operation until the next shutdown, and a low false call rate (FCR) is desirable to avoid unnecessary maintenance work. Accurate and highly reliable data result in a reliable prediction of the remaining service time.

The number and methods of inspection are always a trade-off between the minimum requirements for safe operation and the amount of information needed for optimum maintenance management. In a baseline inspection, it is important to establish, as accurately as possible, the zero condition of the component. It is often good to use several complementary techniques, but one should definitely be the technique that will be used to monitor the equipment. For example, if ultrasonics is the method to be used for monitoring a region, the baseline condition might be obtained using radiography as well as ultrasonics. Care should be taken to precisely establish coordinate systems used for the component and to precisely mark and notate any anomalies that are found.

12.2.1 Ultrasonic monitoring technologies

Ultrasonic monitoring for corrosion/erosion and general wall-thinning processes can be conducted by collecting periodic 0-degree L-wave wall thickness measurements using individual probes, a single probe that is scanned or an array of probes. Ultrasonic corrosion mapping can provide quantitative information on corrosion of vessel plate and welds, and even the condition of certain internal coatings, and is an invaluable and widely used tool. Ultrasonic thickness is based on the time required for the ultrasonic pulse to travel from the front surface of the component to the back and return, based on the known ultrasonic velocity in the component. The limitations of the method are the maximum temperature at which inspection may be carried out, usually lower than approximately 94°C, unless methods for keeping the ultrasonic sensor cool and maintaining fresh and effective couplant are part of the process [2]. One approach is to have a constant supply of couplant applied to the probes, but this could also lead to variations in coupling efficiency. However, the thickness measurement is primarily a time measurement, not an amplitude measurement. In addition, coatings on the probe surface can attenuate the ultrasonic energy going into the part. This is especially true if the coating is thick and possibly delaminated. Also, if the corrosion occurs on the front surface, the ultrasonic beam is scattered by the front surface corrosion, making this approach ineffective. Thus, ultrasonics should primarily be used when the corrosion process is on the inside surface of the component being monitored. Ultrasonic weld monitoring can be done as well, using fixed probes or arrays for shear wave and/or time-of-flight diffraction (TOFD) data to detect weld cracking. Again, the primary limitations of these techniques include providing sufficient long-term couplant as well as access and temperature [1].

One example of a corrosion monitoring system that uses a 0-degree longitudinal mode thickness measurement transducer to measure wall thickness over a period of time under a transducer that is clamped in place is the MIST UT Monitor developed by CC Technologies, located in Dublin, Ohio, USA. The device is fairly simple, but requires that the exact location be known where monitoring is required. This is best used when an inspection reveals a corrosion pit or defect that is at an acceptable level, but is expected to grow. The MIST UT Monitor can be clamped onto the pipe or component at that location and ultrasonic thickness measurements collected periodically.

More complex units have also been developed because recent developments in the fields of electronics, computer systems, and communication technology have enabled the use of new concepts for inspection and monitoring. One such development is the ULTRAMONIT system developed by Cole and Gautrey [2], where an array of 0-degree ultrasonic sensors is placed permanently onto the pipe. The location of the sensor array is usually at a field joint directly at or close to a weld so that the array can monitor for wall-thinning corrosion and cracks. Permanently installed sensors are advantageous because exact position, orientation, and acoustic coupling are maintained between surveys. Compared to conventional ultrasonic inspection, this concept provides significantly improved possibilities for high-resolution data trend analysis. This technology has been applied for both subsea and land facility use.

In the subsea embodiment, the array of 0-degree transducers is implemented as an instrumented pipeline clamp, where the required electronics and sensors are placed and protected in a rugged steel structure as illustrated in Fig. 12.1 [3]. Reliability was a key objective for this system and has been addressed by keeping the permanently installed electronics as simple as possible, while making the more complicated parts of the instrumentation accessible in a remotely operated vehicle (ROV)-carried unit. This ROV is used for interrogation with the sensors by the use of a specially developed inductive coupler, which provides power, analog signal for the ultrasonic transducers, and digital communication for the signal multiplexer. This unit is controlled digitally, and the number of channels is limited only by signal degradation due to total spanned length of analog cable, as the channel count gets very high. For the land facility version, the same system has been installed on a 30-cm pipe at the Kårstø gas processing plant located near Stavanger, Norway. This unit has run since December 2003.

It is important to understand that the application of this technology is very localized and requires that the location of corrosion attack be known so that the sensors can be placed in that area to effectively monitor for potential corrosion failure. If the corrosion begins to affect other parts of the component where the sensor is not located, corrosion could go undetected.

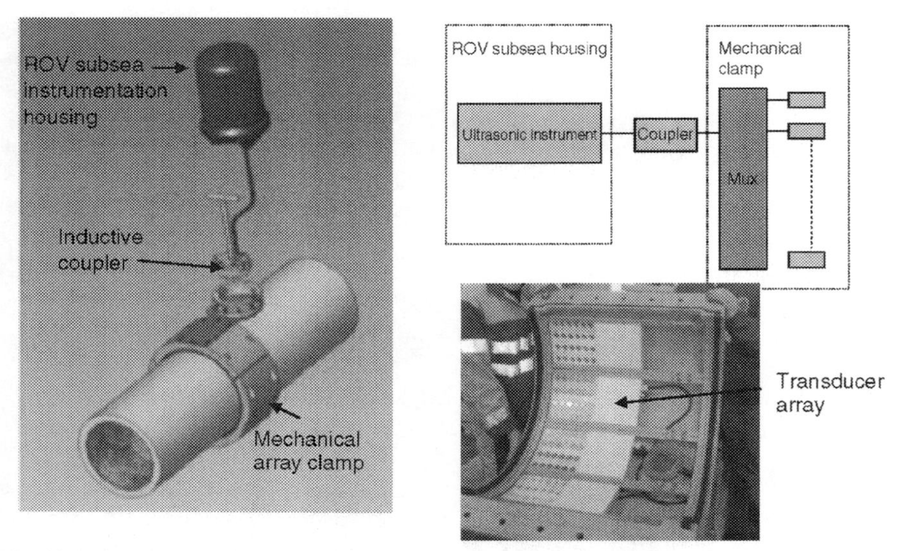

Fig. 12.1 Illustration of the ULTRAMONIT array system used on subsea pipe applications with subsea sensor clamp, inductive coupler, and main instrumentation subsea housing that contains the ultrasonic instrument and multiplexer (MUX).

12.2.2 Eddy current

Eddy current inspection can also be a very useful method for monitoring external corrosion in components by measuring the liftoff between the component nominal surface and the surface of the pit on the metal. This requires either a localized scanning method or an array of probes. An example of an eddy current array is the conformal array shown in Fig. 12.2. The array is flexible; however, to effectively use this to monitor, a means to locate the array precisely each time is needed [4]. Temperature and access to the surface are the main limitations of eddy current.

Pulsed eddy current can operate without having close proximity to the surface. A pulsed eddy current technique that uses a stepped or pulsed input signal for the detection of corrosion under insulation (CUI) has been evaluated. This allows the detection of wall-thinning areas in various types of pipe, including riser pipe, without removing the outside coatings. In addition, it is found that filmless, real-time, and digital radiography can be used to find internal and external corrosion defects in an insulated splash zone, while the riser remains in service. A survey of NDE manufacturing companies, NDE inspection companies, and operating companies was completed to collect information about current instrumentation and inspection/operators' experience for riser inspection. Examples of advanced riser inspection instrumentation and field results were included. The ability of the candidate technologies to be adapted to riser variations, the stage of standardization, and costs was also discussed.

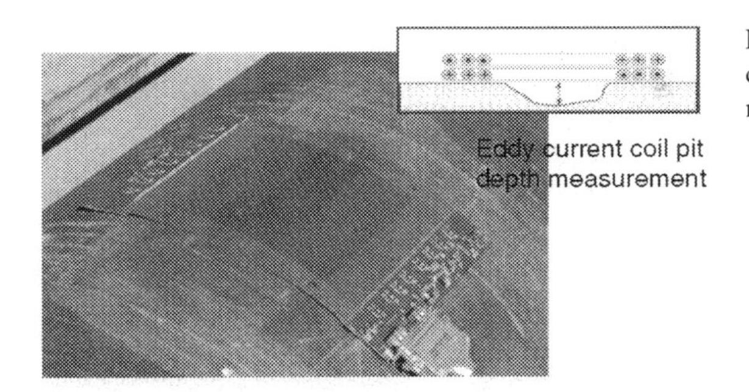

Fig. 12.2 Photograph of the conformable array on top of a pipe monitoring the pit depth.

Pulsed eddy current equipment has been successfully applied in corrosion detection for several years. Whereas field experience on insulated objects has grown significantly, the technique's characteristics also make it highly suitable for other field situations where the object surface is rough or inaccessible. Because (surface) preparations can be avoided, the tool provides a fast, cost-effective solution for corrosion detection.

Röntgen Technische Dienst (RTD) developed a pulsed eddy current tool (called the RTD-INCOTEST) that has been used by the refinery industry for the detection of corrosion under insulation (CUI). It allows the detection of wall-thinning areas without insulation removal. However, this is a point measurement. Currently, a network of 12 companies worldwide operate a total of 32 systems. RTD-INCOTEST applies pulsed eddy currents for the detection of corrosion areas. A pulsed eddy current technique uses a stepped or pulsed input signal, whereas conventional eddy currents use a continuous signal. The advantages of the pulsed eddy current technique are its larger penetration depth, relative insensitivity to liftoff, and the possibility to obtain a quantitative measurement result for wall thickness. This leads to the characteristic that makes it suitable for the detection of CUI: no direct surface contact between the probe and the object is necessary. Also, this tool can be employed in other field situations where the object surface is rough or inaccessible.

The applied operating principle of pulsed eddy current systems can vary from system to system. In order to obtain a quantitative reading for wall thickness, the RTD-INCOTEST uses a patented algorithm [5] that relates the diffusive behavior in time to the material properties and the wall thickness. It operates on low alloy carbon steel, as illustrated in Fig. 12.3.

A pulsed magnetic field is sent by the probe coil. This penetrates through any nonmagnetic material between the probe and the object under inspection (e.g., insulation material). The varying magnetic field will induce eddy currents on the surface of the object. The diffusive behavior of these eddy currents is related to the material properties and wall thickness of the object.

The detected eddy current signal is processed and compared to a reference signal. The material properties are eliminated, and a reading for the average wall thickness within the magnetic field area results. One reading takes a couple of seconds. The

Fig. 12.3 Principle of operation RTD-INCOTEST.

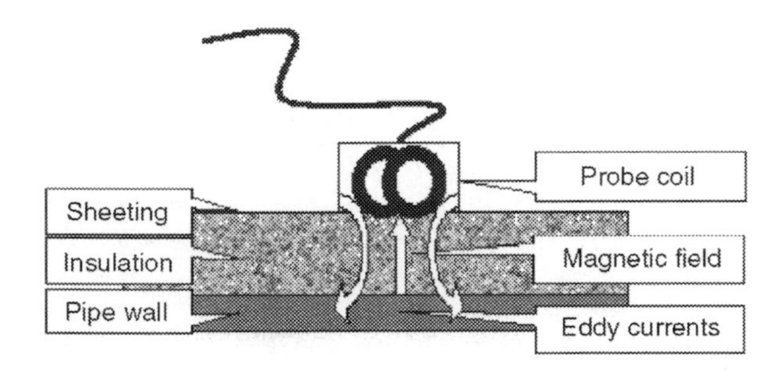

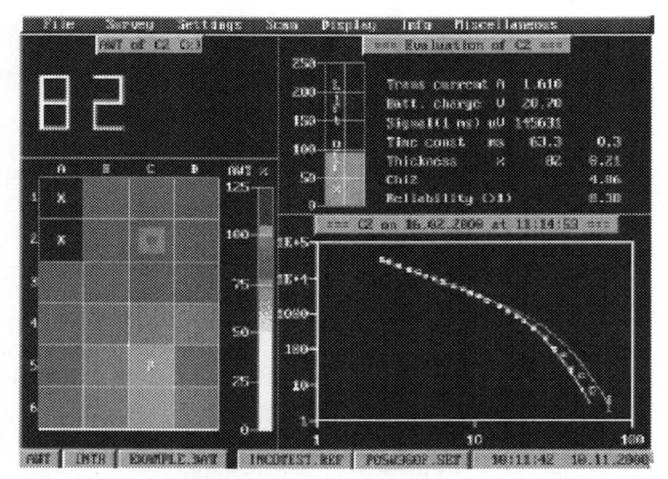

Fig. 12.4 Display of RTD-INCOTEST showing AWT (Average Wall Thickness) reading *(top left)*, logged inspection grid *(bottom left)*, and the decay of the eddy currents *(bottom right)*.

signal is logged and can be retrieved for later comparison in a monitoring approach. An example of the RTD-INCOTEST data output is shown in Fig. 12.4.

The area over which a measurement is taken is referred to as the footprint. Probe design is such that the magnetic field focuses on an area on the surface of the object. The result of the measurement is a reading of the average wall thickness (AWT) over this footprint area. The size of this area is dependent on the insulation and object thickness, as well as the probe design. Roughly, the footprint can be considered to be in the order of the insulation thickness. Due to the averaging effect, detection of highly localized defect types like pitting is not reliable with this tool. Although the AWT reading is not a direct replacement of the commonly used ultrasonic testing (UT)-obtained minimum wall thickness, a quantitative result is obtained that can be interpreted unambiguously.

The outer application ranges of the RTD-INCOTEST tool can be described by:

- Low alloy carbon steel
- Pipe diameter $>$ 50 mm or 2 in.
- Nominal wall thickness between 6 and 65 mm

- Insulation thickness $\leq$ 150 mm
- Sheeting thickness $\leq$ 1 mm stainless steel, aluminum, galvanized steel
- Object temperature $> -100°C$ to $<+500°C$

These ranges are determined on condition that a reliable signal can be obtained under regular field conditions.

12.2.3 Acoustic emission and equipment

Acoustic emission (AE) is a passive NDE technique that makes use of the high-frequency acoustic energy emitted by an object that is undergoing stress, such as when corrosion products formed on a corroding rebar push out on the concrete surrounding it. The primary advantage AE offers over more conventional NDE techniques is that it results directly from the process of flaw growth. Slow crack growth in ductile materials produces few AE events, whereas rapid crack growth in brittle materials produces large numbers of AE events [6–17].

Acoustic emission can also be created as a result of corrosion-induced phenomena. In concrete, for example, corrosion can produce cracking that generates acoustic emissions. In stainless steel, primary AE was produced by the falling-off of grains due to the mutual actions of anodic dissolution and mechanical fracture along a chromium-depleted zone in the grain boundary, and secondary AE was produced by hydrogen gas evolution and by the cracking of corrosion products [6]. AE is able to test/monitor 100% of a structure using relatively few stationary sensors mounted to the structure. AE techniques are primarily sensitive to active defects caused by stress. AE may also be generated by cracking of corrosion products. Unlike most methods, there is no probe temperature limitation with AE, as sensors may be coupled through metal acoustic waveguides welded to the plant, the sensor being outside the insulation.

This technology has been used for detection of corrosion in steel rebar encapsulated in concrete on bridges. AE monitoring of concrete was used to detect rebar corrosion, film cracking, gas evolution, and microcracking. Although attenuation of the AE signal in the concrete was a concern in the past, unique placement of the AE transducers on the reinforcing steel and using the steel as the sound propagation medium allowed the onset of steel corrosion to be detected. In addition, it was possible to use the AE signal to calculate the location where the steel corrosion was occurring, allowing bridge inspectors to determine the extent of corrosion damage. Thus, AE monitoring appears to be a promising technique that can be used for bridge inspection to quantify the condition of steel-reinforced concrete, where corrosion is occurring and where repair is needed.

A typical AE monitoring system uses piezoelectric sensors acoustically coupled to the test object with a suitable acoustic coupling medium (grease or adhesive). The output of the sensors is amplified and filtered by preamplifiers and then fed to the monitor via shielded coaxial cables. The monitor further filters and amplifies the AE signals, processes the data, and displays the results. Both results and raw data are typically recorded for archival purposes or for posttest analysis; for instance, to determine location of the AE signal.

The disadvantage of AE is that commercial AE systems can only estimate qualitatively how much damage is in the material and approximately how long the components will last, so other NDE methods are still needed to do more thorough examinations and provide quantitative results. Service environments are generally very noisy, and the AE signals are usually very weak. Thus, signal discrimination and noise reduction are very difficult, yet extremely important for successful AE applications.

12.2.4 Guided waves and equipment

One of the fastest growing techniques for monitoring corrosion is the use of long-range ultrasonics, primarily using guided waves. Guided waves refer to mechanical (or elastic) waves in ultrasonic and sonic frequencies that propagate in a bounded medium (pipe, plate, rod, etc.) parallel to the plane of its boundary. The wave is termed "guided" because it travels along the medium guided by the geometric boundaries of the medium.

Since the wave is guided by the geometric boundaries of the medium, the geometry has a strong influence on the behavior of the wave [18–20]. In contrast to ultrasonic waves used in conventional ultrasonic inspections that propagate with a constant velocity, the velocity of the guided waves varies significantly with wave frequency and geometry of the medium. In addition, at a given wave frequency, the guided waves can propagate in different wave modes and orders.

The properties of guided waves and examples of their dispersion curves (which refer to the relationship between the velocity and the wave frequencies) are given in Figs. 12.5 and 12.6 for pipe and plate geometries, respectively. In pipe, the guided waves exist in three different wave modes: longitudinal (L), torsional (T), and flexural (F). In plate, they exist in two different wave modes: longitudinal that is generally called "Lamb" waves and exists in symmetric (S) and antisymmetric (A) modes; and shear horizontal (SH).

Although the properties of guided waves are complex, with judicious selection and proper control of wave mode and frequency, the guided waves can be used to achieve 100% volumetric inspection of a large area of a structure from a single sensor location.

Guided waves can be developed at a single location on a component and then travel long distances from the source point to monitor the condition of that component. The guided wave fills the entire volume of the component, and any change in cross-section of the component, such as a weld or corrosion/erosion, usually reflects or scatters the guided wave so that the change in cross-section is detected. Depending on frequency and material conditions, guided waves can travel as far as 150 m or more from one source, and defects such as corrosion scatter back the guided waves, thus giving an indication of corrosion locations.

There are two types of guided-wave systems based on the physics of generating the guided waves. One type uses a large number of piezoelectric transducers that are pressure-coupled to the pipe. This technology was developed and distributed by Guided-Wave Ultrasonics Limited [21] and requires a mechanical fixture to squeeze the transducers onto a pipe. This fixture can be expensive and can only be used on

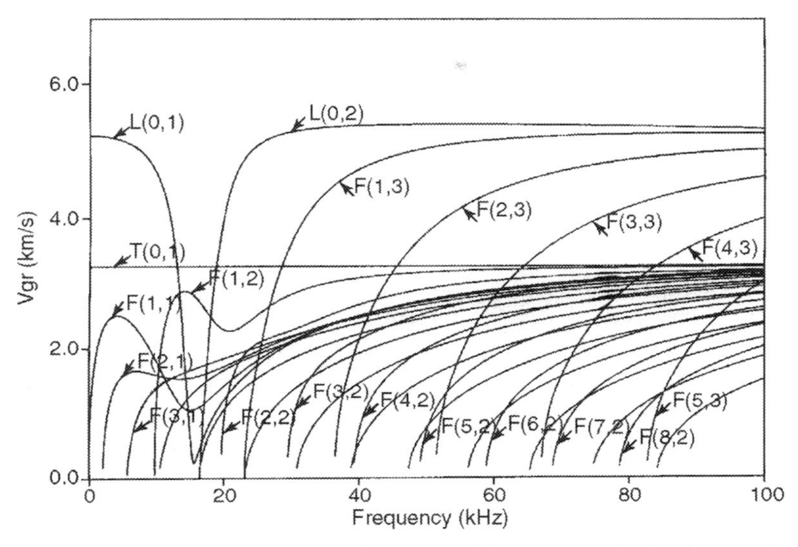

Fig. 12.5 Examples of dispersion waves of various guided-wave modes in pipe (for 114-mm-OD, 8.6-mm-thick pipe). The numbers in parentheses indicate the order of the wave mode.

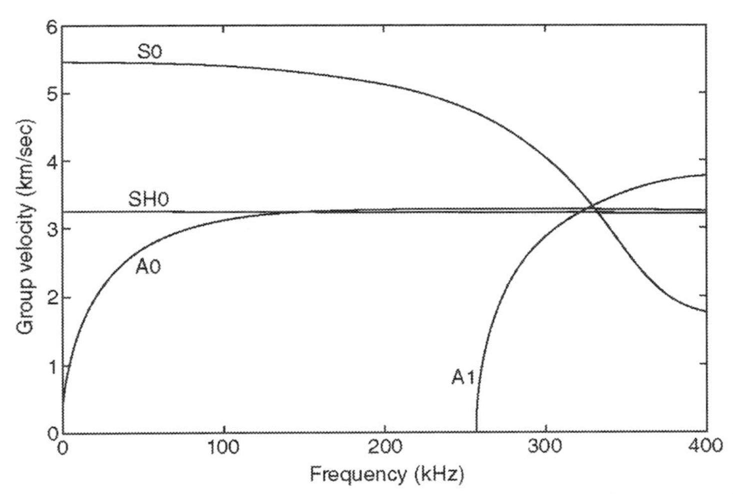

Fig. 12.6 Example of dispersion curves of various guided-wave modes in plate (for 6.35-mm-thick plate). The numbers after the letter (0 and 1) indicate the order of the wave mode.

cylindrical-type geometries such as pipe. The defect detection sensitivity is estimated to be 5%–10% of the pipe cross-section.

The other type uses a thin ferromagnetic strip (approximately 0.15 mm) that has a high magnetostrictive property and is bonded to the pipe [21–28] and is referred to as magnetostrictive sensor (MsS) technology. The ferromagnetic strip and excitation coil are relatively inexpensive and can be used on piping, plate, bar, and a wide range of geometries. However, time for bonding the ferromagnetic strip onto the geometry is required.

Generally speaking, the MsS approach can be used as an inspection mode to detect defects on the order of 3%–5% of the component cross-section, but if the sensor can be left in place for long periods of time (most feasible with the ferromagnetic strip concept) and data are collected at various times during its life, defects as small as 0.5% can be detected. However, guided waves tend to be poor at quantifying the corrosion and should really be used as more of a screening and locating tool.

The piezoelectric sensor system is limited in temperature applications of 70–120° C. The MsS technology has been used on components ranging from approximately −150°C to 300°C.

Guided waves have been used to inspect various types of pipeline, on land as well as offshore. An example of an offshore application is the inspection of risers. Offshore pipeline failure statistics have been collected for more than 30 years and illustrate that the riser predominantly fails as a result of corrosion. The consistent wetting and drying in the splash zone, combined with defects in the coatings, are the usual contributors to the problem. A guided-wave approach using torsional waves can be deployed to detect significant reduction in wall thickness. This allows the detection of wall-thinning areas in the riser without removing the outside coatings. For detecting and monitoring corrosion under insulation, the MsS can be placed on a pipe with many welded sections as shown in Fig. 12.7, with the data shown in Fig. 12.8. The directionality of the guided wave from the MsS is controlled so that the wave can be transmitted to the right of the sensor or to the left.

Presently, the MsS technology uses the T-wave mode primarily for piping inspection that is generated and detected with the thin ferromagnetic (typically nickel) layer approach. Reasons for this practice include: (1) the fundamental T-wave mode is not dispersive and, therefore, no consideration is necessary for possible dispersion effects that exist in the L-wave mode; (2) the T-wave MsS has fewer effects than other extraneous wave modes and, therefore, it gives better signal-to-noise ratio and its data are easier to analyze; (3) the T-wave does not interact with liquid inside the pipe and, therefore, is far superior to the L-wave [24] for inspection of liquid-filled pipes; and (4) the T-wave MsS does not require heavy bias magnets and, therefore, is much easier and safer to handle than the L-wave counterpart. The disadvantage of the

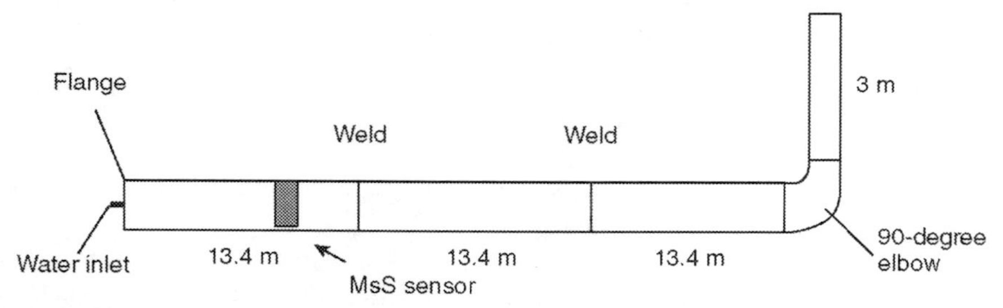

Fig. 12.7 Illustration of MsS guided-wave sensor attached to pipe.

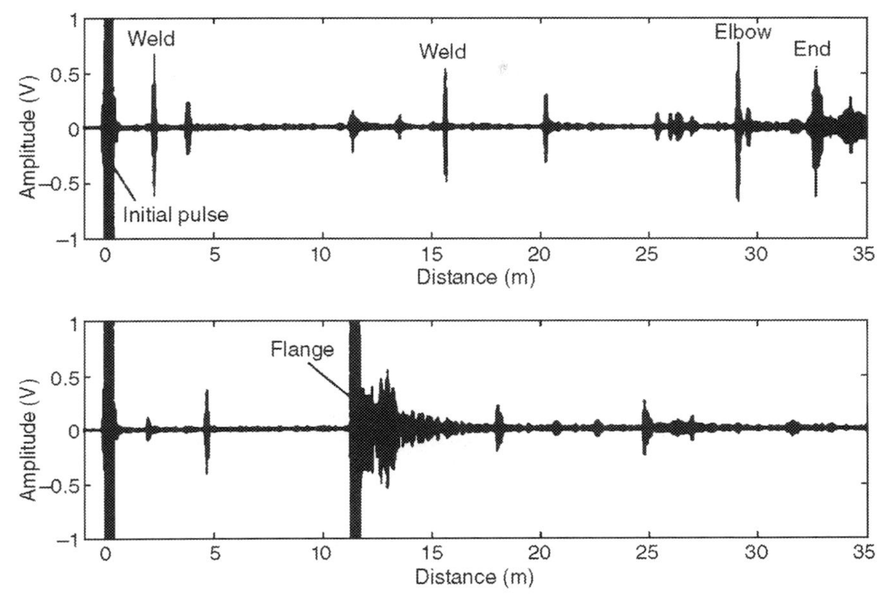

Fig. 12.8 Illustration of guided-wave data collected by MsS system. The top waveform is directed to the right of the MsS sensor shown in Fig. 12.7 and the lower waveform to the left of the MsS sensor shown in Fig. 12.7.

T-wave MsS is the requirement for direct physical access to the pipe surface for bonding of the thin ferromagnetic layer. Therefore, to apply the T-mode MsS for bitumen-coated piping, for example, the coating must be removed beforehand, whereas the L-mode MsS can be applied without removing the coating. However, the advantage of the T-mode MsS greatly outweighs the disadvantage and, consequently, the T-mode MsS is used primarily for long-range piping inspection.

Table 12.1 summarizes the capabilities and limitations of the present guided-wave technology for long-range piping inspection in bare pipes. The effects of pipeline geometric features and other conditions such as coating and liquid inside pipe on the inspection capabilities are also summarized in Table 12.2.

Corrosion monitoring on other components, such as plate, has been demonstrated in the laboratory using MsS technology. For this application, the ferromagnetic sensor (Fig. 12.9) is permanently bonded to the surface of the structure with appropriate adhesive, such as epoxy, and then magnetically conditioned. The coil is used to generate a guided wave in the ferromagnetic strip and also to receive any guided waves scattered from defects. Because the sensor is fixed to the surface, the data collected periodically can be carefully compared to the baseline data established at the time of sensor installation. This allows trending of the structural condition changes as a function of time. An effective determination of damage and its location can be obtained through the monitoring mode and used in a suitable structural management decision code.

Table 12.1 Capabilities and limitations of guided-wave systems for pipe inspection.

Item	Capabilities/limitations
Spatial resolution	3–7 in. (7.5–17.8 cm), depending on frequency and mode
Pipe material	Any material
Pipe size	Up to 60 in. (152 cm) diameter for MsS technology, approximately 32 in. (81.3 cm) for piezoelectric system and less than 0.75 in. (19 mm) wall thickness
Inspection range	100 ft (30 m) or greater, depending on coating and for aboveground piping
Detectable defect type	Isolated corrosion pits and circumferential cracks
Minimum detectable defect size	2% to 5% of pipewall cross-section
Defect location	Axial location within ±2 in. (5 cm), depending on operational frequency and mode
Defect characterization	Limited to rough estimation of circumferential cross-section
Long-term monitoring	Only with MsS system because costs of MsS sensors are much less than piezoelectric sensor belt

Table 12.2 Effects of pipeline geometric features and other conditions on inspection capabilities.

Features/ conditions	Effects
Flange/valve	Prevents wave propagation; forms end point of inspection range
Tee	Causes a large disruption in wave propagation and limits inspection range up to that point
Elbow	Causes a large disruption in wave propagation and limits inspection range no farther than the elbow region
Bend	Has negligible effect if the bend radius is greater than three times the pipe OD; if the bend radius is less than the above, behaves like an elbow
Side branch	Causes a wave reflection and thus produces a signal; no significant effects on inspection capabilities
Clamp	Causes a wave reflection and thus produces a signal; no significant effects on inspection capabilities
Weld attachment	Causes a wave reflection and thus produces a signal; if the attachment is large (such as pipe shoes), can reduce inspection range
Paint	Has negligible effects
Insulation	Has no effects unless the insulation is bonded to the pipe surface, in which case the inspection range will be shortened due to higher wave attenuation
Coating	Has negligible effects if the coating is thin (e.g., fusion-bonded epoxy coating); thicker coating (e.g., bituminous coating, polyethylene coating) increases wave attenuation and shortens inspection range
Liquid in pipe	No effect on torsional wave; significant degradation on longitudinal wave
General surface corrosion	Increases wave attenuation and shortens inspection range
Soil	If pipe is buried, the surrounding soil greatly increases wave attenuation, and the inspection range is significantly shortened

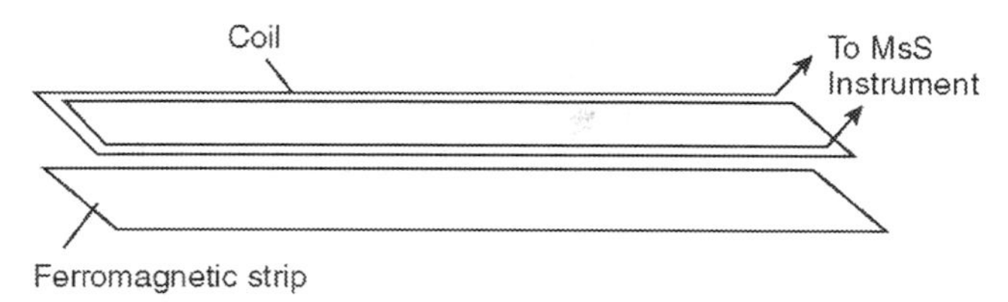

Fig. 12.9 Illustration of the flat MsS guided-wave probe for structural health monitoring. The coil is placed directly on the nickel layer.

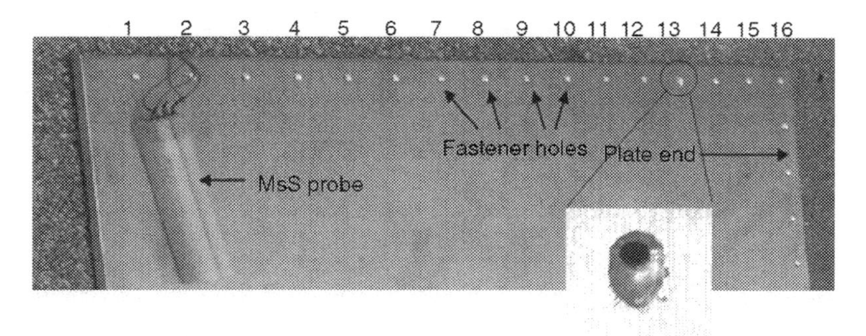

Fig. 12.10 Configuration of test panel and laboratory MsS probe and photograph of fastener hole showing simulated corrosion.

An example application is on aircraft structures such as fuselage and wing skins that are assembled with fasteners. Corrosion and cracking that occur around and under fasteners are major concerns for assuring structural safety of the aircraft. Because of the complicated geometry of the fastened structure and the large number of fasteners, inspection of such structures is time-consuming and difficult. Cost-effective and economical maintenance of fastened aircraft structures could be achieved by applying a suitable structural integrity monitoring (SIM) method.

An example of this application is shown in Fig. 12.10 for an aluminum test panel that was a 1/4 in. thick (6.3 mm), 3 × 4 ft. (0.9 m × 1.2 m) aluminum plate and had a large number of 1/4 in. (6.3 mm) diameter tapered fastener holes along two edges of the plate. The center of each hole was located at approximately 1 in. (25.4 mm) from the edge of the plate, and the distance between the centers of the adjacent holes was approximately 3 in. (76.2 mm). As the simulated corrosion was increased from approximately 0.05 in. (1.3 mm) on the side to 0.2 in. (5 mm) on the side, the monitoring showed that the changes in the corrosion could be detected and monitored, as illustrated in the data shown in Fig. 12.11.

12.2.5 Infrared thermography

Infrared thermography is an NDE tool that images a range of the infrared spectrum given off by a component that is heated. The imaging technology is most useful to detect thermal radiation differences in a component. For example, in the case where

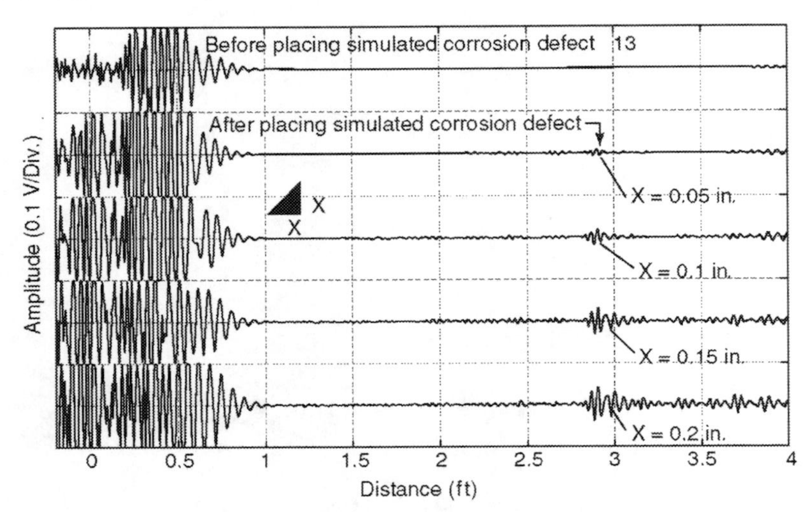

Fig. 12.11 Monitoring data showing the defect signal increasing as the simulated corrosion increases from no corrosion to 0.2 in. by 0.2 in. (5.1 × 5.1 mm) by approximately 0.05 in. (1.3 mm) deep.

a component might have wall thickness changes caused by corrosion or erosion, discontinuities in solid materials can change the heat flow condition, which can result in the fluctuation of the temperature on the surface of the materials. Both infrared testing and thermal image testing use this principle to measure the change of the surface temperature and then to deduce the discontinuity condition in the materials. Care must be taken to properly register images that are periodically collected. This technology has been applied to high-temperature and high-pressure pipelines used in petrochemical plants and power stations. Infrared thermography techniques have advantages over some other NDE technologies because a full-field image is obtained very quickly and there is no harmful radiation required and no contact with the part under inspection.

In the aspect of NDT for industrial equipment, infrared thermography testing is applied to operation condition monitoring of electrical equipment, power plant machinery, and high-temperature equipment [29–38]. Though metals have very high thermal conductivity and detection of wall loss defects for steel pipes by use of infrared thermography has been difficult, this work has shown that wall thinning can be detected. An example of this was the work by Shen [29] in which a series of infrared thermography experiments were performed for four kinds of stainless steel and carbon steel pipes during heating and cooling. The pipes had different size holes drilled on inner surfaces. The experiments were performed using the TVS-2100 Thermal Video System. The infrared camera head of this system is an optically mechanical scanning type. The detector is InSb with 10 × 10 cell arrays. The detecting wavelength is about 3–5.4 μm. The operating range of temperature is about −40°C to 950°C. The minimum detecting temperature difference is 0.1°C at 30°C, and the sensitivity is 0.01°

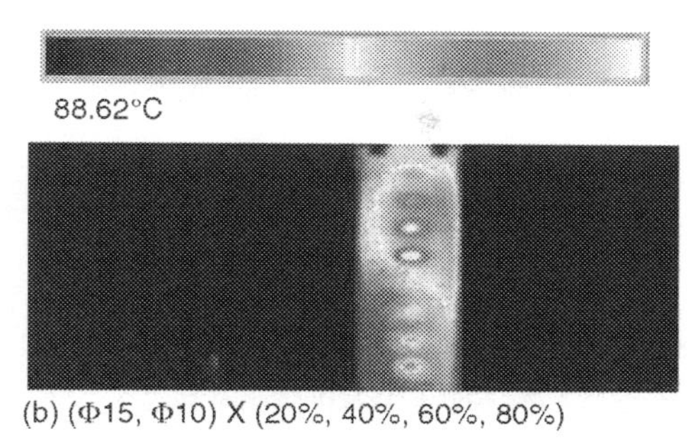

Fig. 12.12 Infrared image of pipe with simulated corrosion defects.

C. The field of view is 10° (V) = × 15° (H). The field resolution is 2.2 mrad. The images were taken at 30 frames per second. For the test, steam at a temperature of 150°C was passed through the pipe to heat from normal temperature. The thermal images of the defects obtained are shown in Fig. 12.12.

This work showed that thermography can be used to detect and monitor corrosion if a thermal gradient can be observed. In many cases, the components being monitored are at a thermal equilibrium, and for these cases, thermography may not be the best approach. However, under the conditions of thermal gradient, the technique is sensitive to small defects. Furthermore, a number of factors affect the potential capability of thermography such as:

1. The thermal conductivity of materials is a key factor affecting the sensitivity of infrared thermography. The lower the thermal conductivity, the higher the sensitivity and the longer the duration of defects emerging.
2. The shape and size of defects is another key factor affecting the sensitivity of infrared thermography. The larger the area of defect, the higher the sensitivity of wall loss.
3. The thickness is also a key factor affecting the sensitivity of infrared thermography. The thicker the material, the lower the testing sensitivity, but the longer the duration of defects emerging.

12.2.5.1 New inspection technology

As mentioned previously, the most common inspection technology is 0 degree longitudinal wave for detection of corrosion and wall loss. Most often, the practice involves the use of gridded UT or UT thickness measurements taken at grid points marked on the surface. This often leads to poor reproducibility of locating the inspection transducer on the grids, and consequently, the erroneous reporting of wall thickness growth instead of loss.

There are two types of Ultrasonic Testing (UT): manual and automated. Manual UT (Non-Encoded) is when only an ultrasonic waveform is displayed. It can easily be performed using a relatively economical device, so it is widely used for general purposes. However, this type of UT requires the inspector to check and record information obtained during inspection (such as the ultrasonic waveform and defect positions) on-site, thus posing a heavy burden on the inspector.

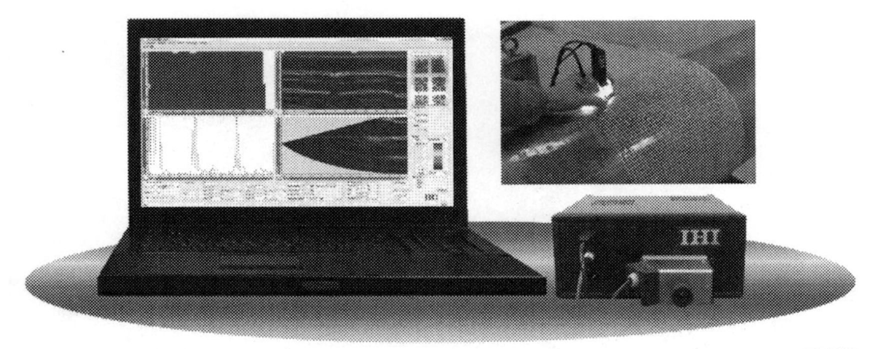

Fig. 12.13 Photograph of the ARMUT^{UT} laptop and control unit showing how the transducer is used on the pipe.

There is a new technology recently developed by IHI Corp that incorporates the use of a transparent Mylar film embedded with bar codes that can be read with a special camera mounted on an ultrasonic search unit to record the search unit location. The film is placed on the pipe or surface and an ultrasonic examination is conducted as-normal with manual manipulation of the search unit (phased array ultrasonics or a single element transducer can be used). The search unit position is determined allowing optical alignment so that a precise A-, B-, and C-scan presentation can be accomplished without a mechanically encoded scanner. This system is shown in Fig. 12.13 and is called the Absolute Recordable Manual Ultrasonic Testing (ARMUT).

The automated (Encoded) UT has the function of automatically saving to the computer all data obtained during inspection, including the ultrasonic waveform. Therefore, data can be recorded and reviewed afterwards.

The recently developed ARMUT (Absolute Recordable Manual Ultrasonic Testing system) incorporates the best of both types—in other words, it leverages the advantages of both manual UT and automated UT—and it can be described as a state-of-the-art ultrasonic testing system that can supersede the conventional grided UT technology. The ARMUT is a small-sized, lightweight system, allowing the technician to carry the device to the inspection site by hand and, after placing the ultrasonically coupled and transparent Mylar film, conduct manual UT. The inspector then performs ultrasonic testing following a procedure that is virtually the same as for manual UT. As the technician is familiar with the procedure, he/she can perform this task quickly.

The only difference is that the ARMUT involves a step in which a sheet with a printed code representing the positional information is attached to the surface of the object being inspected. This one simple step allows all data, including the ultrasonic waveform and defect positions, to be automatically saved to the computer; as is the case with automated UT, this provides a simplified solution during the inspection that can be reproduced at any given time. As a result, the transparency of inspection can be maintained.

The ARMUT's principle of operation is very simple, making it possible to achieve good on-site usability and high reliability. Moreover, functions that are not available in the conventional manual or automated UT have been successfully added by leveraging this principle. For instance, the ultrasonic sensor has to be kept in close contact with the sheet with printed positional information code while scanning the code, otherwise the data obtained will be incomplete. The ARMUT operates in such a way so as to successfully eliminate in principle the possibility of overlooking a flaw detection. Furthermore, by using a camera for reading the positional information code integrated with the ultrasonic sensor, a function has been successfully incorporated to automatically halt data collection when there is insufficient contact between the ultrasonic sensor and the surface to be inspected, or when the incident direction is incorrect because the ultrasonic sensor is skewed excessively. As such, the ARMUT is a reliable system that can be used easily at inspection sites and can automatically prevent the omission of inspection and the failure of data collection.

This system has already been used for piping inspections at Japanese fossil fuel power plants, and it will be used for inspections of multiple units of nuclear equipment outside Japan as well. In addition, public relations activities to provide this system to Japanese nuclear power plants have been carried out with favorable responses.

This technology can be used for inspection and monitoring of corrosion as well as cracking. The grid mesh is positioned relative to a weld or other fiducial mark. Because a map of the area is made, there is less likelihood that the actual position of the data collected will not be reconcilable with data collected previously, thus making monitoring much more realistic.

In addition, the technology can be used with phased array technology.

12.3 Future trends

The industry will tend to use monitoring technologies that leave the inspection sensors in place and to retrieve data as required. To achieve this goal, monitoring technologies will have to incorporate wireless communication technologies so that sensors can be left on a component and be queried at an appropriate time by a base control station. This will require that the sensors and associated instrumentation be inexpensive yet effective. This also will require computerized data acquisition and analysis systems that autonomously detect and analyze acquired data and alert the plant operators when a defect is determined to be growing and possibly heading toward ultimate failure. This is, in effect, diagnosis.

Further, the monitoring technology will be required to handle the operational environment. In most cases, the components that encounter significant corrosion issues are operated at temperatures around 150–650°C. Only a few sensor technologies can continually operate at temperatures above 250 °F.

The ultimate trend is to head toward prognosis. This means that the monitoring data have been fed into the diagnosis system to determine the condition of the component and then the results of the diagnosis put into the prognosis system to determine the remaining life of the component being inspected. Prognosis will incorporate the present condition of the part determined by the monitoring system with the expected operational conditions that the component will be subjected to and, with the understanding of how the material behaves under that operational condition, predict remaining life of the component. Work has been reported [39–42] primarily in high-value components like aircraft rotors and blades. Work [41] has been completed demonstrating that MsS technology, when used in a thin-film, multilayer format, has potential for use in the harsh thermal environments encountered in jet engines.

Although more work and more collaboration between sensor and material technologies are needed, it appears that the ability to continually monitor high-value components during operation will be available in perhaps 5 years. This means that periodic inspection requiring shutdown of a plant operation may be a thing of the past.

More information can be obtained primarily from the internet as well as the American Society of Nondestructive Testing. The following websites can provide additional information: www.swri.org; www.rtd.nl; www.asnt.org; www.pacndt.com; www.ewi.org; www.statoil.com; www.ndt.net.

See also Roderic K. Stanley, Patrick O. Moore, and Paul McIntire (eds), *Nondestructive Testing Handbook, Volume 9*, "Special Nondestructive Testing Methods," American Society for Nondestructive Testing, 1995.

References

[1] M.B. Hoppenbrouwers, Advanced Ultrasonic Methods for in Service Condition Assessment of Industrial Process Installations, Röntgen Technische Dienst BV, Rotterdam, 2000. *http://www.rtd.nl*.

[2] P.T. Cole, S.N. Gautrey, In-service testing and monitoring of process plant, in: 2nd MENDT Proceedings, NDT.net, 2004. 9 (6).

[3] Ø. Baltzersen, A. Solstad, A. Daaland, J.K. Holberg, Multichannel Ultrasonic Monitoring of Corrosion on Subsea Pipelines, 2005, NDT.net, 10(6) *http://www.ultrasonic.de/article/v10n06/baltzers/baltzers.htm*.

[4] T. Goyen, Conformable array for characterization of surface corrosion, in: Proceedings of ASNT 14th Annual Research Symposium, Albuquerque, NM, 14–18 March, 2005, 2005.

[5] M.A. Robers, R. Scottini, Pulsed Eddy current in corrosion detection, in: Proceedings of 8th ECNDT, Barcelona, 2002.

[6] A. Yonezu, H. Cho, M. Takemoto, Monitoring of stress corrosion cracking in stainless steel weldments by acoustic and electrochemical measurements, Meas. Sci. Technol. 17 (2006) 2447–2454, https://doi.org/10.1088/0957-0233/17/9/011.

[7] A.D. Zdunek, D. Prine, Z. Li, E. Landis, S. Shah, Early detection of steel rebar corrosion by acoustic emission monitoring, in: Paper No. 547, CORROSION95, the NACE International Annual Conference and Corrosion Show© 1995 by NACE International, 1995.

[8] M. Huang, L. Jiang, P.K. Liaw, C.R. Brooks, R. Seeley, D.L. Klarstrom, Using acoustic emission in fatigue and fracture materials research, JOM 50 (11) (1998).

[9] A. Almeida, E.V.K. Hill, Neural network detection of fatigue crack growth in reveted joints using acoustic emission, Mater. Eval. 53 (1) (1995) 76–82.

[10] J. Baram, Fatigue-life prediction by an order statistics treatment of acoustic-emission signals, Exp. Mech. 33 (1993) 189–194.

[11] B.A. Barna, J.A. Johnson, R.T. Allemeier, Determination of acoustic-emission sites using a digital nondestructive-evaluation workstation, Exp. Mech. 28 (1988) 210–213.

[12] S. Barre, M.L. Benzeggagh, On the use of acoustic emission to investigate damage mechanisms in glass-fibre-reinforced polypropylene, Compust. Sci. Tech. 52 (1994) 369–376.

[13] R.B. Clough, J.C. Chang, J.P. Travis, Acoustic emission signatures and source microstructure using indentation fatigue and stress corrosion cracking in aluminum alloys, Scr. Metall. 15 (1981) 417–422.
[14] J.E. Coulter, S. Gehl, J.R. Scheibel, D.M. Stevens, Acoustic emission monitoring of fossil-fuel power plants, Mater. Eval. 46 (2) (1988) 230–237.
[15] D. Fang, A. Berkovits, Fatigue design model based on damage mechanisms revealed by acoustic emission measurements, Trans. ASME 117 (1995) 200–208.
[16] K.F. Graff, Wave Motion in Elastic Solids, University Press, Columbus, OH, 1975.
[17] S. Yuyama, Fundamental aspects of acoustic emission applications to the problems caused by corrosion, in: G.C. Moran, P. Labine (Eds.), Corrosion Monitoring in Industrial Plants Using Nondestructive Testing and Electrochemical Methods, ASTM, Philadelphia, PA, 1986, pp. 43–74. ASTM STP 908.
[18] M. Redwood, Mechanical Wave-Guides, The Propagation of Acoustic and Ultrasonic Waves in Fluids and Solids with Boundaries, Pergamon, New York, 1960.
[19] J.L. Rose, Ultrasonic Waves in Solid Media, Cambridge University Press, Cambridge, 1999.
[20] M.G. Lozev, R.W. Smith, B.B. Grimmett, Evaluation of methods for detecting and monitoring of corrosion damage in risers, J. Press. Vessel. Technol. 127 (3) (2005) 244–254.
[21] D.N. Alleyne, B. Pavlakovic, M.J.S. Lowe, P. Cawley, Rapid long-range inspection of chemical plant pipework using guided waves, in: 15th World Conference on NDT, Rome, 2000.
[22] H. Kwun, S.Y. Kim, G.M. Light, Development of a torsional guided wave probe for heat exchanger tube inspection, in: Paper Summaries for the ASNT Fall Conference, Las Vegas, NV, 15–19 November 2004, 2004, pp. 71–73.
[23] G.M. Light, H. Kwun, S.Y. Kim, R.L. Spinks, Health monitoring for aircraft structures, Mater. Eval. 61 (7) (2003) 844–847.
[24] G.M. Light, H. Kwun, S.Y. Kim, R.L. Spinks, Deeply probing, World Pipelines (2003).
[25] H. Kwun, S.Y. Kim, G.M. Light, The magnetostrictive sensor technology for long-range guided wave testing and monitoring of structures, Mat. Eval. 61 (1) (2003) 80–84.
[26] G.M. Light, H. Kwun, S.Y. Kim, R.L. Spinks, Magnetostrictive sensor technology for monitoring bondline quality of and defect growth under adhesively bonded patches on simulated wing structure, in: Proceedings of the Aging Aircraft Conference, San Francisco, CA, 16–19 September, 2002, 2002.
[27] H. Kwun, G.M. Light, S.Y. Kim, R.H. Peterson, R.L. Spinks, Permanently installable, active guided-wave sensor for structural health monitoring, in: Proceedings of Structural Health Monitoring Conference, Paris, France, 10–12 July, 2003, 2003.
[28] H. Kwun, G.M. Light, S.Y. Kim, C.J. Schwartz, C.P. Dynes, R.L. Spinks, Cost effectiveness of magnetostrictive sensor technology for inspection of corrosion under insulation, in: Proceedings of the 1st Middle East Nondestructive Testing Conference & Exhibition, 24–26 September, 2001, Bahrain, 2001, pp. 79–92.
[29] G. Shen, G. Chen, Infrared thermography test for high temperature pressure pipe, in: 10th APCNDT, Brisbane, Australia, 2011.
[30] X. Maldague (Ed.), Nondestructive Testing Monographs and Tracts, Infrared Methodology and Technology, vol. 7, Gordon and Breach Science Publishers, New York, 1994.
[31] R.K. Stanley, P.O. Moore, P. McIntire (Eds.), Nondestructive Testing Handbook, Special Nondestructive Testing Methods, vol. 9, American Society for Nondestructive Testing, Columbus, OH, 1995.

[32] T.-M. Liou, C.-C. Chen, T.-W. Tsai, Heat transfer and fluid flow in a square duct with 12 different shaped vortex generators, Trans. ASME J. Heat Transf. 122 (2) (2000) 327–335.
[33] R.N. Wurzbach, D.A. Seith, Infrared monitoring of power plant effluents and heat sinks to optimize plant efficiency, in: Proceedings of SPIE—The International Society for Optical Engineering Thermosense XXII, 24–27 2000, Orlando, FL, USA, 2000.
[34] R. Rozenblit, M. Simkhis, G. Hetsroni, D. Barnea, Y. Taitel, Heat transfer in horizontal solid-liquid pipe flow, Int. J. Multiphase Flow 26 (8) (2000) 1235–1246.
[35] S. Yamawaki, T. Yoshida, M. Taki, F. Mimura, Fundamental heat transfer experiments of heat pipes for turbine cooling, in: American Society of Mechanical Engineers (Paper), Proceedings of the International Gas Turbine & Aeroengine Congress & Exposition, 2–5 June, Orlando, FL, USA, 1997.
[36] Y. Qin, N. Bao, Thermographic nondestructive testing technique for delaminated defects in composite structure, in: Proceedings of SPIE—The International Society for Optical Engineering Thermosense XVII, 1995, Orlando, FL, USA, 1995.
[37] K.R. Maser, M.S. Zarghamee, Leakage evaluation of a buried aqueduct, in: Proceedings of the Specialty Conference on Infrastructure Condition Assessment, 25–27 August, 1997, Boston, MA, USA, 1997.
[38] X. Maldague, Pipe inspection by infrared thermography, Mater. Eval. 57 (9) (1999) 899–902.
[39] J.M. Larsen, L. Christodoulou, Integrated damage state awareness and mechanism-based prediction, J. Miner. Met. Mater. Soc. (2004) 14.
[40] L. Christodoulou, J.M. Larsen, Using materials prognosis to maximize the utilization potential of complex mechanical systems, J. Miner. Met. Mater. Soc. (2004) 15–19.
[41] S.J. Hudak Jr., M.P. Enright, R.C. McClung, H.R. Millwater, A. Sarlashkar, M.J. Roemer, Potential Benefits of Adding Probabilistic Damage Accumulation to Prognosis of Turbine Engines Reliability, SwRI Final Report to AFRL/DARPA, Contract No. F33615-97-D-5271, 2002.
[42] S.J. Hudak Jr., B.R. Lanning, G.M. Light, J.M. Major, J.A. Moryl, M.P. Enright, R.C. McClung, H.R. Millwater, The influence of uncertainty in usage and fatigue damage sensing on turbine engine prognosis, J. Miner. Met. Mater. Soc. (2005) 157–166.

Acoustic emission

13

Miguel González Núñez (Angel) and Hossain Saboonchi
MISTRAS Group, Inc., Princeton Junction, NJ, United States

13.1 Introduction

Acoustic emission testing is defined as a "passive" testing method for monitoring the dynamic internal stress redistribution within a material that occurs when an external stress is imposed on a component.

Acoustic emission is the term given to the transient elastic mechanical waves that release energy suddenly within a material as a result of a stress condition. This energy release caused by the stress condition is due to the application of a load in the form of pressure, temperature, or due to physical or chemical changes affecting the internal and/or external inherent stresses of the material's microstructure.

Discontinuities and flaws in the materials are the main sources of acoustic emission. The quantity and intensity of how those sources release energy and produce an acoustic emission signal will rely on the mechanical properties of the material to withstand such stresses and the amount of matter involved during the release of energy derived from the stress condition.

As the structure is stressed, the AE instrumentation collects data. Parameters from each emission are measured and then stored within a system data logger. Data from each sensor is stored on separate channels along with the exact time when events are detected. These data are analyzed both during and after the test.

Acoustic emission serves an important function in many industries. It is a global nondestructive examination technique that provides a measure of the structural severity of a defect. As such, it is complementary to other nondestructive examination methods such as radiography, ultrasonics, eddy current, visual dye penetrant, and magnetic particle testing, which are used for examination of local areas.

Research has proven the ability of acoustic emission to detect active corrosion. The chemical and petrochemical industry implemented, since mid-90s, the technology to detect, in particular, stress corrosion cracking (SCC) and pitting of stainless-steel alloys. Phenomena such as crevice corrosion, galvanic, erosion-corrosion, and uniform corrosion have been also detected [1-5].

In early studies, attention was driven by localized corrosion because it is an important damage phenomenon in the chemical, petrochemical, and nuclear industries. Localized corrosion promotes unpredicted in-service failures, with economic, safety, health, and environmental consequences.

Techniques for Corrosion Monitoring. https://doi.org/10.1016/B978-0-08-103003-5.00013-8

13.2 Principle of method

In the traditional AE tests, the material needs to be stimulated through stress, such as thermal or mechanical.

Acoustic Emissions are transient elastic waves generated by the rapid release of the energy from localized sources within a material [6].

Sources could be discontinuities and flaws such as: Corrosion films, cracks, leaks, and delamination.

When dealing with active corrosion, its detection has to do with the propagation of the phenomenon due to localized corrosion as it was postulated by scientists before it was demonstrated in the laboratory [7,8]. When corrosion is active, formation of oxide films both anhydrous and hydrated occur on the metal surface. The crystalline nature of the corrosion products differs from the crystalline structure of the metal and micro stresses are developed naturally. In dry air, oxygen molecules adsorb physically to the metal surface, forming one or more monolayers of oxide with no activation energy requirements [9,10]. In the presence of water (ranging from condensed films deposited from humid atmospheres to bulk aqueous phases), further thickening occurs as partial hydration increases the electron tunneling conductivity through the stable oxide film [11]. Film rearrangement results in the formation of oxide subgrain and grain boundaries; these easy ion migration paths promote the formation of oxide islands and result in an increase in the growth rate of oxide films. Other components in contaminated atmospheres may become incorporated (e.g., H_2S, SO_2, CO_2, Cl^-) which may introduce more stresses to the crystalline oxide lattice, which can go from thin transparent oxides such as passive films on Al, Cr, Ti, and Fe—Cr alloys, or thin visible sulfides on Cu and Ag to thicker visible films, which may be compact, adherent, and protective (oxide films on Al and Ti, $PbSO_4$ films on Pb, etc.) or bulky, poorly adherent, and nonprotective (rust on steel, "white rust" on Zn). Although some of the corrosion products may be adherent, most of it are not. In practice, thermal cycling rather than isothermal conditions more frequently occurs, leading to a deviation from steady state thermodynamic conditions and introducing kinetic modifications. Lattice expansion and contraction, the development of stresses, and the production of voids at the alloys-oxide interface, as well as local pH, temperature-induced compositional changes can all give rise to further complications. The resulting loss of adhesion and spalling may lead to breakaway oxidation in which linear oxidation replaces parabolic oxidation.

Acoustic emission sensors are mounted on the test/monitoring object's surface to detect dynamic motion resulting from acoustic emission signals and convert the detected motion into a voltage vs time signal.

In an acoustic emission test, a network of piezoelectric transducers is attached to a structure. These sensors convert mechanical energy such as elastic waves into an electric impulse that is transmitted by cables to the AE Instrumentation.

The AE processing chain has four main links: the *source* where AE is generated will cause a microdisplacement generating a moment tensor which will be dependent on the local size, amount of matter, physical and mechanical properties; the propagation of the mechanical micromovement throughout the structure until reaching *the sensor* which will vibrate at a particular frequency depending on the movement detected under its surface, which will transform the mechanical signal into an electrical voltage that will be conditioned to be sent to the AE instrumentation.

13.2.1 Equipment

AE equipment is typically conformed by (1) a passive AE sensor, (2) a preamplifier for signal conditioning and enhancing to minimize losses due to signal transmission either through a coaxial cable or wirelessly, (3) the AE instrumentation containing the hardware electronics for detection and feature extraction, and (4) AE software for displaying and analysis of the signals detected. A signal processor block diagram is shown below (Fig. 13.1).

13.2.1.1 Typical sensor

A sensor in general is something that gets information from a specific environment and produces a useful output. As an example, human sensors like the eye and the ear generate nerve impulses that are received by the brain with information related to the surrounding environment. As such, an acoustic emission sensor is an extension of the ear. The difference is that the AE sensor works at much higher frequencies that are undetectable by the human ear. Usually, the AE sensor works in contact with solids rather than in contact with the air and the AE sensor sends its signal to a piece of electronic equipment rather than to the brain.

A transducer is a device that converts energy from one form to another. The sensors used in AE contain transducing elements that convert energy motion into electrical signals. These transducing elements operate under a special property called piezoelectricity.

The word "piezo" comes from the Greek "piezein" meaning "to squeeze"; so piezoelectricity is "squeeze electricity" that is the generation of electric charges by

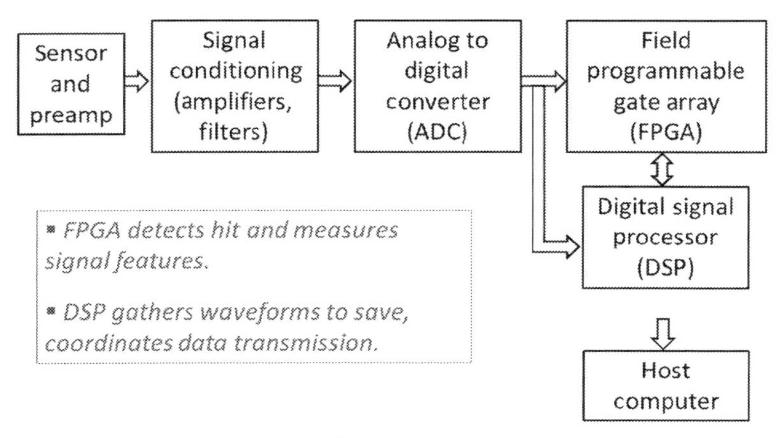

Fig. 13.1 AE signal equipment—block diagram.

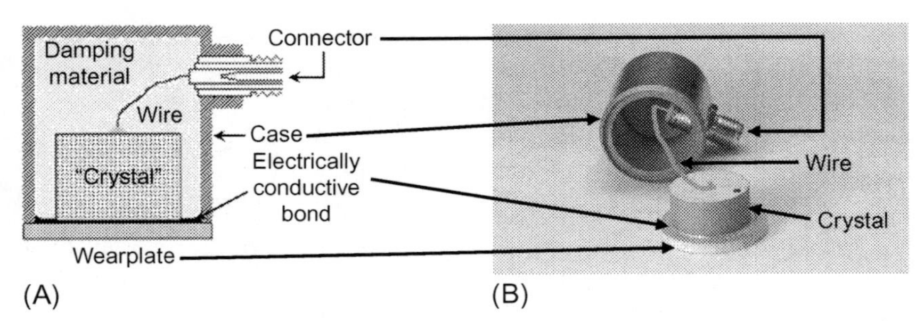

Fig. 13.2 (A) Basic construction of an AE sensor. (B). Internal physical appearance of an AE sensor.

mechanical pulses, or vice versa. The material used most widely in NDT transducers is a ferroelectric ceramic, lead zirconatetitanate, or PZT, developed in World War II for submarine sonar systems.

Fig. 13.2A and B shows the basic content and appearance of an AE sensor.

Typical AE sensors in their simplest form consist of:

1. A piezoelectric transducer "crystal" which is the heart of the sensor.
2. Mounted on a wear plate.
3. Sealed into a protective case for easy handling and electrical shielding, and
4. Wired to an electrical connector.

From the connector, a cable carries the electrical signal on its way to the measurement circuitry.

Piezoelectric sensors can have good fidelity so that the electrical signal output follows very closely the actual surface displacement. The typical minimum displacement detected by a piezoelectric sensor is on the order of picometers (10^{-12} m) or 10^{-11} in.

13.2.1.2 Typical electronics

Back in the 80s, AE systems had some limitations related to large size of electronics for a multichannel AE system, especially if the system was required to take it on board of a ship or aircraft. The first digital systems available had limited on-board processing and speed-difficult to deal with the amount of data produced during an AE test. The automation of data trend analysis was in its infancy; near real-time data analysis was limited to two or three AE features, typically Hits, Counts, and Energy.

With the 21st Century Revolution in the electronics' development, it is possible to have AE systems with powerful and small electronics with a faster software, better data trending, and internet communications capabilities with better structural health monitoring (SHM) capabilities.

Typical AE systems hardware boards use compact, surface mount techniques, in contrast with the early days of AE equipment development, when the systems were limited to acquire time-based features (such as root mean square) and a few hit-based features (such as amplitude, counts, duration, and rise time).

Digital signal processing technology has advanced to the point where it is feasible now to process AE signals in real time in the digital domain. This has huge advantages

Fig. 13.3 Two-channel acoustic emission system on peripheral component interconnect bus plug-in card.

Fig. 13.4 Eight-channel acoustic emission system on peripheral component interconnect bus plug-in card.

over the old instrument designs with their heavy use of analog circuitry. Figs. 13.3 and 13.4 show a two-channel and an eight-channel, digital acoustic emission system on a peripheral component interconnect card with 18-bit analog-to-digital conversions up to 40 and 10 M samples per second waveform conversion, respectively.

Nowadays, the acoustic emission system capabilities have been expanded to have multichannel systems on a small single card with 18 bit Analog-to-Digital conversion capabilities and the ability to acquire multiple Parametrics Inputs (such as load, temperature, RPM, strain, and, pH) that can be correlated with the AE signals. AE systems also have multiple FPGAs (Field Programmable Gate Arrays) for data processing and feature extraction in time and frequency domain in real-time and built-in waveforms streaming that provide enough data for Neural Network and Pattern Recognition techniques on a posttest and/or real-time basis to develop classifiers that can separate and recognize multiple source mechanisms during on-site testing or remote monitoring through the internet.

13.2.1.3 Typical software

Software-based AE systems record three kinds of data which are written to disk. Hit data which are written every time an AE signal crosses the threshold set during the acquisition and information related to the time of detection; AE signal features (such as Amplitude, Energy, and Duration) and parametric information at the moment of the detection of the AE signal are recorded and written for each hit detected. The second kind is the time-driven data which are written at a regular time intervals (such as time of measurement, parametric, RMS, ASL, and Absolute Energy). The third kind of data information recorded is processed by the Analog-to-Digital Converter which digitizes the hit information and shows the waveform associated with every hit detected.

The information in the data records is used to construct the displays, both during the acquisition in real-time or in posttest analysis by replaying the data file. By using the posttest analysis, the data file is always there and it gives complete flexibility to create any display required. The displays can also be modified in real time during the test, although this usually requires a reset and the display will start building again from blank. There are several types of display as summarized later.

AE Data display as the ones shown in Fig. 13.5, provide valuable information about the detected signals, and can be classified into five categories: Histograms, Scatter Plots, Waveforms and Spectra, Location Plots, and Felicity Plots. All these display types, except Waveforms and Spectra are constructed from the hit data and the time-driven data information.

Usually, when performing a standardized test, most of the graphs are set up beforehand and saved in a layout file. For the case of the location plots, the sensor coordinates must be keyed in after the sensors have been mounted and their positions measured.

The function of the AE graphs displayed by the software provides useful information about the activity, intensity, correlation, or source location. The activity graphs have usually time on the x axis, or load as in the case of a Felicity plot and in the y axis an AE feature. Intensity graphs have amplitude or energy on the x axis and in the y axis an AE feature. Correlation graphs can have several forms, but the commonest is a scatter plot with amplitude on the x axis and duration on the y axis. Finally, source location graphs will have channel or x position on the x axis. Examples of the software display graphs are shown later.

13.2.2 Typical tab data for real-time monitoring

Acoustic emission sources from mechanical damage propagation are well-understood; the mechanisms that generate elastic waves from the corrosion process are more complex. It has been possible to identify different corrosion reactions on different alloys.

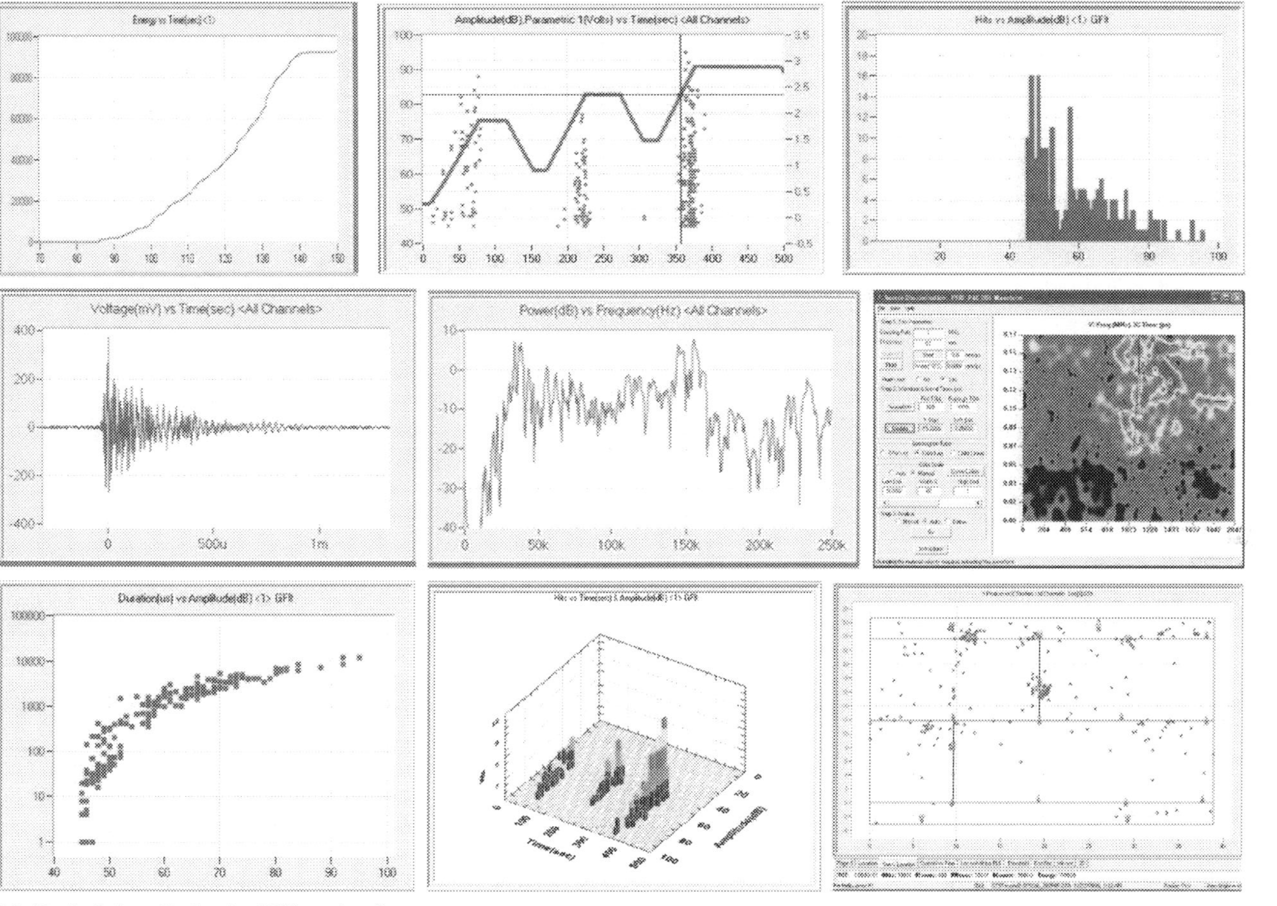

Fig. 13.5 Typical data display in AE-based software.

Uniform corrosion, pitting, crevice, and SCC have been identified and their association with cathodic reactions like the evolution of hydrogen gas bubbles in acidic environments [12]. Monitoring and inspection of storage tanks, corrosion in reinforced concrete, and online monitoring of chemical processes are just a few examples of corrosion monitoring with AE [13]. With the development of neural network classification, it has been possible to characterize the corrosion signals, especially in high background environments, and it has been possible to obtain real-time monitoring information. Information about the activity and intensity of the corrosion process in real-time applications has been obtained to identify what is the cause or condition in the process that triggers the corrosion process. Examples of real-time monitoring tab data are shown later.

Figs. 13.6–13.9 show some of the Acoustic Emission features related to the active corrosion process and an example of the tabular data that is processed and analyzed to determine the severity of the corrosion based on the Activity, Intensity, and Distribution of the AE signals detected and the conditions of the background noise during the acquisition of the data (Table 13.1).

The use of acoustic emission (AE) technique for the detection and location of corrosion damages in metallic structures is well-known. Systematic tests have been conducted in the laboratory and in the field. The AE method can be used for online monitoring to periodically detect corrosion growth during the active corrosion stage or off-line for detection of active corrosion at a specific time. AE signals from the corrosion process on carbon steel are combined with high and low frequency signals related to the breakdown and spalling of the oxide film formed during the corrosion process. Although the signals from the corrosion process are low in amplitude compared to other damaging emission mechanisms such as crack growth, they still may be detected depending on the background noise conditions. AE corrosion signals under particular conditions can be of similar amplitudes or much higher than background noise. Examples of early success applications and recent developments are presented in the following sections.

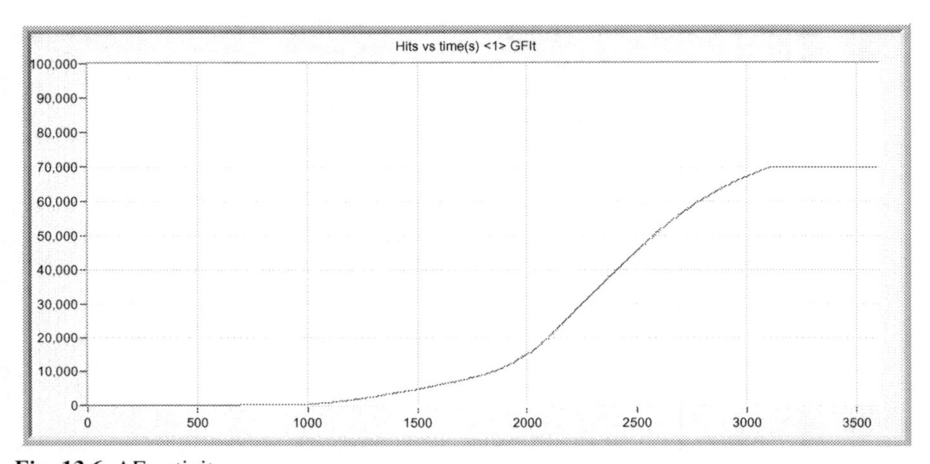

Fig. 13.6 AE activity.

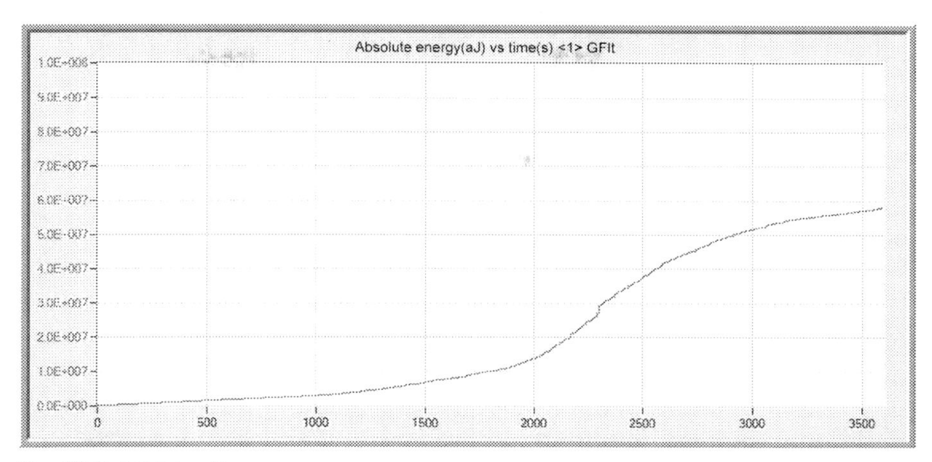

Fig. 13.7 AE intensity.

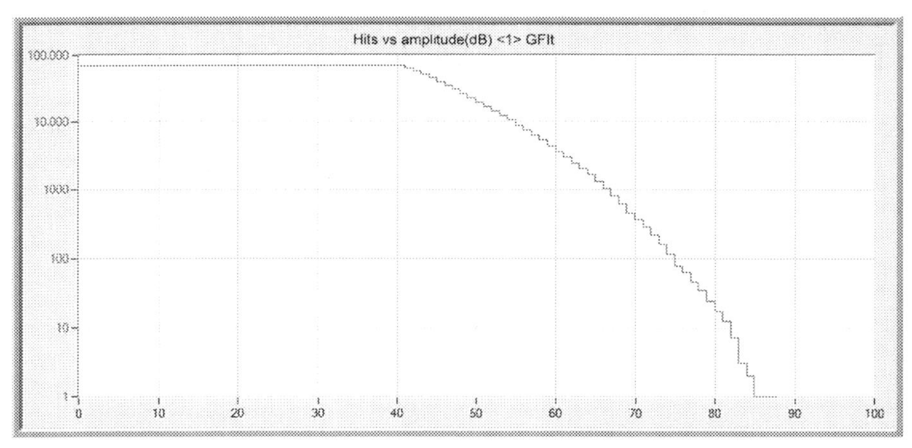

Fig. 13.8 AE activity distribution.

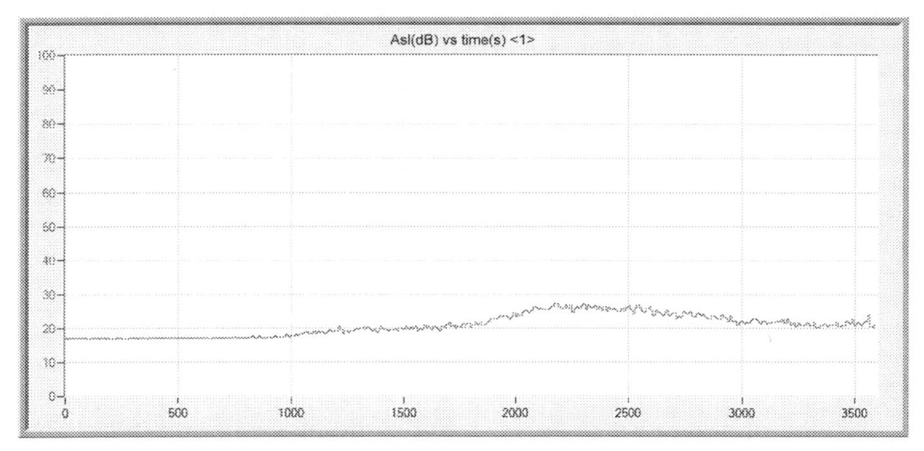

Fig. 13.9 Background noise level.

Table 13.1 Example of acoustic emission characteristic features data from corrosion-related applications, especially in the field that has been used as part of the corrosion mitigation program.

ID	MM/DD/YY	HH:MM:SS:mmmuuun	CH	Rise	Count	Duration	AMP	ABS-Energy
1	06/13/16	12p:13:48.9470183	1	0	1	0	32	0.000E+00
1	06/13/16	12p:14:13.6791950	1	13	2	13	33	924.483E-03
1	06/13/16	12p:14:45.8302052	1	4	3	22	39	5.445E+00
1	06/13/16	12p:14:45.9971055	1	21	13	102	44	31.130E+00
1	06/13/16	12p:14:54.0509607	1	11	2	18	34	1.563E+00
1	06/13/16	12p:14:54.3128190	1	0	2	23	33	1.338E+00
1	06/13/16	12p:15:24.3996430	1	17	3	28	37	4.491E+00
1	06/13/16	12p:15:24.5137607	1	5	2	14	38	2.580E+00
1	06/13/16	12p:15:34.7045252	1	3	3	17	38	2.675E+00
1	06/13/16	12p:15:34.7622600	1	0	1	0	33	107.996E-03
1	06/13/16	12p:15:37.6768617	1	4	2	10	35	1.156E+00
1	06/13/16	12p:15:42.0671593	1	42	9	78	42	20.681E+00
1	06/13/16	12p:15:49.7414938	1	0	1	1	33	200.165E-03
1	06/13/16	12p:15:49.9624350	1	20	27	277	57	415.943E+00
1	06/13/16	12p:15:50.0756380	1	76	3	77	33	2.654E+00
1	06/13/16	12p:15:51.0297763	1	0	1	0	32	2.654E+00
1	06/13/16	12p:15:52.7851043	1	4	5	30	40	6.567E+00
1	06/13/16	12p:15:55.4606187	1	0	1	0	32	6.567E+00
1	06/13/16	12p:15:56.1725012	1	3	2	10	36	1.771E+00
1	06/13/16	12p:16:01.2299098	1	0	2	13	36	1.350E+00
1	06/13/16	12p:16:01.2864355	1	3	4	29	39	3.961E+00
1	06/13/16	12p:16:03.3027662	1	0	1	1	34	191.786E-03
1	06/13/16	12p:16:03.3597927	1	6	4	23	38	4.960E+00
1	06/13/16	12p:16:03.4233745	1	0	1	0	33	46.550E-03
1	06/13/16	12p:16:05.6012993	1	0	1	13	36	1.182E+00

13.2.2.1 Aboveground storage tank inspection

Internal inspection of tank floors is the most expensive aspect of tank maintenance. The costs of removing a tank from service and cleaning it just to inspect the floor are wasted if the floor is then found to be in good condition; by listening first for active corrosion, it is possible to decide if inspection and repair are really necessary. Conversely, tanks that are in an advanced state of corrosion may be removed from service before a costly failure. There are many published papers on this method [14]; the most established procedure is TankPAC from Physical Acoustics (dba MISTRAS Group, Inc.), which has been used on more than 12,000 tanks and has been verified by third party internal inspection on more than 1000 tanks [15–18]. The use of TankPAC is corporate policy for a number of companies and is widely approved by authorities and professional organizations. Emission from corrosion of the floor can be confused with emission from other sources such as internal sacrificial anodes or corrosion of the shell under insulation; experience is required to differentiate between these different sources, as well as between nonrelevant noise sources such as condensation and noise from the environment (Fig. 13.10).

Additionally, aboveground storage tank (AST) inspection and monitoring has been improved by detecting and locating the AE activity related to the corrosion process in difficult access areas such as the annular ring zone of the AST. An example of this improved monitoring/testing application is shown in Figs. 13.11 and 13.12.

13.2.3 Corrosion monitoring of pipe wall thickness

Oil and Gas, Power and Chemical facilities have in their inventory a lot of pipelines and piping systems that need to be monitored periodically or continuously to keep the optimum processing and operation activities uninterrupted.

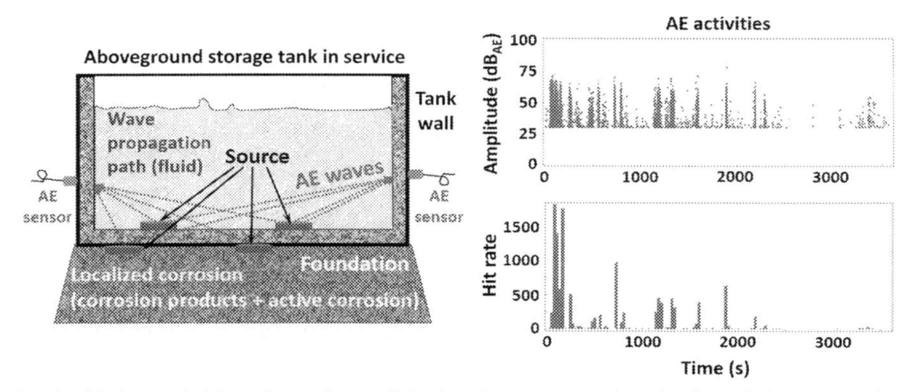

Fig. 13.10 Typical AE activity detected during the inspection/monitoring of aboveground storage tanks.

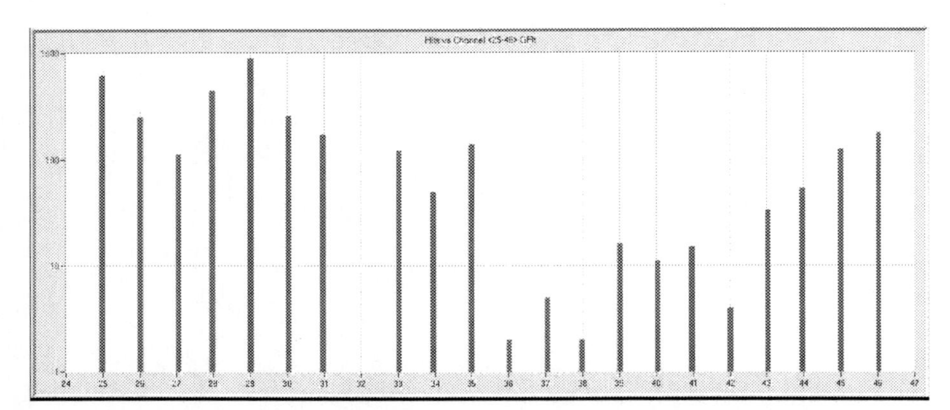

Fig. 13.11 Hits vs channel for detecting corrosion on the annular ring of the tank.

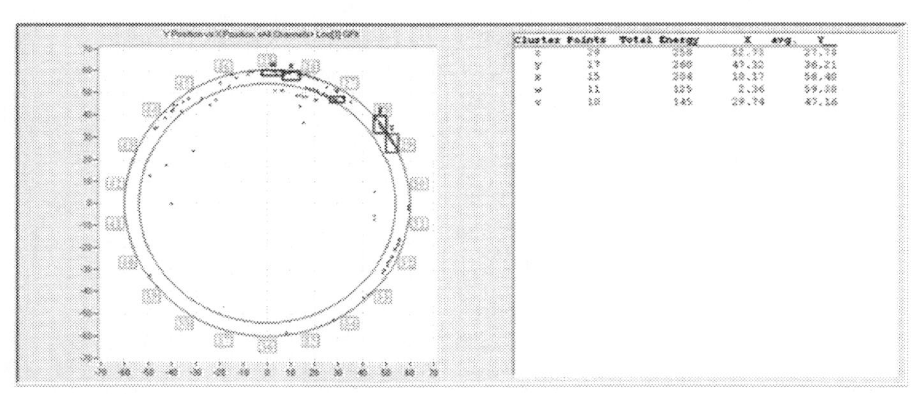

Fig. 13.12 Located accumulated AE activity related to active corrosion of the annular ring of the AST.

Typical inspection/monitoring methods include but are not limited to Metal Loss Coupons, Electrical Resistance (ER), and Ultrasonic Testing (UT). The Vibro-Acoustic metal loss detection technology is one of the newest available methods for this application.

Coupons are selected based on the material specification of pipeline and piping material to try to correlate the loss of material in the coupon with the corrosion rate of the pipe wall. Coupons are inserted in the flow of the process fluid and need to be removed periodically to measure the loss of mass. Specific methodologies are recommended for specific material to clean the corrosion products of the coupon (e.g., ASTM G1, G16, G31).

ER probes rely on the change in electrical resistance of exposed resistor of the probe due to the mass loss caused by corrosion. The resistor material is selected based on the characteristics of the pipe wall and the process fluid. Similar to coupons, they need to be removed and replaced periodically. The benefit of using the

ER probes is that they provide continuous measurement; however, if the process fluid is conductive, they cannot be used.

The need for periodic exchange of coupons and ER probes is a limitation, based on the aggressivity of the corrosive fluid. Typically, they need to be replaced every 3–6 months.

UT transducers are nonintrusive unlike coupons and ER probes and they measure pipe wall thickness, either periodically or continuously. While most UT transducers have lower temperature tolerance, since they are nonintrusive, they have unlimited lifetime.

Acoustic Metal Loss Detector is based on a vibrating tuning fork that changes its frequency as its metal is lost due to corrosion. As a response to mass loss, the sensor decreases its resonance frequency. The insertion of a small corrodible element in the base of the otherwise corrosion-resistant tuning fork tines causes the base to corrode, and as it gets thinner, its spring constant gets smaller and its resonant frequency decreases. This system would need two tuning fork probes in the pipeline circuit, one made of a highly corrosion-resistant material as the reference probe, and one very similar to the first with corrodible components as the measurement probe. This allows to account for the process fluid parameters like temperature, viscosity, pressure, and flow rate.

Other limitations when using corrodible sensors (i.e., coupons and ER probes) to mimic the corrosion behavior of the structure are their useful life, robustness to withstand extreme operation conditions and getting meaningful corrosion results in spite of the effects of scale formation, period of time for corrosion calculation, and sensitivity due to temperature adjustment. The Vibro-Acoustic corrosion probe shown herein overcomes and exceeds all these limitations by combining a reliable sensor design and sensitivity for measuring fluid corrosivity and providing direct information on pressure boundary metal loss for asset monitoring.

The Vibro-Acoustic method has been developed and patented by Exxon Mobil Research and Engineering [19–23]. Then it was turned into a commercial product by Mistras group, Inc. [24] under the marketing name of CALIPERAY Corrflow. Fig. 13.13 shows the operation concept of this method, and Figs. 13.14 and 13.15 show the Metal Loss Detector (MLD) and Metal Loss Probe (MLP) used for this application.

13.2.4 Advantages, limitations, cost, maintenance, ease of use, online and real-time vs offline, effect of environmental factors

A summary of relevant aspects of using different corrosion measurements methods is presented in Table 13.2, including typical application characteristics such as periodic testing or continuous monitoring capabilities, response time, local vs global screening capabilities, if they are intrusive or nonintrusive, temperature and pressure of application, and lifetime.

With respect to the cost and maintenance, these will vary for each particular application which takes into account the frequency of replacement, logistics involved during the maintenance and replacement activities, among others.

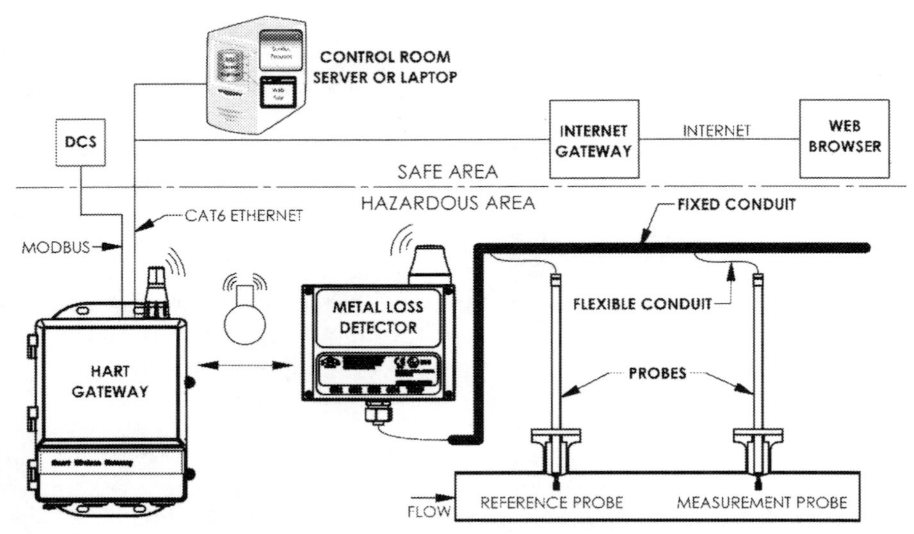

Fig. 13.13 Schematic drawing of Mistras Group Inc. CALIPERAY Corrflow installation.

Fig. 13.14 Tuning fork of metal loss probe.

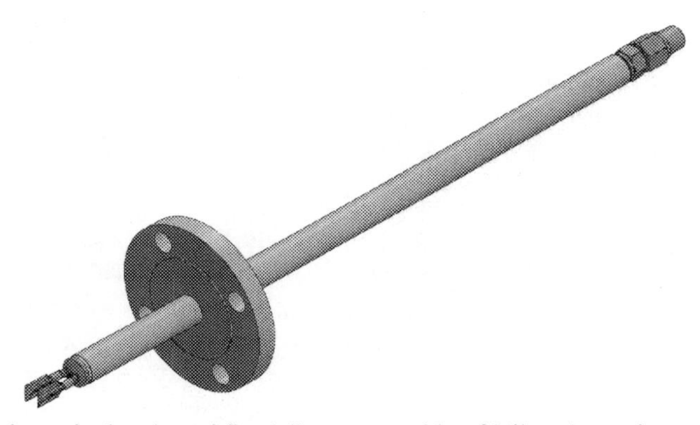

Fig. 13.15 Schematic drawing of fixed flange assembly of Vibro-Acoustic metal loss probe.

Table 13.2 Comparison of nondestructive corrosion measurement methods.

Technique	AE	Coupons	ER	UT	MLD
Periodic Test	Applicable	N/A	N/A	Applicable	N/A
Monitoring	Applicable	Applicable	Applicable	Applicable	Applicable
Response Time	1 day to continuous	2 month or longer	Real time	Periodic/Real time	Real time
Local vs Global	Global	Local	Local	Local	Local
Intrusive/Non-Intrusive	Non-Intrusive	Intrusive	Intrusive	Non-Intrusive	Intrusive
Temperature	175 °C	Material dependent	260 °C	175 °C (Typical)	425 °C
Pressure	N/A	N/A	6,000 psi	N/A	Up to 600 # Flange Class
Lifetime	Unlimited	Limited	Limited	Unlimited	Limited (6 years at 10 mpy)

In terms of ease of use, all of them require some degree of expertise, for the installation of the probes, recovery, data analysis, interpretation, and trouble shooting.

13.2.5 Continuous development and future trends

The continuous evolution of electronics and communication protocols has been the driven factors to deploy more and more ad-hoc AE systems depending on the specific field of applications and the frequency needed to know the actual integrity of structures. Whether the structural integrity is for a specific component in a process or, in general, to know the overall condition of a facility, AE has shown to be very reliable screening method to determine the condition of components/structure for its continuing use/operation or to recommend additional NDE methods to be used to assure a safe use/operation.

Periodic testing vs continuous monitoring have their advantages and disadvantages; however, when dealing with high-risk facilities such as refineries and chemical plants, continuous monitoring has been used more preferably with less and less human intervention.

This necessity has led to the design and development of AE systems that can operate in hazardous areas, minimizing human intervention, and can communicate remotely either if they are in an industrial facility or in a bridge/pipeline in a remote location.

Additionally, the easiness to install AE sensors and deploy the AE systems has led to the development of the so-called autonomous systems (nodes), which in turn can be powered in a daisy chain manner with power and communication driven in the same cable. Wireless communication is used for data transmission from sectors or regions in a structure to a base station; however, all data processing is carried out in each individual node and the processed data transmitted to the base-station.

The next step in this continuous development is the integration of miniature AE sensors to be part of the structure to be continuously monitored and to provide the least intrusive NDE method for SHM applications.

References

[1] Y.P. Kim, M. Fregonese, H. Mazille, D. Feron, G. Santarini, Ability of acoustic emission technique for detection and monitoring of crevice corrosion on 304L austenitic stainless steel, NDT&E Int. 36 (2003) 553–562.

[2] F. Ferrer, T. Faure, J. Goudiakas, E. Andres, Acoustic emission study of active-passive transitions during carbon steel erosion-corrosion in concentrated sulfuric acid, Corros. Sci. 44 (7) (2002) 1529–1540.

[3] F. Bellenger, H. Mazille, H. Idrissi, Use of acoustic emission technique for the early detection of aluminum alloys exfoliation corrosion, NDT&E Int. 35 (6) (2002) 385–392.

[4] F. Ferrer, T. Faure, E. Schille, J. Goudiakas, On the Interest of Acoustic Emission to Detect, Localize and Monitor Corrosion Phenomena in Aqueous Media, Corrosion 2002, Paper 02007, ISBN: 02007 2002 CP.

[5] C. Jirarungsatian, A. Prateepasen, Pitting and uniform corrosion source recognition using acoustic emission parameters, Corros. Sci. 52 (2010) 187–197. https://doi.org/10.1016/j.corsci.2009.09.001.

[6] ASTM E1316-17a, Standard Terminology for Nondestructive Examinations, ASTM International, West Conshohocken, PA, 2017. www.astm.org.

[7] S. Yuyama, T. Kishi, Y. Hisamatsu, AE analysis during corrosion, stress corrosion cracking and corrosion fatigue process, J. Acoust. Emiss. 2 (1–2) (1983) 71–93. Los Angeles, CA, Acoustic Emission Group.

[8] A. Pollock, Acoustic emission capabilities and applications in monitoring corrosion, in: Corrosion Monitoring in Industrial Plants Using Nondestructive Testing and Electrochemical Methods, Special Technical Publication 908, Philadelphia, PA, ASTM International, 1986, pp. 30–42.

[9] J. Benard, Adsorption of oxidant and oxide nucleation, in: Oxidation of Metals and Alloys, Seminar, 1970, American Society for Metals, Ohio, 1971.

[10] H.H. Uhlig, Proceedings of the Third International Congress on Metallic Corrosion, Moscow, 1966, vol. 1, 25, 1969. alsoCorros. Sc., 7, 325 (1967).

[11] F.P. Fehlner, N.F. Mott, Oxidation in the thin-film range, in: Oxidation of Metals and Alloys, Seminar, 1970, American Society for Metals, Ohio, 37, 1971.

[12] A. Arora, Acoustic emission characterization of corrosion reactions in aluminum alloys, Corrosion 40 (9) (1984) 459–465.

[13] P. Cole, J. Watson, Acoustic emission for corrosion detection, in: 3rd MENDT—Middle East Nondestructive Testing Conference & Exhibition, Bahrain, Manama; 27–30 November 2005, 2005.

[14] P.J. Van De Loo, P.T. Cole, Proceedings of the International Conference on Acoustic Emission, Tokyo, Japan, 2001.

[15] S.D. Miller, J. O'Brien, D.L. Keck, Proceedings of 7th European Conference for Non-Destructive Testing, Copenhagen, 1998.

[16] S.N. Gautrey, P.T. Cole, Proceedings of 22nd European Working Group Conference on Acoustic Emission, Aberdeen, 1997.

[17] P.J. Van De Loo, Proceedings of 23rd European Working Group Conference on Acoustic Emission, Vienna, 1998.

[18] P.J. Van De Loo, B. Herrmann, Proceedings of 7th European Conference on Non-Destructive Testing, 1998.
[19] Metal Loss Rate Sensor, Metal Loss Rate Sensor and Measurement Using a Mechanical Oscillator. U. S. Patent No. US7681449B2. 2010.
[20] Wolf, H. A. et al.; Mechanical Oscillator Activated or Deactivated by a Predetermined Condition. U. S. Patent No. 7721605. 2010.
[21] Wolf, H. A. et al.; Determining the Resonance Parameters for Resonant Oscillators.U. S. Patent No. 8676543. 2014.
[22] Wolf, H. A. et al.; Method for Quantifying Corrosion at a Pressure Containing Boundary. U. S. Patent No. 9500461. 2016.
[23] H.A. Wolf, A.M. Schilowitz, N.A. Smeets, New tuning fork corrosion sensor with high sensitivity, Corrosion (2016). Paper 7737.
[24] Saboonchi, H.; Lowenhar, E.; Gonzalez Nunez, M. A. and Carlson, D; Dome-Shape Tuning Fork Transducers for Corrosion Monitoring. U. S. Patent Application No. 62797834. 2019.

Hydrogen flux measurements in petrochemical applications

14

Frank W.H. Dean
Ion Science Ltd., Fowlmere, United Kingdom

14.1 Introduction

Most types of hydrogen damage, including hydrogen-induced cracking (HIC), stress-oriented hydrogen-induced cracking (SOHIC), sulfide stress cracking (SSC), and disbonding, result from the movement of atomic hydrogen, or *hydrogen permeation*, through steel. Hydrogen emanating from the external or *exit* face of a pipe or vessel is known as *hydrogen flux*. Flux measurements provide information relating to cracking risk and also to the causes of hydrogen permeation. In petrochemical operations, hydrogen permeation may be caused by the formation of hydrogen on a steel surface due to some corrosive process, or dissolution of hydrogen into steel at high temperatures, such as in hydrogen plants, or due to release of trapped hydrogen in traps in the steel upon high temperature excursions, such as during welding or in preweld hydrogen bakeouts. In this chapter, we identify situations in which hydrogen flux has been usefully measured and address the interpretation of hydrogen flux measurements to provide an indication of hydrogen damage risk or rate of corrosion.

14.2 Scenarios leading to the detection of hydrogen flux

In petrochemical processing, the detection of hydrogen flux on the external surface of a steel pipe or vessel is usually consequent to the following sequence:

1. Acid corrosion causing atomic hydrogen to substantially enter steel at its internal face
2. Substantial hydrogen permeation through the steel
3. Hydrogen permeation through external surface oxides and coatings
4. Molecular hydrogen desorption from the steel surface
5. Appropriate and reliable flux measurement

For the purposes of this discussion, we assume an adequate and correctly deployed tool. The universally accepted means of hydrogen flux measurement in the lab, the Devanathan cell, which measures the electrochemical current required to oxidize atomic hydrogen at a coated steel surface, precludes the need for step 4 in the sequence above. However, it is severely constrained by practical constraints in the field. In this document, we will assume step 4 is required. It is worth noting that measurement techniques depending on hydrogen exit in air may be subject to substantial hydrogen oxidation, particularly at low hydrogen flux. Hydrogen more freely permeates steel as the

Techniques for Corrosion Monitoring. https://doi.org/10.1016/B978-0-08-103003-5.00014-X

temperature increases, so conditions for flux measurements are nicely defined in terms of temperature, as shown in Table 14.1.

Table 14.1 shows distinct trends with temperature. At low temperatures (<100°C), hydrogen flux sources are restricted to a few "hydrogen promoters," almost exclusively encountered in petrochemical operations upon corrosion of mild steel. At higher temperatures, hydrogen flux may occur as a consequence of corrosion involving naphthenic acid [11] or acidic salts [13]—it does not depend so crucially upon the presence of hydrogen promoters. Molten steel will uptake hydrogen from any hydrogen source. The reason for this trend is that the movement of hydrogen through steel is more facile as the temperature increases:

(a) Internal barriers (e.g., corrosive scale) are less prone to prevent hydrogen access to the steel surface
(b) the kinetic barrier to hydrogen dissolution is more easily overcome

Table 14.1 Required conditions for the existence of a hydrogen flux in typical industrial scenarios.

Temperature (°C)	(A) Hydrogen source	(B) Permeable alloys	(C) Permeable coatings
0–120	H_2S [1–6], amine [3], NH_4HS HCN [1] and HF [4] corrosion. CP[a] [7]. Electroplating	<5% alloy ("carbon" or "mild" steel)	Most [6], not zinc
80–200	Severe acid corrosion, e.g., acetic [8]	<10% alloy	Uncoated. Depending on temperature, alloy oxides present a partial barrier to permeation
200–300	Protonic corrosion, e.g., naphthenic acid [8–11] ...	<13% alloy	
300–400	... also H from traps in preweld H bakeouts [12]	All nonaustenitic	
>400+	H from steam, grease, etc. in high T steel forming	All steels	
>1540	H in molten steel (e.g., welds, any H source)		

[a] CP = cathodic protection.

(c) the solubility of hydrogen, S, and diffusivity of hydrogen, D, through steel increase markedly with temperature of all steels; hence, more alloys permeate a significant flux as temperature increases, step 2, Table 14.1

(d) less hydrogen is retarded or permanently trapped in traps, voids, or preexisting blisters as the temperature increases.

For steel walls of greater than a few millimeters in thickness [14, 15], a well-defined corrosion activity at an entry face will generate a steady state flux through steel inversely proportional to its thickness. Further, the time required for hydrogen to establish a steady state steel increases as the square of thickness: at 300°C (750°F) ½ in. (1.3 cm) thick steel requires less than an hour for flux to reach 90% of its steady state value following a step increase in hydrogen entry at the inside face; 2 in. (5 cm) steel will require half a day. At 20°C (68°F), the corresponding times are about 16 h and a week, respectively. This time delay precludes the use of flux measurements on thick (>1 in.) cold (<30°C) steel, particularly if the interest is in using flux to discern hydrogen damage episodes lasting less than a day.

14.3 A measurement of hydrogen activity based on flux measurement

As indicated above, the hydrogen flux emanating from steel is influenced by both temperature and thickness. Thus, it is preferable to "normalize" flux measurements for steel temperature and thickness, to obtain a more universally comparable parameter indicating HIC risk and corrosive action. It is also desirable for this normalized parameter to have some physical meaning. The activity of hydrogen in steel, $\boldsymbol{a}$, is defined by $\boldsymbol{a} = c/c^0$, where c^0 is the solubility of hydrogen in steel in equilibrium with H_2 gas at 1 bar. Solubility varies with H_2 pressure p according to $c = c^0 \cdot p^{1/2}$. Thus, the hydrogen activity $\boldsymbol{a}$ corresponds to an equivalent gaseous equilibrium pressure of $\boldsymbol{a}^2$ bar of hydrogen. At the hydrogen entry (internal) steel face, hydrogen activity $\boldsymbol{a}^0$, which drives through wall permeation, is the parameter of choice. Physically, $\boldsymbol{a}^0$ is an equivalent hydrogen gas pressure which would be in equilibrium with hydrogen concentration at the entry face necessary to deliver the measurable flux at the exit face.

The conversion of flux, thickness, and temperature to activity is provided by the equation $J_{ss} = \boldsymbol{Pa}^0/\boldsymbol{w}$, where J_{ss} is the steady state flux, P the (T dependent) steel permeability, and $\boldsymbol{w}$ the steel thickness. Permeability data are scarce, but P values obtained by Grabke and Reicke [16] show only a few 10 to 10% variance in P. Interestingly, corresponding D values vary by at least tenfold. Increased trapping in steels causes the solubility S of hydrogen in steel to vary in inverse proportion. Permeability $P = D \cdot S$, so P is fairly invariant.

Hydrogen activities derived in this way are shown in Table 14.2 for a selection of corrosion scenarios based on a review of hydrogen promoters by Dean [17]. A solution saturated with 1 bar H_2S at 20°C can generate $\boldsymbol{a}^0$ values exceeding one million of bar hydrogen equivalent (100 GPa). This can be compared with a typical Young's

Table 14.2 Typical calculated hydrogen activities for a number of scenarios [17].

Scenario	Flux-thickness (at typical temp.)	Activity/bar hydrogen	Comments
15 atm. H_2S	7335 (30°C)	9,100,000	50 cm dia 8.7 mm wall line pipe. Mn 1.0. Ni flash Dev. Cell[a] [18]
NACE TM0284 (H_2S 1 bar, pH 2.7 buff.)	2100 (20°C)	1,500,000	16 mm A516 steel [1]
0.1 mm/year sour corrosion, <50°C, typical	120 (20°C)	6000	Probably dependent on scale type. Co-corrodant dependency
HF acid	1000 (30°C)	2700	Alkylation unit field measurement, typical [4]
Acetic acid	91 (116°C)	6	
H_2CO_3	14 (20°C)	81	0.5 g/L $NaHCO_3$ 5% CO_2 [19]
"Red crude"	864 (417°C)	0.1	TAN 2.3 under reflux
MRO175 "sour" threshold (0.03 bar)	80 (20°C)	2800	Based on NACE MR0175 definition of sour service [20], without low-pressure cut-offs
1 ppm H in ferritic steel, deemed to require PWHT[b]	–	40,000	Assumes $S = 5$ ppb at ambient temperature
0.1 mm/year nap. acid corrosion, typical	200 (250°C)	0.3	Limited data
Ferritic steel Young's modulus, typical	–	2,000,000	=200 GPa

[a] Devanathan cell.
[b] PWHT, postweld heat treatment.

modulus for 200 GPa steel. Not surprisingly then, sour gas corrosion generates atomic hydrogen in steel to associate within steel defects so as to cause hydrogen blistering and cracking of steel, and activity $\boldsymbol{a}^0$, calculable from a steady state hydrogen flux, is a good measure of the crack susceptibility of a corrosion scenario to cause such damage. Fig. 14.1 provides a graphical means by which flux and steel thickness may be used to determine activity $\boldsymbol{a}^0$.

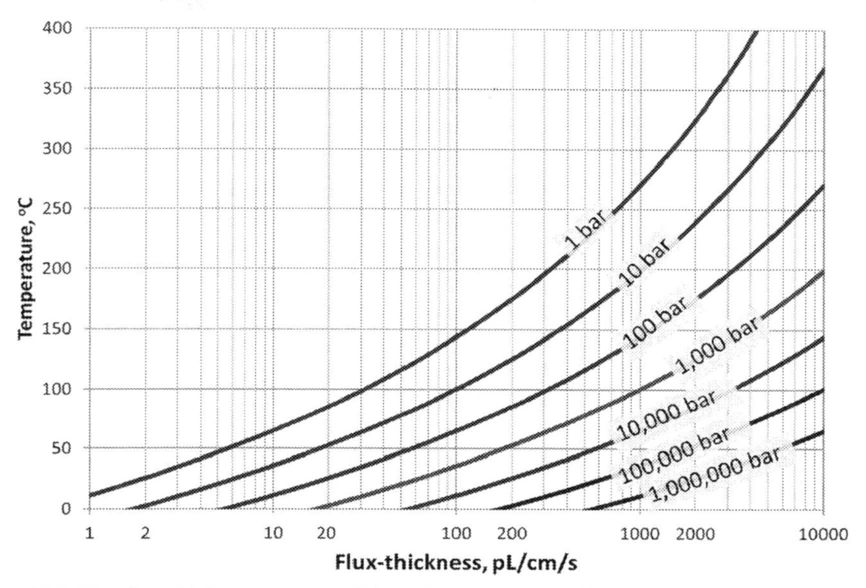

Fig. 14.1 Plot flux-thickness versus mild steel temperature which can be used to estimate hydrogen activity.

There are several flux-sensing technologies that have been used in recent decades, based on the sensing of the hydrogen located at, or exiting from, the steel surface. Fig. 14.2 shows an innovative probe employed using "hydrogen collection method," in which hydrogen flux from a steel of a well-defined surface area is entrained in an air stream and conducted up to 10 m (32 ft) distant to a portable hydrogen analyzer.

14.4 Comments pertaining to particular flux measurement applications

14.4.1 Using flux measurements to assess hydrogen damage risk

Hydrogen damage is never acceptable or anticipated. Generally, it is the result of operating conditions which are outside of the intended operation limits of equipment. Hydrogen damage is usually associated with severe sour corrosion. Field applications of flux monitors in the assessment of crack risk, therefore, chiefly address equipment with a chronic history of unremediable hydrogen damage and determine the effectiveness of process or corrosion control measures intended to mitigate that damage. Laboratory measurements of flux are also extremely valuable in assessing hydrogen damage risk and its remediation in a particular process, with a particular steel.

There is a substantial literature on critical concentrations, c_{crit}, at which hydrogen causes steel cracking. The variance of solubility of hydrogen in steel, S, discussed

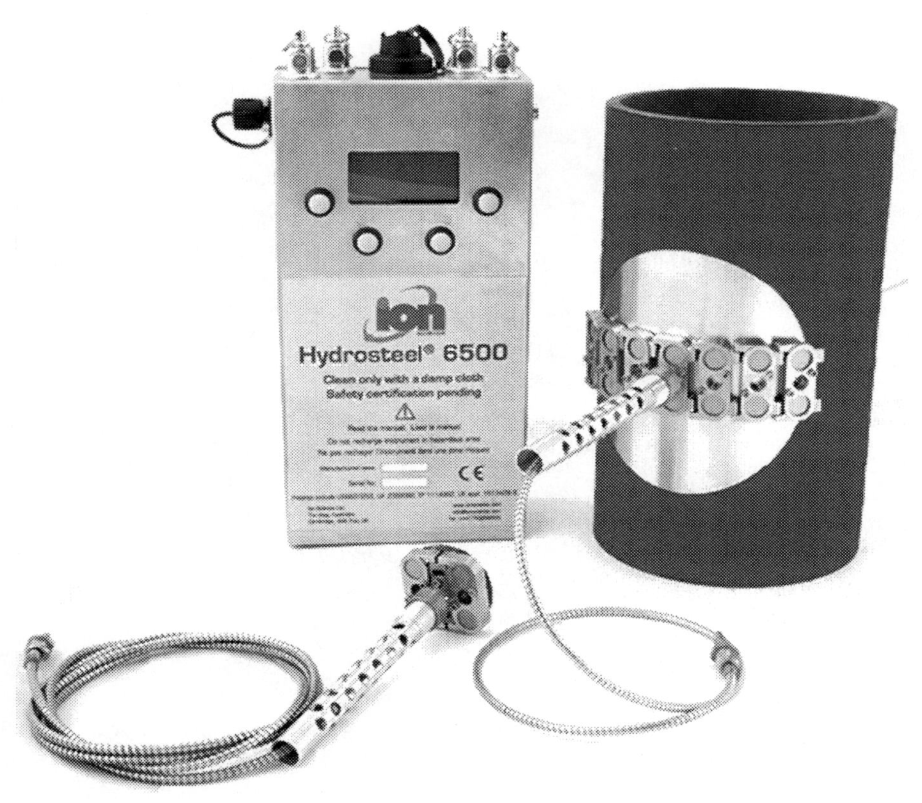

Fig. 14.2 Commercial Hydrogen flux probes, (Hydrosteel 6500, "SR" and "LR" probe, Ion Science Ltd., United Kingdom). An analyzer (not shown) is attached to the gas conduit as described in the text. The spot measurement probes shown are 60 mm (2 in.) and 150 mm (6 in.) diameter, engaged with magnets, and can tolerate temperatures to ~400°C for short periods. Courtesy Ion Science Ltd.

above, as well as the common method of using thin membranes in Devanathan cells to derive the data (from steady state flux and inferred values of D from flux transients), prevents confident comparisons between c_{crit} values. A critical hydrogen activity at which steel cracks, $\boldsymbol{a}_{crit}$, is much to be preferred as a "stand alone" measure of crack susceptibility. Using Fig. 14.1 to obtain activity from flux, thickness, and temperature, Al-Sulaiman et al. [5] present a procedure for evaluating crack risk on the basis of flux measurements on this basis. 10^6 bar is considered sufficient activity to initiate damage in crack susceptible steels; 10^4 bar is a risk to poor steels.

It should be emphasized that the activity conversion assumes steady state hydrogen diffusion through the steel; damage can occur before an appreciable flux exits a steel. Very poor, and previously damaged, steels will not even deliver a substantial hydrogen flux when exposed to hydrogen charging a corrodant, because most hydrogen is taken up by the steel. *Any* activity—any hydrogen flux—is probably symptomatic of damage to steel which is already micro-delaminated due to historical hydrogen damage. Likewise, flux measurements on large blisters tell us more about the pressure (activity) of hydrogen in the blister than at the entry face behind it.

Activity is derived from the permeability P of hydrogen in steel, which is equal to the product of diffusivity D and solubility S. Since both D and S increase rapidly with temperature, so does P. Thus, steel is thousands of times more permeable to hydrogen at say 300°C versus 30°C. Not surprisingly, hydrogen damage at temperatures exceeding 100°C is rare. Correspondingly, diffusible hydrogen in steel in service at high temperatures needs to be allowed to diffuse away before the steel is cooled, lest it increase too much in activity. The use of flux monitoring in this area is likely to increase with increasing acceptance of field worthy flux measurement tools.

14.4.2 Hydrogen bakeouts

Prior to welding steel which may have been subject to hydrogen charging by one of the sources cited in Table 14.1, it is often deemed necessary to raise the steel temperature (typically to 350°C) for a number of hours (typically 2–48 h, depending on the steel type and thickness) to "bakeout" hydrogen which may have become entrapped in the steel. Otherwise, during welding, the hydrogen content of the parent steel is liable to concentrate at the weld's heat-affected zone, causing hydrogen cracking. About 1 ppm hydrogen is often considered an upper acceptable threshold for diffusible hydrogen content of the steel at the elevated temperature. Fick's First Law of Diffusion states that flux

$$J = D \cdot \mathrm{d}c/\mathrm{d}x \tag{14.1}$$

During hydrogen bakeouts, the hydrogen concentration gradient, $\mathrm{d}c/\mathrm{d}x$, at the steel face is expected to be greater than anywhere in the steel interior at all times, while the flux is decreasing from its maximum value. It is possible to conservatively cite a critical, decreasing flux during bakeout as indicating that the steel contains negligible hydrogen. For mild steel at 350°C and <5 cm thickness, the flux is estimated at 500 $\mathrm{pL/cm^2/s}$. This flux provides a real time criterion for extending the time of hydrogen bakeout if necessary, whereby prospective weld damage can be avoided. Correspondingly, if a flux fails to reach, or decreases to below, the threshold flux, the bakeout may be curtailed, and the repair schedule readjusted, whereby the product downtime is shortened. Examples of both of these scenarios are presented by Brown et al. [12].

Steel which has been in high temperature service is occasionally retained at temperature after service, to allow hydrogen in the steel to escape before cooling, sometimes forcibly, under nitrogen. It is possible to forecast a "safe" flux threshold below which the steel may be safely cooled. However, this application of flux monitoring is not reported to date in the literature.

14.4.3 Using flux to indicate corrosion by sour gas and related species

The most well-known cause of hydrogen flux through steel is corrosion by sour gas (H_2S) condensate in pipes and vessels, in both upstream and downstream petrochemical equipment. Different sulfide scales form on steel as a result of sour gas corrosion of steel, depending on sour gas concentration, co-corrodants, pH, ambient temperature, and steel composition. In most cases, corrosion and corrosion-induced flux are completely suppressed by scales. However, since some sulfur scales are semiconducting (in particular, pyrrhotite) [21], thereby providing a means of electrochemical hydrogen formation, not on the steel, but on a scale through which atomic hydrogen migration is unlikely, it is probably best not to assume that the correlation of hydrogen flux with corrosion obtained in one sour corrosion scenario is applicable in another.

Generally, successful deployment of a flux tool depends on the experience of the operator in knowing where and when to use it, particularly as a near real time indicator of sour corrosion and its control. Examples of successful flux measurements at refineries are provided by Addington et al. [3] and Etheridge et al. [2]

More generally, positive flux indications generally occur:

- where there is a history of hydrogen damage
- where inhibitors of corrosion of steel are deemed to be effective
- where corrosion is known to be episodic due to modification, removal, or dissolution of scale, as a consequence of pH change, oxygen, cyanide, or erosion (the latter particularly in relation to ammonium bisulfide corrosion).

Because the permeability of steel is very temperature-sensitive, any attempt to correlate flux with corrosion rate will benefit from conversion of flux to activity, as described above. A more approximate method is simply to compensate for thickness, by multiplying a flux reading (in $pL/cm^2/s$) by thickness (in cm), to obtain readings in pL/cm/s. Field flux due to sour corrosion rarely exceeds 1000 pL/cm/s and values of 10s of pL/cm/s are much more common in the field (cf. values in Table 14.1), typically indicative of corrosion rates of about 0.1 mm/year (a few mil/year).

14.4.4 Using flux to indicate corrosion by HF acid

Hydrogen flux from steel in HF acid alkylation units is more substantial than that from sour systems and more long-lived. This is probably because iron fluoride scales are more soluble and less passivating than sulfide scales. Also, HF corrosion is much less variant than sour corrosion, occurring as it does in a much more well-defined product stream. It is generally not mitigated by inhibitors.

As above, conversion to hydrogen activities is advisable if readings from different sites are to be compared. Flux-thickness of a few 100 $pL^2/cm/s\text{-}cm$ is typical [4], probably indicative of a few tenths of mm/year (about 10 mil/year) corrosion rates.

14.4.5 Naphthenic acid corrosion and sulfidic corrosion

Naphthenic acid is the generic name given to corrosion cyclic carboxylic acids produced from "acid crude" oil. The acids concentrate in distillation column side-cuts at temperatures between approximately 250°C and 400°C (480°F and 750°F). They are corrosive, generating carboxylic iron salts and hydrogen, which, due to the high temperature of the steel, is prone to diffuse through the wall sufficiently to deliver hydrogen flux of up to a few thousand $pL/cm^2/s$ [13]. Thiols and hydrogen sulfide—present as a thermal breakdown product—react similarly to form hydrogen and sulfide salts.

Naphthenic acid corrosion (NAC) is complicated by the plethora of different corrosive constituents in acid crude oil [22], their respective temperature stability and solubility, and interaction with, for example, hydrogen sulfide [23]. Corrosion often depends on the removal of corrosion product due to process stream sheer velocity, as may prevail, for example, at pipe bends or in distillation units. Moreover, in practice, crude blends in refineries are prone to change every few days, which is a challenge for NAC prediction and control. The prevailing high temperatures tend to place limits on corrosion monitoring. For all these reasons, high temperature hydrogen flux monitoring has received considerable interest in refineries. In many instances, flux measurements provide a means of optimizing inhibitor deployment, although it has also been occasionally used to contend with severe episodes of NAC [9, 10].

References

[1] S.J. Mishael, F.W.H. Dean, C.M. Fowler, Corrosion 2004, Paper 04476, Conference Series, NACE, Houston, 2004.

[2] A.M. Etheridge, E.B. McDonald, D.G. Serate, F.W.H. Dean, Corrosion 2004, Paper 04477, Conference Series, NACE, Houston, 2004.

[3] F. Al-Aqeer, F. Addington, Corrosion 2008, Paper 08548, Conference Series, NACE, Houston, 2008.

[4] F.W.H. Dean, P.A. Nutty, M. Carroll, Corrosion 2001, Paper 01636, Conference. Series, NACE, Houston, 2001.

[5] S. Al-Sulaiman, A. Al-Mithin, A. Al-Shamari, M. Islam, S.S. Prakash, Corrosion 2010, Paper 10176, Conference Series, NACE, Houston, 2010.

[6] R.D. Tems, A.L. Lewis, A.L. Abdulhadi, Corrosion 2002, in: Paper 02345, Conference Series, NACE, 2002.

[7] J. Woodward, R.P.M. Procter, R.A. Cottis, The effect of hydrostatic pressure on hydrogen permeation, in: A.W. Thomson, N.R. Moody (Eds.), Hydrogen Effects in Materials, Minerals, Metals and Materials Society, 1996, p. 657.

[8] F.W.H. Dean, S.J. Powell, Corrosion 2006, Paper 06436, Conference Series, NACE, Houston, 2006.

[9] J.M. O'Kane, T.F. Rudd, D. Cooke, F.W.H. Dean, S.J. Powell, Corrosion 2010, Paper 10351, NACE, Houston, 2010.

[10] J.M. O'Kane, T.F. Rudd, J.H. Harrison, S.J. Powell, F.W.H. Dean, Corrosion 2010, Paper 10178, Conference Series, NACE, Houston, 2010.

[11] F.W.H. Dean, S.J. Powell, A. Witty, Corrosion 2014, Paper 14089, Conference Series, NACE, Houston, 2014.

[12] C.N. Brown, M.J. Carroll, F.W.H. Dean, J.H. Harrison, A. Kettle, Corrosion 2004, Paper 04478, Conference Series, NACE, Houston, 2004.
[13] F.W.H. Dean, D.J. Fray, T.M. Smeeton, J. Mater. Sci. Technol. 18 (2002) 851–855.
[14] M.R. Bonis, J.-L. Crolet, Corrosion 2002, in: Paper 2036, Conference Series, NACE, 2002.
[15] F.W.H. Dean, Mater. Sci. Technol. 21 (3) (2005) 347–351.
[16] H.J. Grabke, E. Riecke, Mater. Tehnol. 34 (6) (2000) 331–342.
[17] F.W.H. Dean, Corrosion 2010, Paper 10182, Conference series, NACE, Houston, 2010.
[18] M. Shimitsu, M. Iino, M. Kimura, N. Nakate, K. Ume, Proceedings of Corrosion/87 Symposium, NACE, Houston, TX, 1989.
[19] S. Asher, P.M. Singh, Corrosion 2008, Paper 08411, Conference Series, NACE, Houston, 2008.
[20] NACE, MR0175/ISO 15156-1 International Standard, 'Petroleun and Natural Gas Industries—Materials for Use in H2S-Containing Environments in Oil and Gas Production', first ed., NACE, 2001.
[21] J.C. Ward, Rev. Pure Appl. Chem. 20 (1975) 175–206.
[22] E. Slavcheva, B. Shone, A. Turnball, Corrosion 1998, in: Paper 579, Conference Series, NACE, 1998.
[23] R.D. Kane, M.S. Cayard, Corrosion 2002, in: Paper 02555, Conference Series, NACE, 2002.

Part Three

Corrosion monitoring in particular environments and other issues

Corrosion monitoring in microbial environments

Pierangela Cristiani[a] *and Giorgio Perboni*[b]
[a]Ricerca sul Sistema Energetico, RSE SPA, Milan, Italy, [b]CESI SpA, Piacenza, Italy

15.1 Introduction

In many cases, corrosion damage to industrial equipment cannot be directly attributed to 'chemical' effect. They are often referred to as microbial corrosion, which ascribes the decrease of material performance to the activity of living microorganisms. These kinds of phenomena are ubiquitous in natural environments and are often observed in process waters as well as in cooling circuits, including the facilities exposed to simple air condensates. Nevertheless, it is indeed very difficult to demonstrate exhaustively the mechanism and that these kinds of corrosion are due to a single cause because they usually involve complex systems both temporally and spatially. Taking into account this concern, it is almost always named 'Microbial Influenced Corrosion,' usually abbreviated as MIC. A necessary condition for the development of MIC in industrial plants is the development of a microbial biofilm which can more or less uniformly cover the surface of the materials equipment.

Biofilm is the first layer of biofouling growing on the surface of wetted structures in contact with natural waters, whose thickness can range from a few microns to a few millimeters. It is constituted by microbial cells (bacteria, fungi, and algae) imbued in extracellular polymeric substances (EPS): a complex mixture of macromolecules such as polysaccharides, proteins, nucleic acids, and lipids produced by the cells themselves [1]. EPS facilitate cell adhesion to the substratum and give structural integrity to the biofilm matrix [2]. Depending on the environment condition (Oxygen, pH, conductivity, etc.), several inorganic precipitates enrich biofilm in time. It must be stated that all the bacteria secreting organic acids and EPS or slime are able to play relevant roles in aerobic as well as in anaerobic environments [3–5].

It is well-acknowledged that bacteria synergistically live in multifaceted communities which are able to catalyze electrochemical processes within biofilms and, typically, both the aggressive and inhibitory effects from the metabolism of bacterial population on corrosion reactions are seldom linked to a single bacterial species [6] and the complex interactions between microorganisms are strongly influenced by the presence and nature of corrosion products on the material surface. Nevertheless, some specific types of bacteria better than others were demonstrated to influence corrosion of metals, in particular sulphate-reducing bacteria (SRB) which live in strict anaerobic conditions, sulfur-oxidizing bacteria (SOB), iron-oxidizing/reducing

Techniques for Corrosion Monitoring. https://doi.org/10.1016/B978-0-08-103003-5.00015-1

bacteria (IOB/IRB), and manganese-oxidizing bacteria (MOB) which live in different aerobic condition.

Almost all the metallic materials commonly utilized in plant can be affected by biofilm and MIC. Copper alloys are no exception because the biofilm protects microorganisms from the toxic effect of copper. Indeed, the microorganisms can attack the passive states and inhibit their formation through the excretion of mucopolysaccharides which surround the cells in the biofilm, isolating itself from the harmful effect of copper ions. Bacteria like *Shewanella oneidensis*, *Pseudomonas aeruginosa*, *Escherichia coli*, *Acidovorax delafieldii*, and *Desulfovibrio alaskensis* are just a few of the many microorganisms capable of forming dense biofilms, whose tolerance towards copper ions has been widely demonstrated [7, 8].

The biofilm enhances bacterial resistance against several external factors, including biological competitors and chemical biocides action, forcing the use of intensive antifouling and disinfecting water treatments [9].

The cost of MIC is very high: in the particular case of heat exchangers, 20% of all the corrosion damages were attributed to this type of corrosion [10] and a rough costs of 0.2% GPN for the industrialized world, including increased capital expenditure, energy costs, maintenance costs, cost of production loss, and extra environmental management cost, are generally reported [116]. Studies related to Italian power plants [11, 12] quantified that the majority of the corrosion cases of the condenser tubes might be prevented by better cleaning during the plant operation.

15.2 Biofilm and MIC monitoring

Many techniques are used to monitor and detect MIC in laboratory and some of them can be used also in the field [13]. Some of the methods are very effective for detecting bacteria and monitoring the biofilm growth and others for the corrosion rate estimation, even though none of them can exhaustively quantify the microbial attack in all the possible combinations of material, environment, and metabolism that can actually occur.

A simple technique to document microbial corrosion was proposed by Brenda Little [14], based on a double-chamber electrochemical cell separated by an electrolytic membrane. In this cell, one sample of material to be monitored is posed in one of the chambers kept in sterile condition and another identical sample is posed in the second chamber inoculated by bacteria. When the two samples are electrically shorted, a current detected by a zero ammeter in the circuit can be attributed to the effect of bacteria.

An evolution of this "dual cell" technique was recently proposed, based on the use of microbial fuel cells [15, 16]. In fact, the mechanism generating power in a microbial fuel cells, due to bacteria metabolism, is the same as for microbial corrosion, with the difference that in the first case organic material is oxidized on an un-corrodible anode, while metal is oxidized in the case of corrosion. Connecting the metal under investigation to the anode in the anaerobic compartment of a microbial fuel cell, it is possible to verify if the generated current is affected by the presence of the metal. A current increase can be directly associated to microbial corrosion, if the cathode is not limiting the current increase. In other words, the metal can result in a better anode for bacteria than the un-corrodible anode in the fuel cell. Similarly, other information could be achieved connecting the metal to the cathode in an aerobic chamber. This technique can be versatile to perform preliminary laboratory test, comparing the performance of different materials, and coating, against microbial corrosion. It offers the advantage to have indication in real-time of the "electro-activity" of bacteria on a conductive support without the necessity of microbial analysis or sophisticated electrochemical measurements to be validated. Nevertheless, sophisticated electrochemical analyses and bacteria characterization are still to be carried out to validate the results from this technique, which is in its infancy at the state of the art [16, 17].

It must be underlined that the corrosion rate, especially after the biofilm overcomes the first phase of its development, was found to be independent of parameters such as the biofilm thickness, its roughness, or the number of settled microorganisms. On the other hand, the direct correlation between microbial metabolism and the corrosion rate has been well-documented [18–20] and it is possible to quantify the role of many of the metabolic products of microorganisms inside the biofilm, especially enzymes within the EPS matrix, organic and inorganic acids, and volatile compounds such as ammonia or hydrogen sulphide. The extracellular metabolic components, in particular, significantly alter electrochemical processes at the metal/biofilm interface [6, 21–24] and lead to MIC.

Most of the MIC cases are reported as localized attacks and the localization is connected to nonhomogeneous morphology and composition of the biofilm. In fact, the type of settled microorganisms and the nature/concentration of their extracellular products can change over a distance of a few microns, following the nutrients availability and the environmental condition that can vary at microscale on the surface. Nevertheless, the microbiological corrosion rate detected in nature has so far not been effectively reproduced in the laboratory. Often in the past, it has been attributed only to a few bacterial species because of the measurement techniques and community analyses did not allow to detect most of the microorganisms, being normally not cultivable with traditional culture media.

In recent years, the development of modern molecular biology techniques has allowed us to replicate (amplify) the genetic material or biomolecules produced by microorganisms and to document, in this way, the vast number of microorganisms (archaea, bacteria, and fungae) involved in the corrosion process. First, the technique of electrophoresis on a denaturing gradient gel (DGGE) of DNA fragments [25] and then the techniques of next-generation sequencing (NGS) of the 16S and Internal Transcribed Spacer ribosomal RNA (rRNA) [26], even more comprehensive and precise for the recognition of genes and complex molecules units, have characterized individual species and bacterial groups, allowing to recognize and e-catalog all (virtually 100%) the individual species present on a corroded sample [27].

Generating the 16S rRNA gene amplicons, by domain-level PCR reactions amplifying from genomic DNA and analyzing the data with bioinformatics tools, it was possible to verify the presence of hundreds of bacterial species living together in "corrosive" biofilms, while documenting the predominance of already well-known species, such as the bacteria of the cycle of sulfur and iron. The detection of these species at light microscopy, in fact, is simple as they are able to grow on nutritive soils. Using the new techniques, the presence of recurrent bacterial species on same alloys has been demonstrated in various parts of the world, even in the presence of chlorination treatments [7].

A wide range of sophisticated techniques derived from laboratory experiences can be considered to analyze and monitor MIC phenomena also in field, depending on the requested detail of the study and the application. They are mostly derived from the experience of other types of corrosion monitoring [28, 29]. They are shortly reviewed in the Section 15.2. On the other hand, simple electrochemical tools, developed for direct evaluation of biofilm and MIC risk in industrial water circuits, come into the practice for about twenty years. They are described in details in the Section 15.3.2. Some examples of their application are reported in the section of case studies.

15.3 Corrosion monitoring applied to MIC

The traditional and widely used system to detect and measure corrosion phenomena consists of metallic coupons inserted directly in the plant facility and then extracted for periodical post-exposition evaluation. The advantages and the disadvantages of this method are essentially the same as those regarding the methods to monitor biofilm off-line.

There are essentially no limits to the number and complexity of the possible off-line analyses, from simple weight loss to microscopic observation, corrosion products analyses, etc., so that a detailed information on corrosion can be generally achieved. Biofilm can concurrently be studied at laboratory level and possibly correlated with the morphology and the propagation rate of the corrosion attacks.

Several traditional electrochemical techniques for online corrosion monitoring, such as Redox potential and open circuit potential measurement (OCP), Linear polarization resistance (LPR), Electrochemical Impedance Spectroscopy (EIS), and Electrical Resistance (ER), are used to detect MIC in the laboratory and some in the field [30]. The criteria in the selection of the measurements are mainly related to:

- nature of the metals to monitor (carbon steel, passivable alloys, etc.)
- environmental conditions (aerobic, anaerobic)
- physicochemical characteristics of the water (temperature, flow, conductivity, chemical treatments, pH)
- applications (field, plant, laboratory).

15.3.1 Redox potential and open circuit potential measurement (OCP)

These measurements are very easy to perform both in the laboratory and in the field: they require an electrometer, a reference electrode, a platinum wire for Redox, and a metallic coupon whose corrosion is evaluated for the OCP. The Redox potential can be used in plant mainly to establish whether aerobic or anaerobic conditions are developing in water, or oxidant biocides are decaying in the solution.

Open circuit potential (or free corrosion potential) may give information about the passivation layer of metals and the thermodynamic risk for the integrity of the protective layer. Problems during in-field applications can occur if the reference electrode is not stable in the chosen environment. Zinc and commercial silver/silver chloride reference electrodes were found suitable for seawater; copper/copper sulphate reference electrodes have been found to be stable in soils.

This technique does not give information on the actual corrosion rate and, if not integrated with other techniques, it could provide ambiguous indications about MIC [30]. The following are two examples concerning the use of OCP and Redox potential measurements:

- a high Redox potential suggests the presence of oxygen and aerobic bacteria in the media, but the corrosion of a metal exposed in the same environment can occur under a thick biofilm, at the bottom of which a prevailing anaerobic bacteria population may be present because the oxygen diffused from water is totally consumed in the upper biofilm layers
- a low free corrosion potential measured on an active-passive alloy can indicate corrosion in a passive state, i.e., a low general corrosion rate in an anaerobic environment or, on the contrary, severe localized corrosion in an aerated one.

15.3.2 Linear polarization resistance (LPR)

Several commercial and relatively cheap devices are available for the measurement of the uniform corrosion rate on metallic coupons based on the LPR technique (see Chapter 3). In its basic form, a potentiostat is used to move the potential of the coupon a few mV (normally at maximum of 10 mV) near its free corrosion potential and to concurrently measure the current (ASTM Standard G96-90 (96)). Assuming that in the aforementioned small potential range, the current is proportional to the potential shift, the so-called 'linear polarization resistance', Rp, is evaluated using Ohm's law. Stern and Geary showed that, in case of uniform corrosion in conductive acid solutions, corrosion rate is inversely proportional to Rp.

Taking into account that this technique is relatively easy to perform not only in the laboratory but also in the field and that it is able to provide information in real-time on corrosion rate, the LPR method was applied also in environments different from the (acidic) ones in which it was originally defined. Indeed, this technique was also applied to study MIC on iron in a neutral anaerobic environment, on copper alloys in condensers treated with biocides and so on, in integrated system, as described in the Section 15.4. It should be noted that the data provided by the LPR technique cannot always be accepted without additional information. One of the reasons is that the method refers to uniform corrosion, but localized corrosion often happens in MIC. The other reason is that the presence of corrosion products on samples exposed to a neutral environment can make complicated the interpretation of the data (see [30], for instance); LPR measurements on carbon steel under anaerobic conditions and in the presence of sulphide (5 mg/L or more) create a risk of very large errors due to the formation of porous corrosion products [31]. Also, the potential scanning rate during LPR measurement can play an important role, especially in case of high capacitive current, such as for carbon steel in an anaerobic SRB environment. In that case, LPR technique can overestimate corrosion rate and hours can be necessary attempting to reduce this effect [32]. Actually, the value of the resistance measured with the LPR technique is the sum of the true polarization resistance Rp and of the solution resistance Rs between the coupon and the reference electrode: a large Rs due to a low conductivity of the electrolyte (such as in some potable waters) can result in a large error in the evaluation of the corrosion current. Data provided by the instrument must be corrected when the conductivity of the electrolyte is lower than about 0.5 mS cm^{-1}, depending on the probe geometry [33]. These low values can be found in oligotrophic fresh waters, tap waters, boiler waters, and so on.

The geometry of the electrodes and the finishing of its surface can play a determinant role in MIC monitoring more than any other kind of corrosion for high conductive water, such as seawater, because the biofilm growth is strongly influenced by the value of water flow rate and surface roughness.

Finally, the following suggestions should be taken into account in the installation of the electrochemical cells:

- Use the same kind of alloys that are used in the plant for the electrodes
- Use electrodes with the same surface finishing and corrosion stage development as actually occurs on the surface of the materials operating in the plant
- Use electrodes with the same geometry as the monitoring equipment (tubes of a similar diameter for heat exchangers, tubes or pipelines, or plates for wall or plate exchangers, and so on)
- Make sure that the water flow on the electrodes is representative of the plant condition to be monitored
- Make sure there are no thick or abundant deposits on the electrode surfaces.

Fig. 15.1 shows an example of an LPR probe setup and used in industrial MIC environments [34].

15.3.3 Electrochemical impedance spectroscopy (EIS)

The Electrochemical Impedance Spectroscopy (EIS) instrumentation measures the impedance between a metallic coupon and a reference electrode in a wide frequency range.

In practice, a small sinusoidal potential perturbation (<10 mV in amplitude) at a given frequency is added to the free corrosion potential, joined with the resulting alternating current, and sent to a frequency response analyzer; the result of the analysis is the value of the electrode impedance at the tested frequency; the procedure can be repeated at other frequencies chosen over a wide possible range (e.g., from mHz to MHz, see ASTM G106–89, 1994) so that the impedance spectrum is obtained.

From the impedance values respectively measured at high and low frequencies, the values of the solution resistance (Rs) and polarization resistance (Rp) can be obtained, so that the main disadvantage of the LPR method when applied in a low conductivity solution is eliminated.

By looking at the trend of the impedance values in the intermediate frequency region, one can obtain additional information (double-layer capacitance, diffusion

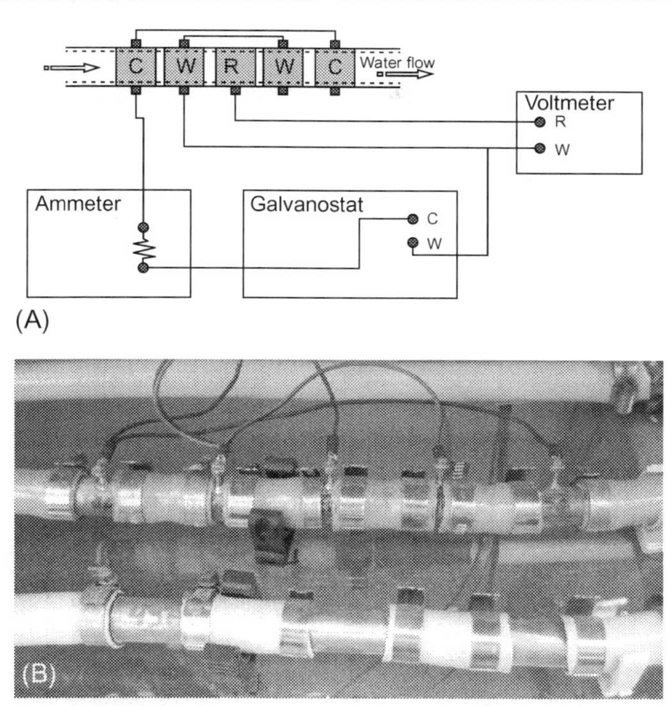

Fig. 15.1 Schematic (A) and photo (B) of samples from a real condenser tube inserted in a hydraulic line for the LPR measurements (coupons with electric connection) and gravimetric determinations (coupons without electrical connection).

phenomena, etc.) about the corrosion mechanism. The analyses of the data are usually complex and must be done by experts, requiring the formulation of an equivalent circuit as model in which every component should have a documented physical meaning. As a result, this technique for mechanistic studies is very useful, but more applicable in the laboratory than in the field. Nevertheless, if a 'model' for data interpretation in the considered system is already known, the analysis of the data is greatly simplified.

Specific tentative studies were made in laboratory experiments with EIS to verify whether the formation of a biofilm affects the value of Rs, measured as a real part of the impedance at high frequencies. Unfortunately, the EIS technique was unable to reveal clearly the first crucial phase of biofilm less than about 200 μm in thickness [35]. Several authors, more recently, reported of equivalent circuits implying a double-layer capacitance of biofilm on un-corrodible electrodes, hypothesizing the possibility of measuring the first stage of biofilm growth by EIS, especially in studies on bacteria adhesion for medicine purposes [36]. Others underlined a pseudo-capacitive behavior of biofilm in a non-turnover food condition (lack of nutrients), indicative of bound redox mediators [37]. Nevertheless, in the case of corrosion of metal alloys, a capacitive behavior of a biofilm made of more than 90% of water seems too speculative and it is more obvious to associate the effect of biofilm in modifying protective corrosion products and consequently their capacitance. As a matter of fact, several authors ([38], for instance) extensively used EIS at laboratory level to study the opposite mechanism of MIC, the microbial influenced corrosion inhibition (MICI). In fact, dense and fat extracellular polysaccharides (EPS) can be produced by specific bacteria such as *Shewanella* strains, which can vary depending on the substrate (electrode) potential and flow condition [39] and can provide excellent corrosion protection in several ways. In some cases, corrosion inhibition appears to be due to reduction of ferric ions to ferrous ions and increased consumption of oxygen, both of which are direct consequences of microbial respiration [40] or biofilm could form an effective protective barrier on the metal surface that can suppress diffusion of corrosion products, resulting in enhanced corrosion inhibition efficiency [41].

15.3.4 Electrochemical noise (EN)

The EN technique consists of the measurement of spontaneous fluctuations (no external signal is applied) of the potential or of current between two identical metal samples as a function of time (see Chapter 5). This technique may detect the initiation of localized corrosion, which is the most important feature of MIC [32], but it does not provide information on the corrosion propagation rate.

Researchers at Argonne laboratories (Tennessee, USA) have developed an online, real-time method to detect sustained localized pitting [42]. This approach includes new software and improved probes to interpret real-time electrochemical noise signals.

Because this technique is relatively easy to apply, EN appears suitable for increasing applications in the field. However, the comprehension of EN signals still remains almost complex [43]. Therefore, the technique is limited by the need for particular instrumentation and expert people for data analysis/interpretation.

Recent application of an advanced technique (the time-resolved instantaneous frequency information of electrochemical current noise (ECN) transients) to MIC studies, combined with monitoring of the OCP and microscopic observations, indicated the development of transients in one of the two electrochemical cells containing SRB with a different instantaneous frequency decomposition as compared to the background ECN signal, which resulted from the anaerobic general corrosion process [44].

Laboratory applications of EN coupled with corrosion rate measurements permitted also the study of the influence of SRB and other organisms in promoting the pitting corrosion of iron in reinforced concrete. In particular, different steps of growth/rupture of the iron sulphide film on the metal surface were documented [45]. Results demonstrate that bacteria are preferentially attracted to iron corrosion products, which determine their spatial distribution [32].

15.3.5 Electrical resistance (ER)

The measure of the Electrical Resistance (ER) of a metallic sample can be used to evaluate the thickness decrease of the sample due to corrosion (see Chapter 11). The measurements give cumulative corrosion, but, if measured as a function of time, they can provide an almost real-time average corrosion rate. Sensitivity depends on the design of the probes and on the instrumentation; probes can be designed with different sensitivities and expected lifetimes. A new, improved highly sensitive ER-technique probe (by MetriCorr, Denmark) has increased sensitivity, despite taking 30 min or more to define a trend if the corrosion rate is high (>0.1 mm/yr), while LPR takes only 10 min or less. The advantage of ER is that this technique can be applied in any media, without limitation regarding the conductivity of the solution [46], and the data analysis is very simple. ER is, therefore, a widely used monitoring technique and has the potential to be useful also for MIC.

Precipitation of semiconducting ferrous sulfides on ER probes in the low-activity SRB media does not seem to affect accuracy in the ER measurements when comparing weight loss. The risk of underestimation is often the major reason for not recommending ER for sulfide media (overestimation is the case with LPR), but it seems that with proper sensor design and by avoiding probes that may short circuit (the same for the case of LPR), the error will be small due to the lower conductivity of ferrous sulfide to that of steel [47]. The main disadvantage of this technique is that the basic hypothesis of uniform corrosion is seldom respected in MIC.

15.3.6 Other techniques

Other various sophisticated electrochemical methods have been experimented: large signal polarization techniques, random potential pulse method, electrochemical relaxation methods (programmed pulse relaxation, sinusoidal AC relaxation), and scanning vibrating electrode techniques (a review in [48]). Various microsensors have also been used [49], but these tools at this moment remain useful mainly for laboratory studies in full controlled 'microcosmos.' In the field, macroscopic MIC cases concern the oil industry pipeline. In a large waxy crude oil pipeline under low flow conditions, MIC corrosion is more of an issue, especially for older fields where reservoir pressure is being supported by water injection. Mitigation methods usually involve periodic biocide treatments of the pipeline.

The monitoring of corrosion in the pipelines requires the use of in-line inspection (ILI) tools often consisting of 'intelligent pig runs' equipped with devices and sensors. ILI tools are forced to pass through the pipeline driven by the fluid flow or towed by a vehicle or cable. The most common technologies used for pipeline inspections are Magnetic Flux Leakage (MFL) and Ultrasonics (UT) associated with geometry measurement tools. The information which can be provided by these intelligent pigs covers a much wider range of inspection and troubleshooting needs; they are not specific for corrosion, but often the MIC monitoring programs refer to the data collected with that facility [50].

The field signature method (FSM, CorrOcean ASA, Norway) is one of the techniques applied to monitor corrosion inside buried pipes and subsea flow lines (see Chapter 25). It is a variant of the ER technique (see Chapters 11): an array of metallic pins (electrodes) is fixed on the external surface of a specially prepared section of pipe in which an aggressive solution flows. A known direct current is applied and it flows through the pipe wall. Voltage drops between the pins are recorded as a function of time and converted into losses of the wall thickness in different areas of the pipe. The sensitivity is claimed to be 1/1000 of wall thickness, or better, in case of general corrosion.

The nonuniformity of the attack can be revealed if the spatial resolution of the pins is high. For example, a resolution of less than 12 mm for the distance between the pins may reveal a pit of 2 mm × 2 mm × 2 mm in a plate 24 mm thick. For corrosion control, the pipe section for FSM is designed to last over the total operational lifetime of the structure and can be engineered for a variety of situations and geometries, e.g., a pipeline on the seabed. This technique is obviously quite expensive and the number of electrodes fixed on the pipe wall can be very high if a good resolution for localized attacks is required. In practice, the sensitivity is not high enough for monitoring initiation of pitting and small attacks.

In addition, the formation of sulphide layers generated by SRB might complicate the signals, as in the cases of ER probe and also the LPR probe.

15.4 Biofilm and bacteria monitoring

In industrial practice, it is obviously more important (and more effective) to prevent biofilm and minimize corrosion risk than monitor the evolution of corrosion phenomena. In this light, specific tools and techniques were implemented to detect and monitor biofilm and bacteria presence. Some of them are used also in the industrial practice.

The traditional way to document biofilm presence involves the use of 'handmade' probes and coupons inserted directly into the plant facility, or in a bypass of a process stream, and periodically collected for off-line analyses [51]. Devices suitably designed for material coupons exposure were early commercialized, for instance, Bioprobe^{TR} (by Petrolite, Texas, USA) and Robbin's Device [52]. Some of these systems can also be used in high pressure. Nevertheless, service companies and professionals often provide self-made assembling devices with studied materials and tubes.

Actually, it is important to make test lines being representative of the real materials and process [53]. Indeed, as the positioning of the probes and the water flow influence the results strongly, they must be chosen to reproduce, at least, the same (or most critical) conditions.

Coupons, in addition, must be made of the representative materials. It has been shown, in fact, that bacteria populations colonization strongly varies on different materials (and different material finishing), even though the planktonic bacterial growth conditions are identical [54].

After the sampling, many analyses can be performed off-line (Table 15.1) to gain better knowledge of the structure and distribution of the biofilm [55]. Several tools and kits are available to detect and count the cultivable bacteria in the biofilm, as described in Table 15.1. Some systems are able to give a rough quantification of cell proliferation biomass measuring the cellular ATP (adenosine triphosphate) activity [56]. The ATP assays are often based on the detection of the light produced by the reaction of ATP with added luciferase and D-luciferin (Britelite^{TR} Perkin Elmer, MA, USA,

Table 15.1 Techniques used to document and study MIC.

Analyses	Technique	Comment
Sampling	Coupons	Always necessary
Deposit analysis	Determination of dry weight, Visual observation of corrosion products, slime, sludge	To document and quantify the corrosion case
Microscopy observations in plant	Stereo zoom microscopy, Optical microscopy, Epi-fluorescence microscopy	Biofilm observation in field, Bacteria counts, ATP, and bacteria count
Microscopy observation in laboratory	Scanning Electron Microscopy (SEM) Environmental SEM (ESEM) for non-fixed and wet samples	Fixed biofilm observation and bacteria count Living biofilm morphology
	X-Ray XPS (X photoemission Spectroscopy), EDS (Electron Diffraction Spectroscopy) microprobe	Chemical analyses of corrosion products and biofilm components
	Atomic Force Microscopy (AFM)	Three-dimensional vision of living biofilm morphology
	Confocal Laser Scanning Microscopy (CLSM)	Section analyses of living biofilm layer allowing the three-dimensional vision of it
Chemical analyses	Colorimetric kits	Analyses of corrosion products in plants
	Laboratory techniques Inductively Coupled Plasma (ICP)	Analyses of corrosion products in laboratory
	Gas Chromatography (GC) Gas Chromatography/Mass Spectroscopy (GC/MS) FT-IR (Fourier Transform Infra Red spectrometer)	Analyses of chemical components of biofilm (fatty acids, chlorophylls, organic compounds, metabolomes) for laboratory
Microbiological analyses	Coupons, dipstick, culture medium, and laboratory equipment.	Plate counts of microorganisms Used in field also
	Plate count of specific bacteria (as anaerobics, iron-bacteria, pathogens, etc.) Kit for luminescent bacteria	
Immunological analyses	Colorimetric assays, ELISA-kits, luminometric assays, laboratory equipment.	Detection of enzymes, used in field also
	Analyses of ATP Analysis of APS (adenosine-5′ phosphosulphate reductase, the enzyme-reducing sulphate to	

Continued

Table 15.1 Continued

Analyses	Technique	Comment
	H_2S in SRB) Analyses of hydrogenase (the enzyme removing hydrogen in SRB).	
Staining	DAPI (4′,6-Diamidino-2-phenylindole) epifluorescent reactive to reveal DNA Acridine orange to reveal adenosine other stains to reveal specific cell components	For epifluorescent, optical microscopy observation
Amplification of genetic components	PCR (Polymerase Chain Reaction) FISH (Fluorescence In Situ Hybridization), laboratory equipment.	Amplification and detection of DNA or RNA genetic matter in microorganisms
Metagenomic analyses	Next-generation sequencing NGS (16S rRNA gene sequence analysis)	Individuation of all bacteria, included the uncultivables at the Genera, Order, Family, and Specie level.
	Transcriptome of specific genes	For the study of gene expression in bacteria

TRACIDE^{TR}, Improchem, South Africa). ATP is present in all metabolically active cells, including uncultivable bacteria, and it declines very rapidly when the bacteria die and the cells decay.

The simplest device for bacteria count contains dipsticks, glass beads, or coupons made of different materials that can be removed for testing. They can be sonicated and the bacteria of detached biofilm counted (plate count) or placed in a specific lysing solution for cellular ATP analysis. Otherwise, the beads may be treated with specific staining and directly observed under optical microscopy. Some commercial kits analyze specific enzymes (Enzyme-Linked ImmunoSorbent Assay, or ELISA test), for instance: the sulphate reductase, common to sulphate-reducing bacteria (SRB) (Rapidcheck^{TR} Strategic Diagnostic Inc., Newark, DE, USA) as well as the hydrogenase responsible for the acceleration of corrosion through the rapid removal of hydrogen formed on the metal surface [57] (Hydrogenase, Caproco International Inc. TX, USA).

Other kits specific to SRB analyze the development of hydrogen sulphide (Sanicheck^{TR}, Biosan Laboratories Inc., MI, USA; BTI-SRB^{TR}, Bioindustrial Technologies Inc., TX, USA).

In addition to the methods listed in Tables 15.1, other analyses are sometimes utilized, especially in laboratories for mechanistic studies [55]. These methods are usually time-consuming, expensive, and complex and require expert people. Due to the indirect correlation of the bacterial number and the corrosion phenomena, the commercial kits for bacteria count are usually used only in the cases where the number of cultivable bacteria has to be strictly controlled or the presence of some specific bacteria has to be mandatorily assessed from plant procedures. However, simple visual observations of coupons and their viscid tactile sensation can provide a 'rough' proof of biofilm presence.

15.4.1 Online biofilm and fouling monitoring

The first approach to the online evaluation of biofilm and fouling growth was the measurement of physical effects such as the decrease of the heat exchange efficiency and/or the increase of the friction factor induced by fouling [58].

Indeed, in a power plant, the measurement of the condensate back pressure, directly correlated with the fouling developed on the condenser tubes, is carefully monitored as it is an important parameter indicating the efficiency of the thermal cycle. The whole condenser can act as a biofilm sensor in this case, biofilm being one of the major concerns for heat exchange efficiency. Heat transfer and pressure drop of condenser tubes can be more precisely measured by suitable devices on a piece of tune mounted directly inside the condensers or in a bypass.

Such devices were put on the market [59] as fouling sensor for cooling circuits, but they were generally used in proprietary equipment for specific studies [60]. Main disadvantages of this kind of devices are:

1. They are not specific for biofilm. Indeed, they signal not only biofilm, but the growth on the condenser walls of the whole fouling, scale, and eventual corrosion products.
2. They are able to signal the presence of a biofilm only if its thickness is $\geq$30–40 μm [22, 61]. As a consequence, they do not provide information on the first phase of the biofilm growth, which is crucial for MIC. Furthermore, they do not reveal residual living microorganisms that have survived the antifouling procedures, leading to a fast regrowth after treatment [62].

Several online fouling sensors based on optical effects have also been developed. Optical fibers detect all kinds of deposit, living or dead biofilm, scale, etc. A simple fiber optic device is based on light reflectance measurement [63]. A differential turbidity measurement device is another optical method based on detection of the deposition of reflecting material.

The optical sensors could be employed particularly to detect fouling in high turbidity solutions (as in pulp and paper mills) and for nonconducting materials. Their sensitivity to the biofilm layer could be bigger than that of pressure drop sensors [64]. The more sophisticated Fourier Transform Infrared Spectrometry (FTIR) flow cell operating in the attenuated total reflectance mode allows the detection of bacterial biofilms as they form on a crystal of zinc selenide or germanium and also gives information about the chemical nature of the deposit. In this case, it could be possible to separate the signal coming from the different fouling components. However, this technique is currently still limited to laboratory studies.

Optical systems are often exclusively licensed by manufacturers to a single service company (as in the case of Nalco Optical Fouling Monitors OFM [65]). Another system proposed by Nalco Water (Ecolab company) for monitoring and online optimization of biocide treatments in cooling towers is a fluorometer (TRASAR™). It detects a fluorescent molecule (biosensor) that interacts with both planktonic and sessile microorganisms and changes its fluorescence spectrum when added to recirculating water [66, 67]. This system is proposed to control the treatment in cooling towers (https://it-it.ecolab.com/nalco-water/offerings/3d-trasar-technology-for-cooling-water) as well other different applications.

There are a number of other methods attempted for the measurement of biofouling. The Laser Light Scattering Device of Wyatt Industries is one of the first commercial systems for bulk phase monitoring, based on the detection of colloidal particles of bacteria by laser light scattering [68, 69]. In another electrochemical method based on a 'Mass Transport Device,' the estimation of the biofilm thickness is obtained by measuring the tracer transport rate through the fouling layer on a gold electrode. Other techniques require a quartz crystal microbalance or various sophisticated ultrasonic devices [70]. These methods have been developed only on a laboratory scale or they were used only in special applications, although they might have the potential for wider use.

15.4.2 Electrochemical biofilm sensors

Bacteria settled on a conductive material like a metal alloy tend to use it at convenience, as electron donor or final electron acceptor and with direct or indirect mechanisms, to sustain their energetic metabolisms [3, 71–74]. Based on the electrochemical phenomena induced by bacteria, specific electrochemical probes were developed in the last twenty years and are now in the industrial practice, extensively used in industrial cooling circuits.

These systems monitor the biofilm growth on a light polarized cathodic electrode of an electrochemical probe suitably inserted in the natural water to be monitored. The polarized electrodes are usually made of stainless steel or titanium.

On a passive alloys such as stainless steel and titanium immersed in aerated water, the microorganisms attempt an electron transfer from the metal to the oxygen (final acceptor), increasing the free corrosion potential of the un-corrodible metal. This phenomenon, often known as 'cathodic depolarization' [75–77], was repeatedly observed on stainless steel by several researchers [73, 78–80] and well-recognized for marine biofilm [81]. In time, the same effect induced by bacteria was confirmed also in other natural waters such as estuarine, river [82, 83], wastewaters [84], and in mineral waters [85]. It is widely accepted that the 'cathodic depolarization' is a common cause of MIC on stainless steel (and similar active-passive alloys) in seawater as a result of a gradually faster oxygen reduction induced by biofilm growth. Indeed, increasing the metal potential, the probability of localized corrosion onset increases, the propagation rate of localized corrosion is faster, and the galvanic current between passive alloys and less noble alloys present in the plant becomes higher [86, 87]. Therefore, a device able to signalize the effect of bacteria in growing biofilm, for the first stages, can be used to signalize the presence on the metal surface of MIC risk.

Although there are variations in the different probes, all the electrochemical devices to monitor biofilm currently on the market reveal biofilm activity as they influence cathodic depolarization. Despite the simplicity of the probes, these devices are extremely sensitive to the effects generated by biofilm during the first phases of growth.

It must be noted, in addition, that these electrochemical sensors are able to signalize the biofilm also in the presence of complex fouling on the metal surface (scale, deposits, mud, etc.)
The main limits of the electrochemical biofilm sensors are:

- they cannot be used if temperature is higher than about 40 °C (at higher temperatures, 'cathodic depolarization' induced by biofilm growth decreases)
- they signalize the biofilm growing on the surface of a stainless steel (or titanium), but the biofilm growing on the surface of different alloys exposed in the same environment can be different.

The first system commercialized was the BIoGEORGE^{TR} (Structural Integrity Associates, USA).

This system includes the probe, integrated electronics, interconnecting cable, display software, user's manual, and product support. The probe is made of an alternated series of stainless steel identic discs (or titanium discs for saline waters), which is polarized at a fixed cathodic potential. The probe can be installed into a piping system or side stream via a 1-in. or 2-in. threaded connection. Special probes can also be built for 'hot tap' type fittings or flow-through probes.

The sensor can work in different kinds of industrial water, including seawater as well as water with the lowest conductivity and in anaerobic conditions [88]. There are many industrial applications of BIoGEORGE^{TR}, for instance, in nuclear and thermal power plants, cooling towers, and in gas field-produced waters, since the 1990s [89–92].

Just after the BIoGEORGE^{TR}, a new and simplified probe BIOX was commercialized by the Italian company Stelar srl, and more recently, also by AMEL srl. The BIOX system was specifically developed to monitor the antifouling treatments of the industrial cooling systems in power plants (Fig. 15.2) [93, 94]. Indeed, it has been frequently used integrated with other corrosion monitoring techniques in order to optimize different kinds of industrial antifouling and anticorrosive and disinfecting water treatments. A different version of the BIOX system was also applied for monitoring biofilm in soil [34].

The last and more recently born from the same research activities was the ALVIM probe [95] that can be suitably adapted to fit different kinds of pipeline plug (Fig. 15.3). Application of ALVIM system recently included the monitoring of biofilm growth in nano-membrane filtration facility [96] and for paper mills [97].

The standard BIOX system generally includes a BIOX tubular electrochemical sensor of internal diameter of 16–20 mm, optionally with a titanium cathode,

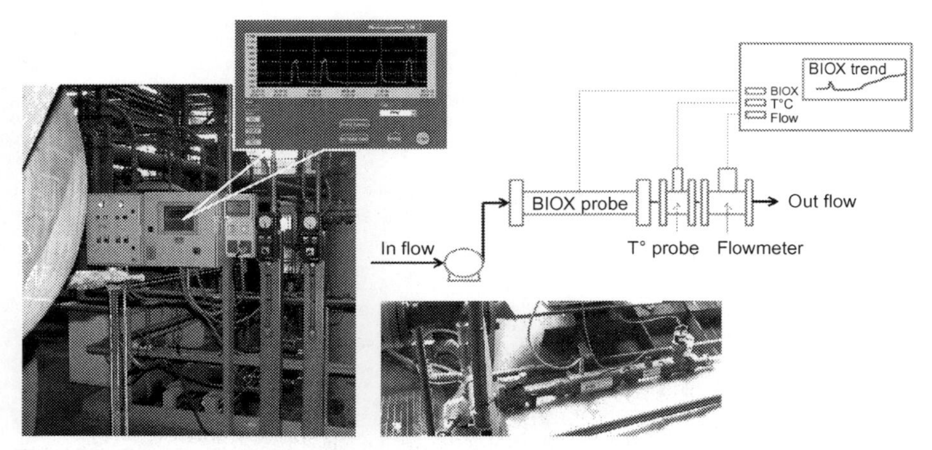

Fig. 15.2 Pictures of a BIOX system (computer and probes set) installed in a power plant. Schematics of the measurement line and of the computer display are also reported.

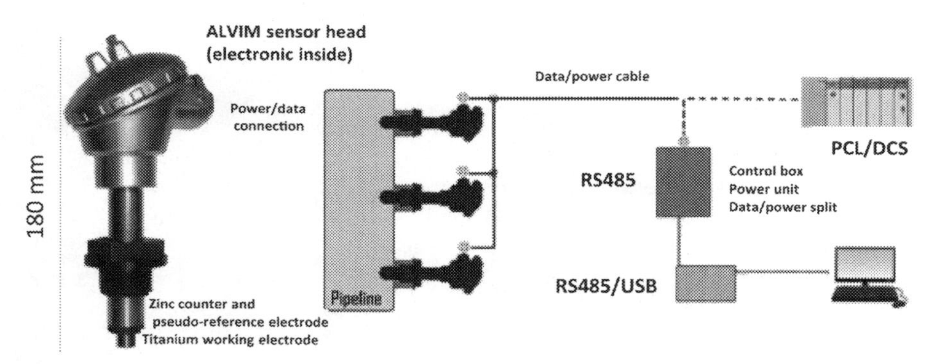

Fig. 15.3 ALVIM probe and schematic of an industrial sensor set.

a temperature meter, and a flowmeter. Both the commercial BIOX and ALVIM systems are usually equipped with dedicated industrial computer and software, with a monitor showing a signals trend in situ or specific electronics for the remote control of the signals coming from one or more probe set.

The BIOX probe consists of a bimetallic couple: a stainless steel sample coupled to a zinc electrode through a high resistor which strongly limits the galvanic current (Fig. 15.4).

This assembly simulates a quasi-intentiostatic technique and the evolution of 'cathodic depolarization' on stainless steel induced by biofilm development is signalized by an increase in the ohmic drop across the resistor. The electrical voltage across the resistor is the BIOX signal.

The ALVIM device also consists of an electrochemical probe whose electrodes are made of Zinc and stainless steel/titanium, but, differently from BIOX, it can switch both in potentiostatic and intensiostatic mode, although its preferred use is in intensiostatic mode, as it is considered the most suitably for industrial applications [95].

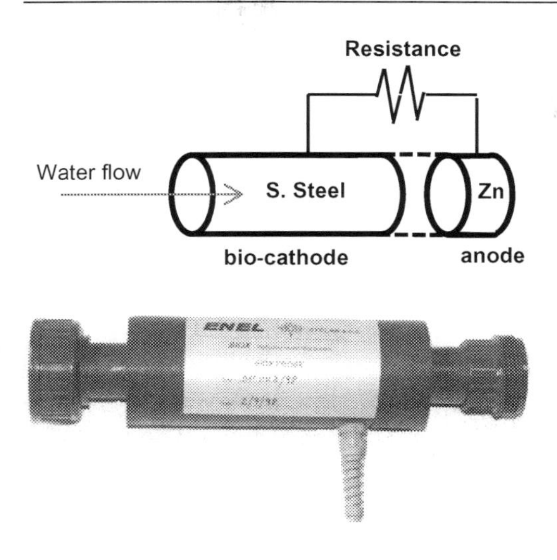

Fig. 15.4 Schematic and image of the BIOX probe.

A different configuration could be suggested for anaerobic water environments, where the BIOX probe configuration can be critical.

In aerated cooling circuit, the specific configuration of the BIOX and ALVIM probes allows the contemporaneous detection of two main phenomena: the aerobic biofilm growth on the working electrode and the cathodic effect of oxidant biocides added to the water (such as chlorine, bromine, chlorine dioxide, hydrogen peroxide, peracetic acid, ozone, and others) [98–100]. A typical trend of the BIOX signal in the presence of chlorine and during biofilm growth in sweater is reported in Fig. 15.5. Two images of the biofilm on the stainless steel cathode at different signal levels

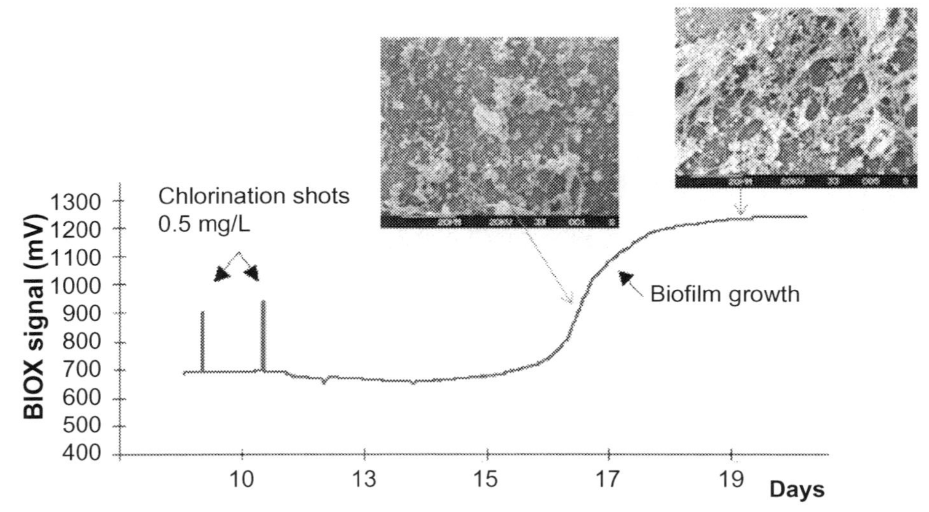

Fig. 15.5 Trend of the BIOX signal during a chlorination shot and when biofilm grows in seawater.

are also reported in the Fig. 15.5. When an oxidant (chlorine in this case) is added into the flowing water, the probe signal increases immediately due to the increase in redox potential of the stainless steel electrode. When biofilm grows on the stainless steel inside the probe, the time response is considerably longer, because several days may be required to completely cover a clean surface with biofilm. By analyzing the signal trend, it is clearly possible to distinguish the contributions of the two phenomena.

Several field tests confirmed that the electrochemical sensors persist to be sensitive in the same configuration in seawater as well as in river water and mineral water with low conductivity [85], even though a shift of both the signal baseline and the saturation level. The signal generally ranges of about 700 mV between 200 and 1400 mV, with a baseline depending on the water conductivity (Fig. 15.6, [85]). A trend of the BIOX signal during the biofilm growth and its destruction (by a shot of 15 mg/L Cl_2) in a test line of mineral water of 250 μS/cm is shown in Fig. 15.7. Early studies in seawater demonstrated that the sensors are able to clearly signalize the presence of a biofilm on the stainless steel cathode starting from a covering of just 1% of the exposed surface area, corresponding to a number of settled bacteria

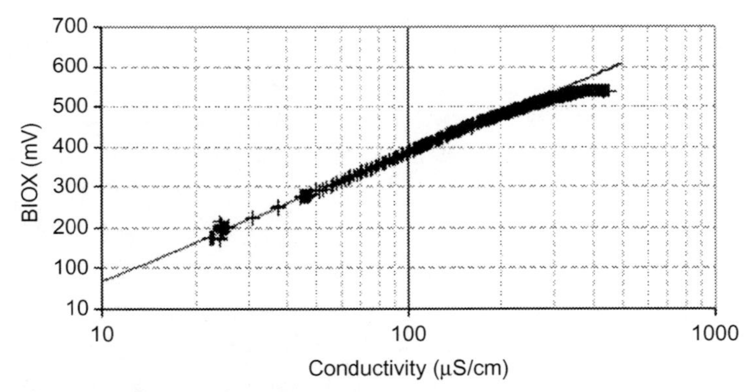

Fig. 15.6 Graphic of the BIOX signal baseline at different water conductivities [85].

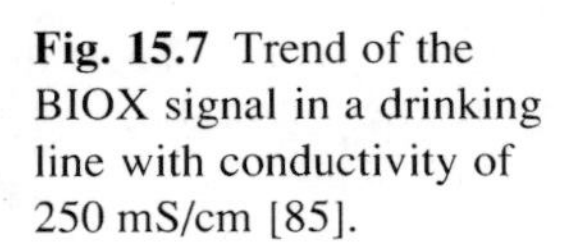
Fig. 15.7 Trend of the BIOX signal in a drinking line with conductivity of 250 mS/cm [85].

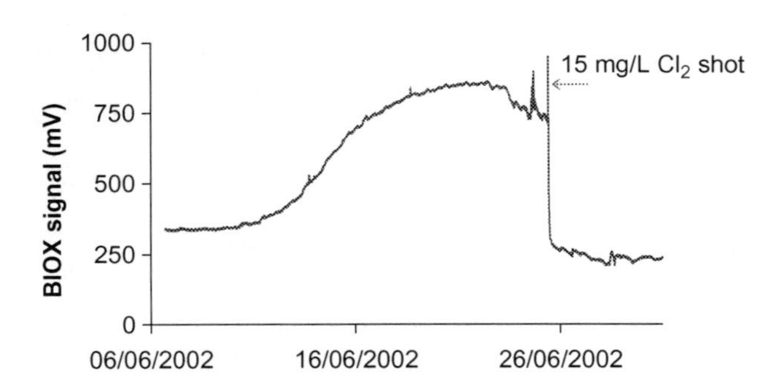

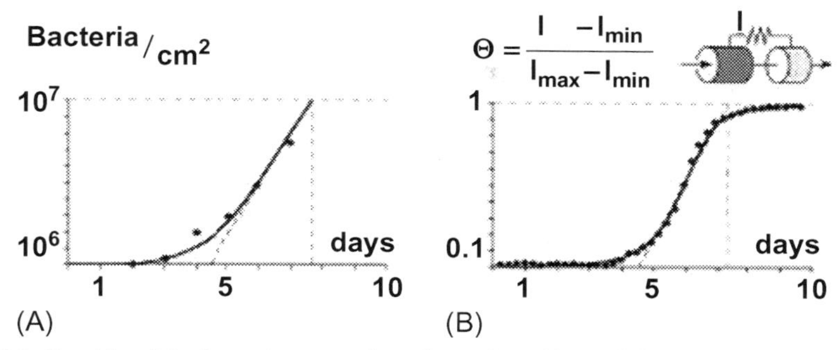

Fig. 15.8 Graphic of the bacteria counted on the surface (A) at different levels of electrochemical signal (B) [94].

in the order of 10^6 bacteria cm^{-2} (Fig. 15.8, [94]). Further biofilm development generates a signal almost proportional to the surface coverage, up to 100% of the surface, with an increase of settled bacteria of about two orders of magnitude. The signal reaches saturation when the number of settled bacteria is in the order of 10^8 bacteria/cm^2, i.e., an additional biofilm growth is no more signalized. The baseline is not significantly influenced by a temperature variation in a range of 5–40 °C or by sunlight. Fig. 15.9 shows the fast signal increase due to biofilm growth after a cleaning on a BIOX probe kept in the dark. Temperature and light variations during the day do not influence the baseline of the other probe exposed at sunlight.

The sensitivity of the probe to the oxidants depends on the nature of the oxidant, but generally it falls in the range of concentration normally useful for antifouling

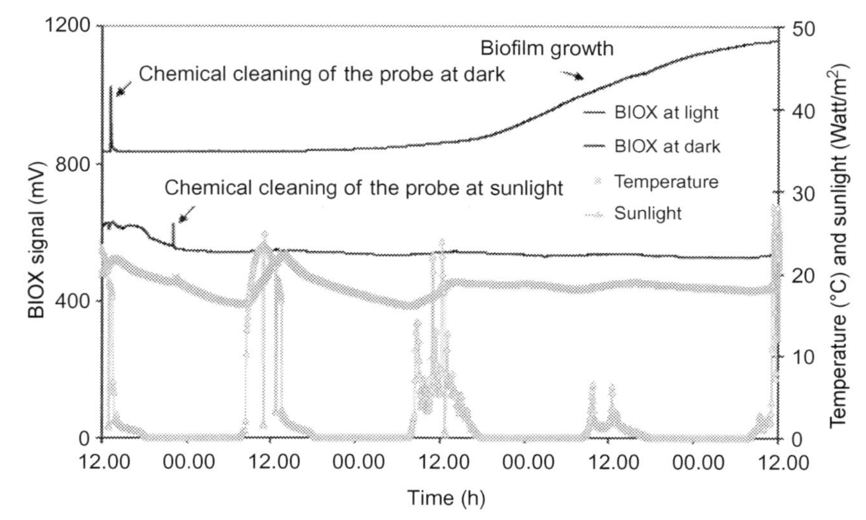

Fig. 15.9 Signal trends of two BIOX probes exposed to sunlight and in the dark in seawater after cleaning. The trends of sunlight and temperature are also reported.

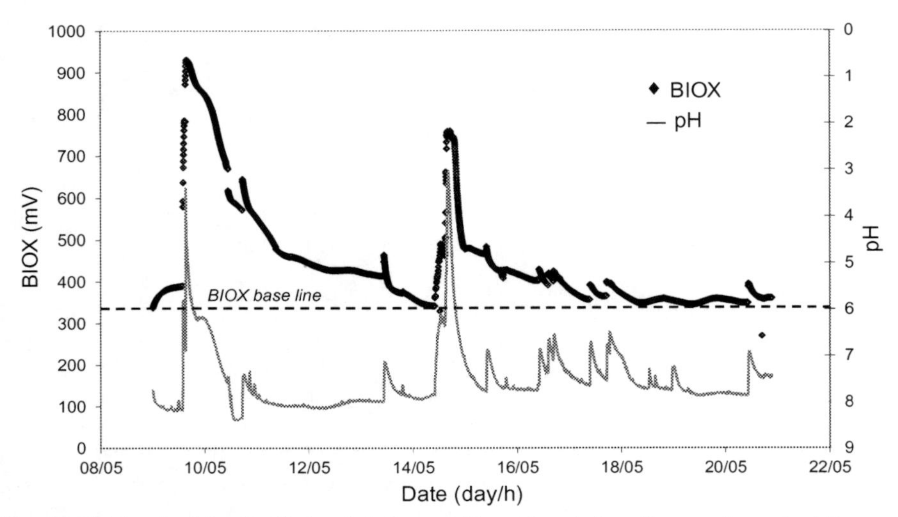

Fig. 15.10 Trend of the BOX signal, influenced by pH variation. Test performed adding hydrochloric acid to demineralized water.

treatments. The range of linearity of the peaks' height in respect to the concentration of biocide oxidants is smaller in fresh water than in seawater. Nevertheless, concentration in the range 0.2–1 mg/L of chlorine can be effectively detected in every kind of water. In seawater, concentrations of 0.02 mg/L of chlorine dioxide have been effectively detected as well as concentrations of 1–10 mg/L of peracetic acid.

Also, the decrease in the pH of the water is clearly detected with an increase of the signal, due to the increase of the redox potential of the stainless steel, as shown in Fig. 15.10. The pH effect is of particular relevance in the case of monitoring in cooling tower circuits.

BIOX and ALVIM probes, differently than the other similar biofilm electrochemical probes, are not affected by carbonate formation on the cathode, because of the very low circulating current. The presence of metal ions (ferrous ion for instance) or the high temperature of the water (more than 32 °C) can promote the growing of inorganic scale on all the surfaces exposed. When metal deposit or thick scale covers the electrodes, the signal baseline falls below its normal value, indicating the necessity to clean the probe. The cleaning operation of the probe can be done by mechanical scrubbing or chemical treatments taking a few minutes, filling the probe with (or circulating) hydrochloric acid 5% for few minutes.

15.5 Integrated online monitoring systems

In industrial facilities, chemical treatments performed against biofilm growth is usually targeted to prevent other correlated and deleterious effects than MIC, such as the reduction of heat exchange efficiency, increased friction factor, or the development of pathogen microorganisms in the water. Nevertheless, it must be taken into account that biofilm enhances bacterial resistance against several external factors, including biological competitors and the action of chemical biocides, which forces operators to use more intensive disinfection programs than those effective in breaking down the sole bioburden of bulk water. Furthermore, biocides could play an active role in corrosion development processes, in particular the oxidant agents which are utilized most. They could also be involved in environmental concerns, because of the discharge of residuals and by-products [101–103].

In all these cases, a higher level of attention is required for the treatment optimization. It means, in practice, that an effective monitoring system must be able to signalize online and in real-time if treatments and procedures applied are effective and, in case, it must provide also information to rationalize and minimize the application of these procedures.

Hence, the biocide dosage, in many cases, is mixed or alternated with other chemicals, likely in the case of industrial process/cooling waters which need chemical treatment against macrofouling growth, scale deposition, and chemical pollution (see Table 15.2). In the light of all these aspects, a reliable monitoring of the water treatment (anticorrosive, antifouling, or disinfecting) must indicate the best way to prevent MIC and the other negative, correlated risks.

Actually, a set of different sensors concurrently working in the plant is necessary to guarantee a reasonable amount of 'cross-validated' information strictly associated with the MIC risk. The number and type of these sensors depend on the specific environment, the metal(s) utilized in the plant, and the nature and sensitivity level of the requested information (see Tables 15.2). Nevertheless, complete or sufficient information is often achievable from few probes operating online and in real-time

Table 15.2 Relevant stages/kinds of fouling for the chemical treatments.

Stage of fouling	Typical application
First stage of living biofilm growth	Wide range of application for MIC risk assessment
Stage of biofilm influencing process parameters	Food industry, pharmaceutical industry, desalination plants, demineralization process (membranes), heat exchangers
Biofilm with microorganisms producing toxics, pollutants	
Clusters of microorganisms in the bulk waters (total viable bacteria, specific bacteria as pathogens, fungae, moulds, microalgae)	Process waters of food industry, pharmaceutical industry, wastewaters, drinking water, pulp and paper industry, cooling towers
Algae, macrofouling growth, larval settlement	Open cooling circuits, cooling towers
Solids, organic matters, oils	Wastewaters, drinking water plants
Scale deposition	Fresh waters, cooling towers, recirculating waters, hot waters

Table 15.3 Online Sensors used for the integrated monitoring of corrosion, biofilm, and industrial water treatments.

Parameter	Sensors	Application
Biofilm growth	BIOX probe, ALVIM probe	Almost always
Corrosion rate	Corrosion-meter LPR cells	When corrosion phenomena or the metal passivation processes have to be monitored
Oxidant biocide concentration (above 0.2 mg/L of chlorine)	BIOX probe Redox probe Amperometric systems Colorimetric systems SensIon probe	For oxidant biocide treatments (chlorination and the others based on bromine, chloramines, chlorine dioxide, peracetic acid, hydrogen peroxide, ozone, etc.)
Oxidant residual concentration (below 0.2 mg/L of Total Residual Oxidant)	SensIon probes Colorimetric systems Amperometric systems	To control very low residual concentration at the discharge. Amperometric systems work best in fresh, clean waters
Nonoxidant chemicals	Colorimetric systems Specific chemical systems	Water treatments alternative to chlorination, such as quaternary ammonium salts or filming amines
pH Redox potential	pH-meter Redox potential-meter	Fresh waters, cooling towers Wastewaters, when the aerobic condition has to be monitored
Temperature	Thermometer	For optimization of biocide dosage according to season; in case of macrofouling risk monitoring; for optimization of scale control; supporting the precision of other measurements
Water flow	Industrial flowmeter	Always in the cases of test line in bypass: to be sure to have the same flow in the measure line as in the plant facility

(see Table 15.3), integrated in a single electrochemical MIC risk monitoring device. This is an important conclusion coming from the experience collected by comparing the data from the different devices employed in practical applications.

The base-integrated system, from the author's experience, includes a single electrochemical probe for biofilm monitoring, a flow meter, and a temperature meter.

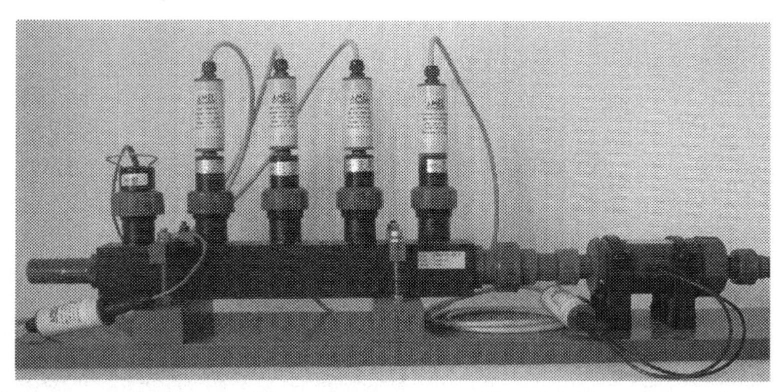

Fig. 15.11 Photos of a commercial integrated sensor set (AMEL). From the left: the rotameter, a LPR probe, a pH electrode, a Redox electrode, a turbidimeter (nephelometric probe), and a BIOX probe.

An additional electrochemical cell, made with material from monitoring equipment using the LPR technique, is necessary when the corrosion rate monitoring is requested. A data acquisition/transmission system (also in remote way) has to complete the monitoring system in any case.

Other online sensors such as a turbidimeter (in open circuits), a pH meter, or Redox measurement (in cooling towers) may be also useful in specific cases. In Fig. 15.11, an integrated electrochemical monitoring system by AMEL factory [12] applied in a power plant is shown. It includes BIOX, LPR pH, Redox, nephelometer, and rotameter (flowmeter) probes. A set of sample coupons for weight loss determination and visual observations of biofouling and corrosion (Fig. 15.1) may be inserted in the test line to collect additional confirmation of the phenomena monitored.

When scale deposition represents the main topic, a scale sensor could be equally useful, but reliable sensors for industrial application still need to be assessed. In the case of chlorination treatments, it must be taken into account that most of the commercially available systems to monitor the TRO (Total Residual Oxidant) online work quite well in fresh and drinking waters (the amperometric ones in particular, commonly used in swimming pools, drinking water plants, and so on), but very few of them were demonstrated to be sufficiently sensitive to give correct and effective measurements in seawater, least of all in cases of concentration less than 0.2 mg/l. Among them can be mentioned the colorimetric instrumentation Analyzer AMI Codes-II CC, base on the DPD-method, for measurement of disinfectants such as free chlorine, monochloramine, total residual chlorine, and combined chlorine, commercialized by SWAN Analytische Instrumente AG, (Switzerland) or the American potentiometric device for the measurement of the TRO (Mod. ORION 1770) by Thermo Scientific. An image of the online monitoring equipment operating in a power plant to monitor and control the chlorination treatment of the condenser cooling circuit, including the AMEL-integrated electrochemical equipment and a colorimetric process analyzer AMI Codes-II CC device, is shown in Fig. 15.12. The pictures show, from the left: the electronic devices for data acquisition, the three chemical reservoirs

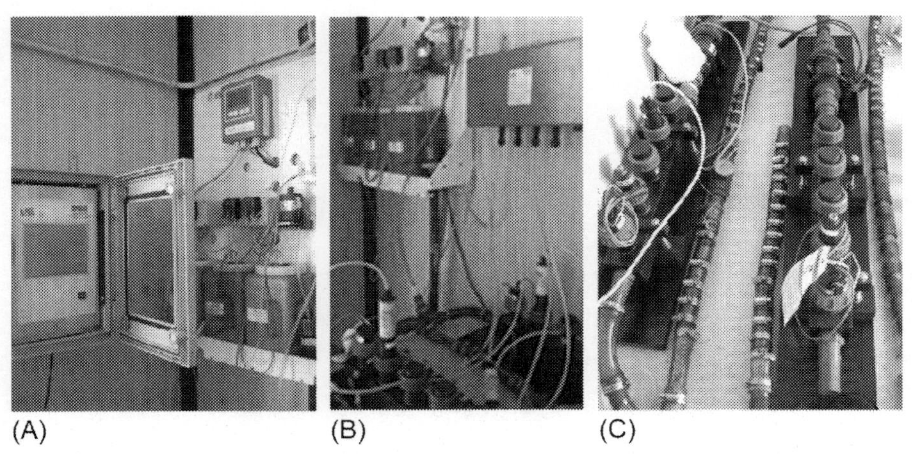

Fig. 15.12 Photos of an integrated electrochemical system installed in a hydraulic bypass of a cooling circuit in a power plant to monitor the antifouling treatment, the condenser corrosion, and residual chlorine: computer, electronics, and tanks of the colorimetric device (A, B), hydraulic lines (C) [53].

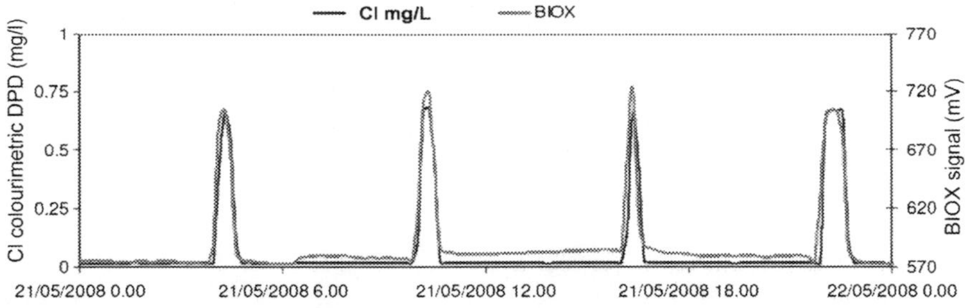

Fig. 15.13 Trends of BIOX signal and residual oxidant concentration measured online with an instrumentation based on DPD colorimetric method [104].

of reactants for the colorimetric device posed near the test line (photo in the middle), and detail of the hydraulic test line including electrochemical probes and specimens cut from real condenser tubes.

In Fig. 15.13 is shown the similar trends of BIOX signal and the concentration of the residual oxidant into the water measured with the online instrumentation AMI Codes-II CC during the chlorination shots [104].

In the case of macrofouling risk (algae, hydroids, mussels, and other sessile organisms), specific panels (concrete, ropes, and plastic supports) can also be immersed at a critical (but reachable) point of the plant. This is typically the case of open marine cooling circuits.

As previously underlined, microbiological kits for bacterial counts could be useful only in the specific cases where the bioburden of bulk water has to be necessarily controlled (as in wastewater or recirculating systems).

15.5.1 Costs of monitoring equipment

The cost of each sensor mentioned above is not very high. The most expensive are the online Total Residual Oxidant Meter for seawater applications, whose cost can be more than 15,000 US$. Those sensors need analytical chemical supply and frequent maintenance (at least weekly). Consequently, the use of those devices is suggested only when a very low concentration of oxidant (chlorine less than 0.2 mg/L for instance) needs to be rigorously controlled, typically at the discharge of open cooling circuits. In these cases, the device must be installed for the other monitoring probes requiring an extra cost to collect the output signal. The computer and the data acquisition system is the most critical and expensive component of the integrated monitoring system, especially when an industrial computer is preferred to solutions with remote control, because the industrial environment (coal dust, atmospheric agents, humidity, high temperature, etc.) is deleterious for the standard electronic components. Commercial software developed for specific applications is already available (e.g., in BIOX, ALVIM, BIoGEORGE, and LPR commercial systems). The most complex ones are able to drive both the LPR electrochemical measures and the acquisitions of all the other probe signals (e.g., the software for the Integrated BIOX system by AMEL). The cost depends mainly on the number of probes and on the application.

The basic electrochemical probes of BIOX, ALVIM, and BIoGEORGE systems cost about the same and they are a bit more expensive than the LPR commercial systems. The electrodes for LPR probes, the coupons of metal samples, and the supports for macrofouling may be prepared in situ using the material coming from the equipment/circuit of the plant. Online sensors and the cases in which they had been frequently integrated into the loop are summarized in Table 15.3 and in Section 15.5.

15.6 Case histories

Some examples of integrated monitoring systems for MIC risk prevention and to optimize water treatments applied in industrial plants are described in this session. Several field tests were conducted to determine whether electrochemical sensors can really be utilized to minimize corrosion problems in plants and if these sensors are reliable for long-term application without maintenance.

15.6.1 Inhibition of crevice corrosion on stainless steel

The results of a test for checking the inhibition of the crevice corrosion on stainless steel [105], lasting one year in 1996, are briefly described.

In the tests, seawater was pumped, once, through a series of electrically insulated tubes of external diameter 25 mm, wall thickness 1.5 mm, made of stainless steel (UNS31600). In order to promote the onset of localized corrosion, the stainless steel tubes were connected using rubber hose which was pressed on the external surface of the stainless steel tubes: in this way, artificial crevices are preformed on the metal under the rubber. An electrochemical probe, in the configuration of a BIOX sensor, was inserted at the end of the test line. It was utilized to evaluate an antifouling procedure to apply to the stainless steel tubes and to the sensor itself.

When the BIOX signal was increased above a prefixed value (as a result of the biofilm growth), a single standard chlorination (at 0.5 mg/L for 30 min) was started automatically and chlorine was added in the middle of the test line, leaving half of the samples upstream unchlorinated. At the end of chlorination, the biofilm on the probe was partially destroyed and, as a consequence, the BIOX signal decreased. When new biofilm was growing, the above-described procedure was automatically repeated. The test started in August 1994 with the seawater temperature close to 28 °C. During the year, the temperature first decreased to about 14 °C in winter and then rose again to 28 °C in the following summer. No maintenance was applied to the probe during the test. The trends of the signal provided by the sensor indicate that, during the two summers, it was judged sufficient to apply a single chlorination approximately every 1.5 days. In winter, the sensor indicated that only one chlorination was sufficient about every six days as a consequence of decreased biological activity.

The tubes exposed to untreated seawater were heavily fouled, whereas the tubes exposed to the automatically treated seawater looked totally clean. In the latter case, observation of the surfaces by the Scanning Electron Microscopy (SEM) revealed only the presence of a few settled bacteria, randomly distributed. Heavy crevice corrosion, leading to perforation of a 1.5 mm thick pipe wall, was observed on the stainless steel tubes exposed for three months to untreated seawater. Less severe and only shallow attack was observed on the stainless steel pipes exposed, for the same time length, to automatically chlorinated seawater (Fig. 15.14). Additional tests showed

Fig. 15.14 Crevice corrosion propagation on AISI 316 pipes after 3 months of exposure to untreated (left side) and automatically chlorinated seawater (right side) [105].

that the application of a weak and inexpensive cathodic protection is also sufficient to avoid the onset of crevice corrosion on stainless steel exposed to automatically chlorinated seawater.

In conclusion, electrochemical sensors were found reliable also for long time application without maintenance and to be able to provide a set of information which can be used to minimize the application of antifouling procedures in seawater, increasing plant efficiency and, concurrently, decreasing corrosion problems on the active-passive alloys.

15.6.2 Optimization of cooling water treatment in power plants

The industrial version of the BIOX system was successfully applied in seawater cooling systems of Italian power plants from 1996 to optimize chlorination treatments [11]. Fig. 15.15 shows typical examples of signal trends provided by the sensor during approximately one month of exposure. The trend of the BIOX signal can be described as composed of several sharp peaks superimposed over a stable baseline at about 500 mV. Peaks were observed when intermittent chlorination (20 min at 0.5 mg/L of residual oxidant) was used in the plant following the program routinely applied during summer (four chlorinations per day); the height of the peaks was found proportional to the residual chlorine concentration in the range 0.2 to 1 mg/L. The sensor suggests that this chlorination program was sufficient to avoid biofilm growth. Indeed,

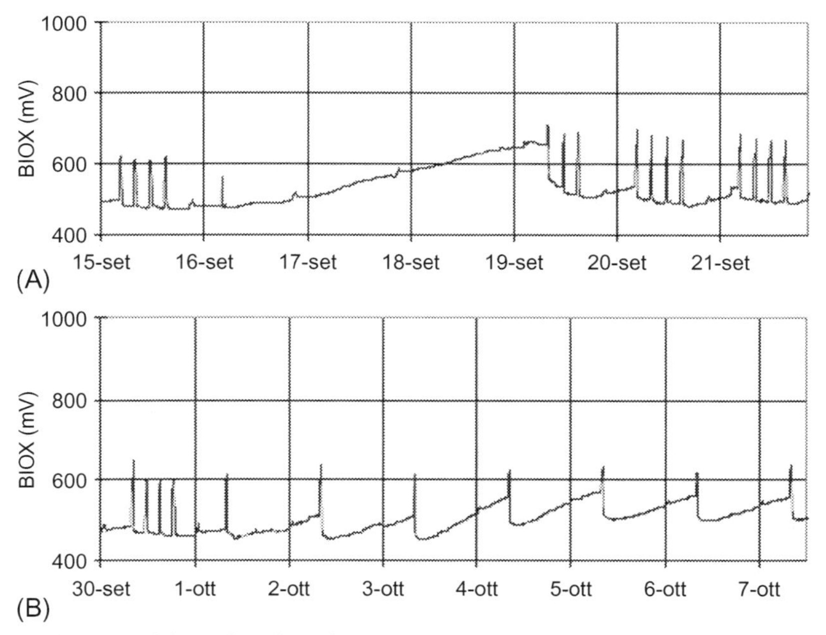

Fig. 15.15 Trend of the BIOX signal in a power plant which treatment varied during the seasons. Treatment is based on 0.5 mg/L dosage of sodium hypochlorite for 20 min (peaks of the BIOX signal), 4 shots/day in September (A), and 1 shot/day in October (B) [93].

the baseline continuously has the same value that was measured at the beginning of the test on the clean probe.

When the chlorination program was not run for three days (from 17 September 1996 to 20 September 1996), the slope of the sensor base signal was positive. The increase of the base signal towards 700 mV means that biofilm was growing, but it was still in the first phase of its development (in this period, no valuable decrease of the condenser efficiency was observed). Once the normal chlorination program was restored, the base value of the signal returned rapidly to the value observed on the clean probe. This suggests that the chlorination program usually applied in the plant was not only able to prevent biofilm formation, but also to rapidly destroy the biofilm once formed.

The antifouling program was changed after ten days and the chlorination frequency was strongly decreased. The graph in Fig. 15.15 indicates that a reasonable, although not complete, control of biofilm growth was obtained with only one chlorination per day, i.e., with 25% of the chlorine previously utilized.

Following this first study, many antifouling treatments for condensers of power plant based on chlorination and other chemicals (chlorine dioxide and peracetic acid included) have been monitored by BIOX system [100]. The results obtained underlined that:

- at the beginning of the BIOX signal increase, the sterilization does not require a huge quantity of biocide, and a continuous, low level of chlorine dioxide results are very effective against biofilm at this stage
- the exponential signal increase, coinciding with the exponential growth of the bacteria community, has very short duration, but it is still easy to chemically control the biofilm at this stage
- after reaching the platform, the steady state could require a stronger dosage of chemicals; the complete disinfection could not be guaranteed

Very low levels of chlorine and peracetic acid were not so effective in controlling biofilm growth as a lower level of chlorine dioxide. However, shot dosage of few mg/L of each chemical was effective enough to destroy mature, untreated biofilms.

Another positive performance of the BIOX system concerned the monitoring of in-service mechanical cleaning by sponge balls. In particular, the monitoring permitted the documentation of different chemical and mechanical force against the biofilm [100]. With a water flow of about 1 m/s in the 20 mm diameter tubular probe, the mechanical treatment was found very effective, as shown in Fig. 15.16: a BIOX signal of about 900 mV, correspondingly fell down to a 'moderate' biofilm growth, after the passage of a single ball. However, with a flow above 2 m/s, a chemical treatment (0.1 mg/L of residual chlorine dioxide), for 1 h, was necessary to destroy the microbiological activity in the thin biofilm.

15.6.3 Detection of biofilm in mineral water plant

During a two-month test conducted with a BIOX system inserted in a bypass in-line with an industrial tank of mineral water (Fig. 15.17) [85], under a flow of 1100 L/h, corresponding to 1.5 m/s in the tubular probe of 16 mm diameter, the probe signal held

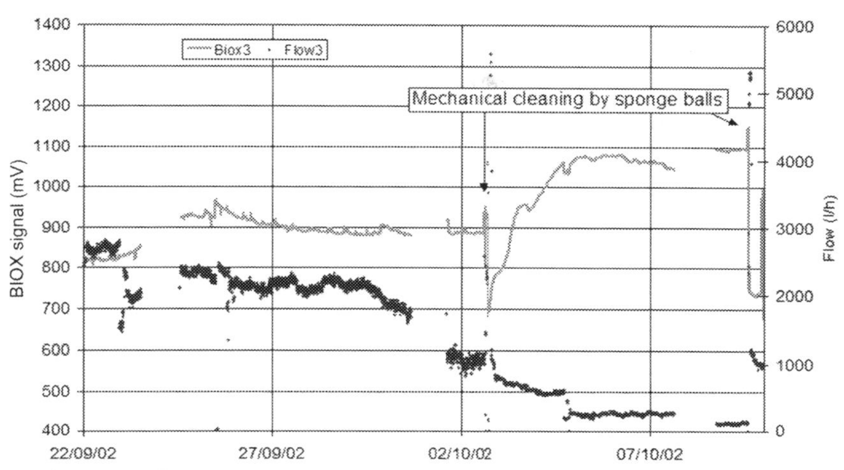

Fig. 15.16 Trend of a BIOX signal and water flow during the passage of a sponge ball from the in-service mechanical cleaning [100].

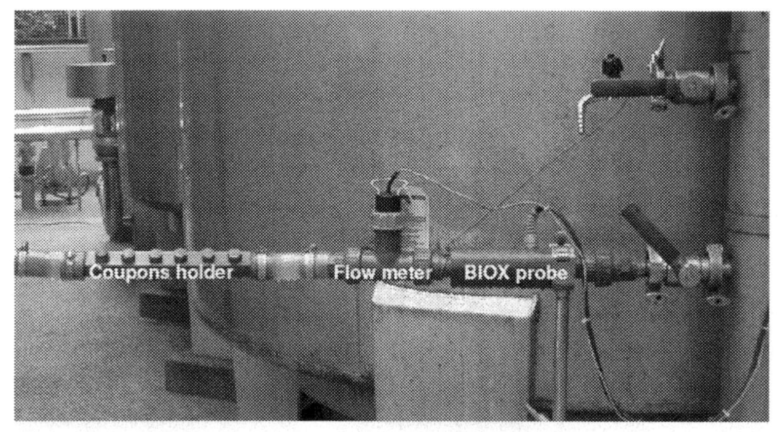

Fig. 15.17 Hydraulic line to monitor online by BIOX system the biofilm absence in mineral water plant [85].

down around 400 mV during all the period of the test, indicating the complete absence of biofilm. The tests confirmed the laboratory results showing that the BIOX system works correctly with water having a low conductivity of 350 μS/cm. In a subsequent test performed with the same water at low flow of about 0.4 m/s, but outlet to a bacteria-contaminated tank, the BIOX probe was able to detect the first stages of biofilm growth reaching the value of 800 mV. The test line was then cleaned and disinfected with a dosage of about 15 mg/L of chlorine for 15 min (Fig. 15.7).

The BIOX signal reached values not more than 900 mV during chlorination (in spite of the high concentration used) and it came back to baseline values after the dosage was complete, keeping the baseline value for a long duration. SEM micrographs of stainless steel coupons sampled from the test line before and after chlorination documented the presence of a thin uniform biofilm on all the surfaces detaching in consequence of chlorination.

15.6.4 Testing of a disinfecting treatment based on peracetic acid in wastewater and seawater

A treatment based on peracetic acid (PAA) to control microbiological growth in a pilot serpentine reactor (10 m^3/h) fed by secondary settled effluent coming from the wastewater treatment plant in Southern Italy was investigated using a BIOX probe with external electrodes submerged into the water flow exposed at sunlight. A number of treatments were carried out during a six-month period at different PAA continuing dosages ranging from 2 to 13 mg/L. PAA decay was determined by monitoring the residual concentration during each treatment using both electrochemical online BIOX signals and standard chemical off-line methods. Data coming from biofilm electrochemical monitoring were integrated and compared with microbiological analytical results. A residual of 3–4 mg/L of PAA was required in order to destroy the biofilm already grown on the electrochemical probes submerged in the final part of the serpentine pilot reactor. PAA dosages lower than 2 mg/L were found sufficient to prevent the early stages of biofilm development on the clean probes. The total microbial charge and amount of *Escherichia coli* in the bulk water of the pilot plant remained under control even at the same lowest tested PAA dosages [99, 106].

Higher dosage of 9–12 mg/L PAA, but intermittent, was necessary to prevent the biofilm and biofouling in a cooling circuit of a power plant on an Italian marine Lagunas. The graphic in the Fig. 15.18 shows the effect of daily dosage of about 10 mg/L of percetic acid monitored by a BIOX probe. The photos in the Fig. 15.19 illustrate supports inserted into the cooling canal for the monitoring of macrofouling settlement, showing that this dosage was effective, although not completely, in preventing the settlement of macrofouling also.

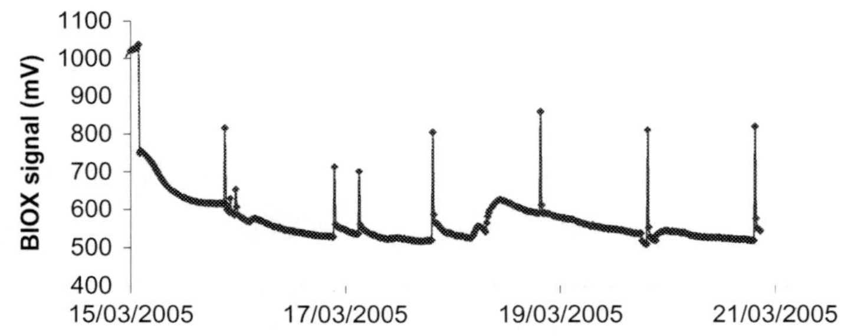

Fig. 15.18 Signal trend of a BIOX probe in seawater cooling canal subjected to a daily dosage of 10 mg/L of PAA.

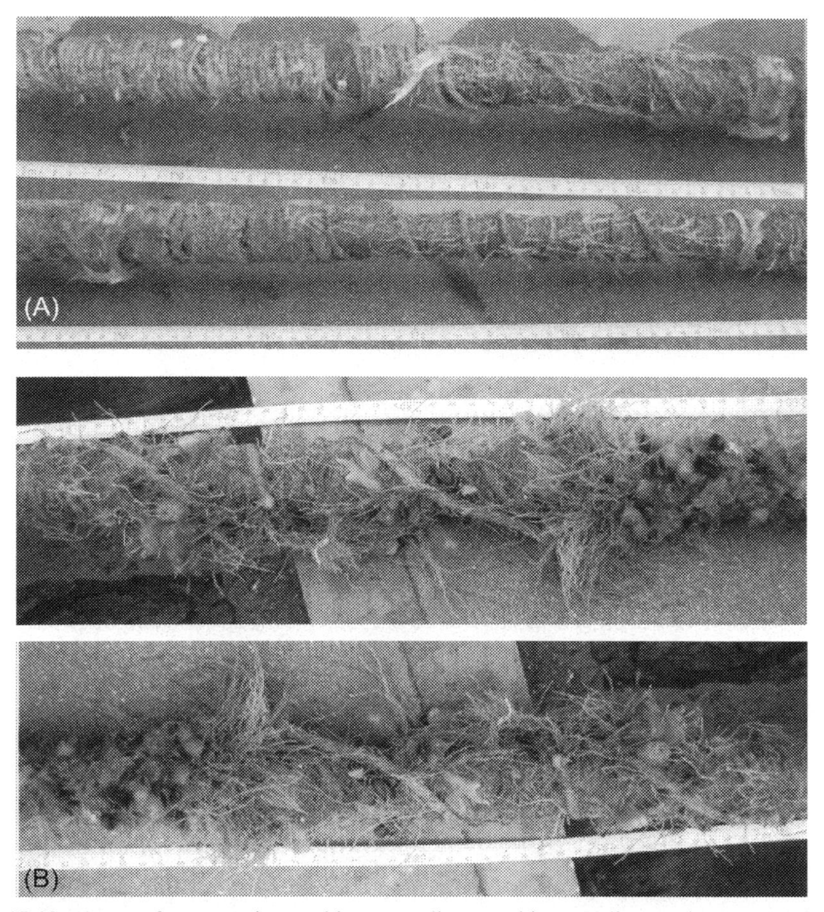

Fig. 15.19 Photos of supports inserted into a cooling canal in an Italian marine Laguna for the monitoring of macrofouling settlement. Cooling water treated with about 10 mg/L PAA (A) and control without any biocide treatment (B).

15.6.5 Monitoring of copper alloy condenser tube passivation in power plants

Microbial activity negatively influences the passivation process of copper alloys, in seawater in particular. In the light of some recorded case histories of early failures in condenser tubes, an integrated monitoring of corrosion, biofilm, and chlorine dosages has been performed on Copper-Nickel 70/30 coupons (UNS C71500, external diameter of 22 mm, wall thickness 1.6 mm) made of the same condenser tubes used in the marine cooling circuit of an Italian power plant.

The following sensors have been chosen for monitoring, inserted in a bypass close to the condenser:

- one BIOX tubular sensor to monitor the biofilm growth and the chlorination treatment;
- one tubular electrochemical cell made of coupons from the condenser tubes to monitor the corrosion rate of CuNi 70/30 by LPR technique;
- one flowmeter, one temperature sensor, and one turbidity meter to monitor the physicalchemical parameters of the water flow.

Weight loss determination and chemical and metallographic analyses have also been done off-line during the monitor program on tubular coupons of the condenser tubes. Gravimetric determinations were used to calculate the conversion coefficient to more precisely estimate the corrosion rate from LPR.

The data collected documented the occurrence of localized corrosion on the surface of the CuNi 70/30 specimens in the case of events causing the enrichment of biologically active mud in sea water (big ships moving or strong rain close to the inlet of cooling circuit) or after the stagnant condition occurring in the tubes. The integrated monitoring permitted an effective chlorination treatment of the condenser cooling water, able to promote the development of a firm 'passive layer' on the CuNi 70/30 tubes.

Fig. 15.20 shows two different conditions for CuNi 70/30 coupons during the integrated monitoring (corrosion rate by LPR technique, biofilm, chlorination, temperature, and water turbidity) in a marine cooling circuit affected by mud incoming (corresponding to the turbidity signal increase). Water was treated with 0.5 mg/L TRO (chlorine-bromine) one hour, 4 times a day, corresponding to steps of BIOX signal. In the graphic above, CuNi 70/30 passivation layer is well established and corrosion rate (CR) is very low. On 5th December, the test line has been moved in the circuit before the chlorination point and samples of the LPR cell have been substituted with new ones; the monitoring is shown in the graphic below. The two graphics show that a significant reduction of the corrosion rate is correlated to an effective chlorination treatment. In fact, corrosion reduction did not occur on new samples (graphic below in Fig. 15.20), untreated with chlorine. This study documented, in conclusion, how a 'little economy' in biocide treatments during the start-up period of exchanger equipment can return severe additional costs because of early failures of the bundle [107].

15.6.6 Monitoring of corrosion-erosion in aluminum brass condenser tubes

Aluminum brass (Al Brass) is another copper alloy (Cu76%–Zn22%–Al2%) widely used for heat exchange pipes of condensers in thermoelectric plants cooled with seawater [11].

In sea water, this alloy shows excellent corrosion behavior after the formation of layers of protective oxides, with corrosion rates at steady state lower than 20 μm/y. However, this alloy may suffer from Flow Accelerated Corrosion (FAC) problems

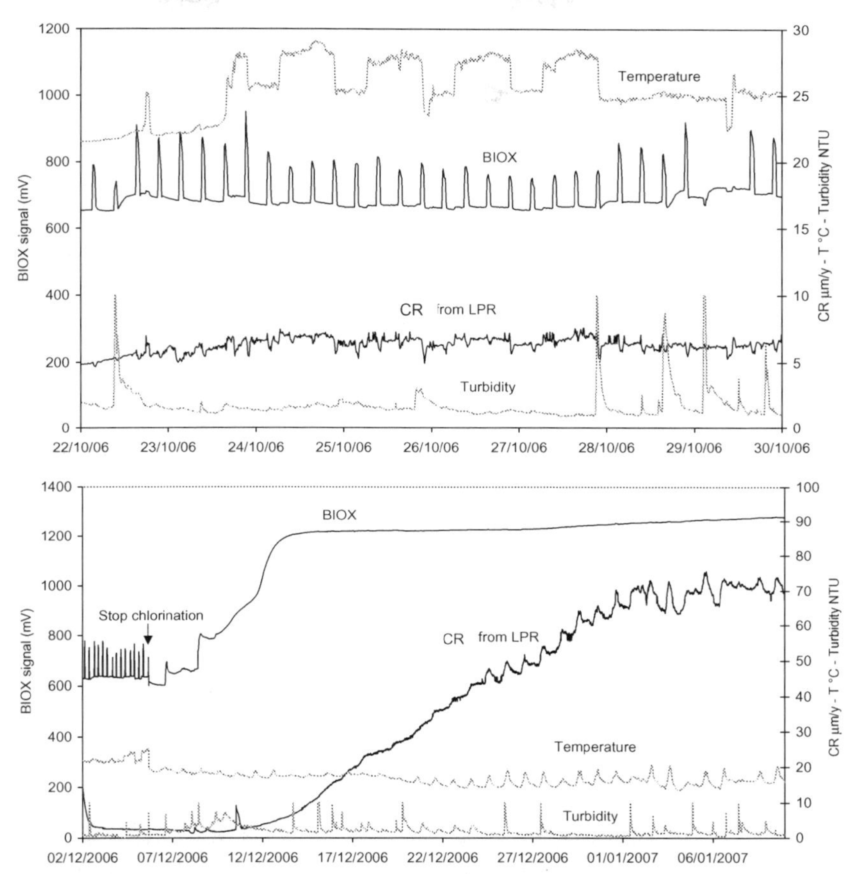

Fig. 15.20 Trends of signals from the Integrated monitoring of corrosion, biofilm, temperature, chlorination, and water turbidity in a seawater circuit rich of mud, treated with 0.5 mg/L TRO (chlorine-bromine) one hour, four time a day (steps of BIOX signal). Vcorr from LPR is applied on electrodes in CuNi 70/30. On 5th December, the test line has been moved before the chlorination point and samples in the LPR cell have been substituted with new ones; the monitoring is shown in the graphic below [107].

if in service with seawater flow velocity > 2.2 m/s, though so high flow could guarantee biofilm control.

An example of corrosion increase due to flow rate is shown in Fig. 15.21.

An integrated monitoring system applied on specimens of this alloy with the online and real-time control of both the metal/cooling water interactions and the antifouling and anticorrosive treatments implemented in a power plant cooled with seawater was conducted for months, using well-passivated condenser tubes specimens and specimens affected by FAC [109]. The monitoring allowed to demonstrate that a treatment with chlorine dioxide (ClO_2) at levels of 0.05–0.1 ppm as free chlorine is able to prevent the formation of biofouling. Furthermore, the dosages on alternate days of ferrous

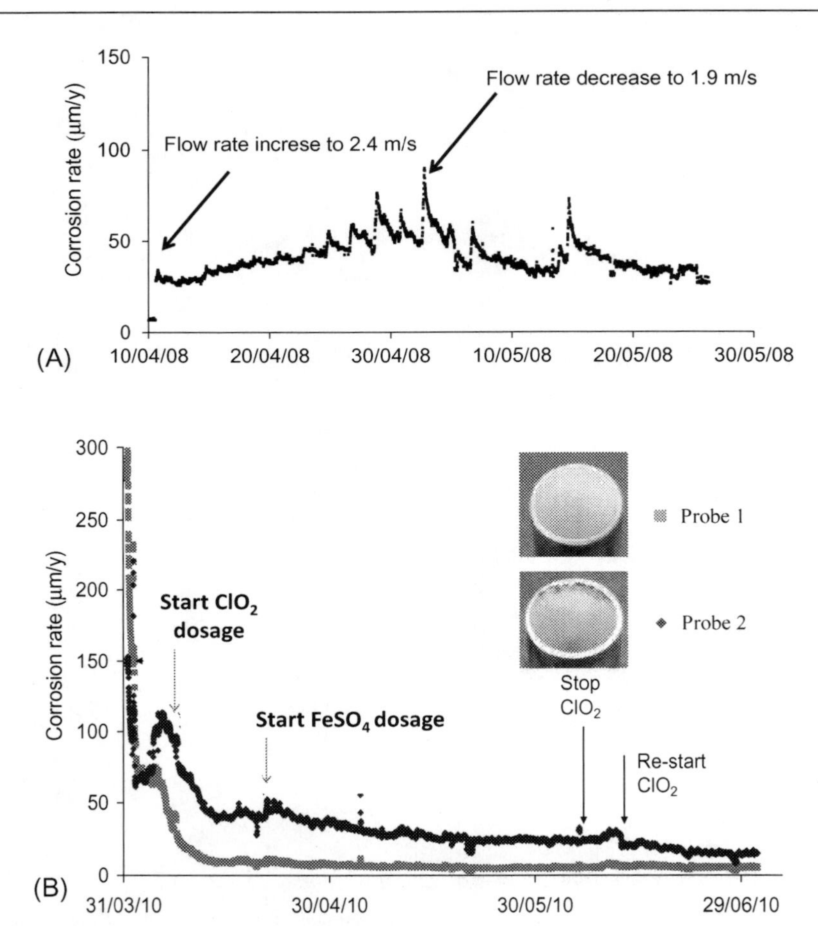

Fig. 15.21 Trends of corrosion rate determined by LPR of Aluminum brass coupons under different flow rate (A) and different passivation condition (B): Probe 1 is the well-passivated coupon treated with ClO_2 and $FeSO_4$; Probe 2 is the coupon subjected to flow accelerated corrosion [108].

sulfate ($FeSO_4$) at concentrations of about 0, 5–1.0 ppm of iron, just at the condenser inlet, were useful because they promote the formation of a protective and almost sterile lepidocrocyte layer on the surface, reducing in this way also the magnitude of the FAC problem (Fig. 15.21).

15.6.7 Monitoring of cooling water treatment in a steel factory

Another integrated monitoring of corrosion, biofilm and water treatment was performed with the metal alloys used (principally carbon steel) in a very large seawater cooling circuit of a steel factory treated with a daily intermittent dosage of chlorine dioxide. The following sensors driven by a computer-data acquisition system were inserted in a bypass close to the condenser:

- one BIOX tubular sensor to monitor the biofilm growth and the chlorine dioxide treatment;
- three tubular electrochemical cells to monitor the corrosion rate by LPR technique; coupons of carbon steel, stainless steel (UNS 31600), and copper-nickel 70/30 UNS C71500, respectively, with diameter of about 20 mm in each case;
- zinc anodes to monitor the free corrosion potential of the three alloys;
- one flowmeter and one temperature sensor to monitor the physicalchemical parameters of the water flow.

Several tubular coupons made of the same three materials were inserted for the off-line visual observations and weight loss determinations.

The treatment in the first period was not effective and a thick layer of iron corrosion products covered the whole surface in the test line. Biofilm growth on corrosion products increased the BIOX baseline. Crevice corrosion was found on stainless steel; the general corrosion rate of the carbon steel was very high; the copper alloy showed localized corrosion. On the basis of the monitoring results, the surfaces were brushed mechanically, and following that, the treatment became more effective to prevent a new corrosion product and biofilm deposition (see Fig. 15.22). In fact, the chlorine dioxide concentration reaching the surface increased a little and the treatment was more effective than before. The responses coming from the electrochemical instrumentation were found to be in a good agreement with the data from the analyses performed on the coupons (visual observations and weight loss) off-line. Subsequently,

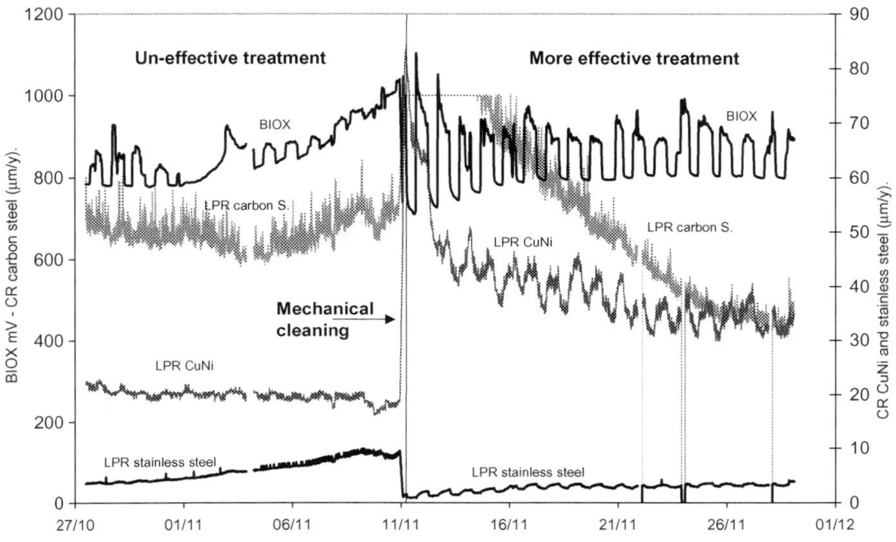

Fig. 15.22 Trends of signals from the integrated monitoring of corrosion, biofilm, and treatment in cooling circuit of a steel factory. Seawater treated with a very low and intermittent residual (<0.1 mg/L) of chlorine dioxide (steps of BIOX signal). LPR techniques have been applied on electrodes made of CuNi 70/30, AISI 316 (Stainless steel), and Carbon steel, before and after the mechanical cleaning of the test line.

good control of crevice phenomena on stainless steel as well as the corrosion of carbon steel and copper alloy was kept with a few small adjustments of the treatment.

15.6.8 Evaluation of a water treatment in a cooling tower

A final example of integrated monitoring of corrosion, biofilm and water treatment refers to a cooling tower fed by river water and treated with short, daily dosages of peracetic acid at different concentrations (Fig. 15.23). The following sensors driven by a computer-data acquisition system were used in the test line:

- one BIOX tubular sensor to monitor the biofilm growth and the peracetic acid treatment
- one electrochemical cell to monitor the corrosion rate by the LPR technique with coupons of carbon steel with diameter of about 20 mm
- one pH meter, one flowmeter, and one temperature sensor to monitor the physicalchemical parameters of the water.

Several coupons made of carbon steel for visual observation and weight loss determination completed the test line.

The increase of the BIOX baseline detected during the first period indicated too light treatment. Increasing the PAA dosages in the range of 8–10 mg/L, the treatment became effective to control biofilm. Furthermore, the measures of corrosion rate remained at the lowest level. Too strong dosages, with concentration of PAA more than 100 mg/L, tested during the final period, associated with a significant decrease of pH, caused an increase in the rate of corrosion.

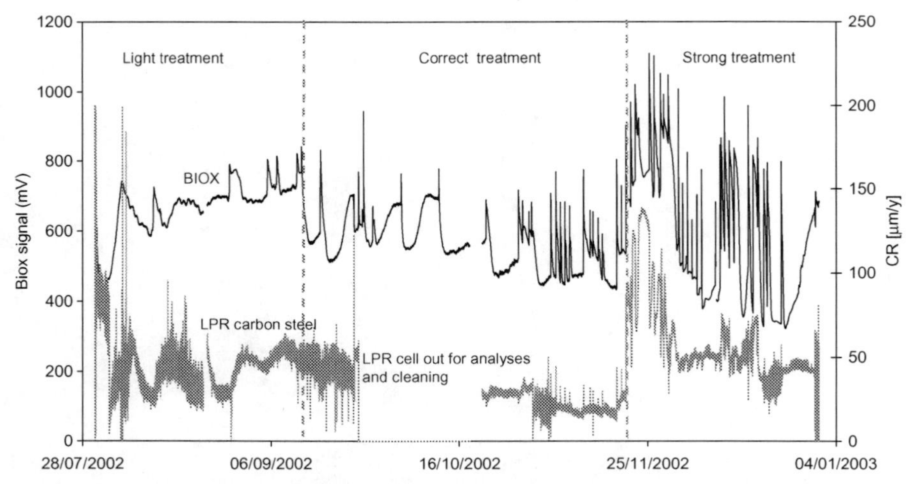

Fig. 15.23 Trends of signals from an integrated monitoring system of biofilm and corrosion in a cooling tower applied to a different treatment based on periodical dosages of PAA: less than 10 mg/l for the light treatment; 10–20 mg/l for the corrected treatment, and 330–100 during the strong treatment. LPR technique has been applied to monitor corrosion of carbon steel (CR). A significant decrease of pH around 25/11/2002 caused the corrosion rate increase. The peaks of BIOX signal correspond to PAA dosages.

15.7 Summary

The biological film growing on wet surfaces is the cause of microbial-induced corrosion (MIC), which strongly can affect the performance of alloys, equipment, and process waters. MIC risk mitigation usually coincides with biofilm risk mitigation.

Biofilm also causes other deleterious effects, reducing heat transfer and promoting the biological risk in industrial facilities; consequently, the biofilm control can prevent several issues, not only MIC. The existing procedures of MIC risk prevention often require the use of biocide treatments, such as chlorination, but standard treatment procedures, field experiences, and macroanalysis of the phenomena are insufficient to build an effective and exhaustive risk assessment program. Valid help, often decisive, was confirmed to come from simple, rugged electrochemical sensors, able to monitor online and in real-time the first stage of biofilm growth. An electrochemical biofilm sensor, applied alone or integrated with a few other traditional sensors and devices, could allow the direct evaluation of MIC risk in several field applications. The monitoring experiences collected with an electrochemical MIC risk monitoring, from several plant and for several years, have demonstrated the possibility to optimize the chemical treatments of cooling water circuits (in particular), guaranteeing a safe condition for the operating materials.

References

[1] J. Wingender, T.R. Neu, H.C. Flemming, What are bacterial extracellular polymeric substances? in: J. Wingender, T.R. Neu, H.C. Flemming (Eds.), Microbial Extracellular Polymeric Substances: Characterization, Structure and Function, Springer-Verlag, New York, 1999, pp. 1–15.

[2] I.W. Sutherland, The biofilm matrix—an immobilized but dynamic microbial environment, Trends Microbiol. 9 (2001) 222–227.

[3] I.B. Beech, J.A. Sunner, Microbially-influenced corrosion: towards understanding of interactions between bacterial biofilms and metallic materials, Curr. Opin. Biotechnol. 15 (2004) 181–186.

[4] P. Chandrasekaram, S.C. Dexter, Factor contributing to ennoblement of passive metals due to biofilm in seawater, in: Proc. of 12th Int. Corr. Cong., NACE, article n. 493, 3696–4707, 1993.

[5] J.N. Wardell, A.H.L. Chamberlain, Bacteria associated with MIC of copper: characterisation and extracellular polymer production, in: A.K. Tiller, S. CAC (Eds.), Microbial Corrosion, European Federation of Corrosion publication N. 15. The Institute of Materials, London, 1995, pp. 49–63.

[6] G.G. Geesey, I.B. Beech, P.J. Bremmer, B.J. Webster, D. Wells, Biocorrosion, in: J. Bryers (Ed.), Biofilms II: Process Analysis and Applications, Wiley-Liss, Inc, London, 2000, pp. 281–326.

[7] M.L. Carvalho, J. Doma, M. Sztyler, I. Beech, P. Cristiani, The study of marine corrosion of copper alloys in chlorinated condenser cooling circuits: the role of microbiological components, Bioelectrochemistry 97 (2014) 2–6.

[8] M.M. Critchley, R. Pasetto, R.J. O'Halloran, Microbiological influences in "blue water" copper corrosion, J. Appl. Microbiol. 97 (2004) 590–597.

[9] P. Cristiani, Electrochemical technologies for antifouling treatments of cooling circuits, in: J. Chan, S. Wong (Eds.), Biofouling, Types, Impact and Anti-Fouling, Nova Science Publishers, Inc., ISBN 978-1-60876-501-0. 2010, pp. 2–24.

[10] H.C. Flemming, G.G. Geesey (Eds.), Biofouling and Biocorrosion in Industrial Water Systems, Springer-Verlag, Berlin, 1991, pp. 47–80.

[11] P. Cristiani, G. Bianchi, Microbial corrosion prevention in ENEL power plants, in: European Federation of Corrosion Publications, NUMBER 22: Aspects of Microbial Induced Corrosion, The Institute of Materials, London, 1997, pp. 137–155.

[12] P. Cristiani, Risk assessment of biocorrosion in condensers, pipework and other cooling system components, in: T. Liengen, D. Féron, R. Basséguy, I. Beech (Eds.), Understanding biocorrosion: fundamentals and applications (EFC 66), Woodhead Publishing Ltd, 2014, Chap. 15. pp. 357–384.

[13] P. Cristiani, Biofilm and MIC monitoring: State-of-the-art, in: Venice, 13/04/2000, Second Workshop of Brite-Euram Thematic Network *BIOCORROSION – MIC of industrial materials*' (BRRT–CT98–5084), UE report 'biocorrosion 00–01', 2000. www.corr-institute.se/english/Web_DT/files/Biofouling1.pdf.

[14] B.J. Little, P.A. Wagner, Application of electrochemical techniques to the study of microbiologically influenced corrosion, in: Modern Aspects of Electrochemistry, vol. 34, Springer, Boston, MA, USA, 2002, pp. 205–246.

[15] P. Cristiani, Metodi innovativi per lo studio e la prevenzione della corrosione microbiologica, in: Proceedings of 36° Convegno dell'Convegno Associazione Nazionale Metallurgia. Parma (I), 21-23 settembre 2016, 2016.

[16] L. Iannucci, J.F. Ríos-Rojas, E. Angelini, M. Parvis, S. Grassini, Electrochemical characterization of innovative hybrid coatings for metallic artefacts, Eur. Phys. J. Plus 133 (12) (2018).

[17] L. Iannucci, M. Parvis, P. Cristiani, R. Ferrero, E. Angelini, S. Grassini, A novel approach for microbial corrosion assessment, IEEE Trans. Instrum. Meas. 68 (5) (2019) 1424–1431.

[18] H.T. Dinh, J. Kuever, M. Mussmann, A.W. Hassel, W. Stratmann, F. Widdel, Iron corrosion by novel anaerobic microorganism s, Nature 427 (2004) 829–832.

[19] G. Ghiara, R. Spotorno, S.P. Trasatti, P. Cristiani, Effect of *Pseudomonas fluorescens* on the electrochemical behaviour of a single phase Cu-Sn modern bronze, Corros. Sci. 139 (2018) 227–234.

[20] A.L. Neal, K.M. Rosso, G.G. Geesey, Y.A. Gorby, B.J. Little, Surface structure effects on direct reduction of iron oxides by *Shewanella oneidensis*, Geochim. Cosmochim. Acta 23 (2003) 4489–4503.

[21] I.B. Beech, Biocorrosion: Role of sulphate-reducing bacteria, in: G. Bitton (Ed.), Encyclopaedia of Environmental Microbiology, John Wiley, London, 2002, pp. 465–475.

[22] W.G. Characklis, K.C. Marshall, Biofilms, John Wiley and Sons, New York, 1990.

[23] W.A. Hamilton, Microbially influenced corrosion as a model system for the study of metal microbe interactions: a unifying electron transfer hypothesis, Biofouling 19 (2003) 65–76.

[24] A.K. Lee, D.K. Newman, Microbial iron respiration: impacts on corrosion processes, Appl. Microbiol. Biotechnol. 62 (2003) 134–139.

[25] G. Muyzer, T. Brinkhoff, U. Nübel, C. Santegoeds, H. Schäfer, C. Wawer, Denaturing gradient gel electrophoresis (DGGE) in microbial ecology, in: A.D.L. Akkermans, J.D. Elsas, F.J. Bruijn (Eds.), Molecular Microbial Ecology Manual, Kluwer, Dordrecht, 1998, pp. 1–27.

[26] S. Sanschagrin, E. Yergeau, Next-generation sequencing of 16S ribosomal RNA gene amplicons, J. Vis. Exp. 90 (2014) Aug 29.

[27] B. Geurkink, S. Doddema, E. Vries, J. Jan Euverink, E. Croese, Value of next generation sequencing as monitoring tool for microbial corrosion—a practical case from biosprophyling to tailor made MMM analysis, in: CORROSION 2016, 6–10 March, Vancouver, British Columbia, Canada, NACE-2016-7764, 2016.
[28] E. Angelini, C.E.A. Posada, E. Di Francia, S. Grassini, L. Iannucci, L. Lombardo, M. Parvis, Indoor and outdoor atmospheric corrosion monitoring of cultural heritage assets, Metall. Italiana 4 (2018) 34–41.
[29] S. Corbellini, E. Di Francia, S. Grassini, L. Iannucci, L. Lombardo, M. Parvis, Cloud based sensor network for environmental monitoring measurement, J. Int. Meas. Confed. 118 (2018) 354–361.
[30] S.W. Borenstein, Microbiologically Influenced Corrosion Handbook, Industrial Press Inc., New York, 1994.
[31] L.R. Hilbert, Monitoring techniques for microbially influenced corrosion of carbon steel, in: Biocorrosion—Monitoring Techniques, Annual meeting of Danish Metallurgical Society, January, Vintermodet, 5–7, 2000.
[32] B. Little, R.I. Ray, P.A. Wagner, J.C. Jones-Meehan, C. Lee, F. Mansfield, Diagnosing microbiologically influenced corrosion, in: COST 520 Biofouling and Materials, Budapest, 31 May–3 June, 2000.
[33] G. Perboni, M. Radaelli, Monitoraggio elettrochimico ad alta temperatura del trattamento ossidante nel generatore di vapore con corpo cilindrico di Turbigo 4, in: Report ENEL-SRI-PDM-CPC-98-003, Luglio 1998, 1998.
[34] P. Cristiani, A. Franzetti, G. Bestetti, Monitoring of electro-active biofilm in soil, Electrochim. Acta 54 (1) (2008) 41–46.
[35] D. Feron, In task 5 final report 'guidelines', in: Brite-Euram Thematic Network MIC of Industrial Materials (BRRT – CT98-5084), 14, 2002.
[36] T. Kim, J. Kang, J.H. Lee, J. Yoon, Influence of attached bacteria and biofilm on double-layer capacitance during biofilm monitoring by electrochemical impedance spectroscopy, Water Res. 45 (15) (2011) 4615–4622.
[37] J.T. Babauta, H. Beyenal, Mass transfer studies of Geobacter sulfurreducens biofilms on rotating disk electrodes, Biotechnol. Bioeng. 111 (2) (2014) 285–294.
[38] A. Nagiub, F. Mansfeld, Evaluation of microbiologically influenced corrosion inhibition (HICI) with EIS and ENA, Electrochem. Acta 47 (2002) 2319.
[39] M. Kitayama, R. Koga, T. Kasai, A. Kouzuma, K. Watanabe, Structures, compositions, and activities of live Shewanella biofilms formed on graphite electrodes in electrochemical flow cells, Appl. Environ. Microbiol. 83 (17) (2017) e00903-17.
[40] M. Dubiel, C.H. Hsu, C.C. Chien, F. Mansfeld, D.K. Newman, Microbial Iron respiration can protect steel from corrosion, Appl. Environ. Microbiol. 68 (3) (2002) 1440–1445.
[41] M.S. Suma, R. Basheer, B.R. Sreelekshmy, A.H. Riyas, T.C. Bhagya, M. Ameen Sha, S. M.A. Shibli, Synergistic action of *Bacillus subtilis*, *Escherichia coli* and *Shewanella putrefaciens* along with *Pseudomonas putida* on inhibiting mild steel against oxygen corrosion, Appl. Microbiol. Biotechnol. (2019) 2019 May:1–15.
[42] J.Y.P. Lin, E.J. St. Martin, J.R. Frank, Monitoring and Mitigation of Sustained Localized Pitting Corrosion, in: Argonne National Laboratory Argonne IL, January 2003 Final Report DOE FEW 49297, 2003.
[43] C. Chandrasatheesh, R.P. George, J. Jayapriya, J. Kamachi, U. Mudali, Detection and analysis of microbiologically influenced corrosion of 316 L stainless steel with electrochemical noise, Tech. Eng. Fail. Anal. 42 (2014) 133–142.
[44] A.M. Homborg, C.F. Leon Morales, T. Tingac, J.H.W. de Witd, M. JMC, Detection of microbiologically influenced corrosion byelectrochemical noise transients, Electrochim. Acta 136 (2014) (2014) 223–232.

[45] A.N. Moosavi, J.L. Dawson, R.A. King, The effect of sulphate reducing bacteria on the corrosion of reiforced concrete, in: S.C. Dexter (Ed.), Biologically Influenced Corrosion, NACE International, Houston, TX, 1986 reference Book N. 8.
[46] L.V. Nielsen, K.V. Nielsen, Differential ER-technology for measuring degree of accumulated Corrosion as well as Instant Corrosion Rate, in: CORROSION2003, NACE International, Houston, TX, 2003 Paper No. 03443.
[47] L.R. Hilbert, T. Hemmingsen, L.V. Nielsen, S. Richter, When can electrochemical techniques give reliable corrosion rates on carbon steel in sulfide media? in: CORROSION/2005 paper no. 05346, Houston, TX, NACE international, 2005.
[48] G. Schmitt, Sophisticated electrochemical methods for MIC investigation and monitoring, Mater. Corros. 48 (1997) 586–601.
[49] Z. Lewandowski, W.C. Lee, W.G. Characklis, B.J. Little, Dissolved oxygen and pH microelectrode measurements at water-immersed metal surfaces, in: Paper No. 93, CORROSION 88, Houston, TX, NACE International, 1988.
[50] G.B. Farquhar, W. Thomas, T.W. Pickthall, J.A. DeCuir, Solving Gulf Coast oil pipeline bacteria-related corrosion problem, Pipeline Gas J. 232 (3) (2005) 28–30. www.pipelineandgasjournal.com.
[51] D.H. Pope, Microbial Corrosion in Fossil-Fired Power Plants, in: EPRI CS-5495 Project 2300-12 Final Report November 1987, Research, 43(4): 1–15, 1987.
[52] J.C. Nickel, I. Ruseska, J.B. Wright, J.W. Costerton, Tobramycin resistance of *Pseudomonas aeruginosa* cells growing as a biofilm on urinary catheter material, Antimicrob. Agents Chemother. 27 (1985) 619–624. www.tylerresearch.com/instr/biofilm/lpmr.shtml.
[53] P. Cristiani, G. Perboni, Antifouling strategies and corrosion control in cooling circuits, Bioelectrochemistry 97 (2014) 120–126.
[54] I.B. Beech, C.L.M. Coutinho, Biofilms on corroding materials, in: P. Lens, A.P. Moran, T. Mahony, P. Stoodly, V. O'Flaherty (Eds.), Biofilms in Medicine, Industry and Environmental Biotechnology—Characteristics, Analysis and Control, IWA Publishing of Alliance House, London, 2003, pp. 115–131.
[55] H.A. Videla, R.A. Silva, C.G. Canales, J.F. Wilkes, Monitoring biocorrosion and biofilms in industrial waters: a practical approach, in: Proceeding of The International Symposium on Microbiologically Influenced Corrosion (MIC) Testing, by ASTM Committee G-1 on Corrosion of metal, Miami, Florida, November 1992, 1992.
[56] R.D. Petty, L.A. Sutherland, E.M. Hunter, I.A. Cree, Comparison of MTT and ATP-based assays for the measurement of viable cell number, J. Biolumin. Chemilumin. 10 (1995) 29–34.
[57] R.D. Bryant, W. Janssen, J. Bolvin, E.J. Laishley, J.W. Corterson, Effect of hydrogenase and mixed sulfate-reducing bacterial population on the corrosion of steel, Appl. Environ. Microbiol. 57 (1991) 2804–2809.
[58] R.E. Hillman, Biofouling Detection Monitoring Devices: Status Assessment, Battelle New England Marine Research Laboratory, 1985 EPRI Report CS-3914.
[59] W. Micheletti, M. Miller, Condensor microbiofouling control handbook, in: EPRITR-102507, 4(1)–4(33), 1993.
[60] C.M. Stuart, M.K. Kaufmann, E.R. Brundage, Practical experience with advanced online monitoring techniques, in: CORROSION/90 Las Vegas, Nevada, April 23–27, Paper n. 360, Houston, TX: NACE International, 1990.
[61] K. Chu, F. Mochizuki, The effect of biofouling on fouling resistance, in: Proceedings of Condenser Biofouling Control: The State of The Art, Menlo Park, CA: CS-4339 April 1985, Electric Power Research Institute, 3-140–3-150, 1985.

[62] R.O. Lewis, The influence of biofouling counter measures on corrosion of seat exchanger materials in seawater, in: Material Performance, September, 31–8, 1982.
[63] T.R. Bott, Fouling of Heat Exchangers, in: Chemical Engineering, Elsevier, Amsterdam, 1995 monograph no. 26.
[64] H.C. Flemming, C. Tamachkiarowa, J.A. Klahre, J. Schmitt, Monitoring of fouling and biofouling in technical systems, Water Sci. Technol. 38 (8–9) (1998) 291–298.
[65] R.L. Wetegrove, R.H. Banks, M.R. Hermiller, Optical monitor for improved fouling control in cooling systems, in: Cooling Technology Institute, Technical paper TP 96–09, 1996. www.cti.org/tech_papers/fouling.shtml.
[66] M. Chattoraj, D.L. Stonecipher, S.A. Borchardt, Demand-based, realcontrol of microbial growth in air-conditioning cooling water system, ASHRAE Trans. 109 (2003) 1.
[67] S.R. Hatch, D.L. Stonecipher, M. Willer, P. Yu, S.A. Borchard, Real word experience with a new cooling water automation system, in: Int. Congr. CORROSION/2003 paper no. 3075, Houston, TX, NACE international, 2003.
[68] P.J. Wyatt, Light scattering in the microbial world, J. Colloid Interface Sci. 39 (1972) 479–491.
[69] P.J. Wyatt, C. Jackson, Discrimination of phytoplankton via light scattering properties, Limnol. Oceanogr. 34 (1) (1989) 96–112.
[70] D. Herbert-Guillou, B. Tribollet, D. Festy, L. Kiéné, In situ detection and characterization of biofilm in waters by electrochemical methods, Electrochim. Acta 45 (1999) 1067–1075.
[71] D.R. Bond, D.R. Lovley, Electricity production by Geobacter sulfurreducens attached to electrodes, Appl. Environ. Microbiol. 69 (2003) 1548–1555.
[72] J.S. Deutzmann, M. Sahin, A.M. Spormann, Extracellular enzymes facilitate electron uptake in biocorrosion and bioelectrosynthesis, MBio 6 (2015) e00496–15.
[73] Y. Ishihara, S.I. Motoda, Y. Suzuki, S. Tsujikawa, Effect of environmental factors on ennobled electrode potential attained in natural sea water for stainless steels, Corros. Eng. 44 (1995) 421–433.
[74] S. Kato, I. Yumoto, Y. Kamagata, Isolation of acetogenic bacteria that induce biocorrosion by utilizing metallic iron as the sole electron donor, Appl. Environ. Microbiol. 81 (2015) 67–73.
[75] Mollica A and Trevis A (1976) Proc. of 4th Int. Cong. Marine Corros. and Fouling, Antibes, France, June 1976, 351–65.
[76] A. Mollica, E. Traverso, G. Ventura, Electrochemical monitoring of the biofilm growth on active-passive alloy tubes of heat exchange using sea water as cooling medium, in: XI Proceedings of International Corrosion Congress, Firenze 2–6 April 1990, Italy, AIM Ed., vol. 4, 333–40, 1990.
[77] A. Mollica, Biofilm and corrosion on active-passive alloys in seawater, Int. Biodeter. Biodegr. 29 (1992) 213–229.
[78] M. Eashwar, S. Maruthamuthu, Ennoblement of stainless alloys by marine biofilm: an alternative mechanism, in: Proc. of 12th Int. Corros. Congr., Houston, TX, NACE International, 3708–16, 1993.
[79] T.S. Lee, R.M. Kain, J.W. Oldfield, The effect of environmental variables on crevice corrosion of stainless steels in seawater, Mat. Perform. 23 (1984) 9–15.
[80] H.A. Videla, M.F.L. De Mele, G. Brankevich, Biofouling and corrosion of stainless steel and 70/30 copper-nickel samples after several weeks of immersion in seawater, in: Corrosion/89, Houston, TX: NACE Paper No. 291, 1989.
[81] V. Scotto, M.E. Lai, The ennoblement of stainless steels in seawater: a likely explanation coming from the field, Corros. Sci. 40 (1998) 1007–1018.

[82] S.C. Dexter, H.J. Zhang, Effect of biofilm on corrosion potential of stainless alloys in estuarine waters, in: Proc. of 11th Int. Corros. Congr., 4 April 1990, Firenze: AIM, 333–40, 1990.
[83] S. Maruthamuthu, G. Rajagopal, S. Sathianarayannan, M. Eashwar, K. Balakrishnan, A photoelectrochemical approach to the ennoblement process: proposal of an adsorbed inhibitor theory, Biofouling 8 (1995) 223–232.
[84] B. Erable, D. Feron, A. Bergel, Microbial catalysis of the oxygen reduction reaction for microbial fuel cells: a review, ChemSusChem 5 (2012) 975–987.
[85] P. Cristiani, Biofilm growth and destruction: electrochemical monitoring for food processing plants, in: Proceedings of Eurotherm Seminar 77 Heat and Mass Transfer in Food Processing, 20–22 June 2005 Parma: ETS Pisa, 2005.
[86] G.J. Licina, Microbial corrosion: 1988 Workshop Proceedings, in: EPRI report ER-6345, April 1989, 1-1/1-14, 1989.
[87] V. Scotto, R. Di Cinto, G. Marcenaro, The influence of marine aerobic microbial film on stainless steel corrosion behaviour, Corros. Sci. 25 (3) (1985) 185–194.
[88] G.J. Licina, G. Nekoksa, An electrochemical method for on-line monitoring of biofilm activity, in: CORROSION/93 Annual Int. Congr. March 1993, New Orleans LA, paper No. 403, Houston, TX, NACE International, 1993.
[89] G.J. Licina, Monitoring biofilms on metallic surfaces in real time, in: CORROSION/01 annual int. congr., Paper No. 1442, Houston, TX, NACE International, 2001.
[90] G.J. Licina, Optimizing biocide additions via real time monitoring of biofilms, in: CORROSION/04 Int. Conference, Paper No. 04582, Houston, TX, NACE International, 2004.
[91] C. Andrade, Real Time Monitoring of Biofilms in Seawater Injection Environments, in: CORROSION06, Paper 06660, NACE International, Houston, TX, 2006.
[92] H. Fallon, G.J. Licina, Plant experience with dosing optimisation of an environmentally friendly biocide, in: Int. Cong. CORROSION 2004, Paper 04086, Houston, TX, NACE International, 2004.
[93] P. Cristiani, U. Giancola, Prevention of fouling and microbial corrosion in power stations using seawater as a coolant, in: H. Muller-Steinhagen (Ed.), Heat Exchanger Fouling Cleaning Technologies, Publio, Essen, 2000 334, 349.
[94] A. Mollica, P. Cristiani, On-line biofilm monitoring by electrochemical probe "BIOX" IWA J. Water Sci. Technol. 47 (5) (2003) 45–49.
[95] G. Pavanello, Alvim srl, (2018). www.alvimcleantech.com/cms/it.
[96] G. Pavanello, M. Faimali, M. Pittore, A. Mollica, A. Mollica, A. Mollica, Exploiting a new electrochemical sensor for biofilm monitoring and water treatment optimization, Water Res. 45 (2011) 1651–1658.
[97] G. Pavanello, A. Mollica, M. Faimali, Wrapped by bacteria: biofilm detection and removal in paper mills, J-FOR, J. Sci. Technol. Forest Products Process. 4 (2017) 6–12.
[98] P. Cristiani, M. Belluato, V. Balacco, L. Bartole, G. Bressan, Chlorine dioxide treatment optimisation by monitoring systems, in: New York Engineering Foundation Conferences: Heat Exchanger Fouling—Fundamental Approaches and Technical Solutions, 8–13 July 2001, Davos, Switzerland, 2001.
[99] P. Cristiani, A. Dell'Erba, D. Falsanisi, M. Notarnicola, L. Liberti, Tecnologie antifouling ad impatto ambientale minimizzato: Valutazione sperimentale dell'effetto della crescita di biofilm e degli inibitori, con particolare riferimento all'acido per acetico, in: Ricerca di Sistema, Report A4524785, 2004. www.ricercadisistema.it.
[100] P. Cristiani, Solutions to fouling in power station condensers, Appl. Therm. Eng. J. 25 (16) (2005) November 2005, Elsevier, 2630–40.

[101] Jenner HA, Taylor CJL, van Donk M and Khalanski M (1997) Chlorination by-products in chlorinated cooling water of some European coastal power stations, Mar. Environ. Res., 43(4), June, 279–93.

[102] J.J. Rook, Formation of haloforms during chlorination of natural waters, Wat. Treat. Examin. 23 (1974) 234–241.

[103] J.M. Symons, S.W. Krasner, M.J. Sclimenti, L.A. Simms, H.W.J. Sorensen, G.E. J. Speitel, A.C. Diehl, Influence of bromide ion on trihalomethane and haloacetic acid formation, in: R.A. Minear, G.L. Amy (Eds.), Disinfection By-Products in Water Treatment: The Chemistry of Their Formation and Control, CRC Press Inc, 1996, pp. 91–130.

[104] P. Cristiani, G. Perboni, Monitoring of the negative influence and the positive effect of chlorination on surface passivation of CuNi70/30 condenser tubes, in: Proceedings of EUROCORR09, paper 7899, 7-10 Nice, September 2009, 2009.

[105] A. Mollica, V. Scotto, Mechanisms and prevention of biofilm effects on stainless steel corrosion: inhibition of crevice corrosion on stainless steel test, European Federation of Corrosion Application no. 19, in: Sea Water Corrosion of Stainless Steels – Mechanisms and Experiences, London, the Institute of Materials, 23–43, 1996.

[106] A. Dell'Erba, P. Cristiani, D. Falsanisi, M. Notarnicola, L. Liberti, Peracetic Acid Treatment for Biofilm Control, in: Engineering Conferences International Water Treatment and Reuse, 11–16 February, Tomar, Portugal, 2007.

[107] P. Cristiani, G. Perboni, A. Debenedetti, Effect of chlorination on the corrosion behavior of CuNi 70/30 condenser tubing, Electrochim. Acta 54 (1) (2008) 100–107.

[108] P. Cristiani, M.L. Carvalho, G. Perboni, Monitoring by an electrochemical integrated system the corrosion and antifouling treatment on aluminum brass condenser tubes, ECS Trans. 50 (31) (2013) 267–274.

[109] M.L. Carvalho, P. Cristiani, Experiences of on-line monitoring of microbial corrosion and antifouling on copper alloys condenser tubes, in: Proceedings of Heat Exchanger Fouling and Cleaning, Eurotherm, Crete, Greece, June 05-10, 2011, 2012.

[116] H. Müller-Steinhagen, M.R. Malayeri, A.P. Watkinson, Recent advances in heat exchanger fouling research, Heat Transfer Eng. 28 (3) (2007) 173–176.

Corrosion monitoring in concrete 16

Till Felix Mayer[a], *Christoph Gehlen*[b], *and Christoph Dauberschmidt*[c]
[a]Sensortec GmbH, Munich, Germany, [b]Technical University of Munich, Munich, Germany, [c]University for Applied Sciences Munich, Munich, Germany

16.1 Introduction

The development and maintenance of the infrastructure is one of the key issues in almost all industrial countries. In Asia, most of the innovative and prestigious infrastructure projects have been completed in recent years (e.g., the Hong Kong-Zhuhai-Macao bridge project) or are still under construction, whereas the largest parts of the infrastructure in Europe and Northern America were constructed in the sixties and seventies of the last century. Especially, the older structures require more intense condition monitoring and refurbishment today and in the next decades. If not performed properly, complete collapse of the structures can be the consequence (for example, the Morandi bridge, Genoa, Italy).

Among the possible deterioration mechanisms, reinforcement corrosion due to chloride ingress plays a dominant part. A recent survey in Germany revealed that approximately 66% of the damage to the major road network bridges was caused by chloride-induced reinforcement corrosion [1]. In 2009, the World Corrosion Organization published a study which stated that the repair of corrosion damage in the industrialized nations annually consumes between 3% and 4% of the nations' GDP [2]. Consequently, a definite need arises for the detection and prevention of corrosion in general and for our reinforced and prestressed concrete infrastructure in particular.

The reinforcement of concrete, with reinforced steel as well as prestressed steel, ensures the load-bearing capacity of most structures such as bridges, parking decks, tunnels, etc. The steel in reinforced concrete is normally protected against corrosion by a thin and impermeably dense passive layer formed in the alkaline medium of the pore solution of concrete ($pH \approx 13.3$). This passive layer can break down in the presence of chlorides at the steel surface or due to a decrease in pH of the pore solution caused by carbonation of the concrete. After the breakdown of the passive layer, corrosion at the reinforcement can be initiated, which can result in severe deterioration of the steel and the concrete, and thus, a significant reduction in load-bearing capacity.

During the initiation phase as well as in the early corrosion phase during which chlorides penetrate the concrete and/or the carbonation depth increases, no visible damage may be apparent on the concrete surface. Only later, when cracks have formed or spalling of the concrete cover occurs, can deterioration be assessed by visual

Techniques for Corrosion Monitoring. https://doi.org/10.1016/B978-0-08-103003-5.00016-3

inspection. However, repair measures are then likely to be more costly and time-consuming compared to measures taken at the earlier stages of damage because repair costs increase exponentially with damage. On this background, corrosion monitoring provides some important tools for the condition assessment of both newly built structures and aging structures, because it may supply additional information on the condition of structures which cannot be provided with common nondestructive testing (NDT) techniques. However, it is obvious that the aims and requirements for monitoring of newly built structures and existing structures can be quite different. While corrosion monitoring of new structures mainly aims at monitoring the ingress of corrosion-promoting substances (e.g., chlorides) towards the passive reinforcement, corrosion monitoring for existing structures usually yields time-dependent changes in corrosion activity, e.g., after a repair measure. As a consequence, different techniques and sensor setups have been developed over the last decades. Those devices are often grouped under the term "corrosion monitoring."

In the following sections the fundamentals of corrosion of steel in concrete, novel approaches for the service life management of reinforced concrete structures, and the basic principles of some of the most important techniques for corrosion monitoring will be discussed. Three case studies at the end of the chapter illustrate the application of corrosion monitoring for real structures and the vast potential that corrosion monitoring offers for structures at almost any time during their service life.

16.2 Deterioration mechanisms for corrosion of steel in concrete

16.2.1 General deterioration model

The time-dependent development of most types of deterioration mechanisms for concrete structures can be modelled by means of a two-phase-curve as illustrated in Fig. 16.1. This is the basis for defining the service life of a structure. Reinforcement corrosion is the most relevant deterioration mechanism in our reinforced concrete infrastructure; the two phases of deterioration are the following [4]:

- *Initiation phase*: During this phase, no noticeable weakening of the material or the functionality of the structure occurs. However, part of the protective barrier is destroyed or overcome by aggressive media. Carbonation, chloride penetration, and sulphate action are examples of such mechanisms. In the case of corrosion, the duration of the initiation period is determined by the thickness and permeability of the concrete cover and the actual exposure conditions, e.g., chloride ingress can be accelerated by cyclic wetting and drying.
- *Deterioration phase*: During this phase, active deterioration develops and a loss of functionality occurs. Fig. 16.1 shows a mechanism of deterioration, leading to an increase in deterioration rate with time. Reinforcement corrosion is such a form of deterioration.

The most common mechanisms leading to corrosion of steel in concrete under moderate climatic conditions are carbonation and chloride penetration in the initiation phase. Both mechanisms are described in further detail in Section 16.2.2.

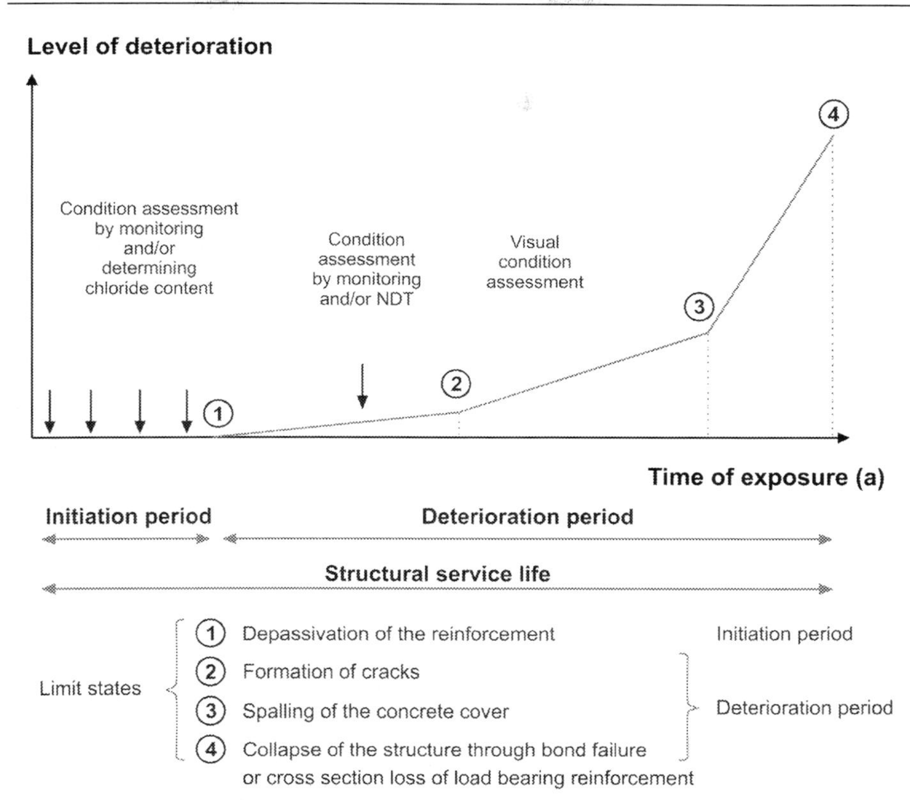

Fig. 16.1 Service life of concrete structures. A two-phase modelling of deterioration with possibilities of assessment.
(Adapted from [3])

16.2.2 Initiation phase

16.2.2.1 Carbonation

Steel in concrete is generally protected against corrosion. Owing to the alkalinity of the pore solution (pH = 12.5–13.5), a microscopic oxide layer is formed at the steel surface preventing the further dissolution of the steel. From an electrochemical perspective, the formation and maintenance of the passive film is also a corrosion process itself; however, the "corrosion rate" of passive steel in concrete is 1 μm/year at most and is, therefore, negligible from a practical point of view.

If the pH of the pore solution drops below approximately 9, the depassivation occurs and, consequently, corrosion protection is lost over a large surface area of the steel. Such a drop in pH can be induced under normal exposure conditions by the carbonation of the concrete. Carbonation of concrete normally involves a chemical reaction between atmospheric carbon dioxide (CO_2) and the products of cement hydration. The normal CO_2 content of the air is 0.0415 vol%, but in particular conditions, such as in tunnels and parking garages, it may easily reach much higher values.

Carbonation starts from the exposed concrete surface and gradually progresses into the concrete structural element. Once loss of alkalinity occurs at the reinforcement surface, the initiation phase ends and—depending on the particular conditions—the deterioration phase commences. The duration of the initiation phase for carbonation will be prolonged by the following measures:

- Increasing the concrete cover of the reinforcement
- Decreasing the permeability of concrete (low water/cement ratios (w/c), sufficient curing) and using a binder which produces a high $Ca(OH)_2$ content
- Increasing the water content of the pore system (water-saturated concrete cannot carbonate)

16.2.2.2 Chloride ingress

Chloride ions may in some cases already be present in the fresh mix, but in most cases they penetrate into the hardened concrete. During the use of the structures, the concrete surface can be exposed to chlorides from various sources. The most important of these are sea water, deicing salts, and PVC (polyvinyl chloride) fires. Chlorides penetrate into the concrete as ions dissolved in partly or completely water-filled pores. A fraction of these chlorides will be bound chemically or physically by the hardened cement paste matrix, leading to the formation of an equilibrium between bound and free chlorides in the concrete [5].

Unlike carbonation, chloride penetration is not associated with a distinct "reaction" front. On the contrary, a chloride profile with a decrease in content from the concrete surface to the interior is usually found for chloride exposure conditions [6]. The resulting chloride profile is generally governed by diffusion, but may show some deviations from pure diffusion behavior close to the concrete surface due to capillary suction and convection combined with periodic exposure to water or chloride solutions.

When assessing the penetration of chlorides into concrete and the associated risk of reinforcement corrosion, it is necessary to distinguish between two basic effects.

- The penetration resistance of the concrete against the ingress of chlorides. This is often described by the diffusion or migration coefficient of chlorides in the concrete. The penetration resistance depends primarily on the pore size distribution of the concrete.
- The binding capacity of the concrete with respect to chloride ions (physical and chemical binding). This affects both the penetration rate and the ratio of bound to free chloride ions in the pore water.

The binding capacity of concrete is particularly important since the risk of corrosion at the reinforcement is determined solely by the free chlorides in the pore water. Consequently, free chlorides in the pore water and the ratio of bound to free chlorides at the steel surface are the decisive factors in judging the reinforcement corrosion risk.

Corrosion at the reinforcement can occur when the so-called "critical chloride content" (i.e., threshold of chlorides inducing corrosion) of the concrete at the surface of the reinforcement is exceeded. In principle, two definitions of "critical chloride content" are conceivable:

1. A critical chloride content at which depassivation of the steel surface commences, whether or not this leads to visible corrosion damage at the concrete surface.
2. A critical chloride content which ultimately leads to a corrosion phenomenon classified as damage.

The critical chloride content may be considerably lower according to definition 1 than for definition 2, because corrosion damage only occurs if, in addition to depassivation of the steel surface, other conditions necessary for an accelerated rate of corrosion (e.g., oxygen supply, humidity) are fulfilled (see Section 16.2.3).

The critical chloride content is not a fixed value, but strongly depends on the quality of the concrete cover (type of cement, water/cement (w/c) ratio, curing, thickness), the quality of the steel-concrete interface, and environmental conditions [7–9]. For concrete permanently subjected to dry environmental conditions as well as for concrete which is constantly saturated with water, the critical chloride content according to definition 2 is much higher than for concrete exposed to frequently changing conditions (e.g., splash water zone). For a first assessment, a chloride content of 0.5 wt% with respect to the cement content at reinforcement level is often considered as a lower threshold value for the critical chloride content. However, under certain conditions, much higher chloride contents may not be critical.

Once the chloride content at the reinforcement exceeds the critical chloride content, the passive film will breakdown locally, thus leading to the end of the initiation phase.

The duration of the initiation phase for chloride penetration can be extended by the following measures:

- Increasing the concrete cover of the reinforcement
- Using a concrete mix with a low permeability (with low w/c ratio and, which is much more decisive, the appropriate type of cement (e.g., blast-furnace slag or fly ash cements), adequate curing)
- Increasing the binding capacity of the concrete, especially by increasing the C_3A content (clinker phase tricalcium aluminate) of the cement
- Reducing the water content of the concrete and the depth of the variable water content of the concrete caused by the actual exposure condition (splash or spray fog zones).

16.2.3 Deterioration phase

16.2.3.1 Corrosion mechanisms

After the pore solution condition for depassivation front has reached the steel reinforcement (either by carbonation of the concrete or by exceeding the critical chloride content), corrosion can be initiated. The occurrence of steel corrosion in concrete can be roughly described by four different processes which need to occur simultaneously.

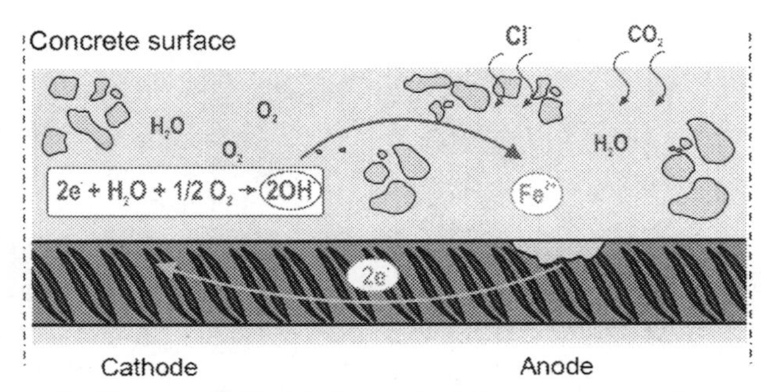

Fig. 16.2 Schematic diagram of chloride-induced reinforcement corrosion.

These processes are: (a) anodic oxidation reaction, i.e., iron dissolution; (b) electric charge transfer through the rebar; (c) cathodic reduction reaction, i.e., oxygen reduction; (d) ionic charge transfer in the pore solution, Fig. 16.2.

For most concrete members subjected to a temperate climate, processes b to d are possible. Under field conditions, mainly two mechanisms are causing the breakdown of the protective passive layer (process a):

- Uniform corrosion. The large-scale breakdown of the passive layer due to chemical reactions of alkaline components of the concrete with carbon dioxide (carbonation) resulting in a distinct drop in pH down to 9 ("carbonation-induced reinforcement corrosion"). This leads to the formation of so-called micro-cells, consisting of pairs of immediately adjacent anodes and cathodes. These are microscopic in size, so that externally they appear to produce uniform dissolution of the steel.
- Pitting corrosion. A local breakdown of the passive layer once the concentration of chlorides (from deicing salts or sea water) at reinforcement level exceeds a critical chloride threshold value ("chloride-induced reinforcement corrosion"). In this case, so-called macro-cells are formed, with the pits acting as anodes and the neighboring sections of the reinforcement as cathodes. Anodes and cathodes are not necessarily adjacent; they may well be situated at a considerable distance from each other.

Since the potential of corroding steel electrodes in concrete is significantly lower than that of passive steel, corrosion initiation is accompanied by a distinct drop in the reinforcement potential in that region.

16.2.3.2 Rate-determining parameters

The corrosion rates at anodic areas in macro-cells are generally much higher than for micro-cell corrosion. This is because the anodic reactions of macro-cells are usually supported by very large areas of the cathodes, leading to high current densities, and hence, high corrosion rates. The size of a macro-corrosion cell is determined largely by the electrolytic resistivity of the concrete, i.e., by its water content, the permeability of the concrete, and the chemical composition of the pore solution. In dry concrete with an extremely low permeability, the rate of corrosion is low. Macro-cells may, however, also occur where the concrete has carbonated over a large area, in regions with an uneven moisture distribution or oxygen availability, or in zones with localized depassivation, e.g., in the regions of cracks or honeycombs. Under these circumstances, corrosion rates higher than those encountered with uniform corrosion of the reinforcement may be expected.

As expected, the chloride content in the concrete at the steel surface has a decisive effect on the corrosion rate. Once the chloride content exceeds the critical corrosion-initiating value, the corrosion rate increases roughly proportional to the increase in chloride content.

According to Fig. 16.2, the corrosion rate of the steel is, besides other resistances, dependent on the polarization resistance at the cathode which is linked to the oxygen diffusivity of the concrete. The diffusivity mainly depends on capillary porosity and the water content of the concrete. Permanently water-saturated concrete structures are unlikely to show significantly high corrosion rates because the resistance to oxygen diffusion is very high. The concrete cover also has a significant effect on oxygen diffusion in concrete mixes with low permeability.

Furthermore, the corrosion rate is dependent on the electrolytic resistance of the concrete. If the water content of the concrete decreases, the electrolytic resistance increases significantly (by up to several decades). Thus, corrosion which causes damage can only occur when the electrolytic resistance of the concrete is below a value in the range of 100–150 Ωm [10], which is normally reached in concrete exposed to an environment with 85%–90% relative humidity or higher. Consequently, technically relevant corrosion can only occur in structures exposed to weather or water, but not in dry indoor conditions. The electrolytic resistance is also dependent on the ion concentration of the pore solution, the temperature, and the permeability of the concrete.

While in uncracked concrete a good quality concrete cover will generally prevent depassivation of the reinforcement for a long period (initiation phase), prevention is practically impossible where cracks cross the reinforcement, since both the ingress of the carbonation front and chloride penetration are much more rapid in the crack zone than elsewhere [11]. The capillary suction capacity of the cracks also plays an important part in the penetration of chlorides.

The corrosion mechanism in the crack zone is of decisive significance for the corrosion rate of the reinforcement. In laboratory investigations, cell current and voltage measurements on cracked reinforced steel beams exposed to chloride attack clearly demonstrate that large macro-cells are formed under certain conditions. The reinforcement in the crack zone acts as an anode and the reinforcement between the cracks at a distance of several decimeters acts as a cathode which often leads to extremely high corrosion rates in the crack zone. The corrosion rates within the cracks are more dependent on concrete cover and composition than on crack width [12].

16.2.4 Service life management

Generally, it is of great importance that a structure or a structural member is able to perform its intended function satisfactorily for its intended service life under the specified environmental and operation conditions. The so-called design service life is the period intended over which a structure or part of it is to be used with maintenance measures, but without major repair work. The design service life is defined by the following:

- The definition of the relevant limit state
- The number of years of the service life
- The level of reliability for not passing the limit state during this period.

The durability of the structure in its particular environment should be such that it remains fit for use during its design service life. This requirement can be fulfilled in one of the following ways:

- Use of materials which, if well-maintained, will not degenerate during the design service life
- Use of construction dimensions to compensate deterioration during the design service life
- The design of protective and mitigating systems
- The choice of a shorter lifetime for structural elements which may be replaced one or more times during the design service life.

All the above approaches are possible in combination with appropriate inspections at fixed intervals or intervals based on the actual condition as well as appropriate maintenance activities.

Designing for a long service life needs some form of verification during the use of the structure. This becomes an integral part of the operation and maintenance activities, the so-called structural service life management. Service life design is based on some assumptions regarding the deterioration mechanisms related to the aggressiveness of the environment. Due to the uncertainties between the assumptions made at the design stage and the real service conditions of the finished structure, it is, at different times during the service life, necessary to verify—or adjust—the initial assumptions and revise the service life forecast accordingly, see Fig. 16.3 (Section 16.3). This need for adjustment, and the benefits of obtaining a more reliable forecast of the service life to be expected, necessitates monitoring the durability performance of the structure in service [4]. The monitoring can be performed destructively or, preferably, nondestructively.

The term "maintenance" is applied to activities that are planned to take place during the service life of the structure in order to ensure the fulfilment of the assumptions in the service life design. A maintenance plan should state the type and frequency of the activities foreseen. The maintenance plan can include activities such as general cleaning, drainage, addition of sealants, replacement of components, etc. Condition control is mainly made to verify the design assumption. The risk of depassivation of the reinforcement can be determined more precisely using a regularly updated condition control (information from monitoring and inspection). Using this information, if necessary, an optimal point in time can be determined for preventive measures, for example, a preventive coating or a cathodic protection.

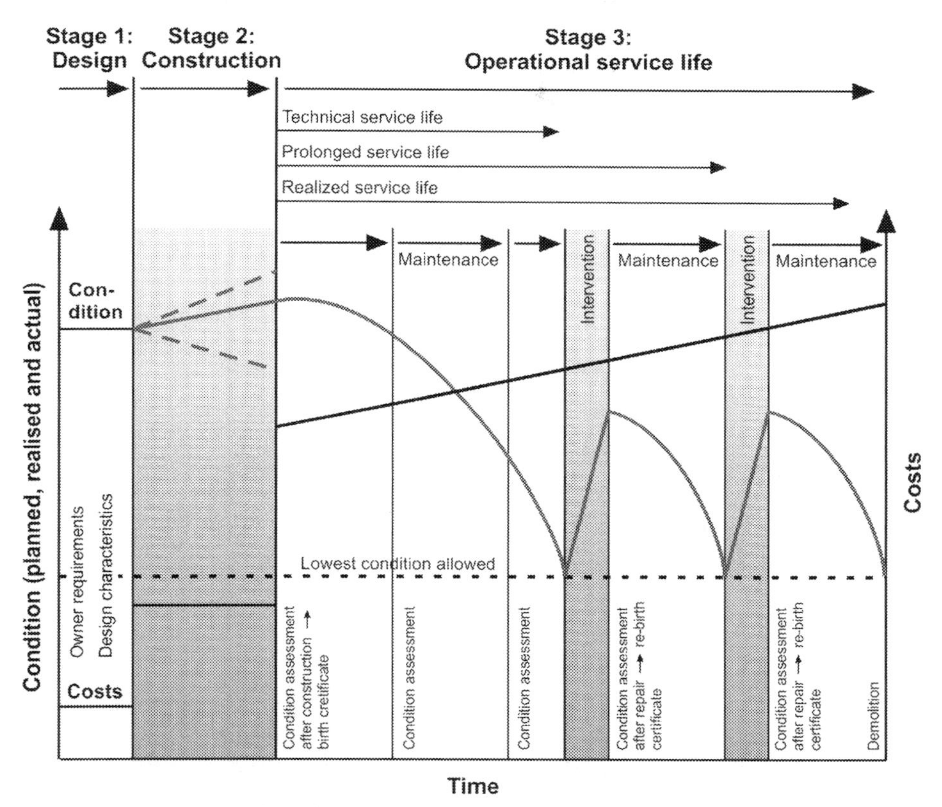

Fig. 16.3 Complete service life from birth to death [13].

16.3 Condition assessment of reinforced concrete structures

Fig. 16.3 shows the service life of a structural component starting with its design, continuing through construction, and followed by the operational service life until the structure is demolished. Each stage may be assessed by appropriate evaluation procedures, the results of which are used to establish appropriate condition control.

16.3.1 Stage 1—Design phase

In the design phase, the specific client requirements, relevant standards specifications, exposure conditions, and implication of the construction materials and methods are considered to identify applicable methods of design, including durability assessment based on deterioration models. The requirements of the client and the standard requirements define the targeted as-constructed-quality and long-term performance of the structure which are expressed in terms of the minimum required condition and reliability.

16.3.2 Stage 2—Construction phase

The construction phase is the most important stage for ensuring that the target service life of a structure is achieved. The conformity of the structure to the design specifications depends highly on the experience and training of the construction team. The actual material properties and geometries on completion of construction are likely to deviate, for better or worse, from those assumed in the durability design. Hence, the "birth certificate" of the structure is intended to reflect the actual properties of the concrete in the new structure, including an update of the service life design to indicate if the maintenance plan has to be modified and to identify critical components/parts of the structure.

16.3.3 Stage 3—Operational service life of a structure

During this period, the structure is in service and should be regularly maintained, assessed, and undergo appropriate intervention measures. Maintenance is commonly applied to limited life items such as gutters and seals, while the actual fabric of the building is designed not to require maintenance throughout its design service life or replacement interval. However, the building does require condition assessment at intervals in order that intervention is possible to avoid unexpected, and potentially costly, deterioration at a later date. These condition assessments are also necessary if the original design service life of the original components and interventions are extended. The results of the measures applied during maintenance are converted into structural reliability by the assessment of the structure using established mathematical deterioration models. Data of a statistical nature obtained by performance testing (condition control) are used to perform the calculation. The calculated reliability level is compared to the minimum level required and additional measures may be necessary if the reliability is less than required. The accuracy of the deterioration model, as well as the available amount of measured data and their quality, has a major effect on the validity of the calculated structural reliability level, and therefore, on the predicted structural condition and further maintenance. Intervention measures may be necessary to prolong the structural life at an acceptable condition level (reliability). Depending on the strategy applied, either proactive or reactive intervention will be necessary. This strategy has to be agreed upon between the persons involved, especially owner/asset managers. The actual condition of the repaired/strengthened structure should be used for the evaluation of the prolonged service life.

As shown schematically in Fig. 16.3, the condition of the structure must be controlled throughout its service life to guarantee an acceptable condition level which lies above a permissible level. The data resulting from inspections on existing structures are used, for example, to estimate the remaining service life of structures, to indicate the time until a further condition assessment, and/or to evaluate the conformity with performance design requirements, i.e., for actions and/or material and/or product properties. In the case of nonconformity, these data are often used as a basis for decisions on the extent/volume of a repair/strengthening measure.

However, for the condition control of existing structures, most of the data obtainable in stages 1 and 2 are not available and alternative data must be supplied at greater cost during the operational service life. Firstly, the condition of the existing structure has to be assessed by a detailed survey. The inspection following construction yields data on the properties at a specific point in time. Secondly, these data are used to determine various aspects of the structure's reliability. Thirdly, the calculated reliability is compared to the required one which is usually specified in standards or specified by the owner. If the reliability is less than required, it is recalculated based on more precise information/knowledge. The structure's reliability is improved by an intervention measure or a lower reliability is chosen if a higher risk level is acceptable.

Following the service life stages in Fig. 16.3, condition assessment is the main basis for decision making, irrespective of whether intervention is necessary or not. In the following sections, some electrochemical methods of assessing the extent of corrosion deterioration are presented. In Section 16.5, case studies illustrate the use of information obtained during surveys for decision making for both newly built structures and existing ones.

16.4 Measurement principles

16.4.1 General

As discussed in Section 16.3, condition control during the use of a structure plays an important part in the service life management of reinforced and prestressed concrete structures. In this context, corrosion monitoring can provide additional and often very useful information. It is evident that the requirements on both the sensor setup and the type of measurements to be conducted will be different for structures which are in the initiation phase as opposed to those with active corrosion. As a consequence, different sensor types and measurement principles have been developed throughout the last decades, which are generally gathered under the term "corrosion monitoring." The four measurement principles which the authors believe to be of the highest practical relevance, i.e., half-cell potential measurements, electrolytic resistance measurements, corrosion current measurements, and linear polarization resistance measurements, will be briefly presented and discussed in Sections 16.4.2 to 16.4.5. Some requirements on the sensor setup for the different measurement methods will also be covered in these sections. In addition, the three case studies in Section 16.5—two for corrosion monitoring during the initiation phase and one for corrosion monitoring during the deterioration phase—contain further information on suitable sensor setups for these applications. A discussion of measurement principles and the presentation of sensor setups provide some help on the design of monitoring systems, but only to a certain extent. The actual choice of measurement principles and sensor setups, the decisions on the number and positioning of sensors, data acquisition, and evaluation always need to be carried out by a corrosion expert with sufficient experience in thorough condition assessment under consideration of the individual boundary conditions of the structure to be monitored.

16.4.2 Half-cell potential measurements

The electrochemical half-cell potential measurement of the steel reinforcement with embedded reference electrodes is a method for assessing the probability of corrosion of the reinforcement in the immediate vicinity of the reference electrodes as well as for controlling the functionality of cathodic protection (CP) systems.

The measurement of the half-cell potential is carried out as a voltage measurement between a reference electrode and the reinforcement. In case of potential mapping, the potential of the steel on the reinforced concrete surface is measured with portable reference electrodes. Furthermore, it is often desirable and useful to make a permanent measurement of the reinforcement steel potential (or the potential difference between the reinforcement steel in the concrete and a reference electrode) with durable, built-in reference electrodes. Typical applications are, for example, difficult or inaccessible reinforced concrete (RC) components, in which the electrodes are installed in critical areas during construction in order to obtain early information about initiated corrosion processes. In addition, potential measurements are used for cathodic protection systems, in which the effectiveness of protection is determined by measuring the steel/concrete potential using embedded reference electrodes. For this purpose, the electrodes are installed at representative positions distributed over the entire structure/or the entire anodic area to be protected.

Regarding the reference electrodes, a distinction is made between so-called thermodynamic "true" reference electrodes, where the potential at the phase boundary electrode/electrolyte depends only on ion concentration and the temperature, and the so-called pseudo-reference electrodes, where there are no well-defined reversible reactions.

There are a whole range of reversible reference electrodes for use in solutions. Only the silver/silver chloride electrodes (Ag/AgCl) and the manganese dioxide electrodes (MnO_2) are currently suitable for permanent use in concrete. A comprehensive field study conducted on 64 reference electrodes of different types installed in the Gimsoystraumen Bridge in Norway rendered that, after 10 years, only 39 reference electrodes were still working, with the lowest failure rate for MnO_2 reference electrodes [14].

Potential measurement must be carried out with a high-ohmic volt meter. The input resistance (impedance) of the measuring device must be at least 10 MΩ. The reference electrode and reinforcement connection must be connected to the measuring device. The resistance between the reference electrode and reinforcement shortly after installation is expected to be $\leq$10 kΩ.

16.4.3 Electrolytic resistance measurements

The electrolytic resistance of the concrete (strictly speaking should be called resistivity) is a geometry-independent material property that describes the electrical resistance as the ratio between applied voltage and resulting current in a unit cell. The dimension of the resistivity is Ωm. The electrical resistivity of concrete may vary over a wide range, from 10^1 to 10^5 Ωm, depending on the moisture content of the concrete (environment) and its composition (material). The electrical resistance of concrete or mortar, thus, represents an essential parameter in the corrosion system "steel in concrete." Therefore, monitoring the resistivity of concrete is important to assess corrosion activity.

Drying building materials always leads to an increase in electrical resistance. This is due to the reduction in the amount of pore water which affects the mobility of the ions responsible for the charge transport. In the case of dry conditions, the electrical concrete resistance can become a controlling factor in the corrosion process, Table 16.1.

To determine electrolytic resistance, low frequency alternating current impedance measurements between two or more electrodes are usually performed. The two or four-electrode arrangement is common, see Fig. 16.4.

Table 16.1 Electrical resistivity and corrosion risk [6].

Resistivity (Ωm)	Corrosion risk
>1000	Negligible
500–1000	Low
100–500	Moderate
<100	High

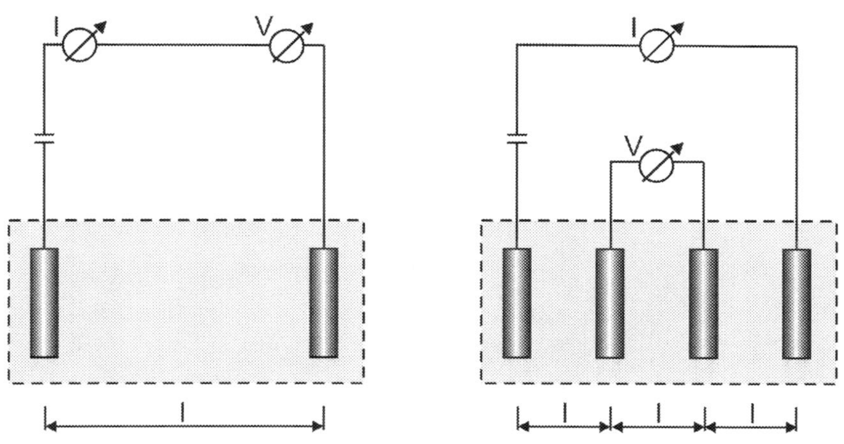

Fig. 16.4 Two (left) and four-electrode arrangement (right) in determining the electrical concrete impedance (DGZfP [15]).

For the measurement, the electrodes are either placed on the concrete surface or embedded during or after construction of the RC-structure in a fixed arrangement. The latter systems are used to permanently determine the electrical concrete resistance for corrosion monitoring.

In the case of two-electrode measurements, the current control and voltage measurement take place with the same electrodes. In addition to the concrete itself, the measuring current has to pass through the contact resistance between the electrode and the concrete twice which results in the so-called transitional resistance. Thus, the two-electrode measurement always measures the sum of concrete and transitional resistance. In case of the four-electrode method, the power feed and voltage measurement are distributed over different pairs of electrodes. The resistance, thus, determined does not contain transitional resistance.

The measured value depends on the geometric shape and arrangement of the electrodes as well as, if necessary, the geometry of the component (for small component dimensions). In principle, a distinction must be made between the measured values.

- Measured electrolytic resistance, R in Ω
- Calculated specific electrolytic resistance, or resistivity, ρ in Ωm,

These are linked by the so-called geometrical factor k (Eq. 16.1).

$$\rho = R \cdot k \tag{16.1}$$

where

- ρ: material-specific resistance in [Ω m].
- k: Geometrical factor (cell constant) in [m].
- R measured resistance in [Ω].

The geometrical factor k results from the electrode and test specimen geometry as well as the electrode arrangement and is to be calculated for the particular structure.

While the assessment of the time-dependent progression of the electrolytic resistance of concrete can be carried out well on the basis of the measurements of R, for further consideration (e.g., quantitative assessment of the water content, influence on corrosion processes, etc.) it is mandatory to use the specific electrolytic concrete resistivity ρ in Ωm.

16.4.4 Corrosion current measurements

The electric current between anodic and cathodic surfaces in a macro-corrosion element is a key factor for the characterization of corrosion activity. In monitoring practice, this type of current measurement is generally referred to as the "corrosion current measurement" (DGZfP [15]). However, it should be pointed out that the measured current is probably only a fraction of the actual corrosion current because cathodic surface reactions may also occur at the anode considered. Nevertheless, monitoring of this current still yields very useful qualitative information on changes in corrosion activity. However, the results should in general not be used to quantitatively determine corrosion rates or losses of cross-section due to corrosion.

Corrosion current measurements can be used during all stages of the service life of a reinforced concrete structure.

- During the initiation phase, corrosion current measurements on anodes placed at different depths between concrete surface and the reinforcement will allow monitoring the movement of the depassivation front into the structure. The time until the initiation of reinforcement corrosion is obtained by extrapolation.
- During the deterioration phase, corrosion current measurements will provide qualitative information on changes in corrosion activity, e.g., after repair measures.

The principle of corrosion current measurements is the measurement of the current between the designated anode and the cathode in a macro-cell corrosion element. This comparatively simple measurement may cause some serious problems for setting up the monitoring system.

- The setup has to be designed in such a manner that the designated anodes and cathodes will actually operate as anodes and cathodes. At the same time, they have to represent the actual corrosion system under observation (i.e., the reinforcing steel in concrete) with sufficient accuracy. Therefore, the materials for both anode and the cathode should have corrosion properties and surface conditions similar to the reinforcing steel. The anodes should be made of reinforcing steel without any modifications of the surfaces. For the cathodes, the use of stainless steel or activated Titanium Oxide (Ti/MMO) may be considered to assure that they will act as cathodes in the corrosion element. It should be kept in mind that this will affect the measured current. In many monitoring applications, the use of the actual reinforcement cage as a cathode was found to be very effective. If the same material is used for anodes and cathodes, the designated anodes should be placed in regions with either higher chloride contents or closer to the concrete surface than the cathodes to make sure that corrosion initiation will first occur on the anodes.
- The anodes should neither be too small nor too large. Depending on the actual setup, anode sizes between approximately 1 cm^2 and 50 cm^2 are recommended.
- Ideally, the anodes are installed during the construction phase. Since this is often not possible, embedding mortars with properties similar to the actual concrete should be used for subsequent installation. In fact, the use of embedding mortars should generally be kept to a minimum because they are normally highly alkaline and free of chlorides which will change the corrosion system dramatically. In some cases, it may be advantageous to use corroding sections of the actual reinforcement as anodes which are taken from the reinforcement cage by cutting or drilling (DGZfP [15]).

Corrosion current measurements can be carried out both continuously ("Stationary current measurement") and discontinuously ("instationary current measurement"). Stationary current measurements, i.e., with a continuous short-circuit between anode and cathode, represent the actual conditions on-site and are, thus, generally recommended for continuously monitoring corrosion activity after corrosion initiation. However, for some applications, e.g., monitoring of the depassivation front movement during the initiation phase, the current measurement does not focus on monitoring time-dependent changes in corrosion activity, but is used to determine whether corrosion initiation has occurred or not. In this case, instationary current measurements may be beneficial because they also enable the measurement of other parameters like the free corrosion potential of the anode shortly before the short-circuit is established and the current measurements are carried out.

To represent the current between the anodes and cathodes that are short-circuited in the RC system, amperemeters that impose near zero voltage, such as the zero-resistance amperemeters, should be used.

16.4.5 Linear polarization resistance (LPR) measurements

The linear polarization resistance (LPR) can be interpreted as the inverse of the corrosion rate at the time of measurement. However, as chloride-induced reinforcement corrosion is a very local corrosion process and the size of the polarized working electrode and distribution of anodic and cathodic regions on this surface is generally not known, the determination of actual corrosion rates from LPR measurements is not recommended. Despite this limitation, LPR measurements can often be very helpful to monitor time-dependent changes in corrosion activity or to identify actively corroding reinforcement surfaces.

For the LPR measurement, the working electrode is polarized starting from its free corrosion potential. Both cathodic and anodic polarization can be selected. In both cases, the actual polarization should be kept very small to avoid significant changes in the corrosion state. As the polarization curve close to the free corrosion potential can be assumed to be linear, the linear polarization resistance can be determined as the quotient of electrode potential shift and resulting current flow. For the actual application of LPR measurements, please refer to [16–18].

For LPR measurements, a three-electrode setup consisting of a working electrode, a counter electrode, and a reference electrode is required. The working electrode is the electrode under consideration, i.e., the reinforcement itself or auxiliary anodes made from reinforcing steel and placed inside the concrete. As counter electrode, either Ti/MMO bars or strips can be used. In case auxiliary anodes are selected as working electrodes, the actual reinforcement cage can also be employed as a counter electrode. The reference electrode is commonly a MnO_2 or, in some cases, an Ag/AgCl reference electrode, cf. Section 16.4.2. A galvanostat or a potentiostat is required for the polarization of the working electrode. It is recommended to carry out LPR measurements in a potentiostatic or a potentiodynamic way. However, galvanostatic and galvanodynamic measurements are possible, too.

The linear polarization resistance strongly depends on the geometry of the corrosion system, i.e., the size of the working electrode, the distance between working electrode and reference electrode, and the position of the reference electrode. In particular, if measurements are carried out using the actual reinforcement as the working electrode, the size of the polarized surface is unknown. In the case of larger working electrodes, the heterogeneity of the surface containing both anodic and cathodic regions will have a major impact on the results. For larger distances between working electrode and reference electrode, the electrolytic resistance of the concrete will also affect the results. However, this effect is more pronounced for actively corroding systems.

For passive systems, it can generally be neglected. Finally, the measurement technique itself and especially the polarization rate will affect the LPR measurements.

Measurements on corroding systems—both corrosion current measurements and linear polarization resistance measurements—allow qualitative or semiquantitative determination on time-dependent changes in corrosion activity. However, conclusions on the actual cross-sectional loss of the rebar in corrosion pits are not possible with both methods or only with large uncertainties.

16.5 Case studies

16.5.1 Case study 1—Corrosion monitoring to avoid chloride ingress

This case study demonstrates a simple yet efficient approach to monitor the functionality of coatings and surface treatments applied as a barrier against chloride ingress into structural concrete.

The condition assessment of a number of highway tunnels in South Germany revealed a very high chloride contamination of the tunnel walls after a relatively short time which was caused by an extensive use of deicing salts during the winter months. As a consequence, comprehensive repair measures had to be carried out in some of the tunnels, leading to significant costs and restriction of use.

To improve the durability of a newly built highway tunnel in 2006, it was decided to apply a hydrophobic treatment to the concrete surfaces close to the tunnel portals, and thus, inhibit the ingress of chlorides. Since experience on the long-time performance of hydrophobic treatment systems is limited, it was agreed to install a monitoring system which would indicate a loss of functionality so that a new hydrophobic treatment could be applied without the need for further repair measures [19]. Since chloride ingress into concrete is always connected with the penetration of water into the concrete pore system, the concept of the monitoring system was to record changes in electrolytic resistivity with time as an indicator for the moisture content of the concrete. A loss of functionality would lead to a significant drop of the electrolytic resistivity of the concrete close to the exposed surface.

The "multiring-electrode" sensor was employed (Sensortec GmbH, Munich/Germany—www.sensortec.de, Fig. 16.5) to monitor the electrolytic resistivity of the concrete. The sensor consists of eight stainless steel rings which are separated by isolating plastic rings in such a way that a distance of 5 mm between two adjacent rings is obtained. The electrolytic resistance between two adjacent rings is measured, thus allowing the user to generate concrete resistance profiles over the sensor depth. Specimens with multiring-electrodes were produced. One half of the specimens were subjected to a hydrophobic treatment and the other half left untreated as a reference. The specimens were then installed close to the tunnel portals under tunnel exposure conditions.

Fig. 16.6 shows the time-dependent development of the electrolytic resistance over the sensor depths for a specimen with hydrophobic treatment (left) and for a reference

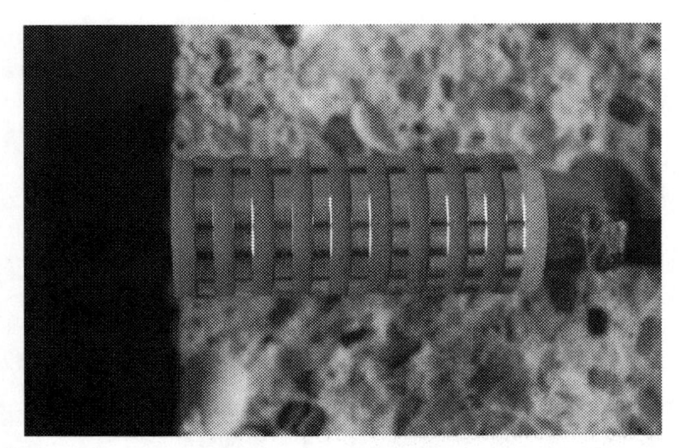

Fig. 16.5 Multiring-Electrode for resistance and moisture monitoring (Sensortec GmbH, Munich/Germany).

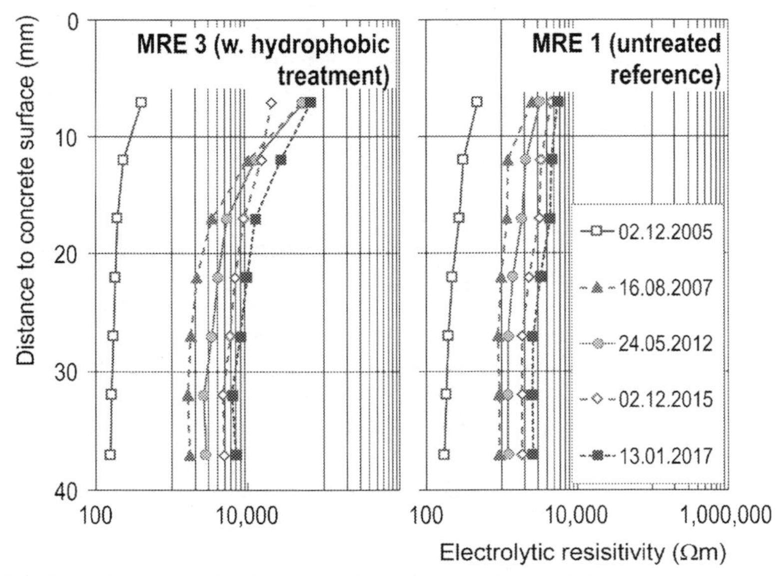

Fig. 16.6 Time-dependent development of the electrolytic resistivity for concrete specimens with (left) and without (right) hydrophobic treatment. Specimens stored under tunnel exposure conditions.

specimen without hydrophobic treatment (right). After twelve years of exposure, the specimens with hydrophobic treatment possess resistivities close to the concrete surface, which are more than an order of magnitude higher than the reference specimens without treatment. Thus, the functionality of the hydrophobic treatment after twelve years of exposure is not impaired [20]. In case future measurements show a distinct decrease in resistivity close to the surface, the hydrophobic treatment will be reapplied in order to avoid future chloride ingress.

16.5.2 Case study 2—Corrosion monitoring during the initiation phase

The parking garage in this case study was completed in 2005 and comprises four parking decks with a total parking deck surface of 270,000 m^2. The parking decks were constructed as prestressed, continuous slabs. Due to the structural system, load-induced cracking at the top surface of the parking decks is expected to occur only in the areas around the columns. Consequently, the application of a crack-bridging coating system to inhibit the ingress of chlorides from deicing salt was limited to these areas. Regarding the adjoining surface areas, it can be assumed that the top surface will be under compression for all relevant load combinations. Thus, a coating system was not applied to these surface areas, but a durability concept was developed which comprises reliability calculations to model the ingress of chlorides from deicing salt applications into the uncracked concrete [21].

For verification purposes (real ingress/predicted ingress), a corrosion monitoring system was installed in the parking decks to monitor the time-dependent ingress of chloride. This system consisted of 25 corrosion sensors ("Anode Ladder", Sensortec GmbH/Germany, Fig. 16.7) which were placed at different locations in the parking area and the traffic lane where the highest chloride loads were expected to occur. The anode ladder consists of six single anodes A1 to A6 placed at different depths inside the concrete cover, i.e., between the concrete surface and the reinforcement (A1 closest to the concrete surface, A6 at rebar level, Fig. 16.7). As chlorides are transported into the concrete, the anodes will corrode once the critical chloride content at the anode level is exceeded, with the anodes closest to the surface corroding first. Since the corrosion initiation of these anodes is monitored continuously, the ingress depths of the critical chloride content can be determined and compared to the actual concrete cover. Extrapolation of the results enables the determination of the remaining

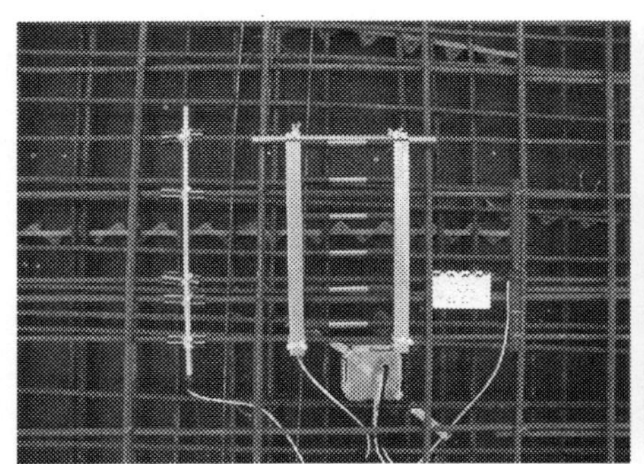

Fig. 16.7 Anode Ladder sensor for corrosion monitoring (Sensortec GmbH, Munich/Germany).

service life until depassivation of the reinforcement. The basic principle of the anode ladder corrosion sensor is illustrated in Fig. 16.8.

The sensors were fixed to the top reinforcement layer before concrete casting and their inclination was adjusted so that the top anode A1 had a concrete cover of approximately 15 mm. Once the parking deck was completed, the actual concrete cover of the top anode A1 was determined nondestructively.

The transport mechanisms leading to chloride ingress into uncracked concrete are comparably slow. Thus, it was decided to carry out sensor readings only twice a year. The measurement scheme consisted of the following sensor readings:

- Free corrosion potential of anodes A1 to A6 vs. the Ti/MMO bar ("cathode")
- Corrosion current between anodes A1 to A6 and the Ti/MMO bar 10 s after the short-circuit was established
- AC impedance between two neighboring anodes (A1 and A2, A2 and A3, etc.)

The corrosion initiation of a single anode can be detected by a distinct decrease in free corrosion potential together with a significant increase in corrosion current when compared to earlier readings. Fig. 16.9 shows the results for one corrosion sensor for which corrosion initiation of the anode A1 closest to the concrete surface was detected in 2010. All other anodes showed no signs of corrosion initiation at that time. In 2015, anode A2 also showed a distinct decrease in the free corrosion potential along with an increase of the corrosion current, indicating corrosion initiation, while all other anodes remained passive. For comparison, Fig. 16.10 shows the results for corrosion sensors for which all six anodes A1 to A6 remained passive throughout the whole time of monitoring.

The evaluation of monitoring data for all 25 corrosion sensors showed that the actual chloride ingress into the uncracked concrete was slower than assumed at the design stage. The sensor data were used to update the original reliability calculations and lead to a relevant increase of system reliability with respect to chloride-induced reinforcement corrosion. Utilizing this alternative approach (model-based prediction of chloride ingress which is permanently monitored and verified instead of a full coating) made it possible to realize significant savings for the owner [22].

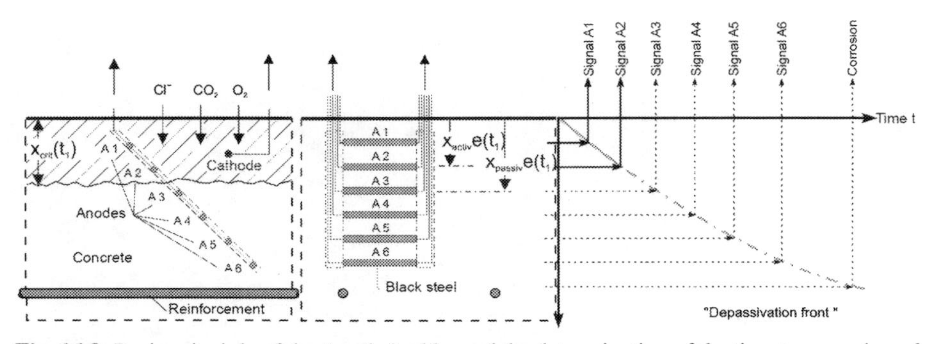

Fig. 16.8 Basic principle of the Anode Ladder and the determination of the time to corrosion of the reinforcement.

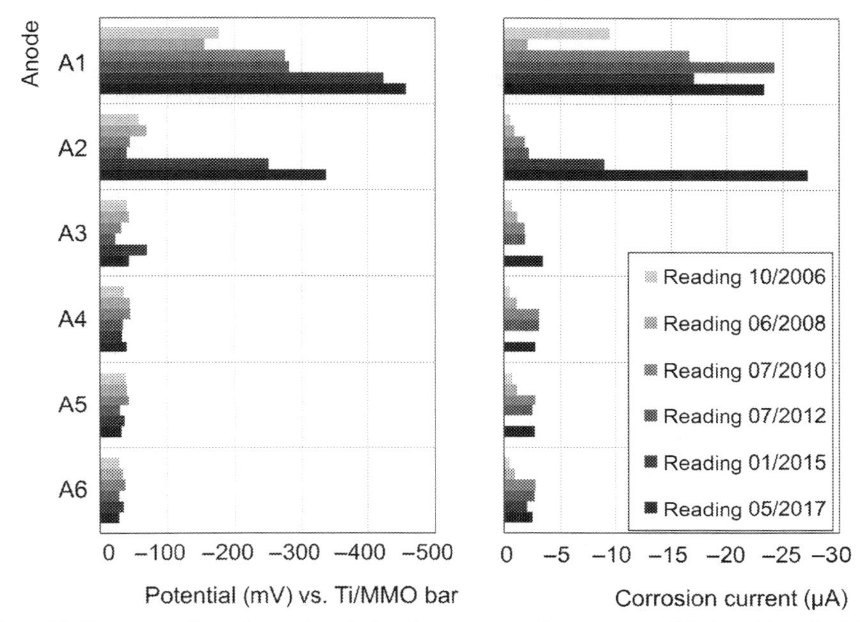

Fig. 16.9 Sensor readings for an Anode Ladder sensor with corrosion initiation of anode A1 in 2010 and of anode A2 in 2015.

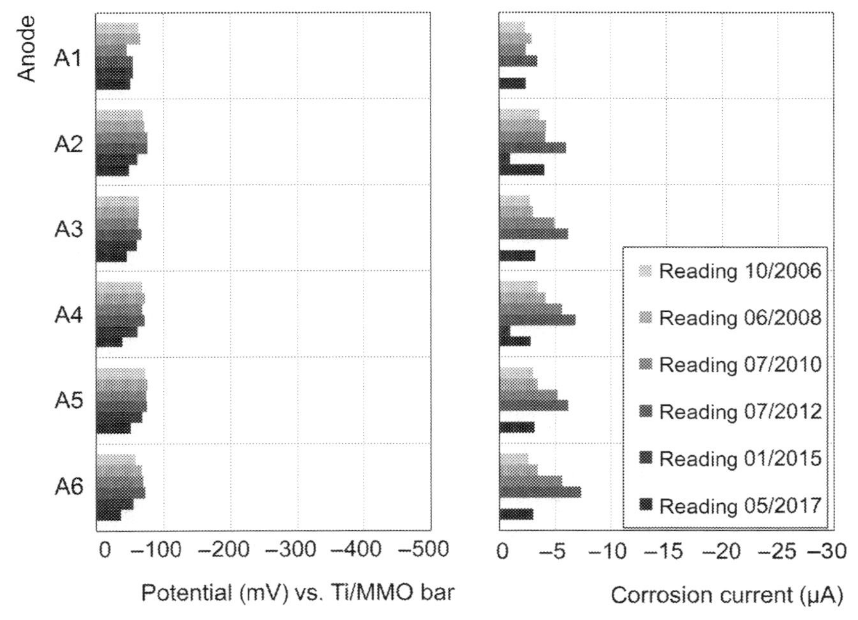

Fig. 16.10 Sensor readings for an Anode Ladder sensor with no corrosion initiation.

16.5.3 Case study 3—Corrosion monitoring during the deterioration phase

The underground car park in this case study was completed in 1998. It has one ground floor with a total area of about 4000 m^2. The floor slab has a varying thickness between 25 and 40 cm and is not needed for load transfer. The upper surface of the floor slab is about 70 cm below ground water level, and therefore, the floor slab forms part of the watertight concrete structure. In order to reduce the ingress of chlorides from deicing salts into the structure, the floor slab was provided with a fairly simple coating system.

A structural assessment approximately 15 years after completion revealed massive crack formation in the floor slab with a total crack length of about 3000 m. The chloride concentration in the cracks at reinforcement level was between 0.50 and 0.90 wt% (with respect to cement) with maximum values at around 2.0 wt%. The uncracked concrete displayed very high local chloride concentrations of up to 3.0 wt% at the surface, but generally possessed uncritical chloride contents <0.5 wt% at reinforcement level. The maximum loss of cross-section due to corrosion was approximately 10% which was considered still to be acceptable from a structural point of view. However, due to the high chloride contents close to the concrete surface in the uncracked concrete, a conventional repair measure would not only include concrete replacement over the whole cross-section in the cracked regions, but also concrete replacement up to a depth of approximately 80 mm in almost 60% of the uncracked regions. It is obvious that this repair strategy would not only cause immense costs, but would also lead to serious use restrictions and noise disturbances. Since the reinforcement in the floor slab is only of minor relevance to load-bearing, an alternative repair approach was developed with the owner which would not require concrete replacement, but the application of a new coating system and crack-bridging bandages in combination with an extensive corrosion monitoring system for changes of corrosion activity. It was agreed with the client that if the corrosion monitoring showed further degradation, a conventional repair procedure would still have to be carried out.

In order to enable a representative and comprehensive assessment of the time-dependent changes in corrosion activity, a total of 40 monitoring positions were selected. Thirty monitoring positions were placed in cracks with high to very high chloride concentrations, 6 in the uncracked concrete with high chloride concentrations close to the surface as well as low concrete cover, and 4 for reference purposes in the uncracked concrete with no increased chloride contamination [23]. On account of the sealed surface of the concrete, half-cell potential mapping of the floor slab surface was not possible. The monitoring positions were selected based on the results of concrete cover mapping, visual signs of corrosion stain and puddles, chloride concentrations, etc.

The installation of the monitoring setup was carried out as described above (Fig. 16.11) and as follows:

- Location of rebar in the crack parallel to the crack and at right angles to the crack
- Isolation of the rebar in the crack ("anode") from the remaining reinforcement cage by means of core drilling

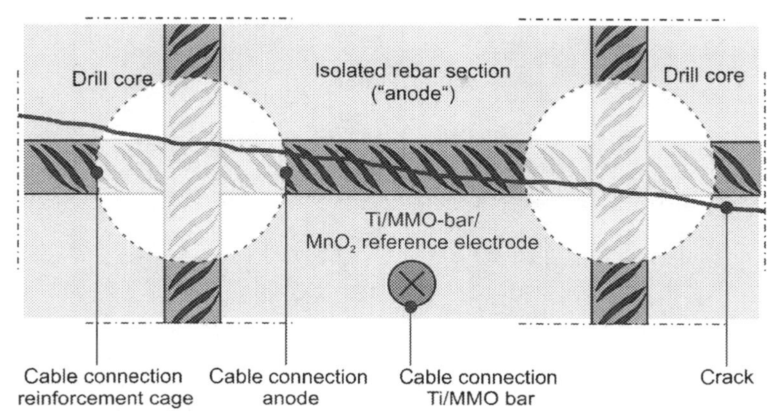

Fig. 16.11 Schematic diagram of corrosion-monitoring setup for the corrosion monitoring in cracked concrete.

- Extraction of drill cores and determination of crack formation over the cross-section, concrete cover, corrosion state of the rebar, and chloride content on the extracted drill cores (Figs. 16.12 and 16.13),
- Fixing of cable connections both to the anode and the reinforcement cage ("cathode"),
- Coating of bore hole walls with epoxy resin to avoid mortar entry into the crack and filling of the bore holes with an appropriate cement-based mortar,
- Installation of a reference electrode and a Ti/MMO bar in drill holes close to the anode.

Once the installation was completed, a regular measurement procedure was defined which comprised measurements of the macro-element current between anode and reinforcement cage and the half-cell potential of the macro-element. Then the anode and reinforcement cage were separated and the depolarization of the anode monitored.

Fig. 16.12 Concrete drill core after extraction. The crack formation over the cross-section and concrete cover can be determined.

Fig. 16.13 Assessment of the corrosion state of rebars inside the drill core.

After a depolarization for two hours, the free corrosion potential of the anode was determined. Once this measurement routine was completed, the anode and reinforcement cage were again short-circuited in order to reproduce realistic corrosion conditions.

Figs. 16.14 and 16.15 show some results of measurements conducted on two sensors that exhibited active corrosion before coating, but the corrosion activity reduced

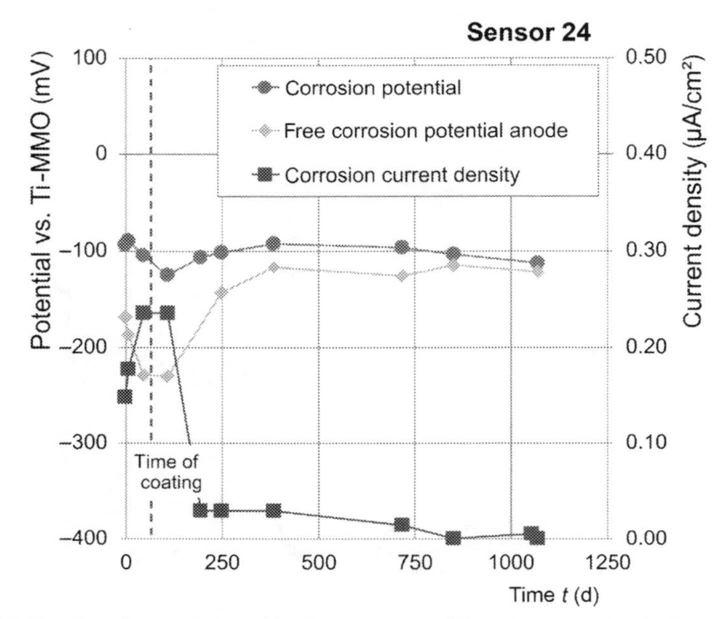

Fig. 16.14 Results of corrosion monitoring on sensor 24 (active corrosion before coating).

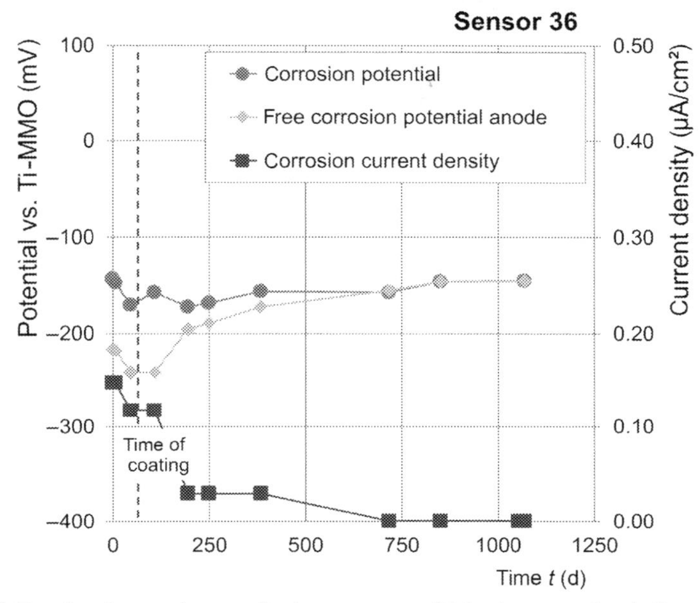

Fig. 16.15 Results of corrosion monitoring on sensor 36 (active corrosion before coating).

significantly within a few months after coating. The reduction in macro-element current is accompanied by a distinct shift of the free corrosion potential of the anode towards more noble values. At the time of publishing, the potential difference between the anode and the reinforcement cage was in the range of only 20 to 30 mV.

Two years after the coating was applied, the macro-element currents for 22 of the 27 corrosion sensors that originally showed active corrosion had reduced to a degree that can be assessed as structurally irrelevant. For most of these sensors, the remaining current was in the same range as that of the passive "reference" sensors. Only five sensors still showed increased element currents, but for three of these sensors the corrosion currents had decreased quite significantly. Until now, an increase in corrosion activity, e.g., due to a redistribution of chlorides, was not apparent for any of the sensors. Based on the current results of corrosion monitoring, no further repair measures will be necessary. Therefore, in this particular case, the combination of coating and corrosion monitoring enabled significant savings of about 80% with respect to both repair costs and downtimes.

16.6 Conclusions

The foregoing sections have addressed the problem of durability in the case of reinforcement corrosion. In order to support the service life management of reinforced and/or prestressed concrete structures, it is crucial to have detailed information about the current condition level of the structure and a reliable assessment of the condition development in the future.

The necessary knowledge can be obtained from extensive sampling or the application of additional inspection techniques, but also from installed corrosion monitoring devices. The principles of some of the most important techniques for corrosion monitoring have been introduced here. Three case studies demonstrate the application of corrosion monitoring for real structures and the vast potential that corrosion monitoring offers for structures at almost any stage during their service life.

References

[1] P. Schießl, T.F. Mayer, Lebensdauermanagement, Heft 572 der Schriftenreihe des Deutschen Ausschusses für StahlbetonBeuth-Verlag, Berlin, 2007.

[2] G. Schmitt, M. Schütze, G.F. Hays, W. Burns, E. Han, A. Pourbaix, G. Jacobson, Global Needs for Knowledge Dissemination, Research, and Development in Materials Deterioration and Corrosion Control, The World Corrosion Organization, New York, 2009.

[3] K. Tuutti, Corrosion of Steel in Concrete, Research Report Nr. Fo 4-82 Swedish Cement and Concrete Research Institute, Stockholm, Sweden, 1982.

[4] fib, Structural Concrete, the Textbook on Behaviour, Design and Performance, vol. 3, International Federation for Structural Concrete (fib), Lausanne, 1999.

[5] L. Li, A. Sagüés, Metallurgical Effects on Chloride Ion Corrosion Threshold of Steel in Concrete, Report WPI 0510806 University of South Florida, Tampa, 2001.

[6] Brite/EuRam, Smart Structures (Integrated Monitoring System for Durability Assessment of Concrete Structures), in: Final Technical Report and Report of Task 2.1: Determination of Key Parameters, Brussels, 2002.

[7] U. Angst, B. Elsener, C.K. Larsen, Ø. Vennesland, Critical chloride content in reinforced concrete—a review. Cem. Concr. Res. 39 (12) (2009) 1122–1138, https://doi.org/10.1016/j.cemconres.2009.08.006.

[8] U.M. Angst, M.R. Geiker, A. Michel, C. Gehlen, H. Wong, O.B. Isgor, et al., The steel–concrete interface. Mater. Struct. 50 (2) (2017), https://doi.org/10.1617/s11527-017-1010-1.

[9] Y. Cao, C. Gehlen, U. Angst, L. Wang, Z. Wang, Y. Yao, Critical chloride content in reinforced concrete—an updated review considering Chinese experience. Cem. Concr. Res. 117 (2019) 58–68, https://doi.org/10.1016/j.cemconres.2018.11.020.

[10] U. Nürnberger, Korrosion und Korrosionsschutz im Bauwesen, Bauverlag GmbH, Wiesbaden, 1995.

[11] P. Schießl, Zur Frage der zulässigen Rissbreite und der erforderlichen Betondeckung im Stahlbetonbau unter besonderer Berücksichtigung der Karbonatisierung des Betons, Heft 255 der Schriftenreihe des Deutschen Ausschusses für StahlbetonBeuth-Verlag, Berlin, 1976.

[12] P. Schießl, M. Raupach, Laboratory studies and calculations on the influence of crack width on chloride-induced corrosion of steel in concrete, ACI Mater. J. 94 (1) (1997) 56–62.

[13] C. Gehlen, S. Matthews, D. Straub, S. Kessler, T. Mayer, S. Greve-Dierfeld, Condition control and assessment of reinforced concrete structures exposed to corrosive environments (carbonation/chlorides): state-of-the-art report, in: International Federation for Structural Concrete (fib), 2011.

[14] R. Myrdal, The Electrochemistry and Characteristics of Embeddable Reference Electrodes for Concrete, Woodhead Publishing Ltd., Cambridge, 2006.

[15] DGZfP, Merkblatt B12—Korrosionsmonitoring bei Stahl- und Spannbetonbauwerken, Deutsche Gesellschaft für Zerstörungsfreie Prüfung, Berlin, 2018.
[16] E. Heitz, W. Schwenk, Theoretische Grundlagen der Ermittlung von Korrosionsstromdichten aus Polarisationswiderständen, Mater. Corros. 27 (1976) 241–245.
[17] M. Stern, A.L. Geary, Electrochemical polarization I: a theoretical analysis of the shape of polarization curves, J. Electrochem. Soc. 104 (1) (1957) 241–245.
[18] C. Wagner, W. Traud, Über die Deutung von Korrosionsvorgängen durch Überlagerung von Elektro-chemischen Teilvorgängen und über die Potentialbildung an Mischelektroden, Zeitsch. Elektrochem. 44 (1938) 391–454.
[19] C. Sodeikat, Feuchtesensoren in der Bauwerksüberwachung. Beton- und Stahlbetonbau 105 (12) (2010) 770–777, https://doi.org/10.1002/best.201000058.
[20] T.F. Mayer, J. Harnisch, G. Ebell, C. Dauberschmidt, Korrosionsmonitoring von Stahlbetonbauwerken. Beton- Und Stahlbetonbau 113 (9) (2018) 632–639, https://doi.org/10.1002/best.201800026.
[21] C. Gehlen, G. Kapteina, T. Mayer, Life cycle management demonstrated on the example of a parking garage, Concr. Australia 40 (4) (2014) 42–49.
[22] C. Sodeikat, C. Dauberschmidt, P. Schießl, C. Gehlen, G. Kapteina, Korrosionsmonitoring von Stahlbetonbauwerken für Public Private Partnership Projekte: Dauerhaftigkeit sichtbar gemacht. Beton- Und Stahlbetonbau 101 (12) (2006) 932–942, https://doi.org/10.1002/best.200600517.
[23] F. Hiemer, F. Wich, C. Gehlen, S. Kessler, T. Mayer, Monitoring von Bewehrungskorrosion in Rissbereichen von Stahlbetonbauwerken, in: Messtechnik Im Bauwesen, pp. 35–43, 2018.

Corrosion monitoring in soil

17

Naeem Khan
NK Consulting, LLC, Fort Collins, CO, United States

17.1 Introduction

Buried metal-loss coupons were used in the past to assess the effectiveness of corrosion control systems for buried structures, in addition to cathodic protection (CP) monitoring. More recently, instrumented probes have been adapted for use in soils. When used in soil side applications, these probes are commonly referred to as "soil corrosion probes" (SCP) or coupon probes.

SCP are connected to the structure negative circuit so that they receive the same level of CP as the structure. Thus, even if the measured potential reading on the structure is not at the minimum required level, the corrosion growth rate measurement (millimeters per year, or mils per year) can be used as a method to assess CP effectiveness. Situations where SCP can be useful include low levels of cathodic polarization in high resistivity soils or erroneous potential readings due to poor soil contact.

17.2 Types of soil corrosion probes

The buried structures corrosion control industry has moved from using buried corrosion coupons for weight loss measurements to modified electrical resistance (ER) probes [1, 2] for use in soil side applications. Additionally, combination CP coupon/corrosion rate monitoring probes have been developed to simultaneously measure corrosion growth rates and CP potentials in soil environments [3–7].

17.3 Electrical resistance probes

Chapter 11 has described the general aspects of electrical resistance probes. This chapter will focus on the application of ER probes in soil environments. As discussed in Chapter 11, ER probes are based on the principle of increase in resistance of a metal strip or wire as a function of metal loss due to reduction in the cross-sectional area of the metal. They indicate the cumulative metal loss of the sample. The measurements taken using an ER probe are evaluated over time to calculate a corrosion growth rate in millimeters or mils per year.

In applying this technique to soil environments, a probe element of known cross-section is exposed to the soil. The electrical resistance of the probe is measured at initial installation and at subsequent time intervals. As metal is lost from the exposed surface of the probe element due to corrosion, the measured electrical resistance will

Techniques for Corrosion Monitoring. https://doi.org/10.1016/B978-0-08-103003-5.00017-5

increase, allowing the amount of metal loss (or corrosion growth rate) to be quantified with an ER measuring instrument.

ER soil corrosion probes are available in various sizes, shapes, and sensitivities. As with CP potential monitoring coupons, the size of the probe element should simulate the size of the possible coating defect sizes of the pipelines or structures under investigation. As the probe is located in the same soil environment as the protected structure, it is assumed that it replicates the effectiveness of CP applied to the structure. The probe element material should match that of the pipeline or structure being monitored, e.g., carbon steel or ductile iron [8].

The sensitivity of an ER probe is a compromise between the response time and the lifetime of the probe. For a probe with 0.635 mm (25 mils) useful life and assuming a sensitivity of 1% of the full scale, the soil corrosion probe system can detect metal thickness losses of approximately 0.005 mm (0.2 mils). The life of a probe connected to the CP system or in corrosion-inhibited soil environments is usually long, because it is protected by the CP current. However, a probe installed unconnected to the CP system to measure unprotected corrosion rates will have a shorter life, depending on the corrosivity of the environment. If there is the presence of stray current, the life of the probe may also be reduced. The probe must be replaced when half of its thickness has been consumed.

17.3.1 Types of ER probes

ER probes fall into two types:

(a) Individual probe
(b) Probe installed as part of a test station

Three designs of individual ER probes are shown below in Figs. 17.1–17.3. The flat surface and cylindrical probes in Figs. 17.1 and 17.2 are typically used in buried pipeline applications or under storage tanks.

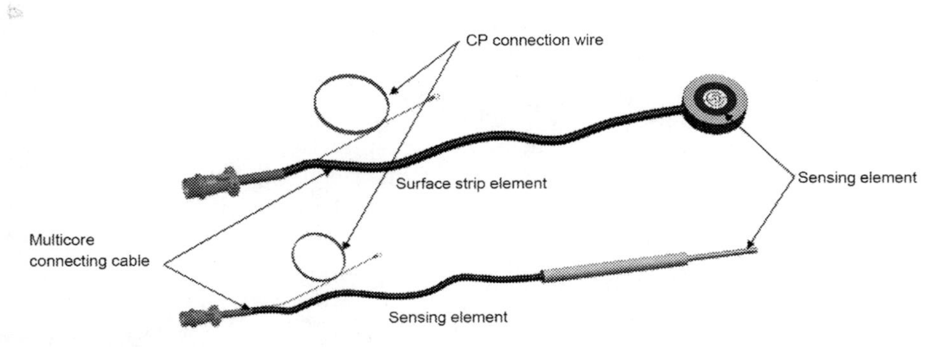

Fig. 17.1 Surface strip and cylindrical ER probes. Courtesy of Metal Samples, Inc.

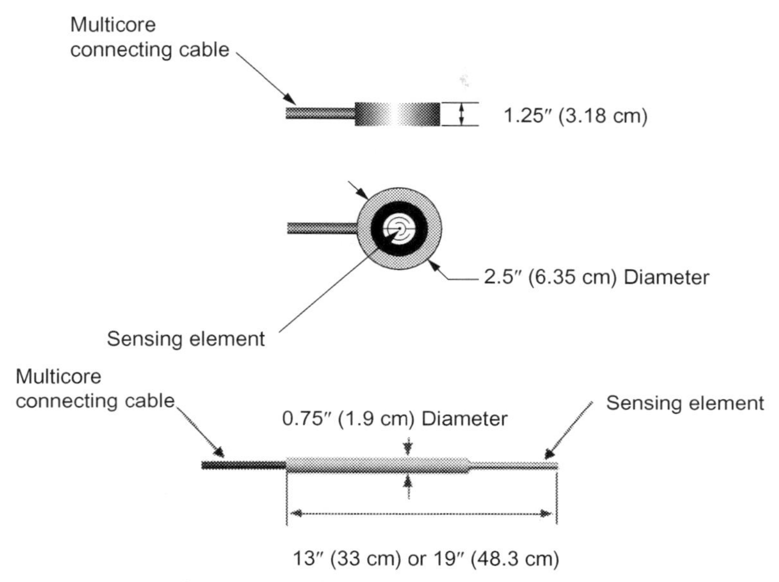

Fig. 17.2 (A) Surface strip ER probe dimensions. (B) Cylindrical ER probe dimensions.
Courtesy of Metal Samples, Inc.

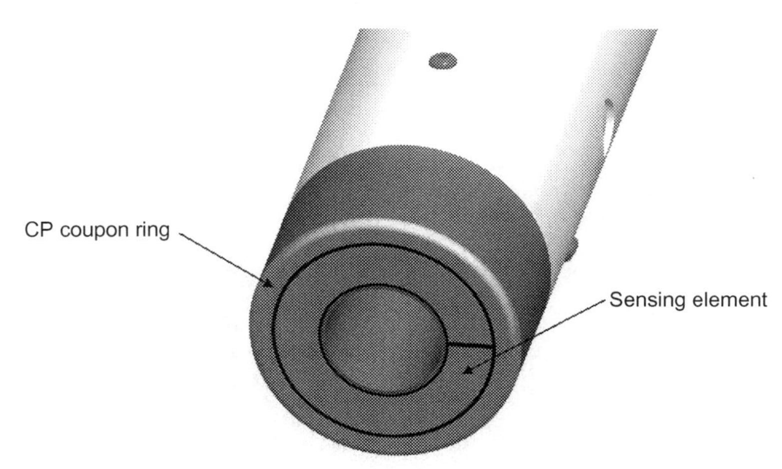

Fig. 17.3 Flush ring ER probe.
Courtesy of Metal Samples, Inc.

The ER probes shown in Figs. 17.1 and 17.2 are installed directly in the soil, or under pipe wrap on pipelines, with the connecting cable brought up to a test/monitoring station. ER probes installed as part of a test station are typically ring type probes, usually installed at the bottom of a combined probe/test station mounting tube, as shown in Fig. 17.3. In this configuration, the probe element has been combined with a coupon ring that can be used as a “native” potential reference.

A multicore connecting cable is terminated on a multipin connector in an above ground test station and is used to connect to the ER instrument. A separate CP connection wire is provided to connect the probe element to the CP circuit by connecting it to the protected structure, thus providing CP to the probe and allowing the effectiveness of the CP system to be monitored.

Both the probe element and the coupon (if used) are wired up through the mounting tube to the test station head (Fig. 17.4), which includes an on/off switch in the circuit connecting the probe element to the CP circuit and a multipin socket for connection of the instrument used to take a probe measurement. The "interrupt" switches allow “off” potentials of the probe and coupon to be measured.

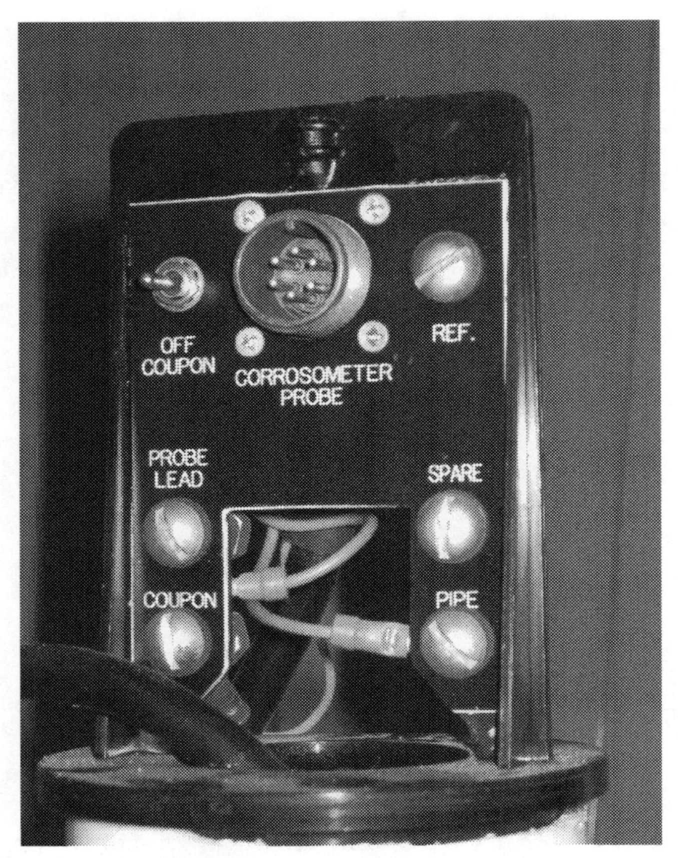

Fig. 17.4 ER probe test station head.
Courtesy of Naeem A. Khan.

The potentials of the probe element can be recorded using a reference cell installed inside the test station mounting tube or in a soil-access tube installed next to the test station. This latter arrangement is used in high-soil resistivity areas, where water-wetting of the reference electrode contact point is typically used to counter the high-soil contact resistance. Pouring water inside the test station mounting tube is not advisable, as water seeping down to the probe element will change the soil environment in the vicinity of the probe, leading to a change in the CP levels, and therefore, erroneous corrosion rate results.

The ER probe is connected to the CP source by connecting to the CP test station lead wire, rather than directly to the pipeline. This simplifies the installation, as a wire connection to the pipe is not typically made. An even more simplified installation routes the cable from a low-profile ER probe (as shown in Figs. 17.1 and 17.2) mounted adjacent to the pipe or under pipe wrap, directly to a terminal in the CP test station.

17.3.2 Typical applications

ER soil corrosion probes are used to determine general corrosion growth rates to determine the effectiveness of CP. ER probes can represent the corrosiveness of a bare (uncoated) pipeline in a particular environment. ER probes are also employed to model the corrosiveness of a coated pipeline at a typically sized defect. This information is typically used for design of CP systems or determination of appropriate CP criteria, for example, in aggressive environments such as high temperatures or areas susceptible to microbiologically influenced corrosion (MIC), in mixed-metal systems, or in areas susceptible to stray current, telluric currents, alternating current (AC) induction, and AC-induced corrosion.

ER probes can be used in areas where potential criteria are difficult or impractical to apply, such as pipeline locations that have proven impractical to meet established CP criteria. These include: pipelines with adjacent buried or submerged metallic structures; locations at which coatings cause electrical shielding; and pipelines with lack of electrical continuity, such as with some forms of mechanically coupled pipe that have not been made electrically continuous through the use of bonding cables or straps welded across each coupling. ER probes can also be installed at locations that are difficult to monitor for CP effectiveness using structure-to-electrolyte potentials, such as the bottom of large-diameter pipelines.

ER probe data can also be used to establish criteria for protection of ductile or cast iron pipe, in mixed-metal systems, systems with cyclical variations in level of protection, and areas with surface conditions limiting access to the electrolyte (such as cased crossings, backfill with significant rock content or rock ledges, gravel or dry vegetation). Additionally, data can also be used to establish criteria for pipelines buried very deep, areas of high contact resistance such as pipe located under concrete or asphalt pavement, frozen ground, or very dry conditions.

Other typical applications of ER probes include installations adjacent to underground storage tanks, installations under aboveground storage tanks, internal surfaces of water tanks, and reinforced concrete structures.

17.3.3 Selection of ER probe installation locations

The ER probe is typically installed close to the pipe or structure in the same soil/electrolyte or environment as the pipe or structure, with the entire surface of the probe element maintaining contact with the soil or environment. Loss of contact due to shrinkage, freezing, or drying out of the soil results in loss of the CP protection, as it would also on the pipe or structure.

The ER probe also needs to be installed at the same distance from the anode as the pipe or structure, so that the probe sees the same potential field as the structure. This is typically not a problem on a pipeline where the anode is some distance from the pipeline. It is much more critical in situations in which the anode is relatively close to the structure, such as in double-bottomed tanks with an anode grid system.

Generally, corrosion probes are installed on well-coated pipelines, with no history of corrosion in the area, but which have low or erratic potentials. They are not recommended for use on pipelines with coatings such as Polyethylene (PE) tapes or three-layer PE coatings, which do not allow CP current to flow to the pipe surface when disbonded from the pipeline (shielding). The installation of probes adjacent to pipelines with poor or disbonded coatings may result in erroneous corrosion rate interpretation, if the size of the probe is much smaller than the exposed surface area of the pipe due to coating defects, or if corrosion occurs under disbonded coating due to CP current shielding. Low-profile type probes (Figs. 17.1 and 17.2) can be used under pipeline coatings, to detect and quantify undercoating corrosion at known disbonded coating locations, or under poly wrap on ductile iron pipes [8].

17.4 Monitoring and data interpretation

Probes are permanently installed, and measurements are to be made on a periodic basis. As a guide, the probe is typically monitored at least once a month for the first 12 months. Thereafter, the monitoring frequency is reevaluated to determine whether more or less frequent readings are warranted. An ER instrument is used to measure the cumulative metal loss on the corrosion probe (see Fig. 17.5). For probe potential measurements, the probe "off" potential is measured by interrupting the current flow to the probe.

The graph in Fig. 17.6 is an example of the data derived from ER probe measurements. The corrosion rate between any two measurement times is the difference in metal-loss measurements divided by the time difference and annualized, i.e., the slope of the metal-loss curve between the selected times. For a series of measurements, the average corrosion rate is calculated using regression analysis from the slope of the trendline as calculated by the least squares method. Most graphing programs, such as an electronic spreadsheet, have built-in capability to calculate slopes.

The average corrosion rate (C) can also be calculated using Eq. (17.1):

$$C = \frac{P \times 365(S_2 - S_1)}{\Delta T \times PSD} \tag{17.1}$$

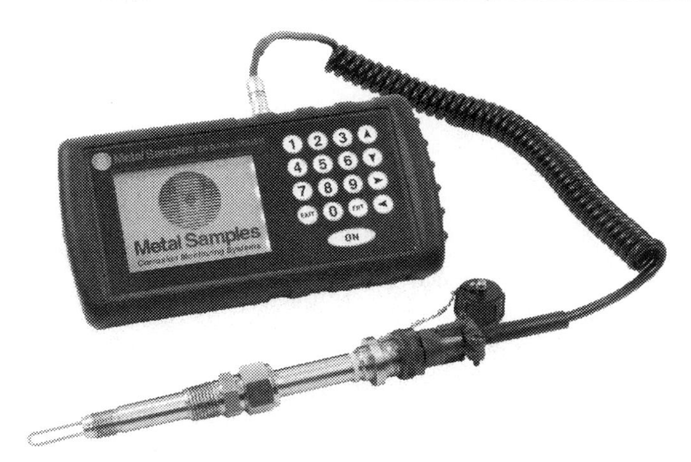

Fig. 17.5 Portable ER instrument.
Courtesy of Metal Samples, Inc.

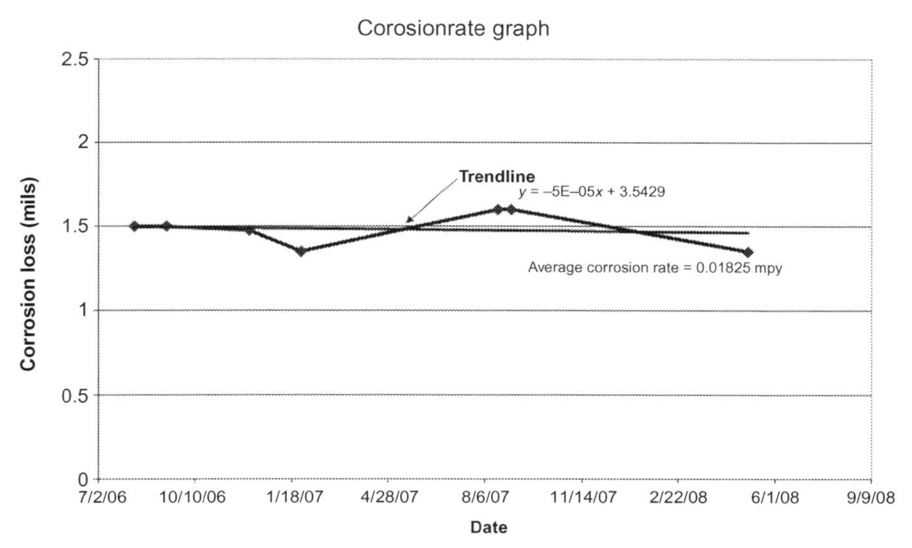

Fig. 17.6 Typical corrosion rate graph.
Courtesy of Naeem A. Khan.

where:

S_1 = the first reading

S_2 = the second (later) reading in divisions (1/1000 of probe span) from typical ER instrument readings

ΔT = the time in days between readings

P = the "probe element span" (probe constant) specified by the probe manufacturer

PSD = ER reading instrument probe span divisions, e.g., 1000th of probe span

If the probe element span is entered in millimeters, the corrosion rate will be in millimeters per year. If the probe span is entered in mils, the corrosion rate will be in mils per year.

17.5 Effectiveness criteria

An average corrosion rate of less than 0.025 mm/year (1.0 mil/year) over a 12-month monitoring period is generally accepted as an indication that CP is effective at the location of the probe. This criterion is often modified based on visual examination of the probe element after a burial time exceeds 12 months and also depending on the condition/integrity of the structure during this time period.

Newer revisions of ISO 15589-1 [9] and NACE International standards [10–12] include the corrosion growth rate criterion as an acceptable alternative to the typical CP potential criteria commonly used to determine CP system effectiveness.

Standards Association of Australia (SAA) AS 2832.19 [13] includes a protection criterion incorporating a corrosion rate not to exceed approximately 0.005 mm/year (0.2 mils/year) in combination with an instantaneous "off" potential of −850 or 100 mV more negative than the depolarized potential.

17.6 New developments in soil corrosion probe monitoring technology

Advances in soil corrosion monitoring standards and specifications have led to the development of newer types of monitoring probes which are able to provide multiple data types in addition to corrosion probe rates. Also, some newer probes utilize technologies other than ER technology to determine corrosion rates in soils. Remote monitoring of probes has also become the norm as part of these new developments.

Two of these newer technologies and their remote monitoring systems are described later.

17.6.1 Multifunction ER soil corrosion probes

One type of a soil side multifunction probe [3] is shown in Fig. 17.7. This probe utilizes 500 or 100 μm thick carbon steel strip as a probe element, with a coupon that simulates the coating defect area of 1 cm^2. The 500 μm thick can detect a corrosion rate of 100 μm/y in 12 h, while the 100 μm thick probe can detect a corrosion rate of 100 μm/y in 0.50 h.

As configured, this probe assembly measures corrosion growth rate (μm/y), coupon DC On and Off potentials, coupon DC current, coupon-induced AC potentials, and coupon AC current.

The remote monitoring system for this probe can be programmed to collect, store, and transmit data per user defined parameters. Measurements are recorded by the data logger on an hourly basis, and then reported to the vendor's website via the cellular

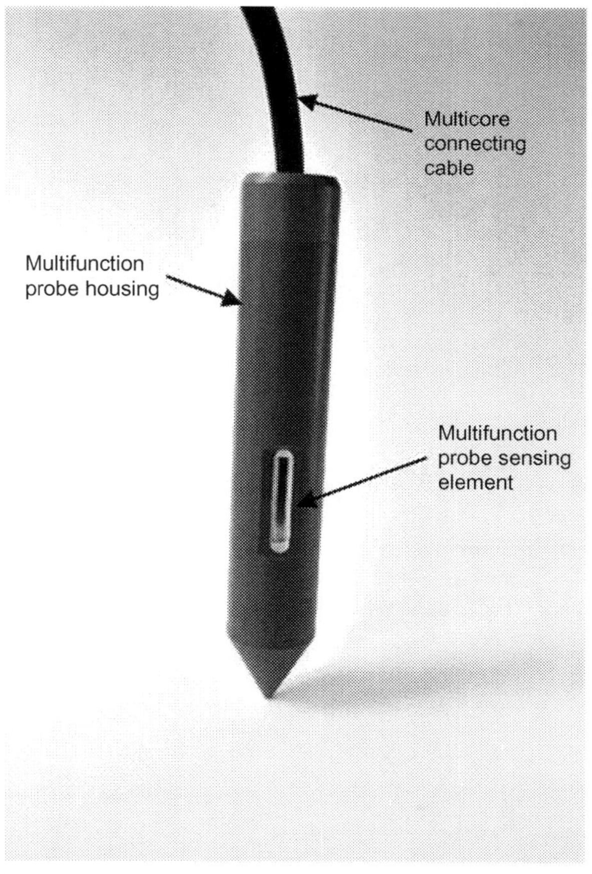

Fig. 17.7 Multifunction ER corrosion probe.
Courtesy of Metricorr.

Fig. 17.8 Multifunction ER corrosion probe remote monitoring terminal. Courtesy of Metricorr.

modem once every 24 h. The data logger has a storage capacity of 80,000 measurements and is equipped with a backup battery. Fig. 17.8 shows a typical corrosion probe remote monitoring terminal.

The vendor's web service server provides complete data charting information for the data types monitored by the probe, which is available from the web site at any time. Data charts are available for up to a 12-month time frame and can be customized for date ranges within the 12-month time frame.

17.6.2 Ultrasound soil corrosion monitoring probes

The use of ultrasound technology (see Chapter 12 for more information) to measure metal loss on coupons is another new development for corrosion monitoring in soils [4]. Both CP potential measurements and metal thickness measurements are acquired with these coupons.

This coupon metal-loss monitoring system utilizes ultrasound transducers, imbedded inside of metal coupons, to directly measure the amount of coupon metal loss. The thickness of the coupon can be measured at any time with a suitable ultrasound pulser/receiver instrument at the test station.

This system uses two independent metal coupons, each imbedded with 2–3 ultrasonic transducers that can be pulsed and interrogated by an operator through a conventional test station. Packaged in the shape of a "hockey puck" (Fig. 17.9), the transducer-coupon assemblies are waterproof and are installed in any configuration on or near the underground structure that is monitored. Ten (10) square centimeters of exposed coupon surface is used to monitor and track metal loss within a DC field, and a one (1) square centimeter of exposed coupon metal is used to monitor and track metal loss within an AC field. The coupon metal-loss data is acquired during typical manual test station measurements (Fig. 17.10), or the data can be transmitted wirelessly with a properly configured wireless communication system (Fig. 17.11).

17.6.3 Coupled multielectrode probes

Coupled multielectrode probes are also used to measure the corrosion rate in soil [6] and the corrosion rate under cathodic protection conditions [7]. Readers are encouraged to read Chapter 8 and Chapter 24 for more information.

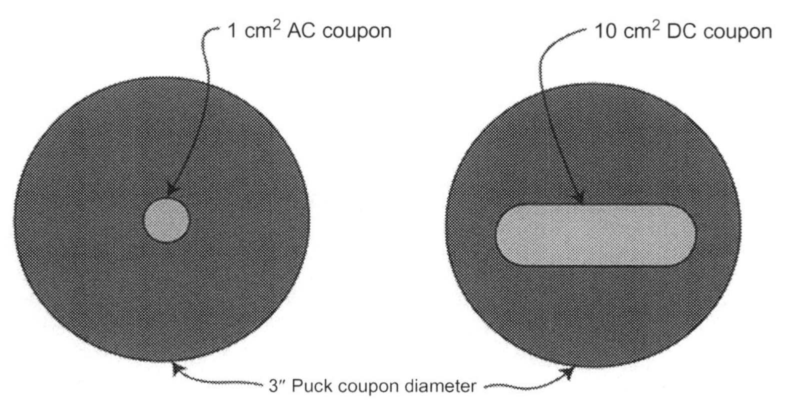

Fig. 17.9 Ultrasound corrosion monitoring probes.
Courtesy of BSI.

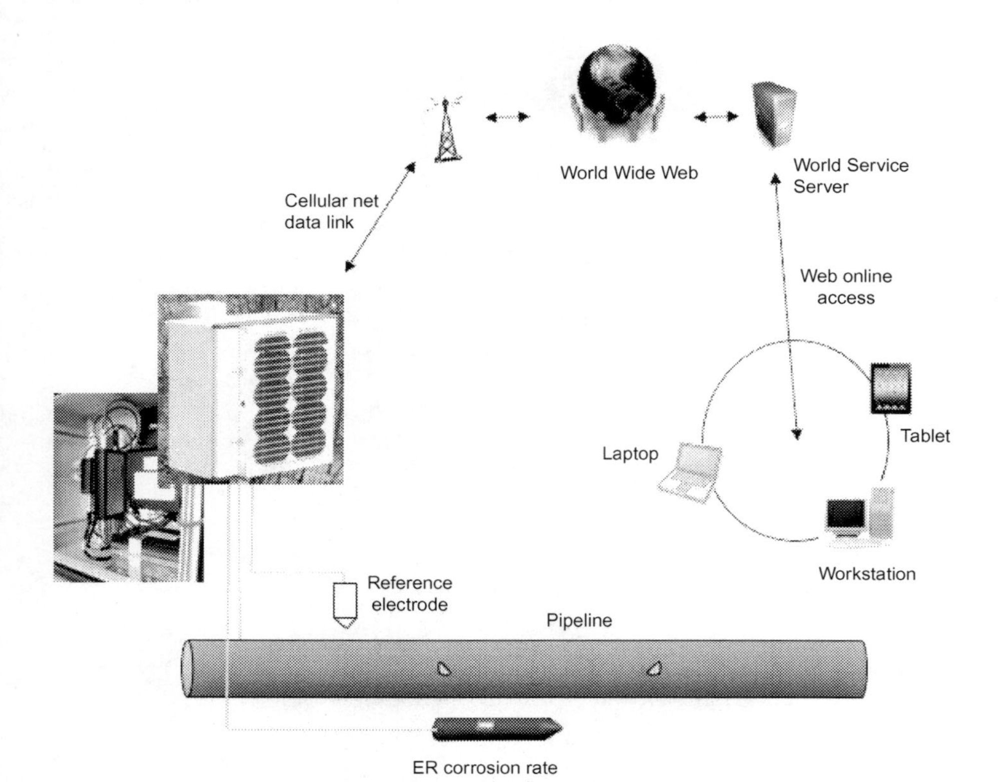

Fig. 17.10 Ultrasound corrosion monitoring probes test station. Courtesy of BSI.

Fig. 17.11 Ultrasound corrosion monitoring probes remote monitoring test station. Courtesy of BSI.

References

[1] J.H. Fitzgerald, P.R. Nichols, R. Niebling, Measuring the Effectiveness of Cathodic Protection on the Exterior Bottoms of New Aboveground Asphalt Storage Tanks Using Corrosion Monitoring Probes, NACE International, CORROSION/99, Paper No. 519NACE International, Houston, TX, 1999.

[2] N.A. Khan, Use of ER Soil Corrosion Probes to Determine the Effectiveness of Cathodic Protection, NACE International, CORROSION/2002, Paper No. 02104NACE International, Houston, TX, 2002.

[3] L.V. Nielsen, L. Galsgaard, Sensor Technology for On-Line Monitoring of AC-Induced Corrosion Along Pipelines, CORROSION/2005, Paper No. 05375NACE International, Houston, TX, 2005.

[4] R. Leary, The Use of Ultrasound for Monitoring the Impact of Induced AC Corrosion on Underground Structures, NACE International CORROSION/2017, Paper No. 9722NACE International, Houston, TX, 2017.
[5] P. Murray, A. Khan Naeem, Rapid Assessment of Station Piping CP Effectiveness in a High Resistivity Environment, CORROSION/2018, Paper No. 10909NACE International, Houston, TX, 2018.
[6] X. Sun, Real-Time Corrosion Monitoring in Soil With Coupled Multielectrode Sensors, CORROSION/2005, Paper No. 05381NACE International, Houston, TX, 2005.
[7] X. Sun, Online Monitoring of Corrosion under Cathodic Protection Conditions Utilizing Coupled Multielectrode Sensors, CORROSION/2004, Paper No. 04094NACE International, Houston, TX, 2004.
[8] G.E.C. Bell, C.G. Moore, S. Williams, Development and Application of Ductile Iron Pipe Electrical Resistance Probes for Monitoring Underground External Pipeline Corrosion, CORROSION/2007, Paper No. 07335NACE International, Houston, TX, 2007.
[9] ISO 15589-1 (Latest Revision), Petroleum and Natural Gas Industries—Cathodic Protection of Pipeline Transportation Systems, International Standard, Switzerland, 2003.
[10] NACE SP0104-2014, The Use of Coupons for Cathodic Protection Monitoring Applications, NACE International, Houston, TX, 2014.
[11] NACE SP0169-2013, Formerly RP0169 Control of External Corrosion on Underground or Submerged Metallic Piping Systems, NACE International, Houston, TX, 2013.
[12] ANSI-NACE SP0502-2008, Pipeline External Corrosion Direct Assessment Methodology, NACE International, Houston, TX, 2008.
[13] SAA AS (Latest Revision), Cathodic Protection of Metals: Pipes and Cables, South Wales Standards Australia Limited, Sydney, New Australia, 2004.

Corrosion monitoring in refineries

18

Kjell Wold
Emerson Automation Solutions, Trondheim, Norway

18.1 Introduction

Petroleum refining is important for the world economy. However, refineries are complex plants, with a variety of chemical reactions, temperatures, and materials—some of them combinations that cause a significant risk of corrosion in the process equipment. According to the Saudi Aramco Journal of Technology, 36% of all maintenance costs in refineries can be linked to corrosion repairs [1].

In addition to maintenance costs, corrosion is also a safety issue and brings risks to plant personnel as well as to the community where the plant is located. Several cases of refinery corrosion accidents are presented in the literature and on the web, discussing reasons for corrosion and its consequences. As an example, it is referred to in the reference list and animation produced by the CSB US Chemical Safety Board [2] describing an accident in a US refinery, not only discussing the accident and how it was handled, but also reasons for corrosion and how better integrity management could have reduced the probability of such an accident.

It is not the scope of this article to dive deep into corrosion-related accidents in the refining industry. However, a lot of interesting information is available. One source of information could be the JRC Scientific and Policy Report—Corrosion-Related Accidents in Petroleum Industries, published 2013 [3]

In addition, refineries are today continuing to see pressures on margins, making opportunity crudes—the discounted, lower price crude oils feedstock purchased on the spot market—a financially attractive option. Simple calculations show that millions of dollars can be saved every year (and tens thousands of dollars every day) by only adding minor amounts of discounted crudes to the feed blend [4].

However, increased use of opportunity crudes adds to corrosion and operational problems in a plant. Increasing the blending of opportunity crudes is, therefore, a balancing game, where the art is to find the best operating window between reduced costs and operational risk. Corrosion monitoring is more and more used as a tool for fine tuning the use of opportunity crudes.

The perspective on corrosion monitoring has traditionally been that it is a required cost to run a plant safely and with integrity. However, corrosion monitoring should, in addition to increasing safety, be considered a cost reduction tool for several reasons:

- Corrosion monitoring contributes to optimized opportunity crude blending
- Increased plant uptime and emergency plant closure due to unexpected corrosion damage can be avoided

Techniques for Corrosion Monitoring. https://doi.org/10.1016/B978-0-08-103003-5.00022-9

- Increased equipment life due to better corrosion mitigation, such as the tuning and verification of corrosion inhibitor programs
- Consumption of chemicals may also be reduced through reliable corrosion monitoring
- Corrosion monitoring may detect failure in other process equipment (like desalters) and help identify needs for equipment repair and better process tuning.

Corrosion monitoring should, therefore, be seen as an opportunity to create safer and more economical refinery operations.

18.2 Types of refinery corrosion

As discussed under Section 18.1, refineries are complex plants with a wide range of corrosion issues.

API RP 571 [5] lists common uniform and localized forms of corrosion taking place in refineries. These are:

- Amine corrosion
  - General or localized corrosion, often pitting.
- Ammonium chloride corrosion
  - General or localized corrosion, often pitting, normally occurring under ammonium chloride or amine salt deposits, often in the absence of a free water phase.
- Hydrochloric acid (HCl) corrosion
  - General or localized corrosion, often pitting.
- Naphthenic acid corrosion (NAC)
  - Primarily in crude and vacuum units and downstream units that process certain fractions or cuts that contain naphthenic acids. Naphthenic Acid Corrosion often starts as localized corrosion, then spreads out to wider attacks, such as "meza" attacks.
  - High temperature and velocities are common triggers for naphthenic acid corrosion (Fig. 18.1).
- High temp H_2/H_2S corrosion

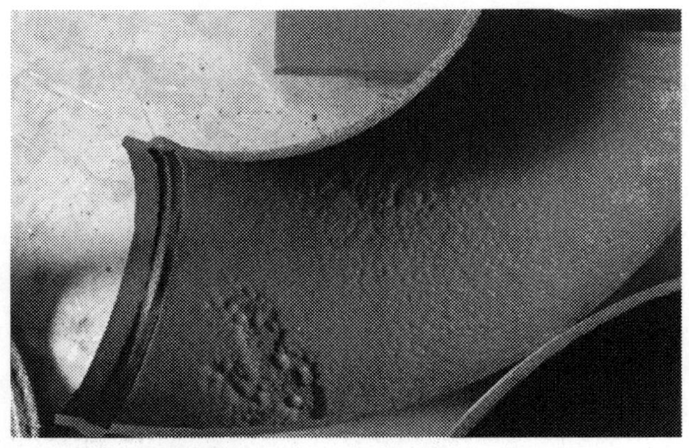

Fig. 18.1 Naphthenic acid corrosion in bend [6].
Courtesy of Emerson Automation Solutions.

- This form of sulfidation results in a uniform loss in thickness associated with hot circuits in hydro-processing units.
- Sour water corrosion (acidic)
  - Corrosion of steel due to acidic sour water containing H_2S, pH between 4.5 and 7.0. Carbon dioxide (CO_2) may also be present. Usually general thinning, but may be highly localized to specific areas of high velocity or turbulence, typically where a water phase is condensing.
- Ammonium bisulfide corrosion (alkaline sour water)
  - General thinning of carbon steel, extremely high localized rates of wall loss at changes in direction or turbulent flow areas above 2 wt% concentration.
  - With low velocities extremely localized underdeposit corrosion, if insufficient water is available to dissolve NH_4HS-precipitated salts.
- Hydrofluoric (HF) acid corrosion
  - High rates of general or localized corrosion. May be accompanied by hydrogen cracking, blistering, and/or HIC/SOHIC.
- Sulfuric acid corrosion
  - Sulfuric acid promotes general and localized corrosion of carbon steel and other alloys.

The mechanisms and reasons for the different forms of corrosion vary, as do the monitoring approaches suggested for tracking corrosion as well as possible strategies for mitigation.

18.3 Corrosion monitoring technologies available

A range of corrosion and inspection methods are available. In this chapter, only monitoring technologies, i.e., methods for continuous monitoring of metal loss or corrosivity at selected monitoring locations, will be briefly discussed.

Generally, corrosion monitoring technologies can be grouped as follows:

	Metal loss-based methods	**Electrochemical methods**
In-line	• Weight loss coupons • Electrical resistance probes	• Linear polarization resistance (LPR) • AC impedance • Electrochemical noise (EN)
Nonintrusive	• Electric field signature method (FSM) • Ultrasonic wall thickness measurements	

In-line (intrusive) methods are based on coupons installed into the flow through an access system. Normally, the access system allows for installation and operation of sensors under full operational pressures. For refineries, the most common arrangement for in-line probes and coupons is the retractable system, as shown in Fig. 18.2.

A general advantage of in-line monitoring solutions is that they provide high-resolution and fast information about changes in corrosion and corrosion rates.

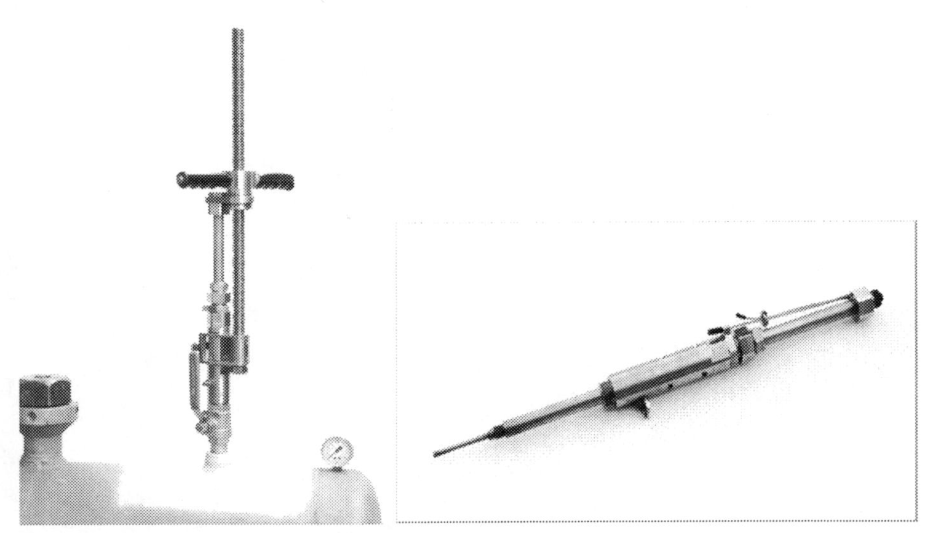

Fig. 18.2 Shows a 1″ retractable Emerson Roxar ER probe with pack box (right). The retractable probe assembly is installed through a valve into the pipe (left). On the side can be seen a retractor tool used for controlled installation and retrieval under full operational pressure [5].
Courtesy of Emerson Automation Solutions.

Nonintrusive methods are based on sensors installed on the external pipe or vessel wall, for monitoring internal corrosion. The methods are normally based on measuring changes in wall thickness, either through ultrasonic wall thickness measurements or electric resistance changes, due to changed wall thickness. Fig. 18.3 shows a nonintrusive FSM system under refinery installation.

The general advantage of nonintrusive systems is that they do not need an intervention through the pipe or vessel wall, which is seen as an operational and safety benefit. On the other hand, nonintrusive systems are less sensitive and have a slower response time compared to in-line monitoring technologies.

Most corrosion monitoring technologies are based on methods for monitoring the *metal loss* over time and calculating a corrosion rate for the actual location.

- *Weight loss coupons* are the oldest and most basic method for corrosion monitoring. Coupons in the same or similar metal to the pipe and vessel material are installed for a defined period (3, 6, 12 months). The coupons are normally preweighed with a sensitivity of 0.1 mg. Upon retrieval, coupons are cleaned and weighed, and from the weight difference, metal loss is calculated and converted to a corrosion rate in mm/year or mils per year (mpy).

  The use of weight loss coupons is simple and reliable and provides information about the form of corrosion, deposits, and possible bacteria activity in addition to corrosion rates. The major disadvantage is that information is slow, not continuous, and difficult to correlate with process changes.
- *Electrical resistance (ER)* probes are probably the most commonly used method for corrosion monitoring today. As discussed in Chapter 11, ER probes are based on measuring the

Fig. 18.3 Shows sensing electrode matrix for electric field signature measurements (Emerson Roxar FSM) installed on refinery bend, with sensing pins connected to cables on the monitored pipe section. The pipe section will also be provided with a cover for temperature and physical protection.
Courtesy of Emerson Automation Solutions.

resistance in the probe element exposed to corrosion and then convert the change in resistance to metal loss and metal loss rates. A reference element is used for compensating for the change in resistivity with temperature.

The major advantage of ER probes is the high-resolution and short response times to changes in corrosion rates. With a resolution in the nanometer range, changes in corrosion rates can be detected within hours or days.

A limitation of ER probes is that they only measure uniform corrosion and are not suitable for detecting localized forms of corrosion. Fig. 18.4 illustrates the fast response to changes in corrosion rates.

- *Ultrasonic wall thickness measurements (UT)* are provided through various concepts, with direct wall thickness measurements using pulse echo technology the most common (there are also concepts using longitudinal lamb waves). With permanently installed sensors, sensitivity and accuracy have been significantly improved.

  A challenge for UT measurements was the temperature rating of the sensors. This has been overcome by an extension arrangement added to the sensor, and UT sensors are now available with a temperature rating of 600°C (1112 F). An array of sensors are used to cover for localized corrosion, e.g., naphthenic acid corrosion (see Fig. 18.5). Readers are encouraged to read Chapter 12 (Section 12.2.1) for additional information on UT sensors.

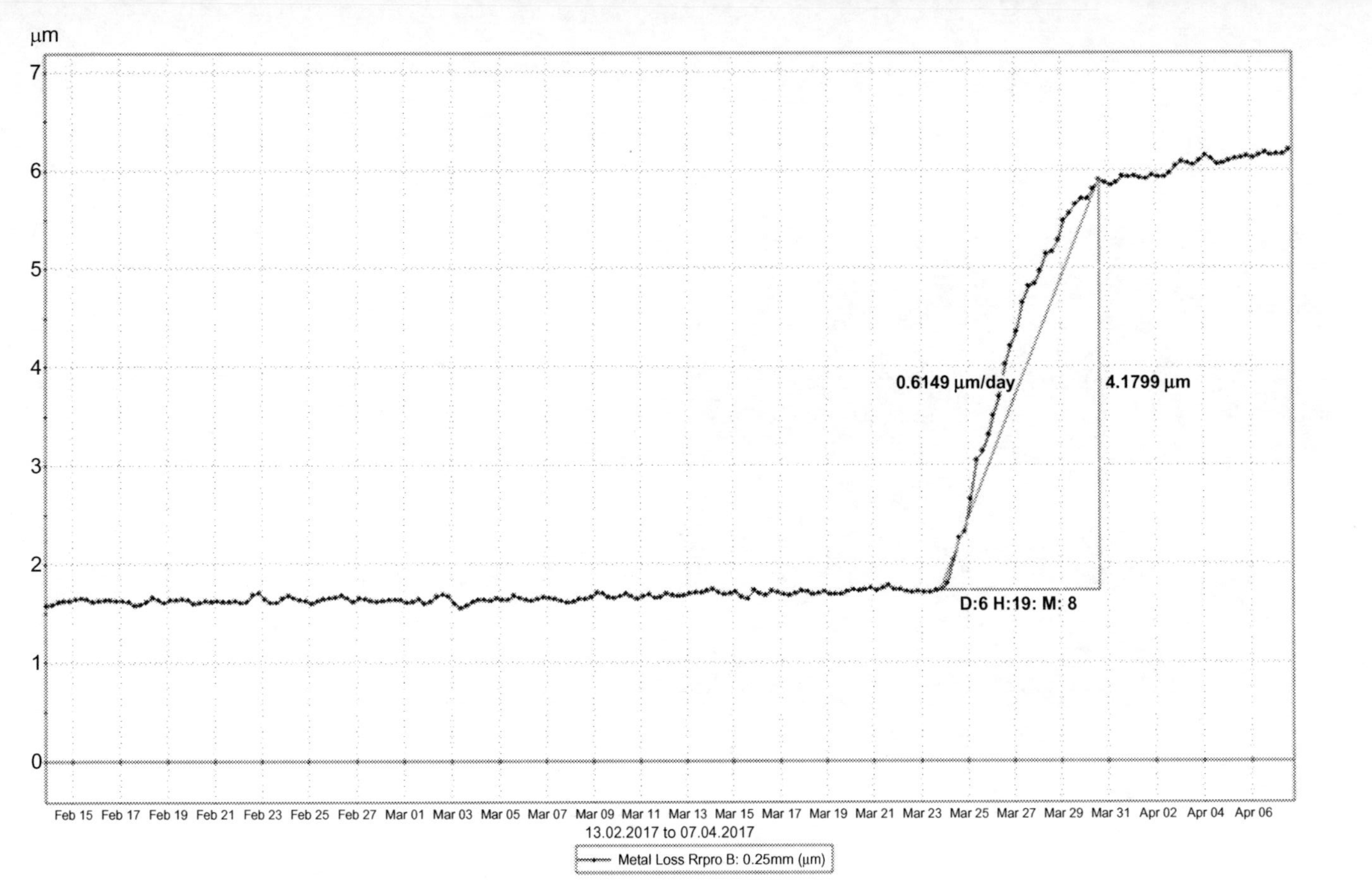

Fig. 18.4 Illustrates the ER probe's sensitivity and response time, showing metal loss vs time. The full scale on the *y*-axis is 7 μm (0.027 mils) and the time scale (*x*-axis) is from 15th February until 6th April same year. With measurement stability in low nanometers, the change in corrosion is detected at nanometer level within less than a day. Total metal loss over the period from 23rd March to 31st March is around 4 μm (0.16 mils), calculated corrosion rate is 0.6 μm/day, or 0.2 mm/year (7.8 mpy). With ER probes, corrosion is detected, mitigated, and the effect of mitigation confirmed before it would be visible with any nonintrusive method [6].
Courtesy of Emerson Automation Solution.

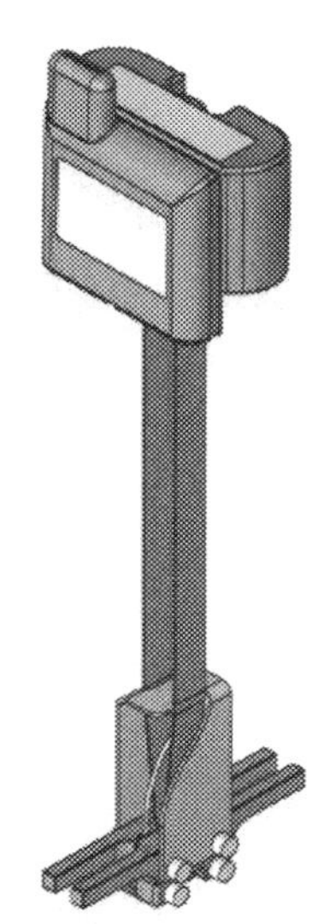

Fig. 18.5 Shows an array of Emerson Permasense UT sensors for monitoring high temperature corrosion (left). Permasense device (right) comprises of a UT instrument sending signal through extension to pipe where the support is welded to pipe (high temperatures). An array of UT devices is used to cover local variations in corrosion. Permasense UT provides online communication via WirelessHART protocol.
Courtesy of Emerson Automation Solutions.

- *Electric field signature measurements (FSM)* are based on feeding an electric current through the monitored pipe section and measuring the resulting voltage drop between an array of sensing pins installed on the external pipe wall. The first reading is called the signature and all other measurements compared to the signature. The technology is described in more detail in Chapter 25 of this book.

  As for UT, the benefit of FSM is that it is nonintrusive and measures changes at the pipe wall. In addition, FSM measures are *between* the sensing pins, meaning that localized corrosion can be detected in addition to uniform corrosion.
- *Corrosion rates for metal loss methods* are calculated from the metal loss over time. Sections 18.4 and 18.5 discuss how metal loss corrosion monitoring should be set up for maximum information value from the corrosion rate calculations, related to sensor sensitivity, and what information it should be used for.

In addition to metal loss-based corrosion methods, monitors based on *electrochemical measurements* are also common. The general advantage of electrochemical measurements is that corrosion rate information is provided instantaneously from one measurement; hence, feedback on changes in corrosivity can be almost immediate.

- *Linear polarization probes (LPR)* are the classic method for electrochemical corrosion measurements. As discussed in Chapter 3, the method is based on Stern-Geary's equations [7]. LPR probes come in 3-electrode or 2-electrode configurations. In a 3-electrode configuration, the probe comprises of a working electrode, a counter electrode, and a reference electrode. The method sets up a small voltage (polarization) on the working electrode (e.g., 20 mV) vs the reference, and the corresponding current response is measured and used to calculate the polarization resistance. For a 2-electrode probe, 2 identical electrodes are used; however, the principle is similar. The polarization resistance is inversely proportional to the corrosion rate.

Electrochemical methods require an electrolyte for the measurements. This means that LPR probes are mostly used in water systems with stable conductivity.

- *AC impedance* is a technology related to LPR, but where polarization is applied at a range of frequencies. Using the frequency spectrum, additional information about corrosion and corrosion forms can be achieved. However, the method is more commonly used in research than in field monitoring.
- *Electrochemical noise (EN)* measurements are also based on the same thinking, but, differently from LPR, an active polarization is not applied, and measurements are based on fluctuations in voltage and current over time. Corrosion rates can be determined, and it is claimed that the initiation of localized corrosion can be detected.

  Since EN measurements are not using active polarization between the electrodes, it is claimed that the conductivity from, e.g., naphthenic acid, would be sufficient for monitoring hydrocarbon systems as well. The author does not have specific knowledge to verify this claim; readers are encouraged to read Chapter 5 and the other literature, e.g., paper by Cox & Al [8]

With the different benefits and limitations of the various monitoring solutions, it is suggested that selecting methods is based on the expected corrosion case, the monitoring objectives, and what the information should be used for. Often, the best solution would be a combination of methods—this is discussed in more detail in Section 18.4 of this chapter. Fig. 18.6 is an example of corrosion monitoring technologies covering different aspects of corrosion types and information needs.

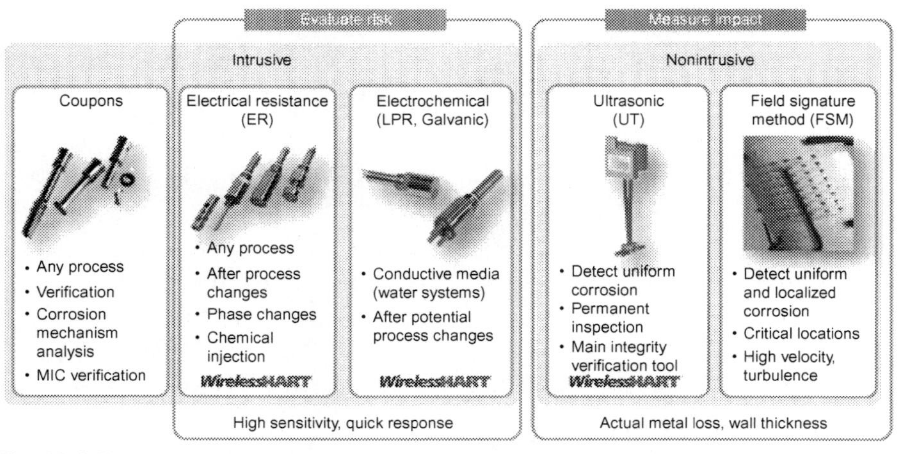

Fig. 18.6 Shows Emerson monitoring technologies and suggested applications [9]. Courtesy of Emerson Automation Solutions.

18.4 Purpose of monitoring—Select monitoring technology and set up accordingly

There are a variety of reasons for monitoring corrosion, which include:

- Alarms: corrosion is a threat to asset life, production uptime, and operational safety. Alarms can be set if the corrosion rate exceeds certain levels related to safety or process issues.
- Process optimization: continuous corrosion monitoring data can be correlated with other process data and provide information on the root causes of corrosion, allowing corrective action before damage takes place. A common application is the tuning of corrosion inhibitors and optimizing crude blends in refineries.
- Integrity management: corrosion monitoring data is important for assessing condition and long-term corrosion rates, contributing to the reduced risk of loss of containment, planned repairs and maintenance, and consequently, extended plant uptime and asset life.

Table 18.1 suggests data management strategies for different corrosion management applications.

18.5 Operational recommendations

Corrosion monitoring has been a part of refinery operations over decades. However, operational experience and trust in corrosion monitoring have not always been good. The reasons for this will be discussed below as well as suggestions for improved performance.

18.5.1 Continuous vs noncontinuous monitoring

Traditionally, corrosion monitoring has been based on methods that do not provide continuous information. The classic monitoring technology is the use of weight loss coupons that are replaced in fixed intervals (3 months, 1 year), cleaned, and analyzed. Corrosion rates are calculated from the weight loss over time.

Weight loss coupons give reliable information about corrosion and also additional information, such as the form of corrosion, deposits, bacteria, etc. However, information is delayed until the coupon is cleaned and analyzed, and the calculated corrosion rate is an average over the exposure period only. Hence, changes in corrosion during the exposure time will not be seen, and changes in process that influence corrosion rates will not be seen. Hence, weight loss coupons are not good for correlation with process changes and for tracking the effect of process changes on corrosion rates.

As a step forward, corrosion probes were introduced, but were normally based on the use of portable instruments for probe readings. Typical measurement frequency was bi-weekly or monthly measurements.

The quality of data trends and corrosion rate calculations depend on the measurement frequency and accuracy of instruments. Traditional, portable corrosion meters had limited sensitivity, and combined with few data points, trends were sometimes erratic and rate calculations not always reliable. Furthermore, with few data points,

Table 18.1 Shows suggested strategy for corrosion data management settings for metal loss-based sensors.

	Alarms	Process optimization	Integrity management
Purpose	Alarm corrosion rates exceed defined limits	Identify root causes and impact of process changes on corrosion Inhibitor optimization and verification Opportunity crude blend optimization	Track longer term metal loss and impact on pipes and equipment—Inspection planning, repairs, maintenance
Suggested technology	High sensitivity, fast response technologies In-line corrosion probes (Nonintrusive) if excessive corrosion possible or in-line probes cannot be used for other reasons	Depending on expected corrosion rates In-line corrosion probes (fast response) Online ultrasound (if high erosion rates expected) Field signature monitoring (If high erosion rates are expected)	Depending on expected corrosion rates Online ultrasound (remaining wall thickness) Field signature monitoring (changes in wall thickness, localized erosion) (In-line probes also used for this application)
Data management settings	Tune settings to exceed variations (avoid false alarms) Short integration periods (hours) for metal loss rates (hours or days)[a] Use differentiated alarms Define alarm verification criteria	Tune settings above random variations (avoid false alarms) Short integration periods (days/weeks) for metal loss rates. Use differentiated alarms Define alarm verification criteria	Longer integration periods give more reliable metal loss rate values. Rate calculations to be based on weeks or months average data
Actions	Explore reason for excessive corrosion and identify corrective measures before significant damage	Change processes that increase corrosion Tune corrosion inhibitor program Tune opportunity crude blending	Use data for integrity and maintenance management purposes—maintenance and repair planning
Comments	Metal loss rates for alarm purposes only, not to be used for integrity evaluations	Metal loss rate calculations focus on tracking changes, values to be used with care for integrity management purposes	Settings not designed to track rapid process changes

[a] See comments and Section 18.5.

correlation with process changes was difficult, and it was challenging to conclude if a changed corrosion rate was due to process changes or random variations in the measurements.

Continuous monitoring, on the other hand, gives a steady flow of corrosion data, providing more reliable trending and the possibility of correlating changes in corrosion rates with process changes. Hence, continuous measurements make corrosion monitoring a more reliable tool that is better suited for corrosion mitigation and control.

18.5.2 Online vs off-line communication

Continuous data can be provided through off-line data loggers or through online communications (wired or wireless). Off-line data loggers do measurements at preset intervals. Hence, corrosion data are continuous and can be correlated with process changes.

The benefit of off-line data loggers is a reduced installation cost (field cabling not required). The main disadvantage is delays in information, since changes in corrosion rates will not be seen in real-time, but only when data is collected and analyzed. Hence, off-line monitors have less value for real-time process control and corrosion mitigation activities.

A refinery is a complex plant, and many of the monitoring locations are difficult to access, scaffolding may be required, and temperatures may be high and make access for measurements difficult. It has commonly been experienced that the motivation for maintaining a manual corrosion program has been getting lower and lower over time, and therefore, corrosion monitoring data have not been collected and used, and the hardware not maintained.

In addition, corrosion monitoring has not always been a priority activity in the refinery, and ownership of the system has not always been clearly defined. The result is that the refining industry has a lot of corrosion monitoring installations that are not maintained, nor actively used today.

Online communication provides several benefits:

- Data are brought to the user in real-time without delays, allowing fast actions when corrosion rates change, increasing the value for system optimization or process changes.
- Online communication does not require operator actions or attention to data collection and processing. Hence, information will be provided with a higher reliability and without loss of data due to lack of attention or ownership.

Traditionally, online systems have required field cabling. When building a new plant, field cabling can be accommodated for and an online corrosion monitoring system can be part of the plant instrument package.

For existing plants, however, running cables is normally more expensive than the corrosion monitoring system itself. Therefore, the upgrading of legacy monitoring systems to online technology has often not been approved due to cabling costs.

The standardization and acceptance of wireless communications (e.g., through the wirelessHART standard) has been a "game-changer" for online upgrades. There is a wide range of applications available with wirelessHART communications today (pressure, temperature, vibration, etc.) that can be combined through the same gateway, providing online communications to a range of data management solutions or control systems. Fig. 18.7 illustrates both in-line and nonintrusive UT corrosion monitoring that are available with wirelessHART communications and can be combined as one complete wireless-based corrosion monitoring solution.

18.5.3 Understanding the data

Another problem is that the data provided from the corrosion monitors are not understood or processed correctly, contributing to reduced trust in the monitoring system, and accordingly, the reduced value of the system.

For electrochemical monitoring solutions (LPR, AC impedance, EN), there are several error sources than needed to be understood when evaluating the data and deciding on possible actions. More advanced electrochemical technologies, in particular, may require an understanding of the method and how information is generated. Electrochemical methods are referred to in more detail in other chapters of this book.

Metal loss-based corrosion technologies measure corrosion either as the metal loss on a probe, or the metal loss directly on the pipe wall (ultrasonic, field signature measurements).

In both cases, *metal loss can be plotted versus time* and gives information on corrosion and corrosion rates. Metal loss rates can be calculated from the slope of the metal loss plot and are defined as the increase in metal loss divided by the period for the change.

Corrosion or sand data management programs are often set up for the automatic calculation of metal loss rates. The *integration period* is the period used in this program for calculating metal loss rates. Shorter integration periods respond faster to changes in corrosion conditions, but reduce the accuracy of the metal loss calculations. Longer integration periods give more accurate metal loss rate calculations.

For all methods, there will be a limit on the sensitivity or resolution of the method. The *response time* is the time needed to have a metal loss exceeding the system sensitivity at a given corrosion rate. Higher corrosion rates give a shorter response time, something that can be used for alarm settings. Fig. 18.8 illustrates metal loss data presentation and calculation of corrosion rates.

A common error in metal loss data management is to set up integration periods shorter than the response time of the system and attempt to use that as a basis for corrosion rate reporting. In those cases, the small variations in probe readings are random variations not related to metal loss. Attempts to extract corrosion rates from changes less than the sensitivity of the system are, therefore, only an amplification of random noise that may lead to misleading conclusions about metal loss and sensor performance (see Fig. 22.9).

Note: Short integration periods may be of value for alarm purposes; however, the metal loss rates generated should not be used for integrity assessments.

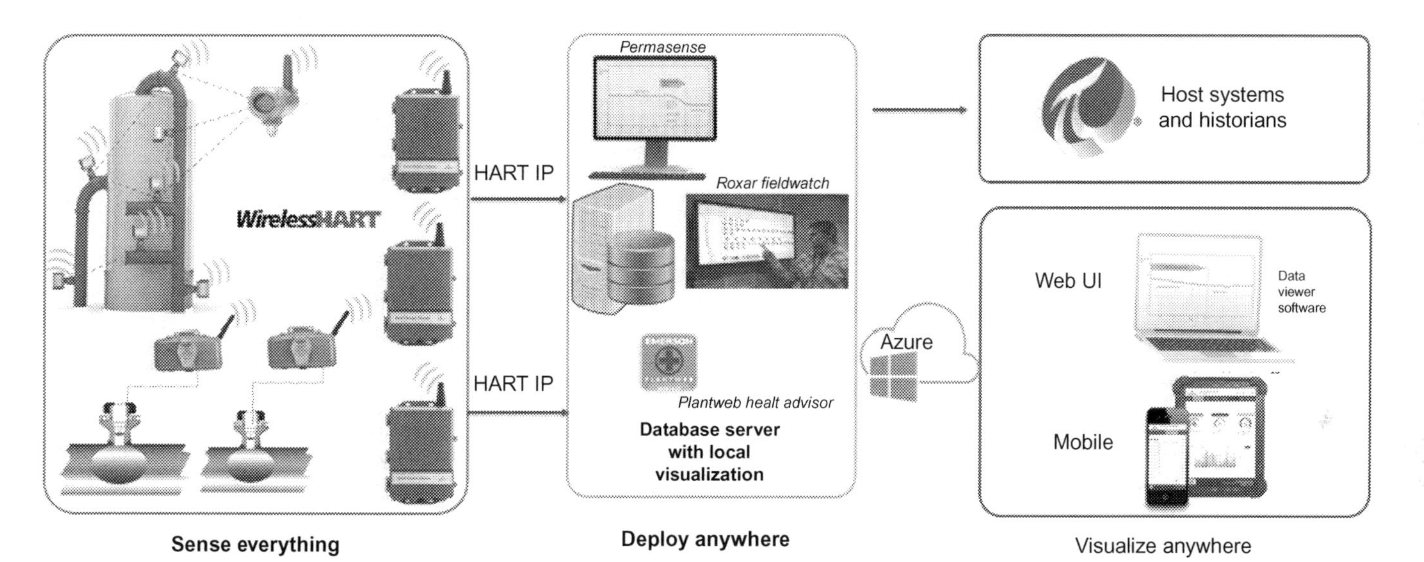

Fig. 18.7 Shows how in-line probes and nonintrusive ultrasonic sensors can be combined into one wireless network, providing complementary corrosion information to user applications [10].
Courtesy of Emerson Automation Solutions.

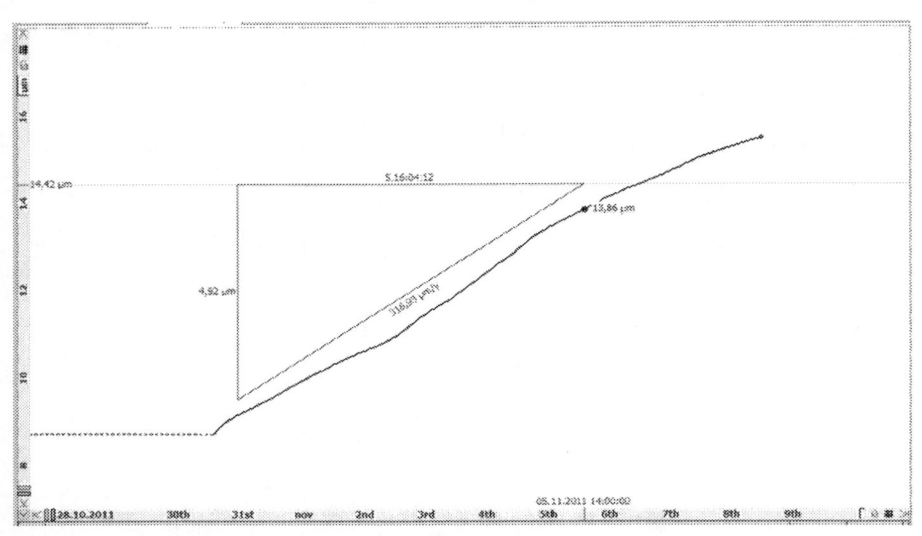

Fig. 18.8 Shows metal loss vs time plot and using "ruler" for metal loss rate calculation. Courtesy of Emerson Automation Solutions.

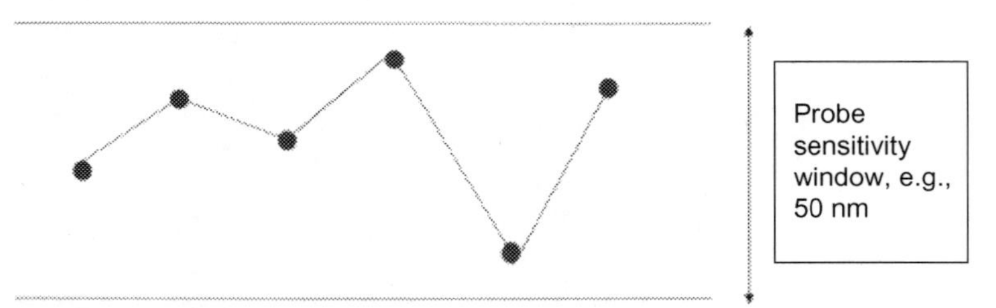

Fig. 18.9 illustrates random data variations when changes are less than the sensitivity window of the method, e.g., 50 nm (0.002 mils). Attempts to calculate corrosion rates over short integration times will give erratic data. It is, however, possible to determine the level of corrosion from the fact that the sensitivity window is not exceeded.

18.5.4 Proactive management of CM system for performance and data management

As mentioned earlier, corrosion monitoring is not a core activity for refinery personnel and system ownership. It is often seen that the success of running a good monitoring program will increase through long-term alliances with the system vendor or other corrosion monitoring specialists.

18.6 Monitoring from tank farm to product

As stated earlier, a refinery is a complex plant with a variety of corrosion mechanisms, operational challenges, and safety issues. Also, as discussed in Section 18.4, there are various objectives for monitoring corrosion, linked to safety, process tuning, cost optimization, and integrity management.

Also, as discussed earlier, there are a range of monitoring technologies available, with benefits and limitations. Selection of a method for each case should be based on corrosion challenge and what the information should be used for. It is the author's general opinion that the best monitoring solution is a combination of methods, to give more reliable information (different information sources) and the best variety of actionable information possible.

It is, therefore, suggested that a corrosion monitoring strategy for a plant is based on a refinery-specific evaluation, where the criteria mentioned below are evaluated.

Note: Table 18.2 is intended as a suggested approach rather than a firm check list for corrosion strategy development.

The next step is then to evaluate the monitoring strategy from tank farm to product. The table below suggests an approach to rate corrosion monitoring for different processes in a refinery, with a suggested spread of technologies. It is recommended that effort is made on this process, e.g., including a site walkdown for existing plants with a subject matter expert, or a similar review for new plants being built.

Ratings are ranked from 1 to 3, where 1 is the primary suggestion for the application.

Note: Table 18.3 is intended as a suggested approach rather than a firm guideline for monitoring technology specifications in a plant.

Also observe that a monitoring strategy for a refinery (or any plant) should not be all about distributing sensors. Successful monitoring also requires the following:

- Efficient communication providing data to the user. Experience shows that online communication (wired, wireless) increases the user value significantly
- A good strategy for data management and reporting
- Ownership of the system in the refinery organization, and skilled personnel understanding the system and the information provided. Since corrosion monitoring is not a core activity within the plant organization, the outsourcing of the management of the corrosion monitoring system may provide more user value.

18.7 Summary and conclusions

The message in this article can be summarized as follows:

- Corrosion is a main challenge in the corrosion industry
- Corrosion monitoring can contribute to
  - Safer operations
  - Reduced downtime of the plant
  - Better planned maintenance and repair
  - Extended equipment life
  - More cost-efficient operation, e.g., due to optimizing opportunity crude blending.
- A range of monitoring technologies are available, and often the best corrosion monitoring strategy uses a combination of technologies. The monitoring technology specification should be based on
  - Intended use of the information
  - How monitoring technology reflects the actual corrosion condition throughout the plant.
- Continuous, online monitoring systems provide better, faster, and more conveniently the information needed by the user.
- Attention to the monitoring system and the understanding of the data are important, but often underestimated by the users.

Table 18.2 Suggesting approach for monitoring strategy development criteria for a refinery.

Criteria	Examples	Comments
Criticality	Probability of corrosion damage and consequences	Highest focus on locations with high probability of corrosion damage, and where the consequence of corrosion would be most critical from a safety and operational perspective Also, focus on locations when monitoring for diagnostics and process tuning
Purpose	Integrity/safety	Moderate sensitivity needed—preference reliable wall thickness monitoring
	Process tuning	Fast response to changes required—preference in-line probe if suitable for application
	Opportunity crude optimization	Will crude be provided from one source, or is the refinery strategy based on blending crudes in the spot market?
Corrosion rate	High or low rate expected or accepted	Low corrosion rate acceptance—preference in-line monitoring High corrosion expectancy—preference wall thickness monitoring
Type of corrosion	General corrosion	Most methods can monitor uniform corrosion. Preference in-line probes for fast response
	Localized corrosion	Preference wall thickness monitoring—area coverage Specialized electrochemical methods—an alternative
	Sour corrosion	FeS deposits may disturb in-line probe readings + safety aspect of interventions—preference wall thickness monitoring
	NAC corrosion	Preference wall thickness monitoring—high temp + localized corrosion High temperature probes option for fast response to changed fluid corrosivity
Environment	Pressure	High-pressure requires nonintrusive or high-pressure access systems
	Temperature	Most methods have temperature limitations, e.g., for particular refinery applications
	Oil/gas/water	Electrochemical methods may require a conductive electrolyte (water)

Table 18.3 Shows suggested approach for selecting monitoring technologies through a plant.

Process	In-line probes	FSM/ EFM	Online UT	Common objective	Suggestion
Tank farm	2	3	1	Integrity of pipe wall	Online UT for tank integrity monitoring In-line probes for associated pipes FSM for possible localized corrosion
Crude blender, crude exchangers, desalter	1		2	Track process changes in desalter efficiency Integrity	In-line probes due to high sensitivity to changes Online UT for integrity
Crude heater	3	1	2	Integrity Opportunity crude tuning	Due to high temperature and severe localized corrosion, FSM/ EFM and/or online UT suggested. High temperature in-line probes back up for tracking rapid process changes
Distillation—overhead, lighter products	1		1	Integrity Opportunity crude tuning	In-line probes and/or online UT
Distillation—heavy products	2	1	1	Integrity Opportunity crude tuning	Online UT and/or FSM/EFM recommended. In-line probes possible, check life expectancy and conditions
Alkylation process	3	3	1	Integrity	With aggressive fluids nonintrusive generally suggested. Online UT Lead solution, in-line probes, and FSM/EFM possible for special applications
Hydrotreating/ amine process	1		2	Integrity	In-line probes and/or online UT combination or choice based on use of info. Probes suggested lead for fast response to changes. Online UT suggested for high H_2S lines
Cooling systems	1		2	Integrity	In-line probes and/or online UT combination or choice based on use of info

The table should not be seen as fixed recommendation, but as a suggested approach to be used with a subject matter expert for system review or walkdown. The distillation area in a refinery is often given the highest priority when designing a monitoring solution.

References

[1] R. Tems, A. Al Zharani, Cost of corrosion in production and refining, Saudi Aramco J. Technol. (2006) 2 Summer.
[2] CSB US Chemical Safety Board, https://www.youtube.com/watch?v=QiILbGbk8Qk; US Chemical Safety Board, April 2013
[3] Wood, Arrellano, Wijk, JRC Scientific and Policy Reports, Corrosion-Related Accidents in Petroleum Refineries, EUR 26331 EN(2013).
[4] K. Wold, T. Olsen, Corrosion Management Strategies for Opportunity Crudes, PTQ, Q4 (2019).
[5] API, Damage Mechanisms Affecting Fixed Equipment in the Refining Industry, API Recommended Practice 571second ed., (April 2011).
[6] M. Carougo, K. Wold, New perspectives on refinery corrosion, integrity management, communications and optimizing opportunity crudes, in: AFPM Annual Meeting, San Antonio, Texas, 2017.
[7] M. Stern, A.L. Geary, Electrochemical polarization, J. Electrochem. Soc. (Jan 1957).
[8] W.M. Cox, W.Y. Mok, R.G. Miller, J.L. Dawson, Corrosion monitoring in refinery overheads, in: NACE Corrosion '90, Las Vegas, Nevada, 23–27 April, 1990.
[9] Stubelj, Vasquez, How Integrated Corrosion Monitoring and Getting Your System Design Right Can Lead to Top Quartile Refinery Performance, Emerson Global Exchange(2018).
[10] Wold, Ruschmann, Stubelj, A new perspective on corrosion monitoring, in: Paper No 10937, NACE Corrosion Conference, Phoenix, AZ, USA, 2018.

Corrosion monitoring undercoatings and insulation

Feng Gui[a] *and C. Sean Brossia*[b]
[a]DNV GL, Dublin, OH, United States, [b]Invista Sarl, Houston, TX, United States

19.1 Introduction

Coatings, including organic and metallic coatings, have been used in many industries as a means to provide protection from a corrosive environment, and thus, extend the lifetime of the protected structure. Although organic and metallic coatings are both used for protection, most metallic coatings usually serve as the sacrificial anode in the coating-substrate couple, such as a zinc coating on steel and the aluminum clad layer on aluminum alloys, in addition to their relative inertness, whereas organic coatings provide protection mainly through the barrier properties of the film. In some industries (e.g., aircraft industry), other coating-like products—corrosion prevention compounds (CPC)—have also been widely used to provide protection for the interior surface of joints due to the wicking and water displacement ability that CPCs possess.

An organic coating system often consists of a conversion coating, primer, and a top coating. Sometimes, inhibitors such as chromate are incorporated into the organic coating to further inhibit corrosion. Some researchers are also working on developing a smart coating system through which inhibitors can be delivered to a corroded site as needed [1].

The protection offered by organic coatings is mainly determined through isolation from the aggressive environment, by adhesion opposing corrosion initiation at the metal/coating interface, and through structural changes taking place during curing of the coatings [2]. Therefore, many times, coatings that have good barrier properties would be better choices for isolating the corrosive environment from the protected structure, although sometimes it is necessary to have a porous coating so that the penetrated electrolyte can trigger the inhibitive action of the pigment in the coating [3]. Frequently, however, corrosion could initiate if aggressive species and water are transported through the defects (pores) in the coating and reach the substrate/coating interface.

In a similar vein, corrosion is also frequently observed under insulation and fireproofing that are to minimize heat loss in production piping and equipment. Corrosion under insulation (CUI) and corrosion under fireproofing (CUF) are typically subject to inspection methods but rarely real-time monitoring, despite being well-known and recognized as a significant integrity risk in process plants, refineries, and offshore platforms. CUI and CUF have been extensively studied and are most commonly encountered where moisture or water can accumulate and buildup on

Techniques for Corrosion Monitoring. https://doi.org/10.1016/B978-0-08-103003-5.00018-7

the external surface of hot, insulated equipment. Carbon steel equipment tends to be susceptible to CUI in the temperature range of 10–350°F (−12–175°C), but has been observed at temperatures as high as 600°F (316°C) when the insulation is subjected to periodic wetting by rain [4]. Equipment constructed using 304 and 316 stainless steel can also experience CUI, but is typically in the form of chloride stress corrosion cracking and is most often observed at temperatures between 140 and 350°F (60–175°C) [4].

The initiation of corrosion and failure of coatings usually can be identified via visual inspection because the failure of protection from organic coatings generally results in the formation of blisters and delaminated regions. CUI, similarly, is most often discovered using visual inspection and is typically included as a part of inspection plans to be conducted every few years depending on risk factors. Blistering and delamination of coatings usually cannot be observed until the later stages of corrosion. Furthermore, visual inspection for CUI can be very labor-intensive as it typically involves removal of protective jacketing and the insulation over large areas that are often not at ground level requiring scaffolding and/or rope access crews. Most times, it is desirable to identify corrosion at an earlier stage to reduce the impact of corrosion and to reduce the overall maintenance that is needed. It is also desirable to identify locations of possible CUI without having to remove significant quantities of jacketing and insulation. Thus, monitoring of coating degradation as well as for CUI is of great importance to the corrosion engineer.

19.2 Corrosion monitoring methods undercoatings

It is usually desired that any technique capable of monitoring undercoating corrosion is not destructive so that corrosion could be detected while maintaining the integrity of the protected structure. Both Electrochemical Impedance Spectroscopy (EIS) and Electrochemical Noise (EN) are nondestructive techniques. In particular, EIS has been successfully used in many cases to detect coating delamination, corroded area undercoatings. During the last decade, many efforts have been made in order to use it for undercoating corrosion monitoring. One typical benefit of EN technique over EIS is that it does not require any perturbation of the system of interest.

The theory behind EN techniques was discussed in Chapter 5 and the scientific literature is replete with discussions around the theory behind EIS. Although the use of these two techniques is not limited to coatings, this section will only review the applications where EIS and EN were used to detect and monitor corrosion under organic coatings. The applications of several other techniques will also be reviewed.

19.2.1 Electrochemical impedance spectroscopy (EIS)

The use of EIS to study the corrosion has been broadened significantly since it was introduced by Epelboin and coworkers [5]. The behavior of a coated metal can be represented with an equivalent circuit model, which has been well-developed [6]. The typical models for a coated metal are shown in Fig. 19.1. An equivalent circuit model

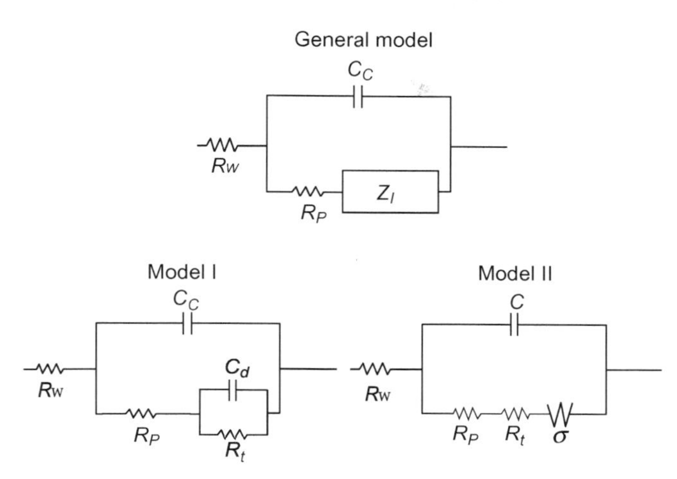

Fig. 19.1 Typical equivalent circuit models for a coating [6].

usually consists of *Rw*, *Rp*, and *Cc* that characterize the solution resistance, coating resistance (or pore resistance), and coating capacitance, respectively. The *Z1* component in the general model in many systems consists of a capacitance that characterizes the double layer behavior and a resistance that reflects the charge transfer resistance (or polarization resistance). In the case that diffusion through the coating is of concern, the Z1 component can be characterized with the charge transfer resistance and a Warburg impedance.

Much information can be obtained by carefully fitting the EIS data to a meaningful equivalent circuit model, such as coating capacitance, pore resistance (coating resistance), double layer capacitance, and charge transfer resistance (polarization resistance). These parameters can be related to the coating performance, such as delamination area and corroded area undercoating, for many coating systems.

Coating capacitance depends on the thickness of the coating (d) and the total sample area (A) that was tested as shown in Eq. (19.1):

$$C_C = \frac{\varepsilon\varepsilon_0 A}{d} \tag{19.1}$$

where ε and ε_0 are the coating dielectric constant and the free space dielectric constant, respectively. Thus, by monitoring the change in the coating capacitance, it is possible to monitor possible water uptake in a coating system. This is an essential step to evaluate the coating performance because water uptake would subsequently cause the loss of adhesion at the coating-metal interface [7–9]. The coating dielectric constant is very sensitive to the uptake of water. The volume fraction (v) of electrolyte absorbed by the coating can be determined based on the coating capacitance [10]:

$$v = \log[C(t)/C(0)]/\log 80 \tag{19.2}$$

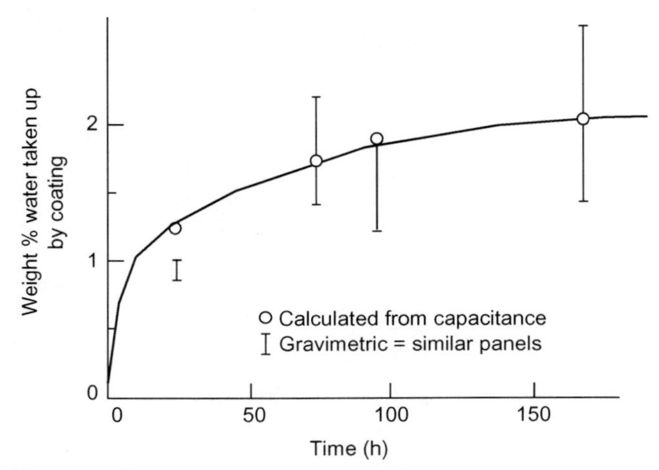

Fig. 19.2 Weight percent water uptake by a polybutadiene coating as determined from the capacitance and as determined gravimetrically [11].

where $C(t)$ and $C(0)$ are the coating capacitance after exposure to time period t and that at time zero, respectively. Water uptake determined from coating capacitance has been found to correlate well with that determined gravimetrically, as shown in Fig. 19.2 [11]. In many cases, a coating that absorbed less water is likely to indicate better performance as the water/electrolyte that diffuses through the coating and reaches the metal/coating interface could often serve as the precursor of localized corrosion. Thus, the ranking of coating performance determined by water uptake during the earlier time period of exposure could correlate well with that obtained by other variables such as coating resistance [12]. Touhsaent and Leidheiser [13] also found that coatings that absorbed less water at earlier times calculated according to coating capacitance performed better than others that showed more water uptake during two years of atmospheric exposure. In some cases, studying water uptake is essential for understanding the protective mechanism through which a coating provides protection. Deflorian and Felhosi [3] found that the low barrier properties of a coating, meaning more water uptake, actually induced the activation of the pigment inhibitive action. Therefore, in this case, the coating provided protection not simply through the barrier properties.

Coating resistance is usually influenced by the presence of defects and conductivity through the pores [3]. Frequently, coating resistance was observed to decrease as the exposure time increased [14, 15]. Haruyama et al. [16] suggested that the coating resistance was related to the delamination area in Eq. (19.3):

$$R_P = \frac{R_P^0}{A_d} \tag{19.3}$$

where R_p^0 is given by the product of the coating resistivity (r) and coating thickness (d)—both of which are assumed to be constant during the exposure—and A_d is the coating delamination area. Thus, caution should be taken when water uptake in the coating resulted in the change in the coating thickness. Mansfeld and Tsai found that there was an excellent agreement between the estimated delamination degree based on the coating resistance and visual observation. And thus, it is possible to determine the delaminated area under the coating from EIS data [15]. Kendig et al. have demonstrated the correlation of coating resistance and the under-corrosion extent as determined based on ASTM D610 [6]. As shown in Fig. 19.3, the corrosion extent (CORR) increased with the decrease in the coating resistance (*Rp*).

After the electrolyte is absorbed by the coating and the metal/coating interface penetrated, a double layer forms and corrosion is initiated. As a result, the double

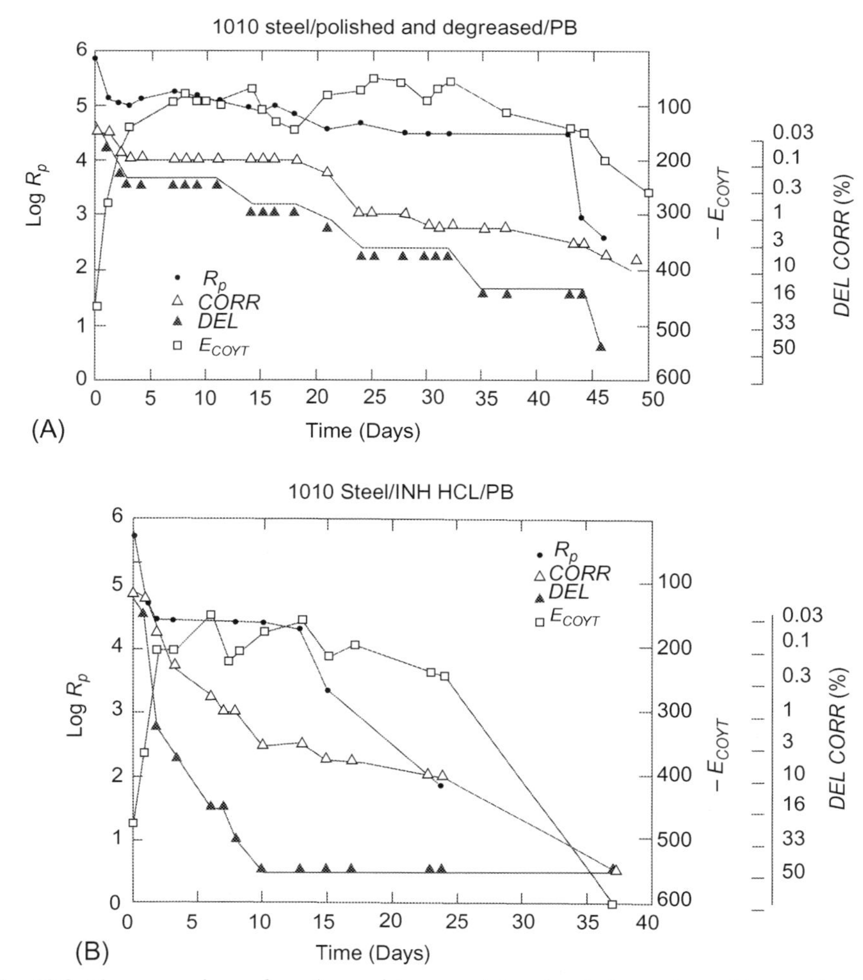

Fig. 19.3 The comparison of coating resistance and the visual observation of the coating performance [6] (E_{corr}, corrosion potential; *DEL*, delamination; *CORR*, corrosion extent).

layer capacitance is increased from zero and can be related to the disbonded area (A_d) in Eq. (19.4):

$$C_d = C_d^0 A_d \tag{19.4}$$

where C_d^0 is the area-specific capacitance that can be determined from bare metal with known area in a simulated underfilm solution [7]. Using Eq. (19.4), the disbonded area could be determined from double layer capacitance. The double layer capacitance has been used by people to calculate the corroded area or delaminated area undercoatings [9, 17–20].

Charge transfer resistance R_t (polarization resistance) describes the electrochemical processes taking place at the metal/coating interface. Similar to the polarization resistance determined via DC electrochemical techniques, such as the linear polarization resistance (LPR) method, a corrosion rate could be obtained, but this rate corresponds with the undercoating corrosion [21]. Frequently, however, the low frequency impedance (sometimes corrected by the solution resistance), which is the summation of coating resistance and sometimes the charge transfer resistance, has been used to evaluate the coating performance. Scully found that the low frequency impedance correlates well with the open circuit potential (OCP) measurement [22]. Specifically, the OCP for an epoxy polyamide-coated steel sample is either unstable or positive when low frequency impedances are high. When impedances decreased below 107 Ω/cm^2, the OCP was measured as −650 mV vs. Ag/AgCl, as shown in Fig. 19.4. Others also found that low frequency impedances could be correlated to the coating performance in various application cases, and thus, a critical value of low frequency impedance could be used to identify the failure of a coating [17, 23].

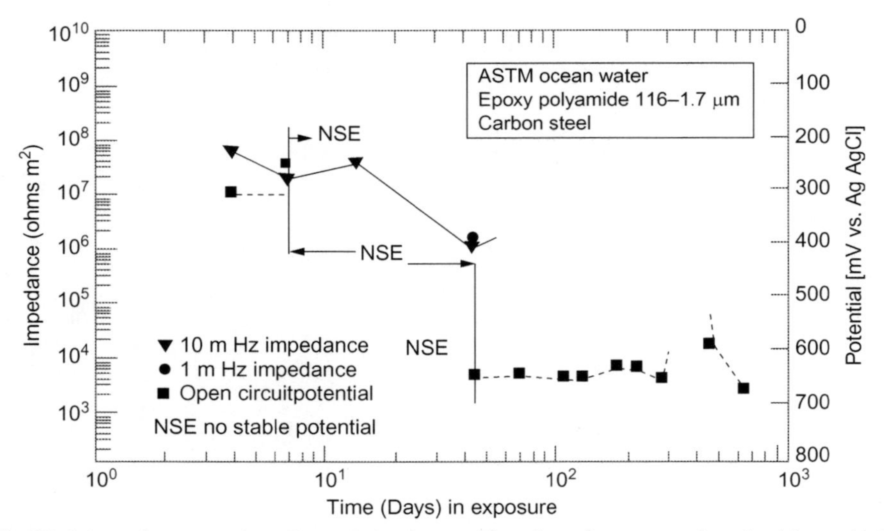

Fig. 19.4 Low frequency impedance behavior as a function of exposure time for 11 μm thick epoxy polyamide-coated steel [22].

Another approach that has been used frequently to detect corrosion under-coatings is the breakpoint frequency method. Breakpoint frequency is defined as the frequency where the phase angle drops to 45° [16]. The breakpoint frequency at high frequency is related to the disbonded area (A_t) in Eq. (19.5):

$$f_h = \frac{1}{2}\pi\varepsilon\varepsilon_0 r = K_f \frac{A_t}{A_0} \tag{19.5}$$

where A_0 is the surface area of the coated sample and K_f is a material constant. Hack and Scully demonstrated the correlation of high frequency, breakpoint frequency, and the ASTM D-610 rating of an epoxy-coated steel sample (Fig. 19.5) [24].

Clearly, the breakpoint frequency was higher as coating damage increased. Since the breakpoint frequency can be determined at high frequency and without running a full frequency sweep, the measurement can be performed rapidly [7]. A few others have also demonstrated the use of breakpoint frequency to evaluate coating performance and detect coating delamination [14, 15, 25].

Note Eq. (19.5) is valid based on the assumption that the electric constant and the coating resistivity did not change significantly during exposure. This situation is usually not true considering that the coating electric constant will change as water uptake occurs. To overcome this problem, Mansfeld proposed to use the ratio of breakpoint frequency and the minimum of frequency (f_{min}) [14] in Eq. (19.6):

$$\frac{f_b}{f_{min}} = \left(\frac{C_{dl}}{C_c}\right)^{1/2} = \frac{C_{dl}^0}{C_c^0} D^{1/2} \tag{19.6}$$

This ratio is independent of coating resistivity. The results by this method agree with the time dependence of R_{po} [14].

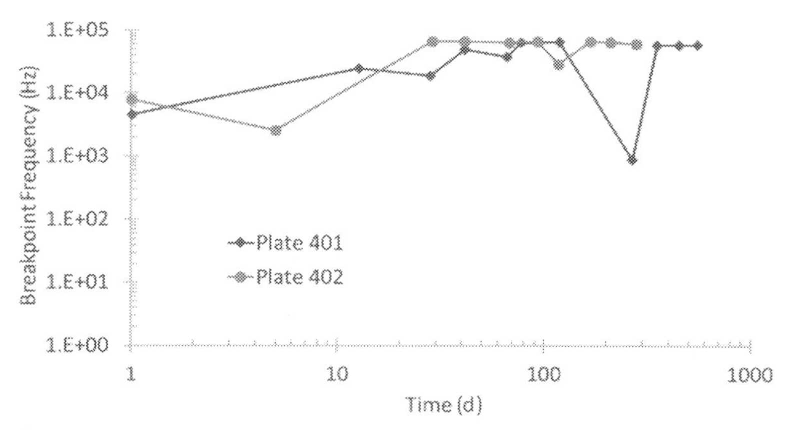

Fig. 19.5 Increase in the higher breakpoint frequency with exposure time for two 55 μm thick epoxy-coated steel (Plate 401 and Plate 402) in ASTM seawater [24].

Like many other corrosion mitigation methods, the challenge of corrosion prevention is to be able to predict when the coating will fail based on short-term test data. To achieve this, many authors have successfully correlated several different parameters determined from impedance to the long-term performance of coatings. Kendig et al. found that the time dependence of polarization resistance for a coated steel observed with five days of exposure in 0.5 M NaCl predicts the failure that occurred at three weeks [6]. With the impedance at 10 mHz and breakpoint frequency determined after an exposure of ten days, Scully was able to predict the coating performance after 550 days in service [22, 26]. Scully also found that the prediction was successful only at times sufficient to permit permeation of specifically H_2, O_2, Na^+, and to a lesser extent Cl^- through the coating to the reacting metallic interface [22]. When using impedance to study the performance of CPC, Gui and Kelly also found that the low frequency impedance at short exposure periods was able to predict the coating performance after six months of exposure [17].

The nondestructive nature of EIS has made it one of the most frequently used electrochemical techniques for corrosion performance evaluation of coating systems. As discussed, the parameters determined from impedance data can generally be related to the coating performance, characterized by water uptake, delaminated or disbonded area, and corroded area under-coating. However, caution needs to be taken when using EIS simply because much information could be obtained from a single measurement. Specifically, the determination of the parameters described in this chapter is not a simple task. Many models could be obtained by fitting the impedance data, but only the one that has physical meanings should be considered as the representative model for the coating behavior. Additionally, measurements of some of the parameters, such as low frequency impedance, could be time-consuming, especially at extremely low frequency range.

Although the bulk of the efforts in EIS so far have been limited to laboratory investigations, some efforts have been undertaken to apply this technology to field measurements of coating performance. Several authors have demonstrated the use of impedance-based sensors for monitoring undercoating corrosion. Although the design of these sensors is different, essentially, they function by measuring the impedance of a two-electrode system under the coating. Simpson et al. demonstrated the use of an atmospheric electrochemical sensor to monitor coating degradation in atmosphere (Fig. 19.6) [27]. The sensor consists of a painted steel coupon on which a sputter-coated electrode has been deposited to serve as a reference/counter electrode combination. Van Westing et al. was able to monitor the curing extent of coatings with a dielectric sensor that consists of a glass plate with a golden grid applied to the surface [28]. With a sensor buried in the structure in question, Brossia and Dunn successfully monitored the coating degradation [29]. In the work by Brossia and Dunn, a parameter was obtained from three different frequencies and the authors were able to identify the quality of the coating during approximately one month of exposure. Similarly, Dacres and Davis also developed an electrochemical in situ sensor for detecting corrosion under-coatings on aging aircraft. Essentially, the low frequency impedance (defined as Near-DC impedance) was monitored with the sensor embedded in the coating system. The results obtained from this sensor were consistent with that obtained by conventional three electrodes measurements [30]. Tsai and Mansfeld demonstrated the

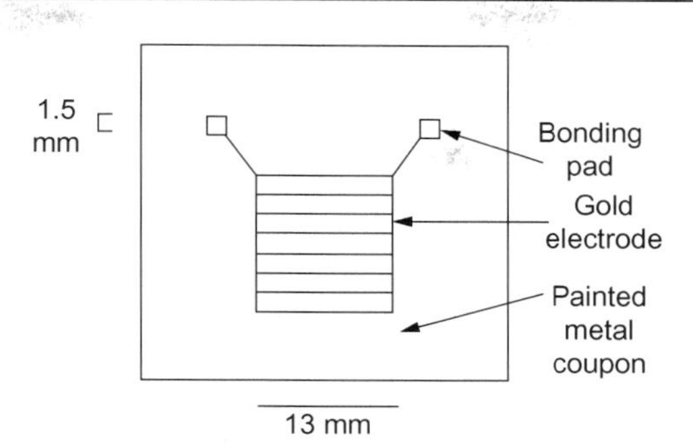

Fig. 19.6 Schematic of the atmospheric electrochemical monitor used by Simpson et al. [27] utilizing an interdigitated array of gold fingers.

use of the impedance at 100 Hz and 10,000 Hz to detect delamination ratio D (the ratio of the delaminated area to the initial sample area) [12]. All of these methods did not require the use of the conventional electrochemical cell or a sweep over a relatively wide frequency range. Therefore, the detection and monitoring of corrosion undercoatings can be achieved in the field with proper design and consideration of probe/element/electrode placement.

19.2.2 Electrochemical noise (EN)

The impedance technique, although powerful and capable of detection undercoating delamination via several parameters, can be complicated. Compared to the impedance technique, electrochemical noise does not introduce any perturbing signal to the system of interest. Instead, the current and voltage fluctuations are obtained at the open circuit potential [31]. Many efforts have been made to use electrochemical noise for coating evaluations [31–38].

Although many different parameters could be determined from the analysis of impedance data, in general the analysis of electrochemical noise only generates the electrochemical noise resistance (R_n), defined as:

$$R_n = \sigma_V / \sigma_I \tag{19.7}$$

where σ_V and σ_I are the standard deviation of the potential noise and the current noise, respectively. Many authors have demonstrated that the electrochemical noise usually decreases with increases in exposure time, which is very similar to that observed from EIS [31, 35–38]. Skerry and Eden also found that good paints exhibited relatively large voltage noise and relatively small current noise compared to poor paint coatings [38]. However, many also agreed that although the general trend of noise resistance changes with time in a similar way to coating parameters (e.g., R_t) determined from impedance, the numerical values were considerably different. As a result, most of the published work has been using EN techniques in conjunction with electrochemical impedance techniques. Further research and theoretical work still need to be conducted to build more confidence in using electrochemical noise to monitor corrosion undercoating.

19.2.3 Other techniques

Besides the conventional electrochemical techniques, some other techniques have also been developed as an effort to provide alternative means to monitor undercoating corrosion.

For field application, online monitoring of corrosion undercoatings is always a challenge. Using coupled multielectrode sensors, Sun monitored the degradation process of a few different coatings on carbon steel [39]. The current signal changed with the increase in exposure time and the signal from the sensors was sensitive enough for detecting the corrosion that had initiated undercoating. The sensitivity makes this technique appropriate for field applications, but the author did not address how to implement the sensors with a protected structure.

Recently, the development of imaging techniques has made it possible for them to be used to study corrosion undercoating. Schneider et al. successfully used confocal laser scanning microscopy (CLSM) as a tool to monitor the corrosion underneath organic coatings [40]. The ability of CLSM to monitor the topography change of the substrate surface and the coating due to the initiation of corrosion undercoating was demonstrated. Souto et al. [41] demonstrated the use of scanning electrochemical microscopy to monitor electroactive species so that the activity undercoating could be identified. Many authors have also used a Scanning Kelvin Probe (SKP) to detect the initiation of corrosion undercoating and to understand the initiation mechanism [42–45]. However, these techniques usually require expensive instruments. Furthermore, some instruments such as CLSM and SKP are extremely sensitive to the environment. The requirement on the samples is also often stricter than that for EIS and EN. Therefore, these microscope techniques are more suitable for using in the laboratory for mechanistic study.

19.3 Corrosion monitoring methods for CUI

As mentioned previously, the most common form of monitoring for CUI and CUF is visual inspection which typically involves removal of the insulation on the outside of the pipe, vessel, or equipment. Visual examination has the advantage of providing 100% examination of the exposed surface; however, it is quite labor-intensive to remove and then reinstall jacketing and the insulation every time. Other inspection tools have also been used to detect CUI, some of which can provide qualitative assessments for possible water accumulation, thereby providing an indirect way to monitor for CUI. Other inspection methods have the ability to provide at least semiquantitative measurement of possible metal loss and/or cracking.

Infrared (IR) imaging and neutron backscattering have been used to help identify possible locations of CUI by detecting water and moisture (and sometimes liquid hydrocarbons) trapped in the insulation. Both methods can be used without removal of the insulation, making them less labor-intensive than direct visual inspection. IR imaging helps identify possible CUI locations by finding areas where the temperature is different than the majority of the pipe or vessel. Typically, this temperature difference needs to be greater than 10°C (18°F) for sufficient resolution. Installation of IR cameras coupled with machine learning has been explored as a possible approach for in situ monitoring of locations where CUI is known to have occurred in the past or is suspected of growing.

Neutron backscattering is based upon how fast neutrons emitted from a radiation source are converted to slow moving thermal neutrons due to interactions with hydrogen atoms present in water and/or hydrocarbons that have accumulated in the insulation. If high energy neutrons are used, they can typically penetrate the jacketing and insulation and impart sufficient energy to the backscattered thermal neutrons to enable detection of these backscattered neutrons. Though this method has been used to identify locations that need additional investigation, it is unclear how this approach could be used as an in situ permanent monitoring method because of the need for a radiation source and the limited field of view of the system which requires scanning/moving the source and detector along the length of area that is of interest. Periodic measurements over time could be measured, but this would need to be performed as a manual process.

Radiation sources have also been used to provide more quantitative information on metal loss. Radiography in the form of traditional profile, film density, real-time [46], computed [47], and digital have all been successfully deployed to find and detect CUI to some extent. In these approaches, a radiation source is placed on the side opposite from the detecting method with the component of interest in between. Each of these methods has advantages and disadvantages in its ability to detect general and pitting corrosion and cracking, its accuracy in sizing the damage detected, and how large a component can be examined. Despite their ability to find CUI, these methods also have the same limitations as the previously mentioned neutron backscattering approach. The need for a radiation source and movement of the source and detector to inspect different locations makes real-time, in situ monitoring using these methods impractical.

Guided wave inspection (see Chapter 12 for more information) has been used to detect CUI on piping systems and can be left in place to conduct in situ measurements. This method uses an array of ultrasonic transducers arranged around the circumference of the pipe. This array can create various UT wave forms that can then be sent in both directions away from the array to detect and size corrosion defects as far away as 30 m (100 ft). The installation does require some amount of jacketing and insulation to be removed, but this amount is typically less than 1 m (3 ft) in length. Except for some limitations at temperatures above 121°C (250°F) and complications if pitting is located near welds, this methodology is suitable for in situ monitoring. An example of test results conducted on a 50 m (164 ft) test section is shown in Fig. 19.7. Gunaltun et al. [48] similarly showed that multiple CUI-type defects of various sizes could be found using guided wave, though some differences were observed in the interpretation of the results by different system operators. They observed that all defects that had depths greater than 3.49% of the cross-sectional area were found by the eight technicians used to conduct the tests. Defects of smaller sizes were not as reliably identified.

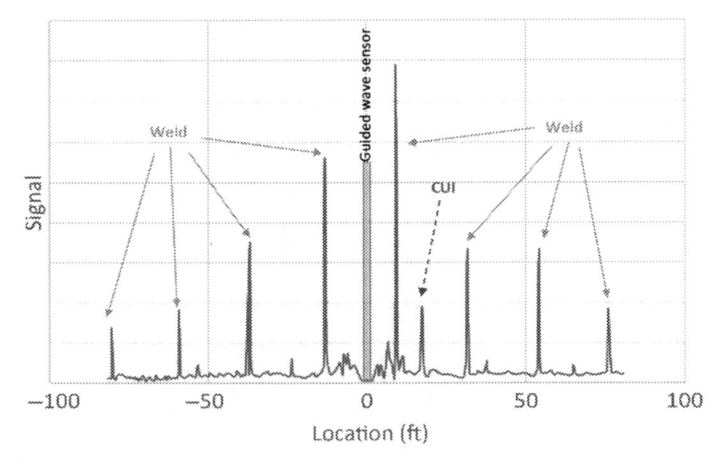

Fig. 19.7 Example results from guided wave inspection for CUI.
(Data adapted from API Recommended Practice 582, 'Corrosion Under Insulation', first ed., American Petroleum Institute, Washington, DC, 2014.)

This would seem to imply that if the system could be automated to a greater extent to eliminate operator-operator differences, a highly robust monitoring system based on this approach could be deployed.

The viability of using pulsed eddy current to detect CUI was discussed by Pechacek [46]. In the pulsed eddy current technique, a coil is used to generate a magnetic field in the steel of interest. This field is pulsed on and off inducing eddy currents in the steel which decay at a known rate, which then decays much faster when the currents reach the other side of the wall of the steel component. By detecting this rate of decay, an average wall thickness of the component can be calculated. The size of the probe used governs the area under which the field is generated and the area of the component that is examined. Provided that the CUI is in the form of general thinning corrosion, relatively accurate determinations of remaining wall loss are possible. Highly localized corrosion, such as pitting, is not as readily detected as the measurement is based on average metal thickness and isolated pitting typically does not affect this average significantly. Though this method does have some merits and can be performed without removal of the jacket or insulation, it still requires moving the device along the length of the component of interest. As a result, it is not well-suited to in situ monitoring, though subsequent measurements could be compared to provide rate information.

Simonetti et al. [49] also developed a method to find wet insulation locations. In their approach, these authors installed a series of small antennas in the insulation and used guided microwaves to identify locations of water accumulation due to differences in microwave transmission/reflectance between the insulation and water. In a series of laboratory tests, it was clearly demonstrated that water intrusion due to seepage and capillary action could be detected over long distances. In their tests, the authors showed that the installed metal jacket acted as a waveguide, significantly extending the range of the generated microwaves. What was not demonstrated, however, was the accuracy of the locating ability of the method and how the method interacts with multiple wet insulation locations over an extended length. As a result, additional testing and development was planned.

Coatings are often used as mitigation scheme to help prevent/minimize CUI and so coating monitoring methods may hold some promise in locating CUI as well. Time of wetness-type sensors and electrochemical noise techniques have been developed and tested that appear to hold some promise along these lines. Funahashi [50] used voltage measurements to indirectly detect the onset of CUI and provide a warning in an approach similar to a time of wetness sensor. In this approach, thin titanium and aluminum tapes were placed on top of a fiberglass tape that in turn was installed on the pipe surface underneath the insulation. The potential between the titanium and aluminum was then monitored using potentiometers that transmitted the data wirelessly to a base station for data collection and analysis. In this relatively simple approach, once water penetrated the insulation at intentionally damaged locations, a voltage difference was measured and transmitted. Though not providing a direct measure of CUI, this approach could be installed in multiple locations and automatically monitored for water intrusion events. Another limitation with this approach is that only the locations where the sensor tape is installed are monitored. If water intrusion takes place a location away from the installation locations, the water intrusion will not be detected unless it spreads to the sensor.

Caines et al. [51] used an electrochemical noise approach to monitor for CUI. In their approach, these authors installed a series of steel bands or circumferential underneath the insulation that were also isolated from the pipe. Because data analysis and interpretation of electrochemical noise can be complicated, a simplified approach using the potential noise differences monitored between the bands was used. The individual voltages of each band were then calculated using the following equations:

$$V_1 = \left(\frac{V_{12}^2 - V_{23}^2 + V_{13}^2}{2}\right)^{0.5}$$

$$V_2 = \left(\frac{V_{12}^2 - V_{13}^2 + V_{23}^2}{2}\right)^{0.5}$$

$$V_3 = \left(\frac{V_{23}^2 - V_{12}^2 + V_{13}^2}{2}\right)^{0.5}$$

where V_1, V_2, and V_3 are the voltages of the individual electrodes and V_{ij} are the voltage differences measured between different electrode pairs. By way of illustration, the calculated voltages for 2 insulated and 1 uninsulated band are shown in Fig. 19.8. Integrating the voltage calculations over time was shown to correlate reasonably well with corrosion rate measurements based on mass loss measurements of the band electrodes. The authors also showed one of the reasons why CUI is so insidious when comparing the insulated and non-insulated electrode readings. During the dry periods, the voltages of the non-insulated electrodes basically went to zero volts (an electrical open

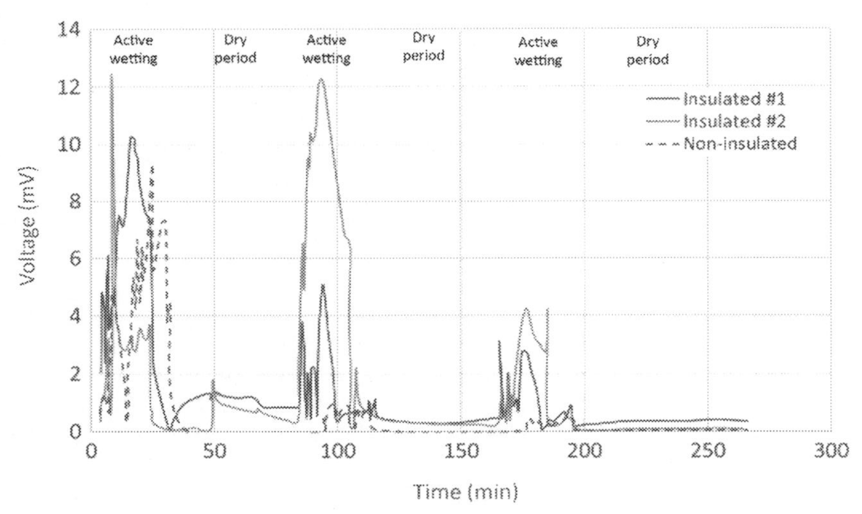

Fig. 19.8 Electrode potential noise measurements for 2 insulated and 1 non-insulated electrode as a function of time.
(Data replotted from S. Caines, F. Khan, J. Shirokoff, W. Qiu, Demonstration of increased corrosion activity for insulated pipe systems using a simplified electrochemical potential noise method, J. Loss Prev. Proc. Ind., vol. 47, pp. 189–202.)

circuit), indicating that no corrosion was taking place. In contrast, the voltage measurements for the band electrodes installed under insulation indicated continued corrosion throughout both the wet and dry periods.Using a completely different approach, Schoon et al. [52] reported on a 2-year study aimed at training five dogs to detect locations of CUI. Dogs were selected to investigate this possibility based on the well-recognized ability of dogs to detect trace explosives, narcotics, and other substances. When the dogs were used individually, the average probability of detecting CUI was approximately 59%, though only a maximum of 3% false positives was observed. When multiple dogs were deployed along the same pipe circuit, the probability of detection increased to nearly 92%. The authors theorized that difference in results obtained when using a single dog compared to multiple dogs was dependent on differences in the responses of the dogs themselves and possibly due to differences in the odors associated with corrosion activity since the dogs were trained to react on slightly different stimuli.

19.4 Summary and conclusions

Several methods have been employed to study coating performance and degradation with the bulk of previous efforts focused on laboratory studies. Some of these approaches have begun to be adopted and applied for field monitoring of coating performance to varying degrees. Of these methods, EIS appears to have shown the greatest promise, but more efforts need to be undertaken to broaden its application in fields. Other techniques, such as EN and impedance-based sensors, have been employed in the field of monitoring, although some more studies have to be performed to correlate EN results with the coating performance. Furthermore, the implementation of the techniques with real structures still needs to be further addressed.

Similarly, monitoring for CUI/CUF has followed a path of laboratory studies with some field applications. Detection of CUI is frequently accomplished via various inspection methods. Unfortunately, most of the methods used are not well-suited to real-time, in situ monitoring for CUI. Though these methods may not be able to be adjusted for in situ monitoring, subsequent measurements over time can provide an indication of the rate of CUI progression. Some methods could conceivably be modified and adjusted to allow for more in situ monitoring. Furthermore, some of the methods and approaches used to monitor coating performance have been applied to monitor CUI and show some promise along these lines. More research and development, however, is needed before these are widely accepted and utilized in the industry.

References

[1] M. Kendig, M. Hon, L. Warren, Smart' corrosion inhibiting coatings, Prog. Org. Coat. 47 (2003) 183–189.
[2] M.I. Karyakina, A.E. Kuzmak, Protection by organic coatings: criteria, testing methods and modelling, Prog. Org. Coat. 18 (1990) 325–388.
[3] F. Deflorian, I. Felhosi, Electrochemical impedance study of environmentally friendly pigments in organic coatings, Corrosion 59 (2003) 112–120.
[4] API Recommended Practice 582, Corrosion Under Insulation, first ed., American Petroleum Institute, Washington, DC, 2014.
[5] I. Epelboin, C. Gabrielli, M. Keddam, H. Takenouti, Alternating-current impedance measurements applied to corrosion studies and corrosion-rate determination, in: F. Mansfeld, U. Bertocci (Eds.), Electrochemical Corrosion Testing, ASTM International, West Conshohocken, PA, 1981, pp. 150–166.
[6] M. Kendig, F. Mansfeld, S. Tsai, Determination of the long term corrosion behavior of coated steel with A.C. impedance measurements, Corros. Sci. 23 (1983) 317–329.
[7] M. Kendig, J. Scully, Basic aspects of electrochemical impedance application for the life prediction of organic coatings on metals, Corrosion 46 (1990) 22–29.
[8] F. Mansfeld, Reply to comments by R. D. Armstrong and A. T. A. Jenkins on the paper 'use of EIS for the study of corrosion protection by polymer coatings' by F. Mansfeld, J. Appl. Electrochem. 25 (1995) 1145.
[9] P.L. Bonora, F. Deflorian, L. Fedrizzi, Electrochemical impedance spectroscopy as a tool for investigating underpaint corrosion, Electrochim. Acta 41 (1996) 1073–1082.
[10] D.M. Brasher, A.H. Kingsbury, Electrical measurements in the study of immersed paint coatings on metal. I. Comparison between capacitance and gravimetric methods of estimating water-uptake, J. Appl. Chem. 4 (1954) 62–72.
[11] M. Kendig, D. Mills, An historical perspective on the corrosion protection by paints, Prog. Org. Coat. 102 (2017) 53–59.
[12] C.H. Tsai, F. Mansfeld, Determination of coating deterioration with EIS. 2. Development of a method for field testing of protective coatings, Corrosion 49 (1993) 726–737.

[13] R. Touhsaent and H. Leidheiser Jr., A capacitance-resistance study of polybutadiene coatings on steel, Corrosion 28 (1972) 435–440.
[14] F. Mansfeld, Use of electrochemical impedance spectroscopy for the study of corrosion protection by polymer coatings, J. Appl. Electrochem. 25 (1995) 187–202.
[15] F. Mansfeld, C.H. Tsai, Determination of coating deterioration with EIS: I. Basic relationships, Corrosion 47 (1991) 958–963.
[16] S. Haruyama, M. Asari, T. Tsuru, M.M. Kendig, H. Leidheiser (Eds.), Corrosion Protection by Organic Coatings, The Electrochemical Society Proceedings Series, Pennington, NJ, 1987 PV 87-2, p. 197.
[17] F. Gui, R.G. Kelly, Performance assessment and prediction of corrosion prevention compounds with electrochemical impedance spectroscopy, Corrosion 61 (2005) 119–129.
[18] F. Deflorian, L. Fedrizzi, P.L. Bonora, Impedance study of the corrosion protection properties of fluoropolymer coatings, Prog. Org. Coat. 23 (1993) 73–88.
[19] F. Deflorian, V.B. Miskovic-Stankovic, P.L. Bonora, L. Fedrizzi, Degradation of epoxy coatings on phosphatized zinc-electroplated steel, Corrosion 50 (1994) 438–446.
[20] R.D. Armstrong, J.D. Wright, Impedance studies of poly ethylmethacrylate coatings formed upon tin-free steel, Corros. Sci. 33 (1992) 1529–1539.
[21] Amirudin, D. Thierry, Application of electrochemical impedance spectroscopy to study the degradation of polymer-coated metals, Prog. Org. Coat. 26 (1995) 1–28.
[22] J.R. Scully, Electrochemical impedance of organic-coated steel: correlation of impedance parameters with long-term coating deterioration, J. Electrochem. Soc. 136 (1989) 979–989.
[23] J.A. Grandle, S.R. Taylor, Electrochemical impedance spectroscopy of coated aluminum beverage containers: Part 1. Determination of an optimal parameter for large sample evaluation, Corrosion 50 (1994) 792–803.
[24] H.P. Hack, J.R. Scully, Defect area determination of organic coated steels in seawater using the breakpoint frequency method, J. Electrochem. Soc. 138 (1991) 33–40.
[25] F. Deflorian, L. Fedrizzi, P.L. Bonora, Determination of the reactive area of organic coated metals using the breakpoint method, Corrosion 50 (1994) 113–119.
[26] J.R. Scully, S.T. Hensley, Lifetime prediction for organic coatings on steel and a magnesium alloy using electrochemical impedance methods, Corrosion 50 (1994) 705–716.
[27] T.C. Simpson, P.J. Moran, W.C. Moshier, G.D. Davis, B.A. Shaw, C.O. Arah, K.L. Zankel, An electrochemical monitor for the detection of coating degradation in atmosphere, J. Electrochem. Soc. 136 (1989) 2761–2762.
[28] E.P.M. van Westing, G.M. Ferrari, J.H.W. de Wit, The determination of coating performance with impedance measurements—I. Coating polymer properties, Corros. Sci. 34 (1993) 1511–1530.
[29] C. S. Brossia and D. S. Dunn, 'Condition Based Coating Degradation Sensor', Paper 02161, Corrosion 2002, NACE, Houston, TX (2002).
[30] G.D. Davis, C.M. Dacres, M. Shookand, B.S. Wenner, C. Ferregut, R. Osegueda, A. Nunez (Eds.), Intelligent NDESciences for Agingand Futurisitc Aircraft, University of Texas, El Paso, TX, 1998.
[31] Q. Le Thu, G.P. Bierwagen, S. Touzain, EIS and ENM measurements for three different organic coatings on aluminum, Prog. Org. Coat. 42 (2001) 179–187.
[32] B.S. Skerry, D.A. Eden, Electrochemical testing to assess corrosion protective coatings, Prog. Org. Coat. 15 (1987) 269–285.
[33] C.P. Woodcock, D.J. Mills, H.T. Singh, Use of electrochemical noise method to investigate the anti-corrosive properties of a set of compliant coatings, Prog. Org. Coat. 52 (2005) 257–262.

[34] G.P. Bierwagen, X. Wang, D.E. Tallman, In situ study of coatings using embedded electrodes for ENM measurements, Prog. Org. Coat. 46 (2003) 163–175.
[35] R.L. De Rosa, D.A. Earl, G.P. Bierwagen, Statistical evaluation of EIS and ENM data collected for monitoring corrosion barrier properties of organic coatings on Al-2024-T3, Corros. Sci. 44 (2002) 1607–1620.
[36] D.J. Mills, S. Mabbutt, Investigation of defects in organic anti-corrosive coatings using electrochemical noise measurement, Prog. Org. Coat. 39 (2000) 41–48.
[37] F. Mansfeld, L.T. Han, C.C. Lee, G. Zhang, Evaluation of corrosion protection by polymer coatings using electrochemical impedance spectroscopy and noise analysis, Electrochim. Acta 43 (1998) 2933–2945.
[38] B.S. Skerry, D.A. Eden, Characterisation of coatings performance using electrochemical noise analysis, Prog. Org. Coat. 19 (1991) 379–396.
[39] X.D. Sun, Online monitoring of undercoating corrosion using coupled multielectrode sensors', Mater. Perform. 44 (2005) 28.
[40] G.O. SchneiderIlevbare, J.R. Scully, R.G. Kelly, In situ confocal laser scanning microscopy of AA 2024-T3 corrosion metrology: II. Trench formation around particles, J. Electrochem. Soc. 151 (2004) B465.
[41] R.M. Souto, Y. Gonzalez-Garcia, S. Gonzalez, In situ monitoring of electroactive species by using the scanning electrochemical microscope. Application to the investigation of degradation processes at defective coated metals, Corros. Sci. 47 (2005) 3312–3323.
[42] B. Reddy, J.M. Sykes, Degradation of organic coatings in a corrosive environment: a study by scanning Kelvin probe and scanning acoustic microscope, Prog. Org. Coat. 52 (2005) 280–287.
[43] P. Nazarov, D. Thierry, Scanning Kelvin probe study of metal/polymer interfaces, Electrochim. Acta 49 (2004) 2955–2964.
[44] B. Reddy, M.J. Doherty, J.M. Sykes, Breakdown of organic coatings in corrosive environments examined by scanning kelvin probe and scanning acoustic microscopy, Electrochim. Acta 49 (2004) 2965–2972.
[45] M. Stratmann, A. Leng, W. Furbeth, H. Streckel, H. Gehmecker, K.-H. Gro-Brinkhaus, Progress in organic coatings, Proc. 20th Int. Conf. Org. Coat. Sci. Technol. 27 (1996) 261.
[46] R. Pechacek, Advanced NDE Methods of Inspecting Insulated Vessels and Piping for ID Corrosion and Corrosion under Insulation (CUI), Paper 03031, Corrosion 2003NACE, Houston, TX, 2003.
[47] S. Nicola, V. Carreto, R.A. Mentzer, M.S. Mannan, Corrosion Under Insulation Detection Technique, Paper 2570, Corrosion 2013NACE, Houston TX, 2013.
[48] Y. Gunaltun, D. Merline, J. Ahmed Al Abri, P. Hivert, and G. Penney, Towards the development of a fitness for service tool for the inspection of corrosion under insulation, Paper 7368, Corrosion 2016, NACE, Houston, TX, 2016.
[49] F. Simonetti, P.B. Nagy, S.M. Bejjavarapu, G. Instanes, and A.O. Pedersen, Long-range microwave detection of wet insulation for CUI mitigation, Paper 5747, Corrosion 2015, NACE, Houston, TX, 2015.
[50] M. Funahashi, Solution to CUI with three layered control and warning systems, Paper 4079, Corrosion 2014, NACE, Houston, TX, 2014.
[51] S. Caines, F. Khan, J. Shirokoff, W. Qiu, Demonstration of increased corrosion activity for insulated pipe systems using a simplified electrochemical potential noise method, J. Loss Prev. Proc. Ind. 47 (2017) 189–202.
[52] A. Schoon, R. Fjellanger, M. Kjeldsen, K.-U. Goss, Using dogs to detect hidden corrosion, Appl. Anim. Behav. Sci. 153 (2014) 43–52.

Cathodic protection and stray current measurement and monitoring

Yanxia Du
University of Science and Technology Beijing, Beijing, People's Republic of China

20.1 Cathodic protection measurement and monitoring

20.1.1 Introduction of cathodic protection mechanism and methods

As an effective electrochemical method, cathodic protection (CP) has been widely used in the industry to protect metal structure from electrochemical corrosion. Electrochemical corrosion is the result of electrochemical reactions driven by the structure-to-electrolyte potential difference between anode and cathode in a corrosion cell.

The mechanism of cathodic protection is reducing the structure-to-electrolyte potential difference between the local anodic and cathodic sites to zero by cathodic polarization, resulting in zero corrosion current flow [1]. This can be accomplished by impressing an external current flowing from electrolyte into steel structure and polarizing the structure-to-electrolyte potentials of cathodic sites in an electronegative direction. As the structure-to-electrolyte potentials of the cathodic sites on the structure are polarized toward the open-circuit potentials of the anodic sites, corrosion current is reduced. When the structure-to-electrolyte potentials of all cathodic sites reach the open-circuit potential of the most active anodic sites, electrochemical corrosion is eliminated on the structure [2–7]. The simplified description on the cathodic polarization process of cathodic protection is given in Fig. 20.1. The values indicated in Fig. 20.1 are structure-to-electrolyte potentials along a buried steel pipeline, which are the potential difference between the metal structure and the electrolyte at the local outer of double layer across the structure-electrolyte interface. The values in "Native corrosion" situation are the open-circuit potentials of the structure at these sites which can be measured if the individual sites are physically isolated from the bulk structure.

There are two kinds of CP methods: sacrificial anode CP and impressed current CP. For sacrificial anode CP, a more active metal is electrically connected to a less active metal forming a galvanic corrosion cell, in which the more active metal corrodes ("sacrifice" itself) to protect the less active metal (see Fig. 20.2). Sacrificial anode CP is generally used for the circumstances with small current requirement, low resistivity of the electrolyte, or local protection for a specific area.

Techniques for Corrosion Monitoring. https://doi.org/10.1016/B978-0-08-103003-5.00019-9

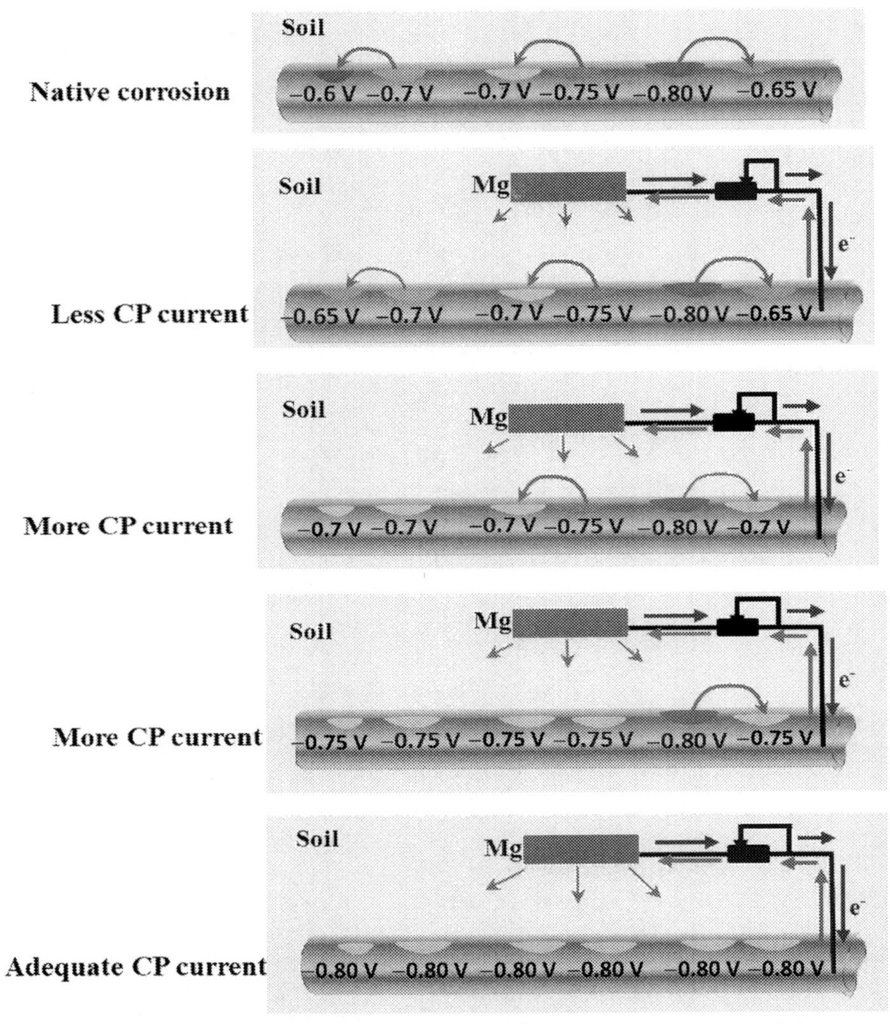

Fig. 20.1 The cathodic polarization process of the structure-to-electrolyte potentials on steel structure under CP Note: *Red* arrow (*dark gray* in print version)—direction of positive current or flow of positive ions in electrolyte or the opposite direction of electron flow in electronic conductor; *blue* arrow (*light gray* in print version)—direction of electrons migration; *green* (*gray* in print version) area—cathodic sites; *red* (*dark gray* in print version) and *orange* (*gray* in print version) areas—anodic areas; potential values—structure-to-electrolyte potentials.

For impressed current CP, the protection current is provided by a DC power supply and it flows from the positive terminal of power supply through cable to auxiliary anode and flows out from the anode, going through the electrolyte and entering the protected structure, and then goes back to the negative terminal of power supply via the protected structure, as shown in Fig. 20.3. Generally, an impressed current

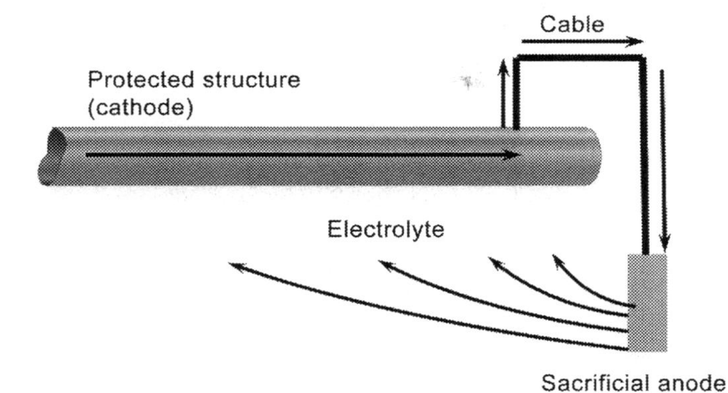

Fig. 20.2 Schematic for sacrificial anode CP. Note: Arrows show the direction of positive current.

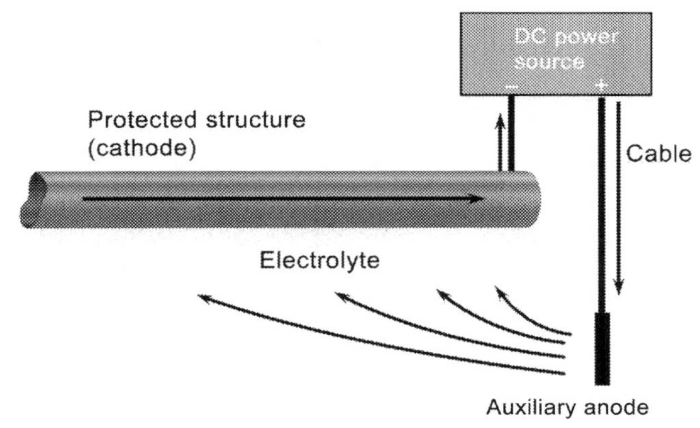

Fig. 20.3 Schematic for impressed current cathodic protection. Note: Arrows show the direction of positive current.

CP system operates at higher current output and is capable of providing voltage levels larger than a sacrificial anode CP system, so it is often used in the circumstance with larger current requirement, without limitation for electrolyte resistivity.

The DC power supply used in an impressed current CP system has several types including rectifiers, potentiostats, galvanostats, and other sources of DC power such as solar power, batteries, wind-driven generators, etc. The most common type of DC power supply used for impressed current CP is a rectifier, which steps the AC power supply voltage to AC output voltage and then converts it into the required DC voltage output. For a potentiostat, the output current and voltage of it is controlled by a preset structure-to-electrolyte potential. The potentiostat constantly monitors the structure-to-electrolyte potential at a certain place and compares it to the preset value. If the monitored structure-to-electrolyte potential changes from the preset value, the potentiostat either increases or decreases current output to bring the structure-to-electrolyte potential back to its preset value. For a galvanostat, the output current is maintained at a preset constant value. If the external resistance changes, the output voltage of the galvanostat is increased or decreased to maintain the preset output current value. Rectifiers, potentiostats, and galvanostats are normally powered by AC power system. When AC power is not readily accessible, other sources of DC power such as solar power, batteries, wind-driven generators, etc. are available.

20.1.2 Cathodic protection criteria and understanding

As introduced before, the true criterion for cathodic protection is polarizing the structure-to-electrolyte potentials of all cathodic sites on the entire structure to the open-circuit potential of the most active anodic site. But in the practice, it is difficult to measure the open-circuit potential of individual anodes and cathodes on a structure. Therefore, the surrogate criteria are needed. It should be noted that CP criteria are different in different countries, and the selection of appropriate criteria according to the protected material and corresponding environment is very important for the successful application of CP.

In NACE Standard SP0169, "Control of External Corrosion on Underground or Submerged Metallic Piping Systems" [8], three criteria for determining the achievement of adequate CP for steel and gray or ductile cast-iron piping are as follows:

> *6.2.1.1 Criteria that have been documented through empirical evidence to indicate corrosion control effectiveness on specific piping systems may be used on those piping systems or others with the same characteristics.*
> *6.2.1.2 A minimum of 100 mV of cathodic polarization. Either the formation or the decay of polarization must be measured to satisfy this criterion.*
> *6.2.1.3 A structure-to-electrolyte potential of −850 mV or more negative as measured with respect to a saturated copper/copper sulfate (CSE) reference electrode. This potential may be either a direct measurement of the polarized potential or a current-applied potential. Interpretation of a current-applied measurement requires consideration of the significance of voltage drops in the earth and metallic paths.*

The selection of appropriate criteria and understanding them are the base for appropriate application of CP [6,7]. Because CP is a polarization phenomenon, the acquirement of polarized potential or polarization for the protected structure is important to determine the level of protection. The potential of interest used to determine the effectiveness of CP is polarized potential. A polarized potential means the potential difference across the structure/electrolyte interface which is the sum of native corrosion potential and cathodic polarization. The most accepted criterion for steel in general soil or water environment is the polarized potential of $-850\ mV_{CSE}$. The evaluation of this criterion to be satisfied or not is based on the direct measurement of the polarized potential or the interpretation of a current-applied potential by eliminating the errors of voltage drops in the earth and in the metallic paths caused by the current (IR drop).

20.1.3 Cathodic protection measurement and monitoring

The measurement of structure-to-electrolyte potential is the most common means for determining if adequate CP has been achieved or not. With the application of CP current flowing from the electrolyte into a structure, the structure-to-electrolyte potential changes. The potential change is a reflection of polarization. Measurements are carried out to determine whether one of the CP criteria is met.

As mentioned in Section 20.1.2, potential measurement often includes errors caused by IR drop, particularly in the electrolyte. According to CP standards, the magnitude of IR drop error must be determined and appropriately corrected during the measurement and evaluation of field data. In some situations, the IR drop error can be neglected if the current and/or the resistance of a current path are very small. However, before neglecting the IR drop, the magnitude of IR drop should be determined in order to verify its insignificance. There have been several widely used methods for determining and correcting the IR drop [6–8]:

- Place the reference electrode close to the structure being measured. On a coated structure, the reference electrode should be placed approximate to the coating defect or holiday.
- Interrupt all the contributing CP current instantly and measure the instant-off potential before significant depolarization occurs to represent the polarized potential.

In the practice, it is difficult or not economical to place the reference electrodes close to the coating holiday in all places; therefore, the first method in the above is only used for specific situations. The "instant-off" method is used more frequently to get polarized potential and evaluate CP effectiveness. This method requires that all the contributing current must be interrupted simultaneously. However, when there are stray current or direct electrically connected galvanic anodes, it is impractical to interrupt all the current. At this time, CP coupons are used to evaluate the effectiveness of CP system.

CP coupons serving as the exposed metal at coating defect (coating holiday) are used to get polarized potential of the structure at specific places and they should be of the same metal as the protected structure and represent the typical size of coating holiday on the structure. The coupons should be placed where they receive the same exposure to CP current as the structure receives and electrically connected to the protected structure through the cables at test stations, as shown in Fig. 20.4.

Advantages of using CP coupons are that polarized potential can be obtained without interrupting multiple power sources, direct-connected galvanic anodes, and stray current. It should be understood that there can be a discrepancy between the polarized potential of the coupon and that of the structure if the coupon cannot represent the coating holiday on the structure completely under certain circumstances as follows:

- To measure the coupon-to-electrolyte potential, a reference electrode is placed in the soil near the coupon, or near the ground. The coupon is instantly disconnected from the structure and the instant-off potential of the coupon is read. The instant-off potential of the coupon is similar to the instant-off potential of the structure if a holiday with the same size were exposed to the soil on the structure at that location. If the coupon meets a certain criterion, then it is inferred that the structure with the same size of holiday would also meet this

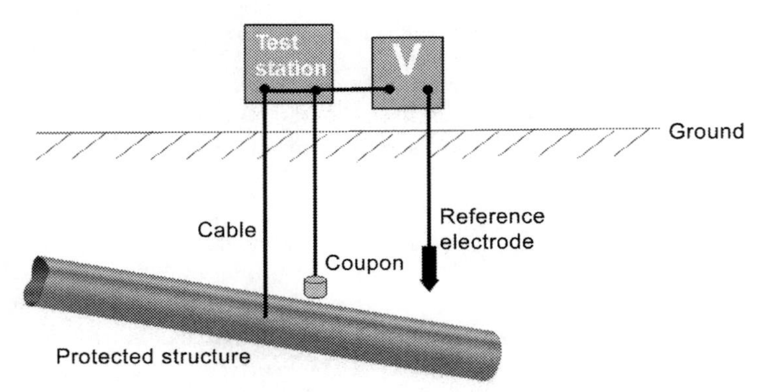

Fig. 20.4 Schematic for test station with CP coupon.

criterion. During the measurement of instant-off potentials of coupons, it is important to know how soon the measurement should be conducted after the disconnection to ensure complete disconnection and avoid depolarization, which is called instant-off delay time and related to environmental conditions, coupon material, size, and polarization characteristics of the coupon. Field experiments showed that, for pipeline steel coupon with enough time of polarization, instant-off delay times of 200 and 300 ms are suggested for the coupons of 6.5 and 15 cm^2, respectively [9].

Sometimes the coupon and reference electrode are assembled as a whole, called CP probe. The CP probe is buried near the protected structure and electrically connected to the pipeline through the test station. The test station with CP probe is shown in Fig. 20.5.

Some of these test stations are available with resistance-type probes to monitor corrosion rate as introduced in Chapter 17.

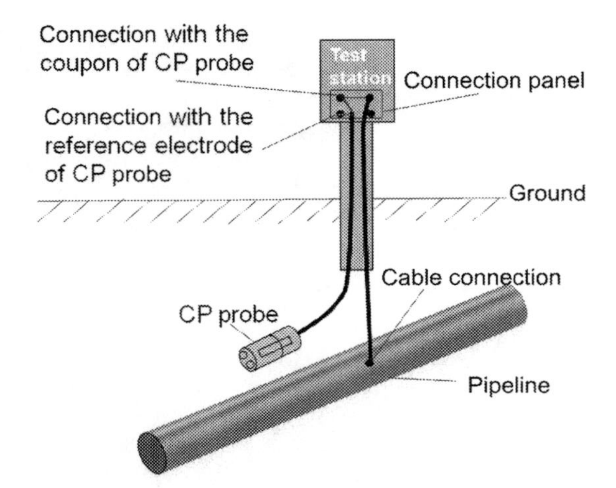

Fig. 20.5 Schematic for test station with CP probe.

As mentioned in Chapter 17, in recent years, a lot of CP monitoring test stations with CP coupons or probes are constructed to realize the remote monitoring of CP system with the help of internet and mobile communication technologies, as shown in Fig. 20.6.

20.2 DC stray current interference detection and monitoring

20.2.1 Introduction of DC stray current interference and types

DC stray current is the current flowing through a metal structure from unintended circuit. The metal structure interfered by DC stray current is called foreign structure or interfered structure and the source of the DC stray current is called interfering structure or interfering source. The DC stray current in an interfered structure is different from the galvanic corrosion current between anode and cathode on the same structure and the current from the CP system of the structure. All the DC stray current picked up by some area of the interfered structure will be discharged from other area of the structure, that is to say there is no net DC stray current accumulation on the whole interfered structure.

DC stray current can be classified as being either static or dynamic. Static DC stray current is defined as the DC stray current that maintains constant value and constant geographical path. Examples of DC stray currents include those from impressed current CP systems and High-Voltage Direct Current (HVDC) ground electrodes [10–13]. Fig. 20.7 shows how DC stray currents from a cathodic protection system

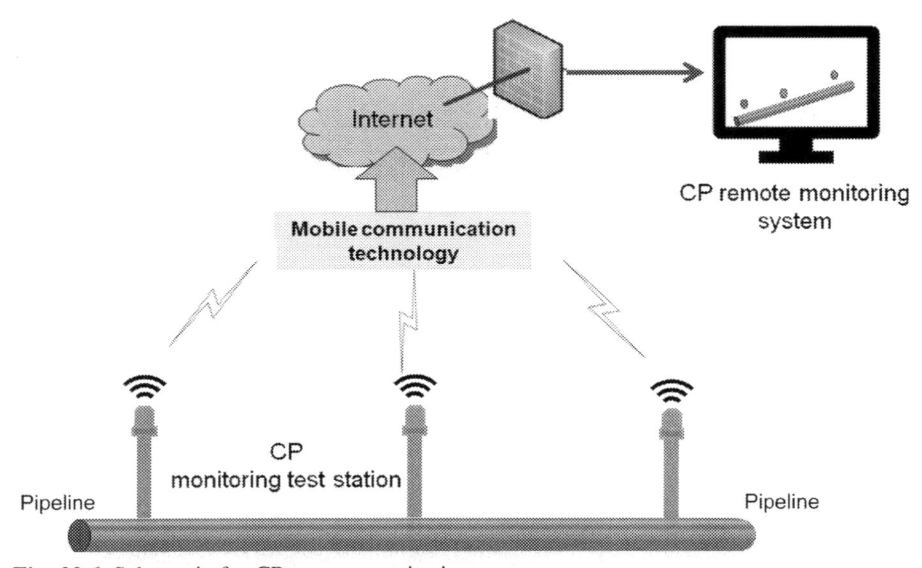

Fig. 20.6 Schematic for CP remote monitoring system.

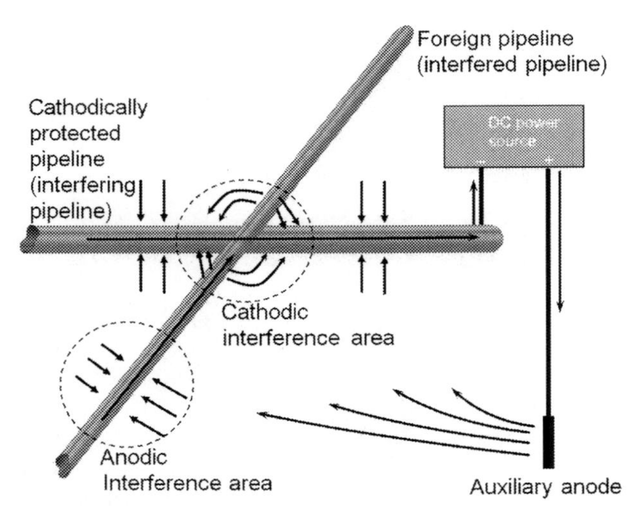

Fig. 20.7 DC stray current interference from an impressed current CP system.

(interfering source) interfere a foreign pipeline in different areas. DC stray current interference includes three types: anodic interference, cathodic interference, or a combination of the two [10–13]. The polarity of the voltage gradients overlapping a foreign structure to remote earth caused by the interfering source determines the type of interference. For cathodic interference, as shown in Fig. 20.7, when a foreign pipeline without CP crosses a cathodically protected pipeline, because of the direction of CP current, a voltage gradient negative with respect to remote earth overlaps the foreign structure and promotes current discharge from the foreign pipeline (interfered pipeline) to the protected pipeline (interfering pipeline). This situation is called cathodic interference and it generally occurs near the cathode of a CP system or other sources. For anodic interference, as shown in Fig. 20.7, when a foreign pipeline without CP is close to the auxiliary anode groundbed of a CP system, a voltage gradient positive to remote earth overlaps the interfered area of the foreign structure and promotes current be picked up by foreign pipeline in this section area. This situation generally occurs near the anode groundbed of a CP system or other sources. When both anodic interference and cathodic interference are obvious in a structure, the situation is called combination interference.

Dynamic DC stray current is the DC current from unintended circuit that varies in amplitude and/or direction with time. One typical example of dynamic DC stray current is DC transit system. The stray current paths originating from a DC transit system are shown in Fig. 20.8.

As shown in Fig. 20.8, although it is the intent that the DC load current (I_L) returns to the substation via the rail (I_R), because there exists a certain resistance along the rail and, at the same time, the rail could not be isolated from the ground completely in most of cases, some of the load current (I_L) will pass through the earth and become stray current (I_e). If there is a steel pipeline buried in the earth in parallel with the rail as

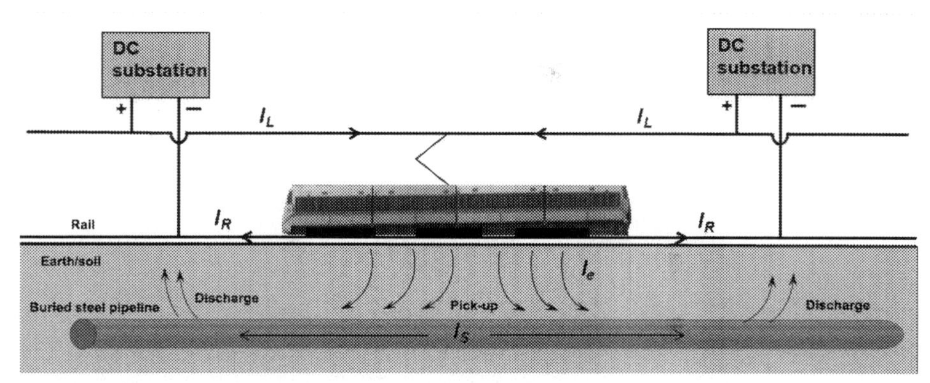

Fig. 20.8 DC stray current paths originating from a DC transit system. Note: I_L—load current; I_R—rail current; I_e—the stray current flowing through earth or soil; I_S—the stray current flowing through buried steel pipeline.

shown in Fig. 20.8, it will carry some of the load current as stray current (I_s). The stray current will be picked up by some area of the pipeline and discharged from other area of the pipeline. Acceleration of corrosion occurs where the current is discharged from the pipeline and undesirable overprotection effects could occur at the location where the current is picked up. For the buried pipeline interfered by DC transit system, the typical characteristic is the dynamic fluctuation of pipe-to-soil potential because of the rapid movement of trains along rail [10, 14–16].

At a particular location on an interfered structure, the presence of stray current could be identified when fluctuating pipe-to-soil potentials are recorded with time. The changes of pipe-to-soil potentials with time have typical characteristics for the pipelines interfered by a transit system. There are considerable potential fluctuations during the daytime when the transit system is running, while light activity from late night to early morning when the transit system is off, as shown in Fig. 20.9.

20.2.2 DC stray current interference evaluation criteria

For static DC stray current evaluation, generally CP criteria are available for the structure with cathodic protection. The European Standard EN 50162 "Protection against corrosion by stray current from direct current systems" [17] describes the criteria used to evaluate DC stray current. For freely corroding steel structures interfered by static DC stray current, it is acceptable when the positive potential shift (excluding IR-drop) is lower than 20 mV. For structures under CP, it is considered unacceptable if the IR-free potential of the structure is outside the potential range specified by CP criteria, i.e., if the IR-free potential of the structure does not match the -850 mV$_{CSE}$ criterion.

However, the criteria used for corrosion evaluation of buried pipelines in the presence of dynamic DC stray current have not reached a consensus as so far in the world. There is no specific NACE standard for corrosion risk evaluation under dynamic DC

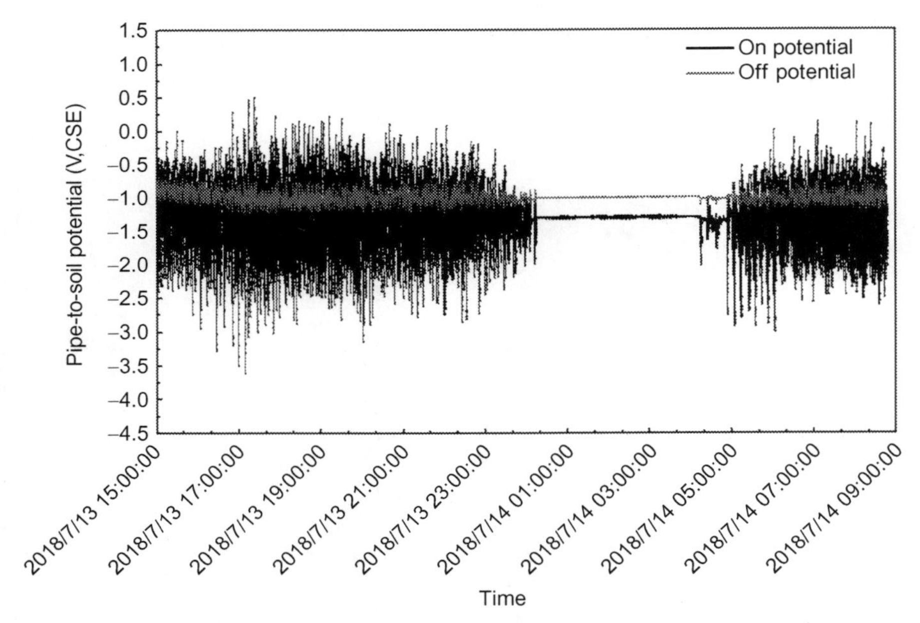

Fig. 20.9 Dynamic characteristic of pipe-to-soil potentials interfered by DC transit system. Note: the measurements were conducted with coupon electrically connected with pipeline interfered by DC transit system and the off-potentials were recorded 200 ms after the disconnection from the pipe.

stray current, and in NACE Standard of SP0169-2013 [8], only the detection and control of stray current are specified, while the corrosion risk evaluation under dynamic DC stray current which is different from CP criteria is not included.

Australian Standard AS 2823.1 "Cathodic protection of metals Part 1: Pipes and cables" [18] specifies the potential and time criteria for structures subject to variable traction stray current, in which the potential more positive than the general protection criterion (−850 mV, referenced to copper-copper sulfate reference electrode, abbreviated as CSE) in a certain time is permitted. It is necessary to record the potential during a sufficient period of time to ensure that the maximum exposure to the stray current effects is encompassed. This period shall include the quiescent traction period and the morning and evening usage peaks and is nominally 24 h with a minimum of not less than 20 h. If a data logger is used for the monitoring of potential, its frequency of sampling shall be not less than 4 samples per minute. The protection criteria for structures subject to traction current effects vary with the structure polarization time as follows:

(a) Structure with short polarization times

Structures with sound coating characteristics, or those have otherwise been proven to polarize and depolarize rapidly in response to stray current, shall comply with the following criteria:

(i) The potential shall not be more positive than the protection criterion for more than 5% of the test period.

(ii) The potential shall not be more positive than the protection criterion plus 50 mV (i.e., −800 mV for ferrous structures) for more than 2% of the test period.

(iii) The potential shall not be more positive than the protection criterion plus 850 mV (i.e., 0 mV for ferrous structures) for more than 0.2% of the test period.

Where such structures do not comply with the above and are characterized by relatively brief anodic excursions interspersed with longer cathodic periods and can demonstrate protection through other means such as resistance probes, coupons or structure corrosion rate history shall still comply, provided they are not more positive than the protection criterion for more than 5% of the test period.

(b) Structure with long polarization times

The potential of structures that exhibit deteriorated coating characteristics, or have otherwise been proven to polarize and depolarize slowly in response to stray current, shall not be more positive than the protection criterion for more than 5% of the test period. The positive excursions shall be brief and have negative excursions between them.

The European Standard EN 50162 "Protection against corrosion by stray current from direct current systems" [17] introduced a method using current probes to evaluate fluctuating stray current interference on cathodically protected structures. In this method, an insulated steel probe with a bare steel surface is pushed into the ground to the depth of the pipeline. The probe is electrically connected to the pipeline. The bare surface of the probe now functions as the steel surface in a simulated coating defect. The recording ammeter is used to determine the direction and the magnitude of the current. The measurement is typically carried out during a period of 24 h. The procedures of measurement and evaluation are as follows: Step1: The probe current corresponding to the cathodic protection potential of the pipeline is measured during a period when the pipeline is not interfered by fluctuating stray current (e.g., at night). This probe current is defined as 100% (reference value). Step 2: The probe current (the resultant of cathodic protection current and stray current) is continuously recorded during a period of typically 24 h. Step 3: For evaluation of the hour with the highest probe current reductions, the hour with the most positive potential fluctuations is identified. Step 4: Probe currents below any of the values given in column 1 of Table 20.1 indicate a high risk of corrosion if their accumulated duration exceeds the corresponding values of column 3 of Table 20.1.

Table 20.1 Current criteria in case of interference due to DC traction systems.

Probe current in % of the reference level	Maximum acceptable occurrence period	
	in % of the worst hour	in seconds
>70	Unlimited	
<70	40	1440
<60	20	720
<50	10	360
<40	5	180
<30	2	72
<20	1	36
<10	0.5	18
<0	0.1	3.6

Note: The figures in table are based on 10 years practical experience.

20.2.3 DC stray current interference detection and monitoring

Static DC stray current interference could be detected by analyses of structure-to-electrolyte potential surveys. The key points to consider in detecting static DC stray current effects on a structure are:

- Potential profiles show abnormally negative or positive potentials.
- The locations with abnormally negative potentials measured on some area of a structure are close to anode groundbed of other CP systems or other current sources.
- The locations with abnormally positive potentials measured on some area of a structure are close to other cathodically protected structures or other sources.
- Unusual currents are measured along the structure.

Dynamic stray current interference can be detected from structure-to-electrolyte potentials and/or line current measurements. A structure-to-electrolyte potential that is changing with time indicates the existence of dynamic stray current. For dynamic stray current detection, the data on the stray current magnitude over a long term (24 h or more) are needed. Dynamic stray current monitoring is important to provide continuous data recording and information on the value of the maximum stray current affecting the structure. A recorder, data logger, or a remote monitoring system could be used for the dynamic stray current monitoring.

Dynamic changes of structure-to-electrolyte on-potentials (including IR drop) only indicate the presence of stray currents. The IR-free potential is required to evaluate the corrosion risk under dynamic stray current. The method to get IR-free potential in the presence of dynamic DC stray currents is to measure the instant-off potential of coupon which is buried in close proximity to the measured pipeline and electrically connected to the pipeline as shown in Fig. 20.10. When the coupon is connected to the pipe, it simulates a coating holiday with similar size on the pipe coating. Upon interruption of the coupon to pipe connection, not only the CP current but also the stray current is interrupted. The use of the coupon also allows for measuring the current passing between the coupon and the protected pipeline.

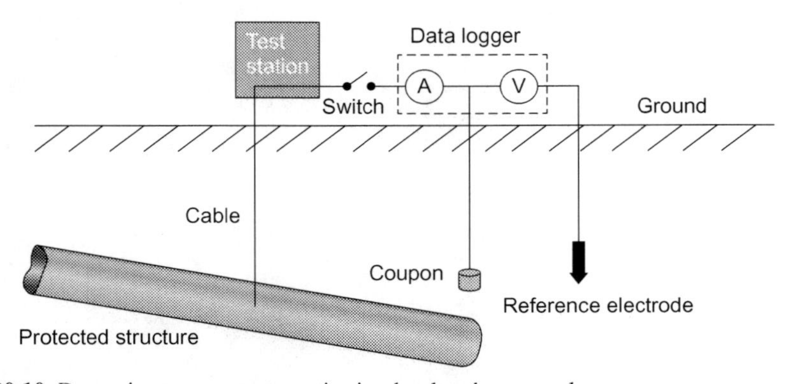

Fig. 20.10 Dynamic stray current monitoring by data logger and coupon.

20.3 AC Stray current interference detection and monitoring

20.3.1 Introduction of AC stray current interference and types

With the development of the power, transportation, and petroleum industries, more and more oil and gas pipelines were buried in parallel with high-voltage alternating current (AC) transmission lines or AC electrified railway, which brings about AC stray current interference (AC interference). In the recent 20 years, AC interference has caused wide attention in the international corrosion community [19–24].

AC interference on pipeline can arise from three possible mechanisms including inductive, resistive, and capacitive coupling between the pipeline and AC carrying structure [24, 25]. The brief introduction for these three mechanisms is as follows:

- Inductive coupling occurs when metallic pipelines are buried parallel to high-voltage AC transmission lines as seen in Fig. 20.11 [25]. In this situation, the buried pipeline acts as the secondary winding of an air-core transformer and the overhead power line acts as the primary winding of the transformer. AC interference could be induced on the pipeline according to Faraday's laws of induction [26].
- Resistive coupling can occur between the grounding of AC interference sources to earth in which a pipeline is buried [25, 26]. It is primarily a concern for fault currents, i.e., a short-circuit of the AC source, as shown in Fig. 20.12; in this case, the fault current is much greater in magnitude than steady state powerline current, and therefore, the resistive coupling can generate very high pipeline-to-soil voltage, resulting in coating damage, melting of the pipeline wall, or a safety hazard to personnel.
- Capacitive coupling can exist between two electrically conductive materials separated by a dielectric medium capable of storing charges. For the pipeline on the ground, capacitive

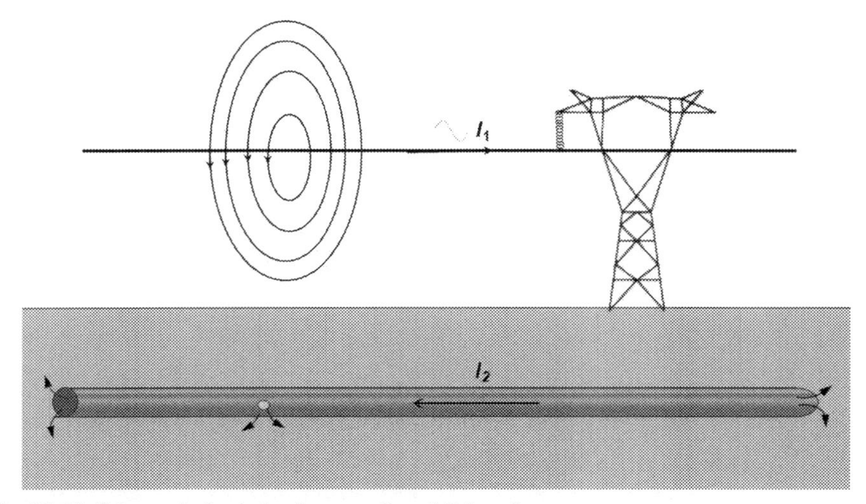

Fig. 20.11 Schematic for inductive coupling AC interference.

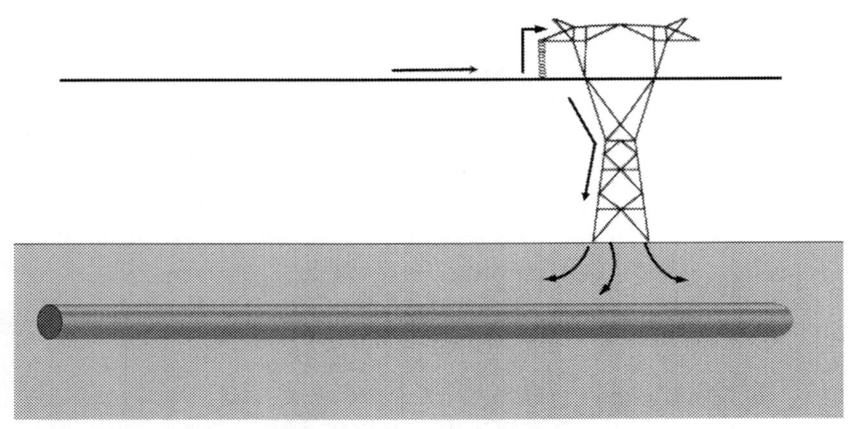

Fig. 20.12 Schematic for resistive coupling AC interference.

coupling could exist between HVAC transmission line and pipeline. However, when the pipeline is buried in soil, there is no such kind of AC interference.

Among the above three kinds of AC interference mechanisms, inductive coupling is the most common form. The AC voltage and current are induced in the pipeline by inductive coupling, which can result in AC corrosion and affect the safety of personnel.

20.3.2 AC interference evaluation criteria

In order to provide reference for the evaluation and mitigation of AC interference in actual production, some standards on AC corrosion risk evaluation have been developed based on existing research results, such as BS EN 15280:2013 "Evaluation of AC corrosion likelihood of buried pipelines applicable to cathodically protected pipelines" [27], ISO 18086:2019 "Corrosion of metals and alloys—Determination of AC corrosion-Protection criteria" [28], and NACE SP21424-2018 "Alternating Current Corrosion on Cathodically Protected Pipelines: Risk Assessment, Mitigation, and Monitoring" [29]. In the first two standards, the provisions on acceptable AC interference levels are similar, and three conditions must be satisfied at the same time. The first is that AC voltage on the pipeline should be decreased to 15 V or less. The second is that the cathodic protection system shall be capable of polarizing all parts of the buried structures to potentials more negative than E_P (IR-Free metal-to-electrolyte potential, $-0.85\ V_{CSE}$ for carbon steels, low alloyed steels, and cast iron in normal soil and water conditions), which are specified in EN 12954:2019 "General principles of cathodic protection of buried or immersed onshore metallic structures" [30] or ISO 15589-1:2015 "Petroleum, petrochemical, and natural gas industries—Cathodic protection of pipeline Systems Part 1: On-land pipelines" [31]. The third one is that at least one of the following items must be met: (a) maintaining AC current density to be smaller than 30 A/m^2, (b) maintaining average cathodic current density smaller than 1 A/m^2 if AC current density is more than 30 A/m^2, and (c) maintaining the ratio between AC current density and DC current density less than 5.

Based on the current studies [20–24], when AC current density is higher than a limited value, AC corrosion cannot be neglected no matter how to adjust CP current density. NACE SP21424-2018 "Alternating Current Corrosion on Cathodically Protected Pipelines: Risk Assessment, Mitigation and Monitoring" [29] specified it as following:

> *Unless effective control of AC corrosion has been shown by a documented corrosion rate less than 0.025 mm/y, the AC current density should not exceed a time-weighted average of:*
> - *30 A/m^2 if DC current density exceeds 1 A/m^2;*
> - *100 A/m^2 if DC current density is less than 1 A/m^2.*

It could be seen that there exists discrepancy between the NACE standard practice and the international ISO or European EN standards. Although a lot of AC corrosion research work has been carried out in the past 20 years [19–24], the mechanism of AC corrosion process has not been completely understood and there still exist a lot of debates on AC corrosion assessment criteria under cathodic protection (CP), especially under high CP level, including the following questions: (1) AC corrosion may be divided into two categories of "high CP" and "low CP" AC corrosion [24], but only the DC threshold for "high CP" is specified in both NACE and ISO standards; (2) In both NACE SP21424-2018 and ISO 18086:2019, the DC threshold is given by the DC current density of 1 A/m^2, but the generally used CP criteria are given by the polarized potential; because the DC current density is theoretically related to the polarized potential, how to determine the thresholds for DC polarized potential under AC interference? (3) Researches have shown chemical environment has great effect on AC corrosion behavior, how to consider the environmental factors during the AC corrosion risk assessment? A lot of research work still need to be carried out to answer the above questions.

20.3.3 AC interference detection and monitoring

To evaluate the AC corrosion likelihood on buried pipeline, detailed detection method has been explained in corresponding AC interference standards. The detected parameters generally include AC voltage, AC current density, DC current density, DC polarized potential, and environmental parameters. The measurement methods for AC voltage and AC/DC current density are introduced in the following portion.

AC voltage: According to CEN/TS 15280-2015 [27] and NACE SP21424-2018 [29], the AC voltage is the ultimate driving force for the AC current density at a coating defect, which may cause corrosion or the AC current density at grounding devices (including galvanic anodes) installed for mitigation purpose. AC voltage measurements are used to determine the level of AC interference. For this purpose, measurements must be made using a data logging device programmed to measure the AC voltage in intervals sufficiently short to capture the steady state long-term interference. The measurement sampling rate shall be consistent with the type of interference.

High-voltage power system interference can typically require a sampling rate in the order of minutes, whereas AC or DC railway interference can require a sampling rate in the order of seconds. AC voltage measurements are usually made with reference to remote earth. However, when continuous anode systems are buried in parallel and close to the pipeline, the AC voltage across coating may be a more accurate measure of the driving force for the AC current density than the measure of AC voltage to remote earth. The measurement of AC voltage to remote earth is described in Annex G of CEN/TS 15280-2015 [27].

AC or DC current densities are very important parameters to evaluate AC corrosion risk under CP, which are generally measured by a coupon or probe used to simulate a coating defect at certain area. The current flowing through the coupon or probe (AC or DC) can be measured by the voltage drop across a resistor in series, as showed in Fig. 20.13. For both AC and DC current measurement, the value of the resistor in series should be sufficiently low to avoid significant disturbance of system. For field measurements, this is typically in the range of 1–10 Ω for a 1 cm^2 coupon. After measurements, the current density is calculated by the current and working area of the coupon.

For AC interference detection, the data on the AC interference magnitude over a long term (24 h or more) are needed because the load of AC source might change in different time. AC interference monitoring is also important to provide continuous data recording and information on the value of the maximum AC interference affecting the structure. A recorder, data logger, or a remote monitoring system could be used for AC interference monitoring.

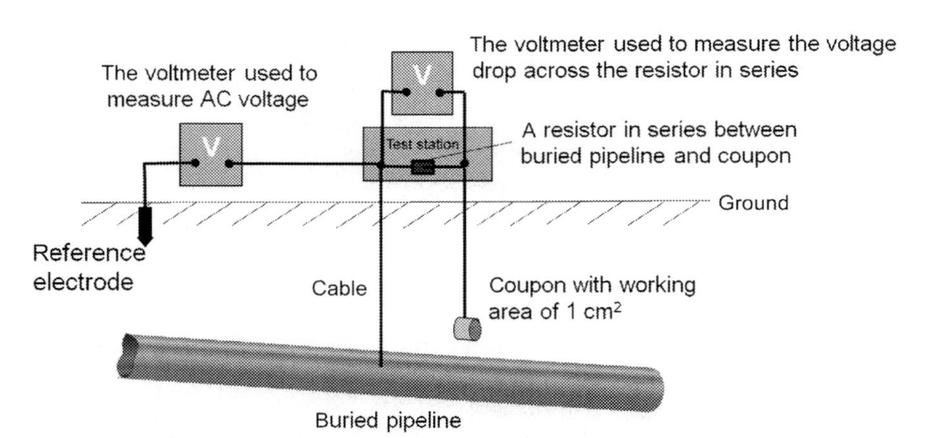

Fig. 20.13 Measurement of AC and DC current by coupon or probe with series resistor.

20.4 Cathodic protection monitoring with corrosion probes

As mentioned earlier, the purpose of cathodic protection is to reduce the corrosion rate of the metal to be protected. Corrosion probes that measure the actual corrosion rate under the cathodic protection conditions may be used to monitor the effectiveness of cathodic protection. Electrical resistance (ER) probe is one kind of probes that can be used for such purpose. During the measurement, the sensing element of the ER probe is connected to the protected metal structure, so the sensing element corrodes at the same potential as the metal structure. If the CP potential is sufficiently low, it will stop the corrosion of the metal being protected and so stop the corrosion of the sensing element of the ER probe. Chapter 11 provides detailed description on the ER probes and Chapter 17 provides more information on the use of ER probes for monitoring pipeline corrosion in soil.

Coupled multielectrode array sensor (CMAS) is another type of probes that are used to monitor the effectiveness of cathodic protection. Chapter 8 provides detailed information on the principle of the CMAS probe. Chapter 24 provides more detailed information on the use of the CMAS probes for monitoring the corrosion of metals in brine solutions, concrete, and soil under cathodic protection conditions.

References

[1] R.A. Gummow, S. Segall, D. Fingas, An alternative view of the cathodic protection mechanism on buried pipeline, in: NACE Corrosion 2016, Paper No. 7531, 2016.
[2] R.B. Mears, R.H. Brown, A theory of cathodic protection, Trans. Electrochem. Soc. 74 (1938) 519–531. Presented October 15, 1938.
[3] W.J. Schwerdtfeger, O.N. McDorman, Potential and current requirements for the cathodic protection of steel in soils, Corrosion (1952) 392.
[4] S.C. Dexter, L.N. Moettus, K.E. Lucas, On the Mechanism of Cathodic Protection, vol. 41, NACE Corrosion, 1985, p. 606 #10, Oct.
[5] N.G. Thompson, T.J. Barlo, Fundamental processes of cathodically protecting steel pipelines, in: Gas Research Conference Proceedings, Government Institutes, Inc, Rockville, MD, 1983.
[6] NACE Standard TM0497 (latest revision), Measurment Techniques Related to Criteria for Cathodic Protection on Underground or Submerged Metallic Piping Systems.
[7] NACE CP Course Material, Chapter 2: Concept of Cathodic Protection, Houston, TX: NACE.
[8] NACE RP0169-2013, Standard Recommended Practice – Control of External Corrosion on Underground or Submerged Metallic Piping.
[9] D. Tang, D. Yanxia, J. Liu, et al., Some considerations on the use of coupon under stray direct traction current interference, in: NACE Corrosion 2018, Paper No. 11028, 2018.
[10] NACE CP Interference Course Material, Chapter 2: DC Interference, Houston, TX: NACE.
[11] F. Brichau, J. Deconinck, T. Driesens, Modeling of underground cathodic protection stray currents, Corros. Eng. 52 (6) (1996) 480–488.

[12] D. Yanxia, J. Wang, M. Lu, Researches on mitigation methods of interference caused by cathodic protection system, in: NACE Corrosion 2015, Paper No. 5884, 2015.
[13] R. Qin, Y. Du, G. Peng, M. Lu, Z. Jiang, High voltage direct current interference on buried pipelines: case study and mitigation design, in: NACE Corrosion 2017, Paper No. 9049, 2017.
[14] S.R. Allahkaram, M. Isakhani-Zakaria, M. Derakhshani, et al., Investigation on corrosion rate and a novel corrosion criterion for gas pipeline affected by DC interference, J. Nat. Gas Sci. Eng. 26 (2015) 453–460.
[15] G. Lucca, Estimating stray current interference from DC traction lines on buried pipelines by means of a monte carlo algorithm, Electr. Eng. 97 (4) (2015) 277–286.
[16] D. Yanxia, D. Tang, H. Qin, M. Lu, Research on parameter fluctuation characteristics and effects on corrosion rates under dynamic DC stray current from metro system, in: Corrosion 2019, Paper No. 13203, 2019.
[17] BS EN 50162-2004, Protection against corrosion by stray current from direct current systems.
[18] Australian Standard AS 2823.1-2015, Cathodic protection of metals Part 1: Pipes and cables.
[19] NACE TG 327 AC Corrosion State-of-the-Art: Corrosion Rate, Mechanism, and MitigationRequirements 2007-02-08.
[20] R.A. Gummow, R.G. Wakelin, S.M. Segall, AC corrosion-a new challenge to pipeline integrity, in: Corrosion/1998, Houston, TX:NACE, paper No. 98566, 1998.
[21] R.G. Wakelin, R.A. Gummow, S.M. Segall, AC corrosion-ease histories, test procedures, mitigation, in: Corrosion/1998, Houston, TX: NACE, , paper No.98565, 1998.
[22] M. Buchler, Alternating current corrosion of cathodically protected pipelines: discussion of the involved processes and their consequences on the critical interference values, Mater. Corr. 63 (1) (2012).
[23] D. Yanxia, Y. Xiao, S. Xie, M. Lu, Research on the effects of environmental parameters on AC corrosion behavior, in: Corrosion 2018, Paper No. 10676, 2018.
[24] A. Junker Olesen, AC Corrosion of Cathodically Protected Pipelines, Technical University of Denmark, 2018.
[25] NACE CP Interference Course Material, Chapter 3: AC Interference, Houston, TX: NACE.
[26] A.R. Hambley, Electrical Engineering—Principdles and Applications, sixth ed., Pearson, 2014.
[27] BS EN 15280:2013, Evaluation of a.c corrosion likelihood of buried pipelines applicable to cathodically protected pipelines.
[28] BS ISO 18086:2019, Corrosion of metals and alloys – Determination of AC corrosion-Protection criteria.
[29] NACE SP21424-2018, Alternating Current Corrosion on Cathodically Protected Pipelines: Risk Assessment, Mitigation and Monitoring.
[30] EN 12954:2019, General principles of cathodic protection of buried or immersed onshore metallic structures.
[31] ISO 15589-1:2015, Petroleum, petrochemical and natural gas industries—Cathodic protection of pipeline Systems Part 1: On-land pipelines.

Remote monitoring and computer applications

21

Rich Smalling, Ed Kruft, Dale Webb, and Leslie Lyon-House
American Innovations, Austin, TX, United States

21.1 Introduction

21.1.1 Why remote monitoring?

The production and transportation of oil and gas is a dangerous business. Pipeline companies are constantly striving to improve their safety records to protect personnel, the general public, and the environment. Data from remote monitoring systems are extremely important for operators of oil and gas-related equipment to help them improve the operation of their assets. Many of the largest and most respected oil and gas companies in the world are using data captured with remote monitoring systems to inform and improve their operations and safety performance.

The United States Pipeline and Hazardous Materials Safety Administration (PHMSA) reports that there are over 2.7 million miles (4.5 million km) of natural gas and hazardous liquid pipelines operated by more than 3400 companies in the United States alone [1]. According to PHMSA, the statistics for 2017 show:

- 415 accidents resulting in $160,383,766 of total cost, 1 fatality, and 1 injury as reported by Hazardous Liquid Operators
- 104 accidents resulting in $72,226,380 of total cost, 16 fatalities, and 34 injuries as reported by Natural Gas Distribution Operators
- 107 accidents resulting in $79,617,650 of total cost, 3 fatalities, and 3 injuries by Natural Gas Transmission Operators [2]

While complying with regulations remains the strongest driver for the remote monitoring of corrosion systems, there are many other benefits of utilizing this technology. In addition to providing monitoring due to seasonal or geographical inaccessibility, remote monitoring systems enable a shrinking workforce to focus on higher value-add activities like analyzing trends, collaborating with other operators, and communicating changes throughout their organizations. Careful monitoring of equipment operation allows for rapid corrective action, which may help prevent the loss of life or product and reduce the expense associated with malfunctioning equipment. Remote monitoring systems also provide valuable data for broader examination of pipeline safety aimed at predicting failures before they occur.

Alternating current (AC) and direct current (DC) interference is a growing issue affecting cathodic protection (CP) management systems. Collocation of high-voltage electric lines and pipelines is becoming more common due to the relatively low cost of

Techniques for Corrosion Monitoring. https://doi.org/10.1016/B978-0-08-103003-5.00020-5

existing right-of-way land and the proliferation of high-voltage electric lines from Solar, Wind, and other energy sources and the proliferation of transmission pipelines resulting from new shale discoveries and production techniques. Unlike most corrosion on pipelines, interference-related failures can happen quickly, especially in areas where high-voltage lines and pipelines have been running parallel and suddenly diverge. This has driven awareness of the threat of AC and DC interference on cathodic protection systems and need for mitigation. Remote monitoring in areas with collocated transmission pipelines and high-voltage electric lines is necessary because manual measurement only provides a snapshot in time versus gathering enough quality data and context to identify trends and changes over long periods. Continued advancements in coverage and bandwidth of wireless networks make it less costly to gather the larger amounts of data required to diagnose this problem than it would have when we first wrote this chapter.

As the capabilities of remote monitoring systems have become more powerful, the systems have come to be utilized for a multitude of applications including: the monitoring of cathodic protection (CP) systems, gas measurement devices, storage tanks, pumps, compressors, and site security. In many of these applications, remote monitors employ alarm features to notify operators immediately when an alarm condition has occurred. Due to their utility, versatility, and dependability, remote monitoring of cathodic protection systems has become standard procedure for pipeline operators.

21.1.2 Chapter scope

This chapter addresses remote monitoring of external corrosion control systems. External corrosion control is one of many remote monitoring applications. It is important to provide a general overview of any remote monitoring system including the critical elements. The goal of this chapter is to help the reader understand the current state of remote monitoring technology, the trade-offs required when deciding upon a particular technology, and the challenges of remote monitoring that are particular to corrosion control systems.

It is also important to point out what will not be covered in this chapter. This chapter considers relatively low-cost systems suitable for the monitoring of corrosion systems. This chapter will not consider Supervisory Control and Data Acquisition (SCADA) or factory control systems as those are typically designed for applications requiring real-time control (e.g., valve control on a petroleum pipeline). The reader should be aware that the lines between SCADA, plant control, and remote monitoring continue to blur with the increased availability of wireless bandwidth at lower prices, the ubiquity of cloud computing, and advances in computing and radio technology.

The main application considered in this chapter is the monitoring of corrosion control systems that protect transportation, distribution, and storage assets for oil, gas, and hazardous liquids. Many other applications for corrosion control systems exist; however, you must consider the cost justification of using a remote monitoring system for these applications. For example, monitoring corrosion in a processing plant is certainly important and can be accomplished remotely, but would most likely be done using the plant's existing process control system and/or manually.

While a process control system can be considered 'remote' monitoring, such systems are beyond the scope of this chapter.

Except in a cursory sense, this chapter does not consider the corrosion sensor or data probe because they are the topics of the other chapters of this book. This discussion assumes that there is an existing sensor or data probe that requires monitoring and that this data source provides the remote monitoring system with a fairly standard industrial input (e.g., digital signal, 0–5 V, 4–20 mA). In the more specific case of corrosion monitoring, this chapter focuses on the particular challenges of monitoring an impressed current cathodic protection system on a pipeline.

Mobile computing technology increasingly intersects with remote monitoring systems. While this is out of scope for this chapter, it is important to note that because of the democratization of mobile devices we will continue to see these used increasingly for configuring and interacting with remote monitors, primarily during field installation. These mobile devices also play a growing role in the larger CP industry for field data collection surveys.

21.1.3 Remote monitoring basics

Remote monitoring systems continuously check the condition of field equipment and communicate this information to a central location where it can be routed to others or accessed by those with appropriate security clearance. The essential parts of a remote monitoring system are: the remote monitoring device, the communications network, and the central data warehouse. The remote monitoring device is often called the remote monitoring unit (RMU); it may also be called the remote terminal unit (RTU), but that term is most often reserved for SCADA systems or other more advanced field electronics with sophisticated sensors and programming capabilities.

The most effective monitoring systems use existing wireless networks to transfer information from an RMU on any remote piece of equipment to the website where it is stored in a secure database that is accessible via the Internet. Using a commonly available, user-friendly Internet browser with proper authorization, information can be viewed from the main office, remote sites, or anywhere in the world 24 hours a day. Data may also be automatically transferred to clients by the website. In some instances, data are transferred from the field to the website, automatically retrieved by the client's information system, and published directly to end-users without any human intervention. Many remote monitoring websites provide users with the ability to set up automated notifications via email or text and allow users to manage communication to and from the RMU. Reliable two-way communication enables users to update firmware and settings and trigger actions on the RMU. Figure 21.1 provides an overview of the essential parts of a remote monitoring system.

The cathodic protection data manager (CPDM) is a data repository with superior capabilities of analysis and reporting used to better manage assets and meet regulations. CPDM manages all sites required for monitoring data collection on cathodic protection systems, whether those sites are automatically monitoring with RMU or checked manually. CPDM is increasingly integrated with the remote monitoring system in order to assess the system as a whole and share corrosion data with other pipeline integrity efforts.

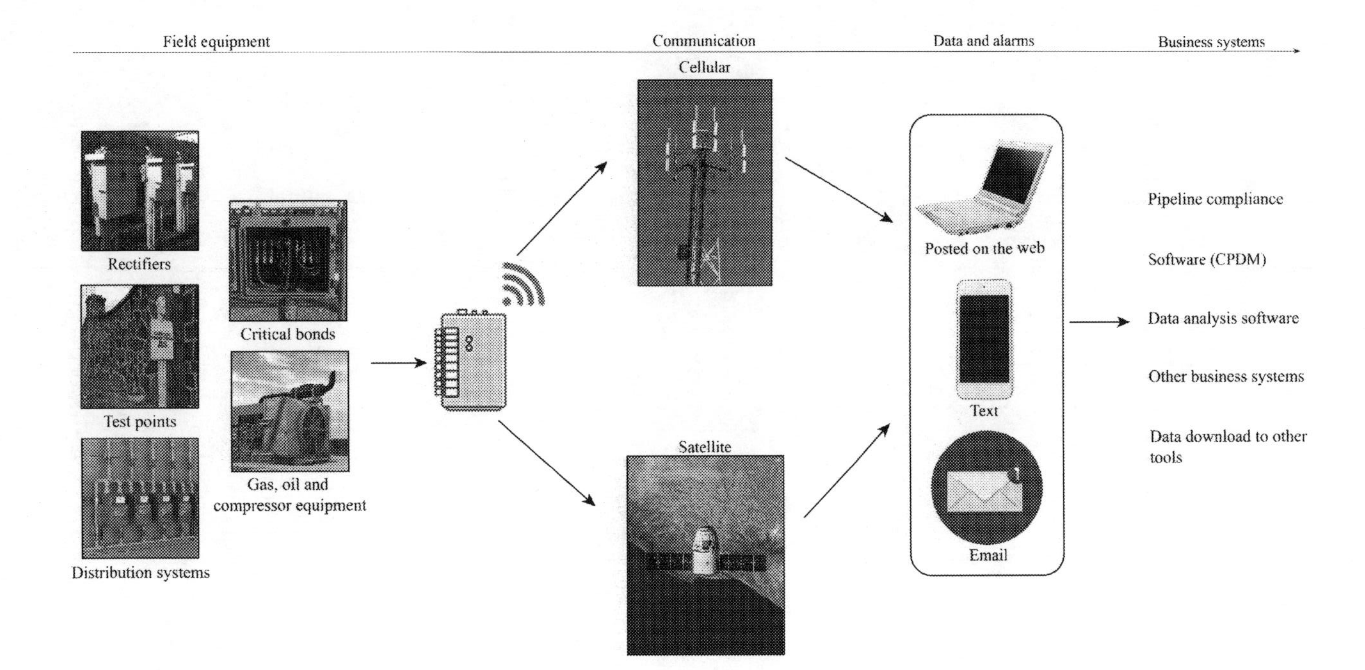

Fig. 21.1 A remote monitoring system.

21.1.4 Critical decisions in a remote monitoring system

There are several important questions you must ask when considering a remote monitoring system. These can generally be grouped as follows:

(1) *Data questions.* What is the nature of the data you wish to collect? How critical is it? How much needs to be transmitted? How often does it need to be transmitted? What are you willing to spend to gather it remotely?
(2) *Network questions.* What communications network(s) are available? Are the sites you wish to monitor clustered within a small geographic area (e.g., one site, one city) or spread across a country, a continent, or the world? Does the equipment you wish to monitor move around or is it fixed in one location?
(3) *Application questions.* Is power available at the remote site or required by the sensor(s)? Do you require two-way communication with the site allowing for requesting current status "on-demand" or remote control? If not, how will you update the RMU firmware over time? What other site or application-specific requirements influence the design of the system?

Each of these questions impacts one or more essential elements of the remote monitoring system. In the following sections, we consider how the answers to these questions guide you through the trade-offs required by the current limits of technology. The data questions help determine whether a remote monitoring system is suitable for the application or whether SCADA or manual data collection is warranted. These questions also help the user narrow the communications network options. The data requirements and characteristics of the site to be monitored determine the communications network(s) to be employed, which then drives the basic design of the RMU. Finally, the RMU may need to be tailored specifically to the application, which may then cause you to reconsider the communications network and/or the suitability of the application for remote monitoring. In the following sections, each of these questions is discussed in further detail. Once you determine what kind of data is required and how communication will be accomplished with the remote sites, the often-overlooked website and the always-underestimated support systems will be discussed.

21.2 Data considerations

As discussed in Section 21.1.2, many applications fall outside of the scope of a remote monitoring system. The characteristics of the data required by the application determine whether a remote monitoring system is technologically or economically viable. The limits of current technology will simply not allow you to consider remote monitoring for every application. Careful consideration of the data requirements will provide a quick answer to the question of viability. Table 21.1 summarizes these considerations.

Table 21.1 Data considerations.

Criticality	**Informational**	**Operational**	**Life support**
Cost of monitoring system failure	Low	Low-moderate	High
Rate of data capture from the sensor(s)	Minutes to days	Seconds	Micro-seconds
Frequency of data transmission	Quarterly-annually	Hourly-monthly	Real-time, <1 min
Amount of data transmitted	Broad range—bits to megabytes	<1 megabyte, typically <100 bytes	Typically >1 megabyte
Maintenance and training required	Low	Low-moderate	Very high
Justifiable system costs	Cost of quarterly or annual site visits	<$5000 equipment <$50/month	>$5000 equipment >$100/month
Typical monitoring system choice	Manual	Remote monitoring	SCADA and/or manual
Example applications	Test points, coupons	Rectifiers, critical bonds	Emergency valves, large compressors

21.2.1 Nature and criticality of the data

Since most corrosion occurs at such a slow rate that instantaneous measurement and communication is not required, data requirements for a corrosion control system are not likely to warrant a SCADA system. Corrosion associated with AC and DC interference of a cathodic protection system is a notable exception. Understanding the scope of that interference requires more frequent data collection. Once the problem is diagnosed and mitigated, data collection can be scaled back by remotely reconfiguring the RMU. As conditions change over time, the RMU can periodically be reconfigured to capture more data as needed, then reset back to infrequent communications to save money.

Even though the characteristics of corrosion are consistent across industries and applications, if control system capacity is available and if the proximity of signal wires in a processing plant environment makes it viable to monitor with the existing control system, it is likely that the corrosion monitoring need will be filled by the existing plant monitoring system. Monitoring rectifiers on incoming oil or gas pipelines supplying an electric power generation facility or other processing plants is a notable exception. In such cases, plant operations consider these rectifiers to be outside of process controls, similar to how oil and gas pipeline operators view the cathodic protection systems, resulting in remote monitoring being deployed. Most other cases requiring instantaneous response to prevent catastrophic results driven by corrosion fit into the category of leak detection and are not, *per se*, corrosion system monitoring.

The nature and criticality of the data for a corrosion control system is not nearly as important a factor as it is in other applications. For example, the types of remote monitoring system contemplated herein would not be suitable for life support or when a failure of the system could lead to loss of human life or other catastrophic results. The data from a corrosion control system on a pipeline is important, and it is important to know when the system is operating outside acceptable parameters. Most RMUs are designed with functionality that ensures messages will be delivered even if the first attempt fails. United States Department of Transportation (DOT) regulations require corrosion system measurements on a bi-monthly basis for pipeline corrosion system rectifiers, and a remote monitoring system that delivers data and alarms on a daily or weekly basis meets the demands for criticality by a wide margin. More often than not, it is the frequency of data transmission required and the geographic locations of the sites that determine the suitability of remote monitoring for corrosion systems. Within North America, improvements in wireless communications networks have enabled the economic collection of virtually any amount of data required to assess corrosion.

21.2.2 Frequency and amount of data transmitted

The amount of data to be transmitted and the transmission frequency required are key factors in determining whether remote monitoring is viable or if another real-time control system is needed. Some applications do not warrant the cost of transmitting large amounts of data over commonly available wireless networks on a daily or hourly basis. This continues to change as consumer demands for bandwidth drive more efficient digital cellular networks at lower costs. There are few applications that require large amounts of data which cannot be economically monitored using available cellular networks. Bandwidth improvements in satellite networks are not following suit, so if cellular networks are not available in your location, then the amount of data required remains a critical factor.

Our appetite for data continues to expand to effectively identify trends and patterns in that data. Monitoring AC interference on transmission pipelines often justifies collecting relatively large amounts of data to allow operators to identify transient events that cause rapid corrosion. Today, data are being used to do more predictive failure analysis and make automated corrections to processing conditions without human intervention.

The frequency of data transmission required for external corrosion monitoring often determines whether remote monitoring is justified. In the United States, DOT regulations contained in the Code of Federal Regulations Title 49, Part 192, Subpart 1 (49 CFR 192.465) require that rectifiers must be inspected 6 times each calendar year, but with intervals not exceeding two and one-half months. Critical bonds (bonds that must be present to ensure structure protection) must be inspected at the same frequency. Test points and noncritical bonds, however, must be tested at least once each calendar year, but with intervals not exceeding 15 months. Due to the quantity of test points, the remote nature of these devices, and the inherent lack of available power, most companies have not found a cost-justifiable way to automatically monitor these points. While technology has advanced, it has not enabled the reduction in cost that would drive significantly more remote monitoring in the corrosion industry. Test points that are in extremely difficult places to reach, such as bayous in Louisiana only accessible by boat, are notable exceptions. As costs continue to come down, this circumstance may change. For financial reasons, most companies choose to manually monitor test points, while automating the monitoring of rectifiers and critical bonds has become the norm.

Of course, regulatory requirements are only one consideration. Monthly or daily monitoring of cathodic protection systems, including rectifiers, test points, and critical bonds, provides operational data that are useful for the safe and efficient operation of a pipeline. In a worst-case scenario, a rectifier or critical bond may fail immediately after a manual visit and remain undetected for up to two and one-half months. Even though corrosion happens at a relatively slow rate, the failure of a corrosion system for a month or more may cause significant damage to the pipe depending upon the general quality of coating and the quantity of holidays (flaws in the coating systems protecting the pipelines) in the specific area of the failure. If this situation occurs, at best, the risk of failure increases, and at worst, the pipe might require mitigation activities that exceed the cost of a remote monitoring system by an order of magnitude. While many monitoring systems are typically configured to report data on a daily or weekly schedule, they might also be configured to monitor the corrosion system continuously and report an alarm condition upon occurrence.

Technology often provides information that users are not prepared to use effectively. Regulations and corrosion professionals are geared toward looking at bimonthly data to ensure all readings are "in compliance." Daily measurements are of little use today because they would provide 60 times more data than people are prepared to handle. In time, however, both the regulations and the technicians may come to see the value of looking for "out of compliance" events within thousands of data points to more effectively predict the risk of failure on a pipeline. The corrosion industry today remains reluctant to gather more frequent data than what is required by regulations, even if that data are "free."

Generally, in today's environment, remote monitoring is difficult to justify when the data are needed any less often than bi-monthly. However, the justification of a remote monitoring system becomes easier when the sites that must be monitored are more difficult to access manually; i.e., when the sites are very remote, geographically spread over a wide area, or in instances where access to the right-of-way (ROW) is behind several locked gates.

21.3 Communications networks

"Once a new technology rolls over you, if you're not part of the steamroller, you're part of the road."—Stewart Brand. (As creator of the Whole Earth Catalog, his ideas were forerunners of the Internet. Brand was one of a group of "futurists" consulted in the planning stage of the 2002 science fiction film *Minority Report*.) Changes in technology are happening at an ever-increasing rate which both enables new solutions and creates new challenges for remote data collection. The most critical and dynamic technology for remote monitoring is related to communication networks.

From the Internet of Things (IoT) and mobile devices to satellite and cellular networks, there are more ways to communicate than ever before. The bandwidth available today comes at a price we couldn't imagine just a few years ago. We can now provide more functionality in a smaller package for the same cost. IoT is beginning to live up to its hype. Twenty years ago, a remote monitoring company had to create all the pieces needed to do the job from end-to-end. Today, things are dramatically different. There are many pieces of the solution that are available as commercial off-the-shelf (COTS) items. Cloud computing means we can quickly and easily scale solutions without the expense and expertise of in-house implementation. Standard mobile and embedded operating systems along with standard protocols and code libraries enable developers to create far more capable systems in less time. The growth of network options available today and promised for tomorrow are changing the way companies look at automation, data, and remote monitoring.

21.3.1 Private networks

Once you have determined that the data requirements are suitable for a remote monitoring system, the next step is to determine the best method for communicating with the site. Site locations have a significant impact on the selection of the communications network(s). Table 21.2 presents an overview of commonly available public communications networks. Of course, you may always elect to build and operate your own communications network. Companies sometimes install a private network to communicate along a pipeline or other transmission system. However, those networks are most often used for SCADA. The operators of those networks do not typically allow

Table 21.2 Network options.

	Cellular	LEOS	GEOS
Coverage	Nationwide	Worldwide	Worldwide
Data payload	Flexible: few bytes to MB	Typically few bytes to KB	Flexible: few bytes to MB
Message latency	<30 s	1–60 min	Typically <3 min
Reliability	Proven reliability over years	Good	Very good
Transceiver cost	US$75–150	US$100–300	US$300–1000
Monthly fees	US$10–50	US$10–50	US$15–100
Risk of obsolescence	Very low	Low	Low
Transceiver size	3 cm × 4 cm	>5 cm × 5 cm	>12 cm × 8 cm
Commercial providers	Kore, AT&T, Verizon	ORBCOMM, Iridium	Inmarsat, Globalstar
Installation, ease of use	Very easy; must determine coverage first	Very easy	Requires clear line of site to satellite

Notes: LEOS, low earth orbiting satellites; GEOS, geosynchronous earth orbiting satellites; MB, megabyte; KB, kilobyte; s, second; cm, centimeter.

traffic from noncritical applications that could potentially affect the SCADA system. Often, the SCADA system does not extend to all of the sites you wish to monitor remotely, and the cost of extending the system is prohibitive.

Some utility distribution companies, whether gas, electricity, or water, install networks for the remote monitoring of utility meters. These networks are typically owned by the distribution company and may be operated by the distribution company or managed by the supplier of the network equipment. These networks may be ideal for remotely monitoring sites that are owned by the distribution company including corrosion systems. However, similar to SCADA systems, these networks are private, primarily used to collect data from the utility meters and designed to exclusively serve the owner of the network, and thus, have limited application in CP. Corrosion control is only recently getting attention from the providers of such "smart meter" networks. The networks typically operate on unlicensed bandwidth in the 800 MHz range. Providers of these networks are beginning to collaborate with corrosion monitoring experts to combine corrosion measurement expertise with private radio communications. It is only a matter of time before utilities can address corrosion monitoring on their existing meter reading network. However, even those systems cannot be economically extended to reach every point requiring monitoring.

Another category of private networks has emerged over the last several years, primarily in cities but also around high-activity areas where there is a concentration of equipment to monitor. These networks utilize a variety of protocols and methodologies that are being deployed for numerous uses. Often called low-power wide-area network (LPWAN), these systems are relatively low-cost and allow fairly long-range communications for power-limited applications. Examples of this type of network include Long Range (LoRA), Long-Range Wide-Area Network (LoRaWAN), Sigfox, and others. For remote monitoring applications, sensors with LoRaWAN capability can connect into existing LoRaWAN networks or operators can set up their own local LoRaWAN gateway to communicate to many devices, allowing only one backhaul for the data of all devices. However, for long-distance, straight-line transmission pipelines, these technologies do not currently provide a compelling solution to operators. We cover WAN in more detail in the next section.

As with data, several important questions need to be answered before an appropriate network can be selected. One such question has already been covered—if you have your own network available that serves all of your sites, skip the remainder of this section.

21.3.2 Wide-area networks

Wide-area networks (WAN) have recently been an area of expansion with a focus on very low-power WAN and IoT solutions like LoRa, SigFox, and Random Phase Multiple Access (RPMA) in unlicensed frequency bands. As mentioned above, some of these networks can be deployed privately by companies operating in cities or other areas where there is a relatively high density of assets to be monitored, like tank farms, gathering, and distribution sites. The landscape is changing rapidly with these technologies and there are competing technologies that have very recently been deployed at a much larger scale based on existing cellular network hardware.

Cellular technology is rapidly changing to address the growth in demand for low-power IoT applications, including remote monitoring of CP systems. Lower-frequency bands built on existing network systems are being deployed as of the writing of this edition, most promising of which are Category M1 (Cat M1) and Narrow-band Internet of Things (NB-IoT). Advantages of using these technologies for remote monitoring applications are longer battery life, longer radio range, penetration below grade and into buildings, and lower overall cost compared to many other traditional communication methods. However, despite the longer range of CatM1 and NB-IoT, cellular technology currently has limited reach outside population-dense areas and corridors, so alternate communication technologies must still be considered for very remote sites.

In the world of CP, especially in transmission pipelines in remote areas, cellular connectivity remains a challenge, even with newer, longer-distance cellular technologies. Many operators must turn to satellite communications in these areas to provide reliable connectivity. Traditionally, these satellite systems have demanded relatively high prices for hardware and data; however, this may change with emerging technology. A variety of companies and universities are promoting lower-cost IoT satellite solutions. While these have been used for educational projects for the last several years, broader deployment has not been fully realized. While this technology shows promise, current system deployment and performance is limited.

No communication system is perfectly suited for all remote monitoring of CP systems. Broad coverage is an area of concern regarding cellular and lower-cost IoT solutions, satellite options can be cost-prohibitive for some applications, and private networks only make sense in certain situations. Remote monitoring system communication must be a balance of performance, capability, availability, and cost.

21.3.3 WAN choice and obsolescence

All of the networks listed in Table 21.2 are suitable for remote monitoring requiring the exchange of hundreds of bytes, up to one megabyte, over the course of a month from fixed points spread over a large geographic area. Other important factors to consider when selecting one or more WANs are coverage area, transceiver cost, monthly fees, network/carrier reliability, network latency, and size of the transceiver. Networks fall into one of three general categories: cellular, low-earth-orbiting-satellite (LEOS), or geosynchronous earth orbiting satellites (GEOS). Each "flavor" of cellular service requires a different transceiver, so you must know what coverage is available for each in order to gage the viability for a given application.

For most applications, it is unlikely that 100% of the sites you wish to monitor will be covered by any one WAN. Even if all of the sites were located within metropolitan areas covered by cellular service, it is likely that there will be one or two sites that cannot connect to the network for some site-specific reason. It is desirable to use the minimum number of WAN possible to cover 100% of the sites for a number of reasons, including: volume discounts on equipment and service, training required on device installation, system configuration, and the cost of maintaining multiple WAN technologies. Consider the specific needs of the application in question and select the network that fits the majority of those sites as the primary network, but be aware that it is likely that a secondary WAN will be required for complete coverage.

Cost is often a primary consideration in selecting a WAN communication type, both for the monitoring device and the service, but is also a factor that changes rapidly. The cost of LEOS transceivers has decreased from US$300 to around US$100 and has gotten smaller—from the size of a novel to a deck of cards. Antennas for LEOS transceivers range from 2 to 3 foot long "whip" antennas to small patch antennas, each having performance and physical form factor trade-offs. GEOS transceivers are still the most expensive and power hungry. While costs have come down, at over US$300, the choice of this network makes the monitoring device significantly more expensive.

The size of the transceiver, while typically of less importance in external corrosion applications, has a similar story as cost. The main benefit of a smaller transceiver is that it allows for a smaller RMU that is less obtrusive and more easily hidden to protect it from vandalism. Cellular transceivers are the smallest in size and getting smaller every day. A small transceiver allows for a smaller package and often reduces the total cost of the RMU. LEOS transceivers are shrinking in size and cost as deployments increase. In most remote external corrosion applications, the half-wave whip antenna commonly utilized with the LEOS transceiver is an acceptable solution because vandalism is less of a concern. GEOS transceivers can be the most difficult to deal with from a size perspective; they are typically the size of a dessert plate and most often attached outside of the monitoring device. On one hand, that is more expensive and makes a good "target" for some people. On the other hand, it means that the radio is easier to upgrade during a field visit. Vandalism is often more of a concern in metropolitan areas where a digital cellular service is likely to be readily available, and integrated antennae are readily available that help make the device less visible. Wireless telecommunication is marked by rapid technological change. As stated earlier, such change can make choosing a technology for fixed point remote monitoring a challenge. Hence, Table 21.2 attempts to focus on networks that appear to be relatively stable and without significant risk of technological obsolescence in the next decade or so. The reader is advised to remain aware of technological shifts in WAN technology and understand that obsolescence is a key risk in deploying any remote monitoring. Most WAN are built for high bandwidth (high dollar) applications, so providers are continually working to provide more efficient service. Providers are switching radios and replacing networks much faster and it is not uncommon for a radio/network selection to become obsolete within 5 years. The rapid replacement of networks creates a strain on both the user and the manufacturer of remote monitoring networks, especially in fixed point monitoring where site visits are required to replace obsolete equipment.

21.4 Application-specific requirements

21.4.1 Power requirements

RMUs require power to operate. Thus, understanding the power requirements of the monitoring site is one of the most important aspects in designing and selecting a system. Many sensors will have their own power supply and will not require power from the RMU, but it is important to make sure that is the case, because an RMU is often designed to provide power solely for itself and some sensors can require a significant amount of additional power.

If the RMU only needs to power itself, the next choice is to determine what type of power is available. In many corrosion applications, AC power is used to power the external corrosion control (CP) system. AC power is rectified to DC power, adjusted to the appropriate level and applied to a deep anode groundbed used to flow direct current into the metal structure to prevent corrosion. Most CP remote monitoring applications today are, in fact, measuring the output current and voltage provided by the rectifier to the ground bed. If there is AC power at the rectifier, the RMU has a ready power source available and thus can avoid the use of a stand-alone power system consisting of batteries and/or solar panels. Batteries, even those that are recharged by solar power, need to be replaced periodically, and thus add maintenance cost. However, few savings in RMU cost are achieved through the use of AC power over batteries, as the electronics in the RMU require additional components to make effective use of the AC and protect the electronics from power fluctuations.

At those sites where AC power is available for remote monitoring, an important advantage emerges in that unlimited power allows for near-immediate two-way communications between the RMU and the website. Most battery-powered designs provide limited power and thus the transceiver is turned off or put in standby mode until a data transmission is required by an alarm condition or scheduled report. Batteries can be sized such that the transceiver may be left on all the time, but such designs are often economically impractical. Transceivers that require less power are constantly being introduced, so it is becoming easier to leave those devices on for at least some portion of each day. Even with these advancements, immediate, two-way access to the RMU is not pragmatic. For most CP applications, instant or on-demand communication is unnecessary. In these circumstances, battery-powered RMU systems that use a queued, two-way communication where messages are only sent and received at discrete time intervals enable the functionality of two-way communication with a far lower impact to battery life. The trade-off with queued two-way communication is latency. Data and commands will not be available within seconds or minutes, instead it will be communicated based on a predetermined schedule which could be hours, days, or longer depending on the application.

Two-way communication between the RMU and the website provides a number of important features. First, it allows the user to check the status at the remote site at any time instead of waiting for the RMU to make a scheduled report. Second, it allows the website to send firmware and/or configuration program updates remotely. For fixed point locations, this feature is particularly attractive, because it allows for troubleshooting, reconfiguration, and new features without a site visit.

Finally, two-way communication also enables remote control. You may wish to take action at the site remotely through the RMU instead of passively monitoring the status of the site. Examples of such actions are activating ancillary equipment such as turning on pumps, adjusting the operating parameters of a compressor, or opening a valve. The need for remote control in corrosion applications will be discussed later in this chapter.

Often, the user will want to know when AC power is off at the site since it means that the site equipment is no longer operational or is relying on limited back-up power supplies. If the RMU is also relying on AC power, it must be designed with a back-up power supply of its own in order to report the outage. This requirement further reduces the cost differential in the AC-powered RMU compared to a battery-powered unit.

Many remote monitoring applications, however, do not have AC power readily available. Most critical bonds and test points are simple test stations inside plastic risers in the middle of a field without any power supply in sight. In those applications, the RMU is typically powered by batteries. In most cases, the RMU will be located where there is enough sunshine to power a solar panel used to recharge the battery. The use of such rechargeable batteries is most economical as they require less frequent replacement and often last for the useful life of the RMU. In cases where the RMU is installed indoors, or where vandalism of the solar panel is a particular concern, lithium-ion batteries are often used. The efficiency of lithium-ion batteries has improved and cost has come down, and the advantages of avoiding a solar panel are many, so lithium-ion batteries have become a staple of RMU. Maintenance of the power supply is another factor that is more important for fixed sites than mobile applications, so in the end, minimizing maintenance is often prioritized over equipment costs.

21.4.2 Environmental requirements

Most remote monitoring applications are outdoors, and the RMU must be protected from the elements to ensure reliable operation of the electronics and transceiver. The environmental requirements and methods used to protect the RMU are similar to those of many other outdoor industrial applications. The RMU must be designed to handle wide variations in temperature and relative humidity. Choosing an RMU with a suitable enclosure for the conditions is important. Many enclosures also provide latches that may be locked to prevent tampering.

An environmental concern more specific to external corrosion monitoring is protection from power surges. Essentially, metal pipelines are huge conductors capable of carrying large surges of current that can damage sensitive electronics if not properly protected. Sources of these surges can include lightning strikes in the vicinity of the pipe, creating large potential energy or other external sources such as faults, surges, and high-voltage AC induction. Various techniques exist to protect the electronics, but as always, there is a cost/benefit trade-off. One technique that is commonly used is to keep the RMU inputs isolated from the pipeline at all times except when a measurement is in process. This technique reduces the probability of failure significantly if adequate materials, creepages, and clearances are designed into the RMU. In lightning-prone areas, it is necessary to add external lightning protection devices such as grounding, isolation, and/or surge arrestors to the rectifier itself and to remote monitoring inputs to protect them from likely close proximity lightning strikes. It is also important to maintain these protection devices over time. Surge arrestors, for example, are rated to protect equipment from a limited number of faults, and this number decreases as fault voltage increases. Some remote monitoring equipment can be outfitted with surge counting devices, so operators can more accurately predict when replacement may be necessary.

There are, of course, many other environmental factors to consider in remote CP monitoring applications. Insect infestation can occur, which can be managed through physical barriers and chemical deterrents. Below-grade installations present a unique set of challenges for electronics and interconnects as moisture ingress, pressure, and soil expansion and contraction all act to shorten the useful life of the system. Vibration can be an environmental stress to consider, particularly when electronics are mounted to rotating machinery like compressor stations.

21.4.3 RMU inputs

An RMU is typically designed to monitor a limited number of I/O from sensors at the site. Most sensors at remote sites provide a standard type of input that can be grouped into three categories: digital, analog, or serial. However, many other types of I/O and control interfaces are often used in CP remote monitoring, as can be seen in the list below.

- AC and DC peak waveforms
- Voltage and current measurements
- Transient protection
- 4–20 mA measurement and power source
- Coupon switching and over current protection
- Instant-off measurements, coupon switching, and synchronized rectifier switching
- Low-power conditions for battery-powered units
- External relay control for interruption or other applications
- RS485 and RS232 serial interfaces
- Electrical resistance (ER) probe interface

Digital inputs are often used to sense a simple state change: ON versus OFF, OPEN versus CLOSED, 1 versus 0. Analog inputs are often used to measure a value and most analog sensors provide a standard range of 0–5 V or 4–20 mA. Some key characteristics of RMU analog inputs are accuracy, auto-calibration, and input impedance. Most RMUs are designed to accommodate several digital inputs and are limited to monitoring state changes. Some RMUs have the added capability of being able to monitor analog inputs along with digital inputs and can therefore be used to monitor a wide variety of applications. Analog inputs are required for external corrosion system monitoring. A general monitoring example of how analog signals are utilized would involve a level sensor on a storage tank that may be calibrated to provide a

4–20 mA current signal relating to the actual level. The signal is converted to a voltage signal acceptable to the RMU such that 4 mA equate to 1 V (representing an empty tank) and 20 mA equate to 5 V (a full tank). The voltage measured by the RMU at any given point in time is assessed for alarm conditions. The voltage value is also potentially transmitted to the website on a scheduled basis and is converted to a volume measurement by knowing the size of the tank being monitored at the site.

Serial connections are often used to allow the RMU to communicate with an intelligent device at the site. Many intelligent industrial devices employ a standard language called Modbus. Modbus is a serial communications protocol published by Modicon in 1979 for use with its programmable logic controllers (PLC).[3] It has become a de facto standard communications protocol in the CP industry and is now the most commonly available means of connecting industrial electronic devices. The main reasons for the extensive use of Modbus over other communications protocols are:

(1) it is openly published and royalty-free
(2) it can be implemented in days, not months
(3) it moves raw bits or words without placing many restrictions on vendors

An RMU with a serial connection and Modbus capability can be used in a wide variety of applications including gas volume correctors, electronic flow measurement systems, gas pressure monitors, or more sophisticated level sensors. Thus, an RMU equipped with a serial connection may be able to determine not only the level of the tank, but also its contents, temperature, pressure, and the last time the valve was opened or closed.

Basic corrosion monitoring typically involves reading two analog inputs from a rectifier. Typical measurements at the rectifier include DC output voltage and DC output current from the rectifier. Current is determined by taking a voltage reading across a 'shunt' of known resistance. That voltage level is then scaled to represent the appropriate current level. Required rectifier output voltage levels are determined by a number of environmental and pipeline construction factors. The DC output voltage of the rectifier can be sensed and converted to a range that the RMU accepts, and then, the voltage measured by the RMU at any given point in time after transmission to the website can be scaled to display the actual DC voltage level at the rectifier. Like the current reading, this value is scaled for display at the website. Sometimes, additional measurements like test point and critical bond voltage levels might also be taken by the RMU at the rectifier.

As discussed earlier, test point and noncritical bonds (places where two or more pipelines cross) may not warrant the cost of remote monitoring systems because of the relatively infrequent need for the data. However, when these points are located in close proximity to the rectifier, the RMU must also have the capability to read them. Regardless of whether these test points and noncritical bond measurements are taken along the pipeline or at the rectifier, the characteristics are the same. These measurements require bi-polar analog inputs since these voltages are typically negative in polarity. They are also usually at a relatively low level; i.e., usually 0–2 V in magnitude. It is important to obtain a clean signal, and thus, AC filtering is a requirement.

The problem with fixed points is that any maintenance or repair requires a site visit. Remote monitoring is most valuable at the most remote sites. Therefore, any visit for repair or maintenance at such a site is likely to be very expensive. In extreme cases, the visit may require transportation by boat or aircraft. Reliability, therefore, becomes an important factor in the selection of a WAN and the design of the remote monitoring device for fixed sites. It is becoming increasingly easier to perform remote diagnostics, repair, or firmware upgrades on fixed sites as more bandwidth becomes available at increasingly reasonable prices. More cost-effective, two-way communication makes it much easier to upgrade field units without a visit—whether to fix a known problem, initiate an action, or add new functionality. As noted earlier, this is still much easier, cheaper, and more reliable on cellular networks than on satellite.

21.4.4 Remote control; output requirements

In Section 21.4.1, the importance of two-way communication was briefly discussed. In corrosion monitoring, two-way communication is needed primarily for remote control to support two specific needs: (a) turning the rectifier on and off at specific intervals and (b) remotely adjusting the rectifier to tune the corrosion control system. However, two-way communication for remote control has many other uses:

- Active, two-way communication with low latency
- Normally asleep units with queued message next wake up retrieval
- Firmware over the air (FOTA) updates to prevent site visit
- Security—user authentication and limited permissions
- Field upgradeable components and self-configuration

While basic monitoring of rectifier voltage and current is employed in the majority of remote sites, there is increasing demand for remote synchronized current interruption (i.e., using relays to cycle rectifier output voltage at specific timing intervals) for the measurement of the instant-off potential. The RMU typically accomplishes this by enabling or disabling a program within an optional current interrupter that uses the Global Positioning System (GPS) to provide timing synchronization. Each current interrupter is preprogrammed with a variety of interruption schedules. The user can remotely trigger a specific interruption program through the RMU. Significant savings are available due to the elimination of a costly manual interruption setup. When a series of rectifiers are interrupted in a concurrent, synchronized manner, additional data can be gathered on the operation of the system via manual survey techniques using mobile devices with the same GPS time synchronization techniques. Mobile devices can be equipped with integral and external digital voltmeters (DVM) to allow for collecting voltages along the expanse of the pipeline at close intervals for later analysis of the viability of the impressed current protection. Once collected, this data is moved electronically into a historical database for analysis, remediation, and regulatory audit purposes.

In the United States, the DOT has established rules regarding External Corrosion Direct Assessment (ECDA) as part of the holistic pipeline integrity management requirement. Pipeline companies are expected to utilize at least two techniques to assess whether their pipelines are properly protected from corrosion. Generally, one chosen technique is focused on finding and sizing holidays. The other technique described directly above is intended to assess where the pipeline is being protected by the CP system and where it is not, so remediation may be achieved. This technique is commonly known as Close Interval Survey (CIS) or Close Interval Potential Survey (CIPS). Readings are taken every few feet while the power to the pipeline is cycled so that a pair of readings at each point (Instant-On and Instant-Off) might be obtained. The need to collect this information from at least two ECDA techniques and analyze the results concurrently is a DOT requirement.

The other important use of two-way communication in corrosion monitoring is to allow for adjustment of auto-potential rectifiers. By using an analog output from the RMU, the user can command that the set point of the rectifier change to support changing environmental conditions. Most often, the need for this type of adjustment is driven by extreme variation in ground moisture that will result in a need for more or less power output from the rectifiers.

21.5 Website and supporting systems

21.5.1 Website basics

The website is an often-overlooked piece of the remote monitoring system. The RMU and WAN, most often, get lots of attention because that is where specific design questions must be answered prior to implementation and is often where the majority of the monitoring cost occurs. However, the RMU needs a website to gather, decode, store, and disseminate the information gathered remotely. The point of remote monitoring is to improve the speed of collection, accuracy, and value of the data measured at the remote site. The website is responsible for delivering that information. Websites typically include significant investments in hardware and software—machines that are connected to WAN and the Internet 24 hours a day and 7 days a week, a database to store the data, and software to distribute information. Another overlooked aspect of the website is the connection to billing. Most remote monitoring service is billed based on the amount and frequency of data transmission, and the website contains that information. The website also tracks which units are active or inactive, both for billing and operations, and often allows the user to change billing plans.

21.5.2 Data security and redundancy

In some remote monitoring applications, security of the data during transport can be a very important feature. For example, if detailed gas measurement data could be intercepted, a competitor might gain an important advantage. Similarly, the security of the data from CP remote monitors is essential to the secure operation of oil and gas infrastructure. Even if the data packet sent from the RMU was readily transparent, the WAN employed often has extensive data encryption to provide security during transport.

In many remote monitoring applications, as with CP, website uptime is critical. Designing an RMU and choosing a WAN for maximum reliability will be efforts made in vain if the website is not online because of power outages, software defects, or lost connectivity. For critical data needs, it is important to have a back-up website that is automatically set in operation, with a current copy of all data collected, whenever the primary website is down for any reason. Some applications rely on the website to be the primary storage site for operational data. In some remote monitoring applications, it may be critical that the website database maintains a long history of the data collected from RMU over time. However, most applications, including corrosion monitoring, allow the user to export the data from the website database to operational software for further analysis and warehousing. In the case of corrosion monitoring, the data is often exported to a package such as the Pipeline Compliance System (PCS) software where it is archived for regulatory reporting and analyzed against other corrosion data, such as manually collected test point readings to identify system defects. Storage capacity, therefore, is not typically an issue with the website database unless one is relying on it to be the primary data warehouse.

21.5.3 Data export, analysis, and grouping

Once collected, decoded, and effectively stored within the website database, users must be able to manipulate the data. As discussed above, data are most often exported from the website to other operational software for Integrity Management (IM) and risk analysis. The website software must make data export possible in a variety of formats. The optimum system allows for the data to be automatically exported and electronically delivered to client business systems. It is important to understand how you want to use the data collected by the remote monitoring system and ensure that the website can accommodate effective and correct delivery of the data.

Some companies may not have access to operational systems for analysis of the data collected from the remote monitoring system. In that event, it is important to ensure that the website software provides analytical tools. Those tools may be as simple as providing for export to a common commercial package like Microsoft Excel software or as complex as providing route scheduling features and Graphical Information System (GIS) support to map locations.

Users often wish to group remote sites according to departments or regions formally responsible for manually collecting data. For example, a pipeline may be separated into different operating areas with different management, personnel, and resources responsible for maintaining the corrosion system on that segment of pipe. The website software must be capable of grouping RMUs for appropriate visibility by the supporting personnel. The optimum system provides security features that allow for varying levels of access to the data according to functional need. Some users might be allowed to make changes, while others are only allowed to view data. Finally, visibility might be controlled according to breadth of responsibility, so a regional manager might have access to a broader range of data than a local technician.

21.5.4 Alarm notifications

Some remote monitoring applications exist solely to screen for alarm conditions—to provide an indication that operating characteristics are outside normal conditions. Some remote monitoring systems allow for the periodic collection of data and the essential capability of immediately notifying users that an alarm condition exists at the site. Whether some corrosion monitoring applications are not particularly time-critical, other applications like AC corrosion can be especially time-critical. It is important to know that a problem exists at the remote site as soon as it occurs. Users should be able to program any number of notification methods including app notification, text or email. The message should provide the recipient with basic facts about the alarm condition so that effective action may be taken. This information should be delivered reliably. Clients should be offered the option of escalating the notifications—if the recipient does not acknowledge receipt of the alarm notification within a certain amount of time, the website sends additional notifications to ensure that the alarm condition is communicated and addressed.

False alarms can be a significant problem in remote monitoring systems and can result in excessive communication costs if not managed properly. The RMU must have the ability to distinguish a true alarm from a false alarm. This is often accomplished by requiring that the alarm condition persists for a given time period. Likewise, false alarms can also be controlled by using a technique called "deadband," whereby an input is not allowed to alarm again until certain characteristics have been met.

21.5.5 Supporting systems

If the website is often ignored, supporting systems are almost always ignored. An effective remote monitoring system includes a software system that supports production of RMU and ensures that the website knows what devices may be reporting to it. Each RMU must be married to its transceiver and stored in the website database before it can report from the remote site. Its transceiver must also be activated on the WAN so that it can be tested end-to-end prior to shipment. These processes must be accomplished correctly and efficiently prior to installation at the site. Such systems also must provide for preprogramming of parameters according to client specifications so that the user does not have to do so on-site.

Due to environmental stress and other factors, products exposed to the elements can sometimes fail and require repair. An effective monitoring system is tied to a return processing system that ensures returns are processed effectively and network fees are properly assigned while the RMU is in repair. You should be able to tell which of your RMUs are in repair, how long they have been there, when to expect them back, and be confident that it will not be charged for communications fees while the RMU is in repair.

While the production and transportation of oil and gas will likely always remain a dangerous business, advances in technology facilitate the ability to come up with more effective, lower-cost solutions for corrosion control. Cloud computing is making data more accessible, and that facilities the comparison and contrast of diverse data to help operators make more informed business decisions which will ultimately contribute to safer operations.

References

[1] Office of Pipeline Safety: Pipeline Safety Initiatives (Published January 19, 2007), currently downloadable at http://ops.dot.gov/init/init.htm#safety.
[2] Pipeline and Hazardous Materials Safety Administration (Published June, 07, 2019), currently downloadable at https://cms.phmsa.dot.gov/data-and-statistics/pipeline/data-and-statistics-overview.
[3] Wikipedia, Modbus, Available at https://en.wikipedia.org/wiki/Modbus (Updated 25 September 2020, Visited 14 October 2020).

Further reading

C. Argent, K. Prosser, D. Norman, P. Morgan, R. Weatherhead, Macaw's Pipeline Defects, Yellow Pencil Marketing, Andover, 2003.
R. Baboian (Ed.), NACE Corrosion Engineer's Reference Book, third ed., NACE Press, Houston, TX, 2002.
P. Bedell, Wireless Crash Course, third ed., McGraw Hill, New York, 2005.
P. Gralla, E. Lindley, How Wireless Works, second ed., Que, Indianapolis, PA, 2005.
P.R. Roberge, Corrosion Inspection and Monitoring, Hoboken, NJ, John Wiley & Sons, 2007.
M. Schwartz, Mobile Wireless Communications, Cambridge University Press, Cambridge, 2005.

Part Four

Applications and case studies

Corrosion monitoring in cooling water systems using differential flow cell technique

*Bo Yang**
Champion Technologies Inc., Fresno, TX, United States

22.1 Introduction

In this chapter, examples of laboratory and field applications of the differential flow cell technique for corrosion monitoring in cooling water systems are described. The differential flow cell technique is an online measurement method capable of real-time determination of localized and general corrosion rates of metals in aqueous solutions. The principles of the method, typical designs of the monitors, and guidance on data interpretation are described in Chapter 7.

22.2 Corrosion inhibition program selection and optimization

This study was conducted in a two-unit pressurized water reactor (PWR) nuclear power plant in the Midwest of the United States [1]. The plant uses water from an impounded lake for once-through cooling. Water from the lake is made-up and blowdown from a river. Condenser circulating water design flow is 720,000 gpm (2730 m^3/min) per unit (with three circulating water pump running). The total essential service water flow is 24,000 gpm (90.8 m^3/min) and the total nonessential service water flow is 70,000 gpm (265 m^3/min). The service water systems were experiencing an increasing trend in corrosion rates based on corrosion coupon results (about 12 mpy (305 μm/year) for carbon steel coupons with >90 day exposure) and pipe inspection/repair frequency. Considering the large volumes of water to be treated (for this reason, no corrosion inhibitor was in use), it was recognized that economics and environmental concerns would be major drivers in the development of a corrosion control program. Traditional treatments known to be effective would cost about $3 million per year to implement. Research was conducted in the laboratory to develop a treatment program that could deliver the desired results. Based on lab results, a combination of blended deposition corrosion inhibitors and a novel feed approach showed the greatest promise. A sidestream system was developed to model the cooling water system at the plant.

* Current affiliation: Prestone Products Corporation, Danbury, CT, United States.

Techniques for Corrosion Monitoring. https://doi.org/10.1016/B978-0-08-103003-5.00024-2

Several trials were run with the localized corrosion monitor (LCM) based on the differential flow cell described in Chapter 7 with varying degrees of success. Results from the treatment program that yielded the greatest potential are shown in Fig. 22.1. Analysis results of the lake water used as the once-through cooling water in the studied period are shown in Table 22.1. In this situation, it was shown that intermittently feeding a relatively high dosage of inhibitors provided the greatest overall corrosion control. This was compared to continuous feed of inhibitors at a medium dosage, which resulted in higher corrosion rates [1]. Not only did the continuous corrosion inhibitor feed give poorer results, it delivered more total amount of chemicals to the system, resulting in a higher costing program.

In Fig. 22.1, the general corrosion rate was shown to drop steadily and stay low following the first slug of inhibitor. The localized corrosion rate initially decreased by about half the original value after the first slug of inhibitor, but shortly thereafter the rates leveled off at a minimum value. At this point, inhibitor feed was discontinued and the localized corrosion rates rapidly increased to their original level. This cycle was twice repeated, showing similar responses in the localized corrosion rate when the corrosion inhibitor was intermittently fed.

From this case study, it is important to note that it is the area under the curve that relates to the total corrosion in the system. In this case, the general corrosion was maintained and the localized corrosion level was intermittently reduced, giving an overall improved performance than could be achieved with a continuous feed approach. Thus, the sensitivity of the LCM allowed for close monitoring of system corrosivity changes, resulting in a novel intermittent inhibitor dosage strategy. Also, during this evaluation, it was shown that temperature and make-up water chemistry have a great impact on corrosion rates. These factors are intuitive to most water

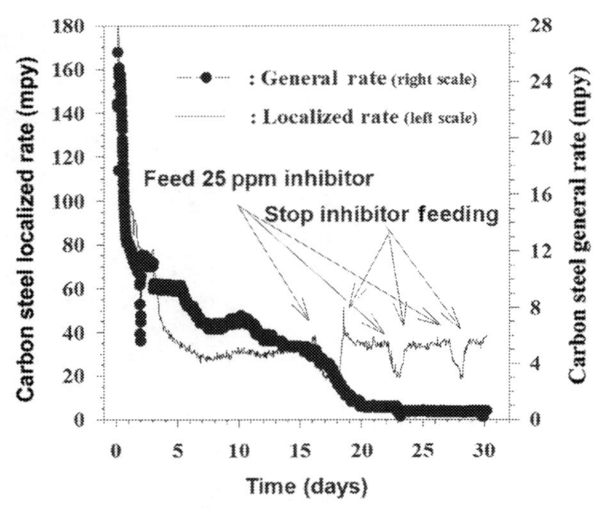

Fig. 22.1 Effect of intermittent inhibitor feed on carbon steel corrosion in a once-through cooling water system.

Table 22.1 Midwest nuclear power plant—Once-through cooling lake water analysis.

Date	26-Feb	12-Mar	13-May	18-Jun	28-Jul
Water analysis					
ICP or IC results (ppm)	**Total**	**Total**	**Total**	**Total**	**Total**
Ca^{2+} ($CaCO_3$)	210	190	190	150	160
Mg^{2+} ($CaCO_3$)	330	310	310	290	330
Iron (Fe)	0.2	0.3	1.9	0.7	0.3
Sr^{2+} (Sr)	0.3	0.3	0.3	0.3	0.3
Na^+ ($CaCO_3$)	100	99	92	84	98
K^+ (K)	7.8	7.1	7.2	6.8	8.1
Aluminum (Al)		0.1	0.1	0.3	0.2
Phosphorus (PO_4)	<0.3	0.5	0.5	<0.3	<0.3
Silica (SiO_2)	3.1	1.5	1.5	3.9	9.8
Sulfur (SO_4)	260	260	270	250	290
Boron	0.2	0.2	0.2	0.2	0.5
Cl^- ($CaCO_3$)	110	110	110	85	120
Sulfate ($CaCO_3$)	230		260	230	270
M-alkalinity ($CaCO_3$)	238	217			
Nitrite (NO_2)		<2.9	<2.9	<2.9	<2.9
Nitrate ($CaCO_3$)		4.8	5.7	4.3	<2.9
Other					
Conductivity (μS/cm)	1130	1140			
pH	7.5	8.6			

Note: *IC*, ion chromatography; *ICP*, inductively coupled plasma atomic emission spectroscopy.

treatment professionals. However, given the sensitivity of the LCM, it is possible to adjust the treatment program based on these clearly identifiable effects.

It should be noted that good correlation between LCM time average rate readings and coupon general rate and pitting rate results are obtained in the same exposure periods in this study [1].

22.3 Program optimization at a chemical processing plant

This case study was conducted in the chlor-alkali cooling water system (system capacity: ~650,000 gallons or 2460 m^3) of a large integrated plastics producer in the Southwest of the United States [2]. The cooling water system had a very long holding time index (HTI, up to 6–9 days; 5–15 cycles of concentration) with variable calcium ions (150–300 mg/L as $CaCO_3$ or 1.5–3 mM in the tower) and used condensate water as partial cooling water make-up. Other challenging conditions in the system included:

- Frequent leakage of 30% and 50% sodium hydroxide from the nickel plate and frame exchangers to the cooling water due to pitting corrosion, crevice corrosion, and plastic deformation at the plate and frame contact points

- Low flow rates in certain shell-side carbon steel heat exchangers due to process demand variations (one shell-side carbon steel column condenser with double segment baffle design experienced a localized corrosion-related failure after 2 years of service, indicating that a time-averaged penetration rate of 40 mpy or 1 mm/year)
- Low flow rate in the ethylene dichloride reactor jackets, resulting in high corrosion rates and plugging with corrosion products and other deposits (consisting of mostly iron oxides and hydroxides)
- Desire to operate at even higher cycles of concentration to conserve water

The system was using an Alkaline Stabilized Phosphate program and chlorine gas at the start of the study. A Stabilized Phosphate program was used before, with unsatisfactory results. High carbon steel general corrosion rates were obtained in the bulk cooling water by coupons and the linear polarization resistance (LPR) probe in the LCM. Severe fouling occurred at the ethylene dichloride reactor requiring acid cleaning once per year.

A newly developed Alkaline Stabilized Phosphate treatment program containing a new stable cathodic localized corrosion inhibitor was judged to be especially suitable for this system [3]. The LCM was used to conduct online optimization of this new treatment program to minimize localized corrosion, the most critical concern at this plant. Several combinations of inhibitors, dispersants, and biocides were used and an excellent program was developed. The results are shown in Fig. 22.2. Chemical analysis of several key parameters of the cooling water in the system (return water after flowing through all the heat exchangers) is shown in Table 22.2.

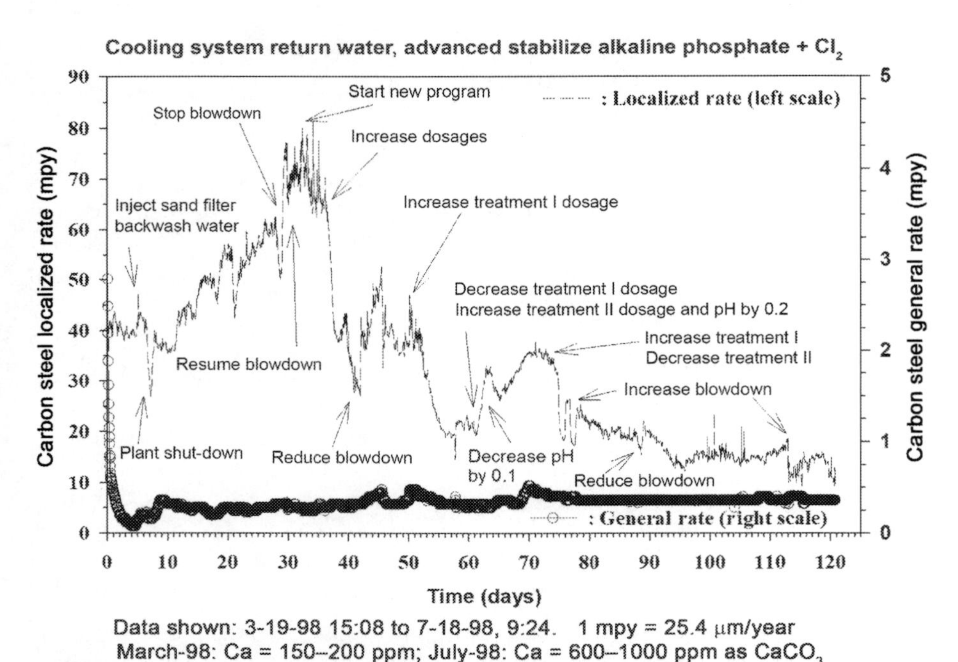

Fig. 22.2 Chemical process plant: chemical treatment optimization using the LCM.

Table 22.2 Chemical processing plant cooling water return analysis—Stabilized alkaline phosphate treatment + Cl_2.

Date	Conductivity uS/cm	Total hardness ppm $CaCO_3$	Ca^{2+} ppm $CaCO_3$	M-alk ppm $CaCO_3$	pH
3/9		212	190		
3/19	1170		203	136	8.4
4/14	1820		205	135	8
4/15	1880	220	166	132	8.41
4/16	2040	250	194	148	8.47
4/17	2230	266	212	156	8.46
4/18	2390	280	220	164	8.33
4/20	2220	248	204	148	8.57
4/21	2250	244	208	156	8.42
4/22	2150	260	188	144	8.22
4/23	2100	270	200	140	8.42
4/24	1910	270	210	132	8.37
4/25	1808	285	220	152	8.4
4/27	1910	350	280	140	8.53
4/28	1970	370	300	150	8.49
4/29	2090	412	324	160	8.47
4/30	2130	460	360	168	8.44
5/1	2140				8.5
5/4	2120	448	400	136	8.33
5/5	2140	470	390	126	8.37
5/6	2100	480	410	140	8.43
5/7	2135	500	412	144	8.5
5/8	2180				8.5
5/11	2360	600	488	144	8.37
5/12	2402	620	500	140	8.4
5/14	2410	624	508	148	8.43
5/16	2620	628	510	148	8.32
5/18	2643	680	552	160	8.5
5/19	2730	710	600	160	8.48
5/20	2930	716	600	140	8.33
5/21	3093	728	608	140	8.34
5/22	2917	716	600	120	8.38
5/23	2990	712	600	120	8.4
5/24	2940			120	8.4
5/25	3040			126	8.46
5/26	3070	736	612	122	8.41
5/27	3091	736	614	168	8.59
5/29	3190	832	692	172	8.57
6/1	3790	824	712	188	8.61
6/2	3850	820	732	204	8.64
6/3	3910	1000	780	200	8.6
6/4	3910	880	760	200	8.46

Continued

Table 22.2 Continued

Date	Conductivity uS/cm	Total hardness ppm $CaCO_3$	Ca^{2+} ppm $CaCO_3$	M-alk ppm $CaCO_3$	pH
6/5	3670	1120	810	160	8.48
6/6		380	340	180	8.4
6/7		820	760	180	8.4
6/8	2710	970	568	180	8.51
6/9	2730	920	552	180	8.57
6/10	2650	870	588	160	8.59
6/11	2770	860	640	180	8.5
6/12	2810	690	580	180	8.69
6/13	2810	680	640	180	8.4
6/14	2870	660	640	180	8.5
6/15	2840	700	668	184	8.58
6/16	2961	720	688	188	8.63
6/17	3090	768	724	196	8.64
6/18	3100	760	640	172	8.6
6/19	3160	920	850	180	8.54
6/20	3200	880	860	180	8.4
6/21	3140	950	880	184	8.4
6/22	3190	860	850	184	8.57
6/23	3260	910	860	180	8.65
6/24	2790	830	720	196	8.5
6/25	3120	900	770	192	8.4
6/26	3310	830	740	180	8.63
6/27	3040	870	750	170	8.62
6/28	3220	810	730	170	8.64
6/29	3120	820	620	176	8.8
6/30	3210	820	640	186	8.5
7/1	3310	840	710	176	8.4
7/2	3420	880	800	168	8.4
7/3	3520	950	850	164	8.45
7/4	3480	850	800	176	8.5
7/5	3550	860	850	160	8.42
7/6	4440	830	640	160	8.51
7/7	4560	860	780	168	8.53
7/8	4880	820	730	185	8.36
7/10	4370	920	800	148	8.46
7/11	4360	880	800	144	8.41
7/12	3960	840	720	266	8.69
7/13	3790	860	820	185	8.5
7/14	3570	720	600	152	8.45
7/15	3560	620	600	160	8.35
7/16	3550	754	612	160	8.49
7/17	3290	670	660	152	8.4
7/18	3680	810	650	160	8.4

Over the 4 months period of this study, the cycles of concentration were steadily increased due to the need of reducing make-up water usage, ultimately rising by a factor of 4 at the end of the study. By taking a macroscopic view of Fig. 22.2, one can see that the LCM was proven to be extremely useful in optimizing treatment performance for localized corrosion, decreasing the localized corrosion rate from 70–80 to 10–15 mpy (1780–2030 to 250–380 μm/year).

At the beginning of the study, the localized corrosion rate was increasing with time after the initial 13 days. Since the direction of the changes is opposite to the expected reduction of corrosion rate with time (see the previous example in Fig. 22.1), this indicates that the treatment performance against localized corrosion in the system is worsening.

Application and optimization of the new program resulted in steadily decreasing localized corrosion rates, despite the approximately fourfold increase in cycles (conductivity changed from ~1200 to 4000–5000 μS/cm and calcium concentration changed from 150–200 to 600–1000 mg/L as $CaCO_3$) during the study. A more microscopic view of the data shows that the LCM identified upset or nonideal operating conditions and treatment program changes. The following are some of the specific results obtained:

Identified upset or changes in operating condition:

- Injection of sand filter backwash water to the cooling tower at about the 4th day;
- A plant shutdown started at about the 7th day;
- Stopped blowdown at about 29th day and resumed blowdown at about the 32nd day;
- Slug dose of new inhibitor treatment at about the 33rd day.

Treatment optimization:

- Increase product feed rates at about the 37th day;
- Increase polymer feed rate at about the 50th day;
- pH was raised, polymer feed rate was decreased, and inhibitor feed rate was increased at about the 62nd day;
- pH was decreased at about the 63rd day;
- Polymer feed rate was increased and inhibitor feed rate was decreased at about the 77th day;
- Final blowdown rate adjustments through the end of the test.

The results from this study show that increasing the dosage of dispersant polymer under the plant conditions was the most effective way to reduce the carbon steel localized corrosion rate. Additionally, increasing the pH and pyrophosphate (corrosion inhibitor) feed at the high calcium conditions and at the same time reducing dispersant polymer feed rate resulted in an increase in localized corrosion rate. This case study demonstrates the detrimental effect on localized corrosion by an excess dosage of deposition corrosion inhibitor and an insufficient dosage of dispersant polymer. The polymer is needed to optimize the interaction of the phosphate and phosphonate corrosion inhibitors with the metal surface for optimum passive film formation. In this case, it is likely that the insufficient polymer dosage resulted in a porous, nonpassive film of phosphate salts on the surface, promoting underdeposit (localized) corrosion. It also should be noted that the averaged localized corrosion rate detected by the LCM before the start of the new treatment program was consistent with the carbon steel shell-side heat exchanger failure frequency observed in the plant (i.e., failed after 2 years in service due to corrosion).

The new optimized treatment showed much better performance in controlling localized corrosion than the old Alkaline Stabilized Phosphate program, despite operating under a much more stressful scale and corrosion environment (higher cycles). Note that the general corrosion rate was much less sensitive to the corrosivity changes in the system and could not have been used to achieve this program optimization. The improved localized corrosion results were confirmed by plant inspection results, showing that no unacceptable carbon steel corrosion problems were found on the plant equipment inspected. The customer had since used the LCM to optimize the treatment programs in two different cooling water systems in the plant [3]. Thus, the LCM provided valuable information for developing a successful new treatment program in this customer application.

22.4 Program optimization using pilot cooling tower tests

To further optimize the chemical treatment program for the Southwest chemical processing plant described in the previous case history, a laboratory pilot cooling tower study was performed [4, 5]. Over the course of several months, many scale and corrosion control programs were used and the program was optimized under long holding time conditions to a localized corrosion level of 10–20 mpy (254–508 μm/year). At one point during the test, the localized corrosion rate increased significantly. Fig. 22.3 shows this increase. Adjustments in the scale and corrosion treatment programs had little effect on the increasing localized corrosion rate. It was also noted that a biofilm had developed on the heat exchanger surface and on the LCM anodes. Continuous bleach feed (maintaining 0.1 ppm free chlorine residual) was used throughout the test.

It is well-known that hypochlorous acid is not as effective a biocide as hypobromous acid at pH values above 7.5. The pH of this test was ~8.4; therefore, the observed results suggested that current program was probably not sufficiently effective in controlling microbial growth. Following a slug dose of a stabilized liquid bromine biocide (STA•BR•EX[a]) [1 ppm TRO (total residual oxidant) as Cl_2], the localized corrosion rates dropped more than half (Fig. 22.3). Localized corrosion rates remained low for ~1 week before they started to increase. A second dose of stabilized liquid bromine (STA•BR•EX) again quickly lowered localized corrosion rates for a period of about 1 week. When localized corrosion rates increased again, a batch of chlorinated water was added to the tower to refill the basin (dosage not recorded). This also showed a beneficial effect on reducing the localized corrosion rate. However, the effects were not longstanding. After allowing time for the excess bleach to be depleted, a 1 ppm slug of bleach was applied, showing little or no effect. Subsequent doses of unstabilized bromine at 1 and 2 ppm TRO as Cl_2 were also effective at decreasing localized corrosion rates by half. Since stabilized bromine (STA•BR•EX)

[a] Trade name of Nalco, IL, United States.

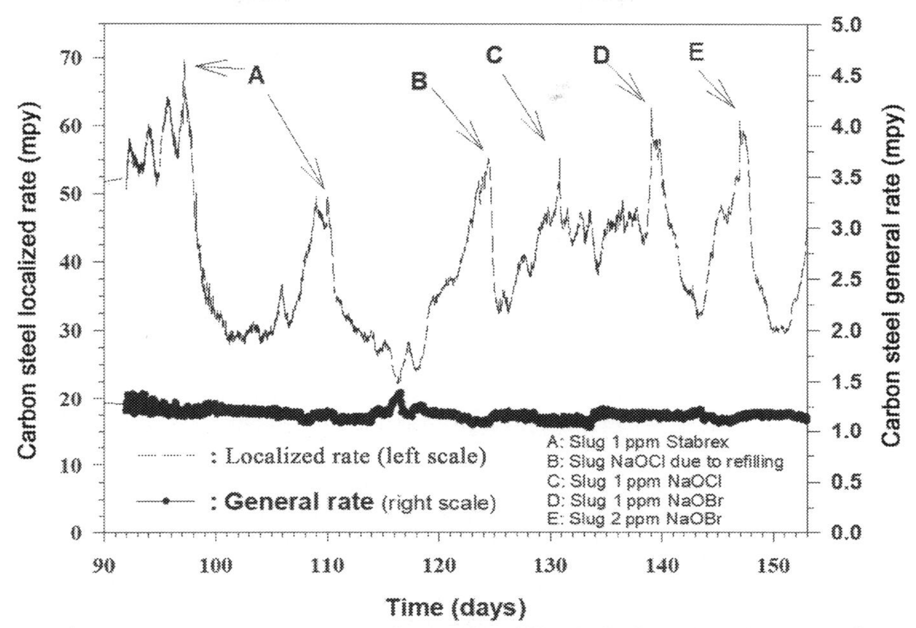

Cooling system return water, advanced stsbilized alkaline phosphate + NaOcl
HTI = 8 days, 260 ppm Ca^{2+}, pH = 8.4, 57.2°C. 1 mpy = 25.4 μm/year
Continuous NaOCl feed. Free Cl_2 = ~ 0.1 ppm, Total Cl_2 = ~ 0.35 ppm
Total aerobic bacteria in the bulk water: < 10^3 cfu/mL

Fig. 22.3 Effect of oxidizing biocides on MIC control.

and the unstabilized bromine are strong oxidants, they would normally be expected to increase corrosion rates. This effect was noticed immediately following the addition of the biocide. However, shortly thereafter, the localized corrosion rate dropped substantially. The corrosion monitor was found to be functioning properly during this test. Furthermore, sulfate reducing bacteria [100 colony-forming unit per square centimeter (cfu/cm^2)] and aerobic bacteria (three orders of magnitude higher than elsewhere in the system) were measured on the corrosion monitor anode, at the end of the test [4, 5]. A likely explanation for these observations is that the high localized corrosion rates were a result of microbial processes. Secondly, bromine-based biocides were very effective at lowering the localized corrosion rate. Bromine chemistry most likely worked better than chlorine as result of the pH of the cooling water used in the tests. Localized corrosion rates did remain low for a longer period following stabilized bromine (STA•BR•EX) application as compared to unstabilized bromine treatments (Fig. 22.3), suggesting that stabilized bromine may be more effective at reducing this type of corrosion under the test conditions.

Microbial control monitoring during this test was primarily done on bulk water samples. Throughout the experiment, bulk water counts of aerobic bacteria did not exceed 10^3 cfu/mL. These results suggest that while the low-level continuous bleach feed

(0.1 ppm as Cl_2) may be effective in controlling microbial growth in the bulk water, it may not be sufficient to control bacteria activity at all the corrosion sites. Surface microbial populations were not monitored regularly during this test. The reason for periodic monitoring is not to measure specific microorganisms, which may be contributing to the increased corrosion, but rather to determine how effective biocide treatments are in reducing the general surface population viability. Rarely does presence, or for that matter absence, of specific groups of microorganisms correlate with localized corrosion rates. This test does demonstrate that different biocides have distinguishable effects on the localized corrosion rate and can be measured using the LCM. Posttest metallurgical examination of the carbon steel cathode tube concluded [4] that the corrosion attack was consistent with microbiologically influenced corrosion (MIC), most likely by acid-producing bacteria (see Fig. 22.4). Biofilms were also observed on the surfaces of the cathode and anodes used in the LCM at the end of the test (see Fig. 22.5).

Fig. 22.5 photo of the anodes and cathode during the pilot cooling tower (PCT) test, showing the likely presence of biofilms on the electrode surfaces.

A second PCT test also yielded similar results (as shown in Figs. 22.6 and 22.7).

In this test, the bleach feed rate was much higher, yielding a free residual oxidants (FRO) as high as 1.84 ppm as Cl_2 and total residual oxidants (TRO) as high as 2.4 ppm as Cl_2. In addition, a different halogen stable phosphonate was used as cathodic corrosion inhibitor and $CaCO_3$ scale inhibitor. FRO and TRO were found to be reduced to close to zero after the addition of glutaraldehyde. The results show that when biofilms were observed on the LCM anodes and carbon steel tubes under the test conditions, slug doses of both oxidizing biocides (Br_2 based) and nonoxidizing biocides (glutaraldehyde and isothiazolinones) (from Katon) in the presence of continuous bleach feed could lead to reduction of the localized corrosion rates. Fifty ppm active glutaraldehyde was found to have no effect on the localized rates. Stopping bleach feed (and turn-off heaters) for a few hours could lead to a substantial increase in corrosion rates when heaters and bleach feed were turned on again. In this test, acid-producing bacteria (10 cfu/cm^2) were detected on a swab sample taken from a Cu exchanger surface

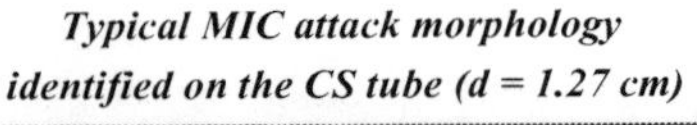

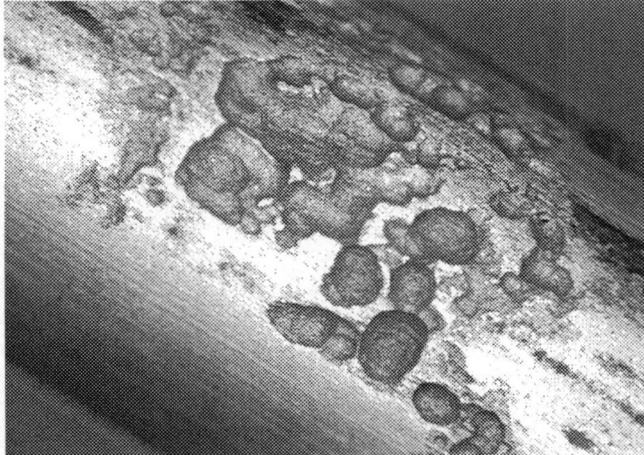

Fig. 22.4 Typical MIC attack morphology identified on the carbon steel tube. Photo of a section of the carbon steel tube (diameter = 1.27cm) used as the cathode in the LCM to obtain the results shown in Fig. 22.3.

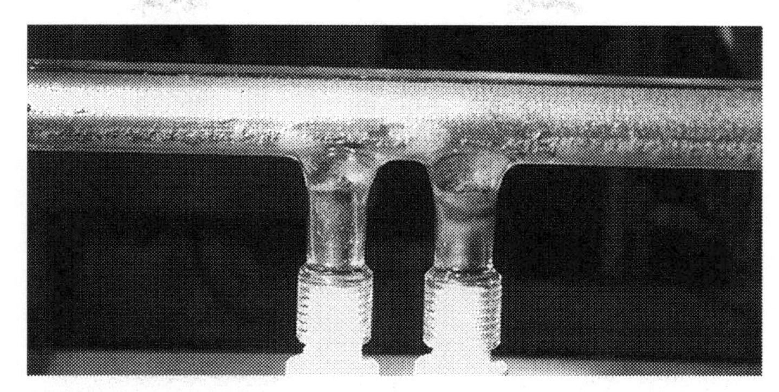

Fig. 22.5 Photo of the anodes and cathode during the PCT test (see Figs. 22.3 and 22.4), showing the likely presence of biofilms on the electrode surfaces.

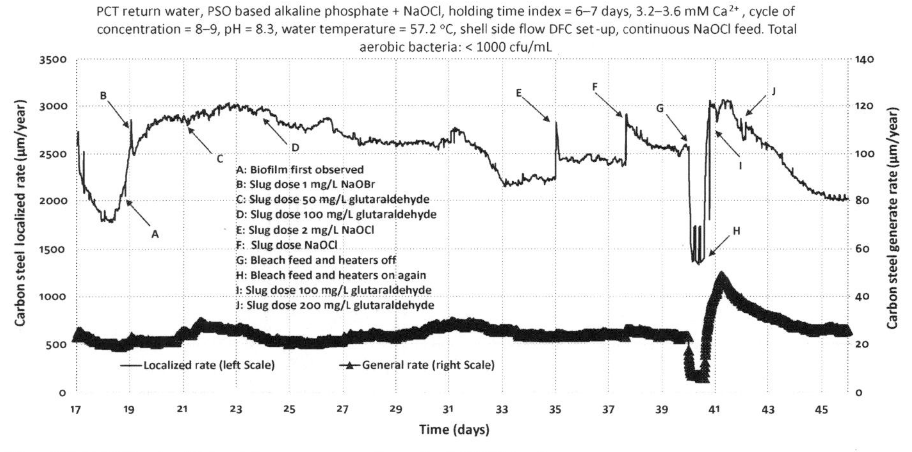

Fig. 22.6 Effect of biocides on corrosion. 2nd PCT test.

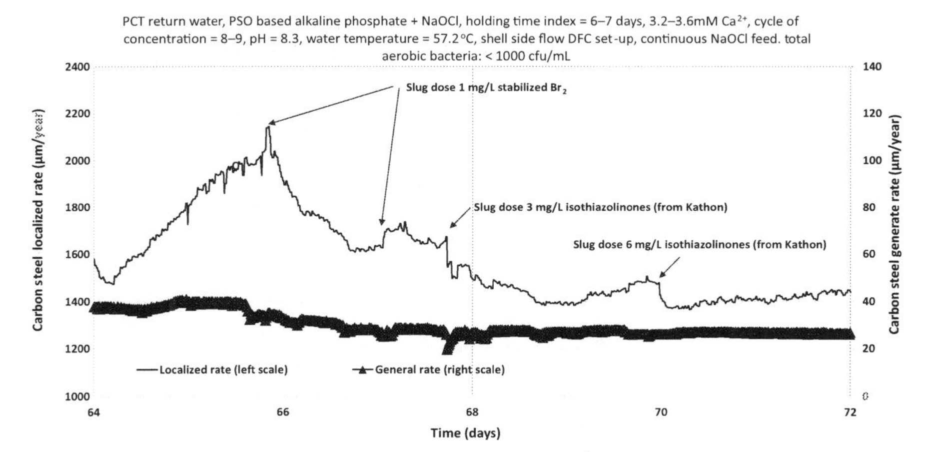

Fig. 22.7 Effect of nonoxidizing biocides on corrosion. 2nd PCT test.

shortly after noticing the appearance of biofilm on the LCM anodes. At the end of the test, 300 cfu/cm^2 of acid-producing bacteria and up to 4.4×10^5 cfu/cm^2 aerobic bacteria (or more than two orders of magnitude higher than elsewhere in the system) were detected on the anodes used in the tests. Corrosion attack morphology similar to the ones shown in Fig. 22.5 was also seen on the carbon steel tube covered with biofilm.

During these tests, bulk water counts of total aerobic bacteria did not exceed 1000 cfu/mL. These results suggest that while the continuous bleach feed may be effective in controlling microbial growth in the bulk water, it may not be sufficient to control bacteria activity at all the corrosion sites.

It is generally known that microbiologically influenced corrosion is difficult to detect and measure. This is because presence of species commonly associated with MIC may not be contributing substantially to the corrosion. Conversely, observation of high corrosion rates may not necessarily indicate that the corrosion is a result of MIC. The results obtained here indicate that the LCM may be a useful tool to monitor and control MIC.

22.5 Refinery hydrocarbon leak detection and control

A corrosion study was performed in a large U.S. Midwest refinery with no known carbon steel heat exchanger failures in the past 5–6 years before the study [2, 6]. The plant has 6 cooling towers. No. 2, 4, and 5 towers normally use the blowdown water from the largest tower (No. 3) as make-up water. Thus, tower No. 3 usually runs at a lower cycle of concentration than the other towers. The carbon steel heat exchangers cooled by the No. 3 tower are all tube-side exchangers (water in the tube side). The system has excess cooling capacity. Hence, some of the exchangers may be idle (i.e., no water flow) in the winter. A low Zn treatment and bleach were used as the cooling water treatment in the system. Some water analytical results of the tower water are shown in Table 22.3. The major concerns of the refinery before the study were:

- Admiralty exchanger stress corrosion cracking due to high NH_3 and SO_4^{2-}
- Carbon steel exchanger fouling
- Effect of a make-up water change in the winter (from lake water, which would freeze in the winter, to a more aggressive plant wastewater)
- Possible options to minimize the potential damaging effects associated with the increased corrosivity

The LCM was installed in a sidestream of a tower return line for the purpose of treatment optimization and control and to help manage problems that may arise due to the make-up water change in the winter. The results obtained from 3 weeks in the winter are shown in Fig. 22.8. Upon immersion, the localized corrosion was initially very high and then the rate decreased drastically (1–2 days). This effect is due to the initial flash corrosion that any fresh metal experiences when exposed to corrosive waters. The localized corrosion rate stabilized for a short period and then rose throughout the remainder of this test. Note that the general rate obtained by the LCM's LPR probe also shows similar pattern, i.e., it was first reduced to a low value upon immersion, followed by a steady increase. However, the increase in general rate occurred

Table 22.3 Midwest refinery Tower no. 3 cooling water return analysis results—Low Zn + bleach.

Date	**11/10/97**	**1/6/98**
Water analysis	**Filtered (0.45 μ)**	**Filtered (0.45 μ)**
ICP and IC results	**(ppm)**	**(ppm)**
Ca^{2+}	69	172
Mg^{2+}	24	48
Zn^{2+}	0.1	0.7
Iron	0.2	0.7
Cu^{2+}		
Sr^{2+}	0.2	1.4
Na^{+}	120	782
K^{+}	2.7	23
Phosphorus	1.9	2.3
Silica (SiO_2)	2.5	22
Boron		0.5
Mn^{2+}		0.1
Cl^{-}	230	994
Sulfate	110	682
M-alkalinity ($CaCO_3$)	144	243
Nitrate ($CaCO_3$) and nitrite (NO_2)	<2.9	<2.9
Other		
Unfiltered organic phosphate (PO_4)	1.4	0.6
Ortho-PO_4 (PO_4)	5.1	1.4
Conductivity (μS/cm)	1090	4280
pH	7.6	8.3

2–3 days after the corresponding increase in the localized corrosion rate, showing the increased sensitivity of the localized rate over that of the general rate. One week later, the LCM's LPR probe failed due to short-circuiting by biofilm deposits. This occurs because the LPR probe tips are in close proximity. Thus, no general corrosion rate information was obtained after that time. About 2 weeks into the test, a leak in a carbon steel heat exchanger served by this tower was reported. The leakage was confirmed by noticing hydrocarbon foaming in the tower. The leaking exchanger was identified and isolated. It was not determined whether the failure was from the waterside or the process side. However, it is known that the heat exchanger had been idled for some time in the previous winter (stagnant water can be a very corrosive condition). It should be noted that after the Light Cat. Gas Oil leak was confirmed, the leaking heat exchanger was identified and isolated from the system. Blowdown from Tower no. 3 to other tower was stopped. Tower no. 3 was blown down and then allowed to cycle up. The chemical analysis of the cooling water under this operating

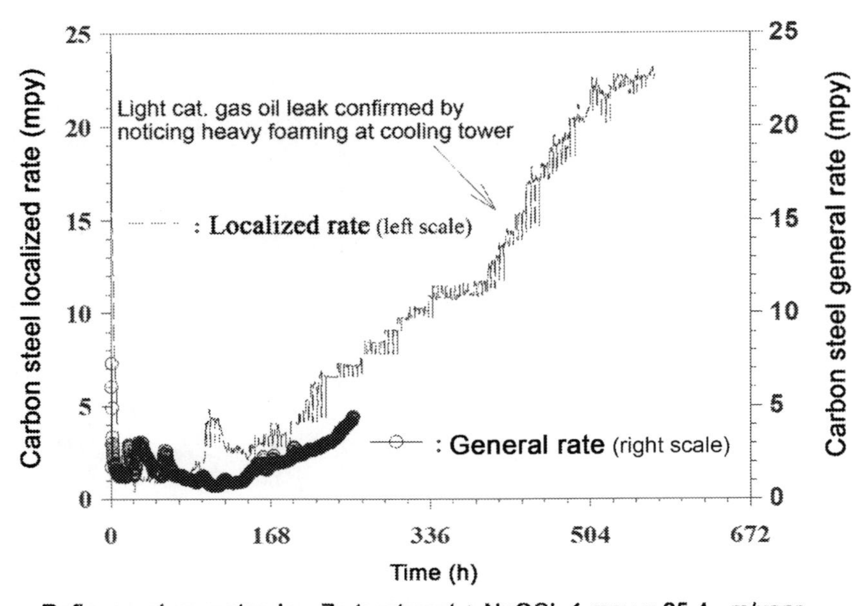

Fig. 22.8 Effect of hydrocarbon leaks on cooling water corrosion.

condition (i.e., sampled on 1/6/98) is shown in Table 22.3. Since there was excess cooling capacity in the refinery, there is no need to stop production.

The results in Fig. 22.8 show that the LCM was able to provide a warning on the heat exchanger failure with the sudden increase in localized corrosion rate 2–3 days before LPR responded, and more than 2 weeks before being confirmed by other methods (cooling tower hydrocarbon foaming). In addition, the LCM electrodes did not short-circuit under heavily fouled conditions, unlike the LPR electrodes. This is due to the physical separation of the LCM anodes into their respective wells, while the LPR electrodes are very close to each other with no barrier in between. It should be noted that the LCM uses a ZRA to measure the galvanically couple current between the electrodes. Thus, even in cases where there is low electrical resistance bridging between the electrodes, the LCM instrument would still likely to be able to detect a major portion of the galvanically coupled corrosion current flowing between the anodes and cathode.

The same LCM study continued for several more months without the LPR general corrosion rate information and is shown in Fig. 22.9.

The spike at about the 24th day was due to a change of LCM anode configuration to better simulate different operating conditions encountered in the plant. However, the localized corrosion rate continued to rise steadily after the "electrode change" spike had receded. Due to a dramatic localized corrosion increase started at about the 55th day, it was suspected that the LCM was malfunctioning. However, at the 80th day, a complete flow and electronics inspection found that the LCM was working normally. A light oily odor in the return water and the cooling tower water was noted after the

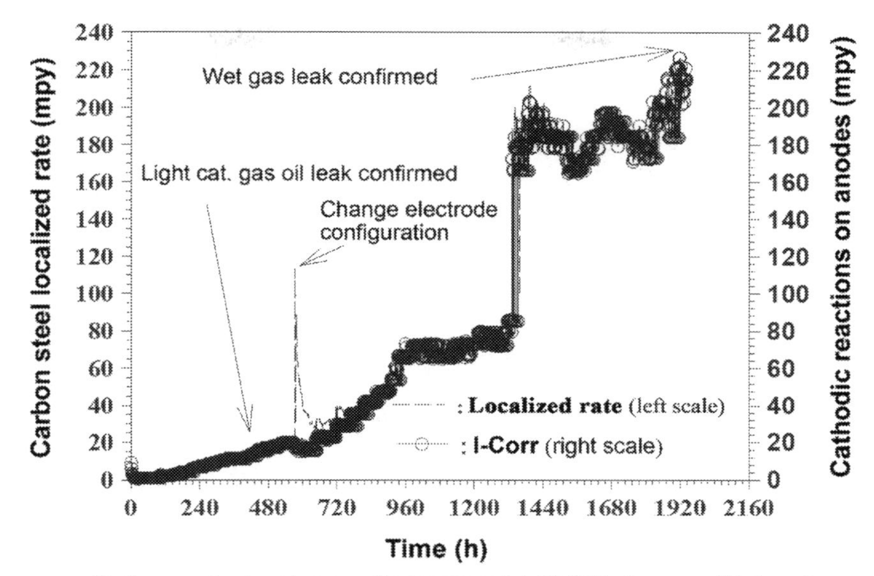

Fig. 22.9 Effect of hydrocarbon leaks on carbon steel corrosion in cooling water.

inspection (make-up water had no hydrocarbon odor and basin only had a very small amount of foaming). Thus, a new hydrocarbon leak was suspected. Indeed, another exchanger leak was confirmed on a different carbon steel heat exchanger later on the same day. Analyzing the biofilm from the tower showed very high aerobic bacteria and sulfate reducing bacteria (SRB) counts (2.9×10^8 cfu/mL total aerobic bacteria and 3.9×10^5 cfu/mL SRB in the biofilm sample), suggesting that the high pitting rate observed may have been due to MIC. Thus, the LCM had detected the effects of a second heat exchanger leakage more than 3 weeks before being confirmed by other methods.

Since the second leak was very likely a direct result of the damaging cascading effect of the corrosive contaminants from the first leak, the ability to provide an early warning to these conditions could be very useful in minimizing further damage to the refinery system.

After the confirmation of the second hydrocarbon leak, various standard damage control measures were taken, including isolating the leak, increasing blowdown rate, and increasing the bleach feed to oxidize the hydrocarbon and sulfide species. However, recommendations for use of more effective corrosion control treatments were not accepted by the refinery, due to concerns about treatment cost. Subsequently, the third heat exchanger leaked 1 month later (detected by LCM, confirmed soon afterwards). Carbon steel localized corrosion rate as high as 400 mpy (10.2 mm/year) was detected by the LCM. The general corrosion rate detected by the LCM was 10–

12 mpy (250–305 μm/year) during the same period. Multiple heat exchanger leaks occurred within the next 2 months before the refinery was shutdown for repair.

It should be noted that the heat exchanger leakage frequency observed in this case study was consistent with the localized corrosion rate results obtained from the LCM. When the localized rate is as high as 300–400 mpy (7.6–10.2 mm/year), leakage of carbon steel exchanger tubes (even for new tubes, typical thickness of 2.08 mm) is predicted to occur in 2–3 months. Higher pressure in the process side would tend to lead to an earlier leakage of the exchanger.

22.6 Refinery leak detection and program optimization

This case study was conducted in a large U.S. West Coast refinery [2, 6, 7]. In this case, the refinery wished to use California Title 22 Water (treated reclaimed wastewater) as make-up water for two of its cooling water systems. Information about the cooling water system, water chemistry conditions, and typical LCM and other corrosion measurement results before and after the introduction of Title 22 water as make-up water can be found elsewhere [7]. The effects of daily temperature fluctuation, due to sunlight and weather conditions, and the introduction of the more corrosive Title 22 water on the localized and general corrosion rates can also be observed in Fig. 22.10. The fact that the localized corrosion rates increased due to the introduction of Title 22 water (initial peak at the beginning of the Figure), while the general rate did not, demonstrates the excellent sensitivity of the LCM toward corrosivity changes in the system.

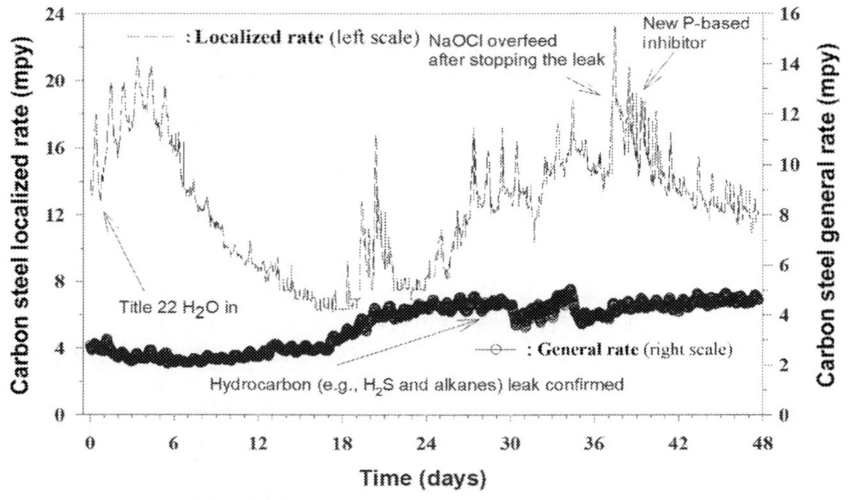

Fig. 22.10 Hydrocarbon leak control.

At about the 18th day into the study, the LCM showed a dramatic increase in the localized corrosion rate. The general corrosion rate also showed an increase. A hydrocarbon leak was confirmed about 9 days later (at about the 27th day) by operators noticing a temperature rise in the cooling water outlet. The leak was further verified by using a leak detection bubbler pot. After discovering the leak, bleach feed was increased to oxidize the H_2S in the leaked hydrocarbon and regain microbial control. The bleach overfeed after stopping the leak (due to less demand by H_2S) was detected by the LCM at about the 37th day. The subsequent effective control of localized corrosion after the leakage situation was corrected is also clearly demonstrated by the LCM results.

Results obtained in this study demonstrate that the damaging effects of a hydrocarbon leak (overhead gases from a column containing H_2S, CH_4, C_3H_8, C_4H_{10}, etc.) can be controlled quite well with timely identification and stoppage of the leak, followed by appropriate chemical treatment procedures, in this case using an optimized Stabilized Phosphate program. The mitigation procedure entailed additional bleach feed to oxidize H_2S and control microbial growth, switching to a more effective cathodic inhibitor and stopping the leak in a timely manner. The polymer dose and blowdown rates remained constant during the hydrocarbon leak. Comparing these results with those obtained from the Midwest refinery case study and considering the similar nature of the leaks, it is reasonable to assume that this West Coast refinery was able to maintain its acceptable level of operation because of the way it responded to the initially identified leak. On the other hand, the Midwest refinery was unable to maintain effective cooling tower operation because it did not effectively respond to the initially identified leaking exchanger. This led to subsequent leaks and, ultimately, an unscheduled unit shutdown.

From another viewpoint, the early response to the hydrocarbon leak also had a great impact on the corrosion performance of the treatment program. The West Coast Refinery was able to address the hydrocarbon leak earlier (about 1 week) than the Midwest Refinery described in the previous case history (2–3 weeks). This earlier response could be the reason that the West Coast Refinery could bring corrosion under control while maintaining blowdown rates, while the Midwest Refinery could not. In addition, it is difficult to say if the nature of the hydrocarbon in one cooling system was more corrosive than in the other. Certainly, the types and quantities of corrosive contaminants present in different hydrocarbon process streams can vary greatly.

22.7 Admiralty brass corrosion control in cooling water system using brackish water as make-up

This study was conducted in a 2-unit 180 MW conventional power plant located in Southwestern United States [6]. One unit (#6, 100 MW) of the power plant started operation in 1968–69 and the other unit (#5) started operation in mid-1950s. Each unit is cooled with a five-cell cooling tower using well water as make-up water. The cooling system volume for unit #6 is 515,550 gallons (1951.6 m^3) with a holding time index of about 30 h. The cooling treatment programs used before 1995 were consisting of tolytriazole, sodium hexametaphosphate, chlorine, and bromine. Since August

1995, a cooling water treatment program consisting of: zinc, phosphate, and tolytriazole as corrosion inhibitors, a phosphonate scale inhibitor, and polymeric dispersants had been used in the system. The biocide treatment consisting of activated bromine (mixing bleach and NaBr at 1:1 mole ratio) had also been used since 1995. Biocide was normally fed for a maximum total time of 1 h per every 2 days due to discharge permit restriction. The goal was to maintain a ~0.2 ppm as Cl_2-free halogen residual at the condenser outlet during the slug feed of the biocide. The feed dosage of corrosion and scale treatment was controlled via online measurement of an inert fluorescence tracer. The residual concentration of tolytriazole (yellow metal corrosion inhibitor) in the cooling water was controlled separately at 1.5 ppm by an online fluorescence monitor/controller. The chemical treatment programs and the state-of-the-art chemical feed control technology were selected to address the high admiralty brass heat exchanger pitting corrosion rate (e.g., frequent admiralty condenser leaks due to cooling water side corrosion. ~3% of the #6 unit condenser tubes were plugged at the start of the study, see Fig. 22.11) and excessive $CaSO_4$ scale formation on the condenser surfaces and tower fill encountered previously in the system. Typical water analysis results of tower #6 using the new chemical treatments are shown in Table 22.4. The corresponding results obtained from corrosion coupon measurements in tower #6 are shown in Table 22.5. These results appear to show that the chemical treatments employed in the system are providing very good corrosion and scale control performance in spite of using a brackish well water as the make-up water in the system. Coupon corrosion rates were very low and well within the acceptable limits. Water analysis results showed that the corrosion product ion concentrations were also very low. In addition, water analysis results also showed that the cycles of concentrations of the deposit-prone ionic species and the nondeposit-forming ionic species were the same in the system, indicating there was little scale formation in the system. Indeed, the good scale and corrosion control performance of the new treatments were confirmed by borescope inspection of the heat exchangers. In addition, admiralty brass (UNS number C44300) heat exchanger leakage had been stopped after the start of new treatment for about 3 years. Admiralty brass corrosion coupon results

Fig. 22.11 A photo of unit #6 condenser tubes showing the presence of tube plugging and debris blockage in several tubes.

Table 22.4 Southwestern U.S. conventional power plant #6 unit cooling water analysis results.

Date	2/17/00				5/7/99		12/1/98
Water analysis	**Filtered (0.45 μ)**		**Total (unfiltered)**		**Filtered (0.45 μ)**	**Total**	**Total**
ICP and IC results (ppm)	**Make-up**	**Tower #6**	**Make-up**	**Tower #6**	**Tower #6**	**Tower #6**	**Tower #6**
Ca^{2+}	260	1300	270	1300	1300	1300	1400
Mg^{2+}	60	300	61	300	300	300	336
Zn^{2+}	<0.01	0.47	<0.01	0.47	0.7	0.86	0.2
Iron	0.04	0.4	0.21	0.78	<0.25	0.66	1.2
Cu^{2+}	<0.01	<0.1	<0.01	<0.1	<0.25	<0.25	<0.1
Sr^{2+}	4.4	22	4.5	22	23	23	24
Na^{+}	190	1000	190	1000	900	900	1012
K^{+}	7	38	7.3	38	43	43	50
Phosphorus	<0.1	1.7	<0.1	1.6	<2.5	<2.5	3.2
Silica (SiO_2)	12	62	13	62	50	50	70
Boron	0.3	1.6	0.3	1.6	<2.5	<2.5	1.6
Li^{+}	0.08	0.37	0.08	0.37	0.51	0.53	0.5
Cl^{-}			330	1700		1600	1917
Sulfate			620	4300		4100	4512
M-alkalinity ($CaCO_3$)			210	110		74	
Br^{-}			<2.9	13		7.8	12
Nitrate ($CaCO_3$) and nitrite (NO_2)			<2.9	<2.9		<2.9	<2.9
Other							
Phosphate (PO_4)	0.2	4.4	0.2	4.7	3.5	3.7	
Inorganic phosphate (PO_4)	0.2	2.1			1.9	1.9	
Ortho-PO_4 (PO_4)			<0.1	2.1	1.4		
Conductivity (μS/cm)			2400	10,000		9600	
pH			8.0	8.0		7.34	
Turbidity (NTU)							0.2

Table 22.5 Southwestern U.S. power plant—Tower #6 corrosion coupon results.

Corrosion coupon	Exposure	General rate (mpy)	Pitting	Exposure length (days)
Admiralty brass	9/23/99 to 12/21/99	0.1	N.D.	89
Admiralty brass	1/5/99 to 3/3/99	0.1	N.D.	57
Admiralty brass	10/14/98 to 1/5/99	<0.1	N.D.	83
Admiralty brass	5/6/98 to 7/30/98	0.2	N.D.	85
Admiralty brass	2/16/98 to 5/6/98	<0.1	N.D.	79
Carbon steel (C1010)	9/23/99 to 12/21/99	0.7	12.7 mpy	89
Carbon steel (C1010)	1/5/99 to 3/3/99	2.0	N.D.	57
Carbon steel (C1010)	10/14/98 to 1/5/99	2.8	N.D.	57

Note: *N.D.*, denotes not determined.

were also improved from 0.4 to 0.1–0.2 mpy (10.2 to 2.5–5.1 μm/year). However, leakage of admiralty brass exchangers reappeared in 1999 (six leaks occurred in the first 7 months of 1999). The condenser leaks were occurring at an increasing frequency (one leak occurred in Jan. 2000). The power plant had little confidence on the reliability of the results obtained from corrosion coupons since they did not simulate the actual corrosion conditions in the plant.

Metallographic analysis of condenser tube section removed from unit #6 in April 2000 shows that the internal surface of the condenser tubing suffered from numerous pitting corrosion under a layer of deposits consisting of mostly calcium, copper, iron and zinc phosphates, and calcium carbonate (see Figs. 22.12–22.14, and Table 22.6). The external surface (i.e., steam side) of the condenser tubes was covered by a thin layer of black oxide. No localized attack on the external surface was observed.

The metallographic analysis suggests that underdeposit corrosion in the water side may be the cause of the recent condenser leakages. It also revealed that the calculated time average maximum localized corrosion would be <1 mpy (25.4 μm/year) since the condenser was put into service ~31 years ago. However, the metallographic analysis did not provide information on the propagating underdeposit corrosion rate under the current operating conditions. Since operating conditions and treatment chemicals had been changed drastically in the long exposure period, one would not expect that the corrosion rates would be constant in the period.

To gain a better understanding of the effects of current operating conditions on condenser tube underdeposit localized corrosion and to optimize the chemical treatment to minimize the localized corrosion rate, a localized corrosion monitor based on the

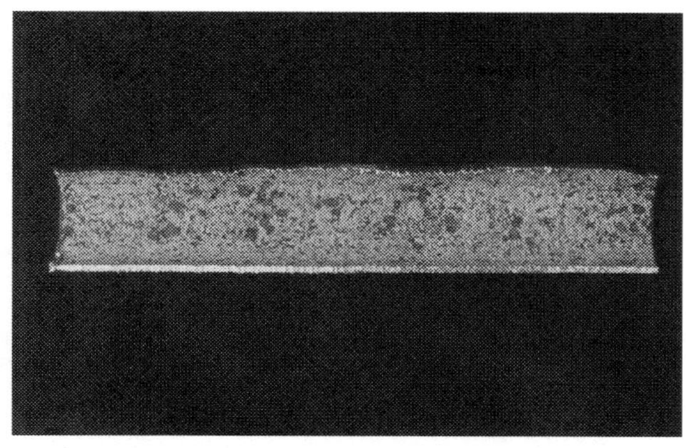

Fig. 22.12 A section of condenser tube (outer diameter = 7/8″ or 2.22 cm, thickness = 0.0046″ or 116.8 μm) removed from unit #6 in April 2000.

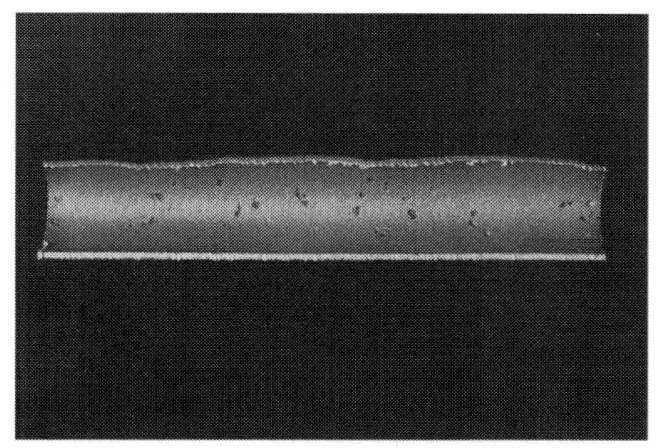

Fig. 22.13 Section of the condenser tube shown in Fig. 22.9 after cleaning via bead blasting, revealing underdeposit corrosion attack sites. The deepest observed attack site penetrated as much as 0.026″ or 66 μm into the tube wall.

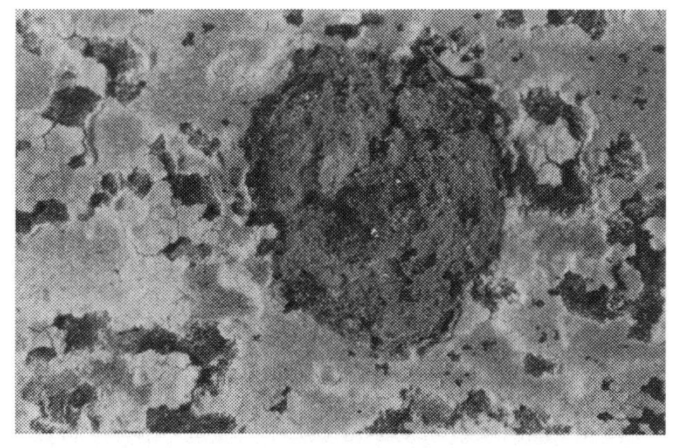

Fig. 22.14 Closed-up view of a corrosion site on the condenser tube before cleaning. Magnification: 12×.

Table 22.6 EDS analysis of material on internal surfaces of the unit #6 condenser tube section.

Element	Tan deposit	Green nodule	Corrosion site interior
Copper (as CuO)	24.0	15.2	83.8
Zinc (as ZnO)	5.3	5.5	4.1
Calcium (as CaO)	37.3	38.5	0.6
Phosphorous (as P_2O_5)	17.3	17.0	0.8
Iron (as Fe_2O_3)	7.2	9.9	0.2
Silicon (as SiO_2)	2.9	5.3	0.9
Magnesium (as MgO)	2.2	2.3	0.1
Aluminum (as Al_2O_3)	1.2	2.0	1.0
Chloride (as Cl)	0.4	0.4	7.0
Sulfur (as SO_3)	2.2	4.1	1.6

Note: Values are in wt% as oxides unless specified otherwise.

DFC method was installed in a sidestream of the outlet of unit # 6 condenser. A tube-side heat exchanger flow setup (see Fig. 6.2 in Chapter 6) was used. Admiralty brass (UNS C44300) was used for both the anodes and cathode. The flow rate was 2.3 gpm (8.7 L/min) or a flow velocity of 0.91 m/s near the cathode tube surface. To simulate the underdeposit corrosion conditions shown in Figs. 22.14 and 22.15, the anodes were occluded (or pulled down) by 3 cm from the flow channel glass tube surface. The exposed anode surface (0.317 cm^2) was precorroded in the plant cooling water containing $>$100 ppm NaOCl for $\sim$1 h before immersion. Laboratory tests (both bench-top and pilot cooling tower) showed that the electrode pretreatment method and the flow cell setup would simulate realistically the underdeposit corrosion attack observed on the plant condenser tube.

Fig. 22.15 Bowl-shape depressions (i.e., underdeposit corrosion sites) beneath the red deposit layers on the condenser tube internal surface. Magnification: 50 $\times$.

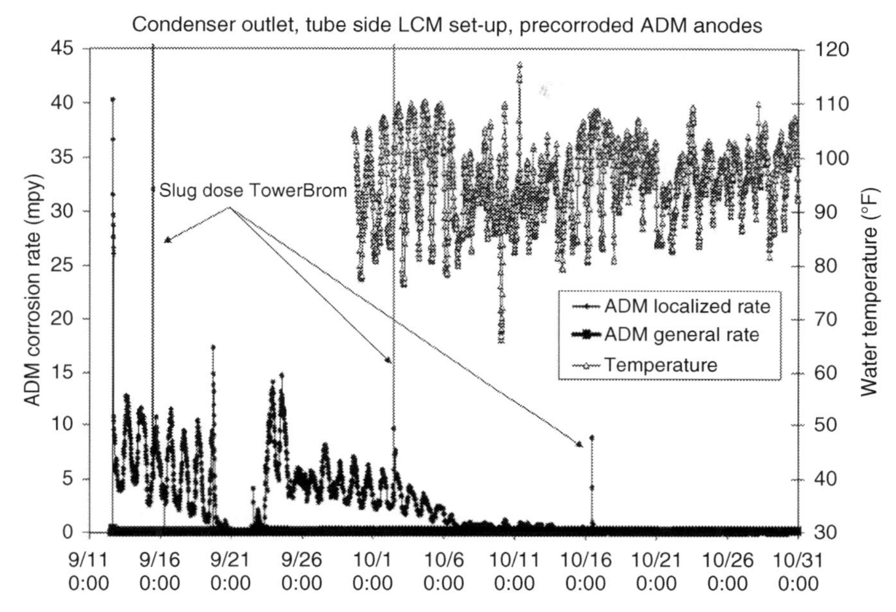

Fig. 22.16 Effect of oxidizing biocide feed on admiralty brass underdeposit corrosion in a conventional power plant cooling water system using brackish water as make-up.
1 mpy = 25.4 μm/year. °C = 5/9 × (°F − 32).

The field study lasted nearly 2 years. Many useful results were obtained. Fig. 22.16 shows the LCM results obtained from 9-12-00 to 10-31-00. Based on laboratory and field studies, the scale inhibitor feed dosage was increased before 9-12-00 to reduce the potential $CaSO_4$ scaling problem on the tower fill (A test tower fill unit installed on-site showed some $CaSO_4$ scaling in early 2000). As shown in Fig. 22.16, the admiralty brass localized corrosion rate was generally quite high in the 2 weeks of immersion (time average localized rate = 3.3 mpy or 83.8 μm/year in the first 2 weeks in Fig. 22.16). The localized corrosion rate also changed with water temperature and showed daily fluctuations in response to the daily water temperature fluctuations associated with weather and operating heat load conditions. These results are similar to the ones observed from LCM field applications using carbon steel as the anodes and cathode. They demonstrate that admiralty brass LCM localized corrosion rate readings are very sensitive to corrosivity changes in the system. It should be noted that due to a data-logger wiring error, the water temperature readings in the initial 17 days of the test were not recorded. The time average general corrosion rate obtained from the LCM during the same time period was very low, i.e., 0.10 mpy (2.5 μm/year), in agreement with the mass loss results obtained from corrosion coupons. The LCM results show that although the admiralty brass general corrosion had been under excellent control by using the selected modern cooling water chemical treatments and the state-of-the-art active and inert fluorescent tracer-based monitor and feed control instruments, the localized corrosion control had substantial room for improvement.

On-site investigation suggested that microbiological fouling control had substantial room for improvement. Tower basin water was turbid and had substantial amount of algae growth, especially during the summer months. High surface aerobic bacteria counts were detected on corrosion coupon swap samples (total count of 1.3×10^8 cfu/swab on both carbon steel and admiralty brass coupons) and on tower fill samples (total count of 500,000 cfu/g) in Feb. 2000. Significant sulfate reducing bacteria activity was also detected in both the tower fill sample (200 cfu/g) and in a cooling tower water sample. In addition, analysis of the dried deposit sample from the condenser in #6 unit showed a high organic content (i.e., 34 w% loss at 925°C), indicating the possibility of fouling by biofilm. Thus, a once every 2 weeks slug dose (i.e., ~50 lb or ~22.7 kg) of high concentration of TowerBrom 60M[b] (i.e., sodium dichloroisocyanurate + NaBr) was initiated in September 2000. In Fig. 22.13, the slug dosing of TowerBrom 60M was detected on 9-15-00 [150.2 mpy (3815 μm/year) at 11:06 a.m.], 10-2-00 [99.5 mpy (2530 μm/year) at 11:21 a.m.], and 10-16-00 [8.8 mpy (224 μm/year) at 11:06 a.m.] as sudden spikes of admiralty brass localized corrosion rate. The general rates were only increased slightly, e.g., 0.04–0.05 mpy (1.0–1.3 μm/year), during the same time periods and the increases were not visible in Fig. 22.13. As shown in Fig. 22.13, the additional slug dose feed TowerBrom 60M was very effective in reducing the admiralty brass underdeposit localized corrosion rate rapidly, especially noticeable after the slug dose on 10-2-00. By 10-15-00, the admiralty brass localized corrosion rate was reduced to <0.2 mpy (5 μm/year). Inspection of the tower basin indicated that the additional TowerBrom 60M biocide slug dose had cleaned up the biomass in the system. The tower water had also become clearer. Furthermore, the feed dosage of tolytriazole to maintain the residual concentration of 1.5 ppm via online monitor/controller was reduced significantly. Roughly, the same localized corrosion rate was observed up to Dec. 2000. The admiralty brass general corrosion rate detected by the LCM remained largely constant during the same period. Results obtained later at the site confirmed that the optimized cooling water treatments (i.e., corrosion and scale inhibitors, biocides) via the help of the LCM were very effective in reducing the high admiralty brass localized corrosion rate problem in the system.

The results obtained in this study suggest that high microbial activity in the system may be one of the major factors in having high underdeposit localized corrosion rates being observed on the condensers. Once every 2 week, slug dose of the bromine-based biocide in the summer months, and the continued use of existing corrosion, scale inhibitors, and biocide treatments, and the associated modern feed control technology could be used effectively to meet the needs of the cooling water system.

[b] Trademarks of Occidental Chemical Corporation, TX, United States.

References

[1] E. Hale, B. Yang, Evaluation of various application regimes to control mild steel corrosion in a nuclear power service water system using a localized corrosion monitor, in: EPRI Service Water System Reliability Improvement Seminar, July 13, Biloxi, MS, 1999.
[2] B. Yang, Minimizing localized corrosion via new chemical treatments and performance based treatment optimization and control, in: NACE Corrosion/99, Paper no. 307, 1999.
[3] D.A. Meier, E.B. Smyk, B. Yang, Advances in zinc free alkaline cooling water treatment, in: Corrosion/99, Paper no. 304, NACE, Houston, TX, 1999.
[4] B. Yang, Advances in localized corrosion control in cooling water systems, PowerPlant Chem. 2 (6) (2000) 321 GmbH.
[5] M. Enzien, B. Yang, Effective use of monitoring techniques for use in detecting and controlling MIC in cooling water systems, Biofouling 17 (1) (2001) 47.
[6] B. Yang, Corrosion control in industrial water systems, in: Presented at NACE Central Area Conference, Corpus Christi, TX, 7–10 October, 2001.
[7] B. Yang, Real-time localized corrosion monitoring in industrial cooling water systems, Corrosion 56 (2000) 743.

Advanced corrosion control at chemical plants using a electrochemical noise method

Takao Ohtsu
Mitsubishi Chemical Corporation, Tokyo, Japan

23.1 Introduction

Chemical plants manufacture products by chemical reactions and the decomposition of feedstock materials. Chemical plant facilities not only face severe environmental conditions, but also the danger of material deterioration, especially corrosion. However, due to the multiple ingredients in the fluids present, it is difficult to pinpoint by analyzing the substances that cause corrosion. To solve this serious problem, it was considered possible to measure chronologically the state of the environments and evaluate the corrosion activity that is brought about by such environments. Such an analysis requires a device that can measure simultaneously the corrosion data and the operating conditions. To do this, a device is required that can monitor the state of the corrosion during the operation of the chemical plant facilities in real-time. Chemical plant corrosion damage has been ascribed to raw material quality, operational conditions, and to other process factors; so, in the past, it has been difficult to propose changes to the operational conditions or other measures to prevent corrosion, unless the causes of corrosion were identified. This holds true even with changes of the service conditions, which have become increasingly sophisticated.

As discussed in Chapter 5, one method to identify the causes of corrosion and formulate preventive measures is the use of electrochemical noise (EN) measurements to measure general corrosion. Recently, developed electrochemical noise measurements are gaining attention as a means to measure real-time corrosion of facilities. This chapter focuses on the issues and countermeasures involved in applying the measurement method to actual plants. Also reported is an example where a combination of laboratory tests and corrosion monitoring was applied to an actual chemical plant, in order to identify the causes of corrosion and to consider and verify measures for the prevention of corrosion.

Techniques for Corrosion Monitoring. https://doi.org/10.1016/B978-0-08-103003-5.00025-4

23.2 Investigation

23.2.1 Principles of the three-electrode electrochemical noise measurement

The development history and principles of electrochemical noise methods have been described in detail in Chapter 5. This section presents a summary of the working principle for the electrochemical noise (EN) measurement methodology, which has been applied to investigate general corrosion using a sensor with three electrodes. Mansfeld and Xiao [1] examined this electrochemical noise method. Eden et al. [2] proposed methodologies for the measurement of electrochemical noise using three electrodes, calculating the corrosion rate and identifying the type of corrosion. Based on this idea, the present author and coworkers at Mitsubishi Chemical Corporation (Japan) conducted research using the data obtained from experiments and observations and, as a result, revised some aspects that had not been clarified or questioned in the earlier research and improved the measurement method [3, 4].

Fig. 23.1 is a schematic representation of the devices used for the three-electrode electrochemical noise measurement. Three metal electrodes of the same material were used with a zero-resistant ammeter installed between two electrodes and an electrometer connected to the remaining electrode, which was used as reference electrodes.

One characteristic of this method is that no polarization current is required to be externally applied to the electrodes while measuring. The corrosion that appears on the electrodes can be measured without generating any electrochemical disturbance to the electrodes.

On the electrodes, multiple anodes and cathodes work together during corrosion activity. Generally, it is impossible to measure externally what is occurring at the

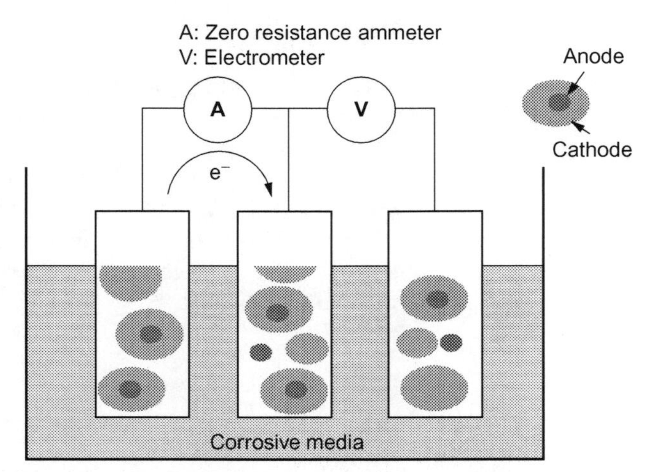

Fig. 23.1 A schematic representation of the devices used for the three-electrode electrochemical noise measurement.

individual anodes and cathodes under natural conditions. However, when the two electrodes are connected by the ammeter, some of the electrons generated by corrosive reactions, which occur at the anodes and cathodes in the proximity of the connection, interact with the impedance of the electrochemical double layer, thereby causing the corrosion potential to fluctuate and this perturbation is monitored by the electrometer. The electrode connected by the electrometer on the right-hand side of Fig. 23.1 functions as the reference electrode. Fig. 23.2 shows an example of measurement data obtained with this method.

Examination of the chronological changes in corrosive reactions at an anode and cathode pair, at a microscopic level, reveals subtle variations in the rate of the corrosive reactions depending on factors that include the supply rate of the oxidizer within the solution and the speed of mobile ions. These microscopic changes within a very short period of time are called fluctuations of corrosion. It was supposed that, within the very short period of time, the ratio of the changes in the electric current to that in the potential remains the same, unaffected by the changes in corrosive reactions occurring at the anode and cathode. The three-electrode electrochemical noise measurement of general corrosion rates originated with consideration of such fluctuations of corrosion. While corrosion progresses on the macro scale, measurement and collection of data at very short intervals should result in measurements unaffected by the corrosion cycle at the anode/cathode pair.

Based on this basic concept, and in an analogy to Ohm's Law, the ratio of change in the current, δI, and the corresponding change in the potential, δE, can be used to calculate the noise resistance, R_n, relevant to the corrosive reaction rate:

$$R_n = \delta E / \delta I \tag{23.1}$$

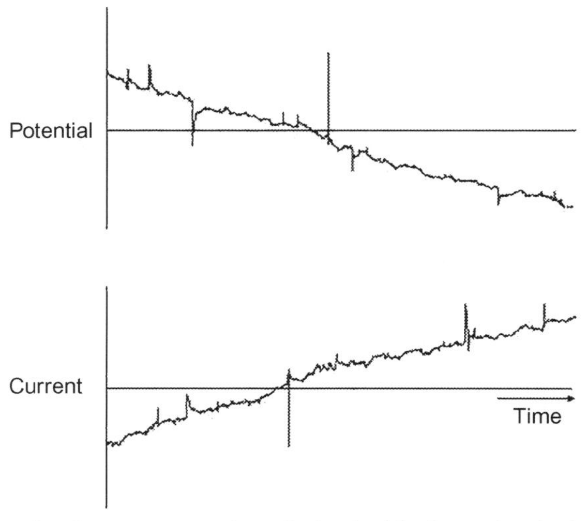

Fig. 23.2 An example of measurement data obtained with electrochemical noise method.

However, during actual measurement, the R_n differs significantly among different measurements taken. Therefore, the standard deviations of the data measured during a specified measurement time (σI: standard deviation of the current, σE: standard deviation of the potential) were computed and used to obtain an average value for R_n. Different time periods of measurement were investigated in order to find one with the fewest fluctuations in the measurement.

$$R_n = \sigma_E / \sigma_I \tag{23.2}$$

The current indication from the zero resistance ammeter should be a representative sample of the electron flow rate between the anodes and the cathodes, generated in the proximity of the two electrodes. However, since the actual areas occupied by the anodes and the cathodes are unknown, in addition to the ratio of the anode to the cathode divided between the electrodes, we are unable to evaluate the values related to the measured current in terms of corrosion current density.

For this reason, the coefficient of corrosion, G, was introduced as a constant of proportion between the corrosion rate (CR) and $1/R_n$:

$$CR = G / R_n \tag{23.3}$$

where CR is the corrosion rate (mm/year), G is the coefficient of corrosion, and R_n is the noise resistance. The corrosion coefficient is not the same across all the areas of measurement. Even within a solution containing the same substances, if the density of the solution and/or other factors change and alter how the corrosion proceeds, the corrosion coefficient also must be altered. Eden et al. [2] employed a constant of proportion, obtained from the Tafel gradient, as the constant of proportion between CR and R_n. This was somewhat similar to the common polarization resistance method as described in Chapters 3 and 4. Still, as mentioned above, the area occupied by the anode may not be constant, and therefore, the use of such a constant of proportion cannot be considered appropriate.

Fig. 23.3 summarizes the measurement device that was developed in accordance with the principles of the three-electrode electrochemical noise measurement. This measurement device consisted of a data analyzer, a measurement section, and a tank containing the test solution into which the sensor was introduced. The data analyzer has two functions: one that outputs analysis results of chronological changes in the electrode potential and current measured, and a second that sets the data collection interval, a crucial feature of the three-electrode electrochemical noise measurement methodology. The measurement section had a zero resistance ammeter (A) between two of the three electrodes and an electrometer (V) connecting the remaining electrode. The electrodes were placed within the solution while measurements were taken.

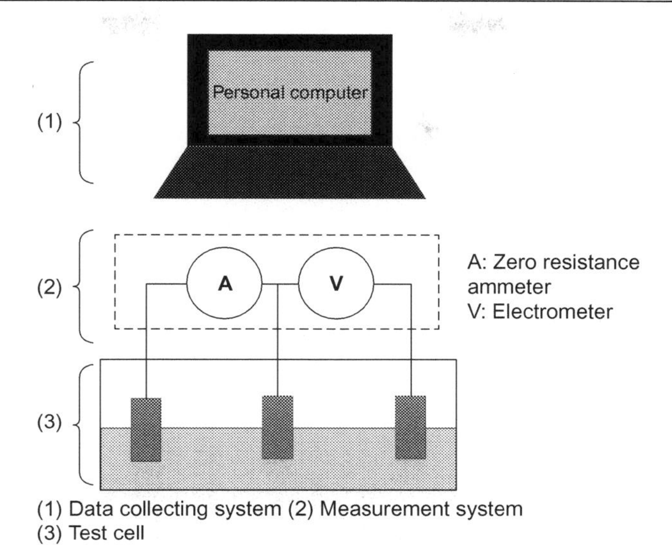

Fig. 23.3 The principles of the three-electrode electrochemical noise measurement system.

23.2.2 Verification of the three-electrode electrochemical noise measurement

Tests were conducted to verify whether the three-electrode electrochemical noise measurement is applicable to the evaluation of general corrosion behavior. In these tests, we attempted to confirm that a correlation existed between the corrosion rate obtained by the weight loss method and the R_n value obtained by the three-electrode electrochemical noise measurement. We employed solutions of the same substances, but with different concentrations (see below).

The measuring device was a SI 1280B manufactured by Solartron Analytical, and the measurement specifications were as follows:

- number of cell electrodes: compatible with 2, 3, and 4 terminals
- range of current measurement resistance: 0.1 Ω to 1 MΩ
- range of current: 200 nA to 2 A
- maximum resolution: 1 pA.

The test sample, made from stainless steel (SUS304: UNS S30400), was a cylinder of 10 mm in diameter and 32 mm in length. The solutions used were four different concentrations of HCl solutions, each containing 100 cm^3 of distilled water and HCl concentrations of 10.0%, 5.0%, 1.0%, and 0.5%.

Fig. 23.4 shows the values of the corrosion rate calculated using the weight loss method, and the R_n calculated with the three-electrode electrochemical noise

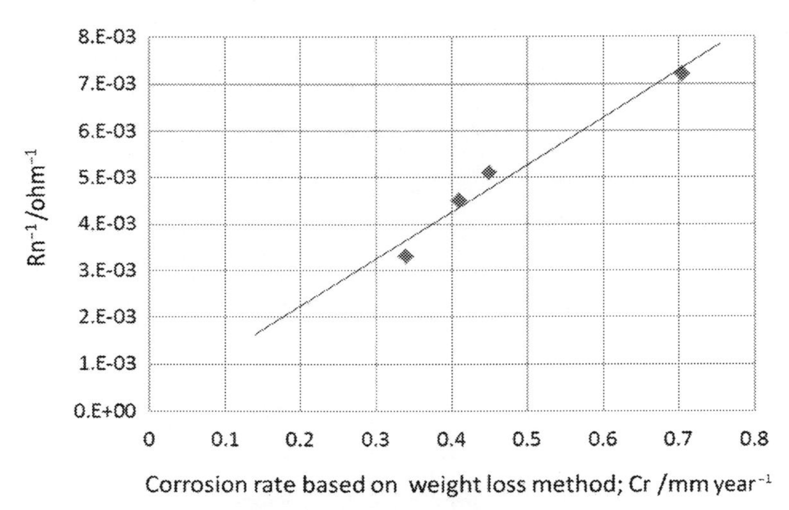

Fig. 23.4 The values of the corrosion rate calculated using the weight loss method, and the R_n calculated with the three-electrode electrochemical noise measurement.

measurement. The figure indicates a proportionality relationship between the corrosion rate from the weight loss method and the inverse of the R_n value from the three-electrode electrochemical noise measurement. Thus, these tests confirmed that the three-electrode electrochemical noise measurement was able to measure uniform corrosion, so long as an appropriate coefficient of corrosion was obtained from laboratory tests.

23.3 Monitoring and corrosion control

23.3.1 Operation

An example case of corrosion control using the corrosion monitor [5] was performed during a reaction process at an organic chemical plant. The controlled corrosion here was inside the fluidized-bed catalytic reactor for gaseous materials. Fig. 23.5 shows the process flow around the reactor. The facility produces ethane dichloride by reacting ethylene and hydrogen chloride as a process reaction. The coil material in the facility is JIS STPA22 (ASTM: P12, BS EN: 13CrMo4-5, DIN EN: 13CrMo4-5), which is a low-alloy steel. During the operation, ethylene and hydrogen chloride, with air as an oxidizer, were fed in, and a flow of catalyst was added inside the reactor to promote the reaction. To start the operation, hot water flowed into the hot-water coil inside the reactor to increase the internal temperature, and hot air was drawn inside to further increase the temperature until it reached 140°C. Then, the catalyst was injected. The reactor was further heated up using the hot-water coil. After it reached 200°C, the raw materials, ethylene and hydrogen chloride, were fed in. The operation was continued by maintaining this condition.

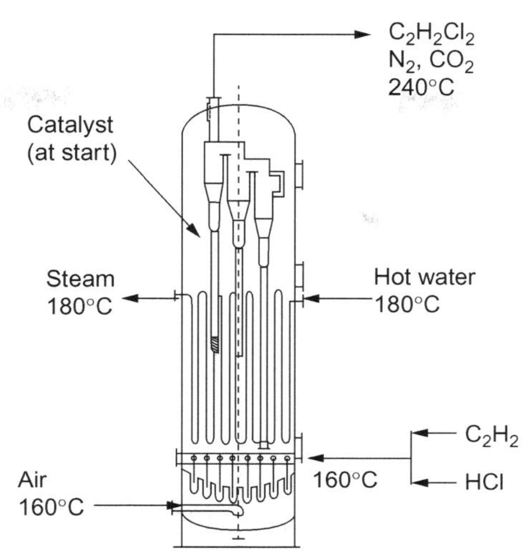

Fig. 23.5 The process flow around the reactor.

23.3.2 Damage

Fig. 23.6 is a photograph of the reactor. The reactor is an independent tower design. Fig. 23.7 shows the state of corrosion damage inside the reactor. As shown in Fig. 23.7, corrosion wastage was evident on the feeding pipe for the material inside the reactor, resulting in a hole in the pipe. In particular, the internal structure at the bottom was severely damaged. Also, the hot-water coil and the feed nozzle were damaged, necessitating repair work during the periodic inspection conducted every 2 years. This situation forced the plant not only to pay the cost of the repair, but also to reduce the time of operation for about a week. The corrosion was initially assumed to be caused by the raw material, hydrogen chloride. However, since this corrosion could not occur because the temperature was higher than its dew point temperature, it was not possible to identify precisely the cause of the damage. Therefore, the erosion by the catalysts, as well as the corrosion caused by hydrochloric acid, was examined as the cause of the damage at the initial stage of the examination.

23.3.3 Monitoring

Under these conditions, the corrosion monitor was installed inside the reactor to take measurements. Fig. 23.8 is a photograph of the electrodes installed inside the reactor. The monitor consisted of a total of nine electrodes. As shown in Fig. 23.9, the electrodes were grouped in these to make three electrodes. Fig. 23.10 is a photograph of the system deployed for measuring the corrosion. The system, except for the sensors shown in the figure, was installed in the control room. The monitor was designed to take measurements continuously for about a year from the time when the reactor started until it stopped. Two extra sensors were installed in case the original sensor

Fig. 23.6 A photograph of the reactor.

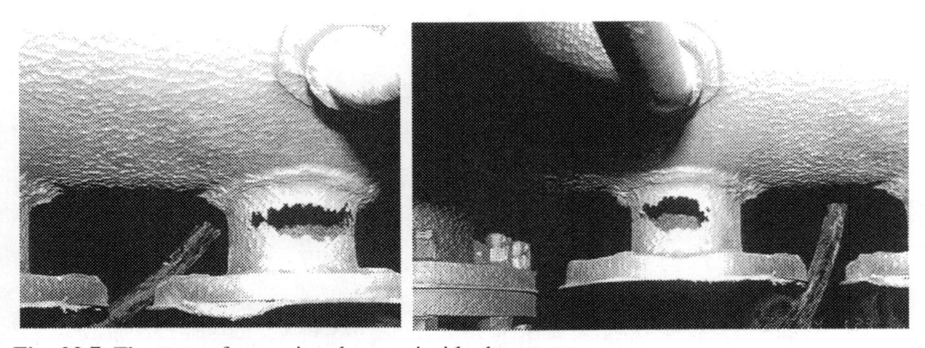

Fig. 23.7 The state of corrosion damage inside the reactor.

failed. One of the two sensors failed in about a month, but the remaining sensor worked for a year taking measurements.

23.3.4 Measurement results

Fig. 23.11 shows the measurement data of the corrosion monitor that was installed inside the reactor. The corrosion damage was not caused by the air feed that was introduced during the initial operation of the reactor, nor was it caused by the increase in temperature of the circulating hot water. The corrosion was caused by the catalyst as it was fed, and the rate of corrosion reached its maximum 2–3 days after the feed was

Fig. 23.8 A photograph of the electrodes installed inside the reactor.

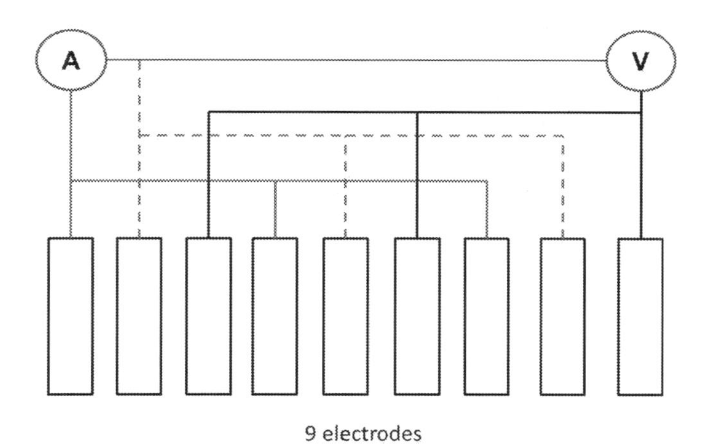

Fig. 23.9 9 electrodes grouped into 3 electrodes.

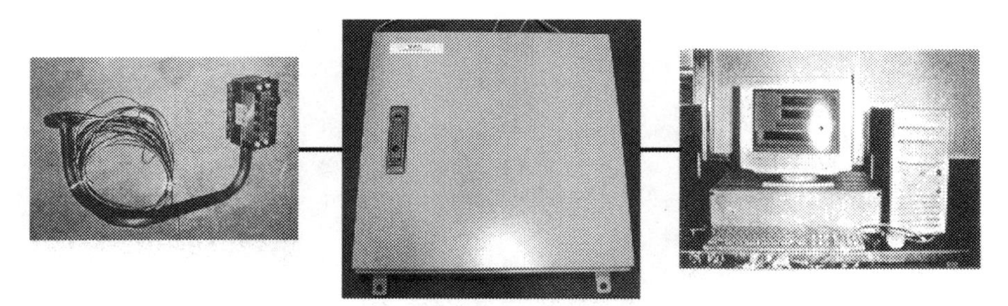

Fig. 23.10 A photograph of the system deployed for measuring the corrosion.

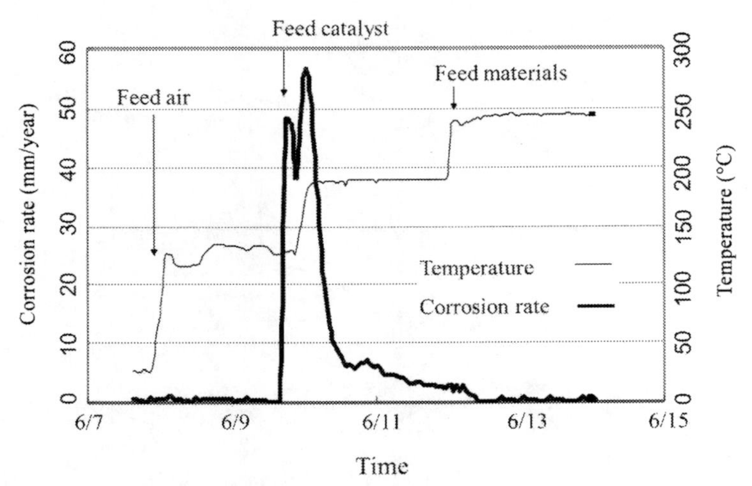

Fig. 23.11 The measurement data of the corrosion monitor.

started. The rate then decreased as the operation continued and did not increase very much even when the raw material was fed in, which caused almost no corrosion during the entire operation. No corrosion was found for a year's operation, except for the corrosion that occurred at the start of the operation. The corrosion in the reactor was considered to be caused because the material gases and/or substances in the catalysts reached their dew point.

23.4 Analysis

The problem of corrosion in the reactor was examined through the results of the corrosion monitoring and the operating procedure. Fig. 23.12 chronologically summarizes the operational items of the reactor and lists the occurrence of corrosion for each corresponding item. The examination results clearly indicate that corrosion began when the catalysts were fed into the reactor. Analysis of the catalyst operation disclosed that the catalysts were sent out of the reactor when the operation stopped, but the hydrogen chloride, a new material, adhered to the surface of the catalysts without any reaction. Furthermore, since the temperature of the catalysts that were being fed in was 100°C, the temperature of the component materials inside the reactor, which were in contact with the catalysts, was almost 100°C, even when the air temperature of the reactor reached 140°C. Because of this, it was found out that the hydrogen chloride emitted from the catalysts combined with water, which condensed and then caused the corrosion. Therefore, the operation needs to be performed in temperature conditions at which dew point corrosion will not occur even when the catalysts are fed in.

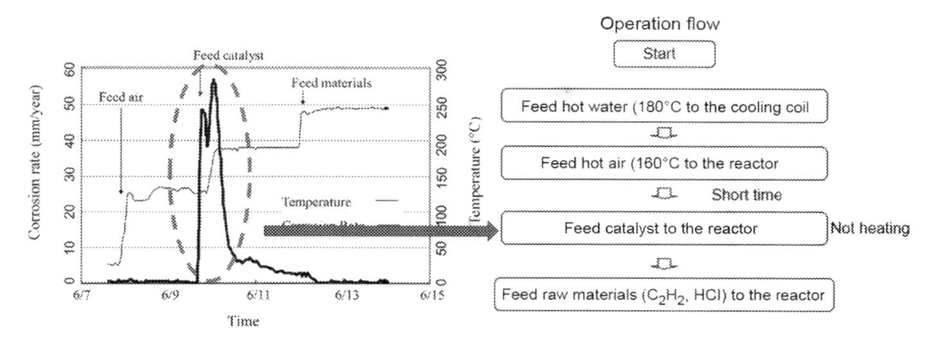

Fig. 23.12 The operational items of the reactor and lists of the occurrence of corrosion for each corresponding item.

23.5 Corrosion control

To implement this measure, the feeding temperature for the catalysts was changed to 140°C and the time allotted for increasing the air temperature was extended to increase the temperature of the component materials inside the reactor. Fig. 23.13 shows the plans for this measure compared with the initial operation. To verify the effectiveness of this corrosion countermeasure, the corrosion monitor was installed to perform measurements after this procedure was implemented. Fig. 23.14 shows the results of measurements taken after the implementation. Almost no corrosion was detected, so the effectiveness of this measure was verified. In addition, Fig. 23.15 shows the result of the overhaul inspection that was conducted at the time of the periodic repair. No corrosion damage was observed from the inspection inside the reactor.

Carrying out corrosion measurements in this way enabled the extention of the interval for the repair of coils and pipes to more than 10 years from an initial interval

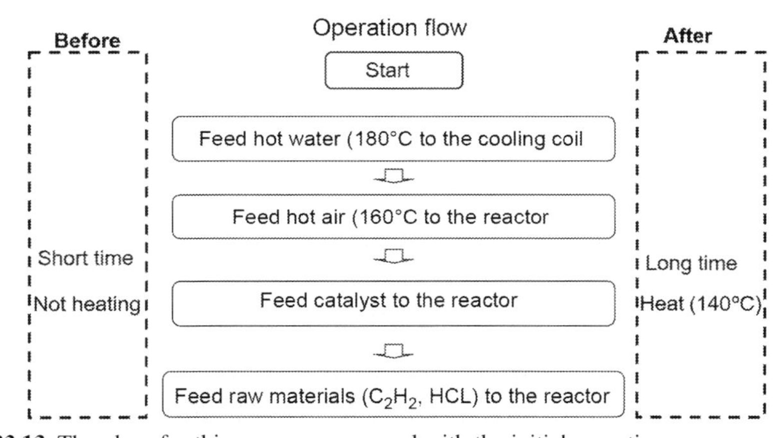

Fig. 23.13 The plans for this measure compared with the initial operation.

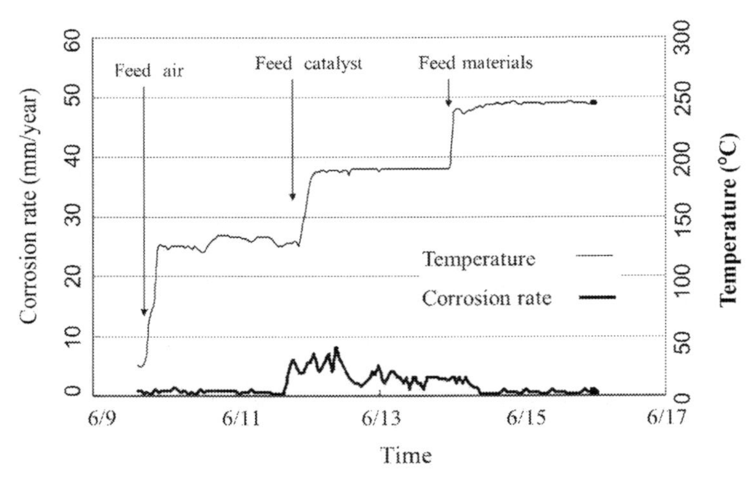

Fig. 23.14 The results of measurements taken after the implementation.

Fig. 23.15 The result of the overhaul inspection.

frequency of 2 years. Furthermore, the reduction in operation stoppages has resulted in an increase in production by 2% or more per year.

23.6 Conclusion

The electrochemical noise method was examined for use in a corrosion monitor, and the method to apply it in an actual plant was established. By using the corrosion monitor, the problem of corrosion in chemical plants, in which several substances are intricately intertwined, was solved, and control measures were successfully formulated and implemented. In addition, it was also possible to verify that the corrosion monitor detected the effect of the implemented measures. The implementation of the corrosion monitor worked efficiently, facilitating the measurement. As a result, it was possible to extend the interval for the repair of the coils and pipes from the initial interval of 2 years to more than 10 years, with savings in maintenance costs at US$200,000 per year and the reduction in operation stoppages resulted in an increase in production by 2% or more per year.

References

[1] M. Mansfeld, H. Xiao, J. Electrochem. Soc. 140 (1993) 2205.
[2] D.A. Eden, D.G. John, J.L. Dawson, European Patent 0 302 703 B1, 1992.
[3] M. Miyazawa, H. Inoue, Zairyo to Kankyo 53 (2004) 317.
[4] M. Miyazawa, H. Inoue, Zairyo to Kankyo 55 (2006) 188.
[5] M. Miyazawa, T. Ohtsu, Zairyo to Kankyo 2001 (2001) 257–258.

Corrosion monitoring under cathodic protection conditions using multielectrode array sensors

24

Xiaodong Sun[a], *Dongmei Sun*[b], *and Lietai Yang*[a]
[a]Corr Instruments, LLC, Carson City, NV, United States, [b]Polytech, Beijing, People's Republic of China

24.1 Introduction

Corrosion monitoring under cathodic protection conditionsCathodic protection (CP) is widely used to protect metallic structures and components in many industries including infrastructure, transportation, oil and gas transmission, and chemical processes. According to a NACE report [1], the total estimated cost for cathodic and anodic protection in the United States alone in 1998 was $2.22 billion, of which the majority cost was associated with the CP. The protection provided by the CP depends on the distribution of the cathodic current density. Due to the variations in geometry of protected structures or components, or due to the location of anodes, certain areas of the protected structure or component may not receive the minimum cathodic current density that is required to provide sufficient protection. Corrosion may take place in these areas and cause catastrophic failures, if not identified and mitigated at an early stage. Because corrosion in these areas is not easily detected, an effective monitoring technique is required to identify the problem at an early stage, in order to alleviate the problem in a timely manner. An online monitoring tool placed near these hard-to-protect and critical locations may provide a real-time indication of the effectiveness of the CP system. Often, the protection potential or current density is set at overly conservative values. An online monitoring tool may make it possible to automate the CP system, to enable the protection current or potential to be set at a value that is just sufficient to protect the critical areas and that, at the same time, can reduce the cost involved in the process.

As discussed in Chapter 20, the true criterion for cathodic protection is to polarize the protected metal structure to the open-circuit potential of the most active anodic site on the structure. Since it is difficult to measure the open-circuit potential of individual anodes and cathodes on a metal, the monitoring of cathodic protection in the pipeline industry has been mostly focused on the pipe-to-soil potential (see Chapter 20 for more details) and the criterion is to maintain this potential between -0.85 and -1.20 V vs $Cu/CuSO_4$ electrode in most cases. This criterion is only a semiquantitative parameter because the potential in this range only indicates that the corrosion rate is most likely below a threshold value (0.01 mm/year for instance), but it does not tell

Techniques for Corrosion Monitoring. https://doi.org/10.1016/B978-0-08-103003-5.00026-6

the operator if the corrosion rate is truly below the threshold value and if the metal is truly protected. In addition, the values of the threshold potentials (−0.85 or −1.20 V) are often determined on the basis of past experiences or experiments conducted under commonly found conditions. These values may be either insufficient or excessive, when the conditions such as the soil pH and temperature in the field change. Therefore, an online corrosion monitor that works under cathodic protection conditions is an ideal tool for the real-time monitoring of the effectiveness of the CP system. Furthermore, the potential criterion does not tell the operator how much margin there is for the degree of adequate protection when the metal is fully protected. A tool that can indicate the CP margin of effectiveness or the degree of adequate CP would be highly useful for optimizing the CP and minimize excessive protections which may cause the blistering of coatings on the metal for painted metals or hydrogen embrittlement.

Coupled multielectrode array sensors (CMAS) have been extensively used as in situ and online monitors for corrosion, especially localized corrosion, in laboratories and industry applications [2–14]. In this chapter, the use of coupled multielectrode corrosion sensors as online monitors under cathodic protection conditions was reviewed. The experimental results of corrosion rates of several cathodically protected metals in simulated seawater, soil, concrete, and drinking water are presented. Furthermore, a parameter that can be used to indicate the CP margin of effectiveness (CPEM) or the degree of adequate protection during the CP is introduced.

24.2 Evaluation of the effectiveness of cathodic protections with CMAS probes

The principle of the CMAS probes has been described in Chapter 8. Fig. 24.1 shows a typical CMAS instrument, nanoCorr analyzer (Model A-50) (nanoCorr is a trade mark of Corr Instruments, LLC., United States). This CMAS analyzer has a current resolution of 10^{-12} A and allows the measurement of coupling currents up to 50 electrodes

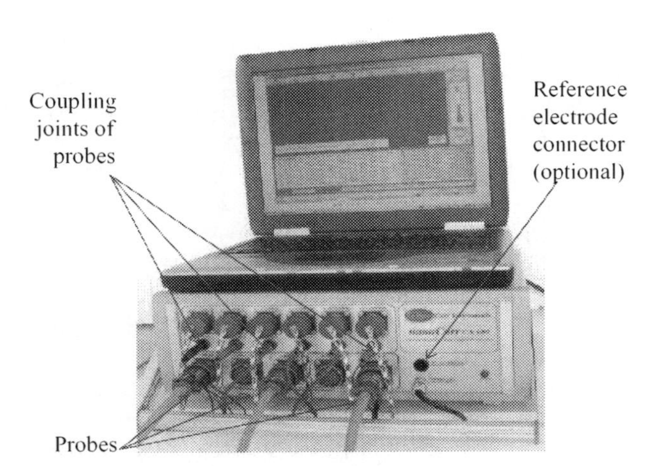

Fig. 24.1 Typical coupled multielectrode array sensor analyzers for corrosion monitoring of cathodically protected systems. Courtesy of Corr Instruments LLC.

in up to 6 separate probes at the same time. Fig. 24.2 shows the wiring diagram between a CMAS probe and the CMAS analyzer when the probe is used to measure corrosion under cathodic protection conditions. In a CMAS analyzer, the coupling joint is essentially at the same potential of the CMAS probe. Therefore, the cathodically protected system (the metal in Fig. 24.2) is connected to the coupling joint of the CMAS analyzer so that the sensing elements (the multiple electrodes) of the CMAS probe would have the same degree of protection as the cathodically protected metal. In Fig. 24.2, the cathodic protection is provided by a power source and this type of CP is often called impressed current CP. This power source is not needed if the anode is a sacrificial anode (often called galvanic anode). The electrolyte can be either a liquid such as seawater, or a wet solid-water mixture such as concrete or soil. The reference electrode is used to indicate the protection potential. It should be noted that the reference electrode is not required for the CMAS to measure the corrosion rate or the CP effectiveness. It is used to gain additional information to understand how the cathodic protection systems work.

Fig. 24.3 shows a typical wiring diagram between a CMAS probe and a CMAS instrument when the CMAS probe is used to evaluate the CP effectiveness for a cathodically protected pipeline in the field. The pipelines in the field are usually protected with protective coating and the access wires that are connected to the pipeline walls are usually available at the test stations (see Chapter 20 for more information). For the coated systems, the multiple electrodes in the CMAS probe simulate the corrosion behavior of the metal at the multiple areas that have coating-defects.

As discussed in Section 24.1 and in Chapter 20, it is difficult to measure the open-circuit potential of individual anodes and cathodes on a metal and a surrogate criterion, the pipe-to-soil potential measurement, has been adopted to evaluate the

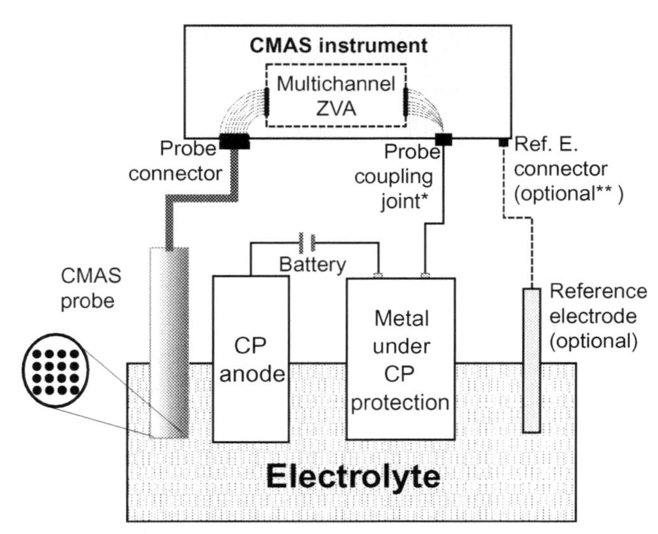

Fig. 24.2 Typical connections of probes and the cathodically protected systems to the CMAS analyzers during the measurement of corrosion of cathodically protected systems.

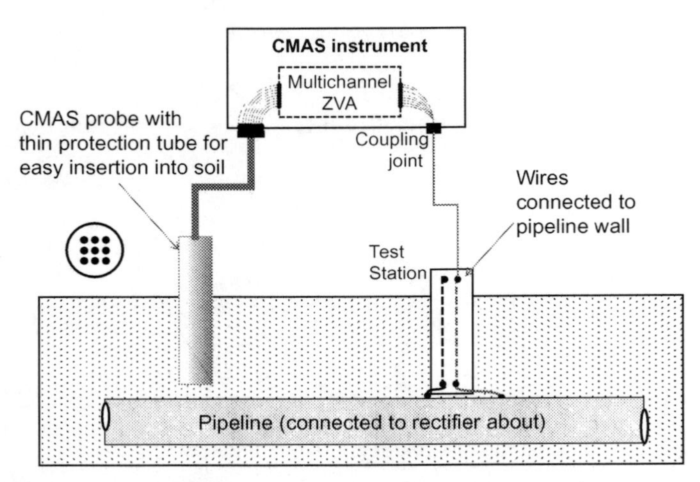

Fig. 24.3 Typical connections between a CMAS probe and a CMAS instrument when the CMAS probe is used to evaluate the cathodic protection of a pipeline in the field.

effectiveness of cathode protection. The concept of the CMAS probe made it possible to measure the open-circuit potential of the probe's individual electrodes that simulate the anodic and cathodic sites of the protected metal. The most anodic electrode on the CMAS probe simulates the behavior of the most active anodic site on the structure. As the polarization of the probe's electrodes increases toward the cathodic direction, the anodic current on the anodic electrodes decreases. When the anodic current on the most anodic electrode dropped to zero, the polarization reached the open-circuit potential of the most anodic electrode or open-circuit potential of the most active anodic site on the structure. So the criterion for using the CMAS probe is to control the cathodic protection such that the anodic current on the most anodic electrode is zero or negative. With this criterion, there is no need for using the reference electrode.

24.2.1 Minimum adequate CP potential, excessive CP potential, and maximum allowable current

Fig. 24.4 shows typical currents from the multiple electrodes and the potential of the coupling joint of a CMAS probe immersed in an electrolyte before and after the CP is applied [15]. The CMAS probe has 9 electrodes that simulate 9 small areas of a corroding metal or 9 small coating-defect areas of a coated pipeline. Before the application of the CP, the potential of the coupling joint was at the free corrosion potential. At the corrosion potential, some electrodes were anodes and some electrodes were cathodes and the current from the most anodic electrode (I^{a}_{max}) represented the maximum corrosion current on the CMAS probe. After the CP was applied, all of the currents

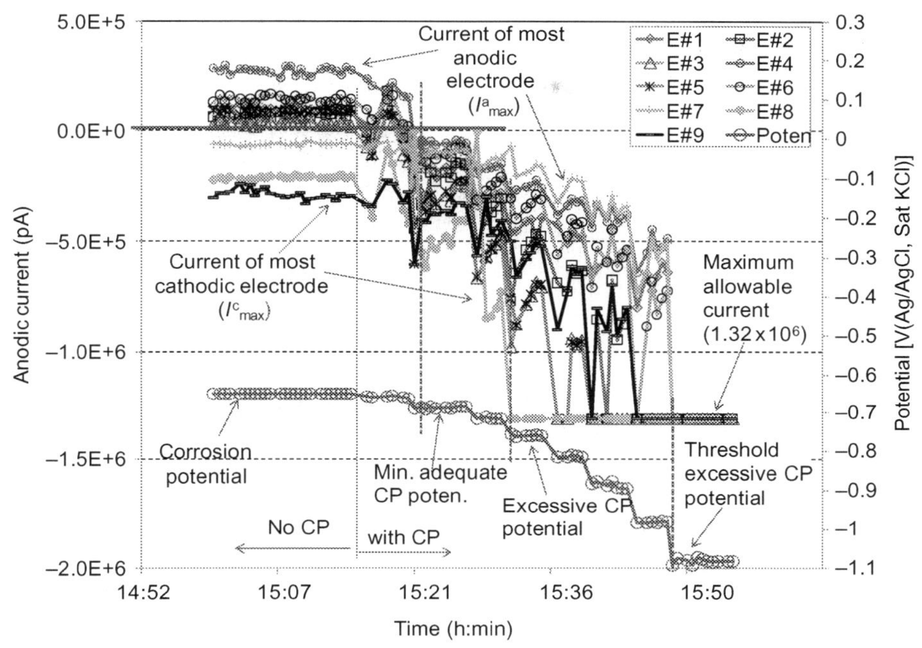

Fig. 24.4 Typical currents from the multiple electrodes and the potential of the coupling joint of a CMAS probe before and after the CP is applied.

started to decrease and the I^a_{max} reached zero when the CP potential reached the minimum adequate CP potential. When the I^a_{max} reached zero, the metal was fully protected because the most anodic electrode (which represents the most vulnerable corrosion site of the metal) is protected.

As the CP potential further decreased, both the I^c_{max} and the I^a_{max} became more and more negative. When the CP potential reached another critical value (the excessive CP potential), I^c_{max} jumped to a large value which usually indicates that excessive hydrogen evolution started on the most cathodic electrode. As the CP potential further decreased and reached the threshold excessive CP potential, I^a_{max} also jumped to the large value which usually indicates that excessive hydrogen evolution also started on the most anodic electrode. Because hydrogen evolution is an undesired reaction and should be avoided, this large value of current is called maximum allowable CP current. The CP potential should be controlled between the minimum adequate CP potential and the excessive CP potential. Under such conditions, the cathodic current from the most anodic electrode (which simulates the site that is the most difficult to protect) is more negative than zero and the cathodic current from the most cathodic electrodes (which simulates the site where excessive hydrogen evolution first starts) is less (more anodic) than the maximum allowable current.

24.2.2 Corrosion rate as an indicator for the effectiveness of CP when the CP is inadequate

Fig. 24.5 shows typical corrosion rate calculated from the I^{a}_{most}, the current from the most corroding electrodes (or the most anodic electrode), of the CMAS probe [15]. The CMAS probe effectively measures the corrosion rate when the cathodic protection is insufficient. As the CP potential becomes more negative and reached the minimum adequate CP potential, the corrosion rate is less than the value specified by the standards (0.01 mm/year) (or simply zero), indicating that the metal is fully protected.

Undoubtedly, the corrosion rate is a very useful parameter for the evaluation of the CP effectiveness. It clearly indicates the quantitative corrosion rate when the CP is inadequate and the quantitative corrosion rate allows the evaluation of the remaining life of the metal structure if the corrosion history is known.

24.2.3 CP margin of effectiveness as the degree of adequate cathodic protection

In practice, CP is usually applied such that the metal structure is slightly more than adequately protected (with the CP potential slightly lower than the minimum adequate CP potential) to guarantee that there is a margin of effectiveness of protection, but not excessive protection which may cause significant evolution of hydrogen and cause damages to the coatings on the metals or hydrogen embrittlement. Since the minimum corrosion rate is zero (corrosion rate cannot be negative), it cannot be used to evaluate

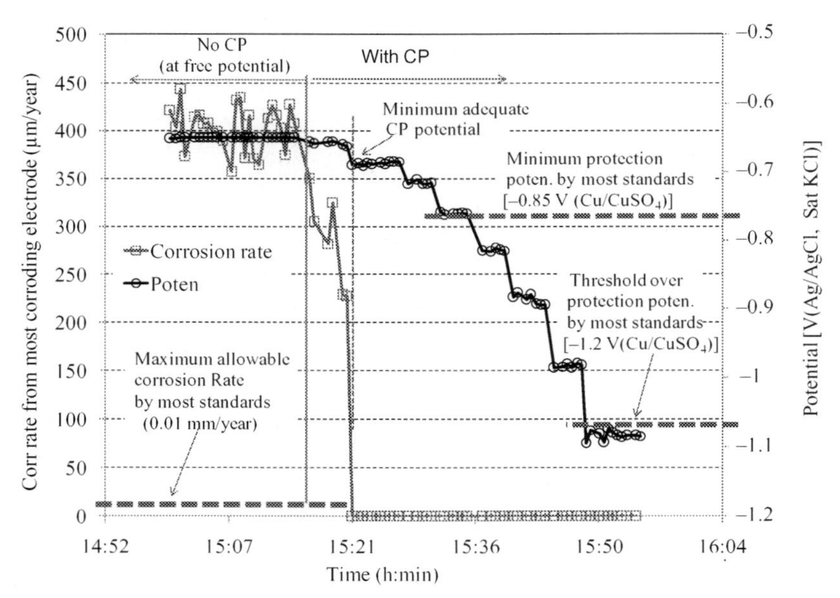

Fig. 24.5 Typical potential and corrosion rate calculated from the current of the most corroding electrodes on the CMAS probe.

the CP margin of effectiveness (or the degree of adequate cathodic protection). Most standards for cathodic protection of carbon steel pipelines consider the pipe-to-soil instant off-potential of −0.85 V (vs $Cu/CuSO_4$) as the minimum protection potential (see Fig. 24.5) and −1.2 V (vs $Cu/CuSO_4$) as the threshold overprotection potential. These threshold values (−0.85 or −1.2 V) are conservatively determined on the basis of past experiences or experiments conducted under commonly found soil conditions. They are not the true threshold potentials at which the metal starts to be adequately protected or the metal starts to experience significant hydrogen evolution. So these values cannot be used to derive a quantitative parameter that can be used to evaluate the degree of adequate cathodic protection.

Fig. 24.6 shows that the ratio between the cathodic current from the most anodic electrode and the large current value at which hydrogen evolution starts to be significant can be used to represent the margin of the degree of cathodic protection or the cathodic protection effectiveness margin (CPEM) [15]. The large current value is also called the maximum allowable current because it corresponds to the threshold overprotection potential (see the below section). When the CPEM is negative, there is still electrode under corrosion which means that the CP is insufficient (if zero is considered to be the maximum targeted current). When the CPEM is equal to or larger than zero, all electrodes are fully protected. The CPEM that corresponds to the excessive protection potential is called the excessive CPEM. When the CPEM is higher than the excessive CPEM, there is excessive hydrogen evolution reaction on at least one of the electrodes (the most cathodic electrode). When the CPEM reaches 100% (the threshold excessive CPEM), even the most anodic electrode starts to experience significant

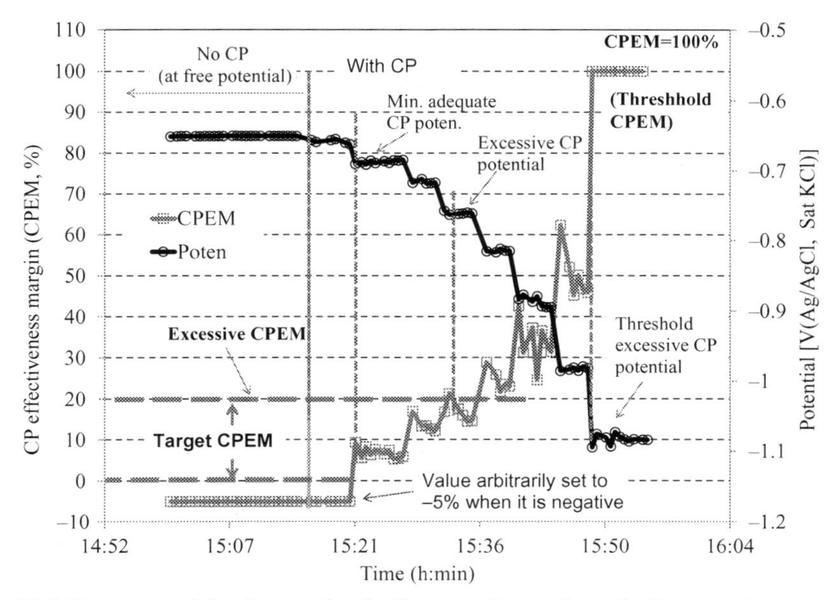

Fig. 24.6 Responses of the degree of cathodic protection or the cathodic protection effectiveness margin (CPEM) to the CP potential.

hydrogen evolution. Therefore, the ideal target range of CPEM value should be between 0 and the excessive CPEM. Because CPEM is an indicator for how safely the metal structures are protected by the CP, this term was originally called CP Safe Margin (CPSM) by the author [15].

The value of CPEM may also be defined such that it equals to zero when the corrosion rate is reduced by the CP to the minimum allowed value by the standard (e.g., 5 or 10 μm/year) and 100% when the CP started to be excessive. For a more reliable result, the cathodic current from the most anodic electrode used to derive the CPEM may be obtained based on statistical analysis of the currents from all the electrodes. For example, the value may be represented by the negative sum of the average value and the standard deviation of the currents from all of the electrodes on the CMAS probe.

24.2.4 Corrosion rate and CP effectiveness margin as the most effective CP parameters from a CMAS probe

Fig. 24.7 shows how the CPEM and the corrosion rate can be used together to effectively monitor the effectiveness of CP [15]. When the CP is insufficient, the CPEM is negative, and the degree of corrosion is shown by the corrosion rate; when the CP is adequate, the corrosion rate reaches zero and lose its effectiveness as the indicator for the degree of protection.

However, the CPEM can be used to guide the operator how to control the CP to the optimum condition when the CP is adequate.

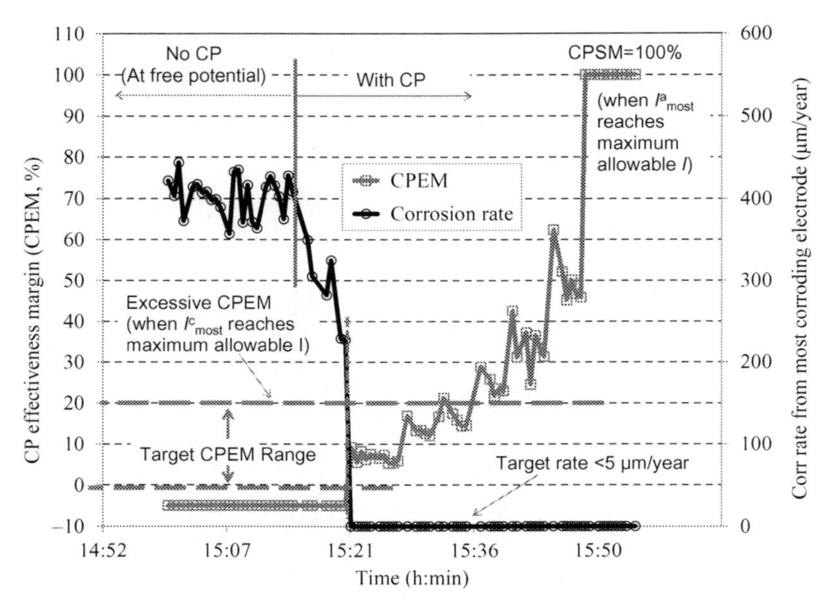

Fig. 24.7 CP effectiveness margin (CPEM) and the corrosion rate used together to monitor the effectiveness of CP.

It is important to note that Fig. 24.7 does not show the potential which means that the CP potential is not required for monitoring the effectiveness of the CP using a CMAS probe. The two parameters solely from the highly robust CMAS probe, the corrosion rate and the CPEM, are all the parameters that are needed to effectively monitor the effectiveness of cathodic protection. In practice, it is highly desirable to use the two parameters from the CMAS probe only without the need for the potential measurements. This is because potential measurements require a reference electrode and the vast majority of the reference electrodes used for CP contain a reference solution and require periodical maintenance or frequent performance verification (every 3 months as specified by some standards for pipeline cathodic protection). In contrast, the CMAS probe is maintenance-free and has a very long life (10–40 years) because the CMAS probe is made of metal electrodes and solid insulators. Eliminating the requirement for the reference electrode means huge savings to the asset owners in terms of maintenance cost and reliable monitoring of the effectiveness of cathodic protection.

24.2.5 Determination of the maximum allowable CP current

Fig. 24.8 shows how the maximum allowable current or current density can be obtained [15]. If it is in an aerated system, the cathodic current is usually due to the reduction of oxygen when the electrodes of the CMAS probe are moderately polarized in the negative direction and the absolute current value gradually increases with the decrease of potential in the negative direction. As the polarization progresses to the more negative direction, the cathodic current starts to be dominated by the reduction

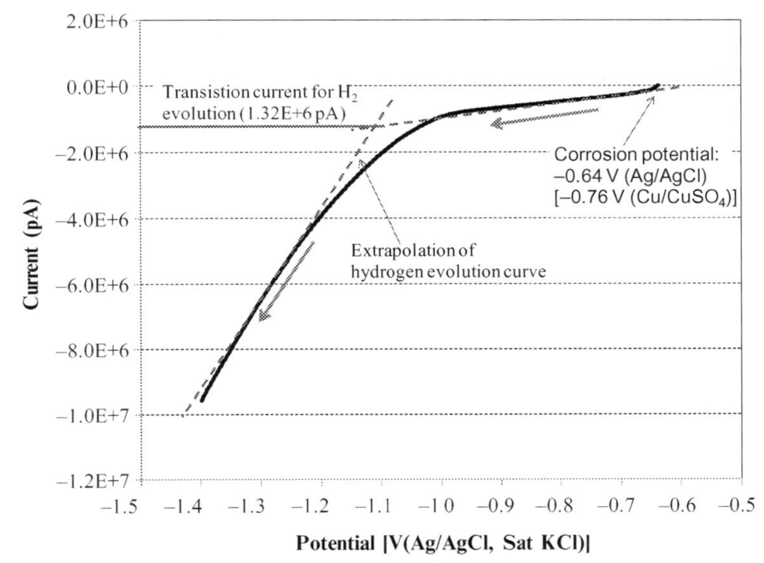

Fig. 24.8 Polarization curve for obtaining the maximum allowable current.

reaction of hydrogen ions and increases more rapidly with the decrease of potential. The inflection point of the curve may be considered as the starting point for the excessive evolution of hydrogen and may be used as the maximum allowable current as shown in Fig. 24.4. The current shown in Fig. 24.8 was the average of all the electrodes on a CMAS probe. The inflection point was 1.32×10^6 pA and this is why the maximum allowable current was set to 1.32×10^6 pA in Fig. 24.4. The electrodes used for Fig. 24.8 were made of Type 1018 carbon steel wire (1 mm diameter) and the exposed surface area for each electrode was the cross-section (0.78 mm^2). The electrolyte used was 0.5 M NaCl solution which simulates the seawater. A simple alternative method to determine the maximum allowable current is to measure the cathodic current on the electrodes of the CMAS probe when its coupling joint is polarized to the threshold excessive CP potential specified by a relevant standard.

24.3 Typical application for cathodically protected carbon steel in simulated seawater

The maximum localized corrosion rate of carbon steel was measured with a nanoCorr A-50 (see Section 24.2) in simulated seawater under different cathodic protection potentials [10]. The sensing electrodes of the coupled multielectrode sensor probes were made from an annealed mild carbon steel wire (concrete rebar wire), 1.5 mm in diameter and 1.77 mm^2 in electrode surface area. Each probe had 16 electrodes flush-mounted in epoxy. Typical probes are shown in Fig. 24.9. Prior to the test, the surfaces of sensing electrodes for each multielectrode probe were polished to 320 Grit and rinsed with distilled water and acetone. The experiments were conducted

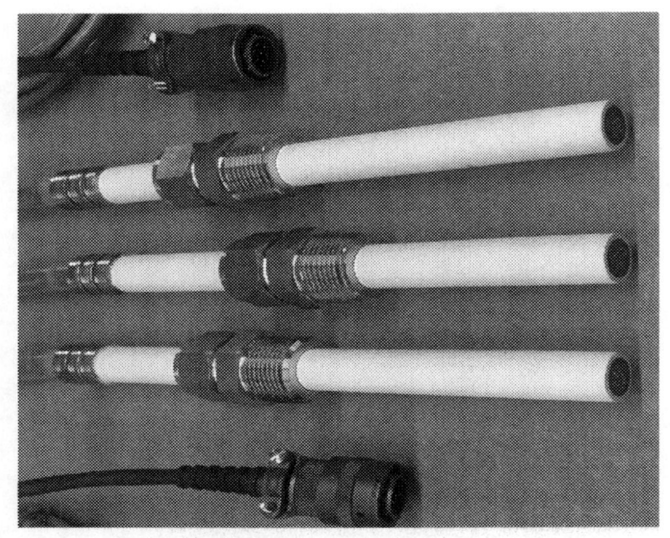

Fig. 24.9 Typical coupled multielectrode array sensor probes.

at 27°C. Typical maximum localized corrosion rates from two carbon steel probes and the cathodic protection potential are shown in Fig. 24.10. The maximum localized corrosion rate for each probe was automatically calculated by the CMAS instrument based on the current from the most corroding or the most active electrode in the probe (see Chapter 8) [2]. So, the maximum localized corrosion rates in Fig. 24.10 have the same meaning as the corrosion rate in Figs. 24.5 and 24.7.

As can be seen from Fig. 24.10, at the start of the test, the potential was at the free open-circuit potential or free corrosion potential. The corrosion rates were approximately 2 mm/year, which is consistent with the result measured in other experiments for carbon steel in the same solution [8].

After a short period of open-circuit measurements, the potential was slowly changed in the cathodic direction. The maximum localized corrosion rates from the two probes gradually decreased with the decrease of the potential and sharply dropped to near zero when the potential was near a critical potential and remained zero when the cathodic protection potential was below this critical potential. This critical potential is the minimum adequate CP potential defined in Section 24.2. All sensing elements (electrodes) of the probe were fully protected when the cathodic protection potential was below this minimum adequate CP potential. The values of the minimum adequate CP potential was −0.72 to −0.74 V(SCE), or −0.793 to −0.813 V ($Cu/CuSO_4$). Because the reference electrode was placed close to the sensing surface of the two probes, the measured potential should not be significantly affected by the IR drop and these measured values were essentially the instant off-potential. Therefore, these values indicate that the −0.85 V($Cu/CuSO_4$) off-potential criterion is valid (i.e., the metal is fully protected when the CP potential is controlled at <−0.85 V ($Cu/CuSO_4$)). However, using the −0.85 V($Cu/CuSO_4$) criterion would not tell the

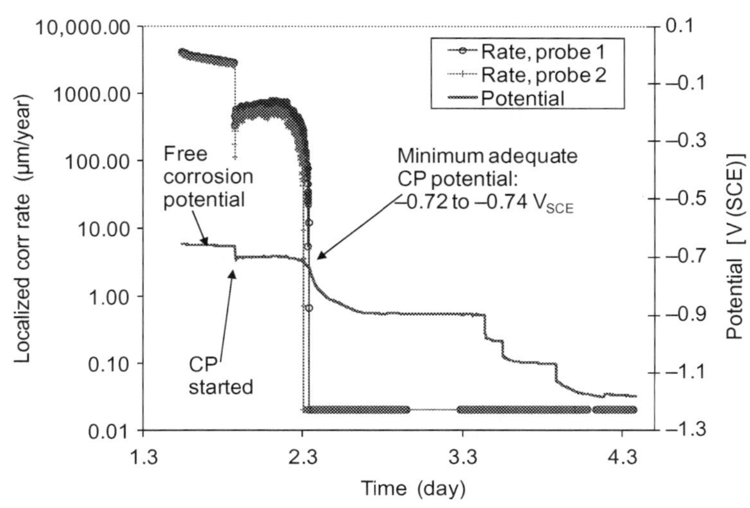

Fig. 24.10 Responses of the maximum localized corrosion rate from two carbon steel probes to the changes of the CP potential in simulated seawater.

operator that the metal was also fully protected when the potential was only −0.813 V ($Cu/CuSO_4$). The signal from the CMAS probe tells the operator the exact potential value at which the metal is adequately protected when the sensing element is conditioned in the same environments as the protected metal.

In addition, the two corrosion rates from the two separate probes behaved nearly identically, showing the good reproducibility of the CMAS probes when used to evaluate the effectiveness of CP.

24.3.1 Measurements of stray current effect and the effectiveness of CP

Fig. 24.11 shows the responses of the corrosion rates of carbon steel probes to the increase of the CP potential caused by stray current in simulated seawater. Before the application of the stray current, both probes were at their free corrosion potentials (about −0.65 V vs SCE) and they experienced a maximum corrosion rate of about 1.1 mm/year. When the stray current effect was applied, the corrosion rates from both probes increased to approximately 20 mm/year (by 20-folds). When the stray current effect was removed, the probes' potentials and corrosion rates dropped instantly and trended toward their original values before the stray current effect. The two corrosion rates from the two separate probes show a good reproducibility of the CMAS probes for the measurement of stray current effect.

Three probes (Probes #1, #2, and #3) were used in another test after they were polished to demonstrate their ability for evaluating cathodic protection and stray current effects. Probe #1 was cathodically polarized to approximately −0.9 V(SCE), which is well below the minimum adequate CP potential. Probe #2 was slightly

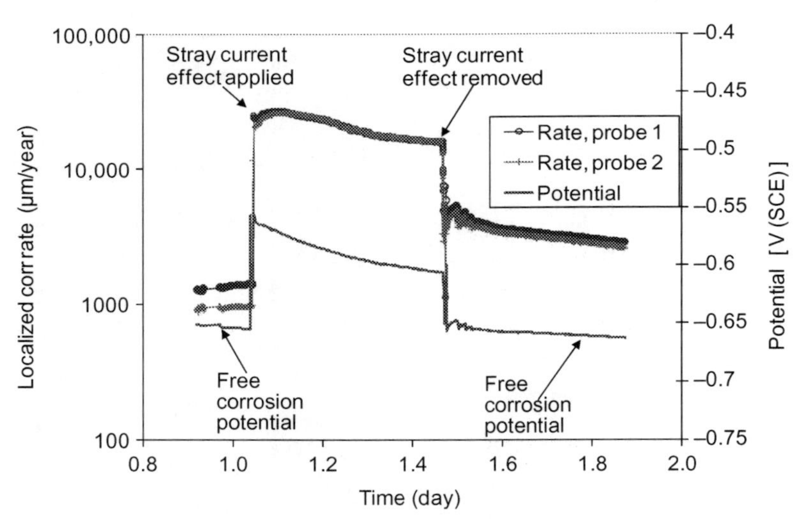

Fig. 24.11 Responses of the corrosion rates of two carbon steel probes to the increase of the CP potential caused by stray current in simulated seawater.

anodically polarized, to simulate the stray current effect. Probe #3 was left at its free corrosion potential, to simulate an inefficient protection from the cathodic protection system or the unprotected condition. Fig. 24.12 shows the measured corrosion current density signals and the potentials of the three probes during the monitoring period. In Fig. 24.12, the corrosion current density signal for each probe was automatically calculated by the software based on the most anodic electrode [2] during the

measurement. The corrosion current from Probe #1 was negative throughout the monitoring period, indicating that there was no corrosion on the sensing electrodes of Probe #1. The corrosion current of Probe #3 was approximately 10^{-4} A/cm^2 during the first 4 days of monitoring and decreased to approximately 10^{-5} A/cm^2 on the 5th day. The corrosion current signal from Probe #2 was approximately 10^{-3} A/cm^2 throughout the monitoring period, indicating that the corrosion increased by more than one order of magnitude, when the potential was raised by approximately 0.07 V from the free corrosion potential. The slight decrease of the free corrosion potential of Probe #3 over the monitoring period was probably due to the pH increase caused by the relatively large cathodic reaction of oxygen on the stainless steel counter electrode used to anodically polarize Probe #2. At the end of the test, a large amount of rust deposit was observed in the electrochemical cell, suggesting a high pH solution.

The maximum localized corrosion rates for the three probes were also available from the software during the measurement (Fig. 24.13). The maximum localized corrosion rate of the unprotected probe—Probe #2—in the first 4 days was approximately the same as the value measured in the previous measurements, which were conducted

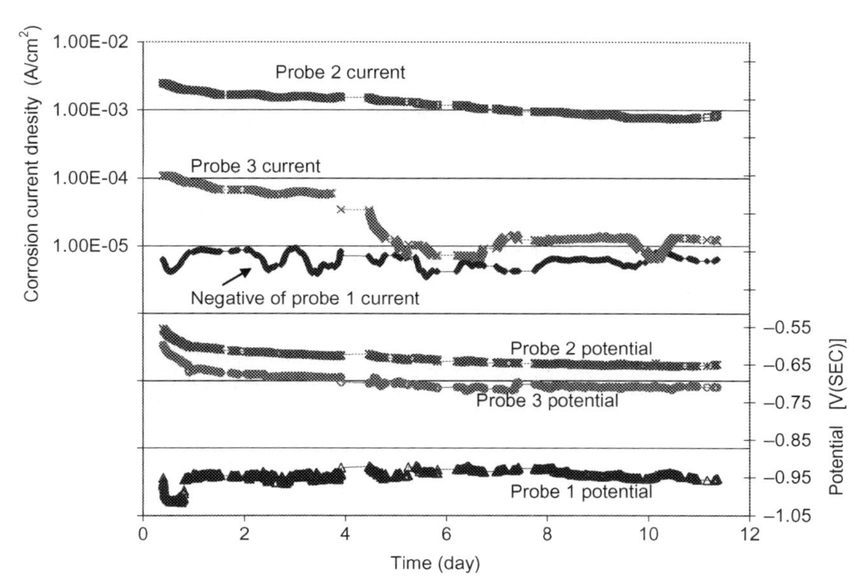

Fig. 24.12 Corrosion current signals and the potentials of the probes during the monitoring period. The corrosion signal from Probe #1 was negative (-6×10^{-6} to -1×10^{-5} A/cm^2), indicating no corrosion was taking place on Probe #1 electrodes.

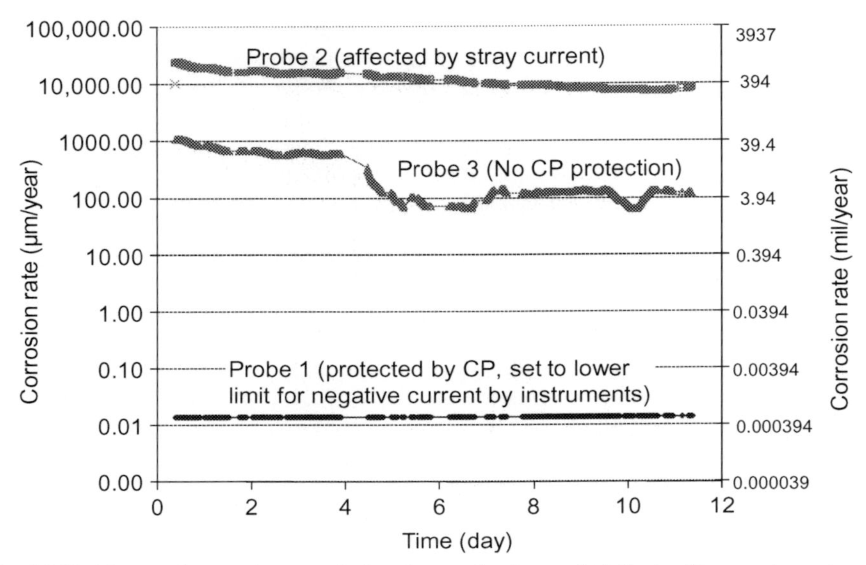

Fig. 24.13 Measured corrosion rate during the monitoring period. Probe #1 corrosion rate was set to the lower detection limit (14 nm/year) by the instrument, because its current was negative, indicating there was no corrosion on the sensing electrodes.

with similar probes in the simulated seawater. However, the rate was slightly lower after the fifth day. The difference between the values measured in the first 4 days and the values measured after the 5th day (Fig. 24.13) is within the expected range of variations for maximum localized corrosion rates in a given environment. The maximum localized corrosion rate for Probe #1 was automatically set to the lower detection limit of the instrument by the software because its corrosion current signal was negative, indicating that there was no corrosion.

Fig. 24.14 shows the appearance of the probes' electrodes after the monitoring period. Clearly, all the electrodes of Probe #2 (except for the two with no connection to the coupling joint) were severely corroded. Slight pitting corrosion can be noticed on some of the electrodes of Probe #3. No significant pitting corrosion was observed on the electrodes of Probe #1. The result, as shown in Fig. 24.14, is in agreement with the measured corrosion rate shown in Fig. 24.13. In addition, Fig. 24.13 shows that, on average, the maximum localized corrosion rate for Probe #2 was 12.7 mm/year. Hence, the total penetration depth for the most corroded electrode on Probe #2 should be 0.38 mm, after the 11 days of exposure. The measured depth of the most corroded electrode on Probe #2, as shown in Fig. 24.14, was approximately 0.4 mm, which is in excellent agreement with the depth estimated by the coupled multielectrode probe.

24.3.2 Measurements of the dynamic stray current effect

As mentioned in Chapter 20, many buried pipelines close to the metro DC transit systems have experienced accelerated corrosion by the dynamic DC stray current [16]. For a metro DC transit system, with the moving of trains along the metro line, both

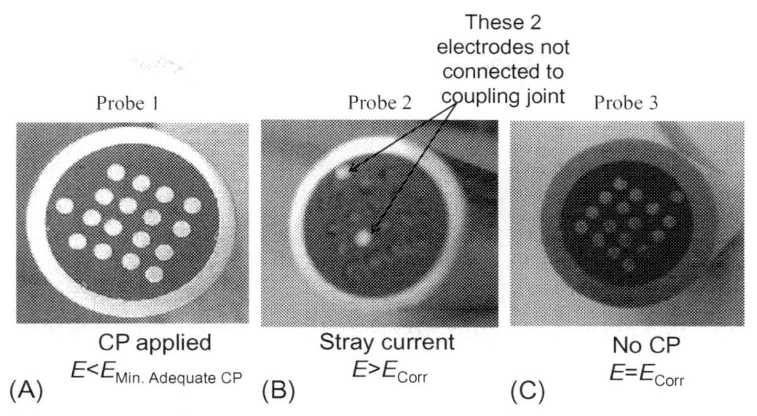

Fig. 24.14 Appearances of the electrodes of the probes after the monitoring period. (A) Probe #1 electrodes were cathodically protected. (B) Probe #2 electrodes were anodically polarized except for the two with no connection to the coupling joint. (C) Probe #3 electrodes were maintained near the free corrosion potential. The electrode outside the 4 × 4 matrix in each sensor was used as a position locator and was not part of the sensing electrodes.

the position and the quantity of the stray current leaked from the metro system vary with time. So, the stray current activity on underground structures arising from the metro system operation is not steady, but dynamic in terms of current and potential amplitude. The stray current flowing into or out of the underground structures often reverses direction. Monitoring the effect of the dynamic stray current on the corrosion of the buried structures is important for mitigating the corrosion accelerated by the dynamic stray current.

The setup of Fig. 24.15 was used in the test for monitoring the effect of the dynamic stray current on corrosion of a carbon steel probe [17]. The electrodes on the CMAS probe were made of Type 1018 carbon steel wire (1 mm diameter) and the exposed surface area for each electrode was the cross-section (0.78 mm^2). The metal was made of a coil of the same carbon steel wire (1 mm diameter and 80 cm length). The anode and counter electrode were made of Alloy 276. The electrolyte was 0.5 M NaCl solution. The cathodic protection power source was a DC power supply. The interfering stray current source was a function generator that supplies a cyclic triangle wave at about 0.01 Hz frequency. The built-in reference electrode was a standard Ag/AgCl electrode filled with saturated KCl (SSC).

Fig. 24.16 illustrates the corrosion rate for carbon steel obtained with and without the effect of the triangle-wave dynamic stray currents [17]. The corrosion rate from the probe was about 200 and 3000 μm/year before and after the application of the stray current, respectively. When the stray current was introduced, the potential of the metal fluctuated between −1.6 and 0.3 V (vs SSC). When the potential was near 0.3 V, the corrosion rate was about 50 mm/year and when the potential decreased to below −0.7 V, the corrosion rate dropped to 0.1 μm/year, which is the lower detection limit of the multielectrode instrument. The rate of 0.1 μm/year is extremely low and can be considered zero.

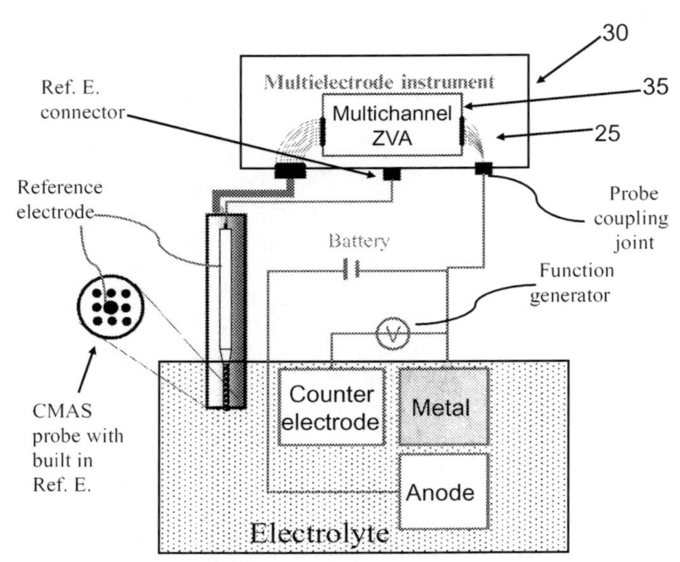

Fig. 24.15 Experimental setup for monitoring the effect of the dynamic stray current on corrosion of a carbon steel probe.

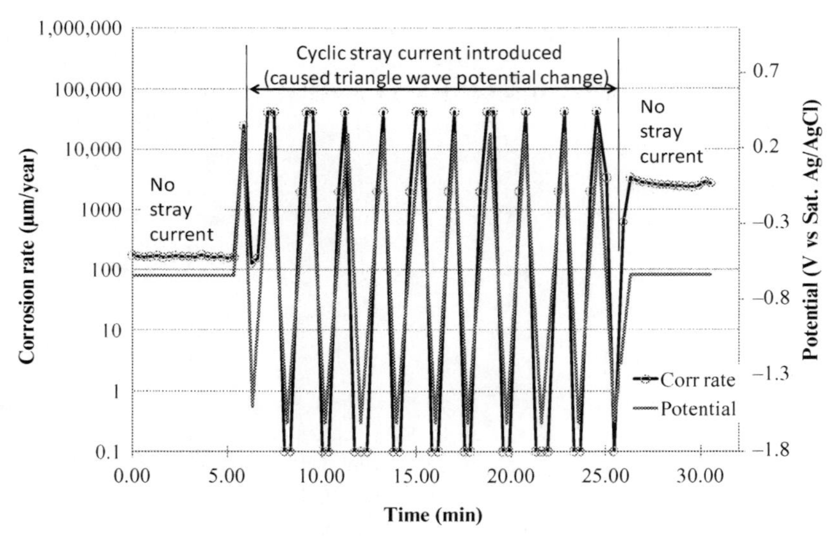

Fig. 24.16 Corrosion rate for carbon steel obtained with and without the effect of the triangle-wave dynamic stray-currents.

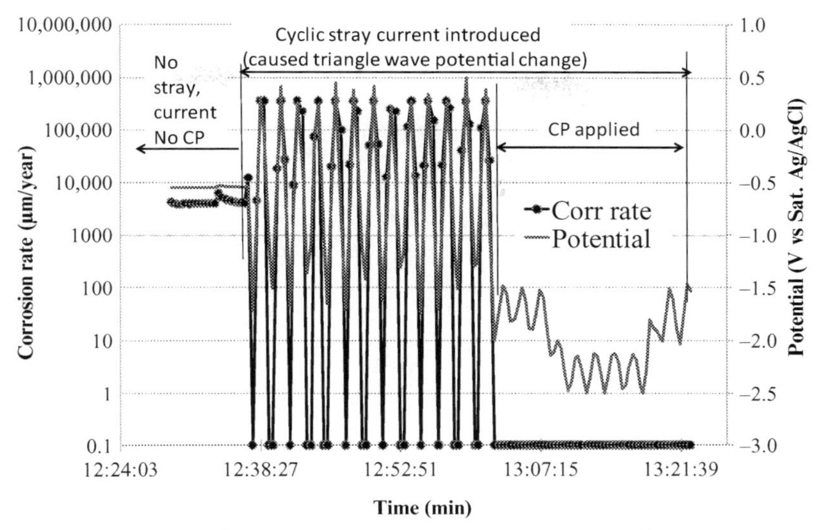

Fig. 24.17 Corrosion rate for carbon steel metal obtained under the effects of triangle-wave dynamic stray-currents and cathodic protection.

Fig. 24.17 illustrates the corrosion rate for carbon steel metal obtained under the effects of triangle-wave dynamic stray-currents and cathodic protection (CP). When enough CP was applied such that the potential of the metal was never above −1.0 V, the corrosion rate remained at 0.1 μm/year, which indicates the effectiveness of the CP under stray-current effect. It should be mentioned that the sampling rate of the multielectrode instrument was designed for DC measurements and could not be increased to values higher than 0.05 Hz. Higher sampling rate may catch even higher corrosion rate near the peak of the potential, but the results in Fig. 24.17 are sufficient to demonstrate the usefulness of the multielectrode measurement for the effect of stray-current.

24.3.3 Summary

It was demonstrated that coupled multielectrode array sensor probes are effective tools for real-time measurement of the minimum adequate CP potential. The probes can also be used as online tools to measure the corrosion rate under cathodic protection conditions with or without stray-current effect, including the dynamic stray-current effect. If placed near the critical locations of engineering components, the CMAS can be used to measure the effectiveness of cathodic protection or the stray-current effect, including the dynamic stray-current effect for cathodically protected system in real-time.

24.4 Measurements of the effectiveness of CP for carbon steel in concrete

Steel reinforcements in concrete structures exposed to aggressive conditions, such as marine environments and deicing salts, are subject to corrosion degradation and significant reductions in their service lives. This section presents the real-time measurements of localized corrosion rates for carbon steel in concrete with and without cathodic protections [11]. This measurement was conducted in an effort to understand the corrosion behavior of the carbon steel reinforcement rebars in concrete.

Fig. 24.18 shows the experimental setup during the measurements. Two CMAS probes were vertically buried into a commercial grade concrete-sand mixture, which was initially mixed with distilled water. The concrete was formed in a plastic container with a dimension of 37 cm (length) $\times$ 25 cm (width) $\times$ 16 cm (height) and cured while it was submerged in distilled water. Corrosion rate measurements were taken shortly before the probes were buried into the concrete and then continued for approximately 7 days, while the concrete was continuously submerged in distilled water. On the seventh day, the concrete was removed from the plastic container and partially immersed in a shallow bath filled with simulated seawater.

At the end of the first month of the test, cathodic protection was applied by connecting the aluminum wire electrodes that acted as the sacrificial anode to the common coupling joint of the CMAS system (see Fig. 24.2). These aluminum wires were vertically buried near the probes.

A saturated calomel electrode (SCE) was dipped into a hole in the concrete, formed during the curing, between the two CMAS probes and was used as the reference electrode for the measurement of the cathodic protection potential of the probes. Distilled water was added to the hole periodically to maintain the conducting path between the reference electrode and the two probes. The sensing electrodes were made from carbon steel concrete rebar wire (1.5 mm in diameter and 1.77 mm^2 in electrode surface area). Each probe had 16 electrodes flush-mounted in epoxy. Prior to the test, the surfaces of the sensing electrodes for each multielectrode array sensor probe were polished to 600 Grit and rinsed first with distilled water and then with acetone. The experiment was conducted at about 27°C.

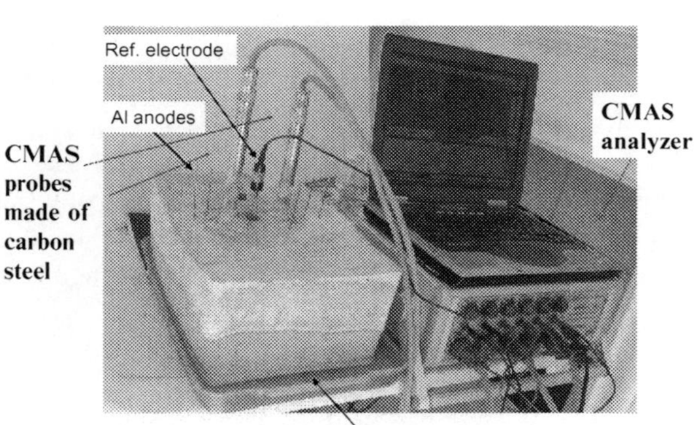

Fig. 24.18 Experimental setup for the measurement of localized corrosion rate in concrete. *Note*: The aluminum anodes were used for cathodic protection during the test.

24.4.1 Localized corrosion of carbon steel in freshly mixed concrete

The maximum localized corrosion rates from the two CMAS steel probes in the first month of the test (with no cathodic protection) are shown in Fig. 24.19. The maximum localized corrosion rates were calculated by the software using the current from the most corroding electrode (also called the most active electrode) among all the sensing electrodes on a probe. Upon placing the probes into the freshly mixed concrete (with distilled water initially), the corrosion rate from the two probes instantaneously increased to approximately 70 μm/year (2.8 mil/year (or mpy)). The corrosion rate decreased rapidly to about 4 μm/year in the first 20 h and slowly to 3 μm/year in the first 7 days, while the concrete was submerged in distilled water.

This decrease in corrosion rate indicates that the carbon steel was passivated by the alkaline environment generated by the concrete. The corrosion rate stayed at about 2 μm/year for Probe #2 and continued to decrease for Probe #1 after the concrete was taken out of the distilled water bath and partially immersed in a shallow simulated seawater bath. The corrosion rate of Probe #1 reached 0.85 μm/year at the end of the first month of the measurements.

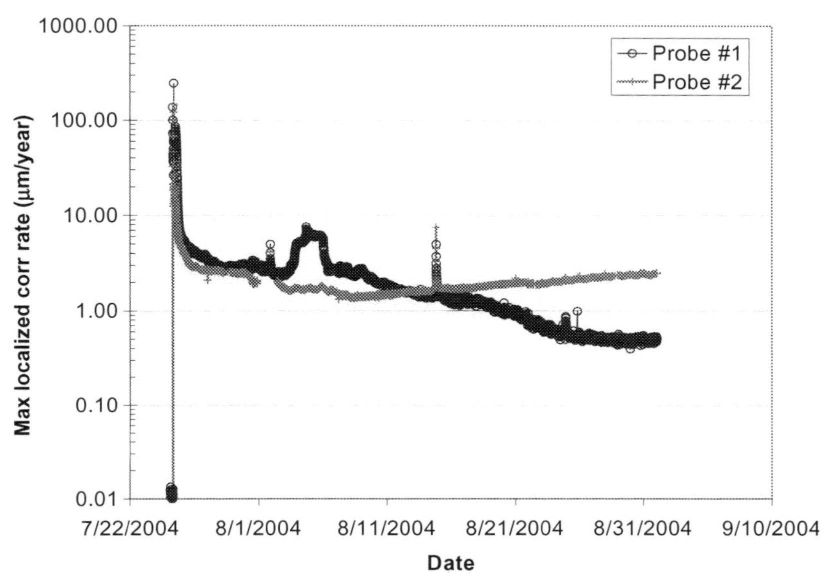

Fig. 24.19 Maximum localized corrosion rates from two independent coupled multielectrode array sensor probes made of rebar material in concrete submerged in distilled water in the first week, and partially immersed in simulated seawater in the remaining test.

24.4.2 Localized corrosion rate during cathodic protection

Fig. 24.20 shows the response of the corrosion rates and the potentials from the two probes to cathodic protection. Upon the connection between the common joints of the CMAS analyzer to the sacrificial aluminum anodes, the electrochemical potential of Probe #1 decreased from −0.735 to −1.27 V(SCE), and the corrosion rates of both probes dropped to 1.3 nm/year, which is below the lower detection limit of the CMAS analyzer (10 nm/year), suggesting that the carbon steel material in the two probes was adequately protected. When the cathodic protection was removed, the potential returned to −1.0 V(SCE), which is significantly lower than the previous value. However, the corrosion rates from both of the probes were significantly higher than those of the previous value. The lower potential and the higher corrosion rate immediately after removing the cathodic protection suggest that the concentrations were low for the corrosion products during the cathodic protection.

24.4.3 Summary

The steady state corrosion rates measured in the concrete partially immersed in simulated seawater were 0.5–2.4 μm/year. When the carbon steel electrodes of the probe were connected to sacrificial aluminum anodes, the corrosion rate decreased instantaneously to a value that is below the lower detection limit of the instruments (10 nm/year). Coupled multielectrode array sensor probes can be used as a real-time tool to measure the effectiveness of cathodic protection for steel reinforcements in concrete.

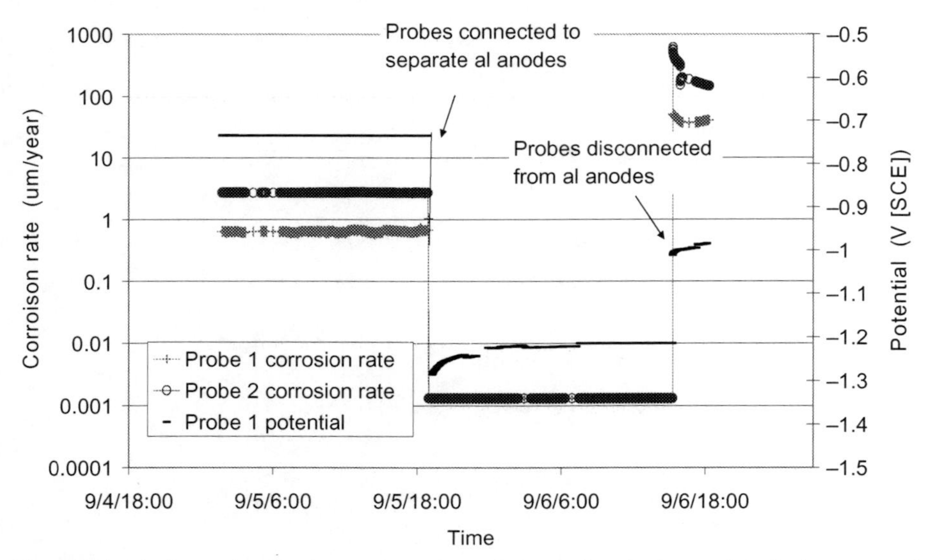

Fig. 24.20 Maximum localized corrosion rates and electrochemical potential of rebar material measured from two coupled multielectrode array sensor probes before, during, and after the probes were cathodically protected by sacrificial anodes.

24.5 Measurements of the effectiveness of CP for carbon steel in soil

Corrosion of metallic components in soil has been a concern in many fields, including the pipeline industry and nuclear waste disposal programs. As discussed in Chapter 17, metal loss corrosion probes based on electrical resistance methods have been used as online tools for corrosion monitoring in soil [12]. However, these probes are not sensitive enough for localized corrosions, such as pitting or crevice corrosion [5]. This section describes the application of the coupled multielectrode array probe as an online and real-time tool for measuring the localized corrosion rate of carbon steel material in soil under cathodic protection conditions.

Fig. 24.21 shows the experimental setup for the measurements of maximum localized corrosion in soil contained in a plastic container (35 cm long × 25 cm wide × 17.5 cm high), which was perforated at a height of 13.5 cm from the bottom to provide drain holes [12]. The coupled multielectrode probes were initially vertically placed in freshly prepared soil that was loose and relatively dry (unsaturated with water). After some initial measurements of the corrosion rate, simulated seawater was added to the container to flood the section below the drain hole. The flooded section is also called the saturated zone. The soil in this saturated zone was sticky and muddy. After the seawater was added, the probe was pushed into the bottom of the soil so that the sensing surface of the probe was in close contact with the sticky and densely compacted soil during the remainder of the test. Distilled water was frequently added from the top to make up for the evaporation loss.

The sensing electrodes of the coupled multielectrode probes were made from annealed mild carbon steel concrete rebar wire (1.5 mm in diameter and 1.77 mm^2 in electrode surface area). Each probe had 16 electrodes flush-mounted in epoxy.

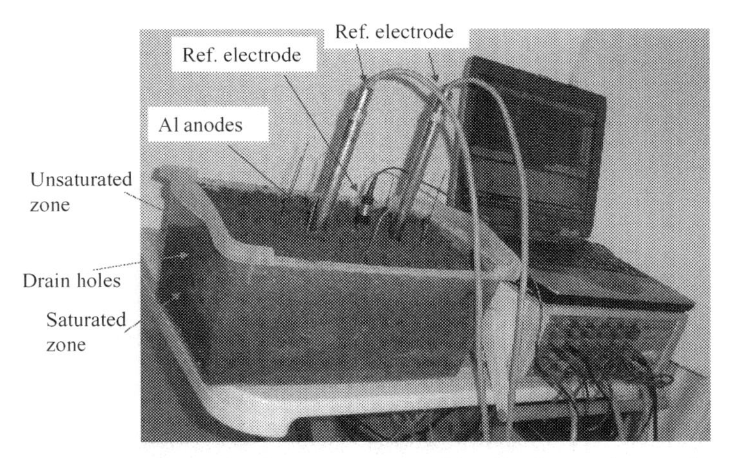

Fig. 24.21 Experimental setup for the measurement of localized corrosion rate in soil. *Note*: The aluminum anodes were used for cathodic protection during the test.

Prior to the test, surfaces of sensing electrodes for each multielectrode probe were polished to 600 Grit and rinsed with distilled water and then with acetone. The aluminum wires shown in Fig. 24.21 were also vertically buried near the probes and were used as sacrificial anodes during the cathodic protection test. A saturated calomel electrode (SCE) was dipped into the saturated zone near the probes and used as the reference electrode for electrochemical potential measurements.

24.5.1 Corrosion rate in soil

Fig. 24.22 shows the maximum localized corrosion rate from the carbon steel CMAS probe in soil before and after the addition of the simulated seawater, but prior to the application of cathodic protection. The corrosion rate was low (0.1–0.2 μm/year) when the simulated seawater was not added and the probe's sensing tip was in the relatively dry and loose soil. This low corrosion rate is expected because the probe was freshly polished (cleaned) and no significant liquid condensate was formed on the sensing surface of the CMAS probe.

After the addition of simulated seawater and the probe was pushed down to the flooded section of the soil, the maximum localized corrosion rate increased to about 600 μm/year, which is close to the corrosion rate of carbon steel in air-saturated simulated seawater [8]. However, a few minutes after the probe was pushed into the densely packed soil saturated with simulated seawater, the rate dropped rapidly and reached 10 μm/year in about 1 h. The rapid decrease in maximum localized corrosion rate was probably an indication that the corrosion process was under mass-transfer

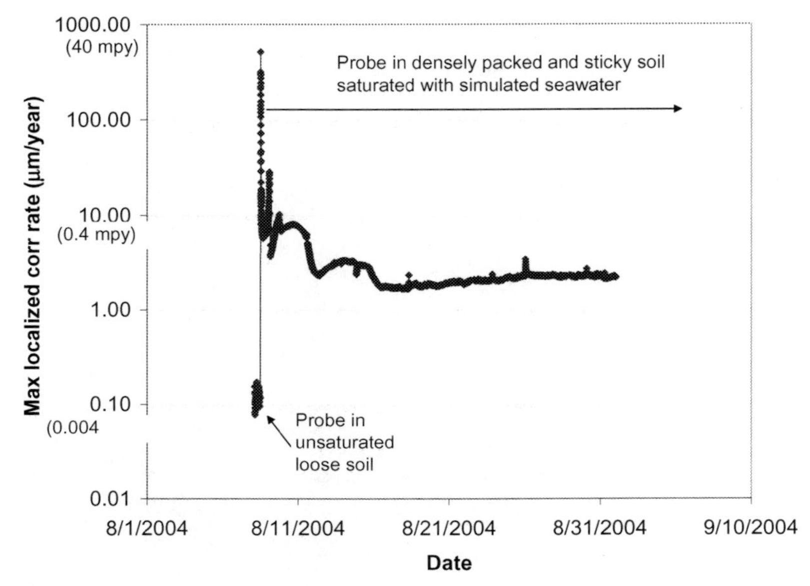

Fig. 24.22 Maximum localized corrosion rates from a coupled multielectrode array sensor probe made of carbon steel in soil soaked with simulated seawater.

control. The migrations of the corrosion products (such as metal ions) away from the corroding sites and the reactants (such as O_2) to the corroding sites are limited by the low diffusion rates of these species in the soil which was densely packed when the probe was pushed into the wet and sticky soil. The maximum localized corrosion rate was about 4 μm/year after 3 weeks in the densely packed soil.

24.5.2 Corrosion rate in soil under cathodic protection conditions

After about 3 weeks in the densely packed and wet soil soaked with the simulated seawater, the carbon steel electrodes of the probe were connected to the aluminum sacrificial anodes (according to Fig. 24.2) to cathodically protect the electrodes. Fig. 24.23 shows the measured corrosion rate and the electrochemical potential of the carbon steel electrodes in soil before, during, and after cathodic protection. As soon as the carbon steel electrodes were cathodically protected, the electrochemical potential decreased from −0.72 to −0.83 V(SCE), and the maximum localized corrosion rate dropped from 3 μm/year to 1.3 nm/year, which is below the lower detection limit of the corrosion analyzer (10 nm/year). This suggests that the carbon steel material was adequately protected. When the cathodic protection was removed, the potential returned to −0.72 V(SCE) and the maximum localized corrosion current returned to approximately 2.2 μm/year.

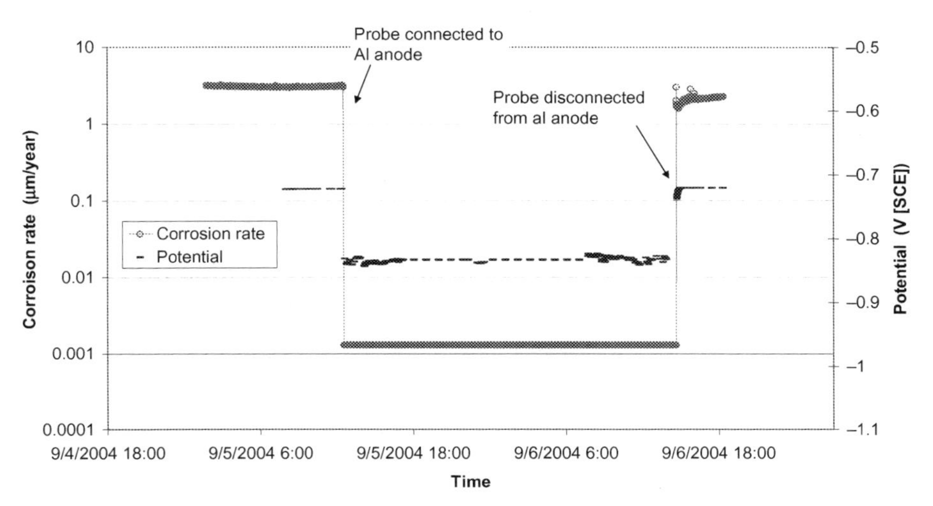

Fig. 24.23 Maximum localized corrosion rates and potential measured from two coupled multielectrode array probes in soil before, during, and after the probes were cathodically protected by sacrificial anodes.

24.5.3 *Measurement of effectiveness of cathodic protection of pipeline in an oil field*

The CMAS probes were tested in an oil field for monitoring the effectiveness of cathodic protection for coated pipelines in the fields. The pipelines were protected by impressed currents that were controlled by potentiostats through anode wells (anode beds) located approximately every 70 km apart along the pipelines. The CP potential near the anode wells were set to approximately −1.20 V(Cu/$CuSO_4$). Two locations were chosen, one was near the anode well and the other one was at the middle point between the anode well (about 35 km away from one of the anode wells) because it is suspected that there was not enough CP at the middle location. The CMAS probes made of 10 carbon steel (Type 1018) electrodes were used in the testing with the setup as shown in Fig. 24.3. A nanoCorr Field Monitor was used in the testing, which measures the maximum localized corrosion rate (the rate from the most corroding electrode) and the CPEM. It also measures the coupling potential of the CMAS probe.

Fig. 24.24 shows the CP parameters from the CMAS probe near the anode well location. After the probe was pushed into the soil, the free corrosion potential was about −0.786 V(Cu/$CuSO_4$) and the corrosion rate from the most corroding electrode was about 0.2 mm/year. As soon as the probe's coupling joint was connected to the pipeline, the probe potential dropped to the "On" CP potential (−1.286 V (Cu/$CuSO_4$)), and the corrosion rate quickly dropped to zero, while the CPEM increased to 100% instantaneously. The 100% CPEM indicates that the pipeline at this location was excessively protected because there were excessive hydrogen evolution on the

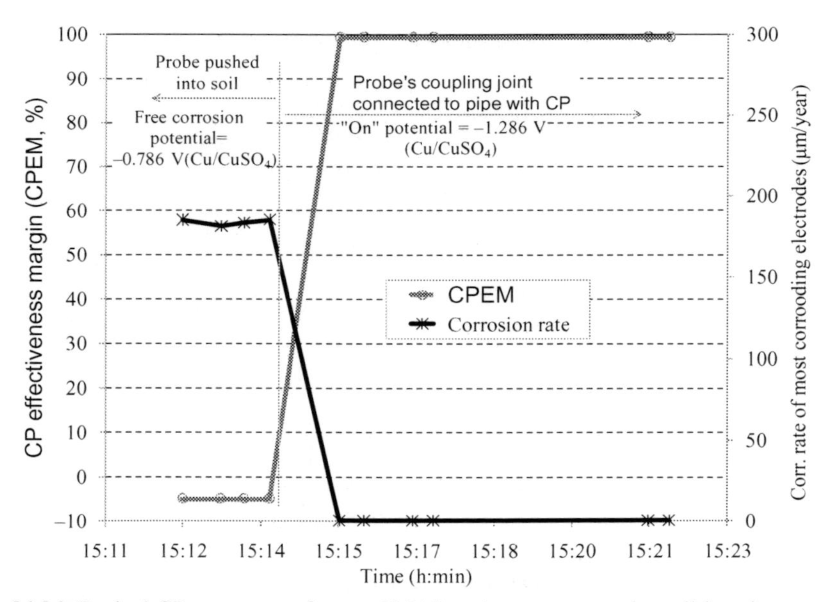

Fig. 24.24 Typical CP parameters from a CMAS probe near an anode well location.

most anodic electrode (the electrode that is the most difficult to protect). The 100% CPEM is also consistent with the CP potential because it is lower than −1.2 V recommended by the standard as the threshold excessive CP potential.

Fig. 24.25 shows the CP parameters obtained at the midpoint between the anode wells. After the probe was pushed into the soil, the free corrosion potential was about −0.65 V($Cu/CuSO_4$) and the corrosion rate from the most corroding electrode was about 0.25 mm/year. As soon as the probe's coupling joint was connected to the pipeline wall, the probe potential dropped to the "On" CP potential (−0.784 V ($Cu/CuSO_4$)), and the corrosion rate quickly dropped to zero, while the CPEM increased about 40% , but did not reach the excessive CPEM according to the value of the current from the most cathodic electrode. The CPEM value of 30%–40% indicates that the pipeline at this location was very well-protected, even though the potential measurements showed that the pipeline was at risk because $|-0.784|$ V is less than $|-0.85|$ V, which is the minimum protection potential as recommended by the standards. The standards do recognize that the minimum protection potential varies with soil conditions and the temperature, but it is difficult for the field operators to determine what the reasonable minimum protection potential is without excessive testing for the soil condition. This underscores the usefulness of the CMAS probe for the evaluation of the CP effectiveness.

In addition, the data in Fig. 24.25 indicate that it took less than 1 min (can be <10 s for dedicated models of CMAS monitors) to obtain the CMAS data. So, CMAS probe may be used as a survey tool for measurements of the effectiveness of CP at different locations.

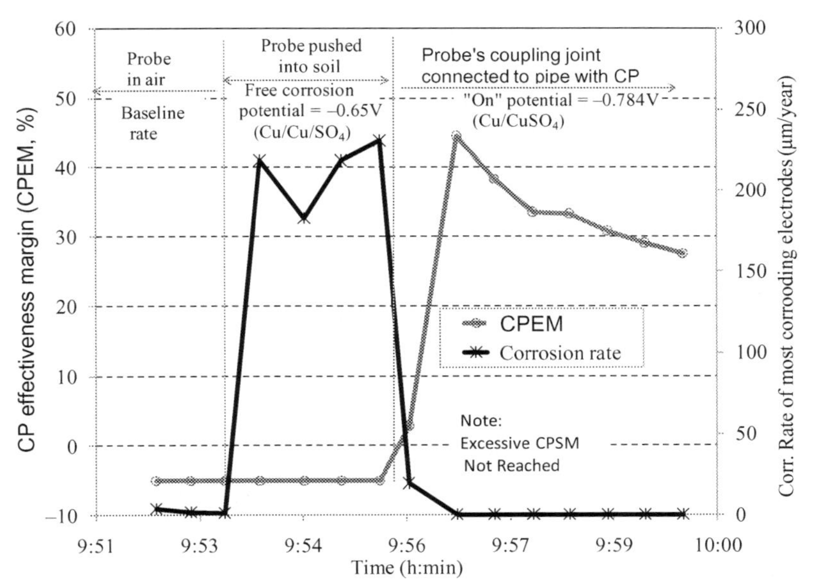

Fig. 24.25 Typical CP parameters obtained from a CMAS probe at the midpoint between anode wells.

24.5.4 Summary

Coupled multielectrode array sensor (CMAS) probes were used for real-time monitoring the corrosion rate of carbon steel material in soil with and without cathodic protection. The steady state corrosion rate measured in the densely packed soil saturated with simulated seawater was found to be approximately 2–15 μm/year. When the carbon steel electrodes of the probe were connected to sacrificial aluminum anodes, the corrosion rate decreased instantaneously to a value that is below the lower detection limit of instrument (10 nm/year), suggesting that the CMAS probe is an effective real-time tool to measure the effectiveness of cathodic protection in soil.

Data were also presented for using the CMAS probe to monitor the effectiveness of CP for pipelines in an oil field. The parameters for CP effectiveness (corrosion rate and CP effectiveness margin) are consistent in most cases with the pipe-to-soil potential criterion recommended by the standards (−0.85 to −1.20 V vs $Cu/CuSO_4$). However, the standard-recommended range of potentials vary with soil conditions and the temperature and it is difficult for the field operators to determine what the reasonable range is. This underscores the usefulness of the CMAS probe for the evaluation of the CP effectiveness. Furthermore, the potential measurements rely on the use of a reference electrode which requires periodical maintenance, while the CMAS probe is maintenance-free. The CMAS method is a more convenient and reliable method to use in the fields.

24.6 Measurements of localized corrosion rates of cathodically protected carbon steel in drinking water

Corrosion of metallic components in drinking water systems has been an ongoing concern. According to a recent report [1], the total estimated cost of corrosion for drinking water systems is $19.26 billion per year, in the United States alone. To effectively control and mitigate corrosion, it is important to measure the real-time rate of corrosion—especially the rate of localized corrosion—taking place in the system. This section describes the application of coupled multielectrode array sensor probes as an online tool for measuring the general and maximum localized corrosion rates of three metals in drinking water systems with and without cathodic protections. These metals include low carbon steel, stainless steel, and brass which are commonly used in drinking water systems.

A nanoCorr model S-50 CMAS analyzer was used in this measurement. In addition to the functions described for the Model A-50 CMAS analyzer, model S-50 also measures the general corrosion rate based on the average anodic currents from the CMAS probe [8]. The experiment was conducted in a beaker filled with natural drinking water (spring water). All coupled multielectrode probes (one carbon steel, one stainless steel, and one brass) and a reference electrode were vertically immersed in water. Prior to the tests, drinking water was placed in the open air to enable saturation with the gases in the atmosphere (e.g., O_2 and CO_2). The water was not agitated during the experiments. One 1-mm-diameter aluminum wire was used as a sacrificial anode for each probe during the cathodic protection test. The reference electrode was a calomel electrode (SCE). The experiments were conducted at a temperature range from 17°C to 23°C. Other parameters such as oxidation and reduction potential (ORP), temperature, and pH are also measured by the CMAS analyzer. These parameters are beyond the scope of this section and users are encouraged to consult the original publication [9] for additional information.

The sensing electrodes of the carbon steel multielectrode probe were made from annealed Type 1008 carbon steel (UNS G10080) wire (1.5 mm in diameter and 1.77 mm^2 in electrode surface area). The sensing electrodes of the brass multielectrode probe were made from Type 260 brass (UNS C26000) wire (1 mm in diameter and 0.785 mm^2 in electrode surface area). The sensing electrodes of the stainless steel multielectrode probe were made from Type 316 L (UNS S31603) wire (1 mm in diameter and 0.785 mm^2 in electrode surface area). Each probe had 16 electrodes flush-mounted in epoxy. Prior to the test, the surfaces of the sensing electrodes for each multielectrode probe were polished to 600 Grit and rinsed with distilled water and then with acetone.

24.6.1 Maximum localized corrosion rates and their usefulness for the evaluation of CP effectiveness

The maximum localized corrosion rates from the three CMAS probes are presented in Fig. 24.26. The maximum localized corrosion rates from the probes were calculated using the anodic current density from the most corroding electrode of the probe. The maximum localized corrosion rate of the carbon steel probe was initially low (90 μm/year or 3.5 mpy) and increased slowly to 330 μm/year (13 mpy) in 2 h prior to cathodic

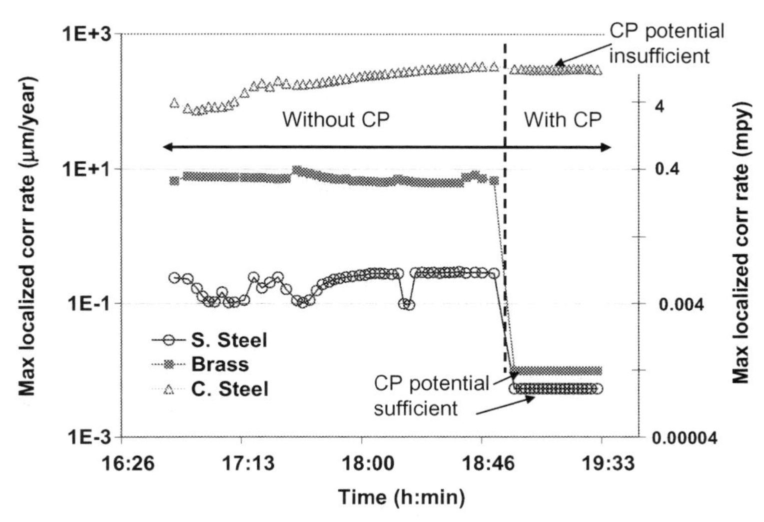

Fig. 24.26 Maximum localized corrosion rates from three CMAS probes in drinking water before and after they were connected to their sacrificial CP anodes.

protection. The maximum localized corrosion rates for the stainless steel and brass were 0.17 and 7.0 μm/year (0.0067 and 0.28 mpy), respectively, in the drinking water and remained unchanged during the 2-h exposure to the water prior to the application of the cathodic protection. When the electrodes of the three probes were connected to their respective sacrificial aluminum anodes, the maximum localized corrosion rates of the Type 316 L stainless steel and the Type 260 brass immediately dropped below the detection limit of the instruments (10 nm/year or 0.0004 mpy), suggesting that these two probes were sufficiently cathodically protected by the potential (see Section 24.6.3) supplied by their sacrificial anodes. The maximum localized corrosion rate of the carbon steel probe, however, only decreased slightly (from 330 to 290 μm/year), when the probe was connected to the aluminum anode, indicating an insufficient cathodic protection. The insufficient protection by the aluminum anode is consistent with the relatively high potential (see Section 24.6.3) supplied by the sacrificial anode, because of its limited surface area.

24.6.2 General corrosion rates

The general corrosion rates from the three probes are presented in Fig. 24.27. The general corrosion rate was calculated using the average anodic current density, which is the total anodic current from all the electrodes of the probe divided by the total surface areas of all the electrodes of the probe. Because this average corrosion rate is similar to the general corrosion rate obtained by weight loss methods or by other electrochemical methods using large electrodes, the use of the average corrosion rate to approximate the general corrosion rate is a reasonable approach. Compared with Fig. 24.26,

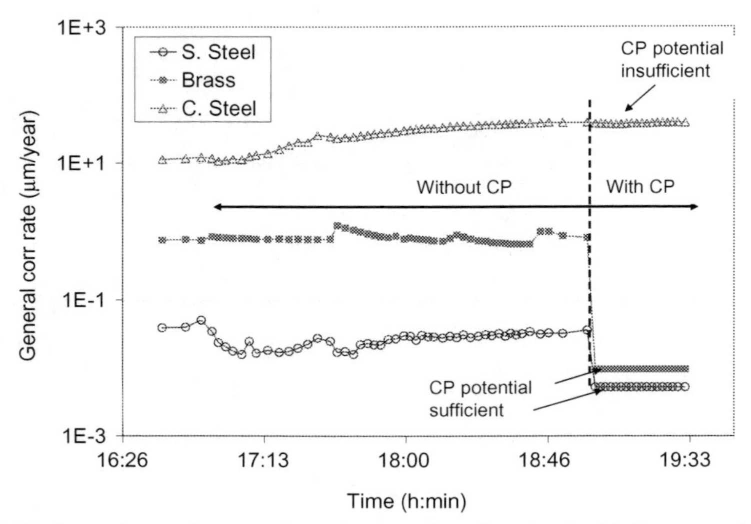

Fig. 24.27 General corrosion rates from the three CMAS probes in drinking water before and after they were connected to their sacrificial CP anodes.

the average/general corrosion rates from these three metals have similar trends with the maximum localized corrosion rates, but the values are much smaller.

The maximum localized corrosion rates were 7.5–8 times higher than their general corrosion rates for the Type 260 brass and low carbon steel during the test. The ratio of the maximum localized corrosion rate to the general corrosion rate for the Type 316 L stainless steel was approximately 9 at the end of the test. These ratios are also called localized corrosion rate factors (see Chapter 8).

24.6.3 Probe potentials

The corrosion potentials of the Type 316 L stainless steel and Type 260 brass in the drinking water prior to the cathodic protection were −20 and −100 mV(SCE), respectively (Fig. 24.28). The corrosion potential of the Type 1008 carbon steel probe was −320 mV(SCE) initially and decreased to −374 mV(SCE) at the time the cathodic protection was about to be applied, indicating that the carbon steel electrode became more and more active, which is consistent with the measured steadily increased maximum localized corrosion rate from the carbon steel probe as shown in Fig. 24.26. When the electrodes of the three probes were connected to their sacrificial aluminum anodes, the potentials of the Type 316 L stainless steel and the Type 260 brass probes immediately dropped by more than 400 and 100 mV, respectively; the potential of the carbon steel electrode, however, dropped only by 10 mV. The variations in the drops of the potentials, during the cathodic protection, were due to the size of the aluminum anodes used as the sacrificial anodes. The surface areas of the aluminum anodes were sufficiently large to lower the potentials of the less active stainless steel and brass probes, but not enough to lower the potential of the more active carbon steel probe.

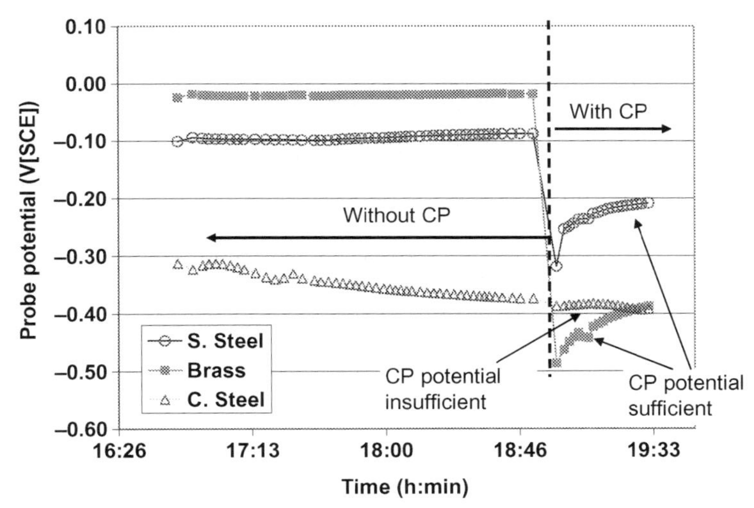

Fig. 24.28 Corrosion potentials from the three CMAS probes in drinking water before and after they were connected to their sacrificial cathodic protection (CP) anodes.

Because of the significant drop in potentials, the stainless steel and brass probes were effectively protected and their corrosion rates were essentially zero (see Fig. 24.26), while the slight drop in potential for the carbon steel probe proved to be insufficient for the carbon steel to be cathodically protected and its maximum localized corrosion rate remained essentially unchanged (from 330 to 290 μm/year) (Fig. 24.26). This underscores the significance of using the coupled multielectrode array sensor to monitor the effectiveness of the cathodic protection.

24.6.4 Posttest visual examination of the probes

Fig. 24.29 shows the appearance of the 260 brass and 316 L stainless steel probe electrodes after an 8-day exposure to the drinking water without cathodic protection. No localized corrosion was seen in the two probes, even though a slight discoloration was observed on some of the electrodes of the Type 260 brass probe, which is consistent with the low maximum localized corrosion rates, as shown in Fig. 24.26.

Fig. 24.30 shows the appearance of the carbon steel probe after the same 8-day exposure to the drinking water without cathodic protection. It is apparent that a few of the electrodes (#5 and #14) were covered by deposits (corrosion products), and most of others were clean. After cleaning off the corrosion products (bottom picture in Fig. 24.30), pitting was observed on the electrodes covered by the deposits. No pitting corrosion was observed on the electrodes that did not have deposits. Therefore, the deposit-covered electrode served as the anodes and the clean electrodes served as the cathode, during the exposure. The pitting corrosion on the carbon steel probe is consistent with the high maximum localized corrosion rate as shown in Fig. 24.26.

It is worth mentioning that no significant corrosion between the electrodes and the surrounding epoxy was observed. The good bonding between the epoxy and the electrode side surface prevented the development of crevice corrosion, which may complicate the calculation of the localized corrosion rate by altering the true surface area.

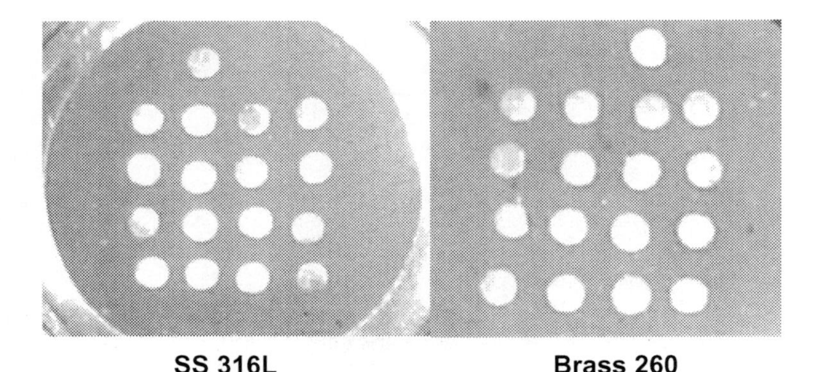

Fig. 24.29 The appearances of the stainless steel and brass probes after 8-day immersion in drinking water without cathodic protection.

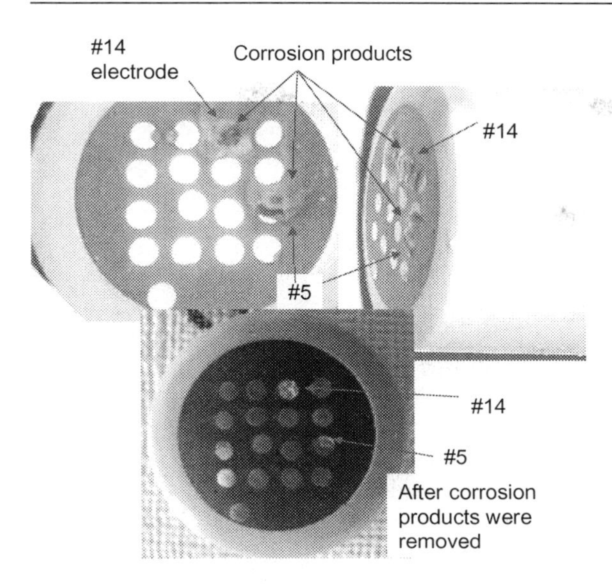

Fig. 24.30 The appearances of the carbon steel probe after 8-day immersion in drinking water without cathodic protection.

24.6.5 Summary

Real-time coupled multielectrode array sensor probes were used to measure the maximum localized corrosion rates of Type 316 L stainless steel, Type 260 brass, and Type 1008 carbon steel materials in drinking water with and without cathodic protection. The maximum localized corrosion rates in the drinking water were approximately 0.3 μm/year (0.012 mpy) for stainless steel and 7 μm/year (0.28 mpy) for brass with no cathodic protection at ambient temperatures. The steady state maximum localized corrosion rate for the carbon steel in the drinking water was found to be 300 μm/year (12 mpy). When cathodic protection was applied to the stainless steel and brass probes, the corrosion rate from these probes dropped below the lower detection limit of the CMAS instruments (10 nm/year). However, the maximum localized corrosion rate and the general corrosion rate from the carbon steel probe remained the same after cathodic protection was applied. The potential measurements indicate that the potentials of the stainless steel and the brass probes were significantly lowered by their sacrificial anodes, but the potential of the carbon steel probe was lowered by only 10 mV. The insignificant lowering of the potential of the carbon steel probe indicates that the surface area of the sacrificial anode for the cathodic protection of the carbon steel probe was not large enough to provide the required cathodic protection potential for the carbon steel. The unchanged corrosion rates from the carbon steel probe after the application of the cathodic protection underscore the significance of the use of the CMAS probes for monitoring the effectiveness of cathodic protection systems.

References

[1] G.H. Koch, M.P.H. Brongers, N.H. Thompson, Y.P. Virmani, J.H. Payer, Corrosion Cost and Preventive Strategies in the United States, NACE Report, FHWA-RD-01-156, NACE, Houston, TX, 2001.

[2] L. Yang, N. Sridhar, O. Pensado, D.S. Dunn, An in-situ galvanically coupled multi-electrode array sensor for localized corrosion, Corrosion 58 (2002) 1004.
[3] X. Sun, Online monitoring of undercoating corrosions utilizing coupled multielectrode sensors, in: CORROSION/2004, paper no. 04033, NACE International, Houston, TX, 2004.
[4] A. Anderko, N. Sridhar, L. Yang, S.L. Grise, B.J. Saldanha, M.H. Dorsey, Validation of a localized corrosion model using real-time corrosion monitoring in a chemical plant, Corros. Eng. Sci. Technol. (formerly British Corrosion J.) 40 (2005) 33–42.
[5] L. Yang, N. Sridhar, C.S. Brossia, D.S. Dunn, Evaluation of the coupled multielectrode array sensor as a real time corrosion monitor, Corros. Sci. 47 (2005) 1794–1809.
[6] L. Yang, N. Sridhar, Coupled multielectrode array systems and sensors for real-time corrosion monitoring—a review, in: CORROSION/2006, paper no. 06681, NACE, Houston, TX, 2006.
[7] X. Sun, L. Yang, Real-time measurement of crevice corrosion with coupled multielectrode array sensors, in: CORROSION/2006, paper no. 06679, NACE, Houston, TX, 2006.
[8] X. Sun, L. Yang, Real-time monitoring of localized and general corrosion rates in simulated marine environments using coupled multielectrode array sensors, in: CORROSION/2006, paper no. 06284, NACE, Houston, TX, 2006.
[9] X. Sun, L. Yang, Real-time monitoring of localized and general corrosion rates in drinking water utilizing coupled multielectrode array sensors, in: CORROSION/2006, paper no. 06094, NACE, Houston, TX, 2006.
[10] X. Sun, Online monitoring of corrosion under cathodic protection conditions utilizing coupled multielectrode sensors, in: CORROSION/2004, paper no. 04094, NACE International, Houston, TX, 2004.
[11] X. Sun, Online and real-time monitoring of carbon steel corrosion in concrete, using coupled multielectrode sensors, in: CORROSION/2005, paper no. 05267, NACE International, Houston, TX, 2005.
[12] X. Sun, Real-time corrosion monitoring in soil with coupled multielectrode sensors, in: CORROSION/2005, paper no. 05381, NACE, Houston, TX, 2005.
[13] L. Yang, R.T. Pabalan, L. Browning, G.C. Cragnolino, Measurement of corrosion in saturated solutions under salt deposits using coupled multielectrode array sensors, in: CORROSION/2003, paper no. 426, NACE, Houston, TX, 2003.
[14] C.S. Brossia, L. Yang, Studies of microbiologically induced corrosion using a coupled multielectrode array sensor, in: CORROSION/2003, paper no. 575, NACE, Houston, TX, 2003.
[15] X.S. Yang, Methods for Monitoring the Degree of Cathodic Rotection for Metal Structutres and Burried Pipelines Using Coupled Mutielectrode Sensors, US Patent Application (Appn No: 16602142), (August 2019).
[16] D. Yanxia, D. Tang, H. Qin, M. Lu, Research on parameter fluctuation characteristics and effects on corrosion rates under dynamic DC stray current from metro system, in: CORROSION/2019, paper no. 13203, NACE International, Houston, TX, 2019.
[17] X.S. Yang, L. Yang, Multielectrode Probes for Monitoring Fluctuating Stray Current Effects and AC Interference on Corrosion of Burried Pipelines and Metal Structures, US Patent Application (Appn No: 16501742), (May 2019).

Corrosion monitoring using the field signature method

Kjell Wold
Emerson Automation Solutions, Trondheim, Norway

25.1 Introduction

A variety of corrosion monitoring technologies are available as of today, with a range of applications, sensitivities, and instrument options. It is the author's opinion that the selection of monitoring technologies should be based on an assessment of the needs in each individual case, and that a combination of monitoring technologies often provides the most extensive and reliable information. This is further discussed in reference [1] to this section.

Field Signature Method (Emerson Roxar FSM) is a nonintrusive technology for monitoring internal corrosion in pipelines, process piping, and vessels. A specific feature of the FSM is that corrosion/metal loss is measured *between* sensors installed at the external surface of the monitored object, which allows a wider area of coverage than most other monitoring technologies and enables the user to differentiate between uniform and localized corrosion. FSM can also be used at high temperatures (generally rated up to 500°C (932 F)). Models of FSM are also certified for installation in hazardous areas (ATEX, IECex).

The FSM technology was first developed in the 1990s, based on a patent from SI (now part of SINTEF, Norway). FSM has been used for many applications—subsea, upstream, pipelines, and in refineries (high temperatures). Various models and applications of FSM have been provided since the first installation, including subsea and off-line/portable systems. Since 2008, commercial FSM systems are based on the use of instruments for "continuous" measurements and, in most cases, online communications to a data server where FSM data are analyzed and reported.

25.2 FSM measurement technology

The FSM technology measurement principle is based on feeding an electric current through a monitored section of a pipe, pipeline, or vessel. The applied current sets up an electric field that is monitored as voltage drop values between a set of sensing pins installed on the external object wall.

The FSM measurement principle is described in Figs. 25.1 and 25.2 below:

- A: When a current is fed into a structure, it spreads out into a pattern which is determined by the geometry of the structure and the conductivity (or electrical resistance) of the material.

Techniques for Corrosion Monitoring. https://doi.org/10.1016/B978-0-08-103003-5.00027-8

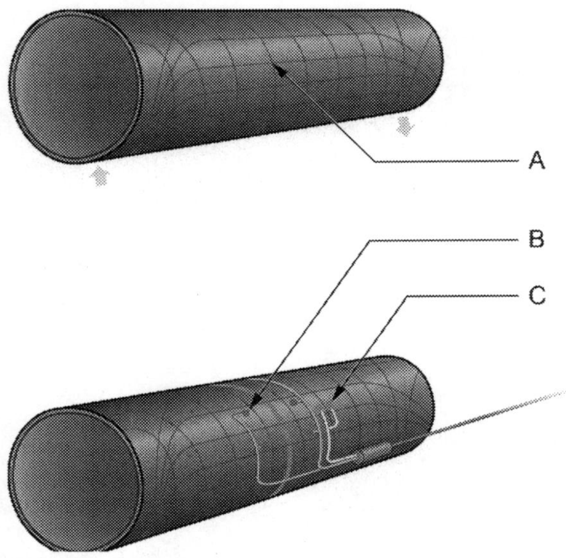

Fig. 25.1 Principle of an FSM measurement. Section A shows electric field lines from current imposed on a pipe section. B illustrates sensing pins installed over a welded section, measuring changes in electric field as a result of corrosion. C illustrates a reference sensing pin pair, to compensate for temperature changes. The reference sensing pins are normally installed on a reference plate that is electrically insulted from the pipe.
Courtesy of Emerson Automation Solutions.

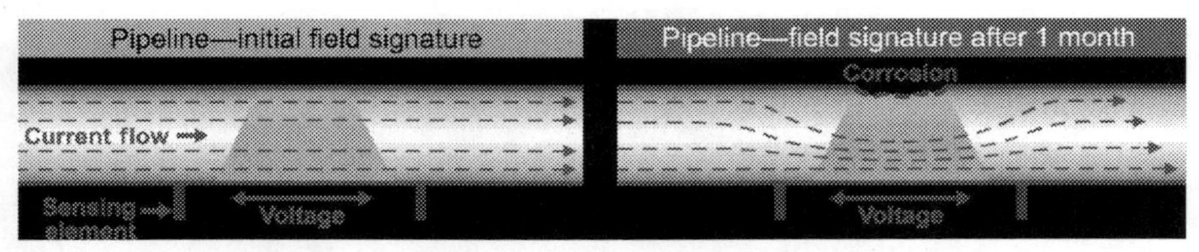

Fig. 25.2 Illustration of how electric current flow is changed by a localized corrosion attack on a pipe, resulting in a higher voltage drop in the area where corrosion takes place.
Courtesy of Emerson Automation Solutions.

The first set of voltage drop measurements is called the Field Signature, and all later measurements are compared to this initial signature.

- B. When corrosion or other types of metal loss takes place, the size and distribution of the electric field changes. For example, if metal loss takes place in the weld zone, a local increase in the FSM measurements can be observed in the actual area.
- C. To compensate for the effect of temperature changes, a pair of reference points are installed outside the monitored area (see Section 25.2.1). Generally, an electrically insulated reference plate is used for this purpose. The current goes through the reference plate and sensing pin matrix to ensure consistency in measurements.

Corrosion is measured *between* pairs of sensing pins, meaning that the complete monitored area is covered, not only at a spot under each sensing pin. Typical sensitivity for

FSM is 0.1% of wall thickness for general corrosion, corresponding to 10–20 μm (approximately ½ to 1 mil) in most cases. For localized attacks, sensitivity is linked to the volume of the localized attack to be detected; the volume must be sufficient to change the voltage drop between the sensing pin pairs by 0.1% to give a "detectable" attack.

25.2.1 Temperature

Temperature changes can interfere with the FSM measurements, since the monitored electrical resistance changes with temperature. Therefore, temperature changes must be tracked and compensated. To compensate for temperature changes, a reference plate and temperature sensors are installed on the monitored object (see Fig. 25.3). The reference plate of the same material as the monitored object is placed next to the pin matrix and is electrically insulated from the rest of the system. The reference is used in the metal loss calculations to distinguish between the changes in resistance caused by temperature and those caused by metal loss.

In addition to the reference plate, temperature sensors may be used to obtain additional temperature information.

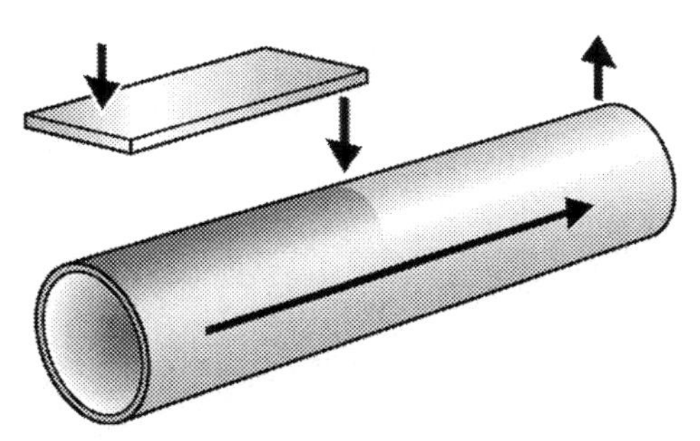

Fig. 25.3 Reference plate arrangement for FSM. Arrows indicate the direction of electric current flow.
Courtesy of Emerson Automation Solutions.

25.2.2 Determining corrosion and erosion data from FSM measurements

The field signature is the basis for the FSM measurements. The signature is based on a series of measurements taken at a time when the wall thickness is known, often when the metal structure is new. The signature should be taken when the system is tuned and at normal operational (temperature) conditions.

For general corrosion, the metal loss will be uniformly distributed over the metal surface. This means that the change in voltage drop between the sensing pin pairs is also uniform. In this case, metal loss can be calculated directly as metal loss proportional to the change in FSM value. As mentioned, a sensitivity of 0.1% of the monitored object's wall thickness is normally claimed for uniform corrosion.

For localized attacks, the FSM readings covering the attack and nearby sensing pin pairs only are affected. This means you can detect localized corrosion, identify the position, and follow through with the propagation. The ability to measure localized corrosion over the total monitored area (between sensing pins) is a unique feature of the FSM technology.

Quantitative interpretation of FSM data for localized corrosion or cracks is more complex, since forms and shapes of localized attacks can vary from wide grooves to narrow pin holes. For cracks, the ability for detection also depends on the crack direction compared to the current direction. Generally, a sensitivity of 0.1% for localized corrosion means that the local attack needs to have a volume sufficient to change the FSM value with more than 0.1% to be significant. Propagation of the localized attack can be followed as a growth/change in FSM readings at the affected local area of the FSM sensing pin matrix.

Observe that even with limitations on quantifying shape and the size of localized corrosion, FSM is still probably the monitoring technology providing the best non-uniform corrosion coverage.

25.2.3 FSM data management

A software application package is needed for running the FSM monitoring system, for data management and data reporting. Fig. 25.3 shows an example window from the FSM application software (Roxar FIeldwatch), where metal loss over time is tracked (left, above), metal loss visualized (right, above), metal loss for each pin pair examined (left, bottom), and a different visualization of metal loss obtained (right, bottom). A variety of tools for data management and analysis are available. It is recommended that the user works with the FSM vendor for the best possible software configuration and setup for each specific application (Fig. 25.4).

25.3 System configurations

A typical FSM configuration is indicated in Fig. 25.5. The sensing pin matrix and reference plate are installed on the pipe and connected via cable to the FSM Log instrument (I). The maximum distance (J) is 10 m.

The FSM Log instrument can be installed in hazardous areas (ATEX and IECEx). When used in hazardous areas, power and communication must be provided by a field interface unit (A) that must be installed in a safe area (or alternatively, in an explosion proof enclosure). If FSM is installed in a non-ex area, power and communication can be provided directly to the FSM Log instrument.

The field interface unit has a range of communication options and communicates with a server with data management software installed. Calculated data can be forwarded, e.g., to Scada/DCS or historians for other reporting purposes.

Power options include solar panels, and communication options include wireless via GSM or radio modems, or fiber-optics.

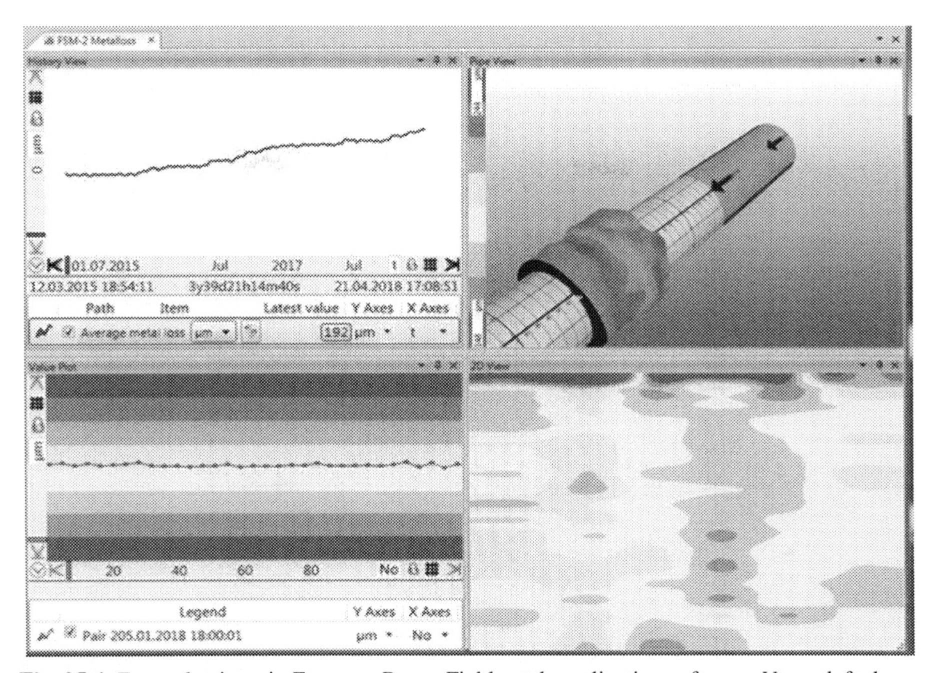

Fig. 25.4 Example views in Emerson Roxar Fieldwatch application software. Upper left shows average metal loss versus time for FSM system (can also be individual pin pairs or clusters of sensing pins). Upper right shows distribution of metal loss on monitored section. Lower left shows distribution of metal loss for individual pin pairs and bottom right different perspective for distribution of metal loss.
Courtesy of Emerson Automation Solutions.

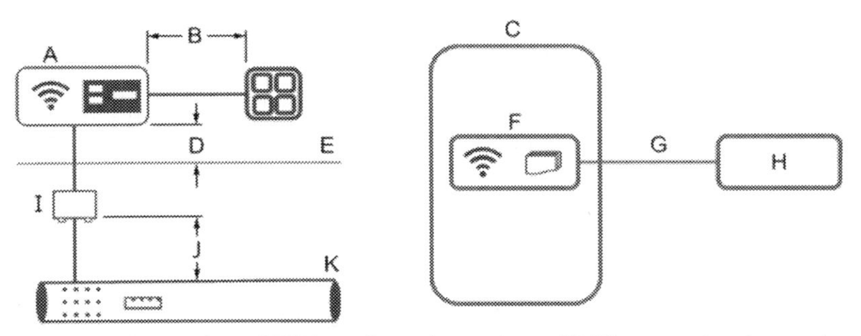

Fig. 25.5 General overview of FSM configuration options. (K) illustrates the pipe section with sensing pin matrix and reference plate installed. (I) represent the FSM Log instrument, with a maximum recommended distance (J) of 7 m from the sensing pin matrix. (A) illustrates cabinet with field interface unit for power and communication with FSM log (required for hazardous area installation)—maximum recommended distance (D) is 500 m (two FSMs on same loop). (B) indicates local power, e.g., from solar panels. Various solutions for communication to control room (C) are possible. Fieldwatch server and software (F) is needed for system control and data management (hazardous areas). (G) is communication from Fieldwatch server to DCS (H) or other locations where FSM information is used.
Courtesy of Emerson Automation Solutions.

25.4 Applications

Corrosion monitoring is generally carried out for three main purposes:

- Tracking changes in process parameters for optimized corrosion management and mitigation. Examples are the tuning of corrosion inhibitors, opportunity crude blending (refineries), or tracking process changes that may influence corrosion conditions.
- Verifying pipeline or plant integrity based on more extensive information about internal corrosion
- Optimizing integrity management programs, e.g., using corrosion monitoring parameters for running more cost-effective inspection programs. An example could be a changing frequency of inspection (e.g., pipeline pig surveys) based on data from continuous corrosion monitoring.

The technology and design selection for a corrosion monitoring solution should be based on a case to case assessment of the objectives of the corrosion monitoring system.

FSM is a relatively costly and complex corrosion monitoring system. The use of FSM is, therefore, focused on applications where the unique features of FSM add value. Such applications are typically (high temperature) applications in refineries, (underground) pipelines, and upstream oil and gas applications above water. FSM is occasionally used also in other industries, e.g., piping in the mining industry.

25.4.1 Pipelines

Pipelines in the following sections refer to all uncovered or underground pipelines, including transportation lines and gathering lines in upstream oil and gas production.

Corrosion in pipelines is normally linked to the presence of water, which is most likely to be in the bottom (6 o'clock) of the pipeline, often at low spots. This makes nonintrusive monitoring, such as FSM, very attractive for pipelines, since sensors are installed directly onto the pipe and can be part of an underground pipeline after installation. (For comparison, to reach the bottom of the pipe, intrusive corrosion probes would either need to be installed through the pipeline from the top of the pipeline, or a pit under the pipeline would be needed for access to corrosion probes.)

25.4.1.1 Value of FSM monitoring on pipelines

Common drivers for using FSM on pipelines are

- Safety
  - Increased information about corrosion reduces the risk of unexpected leaks or accidents.
  - FSM can be installed directly on pipelines where corrosion is most likely to take place (the bottom of the pipe at low spots).
  - FSM responds to localized corrosion as well as general corrosion.
  - FSM is nonintrusive and does not create any additional risk of leaks around the sensors, nor requires online sensor retrieval operations.
- Increased asset uptime, extended life expectancy (pipeline repairs or replacements are costly) [2]
  - Corrosion trend information allows for planned maintenance and repairs and helps avoid unexpected shutdowns.

- Corrosion monitoring data are used to tune corrosion mitigation programs that contribute to extended asset life.
- FSM can be employed at pipe(line) sections, where corrosion thinning is observed, to extend end of life use of the object.

- Reduced/optimized frequency of intelligent pig inspection. Each intelligent pig service campaign may amount to USD 1.5–2 million, depending on pipe length and inspection approach [3].
  - Continuous corrosion monitoring at critical points may justify reduced pig inspection frequencies, and accordingly, reduce operational costs.
  - Alternatively, the findings of increased corrosion rates at monitoring locations may trigger pig inspection services for complete pipeline integrity assessment.

25.4.1.2 Selecting FSM locations along a pipeline

FSM monitors are installed at selected locations along a pipeline only and do not cover the entire length of the pipeline. It is, therefore, important that FSM is positioned at the most critical and representative locations. The criteria for selecting the most relevant monitoring locations are covered below.

There are two main approaches for defining locations for permanent pipeline corrosion monitoring.

Assessment of other data, e.g., intelligent pig data

Intelligent pig data, identifying where internal corrosion is most severe, is often used for selecting FSM monitoring locations. By installing FSM on such locations, the integrity of the critical locations can be monitored on a continuous basis, and the effect of possible corrosion mitigation programs verified.

Pipeline corrosion predictions

If there is no information about the distribution of corrosion along a pipeline, the positioning of FSM must be based on an assessment of corrosion distribution along the pipeline.

The key factor for corrosion predictions along the pipeline is the *presence of water*.

Since water is heavier than oil and gas, a water film would normally be present in the bottom section of the pipeline. Hence, corrosion monitoring should *be focused on the bottom section of horizontal pipelines*.

The accumulation of water will not be even over the length of the pipeline. Accumulation of water will be linked to the topography of the pipeline. A common approach for selecting FSM locations is, therefore, to review the pipeline profile and *pick low spots or locations where there will be an incline in the pipeline elevation* as locations where water accumulation is more likely.

Top of line corrosion is a common problem in gas pipelines, where temperature drops cause condensation of water in the pipeline. Where water condensation takes place, corrosion may be most severe in the top of the pipeline. *If top of line corrosion is expected, top of line sensors* should be considered for an FSM sensor matrix design.

More complex analyses of pipeline corrosion along pipelines can also be done, using available simulation tools including flow models, pH, CO_2, or H_2S models that provide a more comprehensive foundation for FSM monitoring. Such simulation programs can potentially also be linked with a more comprehensive Pipeline Integrity Management programs [2].

25.4.1.3 FSM for pipeline monitoring

FSM sensing pins and reference plates are normally retrofitted to existing pipelines or to new pipelines on location. Sensing pins are stud-welded (pin brazed) in a process that does not harm the pipeline material. Fig. 25.6 shows FSM technician installing sensing pins on a pipeline.

The FSM sensing pin arrangement is then provided with external protection and hooked up with the FSM instrument installed on top of the ground. After installation, the FSM section will be part of the underground pipeline, only the FSM instrument will remain above ground. A range of power and communication options are available, including solar panels for power and GSM/radio communication for long distance wireless communication. Fig. 25.7 shows FSM installed on underground pipeline, FSM Log instrument powered by solar panel.

25.4.1.4 FSM applications on pipelines

Although FSM has been extensively used in pipelines, there are not many papers available referring to the results from pipeline monitoring. Bich and Eng [3] refer to applications with legacy FSM-IT technology (FSM-IT was based on the use of a portable meter and was used up to 2008, when it was replaced by FSM Log technology).

Fig. 25.6 Welding sensing pins to pipeline on location.
Courtesy of Emerson Automation Solutions.

Fig. 25.7 Instrumented FSM section of underground pipeline.
Courtesy of Emerson Automation Solutions.

Fig. 25.8 shows a typical plot from FSM on an underground pipeline, displayed by legacy FSM MultiTrend software. Each bar represents a pair of sensing pins in the sensing pin matrix, and the plot shows the distribution of corrosion in the pipeline folded out, where the middle of the plot shows the bottom of the pipeline (6 o'clock), and the sides top of line (12 o'clock). The plot shows that corrosion in the bottom of the pipeline is more severe than on the sides and the top of the pipeline, which is a common observation for pipeline corrosion.

25.4.2 Refineries

Refineries are complex petrochemical plants, and a variety of corrosion mechanisms and challenges take place through the refinery processes. Corrosion monitoring in refineries is generally covered in a separate chapter of this book (see Chapter 18).

FSM installation in refineries is most commonly linked to high temperature areas (linked to crude and vacuum distillation processes), where severe and often localized naphthenic acid and sulfuric acid corrosion takes place.

25.4.2.1 Value of FSM monitoring in refineries

Common drivers for using FSM in refineries are:

- Safety
  - FSMLog contributes to better integrity management and provides information that reduces the risk of unexpected incidents. FSM's ability to monitor localized corrosion improves safety.

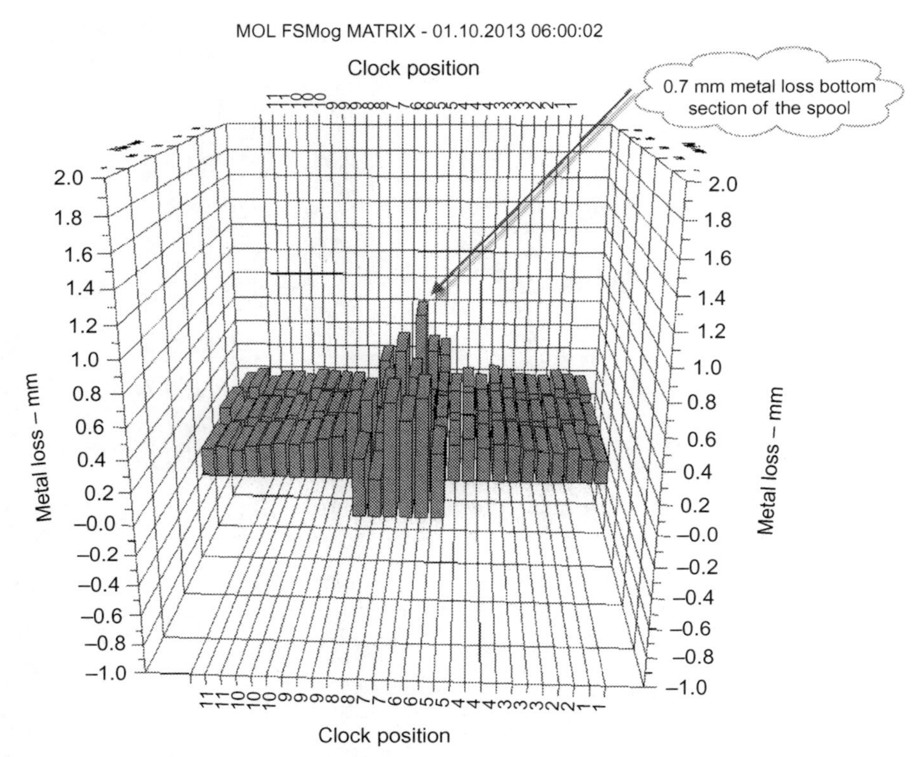

Fig. 25.8 Example data from pipeline, using legacy software version. High corrosion in bottom section of the pipeline.
Courtesy of Emerson Automation Solutions.

- Increased plant uptime
  - Corrosion trend information allows for planned maintenance and repairs and helps avoid unexpected shutdowns and plant downtime. Assuming a profit loss of USD 250,000 per day for a typical refinery, increased plant uptime easily pays for a corrosion monitoring system [4].
- Extended asset life expectancy.
  - Tuned corrosion mitigation (e.g., inhibitors) reduces corrosion and contributes to extended asset life expectancy.
- Optimized inhibitor consumption
  - Efficient monitoring helps verify corrosion protection without excessive inhibitor consumption.
- Reduced crude costs (opportunity crude tuning)—potential savings of multiple million USD/year
  - Corrosion monitoring is used for tuning opportunity crude blending—to find the optimal spot between integrity cost and crude price. FSM adds value for high temperature and localized corrosion.

Aspects of corrosion monitoring in refineries are discussed in more detail in Chapter 18 of this book.

25.4.2.2 Selecting locations for FSM monitoring in refineries

Chapter 18 provides a general discussion on selecting corrosion monitoring technologies and positions through a refinery.

As mentioned, the most common positions for FSM in refineries are linked to high temperature areas in the crude oil and vacuum distillation processes. FSM is often installed on downstream sections of bends in the process, where a combination of temperature and velocity generates preferential positions for naphthenic acid corrosion to take place.

Fig. 25.9 shows FSM sensing pins installed and connected to a refinery pipe section. Each sensing pin is identified with numbers to generate pin pairs to be followed in application software for follow-up. The section shown will later be subject to temperature insulation and mechanical protection.

The sensing pin matrix will be connected to the FSM instrument via cables. Since hazardous areas always apply in refineries, the FSM instrument must be powered and communicate through a field interface unit (see Section 25.3).

25.4.2.3 FSM refinery case stories

FSM is widely used in refineries, particularly in refineries where opportunity crude processing is a key strategy. Monitoring data are, however, the property of the refinery, and access to publishing data is limited.

An application at the world's largest refinery, Reliance's Jamnagar Refinery, was published at the 13th Middle East Corrosion Conference and Exhibition in Bahrain 2010 [5]. The paper discusses the use of FSM in the plant with data examples and

Fig. 25.9 FSM sensing pins installed on refinery pipe section. Observe that all sensing pins identified with number—later used for wiring and identification in Fieldwatch software. Courtesy of Emerson Automation Solutions.

concludes that FSM is important for the integrity monitoring and opportunity crude management of the plant.

Figs. 25.10 and 25.11 show data from another refinery where FSM is installed in both the Crude Distillation Unit (CDU) overpass bend and the Vacuum Distillation Unit (VDU) overpass bend [6].

25.4.3 Upstream

Upstream applications of FSM are mostly linked to import/export lines on oil and gas production platforms, and to pipelines, e.g., gathering lines on land-based productions facilities. Technically, FSM configurations and their use will be similar to what is described under the sections for FSM applications in pipelines (Section 25.4.1) and refineries (Section 25.4.2).

25.5 FSM upgrades

The FSM technology was introduced to the market in the early 1990s, with a variety of technical solutions and applications [7]. The FSM Log technology used today, including online, "continuous" monitoring, was introduced to the market in 2008. The present technology represents many significant improvements compared to legacy FSM versions, with respect to sensitivity and accuracy, convenience of handling data, and data analysis. This means that FSM today provides a significantly better user experience than legacy solutions.

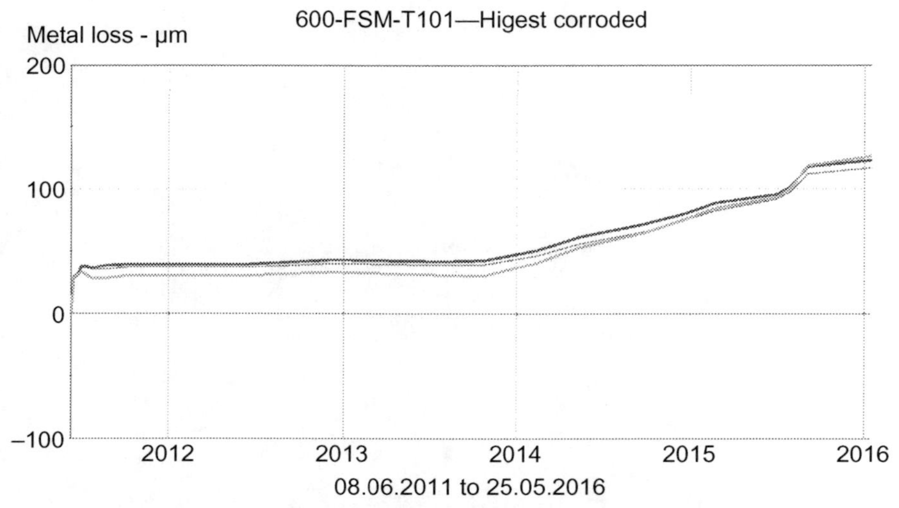

Fig. 25.10 Metal loss versus time plots from FSM, VDU overpass bend, European Refinery. Total scale is −200 to +200 μm, period is from 2011 to mid-year 2016. On lower plot, each line represents metal loss from a cluster of sensing pins. Observe the increase in corrosion rates at the time when opportunity crude processing starts.
Courtesy of Emerson Automation Solutions.

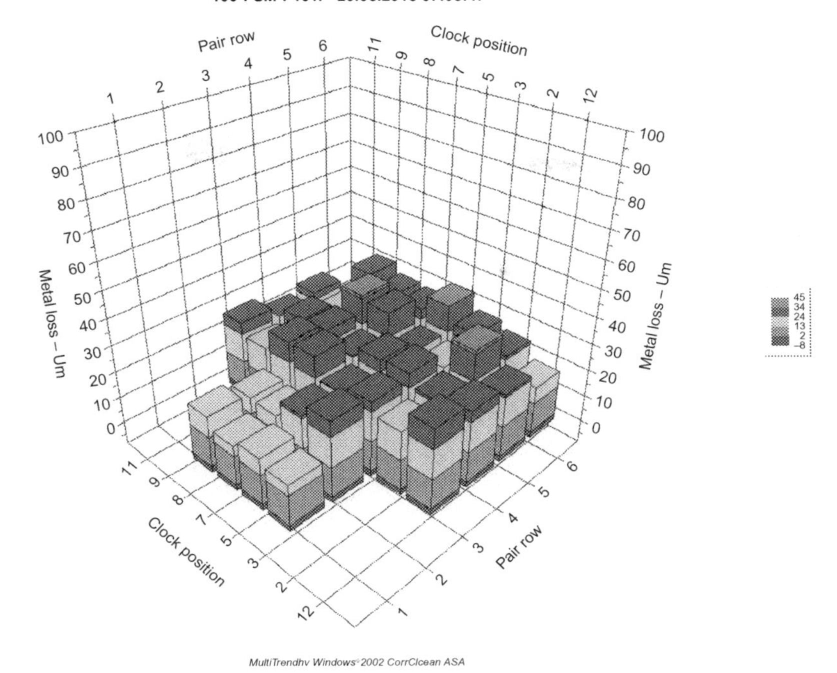

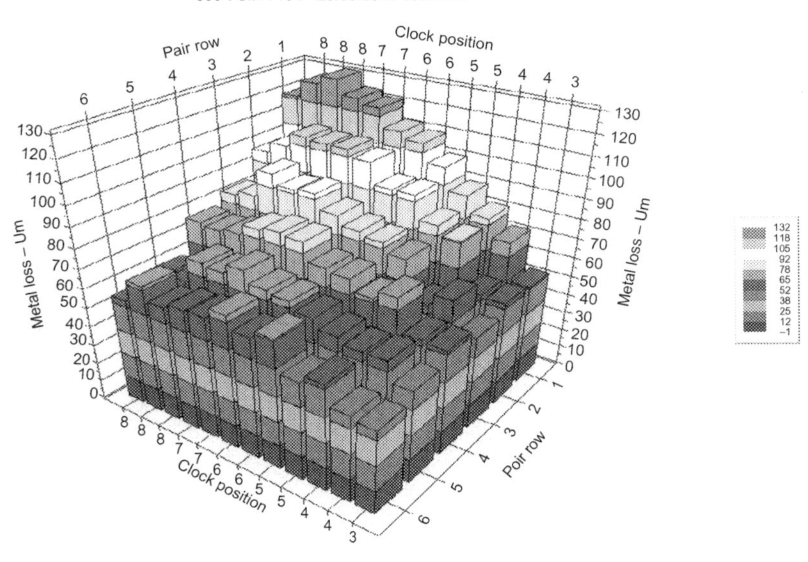

Fig. 25.11 Dimensional plot showing the distribution of corrosion in the same case as Fig. 25.10, CDU overpass line (above) and VDU overpass line (below). Scale micrometers. Observe the distribution of corrosion for the VDU overpass bend where significant corrosion is detected at the end of the matrix. This has been observed on several installations; hence, sensing pin matrixes are now installed in slightly further downstream bends compared to earlier installations.
Courtesy of Emerson Automation Solutions.

For various reasons, there are legacy FSM systems installed on underground pipelines, in refineries, and upstream platforms that are not actively used today. However, the sensing pin matrixes used for legacy FSM systems may still be in good condition and can in many cases be hooked to the FSM Log instrument. Upgrading legacy FSM systems can offer the benefits of present FSM technology at a fraction of the cost of a new FSM system.

25.6 Summary and perspectives ahead

There is clearly a trend towards nonintrusive corrosion monitoring. This trend can easily be understood; nonintrusive monitoring is safer, requires less intervention and reflects corrosion as it happens on the pipe wall. However, as discussed in Section 25.2.2 of this book, nonintrusive technologies still do not have the same high sensitivity and fast response time compared to inline corrosion probes. Therefore, a monitoring solution should be based on the specific corrosion challenges and planned use of data in each case.

FSM is a relatively comprehensive monitoring solution and is used for specific applications where the benefits of FSM add value. These segments are particularly (buried) pipelines and high temperature areas in refineries, particularly when corrosion is not uniformly distributed.

In order to make FSM commercially and technically more attractive, standardization and simplification of the FSM technology has taken place, resulting in a standardized product (FSM Log 48 Area Corrosion Monitor). This standardization has reduced the cost of FSM considerably, both related to hardware and particularly for the time needed for installation and commissioning.

Also, with the complexity of the system, there have been challenges related to interpretation and reporting of FSM data. Focus is now on simplifying the user interfaces in the application software, contributing to a more efficient and easier understood evaluation and of FSM data. It is also seen that FSM is a product where a long-term cooperation between user and vendor adds value and ensures that actionable information is reliable and useful—contributing to increased safety, optimized integrity management, and reduced operational cost

FSM has been in the market since the early 1990s, in various online and off-line versions. Compared to early versions of FSM, the present FSM technology provides a completely different user experience with respect to instrument capabilities, online communication, and data management/application software. In addition to offering a solution for new corrosion monitoring challenges, legacy FSM systems can also be upgraded to latest technology—again providing reliable performance and better user information.

References

[1] K. Wold, H. Ruschmann, I.R. Stubelj, A new perspective on corrosion monitoring, in: Paper no 10937, NACE Corrosion Conference 2018, Phoenix, USA, April 2018.

[2] Gartland, Johnsen, Application of internal corrosion modelling in the risk assessments of pipelines, in: Paper no 03179, NACE Corrosion Conference, Nashville, USA, 2003.
[3] N.N. Bich, P. Eng, Corrosion Monitoring Using the Field Signature Method Inspection Tool (FSM-IT), Technologies for Corrosion Monitoring, first ed., Elsevier, 2007.
[4] M. Carugo, K. Wold, New perspectives on refinery corrosion, integrity management, communications and optimizing opportunity crudes, in: AFPM Annual Meeting, San Antonio, Texas, 2017.
[5] K. Wold, H. Jenssen, R. Stoen, G. Sirness, V. Shinde, U. Anand, On-line, non-intrusive corrosion monitoring based on electric field signature technology—an update on installation experience and typical field data, in: 13th Middle East Corrosion Conference and Exhibition, Bahrain, 2010.
[6] Wold, A holistic perspective on corrosion/erosion monitoring and integrity management, in: Eurocorr Conference 2017, Prague, 2017.
[7] Wold, Sirness, FSM technology, 16 years of field history—experience, status, and further developments, in: NACE Corrosion 07, paper 07331, Nashville, 2007.

Index

Note: Page numbers followed by *f* indicate figures and *t* indicate tables.

F

G

H

N

O

P

Q

R

S

X

Z

Printed in the United States
By Bookmasters